VOLUME SIX HUNDRED AND ELEVEN

METHODS IN ENZYMOLOGY

Intrinsically Disordered Proteins

METHODS IN ENZYMOLOGY

VOLUME SIX HUNDRED AND ELEVEN

METHODS IN ENZYMOLOGY

Intrinsically Disordered Proteins

Edited by

ELIZABETH RHOADES
Department of Chemistry, University of Pennsylvania, Philadelphia, PA, United States

ELSEVIER

ACADEMIC PRESS
An imprint of Elsevier

Academic Press is an imprint of Elsevier
50 Hampshire Street, 5th Floor, Cambridge, MA 02139, United States
525 B Street, Suite 1650, San Diego, CA 92101, United States
The Boulevard, Langford Lane, Kidlington, Oxford OX5 1GB, United Kingdom
125 London Wall, London, EC2Y 5AS, United Kingdom

First edition 2018

ISBN: 978-0-12-815649-0
ISSN: 0076-6879

Publisher: Zoe Kruze
Acquisition Editor: Ashlie Jackman
Editorial Project Manager: Shellie Bryant
Production Project Manager: Divya Krishnakumar
Cover Designer: Mark Rogers

Typeset by SPi Global, India

CONTENTS

CONTRIBUTORS

Sara Abdellaoui
VIB-VUB Center for Structural Biology (CSB), VIB; Structural Biology Brussels (SBB), Vrije Universiteit Brussel (VUB), Brussels, Belgium

Adrian J. Alvarez
Texas A&M Health Science Center, College Station, TX, United States

Perdita E. Barran
The Michael Barber Centre for Collaborative Mass Spectrometry, The School of Chemistry, Manchester Institute of Biotechnology, The University of Manchester, Manchester, United Kingdom

Angela Bekesi
VIB-VUB Center for Structural Biology (CSB), VIB; Structural Biology Brussels (SBB), Vrije Universiteit Brussel (VUB), Brussels, Belgium

Bruno Bellina
The Michael Barber Centre for Collaborative Mass Spectrometry, The School of Chemistry, Manchester Institute of Biotechnology, The University of Manchester, Manchester, United Kingdom

Robert B. Best
Laboratory of Chemical Physics, National Institute of Diabetes and Digestive and Kidney Diseases, National Institutes of Health, Bethesda, MD, United States

Christophe Bignon
CNRS and Aix-Marseille Univ, Laboratoire Architecture et Fonction des Macromolecules Biologiques (AFMB), Marseille, France

Melissa Birol
Department of Chemistry, University of Pennsylvania, Philadelphia, PA, United States

Sarah E. Bondos
Texas A&M Health Science Center; Texas A&M University, College Station, TX, United States

Rebecca M. Booth
Texas A&M Health Science Center; Texas A&M University, College Station, TX, United States

Wade M. Borcherds
Department of Cell Biology, Microbiology, and Molecular Biology, University of South Florida, Tampa, FL, United States

Antoni Borysik
King's College London, Department of Chemistry, London, United Kingdom

Anne Bremer
Max-Planck-Institut für Molekulare Pflanzenphysiologie, Potsdam, Germany

Katrine Bugge
Structural Biology and NMR Laboratory, Department of Biology, University of Copenhagen, Copenhagen, Denmark

Alfredo Vidal Ceballos
Structural Biology Initiative, CUNY Advanced Science Research Center; Ph.D Program in Biochemistry, The Graduate Center, CUNY, New York, NY, United States

Erik C. Cook
Department of Chemistry, The Pennsylvania State University, University Park, PA, United States

Michael D. Crabtree
Department of Biochemistry, University of Oxford, Oxford, United Kingdom

Gary W. Daughdrill
Department of Cell Biology, Microbiology, and Molecular Biology, University of South Florida, Tampa, FL, United States

Stefaan Derveaux
VIB Nucleomics Core, VIB, Leuven, Belgium

Veselin S. Dobrev
Department of Chemistry, Georgetown University, Washington, DC, United States

Shana Elbaum-Garfinkle
Structural Biology Initiative, CUNY Advanced Science Research Center; Ph.D Program in Biochemistry; Ph.D Program in Biology, The Graduate Center, CUNY, New York, NY, United States

David Eliezer
Department of Biochemistry, Weill Cornell Medicine, New York, NY, United States

Aidan P. France
The Michael Barber Centre for Collaborative Mass Spectrometry, The School of Chemistry, Manchester Institute of Biotechnology, The University of Manchester, Manchester, United Kingdom

Lisette M. Fred
Department of Chemistry, Georgetown University, Washington, DC, United States

Monika Fuxreiter
MTA-DE Laboratory of Protein Dynamics, Department of Biochemistry and Molecular Biology, University of Debrecen, Debrecen, Hungary

Kanchan Garai
Tata Institute of Fundamental Research, Hyderabad, India

Kaitlyn P. Gerhart
Department of Chemistry, Georgetown University, Washington, DC, United States

Kris Gevaert
VIB Center for Medical Biotechnology, VIB; Department of Biochemistry, Ghent University, Ghent, Belgium

Stefano Gianni
Istituto Pasteur—Fondazione Cenci Bolognetti, Dipartimento di Scienze Biochimiche 'A. Rossi Fanelli' and Istituto di Biologia e Patologia Molecolari del Consiglio Nazionale delle Ricerche, Sapienza Università di Roma, Rome, Italy

Amy S. Gladfelter
Department of Biology, University of North Carolina at Chapel Hill, Chapel Hill, NC; Marine Biological Laboratory, Woods Hole, MA, United States

Ruth Hendus-Altenburger
Structural Biology and NMR Laboratory, Department of Biology, University of Copenhagen, Copenhagen, Denmark

Vincent J. Hilser
T.C. Jenkins Department of Biophysics; Department of Biology, Johns Hopkins University, Baltimore, MD, United States

Alex S. Holehouse
Department of Biomedical Engineering and Center for Biological Systems Engineering, Washington University in St. Louis, St. Louis, MO, United States

Andrea Holla
Department of Biochemistry, University of Zurich, Zurich, Switzerland

Erik D. Holmstrom
Department of Biochemistry, University of Zurich, Zurich, Switzerland

Natalie Holroyd
King's College London, Department of Chemistry, London, United Kingdom

Jennifer M. Hurley
Department of Biological Sciences; Center for Biotechnology and Interdisciplinary Sciences, Rensselaer Polytechnic Institute, Troy, NY, United States

Madeline M. Keenen
Department of Biochemistry and Biophysics; Tetrad Graduate Program, University of California, San Francisco, San Francisco, CA, United States

Birthe B. Kragelund
Structural Biology and NMR Laboratory, Department of Biology, University of Copenhagen, Copenhagen, Denmark

Erin M. Langdon
Department of Biology, University of North Carolina at Chapel Hill, Chapel Hill, NC, United States

Adam G. Larson
Department of Biochemistry and Biophysics; Tetrad Graduate Program, University of California, San Francisco, San Francisco, CA, United States

Edward A. Lemke
Structural and Computational Biology Unit & Cell Biology and Biophysics Unit, European Molecular Biology Laboratory (EMBL), Heidelberg; Biocenter, Departments of Biology and Chemistry, Pharmacy and Geosciences, Johannes Gutenberg-University Mainz; Institute of Molecular Biology (IMB), Mainz, Germany

Zachary A. Levine
Department of Molecular Biophysics and Biochemistry, Yale University; Department of Pathology, Yale School of Medicine, New Haven, CT, United States

Jing Li
T.C. Jenkins Department of Biophysics, Johns Hopkins University, Baltimore, MD, United States

Sonia Longhi
CNRS and Aix-Marseille Univ, Laboratoire Architecture et Fonction des Macromolecules Biologiques (AFMB), Marseille, France

Anupa Majumdar
Centre for Protein Science, Design and Engineering, Department of Biological Sciences; Department of Chemical Sciences, Indian Institute of Science Education and Research (IISER) Mohali, Punjab, India

Charles J. McDonald
Structural Biology Initiative, CUNY Advanced Science Research Center; Ph.D Program in Biochemistry, The Graduate Center, CUNY, New York, NY, United States

Gabriela Geraldo Mendes
Texas A&M Health Science Center; Texas A&M University, College Station, TX, United States

Steven J. Metallo
Department of Chemistry; Institute for Soft Matter Synthesis and Metrology, Georgetown University, Washington, DC, United States

Lukasz G. Migas
The Michael Barber Centre for Collaborative Mass Spectrometry, The School of Chemistry, Manchester Institute of Biotechnology, The University of Manchester, Manchester, United Kingdom

Andrew D. Miranker
Department of Molecular Biophysics and Biochemistry; Department of Chemical and Environmental Engineering, Yale University, New Haven, CT, United States

Alexander E. Mosier
Department of Biological Sciences, Rensselaer Polytechnic Institute, Troy, NY, United States

Samrat Mukhopadhyay
Centre for Protein Science, Design and Engineering, Department of Biological Sciences; Department of Chemical Sciences, Indian Institute of Science Education and Research (IISER) Mohali, Punjab, India

Geeta J. Narlikar
Department of Biochemistry and Biophysics, University of California, San Francisco, San Francisco, CA, United States

Daniel Nettels
Department of Biochemistry, University of Zurich, Zurich, Switzerland

Rohit V. Pappu
Department of Biomedical Engineering and Center for Biological Systems Engineering, Washington University in St. Louis, St. Louis, MO, United States

Els Pardon
VIB-VUB Center for Structural Biology (CSB), VIB; Structural Biology Brussels (SBB), Vrije Universiteit Brussel (VUB), Brussels, Belgium

Donna L. Pattison
University of Houston, Houston, TX, United States

Jarne Pauwels
VIB Center for Medical Biotechnology, VIB; Department of Biochemistry, Ghent University, Ghent, Belgium

Jacqueline F. Pelham
Department of Biological Sciences, Rensselaer Polytechnic Institute, Troy, NY, United States

Ammon E. Posey
Department of Biomedical Engineering and Center for Biological Systems Engineering, Washington University in St. Louis, St. Louis, MO, United States

Andreas Prestel
Structural Biology and NMR Laboratory, Department of Biology, University of Copenhagen, Copenhagen, Denmark

Elizabeth Rhoades
Department of Chemistry, University of Pennsylvania, Philadelphia, PA, United States

Bankanidhi Sahoo
Tata Institute of Fundamental Research, Hyderabad, India

Benjamin Schuler
Department of Biochemistry; Department of Physics, University of Zurich, Zurich, Switzerland

Sarah L. Shammas
Department of Biochemistry, University of Oxford, Oxford, United Kingdom

Scott A. Showalter
Department of Chemistry; Department of Biochemistry and Molecular Biology, Center for Eukaryotic Gene Regulation, The Pennsylvania State University, University Park, PA, United States

Timir Baran Sil
Tata Institute of Fundamental Research, Hyderabad, India

David Snead
Department of Biochemistry, Weill Cornell Medicine, New York, NY, United States

Wilton T. Snead
Department of Biomedical Engineering, The University of Texas at Austin, Austin, TX, United States

Lasse Staby
Structural Biology and NMR Laboratory, Department of Biology, University of Copenhagen, Copenhagen, Denmark

Jeanne C. Stachowiak
Department of Biomedical Engineering; Institute for Cellular and Molecular Biology, The University of Texas at Austin, Austin, TX, United States

Jan Steyaert
VIB-VUB Center for Structural Biology (CSB), VIB; Structural Biology Brussels (SBB), Vrije Universiteit Brussel (VUB), Brussels, Belgium

Dale Stuchfield
The Michael Barber Centre for Collaborative Mass Spectrometry, The School of Chemistry, Manchester Institute of Biotechnology, The University of Manchester, Manchester, United Kingdom

Piau Siong Tan
Structural and Computational Biology Unit & Cell Biology and Biophysics Unit, European Molecular Biology Laboratory (EMBL), Heidelberg, Germany

Anja Thalhammer
Max-Planck-Institut für Molekulare Pflanzenphysiologie, Potsdam, Germany

Peter Tompa
VIB-VUB Center for Structural Biology (CSB), VIB; Structural Biology Brussels (SBB), Vrije Universiteit Brussel (VUB), Brussels, Belgium; Institute of Enzymology, Research Centre for Natural Sciences of the Hungarian Academy of Sciences, Budapest, Hungary

Francesca Troilo
CNRS and Aix-Marseille Univ, Laboratoire Architecture et Fonction des Macromolecules Biologiques (AFMB), Marseille, France; Istituto Pasteur—Fondazione Cenci Bolognetti, Dipartimento di Scienze Biochimiche 'A. Rossi Fanelli' and Istituto di Biologia e Patologia Molecolari del Consiglio Nazionale delle Ricerche, Sapienza Università di Roma, Rome, Italy

Grace A. Usher
Department of Biochemistry and Molecular Biology, Center for Eukaryotic Gene Regulation, The Pennsylvania State University, University Park, PA, United States

Wouter Van Delm
VIB Nucleomics Core, VIB, Leuven, Belgium

Sławomir Wójcik
Department of Molecular Biophysics and Biochemistry; Department of Chemistry, Yale University, New Haven, CT, United States

Wenwei Zheng
College of Integrative Sciences and Arts, Arizona State University, Mesa, AZ, United States

PREFACE

Historically, our understanding of protein function has been linked to the structure–function paradigm: namely the view that the function of a protein is closely coupled to its folded, globular structure. However, over the past 20 years, this paradigm has been challenged by the identification and characterization of a growing number of proteins which are marked by their absence of stable tertiary structures under physiological conditions. Rather, these intrinsically disordered proteins exist in a dynamic ensemble of interconverting conformations. The physical description of their conformations spans the spectrum from extended, coil-like ensembles to relatively compact ones. Even in the absence stable structure, these ensembles are not entirely random. Many transiently sample secondary structure and/or exhibit conformational biases. Equally broad are the mechanisms by which intrinsically disordered proteins interact with binding partners; while some undergo folding or partial folding upon binding, others remain entirely disordered.

Initially, the idea that intrinsically disordered proteins were functional was met with considerable skepticism. However, there is now a large body of evidence which links intrinsically disordered proteins and intrinsically disordered regions to a broad array of biological functions, including cellular signaling, nuclear transport, cytoskeletal stability, and regulation of transcription and translation. Moreover, bioinformatics analysis of genome data has revealed the abundance of intrinsically disordered protein and regions, with current estimates of significant regions of disorder in 30–40% of eukaryote proteins. Intrinsically disordered proteins have also emerged as key players in a number of devastating human diseases, including many cancers and neurodegenerative disorders, and therefore as important targets for therapeutics. Most recently, intrinsically disordered proteins have been identified as key drivers in the formation of biomolecular condensates, dense dynamic assemblies of proteins and other molecules that are thought to be relevant to a variety of cellular processes and responses. Thus, over the past two decades initial skepticism has been largely overcome and the importance of intrinsically disordered proteins is now firmly established.

While their importance is now widely recognized, intrinsically disordered proteins present challenges to many of the experimental methods developed to study globular proteins. This has spurred methodological development both in established approaches, such as nuclear magnetic resonance, mass spectrometry and computational simulations, as well as in newer methods. To illustrate, single molecule fluorescence spectroscopy, which was in its relative infancy 20 years ago, has grown and reached maturity alongside the intrinsically disordered protein field and has proven itself a valuable tool for their study.

In this volume of Methods in Enzymology I have tried to highlight some of the major experimental and computational approaches to characterizing disordered proteins. Many of the contributions focus on applying a particular technique to a specific intrinsically disordered protein; however, in many cases, the approaches outlined are more generally applicable to the study of other intrinsically disordered proteins, requiring only a few tweaks to be used with the new system. Other articles describe methods used in other fields, such as materials science or membrane biophysics, and which have not yet been broadly adopted for the study of intrinsically disordered proteins.

I thank the scientists who worked so hard to produce the many outstanding contributions to this volume. My hope is that these articles will provide guidance for researchers who are new to the field of intrinsically disordered proteins as well as to motivate established researchers to try new approaches or investigate novel systems. If so, then this volume will contribute to another two decades of growth and discovery in the field of intrinsically disordered proteins.

Elizabeth Rhoades
Department of Chemistry, University of Pennsylvania,
Philadelphia, PA, United States

CHAPTER ONE

Phase Separation of Intrinsically Disordered Proteins

Ammon E. Posey[2], Alex S. Holehouse[2], Rohit V. Pappu[1]
Department of Biomedical Engineering and Center for Biological Systems Engineering, Washington University in St. Louis, St. Louis, MO, United States
[1]Corresponding author: e-mail address: pappu@wustl.edu

Contents

Abstract

There is growing interest in the topic of intracellular phase transitions that lead to the formation of biologically regulated biomolecular condensates. These condensates are membraneless bodies formed by phase separation of key protein and nucleic acid molecules from the cytoplasmic or nucleoplasmic milieus. The drivers of phase separation are referred to as *scaffolds* whereas molecules that preferentially partition into condensates formed by scaffolds are known as *clients*. Recent advances have shown that it is possible to generate physical and functional facsimiles of many biomolecular condensates in vitro. This is achieved by titrating the concentration of key scaffold proteins and solution parameters such as salt concentration, pH, or temperature. The ability

[2] Ammon E. Posey and Alex S. Holehouse contributed equally to this chapter.

Methods in Enzymology, Volume 611
ISSN 0076-6879
https://doi.org/10.1016/bs.mie.2018.09.035

to reproduce phase separation in vitro allows one to compare the relationships between information encoded in the sequences of scaffold proteins and the driving forces for phase separation. Many scaffold proteins include intrinsically disordered regions whereas others are entirely disordered. Our focus is on comparative assessments of phase separation for different scaffold proteins, specifically intrinsically disordered linear multivalent proteins. We highlight the importance of coexistence curves known as binodals for quantifying phase behavior and comparing driving forces for sequence-specific phase separation. We describe the information accessible from full binodals and highlight different methods for—and challenges associated with—mapping binodals. In essence, we provide a wish list for in vitro characterization of phase separation of intrinsically disordered proteins. Fulfillment of this wish list through key advances in experiment, computation, and theory should bring us closer to being able to predict in vitro phase behavior for scaffold proteins and connect this to the functions and features of biomolecular condensates.

1. INTRODUCTION

Intracellular compartments enable the spatial organization of cellular matter and biochemical functions (Diekmann & Pereira-Leal, 2013; Wheeler & Hyman, 2018). Physical barriers formed by membranes delineate compartments such as nuclei, mitochondria, and lysosomes. However, even the earliest ultrastructures of eukaryotic cells indicated the presence of intracellular membraneless entities such as nucleoli and Cajal bodies (Pederson, 2011). Many of these entities are present constitutively while others appear in response to stresses or dynamical events such as the cell cycle. Over the years, membraneless entities have variously been referred to as *protein assemblies*, *cellular bodies*, *membraneless organelles*, and *granules* (Mao, Zhang, & Spector, 2011). Following the pioneering work of Brangwynne et al. (2009) there has been growing recognition that membraneless entities are in fact nonstoichiometric condensates of protein and/or nucleic acid molecules that form either via spontaneous or driven macromolecular phase separation. Consequently, nonstoichiometric membraneless condensates of proteins and/or nucleic acids that are thought to form via intracellular phase transitions are now referred to as *biomolecular condensates* (Banani, Lee, Hyman, & Rosen, 2017; Shin & Brangwynne, 2017).

Biomolecular condensates are multicomponent, nonstoichiometric assemblies that can encompass hundreds, even thousands of distinct types of proteins and nucleic acids (Banani et al., 2017; Hyman, Weber, & Jülicher, 2014; Shin & Brangwynne, 2017). Physical and, in some cases, functional facsimiles of some biomolecular condensates have been reconstituted in vitro (Banani et al., 2016; Elbaum-Garfinkle et al., 2015; Hernandez-Vega et al., 2017; Milovanovic, Wu, Bian, & De Camilli, 2018; Molliex et al., 2015; Patel et al., 2015; Wang et al., 2018; Woodruff et al., 2017; Zhang et al., 2015). So-called *scaffold* proteins and/or nucleic acid molecules are required for the formation of condensates that recapitulate certain physical and functional aspects of their in vivo counterparts (Banani et al., 2017). In contrast, macromolecules that preferentially partition into condensates formed by scaffolds are known as *clients* (Banani et al., 2017, 2016).

Scaffold proteins seem to be identifiable by well-defined sequence/structural features (Wang et al., 2018). These proteins are in fact biological instantiations of *associative polymers* (Rubinstein & Dobrynin, 1997). They are characterized by the multivalency of *stickers*, which are defined as domains or motifs that enable specific, noncovalent interactions known as physical cross-links (Rubinstein & Semenov, 1998; Semenov & Rubinstein, 1998). Associative polymers can either be linear or branched (Fig. 1) and the stickers in both cases are interspersed by *spacers* that appear to be determinants of material properties of condensates (Harmon, Holehouse, Rosen, & Pappu, 2017). Linear multivalent proteins can involve folded domains connected by intrinsically

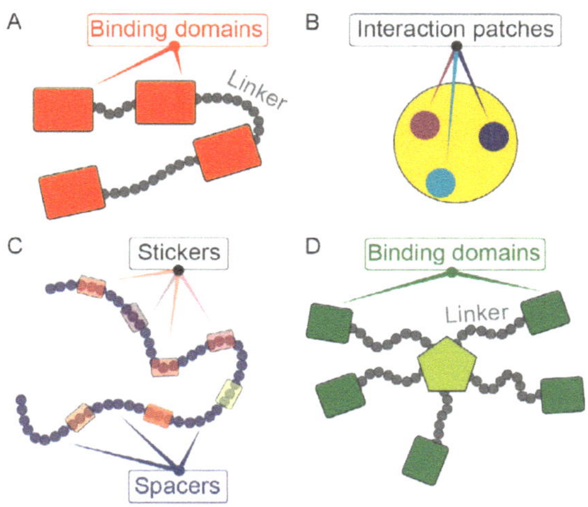

Fig. 1 Different types of biomolecules that can form biomolecular condensates. (A) Biological instantiations of associative polymers that consist of linear multivalent proteins with folded binding domains connected by flexible linkers. (B) Patchy colloids contain distinct patches on the protein surface that act as specific interaction sites. Multivalency arises from the distinct number of interaction patches on the surface. (C) Fully disordered linear multivalent proteins, which are also biological instantiations of linear associative polymers consisting of stickers (motifs) and spacers (linkers). (D) Branched multivalent proteins consisting of a central oligomerization domain that combines multiple folded or disordered domains into a single supramolecular entity.

disordered linkers, or short linear motifs (SLiMs) connected by disordered linkers, or a combination of well-folded domains and SLiMs tethered by disordered linkers. Biological instantiations of associative polymers can also be branched multivalent proteins (Mitrea et al., 2016; Mitrea & Kriwacki, 2016; Mittal, Holehouse, Cohan, & Pappu, 2018) or patchy colloids (Bianchi, Blaak, & Likos, 2011) (Fig. 1). If these molecules lack disordered regions and involve purely folded domains that are either monomeric or oligomeric with defined hot spots that enable specific interactions (Wu & Fuxreiter, 2016), then they are patchy colloids (Ruff, Harmon, & Pappu, 2015). Alternatively, if folded domains have associative polymers tethered to them, then these are instantiations of branched multivalent proteins (Feric et al., 2016; Mitrea et al., 2016; Mitrea & Kriwacki, 2016). There are several examples of linear and branched multivalent proteins that are known to drive the formation of cytoplasmic and nuclear biomolecular condensates (Fig. 2).

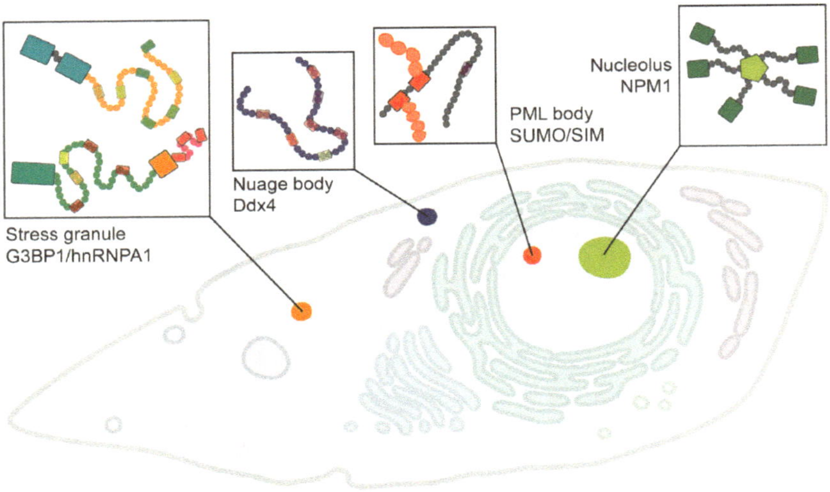

Fig. 2 Specific examples of linear and branched multivalent proteins that drive the formation of distinct cytoplasmic and nuclear biomolecular condensates. In the cytoplasm, proteins G3BP1 and hnRNPA1 (Molliex et al., 2015) serve as scaffolds for stress granules, while Ddx4 is the scaffold for Nuage bodies (Nott et al., 2015). Ddx4 contains globular domains; however, the intrinsically disordered N-terminal domain (depicted here) has been shown to be necessary and sufficient for phase separation. Stress granules and Nuage bodies can also contain RNA, but the scaffold proteins phase separate in vitro in the absence of RNA. In the nucleus, promyelocytic leukemia (PML) protein contains a SUMO interaction motif (SIM) and can also be polySUMOylated, thus providing a means for multivalent interactions that result in PML body formation. The pentameric branched nucleophosmin (NPM1) protein is essential for nucleolus formation (Feric et al., 2016).

Scaffold proteins can undergo distinct types of *phase transitions* (Shin & Brangwynne, 2017). The two types of transitions that are of direct relevance to the formation of condensates are *phase separation* and *gelation* (Harmon et al., 2017). Our focus here is on in vitro phase separation driven primarily by intrinsically disordered linear multivalent scaffold proteins. Of particular interest is phase separation where the end state is the coexistence of two liquid phases, namely, a dilute phase that is deficient in scaffold proteins and a dense phase that is enriched in scaffold proteins. In such a transition, known as liquid–liquid phase separation or LLPS (Shin & Brangwynne, 2017), the relevant *order parameter* is protein concentration (preferred by experimentalists) or protein volume fraction (preferred by theorists) (Brangwynne, Tompa, & Pappu, 2015; Klenin, 1999). Order parameters can be scalars or multicomponent vectors. An order parameter is a quantity with respect to which one can follow the evolution of order vs disorder. The change in order can arise due to a change in symmetry, a change in density, a change in conformation, or some combination thereof (Cheng & Keller, 1998). The key requirement is that the free energy changes for the protein plus solvent system can be written in terms of the order parameter and that the free energy functional is characterized by one or more critical points (Landau, 1936). In vitro experiments are typically performed in closed systems where the total protein concentration is fixed for a given set of solution conditions. Accordingly, the total protein concentration is referred to as a conserved order parameter (Cheng & Keller, 1998).

When the dense phase is a liquid, it implies that the molecular scales are characterized by long-range translational and rotational disorder of the constituent molecules and short-range order that extends out to a few layers defined by the molecular dimensions of the proteins (Shin & Brangwynne, 2017). The dense phase liquids will have characteristic features such as the tendency to round up and form spherical *droplets* as a consequence of the drive to minimize interfacial tension between the dense and dilute phases (Widom, 1988). The effects of minimizing interfacial tension are further manifest in the ability of droplets to fuse with one another such that, at equilibrium, a single large droplet will be thermodynamically favored over multiple small droplets coexisting with one another (Shin & Brangwynne, 2017). Liquid-like organization of proteins within dense phases will have dynamical consequences. This often manifests in the form of a dense phase that behaves like a purely viscous fluid (Taylor et al., 2016), although aging and the onset of viscoelasticity cannot be ruled out. Experiments that probe fluorescence recovery after photobleaching (FRAP) often show that the timescales associated with molecular motions into and out of dense phases as well as within dense phases can be on a par with timescales of typical molecular processes such as protein folding and protein dissociation (Li et al., 2012). Accordingly, protein dynamics probed by FRAP experiments are often used to designate condensates as being liquid-like, although such designations should be treated with caution (Banani et al., 2017).

In the soft-matter physics literature, dense liquid phases formed by associative polymers such as intrinsically disordered linear multivalent proteins are referred to as *network fluids* (Dias, Araújo, & Telo da Gama, 2017). This is because the organization and dynamics of associative polymers within dense liquids are nonrandom, not gas-like, and are governed by physical cross-links among *stickers* (Semenov & Rubinstein, 1998; Tanaka, 2011). These cross-links have characteristic timescales and the extent of cross-linking is facilitated by the properties of *spacers* that modulate the degree of entanglement and overlap among distinct parts of different multivalent proteins (Rubinstein & Semenov, 1998). Therefore, measuring or describing the overall phase behavior of even the simplest associative polymer requires the use of multiple order parameters that go beyond concentration alone (Tanaka, 2011). These parameters include the extent of cross-linking, the stabilities of cross-links, the intrinsic and emergent valencies of stickers, the densities of polymers in distinct phases, and the translational/rotational symmetries of proteins/polymers with respect to one another (Cheng & Keller, 1998). Protein conformational ensembles govern symmetries in the dilute and dense phases and the intra- and intermolecular interactions among folded and disordered regions of multivalent proteins (Harmon et al., 2017; Ruff et al., 2015).

Our focus in this chapter is on the simplest of phase transitions viz., LLPS of associative polymers as a function of changes to either an intensive parameter (temperature or hydrostatic pressure) or the concentration of a ternary component such as salt or ligand. Our narrative will focus on describing phase behavior in terms of *coexistence curves*. We will explain the origins and implications of coexistence curves and describe how comparative assessments of these curves can be used to obtain insights regarding differences in phase behavior of different intrinsically disordered linear multivalent proteins.

2. ANATOMY OF PHASE DIAGRAMS FOR TWO-COMPONENT SYSTEMS

We consider the phase behavior of a simple two-component system comprising of a finite concentration of a specific intrinsically disordered linear multivalent protein dissolved in an aqueous solvent. The free energy of solution ($\Delta\mu_{mix}$) describes the free energy change associated with mixing the protein and solvent in solution by transferring them from their pure liquid forms. The magnitude and sign of $\Delta\mu_{mix}$ are given by the entropy of mixing and by the interplay among protein–solvent, protein–protein, and solvent–solvent interactions. For a polymer of length N, the theories of Flory (1942) and Huggins (1942), which use the device of a lattice model, yield an expression for the mean-field free energy of mixing. Following several simplifying assumptions and approximations, $\Delta\mu_{mix}$ at temperature T may be written in terms of the volume fraction (ϕ) of protein/polymer as (Rubinstein & Colby, 2003):

$$\frac{\Delta\mu_{mix}}{RT} = \left[\frac{\phi}{N}\ln\varphi + (1-\phi)\ln(1-\phi) + \chi\phi(1-\phi)\right] \quad (1)$$

Here, $R = 1.99 \times 10^{-3}$ kcal/mol-K and the mean-field interaction parameter χ quantifies the balance of protein–solvent (u_{PS}), protein–protein (u_{PP}), and solvent–solvent (u_{SS}) interactions such that for a lattice of coordination number z, χ takes the form:

$$\chi = \frac{z}{2}\frac{(2u_{PS} - u_{PP} - u_{SS})}{RT} \quad (2)$$

In a good solvent, protein–solvent interactions are highly favored. Accordingly, χ is negative and the protein will be uniformly dispersed through the solvent for $0 < \phi < 1$ due to a combination of favorable entropies and energies of mixing. For an indifferent solvent, $\chi \approx 0$; the joint effects of protein–protein and solvent–solvent interactions counterbalance protein–solvent interactions. In this scenario, entropy favors mixing for all volume fractions. In a poor solvent, $\chi > 0$ and protein–protein interactions are favored over protein–solvent interactions. Accordingly, depending on the magnitude of χ, there exists a value of ϕ beyond which the well-mixed solution becomes unstable (Fig. 3). Above this volume fraction (or concentration), known as the saturation concentration, the solution separates into two coexisting phases. The dilute phase has a volume fraction ϕ_L and is deficient in protein whereas the dense phase is enriched in protein and has a volume fraction ϕ_H such that $\phi_H > \phi_L$.

In the simplest form of the Flory–Huggins theory, u_{PS}, u_{PP}, and u_{SS} are independent of temperature. Accordingly, since χ is inversely proportional to T, the value of χ will approach zero as T increases and hence ϕ_L and ϕ_H

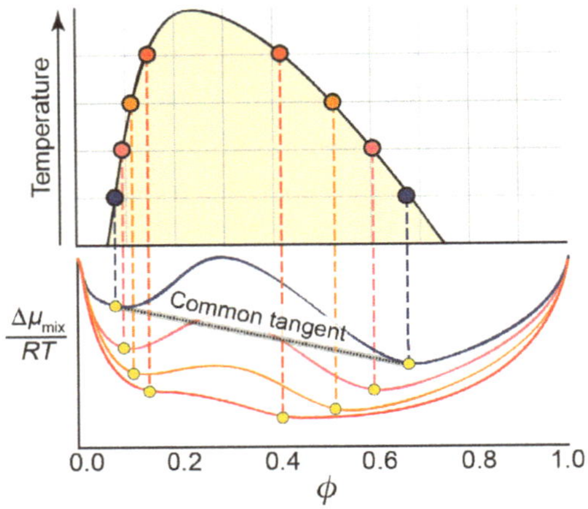

Fig. 3 Mapping coexistence curves from profiles for free energies of mixing. The binodal is constructed by identifying positions at which a common tangent bisects the free energy of mixing curve. The *bottom panel* projects different free energies of mixing curves determined at different temperatures. The *top curve* illustrates how the common tangent intersection points relate to the binodal curve. For clarity, only one common tangent (at the lowest temperature) is shown. These common tangent intersections represent a pair of free energy minima that are accessible to the system should the total concentration lie between them. In this way, the free energy of mixing can be used to construct a phase diagram.

will, respectively, increase and decrease with increasing temperature thus narrowing the gap between ϕ_L and ϕ_H. Above a temperature known as the upper critical solution temperature (UCST) and designated as T_c, phase separation does not occur, regardless of the protein concentration. The critical point, defined by a critical temperature and a critical concentration (ϕ_c), is further identifiable as the point where $(\phi_H - \phi_L) = 0$.

The envelope of temperature dependent ϕ_L and ϕ_H values represents the *coexistence curve* or *binodal* (Fig. 4). For combinations of ϕ and T that lie within (below) the binodal, the protein plus solvent system is in the two-phase regime. In contrast, for any combination of ϕ and T outside (above) the binodal, the system is in the homogeneous (well mixed) one-phase regime where the protein of interest is evenly dispersed in solution. At low protein concentrations, the one-phase regime corresponds to homogenous dilute solutions while at high protein concentrations the one-phase regime will be a viscous, protein-rich solution that can be compared to so-called polymer melts (Rubinstein & Colby, 2003).

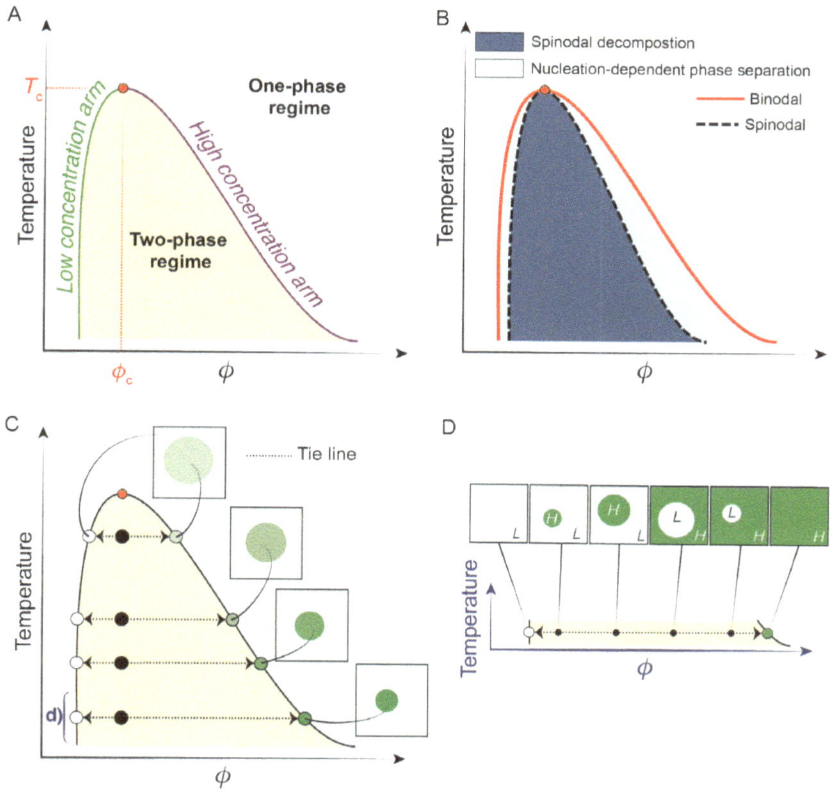

Fig. 4 Phase diagram schematics. (A) Sketch of a binodal for a two-component system drawn in the temperature and volume fraction (ϕ) plane. The binodal is divided into low- and high-concentration arms. Solutions with a temperature and total concentration that fall within the two-phase regime will phase separate, while those that fall outside will remain homogenously well mixed. (B) Binodal and spinodal curves shown for the same system. For the region between the two curves a fully mixed solution can be metastable and phase separation will proceed via a nucleation-dependent mechanism. In contrast, for the region below the spinodal, phase separation will proceed mainly via a diffusion-limited manner (spinodal decomposition). (C) Binodal with a subset of tie lines drawn. The dense phase and dilute phase concentrations obtained depend on the temperature, but the horizontal tie lines illustrate the fact that the temperatures within the dense and dilute phases are the same. Further, at a fixed temperature, the concentrations within the dense and dilute phases will also be fixed. However, the extent to which proteins partition into the dense and dilute phase of fixed concentrations will be governed by the lever rule (see text). (D) The concentration invariance of the dense and dilute phases is further demonstrated for one specific tie line. As the total concentration increases, the fraction of the system occupied by the dense phase increases and the fraction occupied by the dilute phase decreases, but the concentrations within those two phases remain fixed. The dense and dilute phase concentrations will always be equal to the concentration at which the tie line intersects the high and low arms of the binodal, respectively.

2.1 Saturation Concentrations

Coexistence curves feature both low- and high-concentration arms (Fig. 4A). In a space defined by ϕ and T, the low-concentration arm quantifies the temperature dependence of the *saturation concentration* for the system of interest. A solution at a given temperature cannot support more soluble protein than the saturation concentration. If the total protein concentration exceeds the saturation concentration the excess is excluded from the soluble phase by forming a physically distinct second phase, such as a liquid droplet or solid aggregate (Fig. 4C and D). Accordingly, saturation concentrations are temperature and sequence-specific thresholds that are variously designated as c_s (saturation) or c_L (low) in molar units or ϕ_s or ϕ_L in terms of volume fractions.

In the case of LLPS, the dense, protein-rich phase is a liquid as characterized by the fact these phases show liquid-like properties such as wetting, dripping, and fusing of spherical droplets (Shin & Brangwynne, 2017). For proteins that are defined by strong specific/nonspecific interactions, the dense phases may be solids that are amorphous, semicrystalline, or paracrystalline, or they may be precipitates that form via nonspecific interactions (Banani et al., 2017; Posey & Pappu, 2018). Often, solid or semisolid assemblies are collectively described as aggregates. Aggregation is in fact phase separation and many of the same ideas discussed in the phase separation field, including the concept of saturation concentrations are germane to protein aggregation as well (Ciryam, Tartaglia, Morimoto, Dobson, & Vendruscolo, 2013; Crick, Ruff, Garai, Frieden, & Pappu, 2013; Pappu, Wang, Vitalis, & Crick, 2008; Posey & Pappu, 2018). The saturation concentration for LLPS depends on the set of solution conditions under which it is measured. For example, as depicted in Fig. 4, phase separation for a system showing UCST behavior will occur more readily at low temperature than at higher temperatures. Accordingly, the saturation concentration at low temperatures will be lower than the saturation concentration at higher temperature.

How might one compare sequence-specific driving forces for phase separation? We prefer to avoid the use of bioinformatics-based heuristics that attempt to predict or differentiate sequence-specific phase separation propensities (Ryan & Kinsella, 2017). Such measures only have value if one can be sure that the contributions of solution conditions are factored into the estimates. Even then, it is imperative that the comparisons between different sequences be made for identical solution conditions. Otherwise, they lack any meaning or thermodynamic relevance.

One way to compare driving forces for phase separation is to compare saturation concentrations. We shall consider two temperatures T_1 and T_2 such that $T_2 > T_1$ for a protein plus solvent system that has UCST phase behavior. We shall designate the saturation concentrations, in molar units, as c_{s1} and c_{s2}, respectively. The driving force for phase separation is quantified in terms of chemical potentials, which are partial molar free energies. Based on the standard definition of chemical potentials, it follows that $\mu_s(T_1)$ and $\mu_s(T_2)$ are proportional to $-RT_1\ln(c_{s1})$ and $-RT_1\ln(c_{s2})$, respectively. The difference in driving forces for phase separation at T_2 vs T_1 may be written as $\Delta\mu = \mu_s(T_2) - \mu_s(T_1)$ and this will take the form:

$$\Delta\mu = -R\left\{ \ln\left[\frac{(c_{s2})^{T_2}}{(c_{s1})^{T_1}}\right]\right\} \tag{3}$$

Eq. (3) shows that comparative assessments of driving forces for phase separation at two different temperatures, for the same protein plus solvent system, requires the joint consideration of saturation concentrations and temperatures, i.e., changes to solution conditions.

2.2 Tie Lines and High-Concentration Arm of a Binodal

The high-concentration arm of a binodal quantifies the protein concentration within the dense phase. Despite growing interest in systems that drive phase separation, there are, to our knowledge, very few known instances (Brady et al., 2017; Simon, Carroll, Rubinstein, Chilkoti, & Lopez, 2017; Wei et al., 2017) of full binodals for intrinsically disordered linear multivalent proteins. This is because it is challenging to measure the high-concentration arms of binodals. Phase separation results from the coexistence of two distinct phases through equalization of chemical potentials (Rubinstein & Colby, 2003). Accordingly, for a binary solution defined by a combination of fixed ϕ and T that falls below the binodal, the concentrations in the dilute and the dense phases viz., ϕ_L and ϕ_H are fixed with respect to the total protein concentration ϕ (Fig. 4C and D). The value of ϕ_L is fixed for a fixed temperature because more protein cannot be forced into the saturated soluble phase, although such a supersaturated state may be metastable or kinetically stable (Ciryam et al., 2013; Khan et al., 2018). The addition of more protein to the system leads to the excess protein being taken up by the dense phase such that the *total volume* of the dense phase increases. Accordingly, just as the concentration of protein in the dilute phase is fixed at the saturation concentration, the concentration of protein in the dense phase is fixed at ϕ_H.

Hence, as the total protein concentration increases beyond ϕ_L, the *volume* of the dense phase changes to accommodate the extra protein (Fig. 4D). This is manifest as a growth in the sizes of dense phase droplets. However, this growth in volume is not accompanied by a growth in protein concentrations within droplets, which stays fixed. Accordingly, ϕ_H may be thought of as a saturation concentration of the dense phase.

The two-phase regime is also characterized by a series of *tie lines* (Fig. 4C). For an initial value of ϕ and T (the quench depth) that places a system below its binodal, a tie line passes through the point of interest and connects it to points on the low- and high-concentration arms of the binodal. The slope and length of a tie line are informative parameters. In the (ϕ, T)-plane, tie lines will be horizontal because the temperature within the dense and dilute phases should be identical. However, tie lines need not be horizontal if the ordinate corresponds to the pH or salt concentration. Tie lines represent information regarding the relative volumes and compositions of the two phases (Rubinstein & Colby, 2003). Accordingly, if the ordinate measures salt concentration, and the slope of the tie line is positive then the salt concentration will be higher inside the droplet than in the dilute phase. The converse will be true for a negatively sloping tie line. Similar inferences apply if the control parameter is the overall pH of the solution.

The relative volumes of the dilute vs dense phase are determined by the total concentration of the system. Relative lengths of tie lines help quantify these relative volumes. At a fixed temperature T, if the total protein concentration is denoted as ϕ, then the ratio r defined in Eq. (4) will quantify the relative system volumes of the dilute vs dense phases:

$$r = \left(\frac{\phi - \phi_L}{\phi_H - \phi} \right) \tag{4}$$

If the total concentration is half way between the low-concentration arm and the high-concentration arm (the inversion transition point) the relative volumes of dense and dilute phases will be equal. Thus, at total protein concentrations below this point, dense spherical droplets will be surrounded by a dilute solution, and at total concentrations above this point dilute droplets will be encased in a surrounding dense phase—a phenomenon referred to as vacuolization.

2.3 Comparing Sequence-Specific Binodals to One Another

How might one compare driving forces for phase separation of different sequences? This requires the measurement of saturation concentrations for different sequences under identical conditions. In such a scenario, one can use the approach prescribed in Eq. (3). Comparisons become ill-posed problems because differences in phase behavior might arise due to a variety of differences in the sequence-encoded interactions, such that the differences in phase behavior cannot be linked to a particular sequence-encoded driving force. Specifically, the differences in saturation concentrations might be due to different amino acid compositions or differences in sequence patterning that contribute to a spectrum of protein–protein and protein–solvent interactions. A comparison of full binodals to one another enables the linking of specific differences in phase behavior to specific sequence-encoded driving forces because different parts of the binodal may well be influenced by specific properties encoded by sequence, as described later.

Fig. 5A shows hypothetical binodals for three sequences. Here, the binodals are shown as nested curves. One can quantify differences in sequence-specific critical points. Additionally, one can rightly infer that: (a) higher concentrations are required to observe the onset of phase separation for sequences with narrower binodals and lower critical temperatures; (b) concentrations or temperatures that support phase separation for sequence A will not support phase separation for sequence B if the ϕ and T values are above the binodal for sequence B while being below the binodal for sequence A; (c) the dense phases realized for sequences corresponding to the outer binodal will be more concentrated than the dense phases for sequences corresponding to the inner binodals.

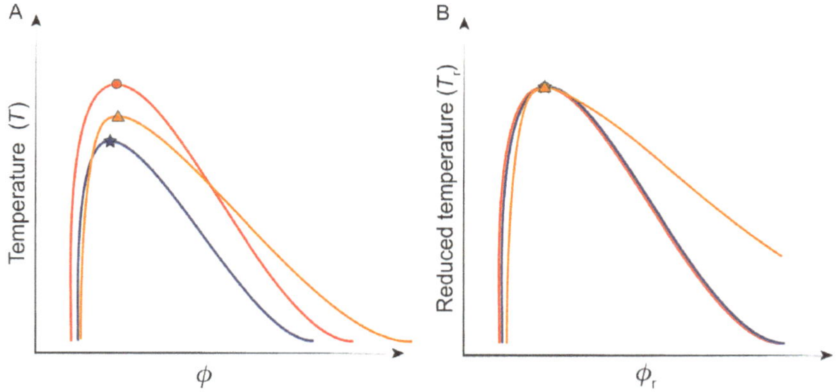

Fig. 5 Rescaled binodals in terms of reduced variables allow for an effective comparison of phase behavior. (A) Three different systems that demonstrate distinct binodal curves. (B) Rescaling of binodals into reduced units demonstrates the *blue* and *red curve* are almost entirely overlapping, implying similar underlying interactions albeit with different strengths. In contrast, the *orange curve* does not overlap, suggesting differences in the underlying interactions.

One can also construct rescaled binodals in terms of reduced concentrations ϕ_r and reduced temperatures T_r (Fig. 5B). For each sequence, if we know the critical points, then one way to define reduced parameters would be as follows: $\phi_r = (\phi/\phi_c)$ and $T_r = (T/T_c)$. If the only difference between a pair of sequences is the valence of stickers, then the rescaled binodals will coincide with one another. This would imply that the chemistries of stickers and spacers are essentially similar if not identical for the two sequences. In direct analogy with the principle of corresponding states that is well established in the field of liquid–vapor phase equilibria (Guggenheim, 1945), the overlap of rescaled binodals drawn in terms of ϕ_r and T_r for a pair of sequences arises from what we shall refer to as the *principle of corresponding sequences*. A simple rescaling of binodals does not have to yield coincident curves. Deviations between pairs of simply rescaled binodals can come in distinct flavors, and these will have distinct origins. For example, the widths of the two-phase regimes could be different from one another. This will imply that the chemistries of stickers and/or spacers are different from one another. Alternatively, the low-concentration arms of rescaled binodals might coincide with one another, but the high-concentration arms might deviate. This would imply that the spacers enable alterations of concentrations within the dense phase either by modulating the extent of cross-linking or by altering the effective valence of stickers. Details of how these inferences can be drawn will be presented elsewhere. If rescaling the binodals by using reduced parameters for pairs of sequences deviate from one another, then we would need more elaborate reduced variables to find a way to rescale the binodals and get them to coincide. The formal procedure one adopts to obtain coincident rescaled binodals will provide insights regarding the differences in sequence-encoded driving forces for phase separation.

The preceding discussion highlights the fact that both raw binodals and rescaled ones enable rigorous comparative assessments of sequence-specific phase behaviors. They also enable the use of other order parameters, providing such assessments are made for equivalent *quench depths* into the two-phase regime. This refers to the joint values of ϕ_r and T_r that lie under the rescaled binodal thereby defining the distances from the left and right arms.

2.4 Coexistence Curves Encompass Instability Curves As Well

For all combinations of ϕ and T that place the system below the binodal, there exists an envelope of points where the second derivative of $\Delta\mu_{\text{mix}}$ with respect to ϕ is zero. This envelope of points is known as the *instability curve* or the *spinodal* (Rubinstein & Colby, 2003). For points that lie between the binodal and the spinodal the well mixed dilute or dense phases are supersaturated although they can be long-lived or metastable (Khan et al., 2018) (Fig. 4B). LLPS in this regime typically proceeds via nucleation whereby a nucleus of the new dense phase forms within the old dispersed dilute phase or nucleation of a dilute phase occurs within a homogenous dense phase. For combinations of ϕ and T that lie below the spinodal, the homogeneous phase is always unstable and phase separation proceeds in a thermodynamically downhill fashion. The dynamics of phase separation can become quite esoteric in this regime. Instead of an organized formation of spherical droplets, there is the possibility of forming worm-like structures that are governed by thermodynamic considerations such as the quench depth, and dynamical considerations such as the motions of chains around one another (Berry, Brangwynne, & Haataja, 2018; Bray, 1994).

3. MAPPING BINODALS USING EXPERIMENTS

The mapping of binodals requires the measurement of points along the low- and high-concentration arms under solution conditions that are identical except for the two parameters of interest (e.g., protein concentration and temperature). This involves measuring protein concentrations in the dense and dilute phases at many distinct temperatures. Alternatively, one can determine the temperature at which a sample of known total concentration shows signs of phase separation. Either of these approaches provide a set of points that will fall on the coexistence curve, allowing a curve connecting the points to be drawn directly. Often, experimental limitations prevent measurement of sufficient points to achieve complete coverage of the coexistence curve. To obtain insights regarding the coexistence curve in regions where data cannot be collected requires some type of theoretical treatment of the data. Flory–Huggins models have been used, as have more advanced approaches, and the theoretical fitting of phase diagrams remains an active area of development (Brady et al., 2017; Simon et al., 2017; Wei et al., 2017).

3.1 The Low-Concentration Arm

Points along the low-concentration arm can be measured in a number of ways (Wang et al., 2018). One method involves centrifugation of a phase-separated sample followed by careful removal of supernatant (dilute phase) for subsequent concentration determination by spectrophotometric or colorimetric assay. The phase separating protein must contain tryptophan if its concentration is to be accurately measured using spectrophotometry. The dilute phase must be at a detectable concentration. This is an important constraint because colorimetry is typically more limited in detection range than spectrophotometry.

An alternative approach for mapping binodals involves systematic titrations of protein concentration until the onset of turbidity (Riback et al., 2017; Simon et al., 2017). A sharp increase in turbidity reflects the appearance of large droplets that scatter light. One can augment turbidity measurements using dynamic light scattering as well (Riback et al., 2017). Direct image analysis is also an option. For fluorescently tagged proteins the onset of phase separation should be accompanied by the formation of bright foci, thus providing a clear marker for dense phases that are easily identified via standard image processing approaches.

For a system showing UCST behavior one can prepare solutions of various known total concentrations of protein at a temperature that is greater than T_c and then lower the temperature for each fixed-concentration sample until the observation of cloudiness or turbidity (Molliex et al., 2015; Petsev, Wu, Galkin, & Vekilov, 2003). Even though the dilute phase is not isolated for direct concentration measurement as in the previously described centrifugation method, this turbidity approach is still reasonable because it leverages the fact that, at total concentrations near the binodal, the concentration of the most abundant phase will be similar to the total concentration. Because of this, it can also be used to measure the high-concentration arm, assuming samples of high enough total concentration can be obtained. Turbidity measurements require large sample volumes to facilitate cuvette-based measurements. In this context, a microfluidic approach requiring much smaller sample volumes seems promising, at least for quantifying temperature-dependent phase separation (Holden & Cremer, 2005; Mao et al., 2004; Mao, Yang, & Cremer, 2002; Simon et al., 2017). Unfortunately, tie line information cannot be obtained using this method. However, for a protein plus solvent with thermoresponsive phase behavior the tie lines will be horizontal, so this becomes less of an issue.

3.2 The High-Concentration Arm

Direct measurements of high-concentration arms are significantly more challenging. The most conceptually simple approach is to generate a system-spanning dense liquid phase (as opposed to a precipitate or some other solid phase), extract a sample of pure dense phase from that droplet, and then measure the absolute concentration after dilution into some condition that favors a single phase, such as denaturant (Brady et al., 2017). The dense phase concentration can then be calculated from the concentration measured in this solution since the amount of dilution is known. While such an approach is appealing, it depends on the accurate extraction of pure dense phase and assumes that the artificially prepared dense phases are exact facsimiles of the dense liquid phase. Consequently, the reproducibility of such measurements can be challenging, although this may be the most straightforward approach other than the turbidity method described earlier. Too often, however, the generation of a suitable dense phase requires the use of a pelleting procedure and this invariably invites the concern that the system is crossing some form of phase boundary into a solid or semisolid phase. The concerns stem from the realization that the underlying structure, density, and organization of molecules are likely to be fundamentally different for network liquids vs solid phases.

Rather than dilution, one could directly measure fluorescence intensity from within the droplet using fluorescently tagged proteins and then compare against a standard/calibration curve. For this to work, one needs to attach fluorescent tags to the intrinsically disordered multivalent protein of interest and these tags can be (relatively large) fluorescent proteins or hydrophobic/aromatic fluorophores that may or may not impact the phase behavior. Moreover, the impact of the droplet environment on fluorescence quantum yields should not necessarily be expected to lend itself to simple comparison with a standard curve in which concentration and intensity are linearly extrapolated. Nevertheless, while it is possible that absolute values may be offset, relative trends obtained via this method should, in principle, be robust. A technically demanding but relatively effective approach is to include a small fraction of chemically labeled protein such that droplets are easily visible, and then perform confocal imaging through/across a sample volume (Wei et al., 2017). The total concentration of protein in solution (c_{total}) can be measured accurately. The dilute phase concentration (c_{L}) can be accurately determined via methods described earlier. Finally, the droplet volumes can be extracted from 3D imaging, thus yielding ϕ_{H}. With this information in hand, and given that $\phi_{\text{L}} = (1 - \phi_{\text{L}})$ because ϕ is a conserved order parameter, one can use the lever rule to determine the dilute phase volume fraction because thus allowing us to estimate c_{H} using:

$$c_{\text{total}} = (\phi_{\text{L}} c_{\text{L}} + \phi_{\text{H}} \phi_{\text{H}}) \tag{5}$$

Beyond these approaches, novel experimental methods have been developed for direct measurement of concentrations within droplets. Fluorescence correlation spectroscopy (FCS) can be used for accurate measurements of protein concentration. However, this invariably requires a standard or calibration curve to extract the actual values. In recent work, Wei et al. implemented a novel ultrafast scanning FCS, in which a 3D confocal volume is known precisely, allowing calibration free measurement of protein concentration to be performed (Wei et al., 2017). This methodology, which should be generalizable to other forms of scanning FCS methods, provides a comprehensive and accurate approach for measuring concentrations in vitro and in vivo.

The dynamics of phase separation can also contribute directly or indirectly to the approaches deployed for mapping system-specific binodals. For example, accurate determination of coexistence curves for slowly equilibrating systems can be a major challenge. These systems introduce limitations on the types of techniques that can be readily applied. For example, rather than preparing a sample, immediately performing a measurement, and then diluting/concentrating the original sample to create the next sample for measurement, one would need to prepare many separate samples and allow them to incubate for the necessary number of hours before making any measurements. In addition to more material, this requires a more extensive sampling of concentration/solution condition space and takes considerably longer.

3.3 Full Binodals Are the Desired Goal, but They Are Difficult to Obtain

It is a tautology that full binodals are important for drawing comparative inferences regarding the details of phase separation. Along with information regarding ϕ_c and T_c one can obtain a complete thermodynamic and comparative description of phase behavior, thereby uncovering the sources of sequence-specific driving forces for phase separation. However, measuring full binodals, specifically the high-concentration arms of coexistence curves remains challenging. Most methods require modifications to the system such as pelleting and indirect inferences via the use of standard curves that are questionable because they impose uncontrolled assumptions regarding the dense phases. Direct methods are either labor intensive thus hindering throughput or they are technically challenging thus limiting their broad applicability.

Clearly, the key challenge to overcome with respect to obtaining comparative assessments of sequence-specific phase behavior is the construction of sequence-specific binodals. A blend of experimental, computational, and theoretical methods might be the way forward. In this blended approach, experiments can be designed to estimate the temperature (or salt/pH) dependent second and third virial coefficients (Bloustine, Virmani, Thurston, & Fraden, 2006; Petsev et al., 2003; Wei et al., 2017). In addition, techniques borrowed from the polymer physics field should be adaptable to estimate ϕ_c and T_c and even probe the vicinity of critical points for different systems (Wu & Li, 2018). Measurements of conformational properties such as average sizes and shapes of protein molecules within dilute and perhaps even dense phases will yield insights regarding the interplay between the effects of fluctuations vs mean-field attributes, thus enabling the utilization of suitable theories (Wei et al., 2017). Alternatively, one can incorporate experimentally derived information into coarse-grained, sequence-specific models to directly calculate coexistence curves from computer simulations. These approaches are being pursued and it is likely that accurate and high-throughput comparative assessments of binodals will be forthcoming in the near future.

4. IMPACT OF CHANGES TO SOLUTION CONDITIONS

Various solution components have been shown to influence the phase behavior of intrinsically disordered linear multivalent proteins. In particular, titrations of salt and the organic compound 1,6-hexanediol have been used to modulate phase behavior and draw inferences regarding the driving forces for phase separation. The typical route is to compare droplet formation by varying solution conditions or mutations without varying protein concentration. This makes it difficult to interpret how changes to droplet formation relate to phase behavior (Fig. 6).

4.1 Impact of Solution Ions

Numerous studies have shown that the concentrations and identities of solution ions can play an important role in modulating phase behavior. Salt sensitivity is frequently used as being suggestive of an electrostatic component to the driving forces for phase separation. While this is certainly reasonable, monovalent and divalent ions can have a range of effects on polypeptides that go beyond screening attractive electrostatic interactions. For example,

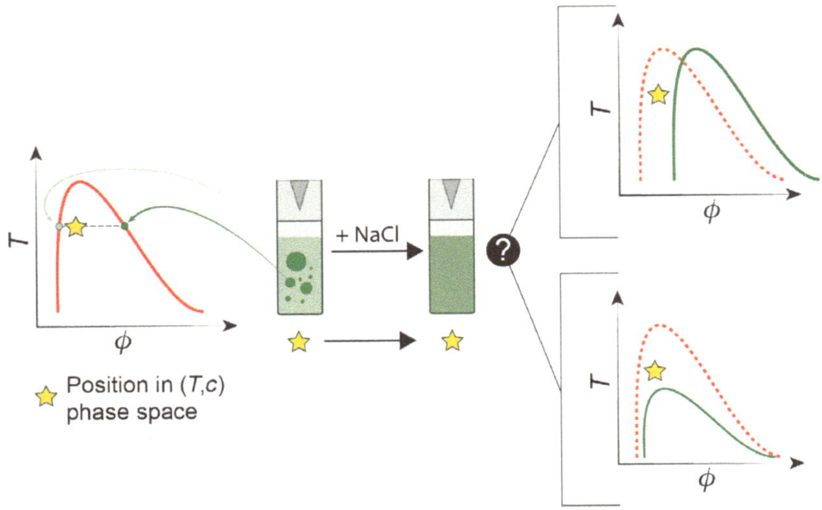

Fig. 6 Measuring phase behavior under a single solution conditions does not necessarily provide insight into how a phase diagram has changed. For a solution of fixed total concentration at a given temperature and (*c*,*T*), interpreting measurements of phase behavior obtained when a third variable is changed (e.g., NaCl concentration) is challenging. The absence of droplets could reflect a systematic shift of the binodal to the right without changing the critical temperature; it could reflect a drop in the critical temperature with minimal impact on the low concentration arms at lower temperature. The former implies phase separation will be obtained if the total concentration is increased, while the latter suggests phase separation will not occur at this temperature. For systems that are well described by mean field theories for long homopolymers in a two-component system, the location of the critical point, and width of the two-phase regime are strongly coupled. However, in complex, multicomponent systems comprising of finite-sized heteropolymers, the heterogeneities of interactions, and the presence of conformational fluctuations can result in a decoupling among the location of the critical point, and the locations of the low- and high-concentration arms of binodals.

in complex coacervation—phase separation of oppositely charged polymers—phase transitions are driven by the entropic release of counterions whereby polymer–polymer interactions replace polymer–ion interactions (Burgess, 1990; Chang et al., 2017; Michaeli, Overbeek, & Voorn, 1957; Sing, 2017; Tainaka, 1980; Tainaka & Yomosa, 1980). This would lead to a nonmonotonic dependence of phase behavior on salt concentration, in which phase separation is suppressed at very low salt concentrations. The chemical potentials of anions and cations must be equalized across the phase boundary. This can lead to nontrivial effects such as preferential partitioning of anions vs cations into dense vs dilute phases. These preferential partitioning effects combined with the modulation of electrostatic interactions by salts can impact the interfacial tensions at phase boundaries. The totality of these effects is rather complex and cannot be reduced to purely *ionic strength* effects, which are restricted to the Debye–Hückel or other mean-field descriptions of electrostatic attractions (Record, Anderson, & Lohman, 1978). To complicate matters further, there are well-defined examples in which phase separation shows a pronounced dependence on monovalent salts despite the absence of any charged residues as well as examples in which salt reduces or enhances phase separation in systems for which electrostatics are not expected to be the main driving force for phase separation (Brady et al., 2017; Burke, Janke, Rhine, & Fawzi, 2015; Molliex et al., 2015; Nott et al., 2015). The identities of anions and cations can also impact phase separation (Boeynaems et al., 2017; Bogaert et al., 2018). Similarly, organic salts may also impart complex effects that are not yet fully understood (Patel et al., 2017). This suggests that the specific chemistries of distinct cations and anions can enhance or suppress peptide–solvent interactions, likely in a sequence-specific manner. As an example, a proline–arginine dipeptide repeat (PR20) undergoes salt-dependent complexation in the presence of 30% polyethylene glycol, with a strong dependence on the anion ($HPO_4 > CO_3 > Cl$) (Boeynaems et al., 2017).

4.2 Impacts of 1,6-Hexanediol

Following the discovery that aliphatic alcohols can dissociate nuclear transport receptors from hydrophobic resins and disrupt the permeability barrier of the nuclear pore complex, 1,6-hexanediol (1,6-HD) has been used extensively to modify phase separation and hydrogel formation in vitro and in vivo (Boeynaems et al., 2017; Dao et al., 2018; Jain et al., 2016; Kroschwald, Maharana, & Alberti, 2017; Kroschwald et al., 2015; Larson et al., 2017; Lin et al., 2016; Peskett et al., 2018; Rog, Kohler, & Dernburg, 2017; Sabari et al., 2018; Schmidt & Gorlich, 2015; Strom et al., 2017; Updike, Hachey, Kreher, & Strome, 2011; Wegmann et al., 2018). Despite its frequency of use, systematic studies of the impact 1,6-HD on phase separation are lacking. While many different proteins that undergo condensate formation show dependence on 1,6-HD, many others do not (Molliex et al., 2015; Peskett et al., 2018; Rog et al., 2017).

In early work on biological hydrogels formed by FG Nups it was noted that the efficacy of aliphatic alcohols on the disruption of the nuclear pore permeability barrier was approximately proportional to their hydrophobicity (Ribbeck & Gorlich, 2002). This result was rationalized by the hydrophobic barrier model, in which the nuclear pore lumen is treated as a hydrophobic sieve. This in turn has led to the oft-stated mechanism of action that 1,6-HD disrupts hydrophobic interactions. However, the breadth of systems that 1,6-HD disrupts is inconsistent with an effect that is mediated solely by hydrophobicity (Kroschwald et al., 2017). Recently, Rog et al. titrated the number of–OH groups and the length of the hydrocarbon chain and found that 1,6-HD was neither the most nor least effective aliphatic alcohol in disrupting the synaptonemal complex (Rog et al., 2017). Interestingly, 1,2-pentanediol and 1,2-hexanediol were more effective than 1,6-HD, suggesting that the mechanism of action is more complex than hydrophobicity alone. These observations suggest the need for an orthogonal titration of small compounds with distinct chemistries that might help elucidate the types of interactions mediating condensate formation, based on their partitioning and/or influence on binodals.

4.3 pH and Temperature

Changes to pH and temperature can have complex effects on biological macromolecules. At low temperatures enthalpic forces can dominate, while at high temperatures entropic effects take precedent. Extremes of pH can modulate the local sequence characteristics of proteins through protonation or deprotonation of labile chemical moieties (Franzmann et al., 2018). Importantly, the local chemical environment can profoundly influence the intrinsic pK_a values of ionizable residues with downshifts/upshifts leading to the potential for novel pH sensing (Franzmann et al., 2018). It is noteworthy that changes to pH and temperature are commonly associated with cellular stresses that lead to the formation of intracellular protein and RNA condensates known as stress granules. While heat-stress has been studied extensively, a growing consensus appears to implicate changes to cytosolic pH as a general stress response that drives condensate formation through a variety of different protein-based sensors (Franzmann et al., 2018; Kroschwald et al., 2015, 2018; Munder et al., 2016; Riback et al., 2017). It remains unclear if pH and temperature are intrinsically coupled or if they are sensed via distinct but related mechanisms. Mapping binodals in a three-parameter space comprising of concentration, pH, and temperature should provide insights into the intrinsic and evolved relationships among the formation of biological condensates and the combination of pH and temperature (Fig. 7). Insights from such studies are likely to be of direct relevance to naturally occurring biological systems.

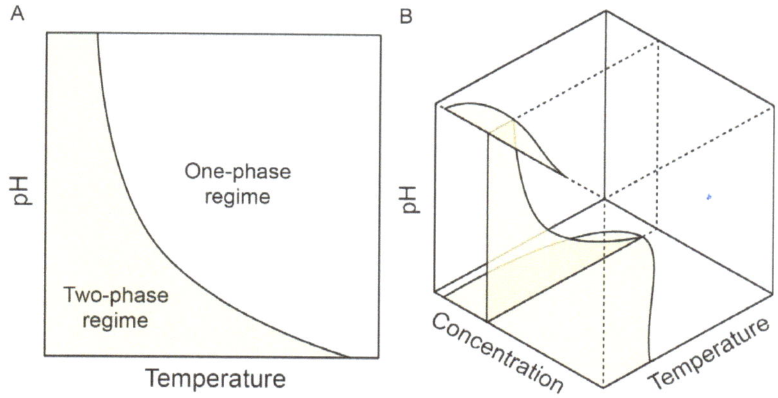

Fig. 7 Example of binodals in pH/temperature space. (A) At a fixed total protein concentration, the temperature of phase separation can be measured as a function of pH (or vice versa) to identify a phase boundary in pH/temperature space. (B) In reality, the projection onto the pH/temperature plane for a fixed protein concentration is a cross-section or slice of the three-dimensional phase boundary that represents a generalization of the two-dimensional representations discussed in Figs. 3–6.

Finally, while we have focused here on phase diagrams that display UCST behavior, systems comprising of intrinsically disordered proteins can also exhibit lower critical solution temperature behavior (Dao et al., 2018; Quiroz & Chilkoti, 2015). For such systems, phase separation occurs if the temperature exceeds T_c and the underlying concepts have been reviewed recently (Ruff, Roberts, Chilkoti, & Pappu, 2018). Here, phase separation is an entropically driven process, in which the release of solvating water drives the relevant transition. Despite the inverted temperature dependence, the concepts discussed here are all directly applicable to such systems, albeit with added nuances.

5. SUMMARY AND CONCLUSIONS

In this chapter, we have focused on a discussion of concepts and techniques that are directly relevant for describing and comparing sequence-specific phase separation of intrinsically disordered multivalent proteins in aqueous solvents. One of the key takeaways is that there are numerous complexities associated with understanding liquid–liquid phase separation even in seemingly simple, two-component systems that comprise of just protein and solvent. Clearly, binodals are important and central for obtaining comparative assessments of the sequence-encoded differences in the driving forces for phase separation. The availability of binodals, with knowledge of critical points, will provide a near complete thermodynamic description of phase behavior in terms of the most directly relevant and physiologically controllable order parameter, ϕ. Unfortunately, despite the massive growth of interest in the topic of intracellular phase separation, serious challenges remain with regard to obtaining full binodals even for the simplest systems. This is sobering when we realize that the condensates of biological interest and relevance are truly multicomponent systems comprising of hundreds—if not thousands—of distinct types of protein and/or nucleic acid molecules. Such systems will be characterized by multiple critical points (Koningsveld, Kleintjens, & Nies, 1987), reentrant phase transitions (Bernardino & da Gama, 2012; Zhang et al., 2008), and phase boundaries that need to be mapped and interpreted in higher dimensional spaces than the simple examples considered in this chapter. The parsing of proteins (and RNA molecules) into scaffolds vs clients (Banani et al., 2017) might be a useful strategy for understanding the phase behavior of multicomponent systems. If it turns out that most biological condensates require only a small number of proteins or nucleic acids to be scaffolds, then one might be able to pursue systematic reconstitutions of functional facsimiles of condensates in vitro and connect the observations of these facsimiles to their biologically relevant counterparts. Additionally, the methods developed in the synthetic polymer literature might prove to be invaluable in efforts to obtain comparative assessments of sequence-specific phase behavior.

Two topics of crucial relevance pertain to the dynamics of phase separation and the interplay between spontaneous vs driven (or passive vs active) processes as determinants of phase separation (Berry et al., 2018). We deliberately ignored these topics because we wished to highlight the importance of thermodynamic descriptions and focus on binodals as descriptors of phase behavior. This in no way diminishes the importance of topics that have been glossed over here. Indeed, the topic of the dynamics of phase separation raises many important and puzzling questions. Many phase-separating systems undergo phase separation rapidly, on the order of minutes, seconds, or faster. Some, however, may phase separate on a timescale of hours or even weeks. These dynamical features are undoubtedly connected to the interplay between binodals and spinodals (Berry et al., 2018) and hence to the metastabilities of homogeneous, albeit supersaturated phases. In addition to posing a challenge to determining coexistence curves, slow dynamics might be a way to divert molecules into arrested states and impose dynamical control over intracellular phase transitions.

Finally, we note that in vitro characterization of phase behavior, blended of course with the use of theory and computation, is invaluable for understanding the sequence determinants of the driving forces for phase separation (Holehouse & Pappu, 2018). Observing phase separation in vivo is important for establishing the biological relevance of a particular phase separating system. However, in order to explore the underlying mechanisms and physical principles that govern phase separation, the cellular environment would appear to be too complex and experimentally cumbersome to allow the extraction of binodals or the quantitative characterization of protein-specific phase behavior—although recent studies directly challenge our seemingly arcane view (Cho et al., 2018; Khan et al., 2018; Shin et al., 2017; Wang et al., 2018). The ability to study phase separation in a controlled in vitro setting does, however, seem to be essential for understanding the phase separation of a given system. The difficulty arises when one must connect in vitro and in vivo observations. Despite the best intentions and efforts, "physiological" buffer conditions are typically far from being physiologically relevant, simply because they lack the complexity of cellular environments. Buffer solutions are often simplified to a single "salt" component such as NaCl or KCl, when the cellular milieu of interest is composed of much more complex mixtures of salts, including but not limited to divalent salts as well as osmolytes. In vitro studies also overlook the contributions of additional biomacromolecules that are either directly involved in phase separation or indirectly involved through crowding or other effects. And yet the power and convenience of a simplified system cannot be denied. Ideally, a range of solution conditions would be explored in vitro, with each of multiple components being titrated independently, in order to draw conclusions about the influence of each on the phase separating system. Inasmuch as this is the case, it appears that a major focus of in vitro investigations should be on the mapping and comparative assessment of sequence-specific binodals. This will enable the development of new theoretical descriptions for the phase behavior of biologically relevant, finite sized intrinsically disordered heteropolymers that fit into the *stickers* and *spacers* architecture thus enabling the extraction of principles that underlie the evolution of these sequences (Brangwynne et al., 2015; Wang et al., 2018).

ACKNOWLEDGMENTS

We are grateful to our colleagues and collaborators Simon Alberti, Louis-Phillip Bergeron-Sandoval, Clifford Brangwynne, Ashutosh Chilkoti, Jeong-Mo Choi, Furqan Dar, D. Allan Drummond, Shana Elbaum-Garfinkle, Mina Farag, Julie Forman-Kay, Titus Franzmann,

Amy Gladfelter, Randal Halfmann, Tyler Harmon, Anthony Hyman, Richard Kriwacki, Erik Martin, Stephen Michnick, Tanja Mittag, Diana Mitrea, Stefan Roberts, Michael Rosen, Kiersten Ruff, Geraldine Seydoux, Andrea Soranno, Lucia Strader, J. Paul Taylor, Jie Wang, and Ming-Tzo Wei for many stimulating and enlightening discussions. Our efforts to understand the concepts underlying protein phase separation are supported by the Human Frontier Science Program (RGP0034/2017), National Science Foundation (MCB1614766 and DMR1729783), the National Institutes of Health (5R01NS056114 and R01NS089932), and the St. Jude Children's Research Hospital research collaborative.

REFERENCES

Banani, S. F., Lee, H. O., Hyman, A. A., & Rosen, M. K. (2017). Biomolecular condensates: Organizers of cellular biochemistry. *Nature Reviews Cellular Molecular Biology*, *18*(5), 285–298.

Banani, S. F., Rice, A. M., Peeples, W. B., Lin, Y., Jain, S., Parker, R., et al. (2016). Compositional control of phase-separated cellular bodies. *Cell*, *166*(3), 651–663.

Bernardino, N. R., & da Gama, M. M. T. (2012). Reentrant wetting of network fluids. *Physical Review Letters*, *109*(11), 116103.

Berry, J., Brangwynne, C. P., & Haataja, M. (2018). Physical principles of intracellular organization via active and passive phase transitions. *Reports on Progress in Physics*, *81*(4), 046601.

Bianchi, E., Blaak, R., & Likos, C. N. (2011). Patchy colloids: State of the art and perspectives. *Physical Chemistry Chemical Physics*, *13*(14), 6397–6410.

Bloustine, J., Virmani, T., Thurston, G. M., & Fraden, S. (2006). Light scattering and phase behavior of lysozyme-poly(ethylene glycol) mixtures. *Physical Review Letters*, *96*(8), 087803.

Boeynaems, S., Bogaert, E., Kovacs, D., Konijnenberg, A., Timmerman, E., Volkov, A., et al. (2017). Phase separation of C9orf72 dipeptide repeats perturbs stress granule dynamics. *Molecular Cell*, *65*(6). 1044-1055.e5.

Bogaert, E., Boeynaems, S., Kato, M., Guo, L., Caulfield, T. R., Steyaert, J., et al. (2018). Molecular dissection of FUS points at synergistic effect of low-complexity domains in toxicity. *Cell Reports*, *24*(3). 529-537.e4.

Brady, J. P., Farber, P. J., Sekhar, A., Lin, Y. H., Huang, R., Bah, A., et al. (2017). Structural and hydrodynamic properties of an intrinsically disordered region of a germ cell-specific protein on phase separation. *Proceedings of the National Academy of Sciences of the United States of America*, *114*(39), E8194–E8203.

Brangwynne, C. P., Eckmann, C. R., Courson, D. S., Rybarska, A., Hoege, C., Gharakhani, J., et al. (2009). Germline P granules are liquid droplets that localize by controlled dissolution/condensation. *Science*, *324*(5935), 1729–1732.

Brangwynne, C. P., Tompa, P., & Pappu, R. V. (2015). Polymer physics of intracellular phase transitions. *Nature Physics*, *11*(11), 899–904.

Bray, A. J. (1994). Theory of phase-ordering kinetics. *Advances in Physics*, *43*(3), 357–459.

Burgess, D. J. (1990). Practical analysis of complex coacervate systems. *Journal of Colloid and Interface Science*, *140*(1), 227–238.

Burke, K. A., Janke, A. M., Rhine, C. L., & Fawzi, N. L. (2015). Residue-by-residue view of in vitro FUS granules that bind the C-terminal domain of RNA polymerase II. *Molecular Cell*, *60*(2), 231–241.

Chang, L. W., Lytle, T. K., Radhakrishna, M., Madinya, J. J., Velez, J., Sing, C. E., et al. (2017). Sequence and entropy-based control of complex coacervates. *Nature Communications*, *8*.

Cheng, S. Z. D., & Keller, A. (1998). The role of metastable states in polymer phase transitions: Concepts, principles, and experimental observations. *Annual Review of Materials Science*, *28*(1), 533–562.
Cho, W. K., Spille, J. H., Hecht, M., Lee, C., Li, C., Grube, V., et al. (2018). Mediator and RNA polymerase II clusters associate in transcription-dependent condensates. *Science*, *361*(6400), 412–415.
Ciryam, P., Tartaglia, G. G., Morimoto, R. I., Dobson, C. M., & Vendruscolo, M. (2013). Widespread aggregation and neurodegenerative diseases are associated with supersaturated proteins. *Cell Reports*, *5*(3), 781–790.
Crick, S. L., Ruff, K. M., Garai, K., Frieden, C., & Pappu, R. V. (2013). Unmasking the roles of N- and C-terminal flanking sequences from exon 1 of huntingtin as modulators of polyglutamine aggregation. *Proceedings of the National Academy of Sciences of the United States of America*, *110*(50), 20075–20080.
Dao, T. P., Kolaitis, R. M., Kim, H. J., O'Donovan, K., Martyniak, B., Colicino, E., et al. (2018). Ubiquitin modulates liquid-liquid phase separation of UBQLN2 via disruption of multivalent interactions. *Molecular Cell*, *69*(6). 965-978.e6.
Dias, C. S., Araújo, N. A. M., & Telo da Gama, M. M. (2017). Dynamics of network fluids. *Advances in Colloid and Interface Science*, *247*, 258–263.
Diekmann, Y., & Pereira-Leal, J. B. (2013). Evolution of intracellular compartmentalization. *Biochemical Journal*, *449*(2), 319–331.
Elbaum-Garfinkle, S., Kim, Y., Szczepaniak, K., Chen, C. C., Eckmann, C. R., Myong, S., et al. (2015). The disordered P granule protein LAF-1 drives phase separation into droplets with tunable viscosity and dynamics. *Proceedings of the National Academy of Sciences of the United States of America*, *112*(23), 7189–7194.
Feric, M., Vaidya, N., Harmon, T. S., Mitrea, D. M., Zhu, L., Richardson, T. M., et al. (2016). Coexisting liquid phases underlie nucleolar subcompartments. *Cell*, *165*(7), 1686–1697.
Flory, P. J. (1942). Thermodynamics of high polymer solutions. *The Journal of Chemical Physics*, *10*(1), 51–61.
Franzmann, T. M., Jahnel, M., Pozniakovsky, A., Mahamid, J., Holehouse, A. S., Nuske, E., et al. (2018). Phase separation of a yeast prion protein promotes cellular fitness. *Science*, *359*(6371), eaao5654.
Guggenheim, E. A. (1945). The principle of corresponding states. *The Journal of Chemical Physics*, *13*(7), 253–261.
Harmon, T. S., Holehouse, A. S., Rosen, M. K., & Pappu, R. V. (2017). Intrinsically disordered linkers determine the interplay between phase separation and gelation in multivalent proteins. *eLife*, *6*, 30294.
Hernandez-Vega, A., Braun, M., Scharrel, L., Jahnel, M., Wegmann, S., Hyman, B. T., et al. (2017). Local nucleation of microtubule bundles through tubulin concentration into a condensed tau phase. *Cell Reports*, *20*(10), 2304–2312.
Holden, M. A., & Cremer, P. S. (2005). Microfluidic tools for studying the specific binding, adsorption, and displacement of proteins at interfaces. *Annual Review of Physical Chemistry*, *56*, 369–387.
Holehouse, A. S., & Pappu, R. V. (2018). Functional implications of intracellular phase transitions. *Biochemistry*, *57*(17), 2415–2423.
Huggins, M. L. (1942). Some properties of solutions of long-chain compounds. *The Journal of Physical Chemistry*, *46*(1), 151–158.
Hyman, A. A., Weber, C. A., & Jülicher, F. (2014). Liquid-liquid phase separation in biology. *Annual Review of Cell and Developmental Biology*, *30*(1), 39–58.
Jain, S., Wheeler, J. R., Walters, R. W., Agrawal, A., Barsic, A., & Parker, R. (2016). ATPase-modulated stress granules contain a diverse proteome and substructure. *Cell*, *164*(3), 487–498.

Khan, T., Kandola, T. S., Wu, J., Venkatesan, S., Ketter, E., Lange, J. J., et al. (2018). Quantifying nucleation in vivo reveals the physical basis of prion-like phase behavior. *Molecular Cell*, *71*(1). 155-168.e7.

Klenin, V. J. (1999). Chapter 1—Stability and phase separation. In V. J. Klenin (Ed.), *Thermodynamics of systems containing flexible-chain polymers* (pp. 1–106). Amsterdam: Elsevier Science B.V.

Koningsveld, R., Kleintjens, L. A., & Nies, E. (1987). Polymers and thermodynamics. *Croatica Chemica Acta*, *60*(1), 53–89.

Kroschwald, S., Maharana, S., & Alberti, S. (2017). Hexanediol: A chemical probe to investigate the material properties of membrane-less compartments. *Matters*, *3*(5), e201702000010.

Kroschwald, S., Maharana, S., Mateju, D., Malinovska, L., Nuske, E., Poser, I., et al. (2015). Promiscuous interactions and protein disaggregases determine the material state of stress-inducible RNP granules. *eLife*, *4*, e06807.

Kroschwald, S., Munder, M. C., Maharana, S., Franzmann, T. M., Richter, D., Ruer, M., et al. (2018). Different material states of Pub1 condensates define distinct modes of stress adaptation and recovery. *Cell Reports*, *23*(11), 3327–3339.

Landau, L. (1936). The theory of phase transitions. *Nature*, *138*, 840–841.

Larson, A. G., Elnatan, D., Keenen, M. M., Trnka, M. J., Johnston, J. B., Burlingame, A. L., et al. (2017). Liquid droplet formation by HP1alpha suggests a role for phase separation in heterochromatin. *Nature*, *547*(7662), 236–240.

Li, P., Banjade, S., Cheng, H. C., Kim, S., Chen, B., Guo, L., et al. (2012). Phase transitions in the assembly of multivalent signalling proteins. *Nature*, *483*(7389), 336–340.

Lin, Y., Mori, E., Kato, M., Xiang, S., Wu, L., Kwon, I., et al. (2016). Toxic PR poly-dipeptides encoded by the C9orf72 repeat expansion target LC domain polymers. *Cell*, *167*(3). 789-802.e12.

Mao, H. B., Li, C. M., Zhang, Y. J., Furyk, S., Cremer, P. S., & Bergbreiter, D. E. (2004). High-throughput studies of the effects of polymer structure and solution components on the phase separation of thermoresponsive polymers. *Macromolecules*, *37*(3), 1031–1036.

Mao, H. B., Yang, T. L., & Cremer, P. S. (2002). A microfluidic device with a linear temperature gradient for parallel and combinatorial measurements. *Journal of the American Chemical Society*, *124*(16), 4432–4435.

Mao, Y. S., Zhang, B., & Spector, D. L. (2011). Biogenesis and function of nuclear bodies. *Trends in Genetics*, *27*(8), 295–306.

Michaeli, I., Overbeek, J. T. G., & Voorn, M. J. (1957). Phase separation of polyelectrolyte solutions. *Journal of Polymer Science*, *23*(103), 443–450.

Milovanovic, D., Wu, Y., Bian, X., & De Camilli, P. (2018). A liquid phase of synapsin and lipid vesicles. *Science*, *361*(6402), 604–607.

Mitrea, D. M., Cika, J. A., Guy, C. S., Ban, D., Banerjee, P. R., Stanley, C. B., et al. (2016). Nucleophosmin integrates within the nucleolus via multi-modal interactions with proteins displaying R-rich linear motifs and rRNA. *eLife*, *5*, e13571.

Mitrea, D. M., & Kriwacki, R. W. (2016). Phase separation in biology; functional organization of a higher order. *Cell Communication and Signaling*, *14*, 1.

Mittal, A., Holehouse, A. S., Cohan, M. C., & Pappu, R. V. (2018). Sequence-to-conformation relationships of disordered regions tethered to folded domains of proteins. *Journal of Molecular Biology*, *430*(16), 2403–2421.

Molliex, A., Temirov, J., Lee, J., Coughlin, M., Kanagaraj, A. P., Kim, H. J., et al. (2015). Phase separation by low complexity domains promotes stress granule assembly and drives pathological fibrillization. *Cell*, *163*(1), 123–133.

Munder, M. C., Midtvedt, D., Franzmann, T., Nuske, E., Otto, O., Herbig, M., et al. (2016). A pH-driven transition of the cytoplasm from a fluid- to a solid-like state promotes entry into dormancy. *eLife*, *5*, e09347.

Nott, T. J., Petsalaki, E., Farber, P., Jervis, D., Fussner, E., Plochowietz, A., et al. (2015). Phase transition of a disordered nuage protein generates environmentally responsive membraneless organelles. *Molecular Cell*, *57*(5), 936–947.

Pappu, R. V., Wang, X., Vitalis, A., & Crick, S. L. (2008). A polymer physics perspective on driving forces and mechanisms for protein aggregation. *Archives of Biochemistry and Biophysics*, *469*(1), 132–141.

Patel, A., Lee, H. O., Jawerth, L., Maharana, S., Jahnel, M., Hein, M. Y., et al. (2015). A liquid-to-solid phase transition of the ALS protein FUS accelerated by disease mutation. *Cell*, *162*(5), 1066–1077.

Patel, A., Malinovska, L., Saha, S., Wang, J., Alberti, S., Krishnan, Y., et al. (2017). ATP as a biological hydrotrope. *Science*, *356*(6339), 753–756.

Pederson, T. (2011). The nucleolus. *Cold Spring Harbor Perspectives in Biology*, *3*(3).

Peskett, T. R., Rau, F., O'Driscoll, J., Patani, R., Lowe, A. R., & Saibil, H. R. (2018). A liquid to solid phase transition underlying pathological huntingtin Exon1 aggregation. *Molecular Cell*, *70*(4). 588-601.e6.

Petsev, D. N., Wu, X. X., Galkin, O., & Vekilov, P. G. (2003). Thermodynamic functions of concentrated protein solutions from phase equilibria. *Journal of Physical Chemistry B*, *107*(16), 3921–3926.

Posey, A. E., & Pappu, R. V. (2018). A first glimpse of nucleation of phase transitions in living cells. *Molecular Cell*, *71*(1), 1–3.

Quiroz, F. G., & Chilkoti, A. (2015). Sequence heuristics to encode phase behaviour in intrinsically disordered protein polymers. *Nature Materials*, *14*(11), 1164–1171.

Record, M. T., Anderson, C. F., & Lohman, T. M. (1978). Thermodynamic analysis of ion effects on binding and conformational equilibria of proteins and nucleic-acids—Roles of ion association or release, screening, and ion effects on water activity. *Quarterly Reviews of Biophysics*, *11*(2), 103–178.

Riback, J. A., Katanski, C. D., Kear-Scott, J. L., Pilipenko, E. V., Rojek, A. E., Sosnick, T. R., et al. (2017). Stress-triggered phase separation is an adaptive, evolutionarily tuned response. *Cell*, *168*(6). 1028-1040.e19.

Ribbeck, K., & Gorlich, D. (2002). The permeability barrier of nuclear pore complexes appears to operate via hydrophobic exclusion. *The EMBO Journal*, *21*(11), 2664–2671.

Rog, O., Kohler, S., & Dernburg, A. F. (2017). The synaptonemal complex has liquid crystalline properties and spatially regulates meiotic recombination factors. *eLife*, *6*, e21455.

Rubinstein, M., & Colby, R. H. (2003). *Polymer physics: Chapter 4*. New York, NY: Oxford University Press.

Rubinstein, M., & Dobrynin, A. V. (1997). Solutions of associative polymers, trends in polymer. *Science*, *5*(6), 181–186.

Rubinstein, M., & Semenov, A. N. (1998). Thermoreversible gelation in solutions of associating polymers. 2. Linear dynamics. *Macromolecules*, *31*(4), 1386–1397.

Ruff, K. M., Harmon, T. S., & Pappu, R. V. (2015). CAMELOT: A machine learning approach for coarse-grained simulations of aggregation of block-copolymeric protein sequences. *The Journal of Chemical Physics*, *143*(24), 243123.

Ruff, K. M., Roberts, S., Chilkoti, A., & Pappu, R. V. (2018). Advances in understanding stimulus-responsive phase behavior of intrinsically disordered protein polymers. *Journal of Molecular Biology*. https://doi.org/10.1016/j.jmb.2018.06.031.

Ryan, B. J., & Kinsella, G. K. (2017). Differential precipitation and solubilization of proteins. *Methods in Molecular Biology (Clifton, N.J.)*, *1485*, 191–208.

Sabari, B. R., Dall'Agnese, A., Boija, A., Klein, I. A., Coffey, E. L., Shrinivas, K., et al. (2018). Coactivator condensation at super-enhancers links phase separation and gene control. *Science*, *361*(6400), eaar3958.

Schmidt, H. B., & Gorlich, D. (2015). Nup98 FG domains from diverse species spontaneously phase-separate into particles with nuclear pore-like permselectivity. *eLife*, *4*, e04251.

Semenov, A. N., & Rubinstein, M. (1998). Thermoreversible gelation in solutions of associative polymers. 1. Statics. *Macromolecules*, *31*(4), 1373–1385.
Shin, Y., Berry, J., Pannucci, N., Haataja, M. P., Toettcher, J. E., & Brangwynne, C. P. (2017). Spatiotemporal control of intracellular phase transitions using light-activated optoDroplets. *Cell*, *168*(1–2). 159-171.e14.
Shin, Y., & Brangwynne, C. P. (2017). Liquid phase condensation in cell physiology and disease. *Science*, *357*(6357), eaaf4382.
Simon, J. R., Carroll, N. J., Rubinstein, M., Chilkoti, A., & Lopez, G. P. (2017). Programming molecular self-assembly of intrinsically disordered proteins containing sequences of low complexity. *Nature Chemistry*, *9*(6), 509–515.
Sing, C. E. (2017). Development of the modern theory of polymeric complex coacervation. *Advances in Colloid and Interface Science*, *239*, 2–16.
Strom, A. R., Emelyanov, A. V., Mir, M., Fyodorov, D. V., Darzacq, X., & Karpen, G. H. (2017). Phase separation drives heterochromatin domain formation. *Nature*, *547*(7662), 241–245.
Tainaka, K. I. (1980). Effect of counterions on complex coacervation. *Biopolymers*, *19*(7), 1289–1298.
Tainaka, K., & Yomosa, S. (1980). Condition for the occurrence of complex coacervation in low polymer concentration. *Journal of the Physical Society of Japan*, *48*(5), 1791–1792.
Tanaka, F. (2011). *Polymer physics: Applications of molecular asssociation and thermoreversible gelation*. Cambridge, UK: Cambridge University Press.
Taylor, N., Elbaum-Garfinkle, S., Vaidya, N., Zhang, H., Stone, H. A., & Brangwynne, C. P. (2016). Biophysical characterization of organelle-based RNA/protein liquid phases using microfluidics. *Soft Matter*, *12*(45), 9142–9150.
Updike, D. L., Hachey, S. J., Kreher, J., & Strome, S. (2011). P granules extend the nuclear pore complex environment in the *C. elegans* germ line. *The Journal of Cell Biology*, *192*(6), 939–948.
Wang, J., Choi, J.-M., Holehouse, A. S., Zhang, X., Jahnel, M., Lemaitre, R., et al. (2018). A molecular grammar underlying the driving forces for phase separation of prion-like RNA binding proteins. *Cell*, *174*, 688–699.
Wegmann, S., Eftekharzadeh, B., Tepper, K., Zoltowska, K. M., Bennett, R. E., Dujardin, S., et al. (2018). Tau protein liquid-liquid phase separation can initiate tau aggregation. *The EMBO Journal*, *37*(7), e98049.
Wei, M. T., Elbaum-Garfinkle, S., Holehouse, A. S., Chen, C. C., Feric, M., Arnold, C. B., et al. (2017). Phase behaviour of disordered proteins underlying low density and high permeability of liquid organelles. *Nature Chemistry*, *9*(11), 1118–1125.
Wheeler, R. J., & Hyman, A. A. (2018). Controlling compartmentalization by non-membrane-bound organelles. *Philosophical Transactions of the Royal Society of London. Series B, Biological Sciences*, *373*(1747), 20170193.
Widom, B. (1988). Note on the interfacial-tension of phase-separated polymer-solutions. *Journal of Statistical Physics*, *52*(5–6), 1343–1351.
Woodruff, J. B., Ferreira Gomes, B., Widlund, P. O., Mahamid, J., Honigmann, A., & Hyman, A. A. (2017). The centrosome is a selective condensate that nucleates microtubules by concentrating tubulin. *Cell*, *169*(6). 1066-1077.e10.
Wu, H., & Fuxreiter, M. (2016). The structure and dynamics of higher-order assemblies: Amyloids, signalosomes, and granules. *Cell*, *165*(5), 1055–1066.
Wu, C., & Li, Y. (2018). Universal scaling of phase diagrams of polymer solutions. *Macromolecules*, *51*(15), 5863–5866.
Zhang, H., Elbaum-Garfinkle, S., Langdon, E. M., Taylor, N., Occhipinti, P., Bridges, A. A., et al. (2015). RNA controls PolyQ protein phase transitions. *Molecular Cell*, *60*(2), 220–230.
Zhang, F., Skoda, M. W., Jacobs, R. M., Zorn, S., Martin, R. A., Martin, C. M., et al. (2008). Reentrant condensation of proteins in solution induced by multivalent counterions. *Physical Review Letters*, *101*(14), 148101.

CHAPTER TWO

Methods and Strategies to Quantify Phase Separation of Disordered Proteins

Alfredo Vidal Ceballos[*,†,2], **Charles J. McDonald**[*,†,2], **Shana Elbaum-Garfinkle**[*,†,‡,1]
*Structural Biology Initiative, CUNY Advanced Science Research Center, New York, NY, United States
†Ph.D Program in Biochemistry, The Graduate Center, CUNY, New York, NY, United States
‡Ph.D Program in Biology, The Graduate Center, CUNY, New York, NY, United States
[1]Corresponding author: e-mail address: shana.elbaum@asrc.cuny.edu

Contents

Abstract

Phase separation has emerged as a new paradigm currently revolutionizing our understanding of cell biology and intracellular organization. Disordered protein domains have recently been demonstrated as integral drivers of phase separation into condensed liquids with emergent material properties. Using in vitro model systems employing purified protein components is necessary to interrogate the molecular mechanisms underlying phase separation; however, these systems pose many experimental challenges. In this chapter we describe general strategies for purifying, handling, imaging, and characterizing the phase behavior of disordered proteins. We further outline methods for the purification of the model P granule protein LAF-1, the construction of phase diagrams, and the quantification of liquid droplet fusion or coalescence.

[2] Contributed equally.

Methods in Enzymology, Volume 611
ISSN 0076-6879
https://doi.org/10.1016/bs.mie.2018.09.037

1. INTRODUCTION

Phase separation of biomolecules into condensed material states has recently emerged as a key principle underlying intracellular organization (Banani, Lee, Hyman, & Rosen, 2017; Boeynaems et al., 2018; Dolgin, 2018; Shin & Brangwynne, 2017). Intrinsically disordered proteins or domains have been identified as major drivers of phase separation into liquid droplets, including the P granule proteins LAF-1 (Elbaum-Garfinkle et al., 2015), PGL-3 (Saha et al., 2016) and MEG proteins (Smith et al., 2016; Wang et al., 2014), the germ granule protein DDX4 (Nott et al., 2015), the nucleolar protein Fib1 (Berry, Weber, Vaidya, Haataja, & Brangwynne, 2015; Feric et al., 2016), and stress granule proteins FUS (Burke, Janke, Rhine, & Fawzi, 2015; Murakami et al., 2015), TDP-43 (Conicella, Zerze, Mittal, & Fawzi, 2016), and hnRNPA1 (Lin, Protter, Rosen, & Parker, 2015; Molliex et al., 2015).

Reconstituting intracellular biomolecular condensates using in vitro model systems of purified disordered proteins or domains can offer significant insight into the mechanisms underlying phase separation and is a rapidly growing area of research. However, phase-separating proteins are often difficult to purify, handle, and characterize. This chapter outlines several methods and strategies for studying phase-separating disordered proteins, including protein purification, hunting for droplets, generating phase diagrams, and quantifying droplet morphology and coalescence.

A liquid–liquid phase separation system in equilibrium can be quantitatively described by a phase diagram—a state diagram that delineates the regions where phase separation will and will not occur (Fig. 1A). The phase boundary, also known as a liquid–liquid coexistence curve or binodal, defines the two-phase or droplet forming region and the single-phase homogenous region where no phase separation occurs. The left arm of the binodal denotes the concentration of the bulk solution outside the droplets, i.e., the saturation concentration, while the right side denotes the significantly higher concentration of protein within the condensed phase droplets. Such diagrams are derived using the Flory–Huggins formalism and mean field theory approximation (Brady et al., 2017; Brangwynne, Tompa, & Pappu, 2015) where the enthalpy of interaction between molecules must be strong enough to counterbalance the entropy of mixing, for phase separation (i.e., demixing) to occur. In this framework, the Y-axis reflects the interaction strength, while the X-axis denotes protein concentration. Therefore, anything

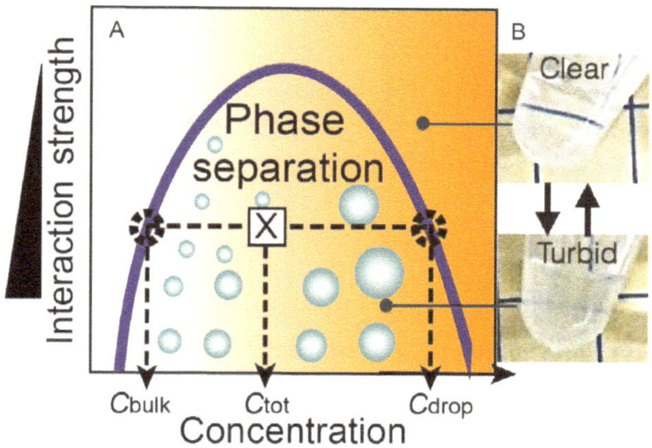

Fig. 1 Phase separation. (A) Phase diagram, binodal or liquid–liquid coexistence curve for a simplified 2-component system delineates the phase-separating region. Any starting total concentration/interaction strength pair that falls within that region—will result in phase separation. At equilibrium, the *left side* of the binodal denotes the concentration outside the droplets or the saturation concentration, while the *right side* denotes the concentration inside the droplets (B) LAF-1 protein solutions reversibly turn turbid upon phase separation.

that tunes the interaction strength between molecules, including temperature, pH, ionic strength, etc., can influence their resulting phase separation behavior. Consideration of potential phase diagram parameters is important when strategizing the purification and characterization protocols for proteins suspected of phase separation.

2. METHODS

2.1 Purification Strategies

Phase separation of protein during the purification process often results in significant protein loss due in part to the increased viscosity of sticky condensed phases. To prevent phase separation all potential phase diagram interaction parameters, such as temperature, ionic strength, pH, or protein concentration, should be considered. Here, we briefly discuss general strategies for optimizing purification protocols of proteins prone to phase separation. These strategies are further illustrated in the detailed purification protocol for the P granule model protein LAF-1 in Section 2.2.

As with proteins that are prone to any concentration-dependent self-assembly, including aggregation and phase separation, care should be taken to avoid high local protein concentrations. During cell growth, such measures include careful monitoring of the OD_{600} of *Escherichia coli* cultures before induction—we find that induction at OD_{600} values of ~0.4 significantly reduce protein lost to the insoluble fraction compared to the more standard OD_{600} values of ~0.6–0.7. Similarly, overnight growth at cool temperatures (~18°C) after induction is often preferred. Additionally, purification column protocols may need to be adjusted to avoid drastic increases in local protein concentration that may induce phase separation on the column. Column modifications include, increasing column resin volumes or performing batch purifications by incubating dilute protein solutions with free resin. Furthermore, protein solutions should be buffer exchanged with extra care, especially if using centrifugal filter concentrators—spin times should be short and solutions should be frequently mixed inside the concentrator so that high local concentrations do not accumulate at the bottom of the centrifugal filter device.

If phase separation is suspected to have occurred buffer conditions should be systematically modified. A classic signature of liquid phase separation is the appearance of turbid solutions (Fig. 1B), resulting from the light scattered by liquid droplets in solution. Sudden and/or mysterious loss of large amounts of protein during a purification could also be a sign of phase separation, as the viscous phase-separated droplets tend to stick to tube walls and surfaces. Temperature is a very strong interaction parameter for phase separation and can be easily modified during purification. For example, while most proteins require purification at 4°C, proteins like LAF-1 more readily phase separate at cooler temperatures; therefore, we perform most of the purification at room temperature. Likewise, ionic strength, pH, and other buffer additives can be tuned to optimize protein yield. When working with a new protein, it is useful to pay attention to buffer changes that significantly alter yield, as this could indicate interaction parameters that will be useful when subsequently characterizing the protein's phase behavior.

2.2 LAF-1 Purification Protocol

2.2.1 Buffers

All buffers, except lysis buffer, are stored and used at room temperature.

Lysis buffer (stored at 4°C)

20 m*M* Tris–HCl, pH 7.4
500 m*M* NaCl
10 m*M* imidazole
10% (vol/vol) glycerol
1% Triton-X

Added day of: 1 mg/mL lysozyme and protease inhibitor mixture one tablet EDTA-free per 50 mL (Roche Diagnostics), 14 m*M* β-mercaptoethanol, 1 m*M* PMSF

Nickel wash buffer

20 m*M* Tris–HCl, pH 7.4
500 m*M* NaCl
10% (vol/vol) glycerol
25 m*M* imidazole
Added day of: 14 m*M* β-mercaptoethanol

Nickel elution buffer

20 m*M* Tris–HCl, pH 7.4
500 m*M* NaCl
10% (vol/vol) glycerol
250 m*M* imidazole
Added day of: 14 m*M* β-mercaptoethanol

Heparin binding buffer

20 m*M* Tris–HCl, pH 7.4
50 m*M* NaCl
1% (vol/vol) glycerol
Added day of: 2 m*M* DTT

Heparin elution buffer

20 m*M* Tris–HCl, pH 7.4
1 *M* NaCl
1% (vol/vol) glycerol
Added day of: 2 m*M* DTT

2.2.2 Protocol

2.2.2.1 Cell Growth

LAF-1 plasmid, which has been codon optimized for *Escherichia coli*, synthesized and inserted into a Kanamycin resistant pET28a backbone with a N-terminal 6x-His tag (Elbaum-Garfinkle et al., 2015) is transformed and expressed in BL21(DE3) cells according to standard protocols. LAF-1 protein yields are very sensitive to OD. We find it is essential to induce at relatively low OD of 0.3–0.4, followed by overnight growth at 18°C with 500 m*M* IPTG. Cell pellets can be flash frozen or immediately lysed and purified.

2.2.2.2 Cell Lysis

(1) Add lysis buffer to pelleted (and/or thawed) cells to resuspend (~15–20 mL/L of culture). Gently swirl buffer over cell pellet until dissolved.

(2) Transfer to 50-mL conical tubes. Incubate immersed in ice bucket for 30–60 min—swirl tube every so often.

Note: this is a good time to prepare the nickel beads for the purification (see later)

(3) Sonicate solutions on ice to further disrupt/lyse cells—pulse on/off 10 s, Amp: 40%, Time: 2 min total.

(4) Centrifuge balanced tubes at 20,000 × *g* for 30 min. Decant supernatant (cleared cell lysate) into fresh conical tube and discard cell pellet (cell debris).

2.2.2.3 Nickel Column Purification: "Batch" (Thermo Fisher Ni-NTA)

(1) Preparing nickel beads:

a. Invert 50% nickel slurry several times before pipetting from the bottle to ensure it is well mixed.

b. Remove desired amount and place in 15-mL Falcon tube. A good starting estimate is ~0.5 mL slurry for each Liter of original growth. (Note: growths are usually several liters; If you are doing a small prep and are only using 0.5 mL or less of slurry, then use an Eppendorf tube instead of a 15-mL conical tube to make sure you can get a nice pellet of the nickel resin).

c. Add 10–12 mL of nickel wash buffer to the 15-mL conical tube. Invert several times to mix.

d. Centrifuge at ~600 × *g* for 2 min to pellet the nickel beads.

e. Carefully pour or pipette off wash buffer into a waste bottle.

f. Repeat steps c–e 3–4 × more times

(2) Combine washed nickel beads with cell lysate. Tape to rocker at 4°C. Allow protein to bind the nickel for ~30 min with gentle rocking.

Note: from this point forward we work at room temperature.

(3) Distribute lysate-bead mixture into several 15-mL Falcon tubes and centrifuge at 600 × *g* for 2 min to collect the beads. Pour/pipette off supernatants and combine into 50-mL Falcon tube labeled "Flow Through." Be very careful not to pour off any of the nickel beads.

(4) Add ~10 mL nickel wash buffer to tubes and mix by inverting tube several times or gentle pipetting. Centrifuge at 600 × *g* for 2 min. Pour/pipette off supernatant and combine into 50-mL Falcon tube labeled "Wash 1." Repeat 2 × for "Wash 2" and "Wash 3." Add additional 1–3 mL of wash buffer to each tube to resuspend nickel beads and combine all beads into a single 15 mL Falcon tube. Collect the beads by centrifugation. Remove supernatant, leaving minimal buffer to transfer beads to column.

(5) Add nickel beads to a prepared, capped, disposable hand-held gravity flow column with filter and let the nickel beads sediment. Do not allow flow out of the column during this time! May use additional Nickel Wash Buffer to remove as much of the nickel beads as possible from the Falcon tube.

(6) Once nickel beads have settled into a packed bed, open stopper and collect flow through. Do not let the bed dry out, stop flow just above top of bed.

(7) Protein elution: Add nickel elution buffer carefully so as not to disturb the nickel bed. Collect elution fractions in Eppendorf tubes (0.5–1.0 mL fractions).

(8) Measure protein concentration via absorption at 280-nm for each tube. Combine the tubes with protein and proceed with Heparin Column to further purify and remove any bound RNA. (Note: may stop here to run a gel).

2.2.2.4 Heparin Column Purification (GE HiTrap Heparin Columns, 1 or 5 mL)

Filter Heparin Buffers before use. Use Lueur lock syringes connected "drop-to-drop" with column to avoid introducing air to the column as per manufacturer instructions.

(1) Equilibrate 1 or 5 mL column with 10× column volumes of Heparin Binding buffer

(2) Dilute protein obtained from nickel resin purification at least 5× with room temperature Heparin Binding Buffer and load onto column.

   a. *Note*: If the protein concentration is sufficiently high, the solution will instantly become turbid as droplets form! Proceed quickly with loading onto the column. The protein still binds and is eluted as a clear solution with high salt buffer.

(3) Load the protein solution on the column and collect flow through.

(4) Wash with 10× column volume Heparin Binding Buffer (i.e., 10 or 50 mL) and collect.

(5) Elute with (~2× column volume) Heparin Elution Buffer and collect in <~0.5 mL aliquots. Check concentrations of each fraction by absorbance at 280 nm. Pool protein-containing fractions, avoiding very dilute fractions so as not to reduce to the final protein concentration.

(6) Add glycerol to pooled protein fractions to 10% (vol/vol), aliquot to desired volumes, and flash freeze with liquid nitrogen for storage in −80°C.

2.2.2.5 Thawing LAF-1 Protein Aliquots

LAF-1 aliquots are thawed by placing them on the bench top at room temperature. We find that both slow thawing on ice or fast thawing at 37°C promote aggregation. It is best to let aliquots thaw, untouched, on the bench top before proceeding. Typically, we perform a 2 min high speed centrifugation step (maximum speed in a small table-top centrifuge) to pellet any aggregates, and then buffer exchange at room temperature into fresh storage buffer (20 m*M* Tris, pH 7.4, 1 *M* Nacl, 1 m*M* DTT).

2.3 Mapping Phase Diagrams

2.3.1 Hunting for Droplets/Identifying an Interaction Parameter

Determining whether or not your protein of interest phase separates, and/or identifying the conditions that promote or deter phase separation can be a laborious and often daunting process. Proteins may or may not require a binding partner to phase separate, including protein partners and/or RNA or other nucleotide analogs/small molecules. Additionally, macromolecular crowding agents may be required and can be sampled as necessary. Thus, the "hunt for droplets" is in many ways analogous to protein crystallography screens, where proteins of various concentrations are combined with a spectrum of buffer conditions and/or additives, incubated at multiple temperatures and screened via imaging or other forms of spectroscopy.

To determine phase separating conditions, we mix protein solutions in an imaging chamber with coverslip bottom ($N = 1.5$) and image over time using a high magnification brightfield, DIC or fluorescence inverted microscope. Different sized imaging chambers or wells can be used in order to accommodate varying sample conditions. For small, quick screens we make glass slide-coverslip chambers, either with rubber gaskets (Grace BioLabs) or with melted parafilm (Fig. 2). Larger culture wells (such as Nunc or Grace BioLabs) can be used for larger volumes (100–200 μL) and are easier to work

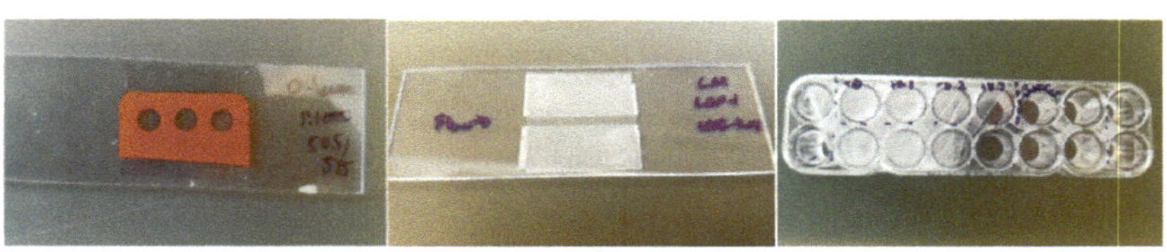

Fig. 2 Imaging chambers glass slide-coverslip chambers made with Grace BioLabs spacers (*left*) and melted parafilm (*middle*). Grace BioLabs CultureWells with coverslip bottom (*right*).

with if solution components are added/mixed in at later times. For larger screens, we also use 96- or 384-well plates.

Additional notes

- One should be very cautious in evaluating the presence of droplets, especially if the protein is unlabeled. Condensation of any kind can give the appearance of droplet artifacts, including immersion oil on the bottom surface of the coverslip! Additionally, any evaporation or leakage, particularly problematic for small volume slides, can confound images and results. Sanity checks or controls for these artifacts include, always including a no-droplet control, cleaning the slide surfaces, and paying close attention to the *Z*-plane as you are imaging.
- Fluorescently labeling protein, either through fusion to fluorescent proteins or via conjugation to fluorescent dyes following purification, can facilitate "droplet hunting" by improving image signal and minimizing nonprotein droplet artifacts. However, controls with unlabeled protein are strongly recommended as the phase behavior can be significantly influenced by fluorescent protein modifications.
- To improve image contrast without fluorescence protein fusion or conjugation, small amounts of free fluorescent dyes can be added to the unlabeled protein solution. Many dyes will partition unequally between the condensed droplet phase and the dilute phase, depending on the dye–protein pair. Even here, care must be taken, as too much free dye, especially if it partitions into the droplet, may affect the phase behavior.
- It is not always trivial to discern between the formation of liquid droplets and other types of assemblies or aggregates by imaging alone, especially if the assemblies are close to the resolution limit. For assemblies at least a few microns in size, liquid droplets appear spherical above the surface and may deform upon wetting the surface (see Section 2.4). Morphology and fusion events may be difficult to assess for assemblies closer to the resolution limit.

2.3.2 Mapping a Phase Boundary

As described above, a liquid–liquid phase diagram or coexistence curve delineates the conditions that do or do not give rise to phase separation. Mapping a full or partial phase diagram can be a useful first step in working with protein condensates, as it provides a framework for experimental setup and strategy. In addition to convenience, phase diagrams provide important information about the interactions that promote phase separation. LAF-1 droplets, for example, are very sensitive to salt concentration suggesting that electrostatics are very important. Other systems that are more driven by hydrophobic interactions may be less sensitive to salt, but may be more sensitive to temperature fluctuations. It is also insightful to see how certain additional perturbations to the system shift the phase diagram. For instance, macromolecular crowding agents often shift the phase diagram to the left, requiring less protein to phase separate. Experiments can be further designed to determine how sequence mutations and various binding partners or competitors influence the phase diagram and thus the interaction strengths of the system.

As depicted in Fig. 1A, a full phase diagram or binodal consists of left and right arms, denoting the concentrations of the bulk or dilute phase and droplet or condensed phase, respectively. Determination of full binodals has been recently described (Brady et al., 2017; Wei et al., 2017). Here, we outline complementary methods to determine the critical concentration for phase separation, or the left arm of the binodal/phase diagram.

2.3.3 Determining the Critical/Saturation Concentration Via Imaging

Optimize an imaging system for small volume samples for high throughput as described above. Prepare samples of varying protein concentration and interaction parameter (i.e., salt). It is best to mix directly within the imaging chamber, or immediately before placing samples into the imaging chamber. If condensed droplets form, they can stick to many surfaces and thus become lost during transfer.

Use a known droplet forming condition to determine the timescale of droplet formation and system equilibrium. For LAF-1, droplets form instantaneously for most conditions and settle down onto the coverslip roughly within 2h. For LAF-1 phase diagrams, mixed samples were observed after approximately 3h. Note: Some protein systems may phase separate on significantly longer timescales and the protocol should be adjusted accordingly.

Beginning with a previously determined droplet forming condition, decrease protein concentration until no droplets are observable. Then narrow down the phase boundary by testing intervening conditions. For any given set of conditions, measurements should be made a minimum of three times to ensure reproducibility. Any set of conditions is scored "yes/no" depending upon whether droplets are observed.

When close to the phase boundary, the system is very sensitive to small pipetting errors—thus conditions close to the boundary may result in inconsistent results for droplet formation upon repetition. A phase boundary can

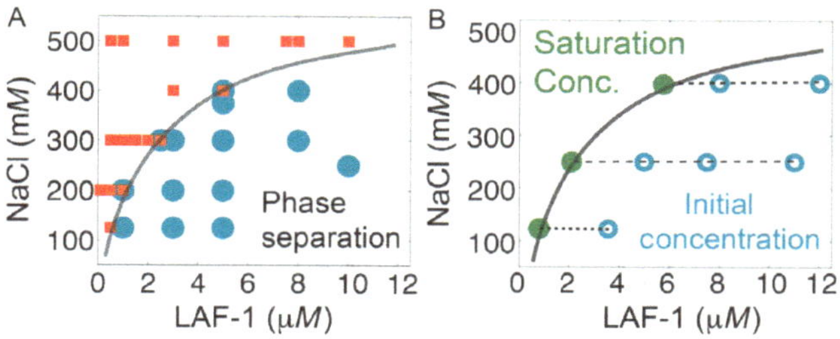

Fig. 3 Mapping the phase boundary. (A) LAF-1/NaCl conditions indicated were mixed, incubated, and scored for droplet formation via DIC microscopy. Positive and negative scoring for droplet formation are indicated by the *blue circles* and *red squares*, respectively. The phase boundary is drawn to distinguish between the two regions. (B) LAF-1/NaCl initial conditions within the phase separation region are incubated (*open blue circles*). The saturation concentration (i.e., critical concentration) is directly measured via absorbance at 280 n*M* of the solution after droplet removal via centrifugation. These values (*green filled circles*) are constant for increasing total protein concentration for a given NaCl concentration and overlay with the phase boundary determined in (A).

be drawn through these points to separate the region for phase separation which roughly corresponds to the saturation concentration (Fig. 3A).

Scoring yes/no for droplets is naturally limited to droplets that are optically resolvable. A high magnification brightfield or a DIC microscope can be used to detect phase separation of unlabeled proteins (see additional notes in Section 2.3.1 for discussion of imaging tips and strategies).

2.3.4 Determining the Critical/Saturation Concentration: Complimentary Methods

Another method to measure the saturation concentration is a centrifugation assay. This method should be considered a complementary method, and droplets should always be confirmed by imaging. For a given interaction parameter value, create samples with increasing protein concentrations. Allow droplets to settle or slowly centrifuge for several hours to collect droplets. Then carefully measure the protein concentration in the supernatant via absorbance at 280 nm. The concentration of the supernatant corresponds to the critical or saturation concentration (Fig. 3B) and should be invariant for increasing total protein concentration at any single interaction parameter value.

One can also use turbidity measurements to report on phase separation. However, just as with centrifugation, many phenomena can result in turbidity increases, including aggregation or polymerization. Therefore, the presence of droplets in contrast to other assembled structures should be confirmed directly by imaging.

2.4 Droplet Shape: Surface Treatments

Pure viscous liquid droplets take on spherical shapes in solution, resulting from the energetic need to relax interfacial tension at the liquid–liquid interface (Hyman, Weber, & Julicher, 2014). Observance of spherical morphologies of protein assemblies has been used frequently to identify the presence of liquid material states (Banani et al., 2017; Shin & Brangwynne, 2017). However, liquids are not always spherical as they can deform upon wetting different surfaces, such as membranes in the case of P granules in the adult *Caenorhabditis elegans* germline (Brangwynne et al., 2009) or glass slides with varying surface treatments (Feric et al., 2016). Furthermore, when droplets approach the resolution limit, accurately resolving droplet morphology becomes increasingly challenging. Additionally, if a system transitions from a liquid to solid state, as has been suggested for Balbiani bodies (Boke & Mitchison, 2017), and centrosomal SPD-5 condensates (Woodruff et al., 2017), it may retain a spherical shape despite no longer being a pure viscous liquid.

While further investigation is necessary to confirm the liquid material properties of a condensed protein phase (i.e., the coalescence assay described below, Fluorescence recovery after photobleaching (FRAP) or micro-rheology (Elbaum-Garfinkle et al., 2015; Zhang et al., 2015), varying surface treatments to promote spherical shapes can greatly assist with initial screening of droplet formation. Common surface treatments include Pluronics F-127, BSA, Sigmacote, and PEG. Protein droplets of varying content may interact uniquely with different surface treatments. LAF-1 droplets significantly wet untreated glass surfaces, but remain spherical upon treatment with Pluronics F-127 (Fig. 4). To treat glass surfaces, we add a filtered 10% solution of Pluronics F-127 to glass chambers/slides and incubate for ~1 h. The chambers are rinsed 4–5 × with MilliQ water—and remain immersed in water until just before an experiment. Such treatments provide reproducible surfaces necessary for the quantification of droplet size and droplet fusion or coalescence.

2.5 Quantifying Droplet Coalescence

We can study the physical properties of our liquid droplets by taking advantage of their liquid-like behavior and spherical shape. Two spherical liquid droplets of length (ℓ) will collapse exponentially over time (τ) into a single spherical form. The time (τ) it takes for two liquid droplets to collapse into a

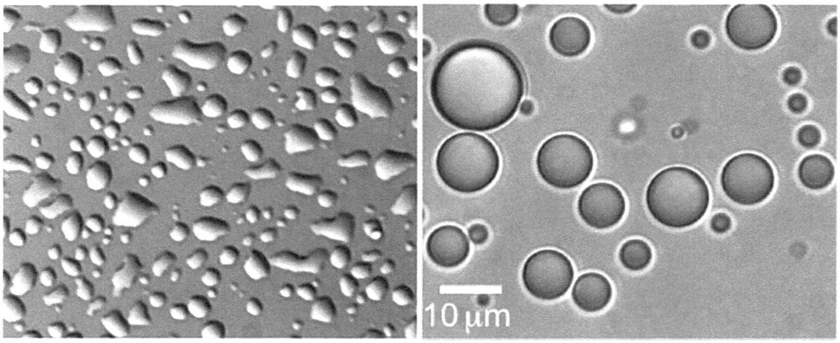

Fig. 4 Droplet shape and surfaces. *Left*: LAF-1 droplets wet the surface of untreated glass. *Right*: LAF-1 droplets do not wet the surface of Pluronics F127 treated glass and remain spherical.

single spherical shape is dictated by the relationship $\tau \approx (\eta/\gamma) \cdot \ell$, in which η is the droplet viscosity and γ is the surface tension (Eggers, Lister, & Stone, 1999). We can quantify the inverse capillary velocity, which is defined as the ratio of viscosity to surface tension (η/γ) by monitoring the coalescence of two droplets into a single sphere (Brangwynne et al., 2009; Brangwynne, Mitchison, & Hyman, 2011; Elbaum-Garfinkle et al., 2015).

2.5.1 Data Collection

To observe the coalescence events of LAF-1 liquid droplets the protein is incubated in CultureWell coverglass wells (GraceBioLabs) and visualized on an inverted microscope. The bottom of the well is treated with a Pluronics F-127 to prevent the liquid droplets from wetting the surface of the well, and to preserve their spherical shape. To visualize the liquid droplets, we mix unlabeled protein with ~1% labeled protein conjugated to DyLight NHS-Ester fluorophores (Thermo Fisher Scientific). We buffer exchange and concentrate the LAF-1 protein into a high salt (1 *M* NaCl) buffer along with reducing agent (1 m*M* DTT). The droplet size may vary depending on protein concentration, with higher concentrations resulting in larger droplets. We induce droplet formation by diluting into buffer with no or low NaCl depending on the experiment. We immediately observe the sample on the microscope, to image droplets and to capture fusion events as the liquid droplets coalesce and sink to the bottom of the well. Be mindful of the acquisition time per frame in each video as it will be important for the analysis.

2.5.2 Analysis

Once data collection is completed, time lapse images are first carefully reviewed in ImageJ. Ideal fusion events involve only two droplets that are similar in size. A single movie may contain several fusion events that can be useful for analysis, so it is important to look through the footage several times. We analyze each fusion event separately, noting the frames of each fusion event and saving each event with at least 10 frames prior to and after the event is completed. Each image sequence is saved as a .tif format file in a separate folder by making a unique "substack" folder for each fusion event to be analyzed.

Note: It is important to save several frames before and after the coalescence takes place to have a reference plateau and make sure the event is captured entirely. We look for liquid droplets in close proximity to each other, but which have not started the coalescence process. In this way we can track the aspect ratio change over time (Fig. 5A). The aspect ratio is defined as the ratio between the major and minor axis, given by the equation $\alpha = a/b$ in which a is the major axis and b is the minor axis. In our analysis, we assume the droplets to have a dimer shape because of their close proximity, with an aspect ratio of approximately 2. We monitor the collapse of the dimers into a sphere of aspect ratio equal to 1 after time (τ), which can be fit into an exponential decay graph. The MATLAB® analysis used to quantify coalescence is summarized in Figs. 5 and 6 and outlined here in detail:

(a) Define parameters in your command window such as the pixels per micron in your images and the timestep in seconds of the video, i.e., the time interval of each image obtained.

Note: As with all time-sensitive imaging, be sure to confirm that the recorded frame rate corresponds to the frame rate set in acquisition settings!

(b) Select the first image of the sequence and convert to double precision `[double(x)]` if the image does not already have this format.

(c) Use `ag=mat2gray(x)` function to convert the intensities of the colors into ranges with values between 0 and 1.

(d) Input this image into the command `T=graythresh(image)` to calculate the mean value of the image threshold. This will improve future visualization of the image.

(e) Finally, convert the values of the image pixels to binary values of 0 and 1 using the function `c=im2bw(ag,T)`, where T is the calculated threshold from step (d). Steps (c)–(e), should be input together and then observed

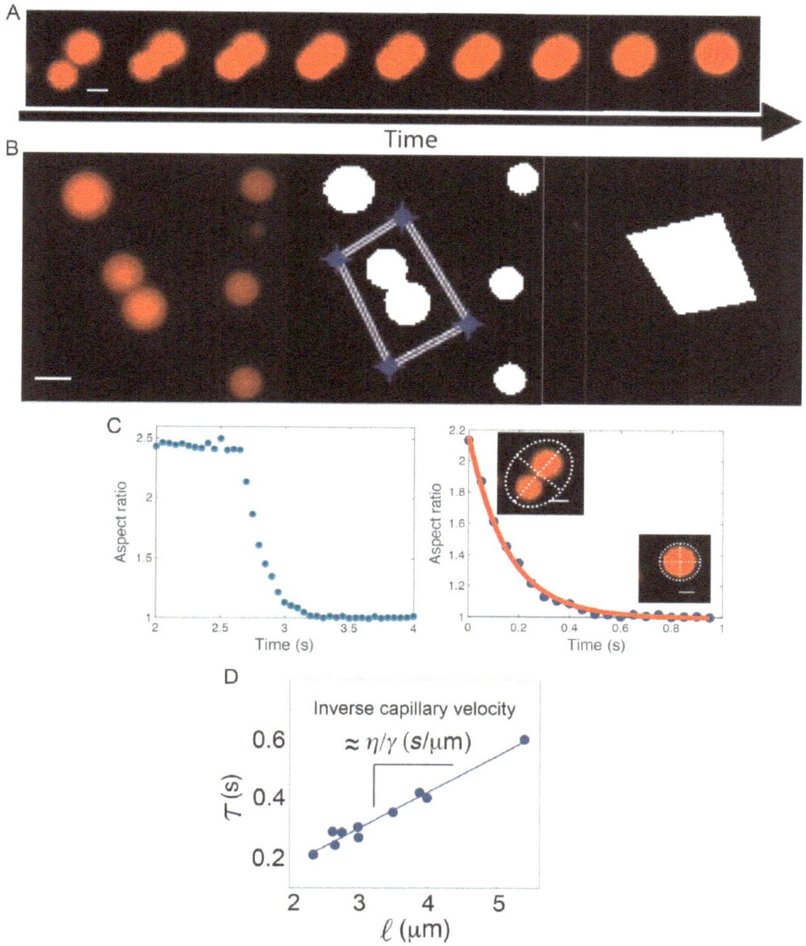

Fig. 5 Coalescence assay. (A) Protein liquid droplets coalesce into a single sphere over time. Scale bar: 2 μm. (B) Single fusion events are selected and tracked over time in MATLAB®. The region corresponding to a single event (*left*) is selected (*middle*) in order to create a mask (*right*) to be applied to each frame in the time lapse. Scale bar: 5 μm. (C) Droplet coalescence raw data (*left*) are fit to an exponential decay (*right*). Inset: droplet ellipse with major/minor axis depicted by *white dashed lines*. (D) The fusion rate and droplet size are plotted and the slope is calculated to determine the inverse capillary velocity.

using the `imtool` or `imshow` functions to see final product. Note that T is an automatic value calculated by MATLAB®. If the image looks too dark or too bright, you can decrease or increase the value of threshold to have a better contrast with the background.

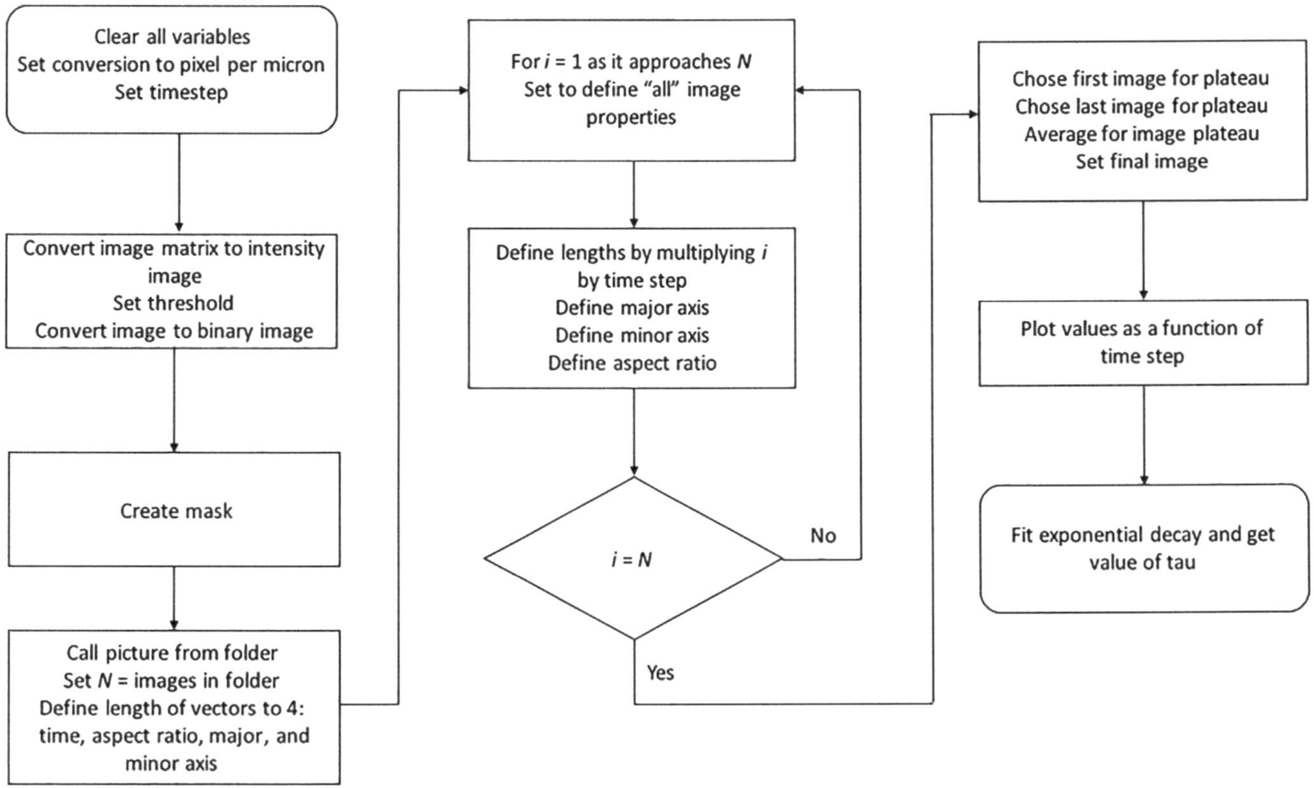

Fig. 6 A flow chart of the image analysis used to quantify droplet coalescence.

(f) Next draw a mask around the droplets to be analyzed. Call the modified image described earlier by using the command `figure, imshow (c,[ ])`. Input the command `h = impoly` to introduce a desired mask surrounding the droplets.

Note: It is important to avoid masking sections of droplets in the surrounding area. Take into consideration that droplets can move into your mask as the video progresses, making your analysis inaccurate. Introduce the commands `BW=createMask (h)` and `figure, imshow (BW,[ ])` to return the mask created with the `h=poly` function. The mask shape is drawn by constructing line segments with the mouse, using a single click for each new line segment and a final double click once the mask shape is complete. A visual representation of this mask is shown in Fig. 5B.

(g) Read in the full image sequence with the command `pics=dir (fullfile('Name_of_file','*.tif'));` Define the number of elements in each folder; this will be the individual pictures in your folder (`N = numel(pics);`).

(h) Run a "For" loop from image 1 to N (total) to: convert and threshold each image in the sequence as in steps (b)–(e); multiply mask created in step f to assign the area for analysis; quantify the major and minor axis and corresponding aspect ratio using the `regionprops` command.

(i) After the loop is completed, create a graphical representation of the information. Set an average plateau for the graph by taking the average values of the matrix before the coalescence event takes place (command: `ARi1=Image Number` (initial), `ARi2=image number` (final), `AR0=mean(AR(ARi1:ARi2))` (average between initial and final). The average will give an accurate starting point for the exponential decay of the droplets. Choose the final frame for the plot. *Note*: The final frame should be a few frames after the droplets have fused to ensure that the fusion event is captured in its entirety. Plot these values along with the timestep per frame. The resulting graph should look like the graph shown in Fig. 5C (left). Set the average plateau to equal the first data point in your graph.

(j) Fit the decay of aspect ratio as it approaches 1 with the exponential decay form $1 + (AR0 - 1) * e^{-t/\tau}$ where AR0 is the initial aspect ratio and τ is the characteristic relaxation time, as shown in Fig. 5C (right). This will give a value for the decay represented by τ, which is the time of exponential decay as the two droplets fuse.

(k) Plot the characteristic relaxation time (τ) against the length scale, or average size of the initial droplets. Since, $\tau \approx (\eta/\gamma) \bullet \ell$, the slope of τ v.s ℓ corresponds to the inverse capillary velocity, (η/γ), (Fig. 5D).

Additional notes

– If the time step of the videos is too long relative to the fusion rate, the exponential decay may not have sufficient points to provide a reliable fit. (A minimum of 10 data points per decay is recommended).
– If the aspect ratio decay cannot be fit to a single exponential, this may indicate divergence from a pure viscous liquid and/or contributions from interactions with the surface.
– Try to avoid droplets that are too close to the surface, as any wetting of the droplets to the surface due to imperfect coating can inhibit fusion rates and compromise analysis.

3. CONCLUDING REMARKS

Here, we have outlined several methods and strategies for working with disordered proteins that are implicated in liquid phase separation. Optimizing reconstituted phase separation systems in vitro will provide a crucial foundation for uncovering the rules that govern disordered protein driven phase separation in and out of the cell.

REFERENCES

Banani, S. F., Lee, H. O., Hyman, A. A., & Rosen, M. K. (2017). Biomolecular condensates: Organizers of cellular biochemistry. *Nature Reviews. Molecular Cell Biology*, *18*(5), 285–298. https://doi.org/10.1038/nrm.2017.7.

Berry, J., Weber, S. C., Vaidya, N., Haataja, M., & Brangwynne, C. P. (2015). RNA transcription modulates phase transition-driven nuclear body assembly. *Proceedings of the National Academy of Sciences of the United States of America*, *112*(38), E5237–E5245. https://doi.org/10.1073/pnas.1509317112.

Boeynaems, S., Alberti, S., Fawzi, N. L., Mittag, T., Polymenidou, M., Rousseau, F., et al. (2018). Protein phase separation: A new phase in cell biology. *Trends in Cell Biology*, *28*(6), 420–435. https://doi.org/10.1016/j.tcb.2018.02.004.

Boke, E., & Mitchison, T. J. (2017). The balbiani body and the concept of physiological amyloids. *Cell Cycle*, *16*(2), 153–154. https://doi.org/10.1080/15384101.2016.1241605.

Brady, J. P., Farber, P. J., Sekhar, A., Lin, Y. H., Huang, R., Bah, A., et al. (2017). Structural and hydrodynamic properties of an intrinsically disordered region of a germ cell-specific protein on phase separation. *Proceedings of the National Academy of Sciences of the United States of America*, *114*(39), E8194–E8203. https://doi.org/10.1073/pnas.1706197114.

Brangwynne, C. P., Eckmann, C. R., Courson, D. S., Rybarska, A., Hoege, C., Gharakhani, J., et al. (2009). Germline P granules are liquid droplets that localize by controlled dissolution/condensation. *Science*, *324*(5935), 1729–1732. https://doi.org/10.1126/science.1172046.

Brangwynne, C. P., Mitchison, T. J., & Hyman, A. A. (2011). Active liquid-like behavior of nucleoli determines their size and shape in Xenopus laevis oocytes. *Proceedings of the National Academy of Sciences of the United States of America, 108*(11), 4334–4339. https://doi.org/10.1073/Pnas.1017150108.

Brangwynne, C. P., Tompa, P., & Pappu, R. V. (2015). Polymer physics of intracellular phase transitions. *Nature Physics, 11*(11), 899–904. https://doi.org/10.1038/Nphys3532.

Burke, K. A., Janke, A. M., Rhine, C. L., & Fawzi, N. L. (2015). Residue-by-residue view of in vitro FUS granules that bind the C-terminal domain of RNA polymerase II. *Molecular Cell, 60*(2), 231–241. https://doi.org/10.1016/j.molcel.2015.09.006.

Conicella, A. E., Zerze, G. H., Mittal, J., & Fawzi, N. L. (2016). ALS mutations disrupt phase separation mediated by alpha-helical structure in the TDP-43 low-complexity C-terminal domain. *Structure, 24*(9), 1537–1549. https://doi.org/10.1016/j.str.2016.07.007.

Dolgin, E. (2018). What lava lamps and vinaigrette can teach us about cell biology. *Nature, 555*(7696), 300–302. https://doi.org/10.1038/d41586-018-03070-2.

Eggers, J., Lister, J. R., & Stone, H. A. (1999). Coalescence of liquid drops. *Journal of Fluid Mechanics, 401*, 293–310. https://doi.org/10.1017/S002211209900662x.

Elbaum-Garfinkle, S., Kim, Y., Szczepaniak, K., Chen, C. C., Eckmann, C. R., Myong, S., et al. (2015). The disordered P granule protein LAF-1 drives phase separation into droplets with tunable viscosity and dynamics. *Proceedings of the National Academy of Sciences of the United States of America, 112*(23), 7189–7194. https://doi.org/10.1073/pnas.1504822112.

Feric, M., Vaidya, N., Harmon, T. S., Mitrea, D. M., Zhu, L., Richardson, T. M., et al. (2016). Coexisting liquid phases underlie nucleolar subcompartments. *Cell, 165*(7), 1686–1697. https://doi.org/10.1016/j.cell.2016.04.047.

Hyman, A. A., Weber, C. A., & Julicher, F. (2014). Liquid-liquid phase separation in biology. *Annual Review of Cell and Developmental Biology, 30*, 39–58. https://doi.org/10.1146/annurev-cellbio-100913-013325.

Lin, Y., Protter, D. S., Rosen, M. K., & Parker, R. (2015). Formation and maturation of phase-separated liquid droplets by RNA-binding proteins. *Molecular Cell, 60*(2), 208–219. https://doi.org/10.1016/j.molcel.2015.08.018.

Molliex, A., Temirov, J., Lee, J., Coughlin, M., Kanagaraj, A. P., Kim, H. J., et al. (2015). Phase separation by low complexity domains promotes stress granule assembly and drives pathological fibrillization. *Cell, 163*(1), 123–133. https://doi.org/10.1016/j.cell.2015.09.015.

Murakami, T., Qamar, S., Lin, J. Q., Schierle, G. S., Rees, E., Miyashita, A., et al. (2015). ALS/FTD mutation-induced phase transition of FUS liquid droplets and reversible hydrogels into irreversible hydrogels impairs RNP granule function. *Neuron, 88*(4), 678–690. https://doi.org/10.1016/j.neuron.2015.10.030.

Nott, T. J., Petsalaki, E., Farber, P., Jervis, D., Fussner, E., Plochowietz, A., et al. (2015). Phase transition of a disordered nuage protein generates environmentally responsive membraneless organelles. *Molecular Cell, 57*(5), 936–947. https://doi.org/10.1016/j.molcel.2015.01.013.

Saha, S., Weber, C. A., Nousch, M., Adame-Arana, O., Hoege, C., Hein, M. Y., et al. (2016). Polar positioning of phase-separated liquid compartments in cells regulated by an mRNA competition mechanism. *Cell, 166*(6), 1572–1584, e1516. https://doi.org/10.1016/j.cell.2016.08.006.

Shin, Y., & Brangwynne, C. P. (2017). Liquid phase condensation in cell physiology and disease. *Science, 357*(6357). https://doi.org/10.1126/science.aaf4382.

Smith, J., Calidas, D., Schmidt, H., Lu, T., Rasoloson, D., & Seydoux, G. (2016). Spatial patterning of P granules by RNA-induced phase separation of the intrinsically-disordered protein MEG-3. eLife, 5. https://doi.org/10.7554/eLife.21337.

Wang, J. T., Smith, J., Chen, B. C., Schmidt, H., Rasoloson, D., Paix, A., et al. (2014). Regulation of RNA granule dynamics by phosphorylation of serine-rich, intrinsically disordered proteins in *C. elegans*. *eLife, 3*, e04591. https://doi.org/10.7554/eLife.04591.

Wei, M. T., Elbaum-Garfinkle, S., Holehouse, A. S., Chen, C. C., Feric, M., Arnold, C. B., et al. (2017). Phase behaviour of disordered proteins underlying low density and high permeability of liquid organelles. *Nature Chemistry*, *9*(11), 1118–1125. https://doi.org/10.1038/nchem.2803.

Woodruff, J. B., Ferreira Gomes, B., Widlund, P. O., Mahamid, J., Honigmann, A., & Hyman, A. A. (2017). The centrosome is a selective condensate that nucleates microtubules by concentrating tubulin. *Cell*, *169*(6), 1066-1077.e1010. https://doi.org/10.1016/j.cell.2017.05.028.

Zhang, H., Elbaum-Garfinkle, S., Langdon, E. M., Taylor, N., Occhipinti, P., Bridges, A. A., et al. (2015). RNA controls polyQ protein phase transitions. *Molecular Cell*, *60*(2), 220–230. https://doi.org/10.1016/j.molcel.2015.09.017.

CHAPTER THREE

Visualization and Quantitation of Phase-Separated Droplet Formation by Human HP1α

Madeline M. Keenen[*,†], **Adam G. Larson**[*,†], **Geeta J. Narlikar**[*,1]
[*]Department of Biochemistry and Biophysics, University of California, San Francisco, San Francisco, CA, United States
[†]Tetrad Graduate Program, University of California, San Francisco, San Francisco, CA, United States
[1]Corresponding author: e-mail address: geeta.narlikar@ucsf.edu

Contents

Abstract

The ability of the heterochromatin protein-1 (HP1) to phase separate into droplets suggests new mechanisms of gene organization in the cell nucleus. An accumulating body of work suggests that other nuclear proteins also display phase separation behaviors in vitro. To understand the mechanistic and biological significance of such droplet formation a rigorous biophysical characterization of this behavior is necessary. Herein we describe procedures for imaging HP1 droplets by brightfield microscopy, and two methods to quantify phase separation.

Methods in Enzymology, Volume 611
ISSN 0076-6879
https://doi.org/10.1016/bs.mie.2018.09.034

1. INTRODUCTION

A cell densely packed with macromolecules must accomplish a broad range of competing molecular processes. Classic compartmentalization, such as the membrane-enclosed nucleus, provides physical separation between transcription and translation and therefore many points of regulation. Within the nucleus, gene poor and transcriptionally silent heterochromatic regions of the genome are known to be more compact than highly expressed euchromatic regions. Heterochromatin also occupies physically distinct territories than euchromatin. Phase separation, or the emergence of a dilute and concentrated phase of macromolecules in solution, has been proposed to help mediate this distinct packaging (Banani, Lee, Hyman, & Rosen, 2017; Larson & Narlikar, 2018). In particular, heterochromatin protein-1 (HP1) is a major mediator of chromatin compaction and has been shown to form protein-rich droplets in vitro (Larson et al., 2017) (Fig. 1A). Correspondingly, in vivo heterochromatin puncta containing HP1 display dynamic properties that are suggestive of a liquid droplet-like state (Strom et al., 2017).

Molecules of unmodified HP1α are typically miscible up to millimolar concentrations in physiological buffers. Upon phosphorylation of the N-terminal extension (nPhos-HP1α) or the addition of DNA, the solution becomes nonhomogenous and partitions into concentrated and dilute populations (Fig. 1A). The high concentration population will often minimize surface tension by forming round droplets. These manifest in a test tube

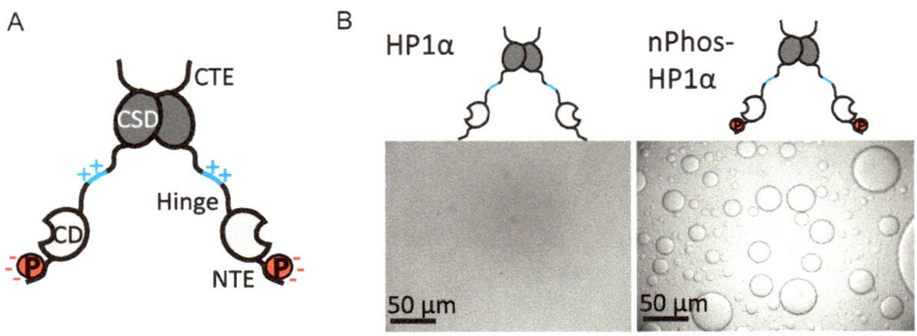

Fig. 1 Phase separation of HP1α. (A) Schematic of nPhos-HP1α containing two globular domains (chromodomain, CD and chromoshadow-domain, CSD), two terminal extensions (CTE and NTE), and a disordered hinge region. Phosphorylation of the NTE binds to the positive patch on the hinge of an adjacent dimer, leading to phase-separated droplets. (B) Brightfield microscopy images of HP1α or nPhos-HP1α visualized in a 384-well plate with a 40 × objective.

as a cloudy, turbid solution. The formation of phase-separated droplets is usually positively correlated with the ability of a protein to make multiple contacts with other proteins in three dimensions (Li et al., 2012). This property, also called multivalency, holds true for HP1α, as it typically exists in solution as a dimer but upon N-terminal phosphorylation forms higher order oligomers. In addition to mutlivalency, the presence of disordered regions has also been shown to correlate with droplet formation (Hyman, Weber, & Jülicher, 2014). In HP1α, the N-terminal region and the hinge region are intrinsically disordered. These regions contribute to oligomerization through electrostatic interactions between the N-terminal phosphates of one dimer and a patch of positive residues on the hinge region of another dimer (Larson et al., 2017; Larson & Narlikar, 2018) (Fig. 1A). Increasing the concentration of monovalent salts increases the concentration of HP1α required to form droplets, consistent with charge shielding of these electrostatic interactions. HP1α droplet formation also decreases as a function of increasing temperature (Larson et al., 2017). Therefore, both the ionic strength of the solvent and the temperature are critical aspects to consider and control when examining HP1α phase separation.

In this chapter we describe methods used to visualize phosphorylated-HP1α droplet formation as well as two methods to quantify phase separation. The first measurement method, termed the spin down assay, utilizes the increase in density of the protein-rich droplets, which pellet to the bottom of the tube with centrifugation. The concentration of protein in the supernatant and pellet can then be determined. The second method is an assay that measures the turbidity of the solution. In this method, light scattering caused by the droplets is assayed by measuring absorbance. Together the techniques can allow reproducible quantification of the phase separation properties of a given protein in a given solvent.

2. PURIFICATION OF nPhos-HP1α

The nPhos-HP1α protein is prepared by coexpressing the kinase CKII and HP1α in bacteria. HP1α contains phosphorylated serine sites in both the N-terminus and hinge domains. To restrict phosphorylation to the N-terminus, we use a HP1α plasmid that has the serine residues in the hinge region mutated to alanines. Further, the construct contains a N-terminal 6×-his tag followed by a TEV cleavage domain that allows for purification via a cobalt column and subsequent cleavage by TEV protease to remove the His tag. Following cleavage, nPhos-HP1α is run over two columns: an anion exchange column (MonoQ) followed by a sizing column

(Superdex 75). The anion exchange step removes the TEV protease and separates HP1α species that are not fully phosphorylated and the sizing column separates leftover contaminants. The resulting nPhos-HP1α protein is ~99% phosphorylated as shown by mass spectrometry (Larson et al., 2017).

Notes

- nPhos-HP1α phase separation is mediated by electrostatic interactions, and therefore is extremely sensitive to the salt concentration. Throughout the purification we use buffer salt concentrations ranging from 150 to 800 m*M* KCl, as described below. We do not lower the salt concentration below 150 m*M* KCl. Salt concentration should be closely monitored as phase separation during concentration steps over a 10K spin concentrator results in protein loss due to permanent sticking of the protein to the concentrator membrane.

Solutions, Reagents, and Equipment

- CKII, in a pRSF-Duet vector (kanamycin resistance)
- HP1α, in a pBH4 vector (carbenicillin resistance)
- *E. coli*. Rosetta (DE3) competent cells (chloramphenicol resistance)
- 2 × LB broth
  - 20 g/L tryptone
  - 10 g/L yeast extract
  - 10 g/L NaCl
- Kanamycin
- Carbenicillin
- Chloramphenicol
- C3 Emulsiflex (Avestin)
- Cobalt-NTA affinity resin (Clontech)
- FPLC
- Mono-Q 4.6/100 PE anion exchange column (GE)
- Superdex-75 16/60 size-exclusion column
- Amicon Ultracel-10K spin concentrator

Lysis Buffer

- 20 m*M* HEPES pH 7.5
- 300 m*M* NaCl
- 10% Glycerol
- 7.5 m*M* imidazole

Elution Buffer

- 20 m*M* HEPES pH 7.5
- 150 m*M* KCl
- 400 m*M* imidazole

Dialysis Buffer
- 20 mM HEPES pH 7.5
- 150 mM KCl
- 3 mM DTT

MonoQ High Salt Buffer
- 20 mM HEPES pH 7.5
- 1 M KCl
- 1 mM DTT

MonoQ Low Salt Buffer
- 20 mM HEPES pH 7.5
- 1 mM DTT

Size Exclusion Chromatography (SEC) Buffer
- 20 mM HEPES pH 7.5
- 200 mM KCl
- 1 mM DTT
- 10% Glycerol

Procedure

Growing Cells

1. Perform a single transformation into Rosetta competent cells with the CKII and HP1α plasmids as per the vendor protocol. As Rosetta cells contain a plasmid with chloramphenicol resistance, cells will be selected on LB plates with 25 μg/mL chloramphenicol, 25 μg/mL kanamycin, and 50 μg/mL carbenicillin. After plating cells, let sit for 12 h at 37°C.
2. The next day make a starter culture by picking a single colony and placing it into 50 mL LB with 25 μg/mL chloramphenicol, 25 μg/mL kanamycin, and 50 μg/mL carbenicillin. Let grow while shaking overnight for 6–12 h.
3. Insert the starter culture media into 1 L 2 × LB supplemented with 25 μg/mL chloramphenicol, 25 μg/mL kanamycin, and 50 μg/mL carbenicillin. Let grow while shaking at 37°C until an OD of 0.4 is reached.
4. Place cells at 18°C and let grow while shaking until an OD of 0.8 is reached.
5. Induce protein expression in cells by adding isopropy-β-D-thiogalactopyranoside (IPTG) to a final concentration of 0.3 mM. Let grow while shaking at 18°C for 14 h.
6. Collect cells by spinning at 4000 × g for 30 min. Remove the media and transfer the cells to a 50-mL conical tube. Cells can be lysed

immediately or frozen in liquid nitrogen and stored at −80°C for future use.

Cell Lysis

1. Resuspend the cell pellet with 30 mL of lysis buffer supplemented with the protease inhibitors 1 m*M* phenylmethanesulfonyl fluoride, 1 μg/mL pepstatin A, 2 μg/mL aprotinin, and 3 μg/mL leupeptin. Vortex and mix with a serological pipette until cells are completely suspended in solution with no residual chunks. A dounce homogenizer is used if vortexing does not result in a homogenous sample.
2. Lyse the cells by passing through a C3 Emulsiflex.
3. Spin down the lysate at 25,000 × *g* for 30 min. Remove supernatant (clarified lysate) into a fresh tube.

Cobalt Column

1. Take 2 mL cobalt bead slurry per liter of culture being purified and transfer into a fresh 50 mL conical. Add water up to top of conical and spin at 1000 × *g* for 3 min. Remove water and repeat one more time.
2. Remove excess water and place 20 mL lysis buffer into tube. Spin at 1000 × *g* for 3 min and remove excess buffer.
3. Add clarified lysate into beads and let incubate at 4°C for 45 min while rotating.
4. Place bead-lysate mixture into a gravity column. Let lysate flow through, but do not let beads run dry.
5. Wash beads with 50 × the bead volume of lysis buffer. Do not add any protease inhibitors to this lysis buffer as this will interfere with the subsequent TEV cleavage step.
6. Collect protein by adding 10 mL of elution buffer to the column. Let slowly flow into a 50-mL conical tube.
7. Add TEV protease into sample at a final concentration of 3 mg/mL.
8. Dialyze sample into 2L of dialysis buffer overnight at 4°C.

MonoQ Column

1. Equilibrate the monoQ column with 15% high salt solution (therefore 150 m*M* KCl; made by mixing 15% High salt MonoQ buffer with 85% low salt MonoQ buffer). Inject sample onto column.
2. Perform salt gradient from 15% high salt (150 m*M* KCl) to 80% high salt (800 m*M* KCl; made by mixing 80% High salt MonoQ buffer with 20% low salt MonoQ buffer) over 15-column volumes. Fully phosphorylated nPhos-HP1α should come off the column at

380 m*M* KCl. There is often multiple peaks prior to the fully phosphorylated species that represent HP1α that is not fully phosphorylated and therefore has a lower net negative charge. We do not collect these species.

3. Collect the nPhos-HP1α fractions and concentrate in a Amicon 10K spin concentrator until the sample is under 500 μL.

S75 Sizing Column

1. Equilibrate the column with SEC buffer.
2. Inject the nPhos-HP1α sample onto the column. The protein normally comes off at 7 mL of a 25-mL column.
3. Collect the fractions containing nPhos-HP1α and concentrate down to a concentration greater than 500 μ*M*.
4. Flash-freeze 50 μL aliquots in liquid nitrogen then store at −80°C until further use.

3. PREPARING nPhos-HP1α SAMPLE FOR EXPERIMENTS

Prior to performing the experiments described below, we dialyze nPhos-HP1α overnight into a salt concentration where phase separation does occur (75m*M* KCl). After dialysis, recovery of protein off the membrane is facilitated by letting the solution sit at room temperature for a few hours. Phase separation is promoted at lower temperatures, and droplets tend to be stickier than miscible protein. Although nPhos-HP1α will phase separate at both 4°C and 22°C, the protein droplets better release from the membrane at room temperature. Concentration measurements can then be taken by an absorbance measurement at 280 nm (A_{280}) on a NanoDrop Spectrophotometer as described below in detail.

Notes

- At room temperature (22°C) droplet formation of nPhos-HP1α is visible at 200 μ*M* in a buffer containing 75 m*M* KCl, 20 m*M* HEPES, pH 7.5, and 1 m*M* DTT.
- If using dialysis tubing with width larger than 10 mm, it is advisable to dialyze a larger volume of protein sample to avoid losing sample. Note that concentration should be maintained above 400 μ*M*.

Solutions, Reagents, and Equipment

- NanoDrop Spectrophotometer
- Dialysis Tubing, 10 kDa molecular weight cutoff, 10 mm width
- Dialysis clips

Dialysis Buffer Components

- 20 m*M* HEPES pH 7.5
- 75 m*M* KCl
- 1 m*M* DTT

Procedure

1. Dialyze phosphorylated nPhos-HP1α overnight at 4°C, preferably 50 μL with a concentration ~400 μ*M* or higher.
   a. Boil dialysis tubing in water.
   b. Squeeze out excess water and immerse in buffer. Squeeze out buffer and repeat a few times.
   c. Clip one end of tubing with dialysis clip. Pipette 50 μL nPhos-HP1α into tubing. Clip other end.
   d. Place into 1 L of dialysis buffer with stir bar at bottom. Let incubate at 4°C overnight while stirring.
2. Let dialyzed protein and buffer sit at room temperature for ~3 h. Take out of dialysis tubing and pipette against bag to resuspend any protein sticking to the sides. Transfer to a fresh tube. Protein solution should look turbid, i.e. cloudy.
3. Dilute and take 280 nm absorbance measurement. To measure the concentration of solution it is important to dilute the turbid solution to a concentration where visible droplet formation does not occur. For this we recommend the following protocol. First, mix the solution with a pipette tip, then take 5 μL of the mixed solution and dilute into a volume of dialysis buffer that results in a clear solution. We typically dilute 1:10, but additional dilution might be required depending on the concentration of your stock solution. Use Beer's law to calculate concentration of solution (extinction coefficient of HP1α is 29,500 M^{-1} cm^{-1}).

4. IMAGING DROPLETS

It is difficult to determine why a solution of purified protein has become cloudy. Common purification issues, such as protein precipitation that occurs when proteins unfold and begin to nonspecifically stick to one another, can also cause turbidity. While these precipitations, unlike droplet formation, are often irreversible, it is important to visualize the solution using a brightfield microscope before making any conclusions. To visualize droplets we either use 384-well plates or chambers crafted out of glass slides, double-sided tape and glass coverslips (Fig. 2). To prevent nonspecific sticking of protein to the glass we first coat the surface as described below.

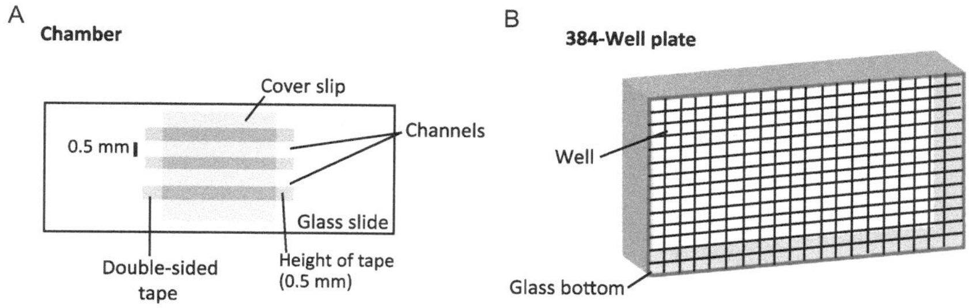

Fig. 2 Tools for droplet visualization. Schematic of chamber setup (A), and 384-well plates with imaging quality glass bottom (B).

4.1 Coating Glass Surfaces With PEG Silane

One common issue with visualizing proteins is the wetting of protein onto glass coverslips due to the dipole moments of silicon dioxide attracting the charged protein. Many procedures are utilized to block the glass surface, such as washing with a relatively neutral protein like BSA or casein before the addition of the turbid protein solution. We utilize a surface reactive PEG derivative, PEG silane that reacts with the silica and blocks the surface from wetting of nPhos-HP1α.

Two different imaging setups are used. The first is a 384-well plate with an imaging quality glass bottom and the second is a set of chambers created from glass slides, cover slips, and double-sided tape.

Notes

- PEG silane is reactive with moisture and light sensitive. Resuspend fresh powder before each use and keep materials in the dark during wash steps.

4.1.1 Well Plate Protocol

Notes

- The thickness of the glass bottom needed depends on the type of microscopy being used. A thickness of 0.15 mm allows visualization with both brightfield and confocal microscopy. However, thicker glass is often more cost-effective and can be easily used with a brightfield microscope. Avoid glass that has surface treatments such as collagen for cell culture.

Solutions and Reagents

- 384-Well plate with imaging quality bottom
- 2% Hellmanex detergent
- 1 *M* NaOH
- Distilled water (dH_20)
- 95% Ethanol
- PEG silane powder

Procedure

1. Pipette 20 µL of 2% Hellmanex into each well. Let sit for 1 h.
2. Wash 3 × with distilled water. Remove Hellmanex, and pipette in 100 µL of dH_2O. Repeat three times with fresh water. Remove excess water.
3. Pipette 20 µL 1 *M* NaOH into each well. Let sit for 30 min. Do not let sit longer, as it might begin to degrade the plastic.
4. Wash 3 × with distilled water, as in step 2.
5. Let sit in hood until residual water evaporates.
6. Dissolve PEG silane in 95% ethanol to a concentration of 20 mg/mL. Pipette 20 µL into each well, cover with parafilm. Let the PEG silane react with the silica overnight (10–14 h) in the dark.
7. Next day wash 3 × with distilled water, as in step 2.
8. Dry the wells, as in step 5.

4.1.2 Chamber Protocol

Solutions and Reagents

- Glass slides 75 × 25mm, 1 mm thickness
- Cover slips 25 × 25mm, .15 mm thickness
- 2% Hellmanex detergent
- 1 *M* NaOH
- Distilled water
- 95% Ethanol
- PEG silane powder

Procedure

1. Immerse glass slides and coverslips in a beaker or slide holder with 2% Hellmanex. Let sit for 1 h.
2. Wash 3 × with distilled water. Hellmanex can be removed and water directly added to slide holders. Repeat three times with fresh water and remove excess water.
3. Immerse glass slides in 1 *M* NaOH. Let sit for 30 min.
4. Wash 3 × with distilled water, as in step 2.
5. Dry slides with nitrogen gas dispensed from a standard nitrogen gas tank. If nitrogen tank is not available the slides can be air-dried in a chemical fume hood for 2–3 h.
6. Dissolve PEG silane in 95% ethanol to a concentration of 20 mg/mL. Pipette 50 µL of solution onto slide and cover with cover slip. Press down to remove any air bubbles. Let the PEG silane react with the silica overnight (10–14 h) in the dark. During this time we prevent the

solution from drying out by placing the slides on the top rack of an empty tip box with half an inch of water in the bottom.

7. Wash 3× with distilled water while keeping track of which side was exposed to PEG silane.
8. Dry slides as in step 5. Drying with nitrogen is preferred as it avoids residue forming on glass surface.

4.2 Creating Chambers

Once glass slides and coverslips are clean, they can be assembled with double-sided tape into chambers with a volume of ~10 μL (Fig. 2A).

Solutions and Reagents

- PEG silane-coated glass slides and cover slips
- Double sided tape, 0.5 mm thick

Procedure

Creating chambers

1. Cut out strips of tape and adhere to the PEG silane-coated side of the slide in parallel fashion, leaving ~0.5 mm in between each strip (Fig. 2).
   a. The length between two strips of tape and the height of tape will determine the volume of sample needed to fill chamber.
   b. Depending on width of tape and channel, you can assemble between 1 and 5 chambers per slide.
2. Adhere cover slip to taped slide. Apply pressure with a pipette tip to remove any pockets of air.

4.3 Imaging Droplets

After preparing chambers or the 384-well plates, the nPhos-HP1α sample can be inserted and visualized with brightfield microscopy.

Notes

- Droplet formation and size is often time sensitive. It is therefore important to measure and standardize the time spent between preparing solutions and imaging on the microscope, especially when comparing different conditions and protein mutants.

Solutions and Reagents

- 384-Well plates OR chambers
- Vacuum grease (optional)

Final Buffer Components

- 20 m*M* HEPES pH 7.5
- 75 m*M* KCl

- 1 mM DTT
- Final protein concentration, 300 μM nPhos-HP1α

Imaging With 384-Well Plate

1. Pipette 20 μL nPhos-HP1α sample into well of plate.
2. Image with a 40 × objective on a brightfield microscope (Fig. 1B).

Imaging With Chambers

1. Pipette nPhos-HP1α sample into space between cover slip and slide. Volume of sample will depend on channel width. Capillary action will pull your sample further into the chamber. If you want to avoid movement of the sample you can seal each side of the chamber with vacuum grease.
2. Image with a 40 × objective on a brightfield microscope.

5. QUANTIFYING PHASE SEPARATION

We utilize two methods to quantitate the phase separation properties of nPhos-HP1α, a spin down assay and a turbidity assay. These two methods quantify two different, though related, properties associated with phase separation. These quantitative measures allow for systematic comparison between mutants and a means to compare the potencies of external ligands that regulate droplet formation.

5.1 Spin Down Assay

This is a simple experiment that utilizes the separation of the solution into two distinct phases. The basis for this assay is akin to a solubility measurement. If the concentration of the protein is greater than can remain soluble, the excess protein separates out in the form of droplets. The droplets, which are denser, will spin down to the bottom of a test tube with centrifugation. The concentration of the supernatant and the pellet can then be determined by measuring absorbance at 280 nm (A_{280}) (Fig. 3). This is a rapid assay,

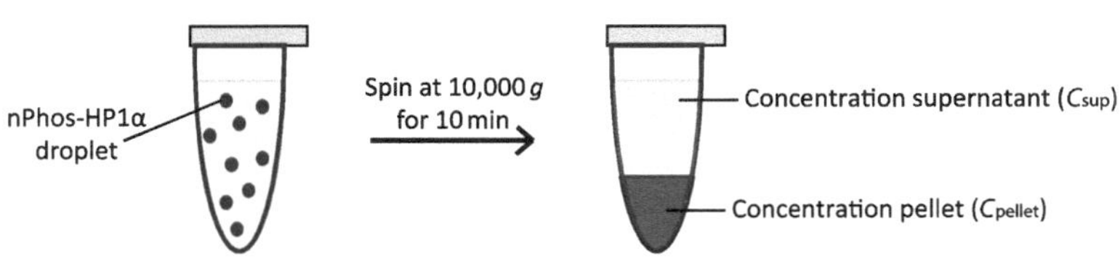

Fig. 3 Spin down assay. nPhos-HP1α forms phase-separated droplets under the conditions listed in Section 4.1. After spinning, the droplets pellet to the bottom and A_{280} reads can be taken to calculate the protein concentration in the supernatant and pellet.

which provides reproducible measurement of the protein concentration that remains in the supernatant. However, the supernatant concentration is dependent on both the density of the droplets and the speed of the centrifuge. Care must be taken to ensure the same rotor speed is maintained across different measurements. Further, it is best practice to visualize the supernatant on a brightfield microscope after centrifugation to see if any droplets remain in the supernatant.

Solutions and Reagents

- Microcentrifuge
- NanoDrop Spectrophotometer
- nPhos-HP1α dialyzed as in Section 4.1

Buffer Components

- 20 m*M* HEPES pH 7.5
- 75 m*M* KCl
- 1 m*M* DTT

Procedure

1. Pipette 10 μL of nPhos-HP1α prepared in Section 4.1 into a new 1.5 mL tube. Let incubate at room temperature for 5 min.
2. Spin solution down at 10,000 × *g* for 5 min in a table top microcentrifuge.
3. Remove 4 μL of supernatant. Take an A_{280} measurement on the NanoDrop Spectrophotometer in triplicate. The average of the three is the supernatant concentration (C_{sup}).
4. The remainder of supernatant can be removed, and A_{280} measurement taken to determine the protein concentration within the phase. However, we typically do not measure the pellet concentration as removing the supernatant can be technically challenging if the pellet is small.

5.2 Turbidity Assay

A classic example of turbidity in biochemistry is the transformation of a solution of tubulin molecules from clear to cloudy upon the induction of polymerization. Longer polymers of tubulin scatter more light than dimers, and this output has often been measured by the absorbance of the solution at 350 nm (Gaskin, Cantor, & Shelanski, 1974; Mirigian, Mukherjee, Bane, & Sackett, 2013). Similarly, nPhos-HP1α is a monomer or dimer at low concentrations but when phase separation occurs, droplets form with much larger radii than an individual dimer. These droplets will scatter light more effectively. We define the $K_{1/2}$ turbidity as the concentration of protein necessary to reach half-maximal turbidity (Fig. 4). Assaying for turbidity

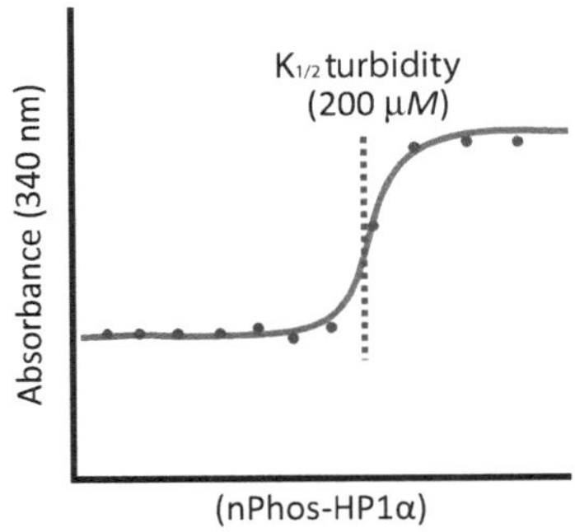

Fig. 4 Turbidity assay. Schematic of turbidity data for nPhos-HP1α, with $K_{1/2}$ turbidity referring to the half-maximal protein concentration.

allows for reproducible, quantitative measurements of phase separation. However, due to the dependence of the scattering signal on droplet size, smaller droplets may be underrepresented. This assay requires a dilution series of nPhos-HP1 and a plate reader that can measure absorbance at 340 nm. If needed the assay can also be adapted to be carried out using a cuvette instead of a plate but this requires larger sample volumes.

Notes

- This procedure can be done with glass 384-well plates treated with PEG silane. After quantitation of 340 absorption the same plate can be taken to the microscope and imaged. Subsequent imaging is best practice, as light scattering at 340 nm is dependent on the size of the droplets.

Solutions and Reagents

- Temperature controlled plate reader that can read absorbance at 340 nm. We utilize the Spectramax M5 plate reader.
- 96- or 384-well plate with clear bottom
- nPhos-HP1α dialyzed in Section 4.1

Buffer Components

- 20 m*M* HEPES pH 7.5
- 75 m*M* KCl
- 1 m*M* DTT

Procedure

1. Beginning at 400 μ*M*, perform a twofold serial dilution using the same buffer as above in a 12-tube PCR strip with a final volume of 25 μL for each sample.
2. Transfer 20 μL of each sample to clear bottom 384-well plate.
3. Let incubate for 5 min.
4. Read absorbance at 340 nm at room temperature on plate reader.

Analyzing Data

The data are plotted and fit with the following sigmoid function, as shown in Fig. 4 schematic.

$$y = A\left(\frac{1}{1 + e^{-k(x - x_C)}}\right) + B$$

A = absorbance value at max concentration; B = absorbance value at min concentration; Xc = inflection point, half-maximal concentration; k = slope factor of sigmoid curve; y = absorbance; x = protein concentration.

6. CHARACTERIZING PHASE SEPARATION OF OTHER PROTEINS

If it is speculated that a protein of interest phase separates, the first hurdle is a successful protein purification, as droplets stick to many common purification tools such as columns and spin concentrators. Analysis of the protein sequence can facilitate this process, and the first relevant question is whether there are sequence stretches predicted to be disordered. If the predicted disordered regions are enriched for charged residues, then purification in a high salt regime (200 mM KCl or greater) is recommended as low salt promotes electrostatic interactions. If instead these regions contain an abundance of hydrophobic residues, purification in a lower salt regime (30–100 mM KCl) may be preferable as increasing salt concentration promotes hydrophobic interactions. Once purification is achieved, some relevant parameters to consider are protein concentration, salt concentration, pH, and crowding conditions. In some experiments researchers have added crowding agents such as PEG and Ficoll, to mimic the macromolecular crowding within cells (Lin, Protter, Rosen, & Parker, 2015). After dialyzing the protein into a buffer space where phase separation is suspected to occur, titrations of the protein and salt concentrations can be performed and imaged as in Section 4. This will allow for visualization of any turbidity and droplet formation, and subsequent quantitation of the two phase separation properties described in Section 5. Care should be taken to match the in vitro phase separation reaction conditions as much as possible to those expected in the cell. Correspondingly, interpretation of the physiological significance of phase separation observed in vitro should take into account potential differences between the in vitro and cellular conditions.

7. CONCLUSIONS

Within this chapter we define our protocol for imaging nPhos-HP1α droplets and two methods to quantitate phase separation properties. The emergence of phase-separated droplets is only one of many causes for the emergence of a turbid solution. Classic examples include protein denaturation and precipitation or tubulin polymerization. Therefore, if presented with turbidity it is essential to image the solution to determine the cause.

Phase separation of proteins in vitro is strongly linked to the ionic strength of the solvent, the temperature and protein concentration used. A thorough investigation of additional buffer parameters like mono or divalent salts and pH while using the above techniques can facilitate teasing apart the mechanisms of multivalency. To assess the physiological relevance of droplet formation in vitro, it is important to (i) match as best as possible the in vitro buffer conditions to those in the corresponding cellular context and (ii) determine how the cellular concentration of a protein compares with the C_{sup} and $K_{1/2}$ turbidity values.

REFERENCES

Banani, S. F., Lee, H. O., Hyman, A. A., & Rosen, M. K. (2017). Biomolecular condensates: Organizers of cellular biochemistry. *Nature Reviews. Molecular Cell Biology*, *18*(5), 285–298. https://doi.org/10.1038/nrm.2017.7.

Gaskin, F., Cantor, C., & Shelanski, M. L. (1974). Turbidimetric studies of the in vitro assembly and disassembly of porcine neurotubules. *Journal of Molecular Biology*, *89*, 737–758.

Hyman, A. A., Weber, C. A., & Jülicher, F. (2014). Liquid-liquid phase separation in biology. *Annual Review of Cell and Developmental Biology*, *30*(1), 39–58. https://doi.org/10.1146/annurev-cellbio-100913-013325.

Larson, A. G., Elnatan, D., Keenen, M. M., Trnka, M. J., Johnston, J. B., Burlingame, A. L., et al. (2017). Liquid droplet formation by HP1α suggests a role for phase separation in heterochromatin. *Nature*, *547*(7662), 236–240. https://doi.org/10.1038/nature22822.

Larson, A. G., & Narlikar, G. J. (2018). The role of phase separation in heterochromatin formation, function, and regulation. *Biochemistry*, *57*(17), 2540–2548. https://doi.org/10.1021/acs.biochem.8b00401.

Li, P., Banjade, S., Cheng, H.-C., Kim, S., Chen, B., Guo, L., et al. (2012). Phase transitions in the assembly of multivalent signalling proteins. *Nature*, *483*(7389), 336–340. https://doi.org/10.1038/nature10879.

Lin, Y., Protter, D. S. W., Rosen, M. K., & Parker, R. (2015). Formation and maturation of phase-separated liquid droplets by RNA-binding protein. *Molecular Cell*, *60*(2), 208–219.

Mirigian, M., Mukherjee, K., Bane, S. L., & Sackett, D. L. (2013). Measurement of in vitro microtubule polymerization by turbidity and fluorescence. *Microtubules, in vitro: Vol. 115*. (pp. 215–229). Elsevier. https://doi.org/10.1016/B978-0-12-407757-7.00014-1.

Strom, A. R., Emelyanov, A. V., Mir, M., Fyodorov, D. V., Darzacq, X., & Karpen, G. H. (2017). Phase separation drives heterochromatin domain formation. *Nature*, *547*(7662), 241–245. https://doi.org/10.1038/nature22989.

CHAPTER FOUR

Probing RNA Structure in Liquid–Liquid Phase Separation Using SHAPE-MaP

Erin M. Langdon*, Amy S. Gladfelter*,†,1

*Department of Biology, University of North Carolina at Chapel Hill, Chapel Hill, NC, United States
†Marine Biological Laboratory, Woods Hole, MA, United States
[1]Corresponding author: e-mail address: amyglad@unc.edu

Contents

Abstract

RNA is an integral component of many biological condensates. A variety of features of RNAs are linked to their function in biological phase separation. Length and negative charge provide fairly generic chemical inputs that drive condensation while sequence has been shown to influence both the molecular identity and biophysical properties of droplets. mRNA sequence guides the secondary structure of the polymers and RNA secondary structure licenses-specific RNA–RNA interactions and the recruitment of RNA-binding proteins. Here, we describe a method for directly probing the structure of mRNAs in the context of RNP-droplets formed via LLPS.

Methods in Enzymology, Volume 611
ISSN 0076-6879
https://doi.org/10.1016/bs.mie.2018.09.039

1. INTRODUCTION

Membrane-less RNA–protein (RNP) bodies are a key mode of intracellular organization in which diverse RNA and protein molecules coalesce for an array of cellular functions. Recent work has shown that these bodies often display liquid-like properties, assemble through disordered protein sequences and RNA, and can be modeled as liquid–liquid phase separations (LLPS) based on in vitro reconstitution (Banani, Lee, Hyman, & Rosen, 2017; Hyman, Weber, & Jülicher, 2014; Shin & Brangwynne, 2017). A growing list of compartments and processes have been shown to be governed by LLPS, including but not limited to the nucleolus, RNA granules, cell signaling networks, chromatin, transcription regulation, and the synaptonemal complex (Banjade & Rosen, 2014; Brangwynne et al., 2009; Li et al., 2012; Lin, Protter, Rosen, & Parker, 2015; Molliex et al., 2015; Rog, Kohler, & Dernburg, 2017; Saha et al., 2016; Smith et al., 2017; Strom et al., 2017; Weber & Brangwynne, 2015; Zhang et al., 2015).

To determine mechanisms for how distinct assemblies may emerge in the cytoplasm, we developed a simplified reconstitution of RNA–protein droplets in vitro with purified protein and mRNAs (Fig. 1). This model LLPS system uses a fungal RNA-binding protein called Whi3 which has an RRM-motif and an extended poly-glutamine (polyQ) tract along with mRNAs that bind to this protein. Our work has focused on mRNAs encoding cell cycle (the cyclin, *CLN3*) and cell polarity (formin, *BNI1*) factors (Lee, Occhipinti, & Gladfelter, 2015; Lee et al., 2013). Purified Whi3 protein can phase separate on its own in the absence of mRNA at

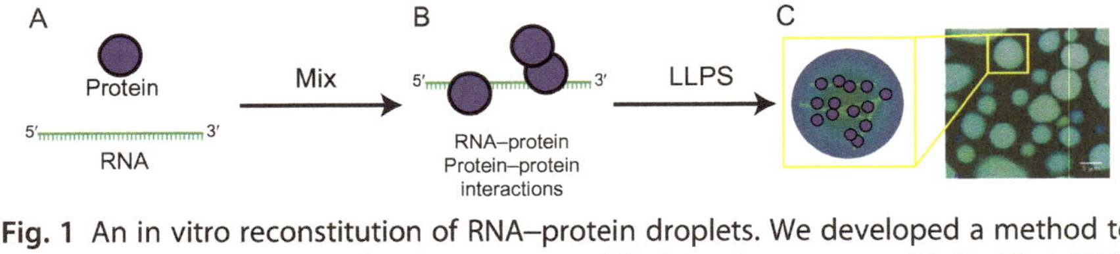

Fig. 1 An in vitro reconstitution of RNA–protein droplets. We developed a method to reconstitute RNA–protein droplets in a simplified in vitro system. (A) Purified RNA-binding proteins and in vitro transcribed RNAs are mixed together. (B) Protein binds the RNA by sensing specific sequences on the RNA. Further protein is recruited through protein–protein interactions. (C) Altogether, the RNA–protein complexes undergo liquid–liquid phase separation, forming large spherical droplets in vitro, which can be imaged using RNA dye and fluorescently tagged protein.

low salt concentrations, but RNA lowers the critical protein concentration needed to de-mix at physiological conditions. Using this method, we also showed that different RNAs produce droplets with different protein densities and distinct biophysical properties (Langdon et al., 2018; Zhang et al., 2015).

It is well established that RNA can modulate the protein-binding behavior of RNA-binding proteins in certain structural contexts (Jain, Lin, Morgan, Harris, & Tolbert, 2017; Taliaferro et al., 2016; Tamayo, Teramoto, Chatterjee, Hall, & Gavis, 2017); however, until recently, it has been poorly understood how certain RNA sequences could affect the assembly or properties of RNPs in LLPS. RNA has been shown to be a driver of LLPS and can modulate the material properties of droplets (Elbaum-Garfinkle et al., 2015; Taylor et al., 2016; Zhang et al., 2015); however, there has been limited understanding how specific RNAs are recruited to and maintained in distinct droplets. We recently found that the structure of mRNAs is a key determinant of the identity of droplets formed by LLPS and is critical for the sorting of different mRNAs into distinct droplets. To identify this mechanism for droplet identity, the droplet assay for Whi3 and target RNAs was coupled with the already established SHAPE-MaP protocol (Smola, Rice, Busan, Siegfried, & Weeks, 2015), which distinguishes between folded and unfolded structures along mRNA sequences. Using this approach, we probed intradroplet RNA structure and identified secondary structure information of mRNAs and determined secondary structure changes on mRNA in the presence of protein (Fig. 2) (Langdon et al., 2018). The method described below should be amenable for use with many combinations of RNAs and RBPs that condense into droplets to identify how RNA structures both influence and are influenced by LLPS.

2. MATERIALS

2.1 Recombinant Protein Expression and Purification

1. BL21 DE3 *Escherichia coli* (New England Biolabs) and plasmid capable of expressing a 6-histidine tag with and without GFP tag.
2. Nickel NTA Resin (Qiagen), gravity columns.
3. Lysis buffer (1.5 *M* KCl, 20 m*M* Tris pH 8.0, 20 m*M* Imidazole pH 8, 1 m*M* DTT, 1 tablet of Roche protease inhibitor cocktail).
4. Elution buffer (150 m*M* KCl, 20 m*M* Tris pH 8.0, 200 m*M* Imidazole pH 8.0, 1 m*M* DTT).

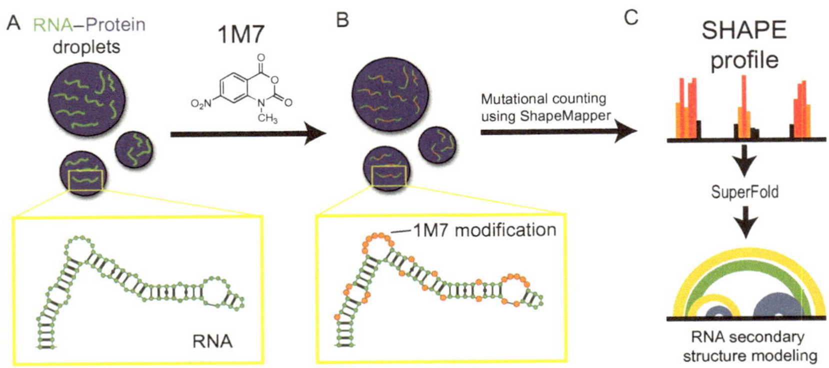

Fig. 2 The DropSHAPE method to determine secondary structure of mRNAs within LLPS droplets. Combining SHAPE-MaP with our in vitro droplet assay allows us to probe RNA secondary structure within droplets. (A) After RNA–protein droplets are allowed to form, we treat with the SHAPE reagent 1M7. (B) 1M7 modifies unpaired nucleotides on the RNA in a structure-specific manner (*orange*). (C) Using the ShapeMapper algorithm, we can determine the SHAPE profile from mutations induced by special reverse transcription coupled with deep sequencing of treated droplets. Using SuperFold, we can further elucidate the secondary structure of the droplet RNA through modeling.

5. Dialysis/storage buffer (150 m*M* KCl, 20 m*M* Tris pH 8.0, 1 m*M* DTT).
6. Dialysis cassettes.
7. Liquid nitrogen.

2.2 In Vitro RNA Transcription and Purification

1. Linearized plasmid containing T7 promoter (5′ TAATACGACTCACTATAGGG) before the gene-coding sequence of interest.
2. In vitro transcription kit and a labeled nucleotide.
3. DNAse reaction buffer and enzyme (NEB), EDTA pH 8.0.
4. RNASE-free reagents (water, TE buffer pH 8.0, Ethanol, Lithium Chloride).

2.3 Droplet Assembly and Imaging

1. Recombinant protein (unlabeled and labeled).
2. Labeled in vitro transcribed RNA.
3. Droplet buffer (150 m*M* KCl, 20 m*M* Tris pH 8.0, 1 m*M* DTT).
4. Bovine serum albumin (30 mg/mL).
5. Chambered cover glass (Grace-Bio).
6. Microscope suitable for fluorescent imaging.

2.4 Probing Intradroplet RNA Structure Using Selective 2′-Hydroxyl Acylation Analyzed by Primer Extension and Mutational Profiling (SHAPE-MaP)

1. Recombinant protein (unlabeled and labeled).
2. Labeled in vitro transcribed RNA.
3. RNASE-free Droplet buffer (150 m*M* KCl, 20 m*M* Tris pH 8.0, 1 m*M* DTT).
4. RNASE-free Bovine serum albumin (30 mg/mL).
5. Chambered cover glass (Grace-Bio).
6. DMSO and 1-methyl-7-nitroisatoic anhydride (1M7).
7. Heat block capable of reaching 100°C.
8. RNA and DNA purification beads (Agencourt) and magnet.
9. Superscript II reverse transcriptase (Invitrogen) RT primer, step 1 and step 2 PCR primers (Integrated DNA Technologies, custom synthesis) (see Note 1).
10. MaP buffer: 5 × (250 m*M* Tris (pH 8.0), 375 m*M* KCl, 50 m*M* DTT, and 2.5 m*M* each dNTP. This solution is stable for months at −20°C, but storage of small aliquots at −20°C is critical. Discard the buffer after five freeze–thaw cycles. 2.5 × (125 m*M* Tris (pH 8.0), 187.5 m*M* KCl, 15 m*M* $MnCl_2$, 25 m*M* DTT, and 1.25 m*M* dNTPs). Prepare this solution fresh by mixing equal volumes of 5 × MaP buffer and 30 m*M* $MnCl_2$.
11. Deoxynucleotide triphosphates (dNTPs; 10 m*M* each nucleotide; New England BioLabs).
12. Illumina sequencing primers and MiSeq instrument.
13. ShapeMapper and SuperFold software available from http://www.chem.unc.edu/rna/software.

3. METHODS

3.1 Production and Maintenance of Protein

1. In the evening, inoculate 25 mL LB culture containing resistance antibiotic with a colony of freshly grown Bl21 (DE3) *E. coli* containing the construct. Incubate shaking at 37°C overnight.
2. Prepare a 1 L Terrific Broth culture containing resistance antibiotic. Dilute starter culture 1:60 and grow at 37°C with shaking for 2 h, then drop to 18°C and grow until culture reaches an OD 0.6–0.7. At this

point, induce culture with IPTG to a final concentration of 1 m*M* and incubate with shaking at 18°C for 20 h.

3. Pellet induced *E. coli* at 10,000 × rpm. Pour off supernatant and immediately begin with lysis or freeze pellet at −80 for purification later.
4. Resuspend pellets in a total of 25 mL lysis buffer, 1.5 mL 10 mg/mL lysozyme, 1 tablet of Roche complete protease inhibitor. Resuspend over the course of 30 min. Transfer pellet to a 50-mL falcon tube and sonicate for 10 s. Allow foam to settle for ~2 min. Sonicate for another 10 s.
5. Pellet lysed and sonicated *E. coli* at top speed (20,000 rpm) for 30 min. After spin, pass supernatant through 0.44-μm filter using a 60-mL syringe. This usually requires changing the filter every 5 mL.
6. Ni-NTA Bed should be made using 2 mL NTA resin for each liter of cells. Equilibrate column with lysis buffer by 2–5 × column volume washes. Load filtered supernatant from cleared lysate onto the equilibrated Ni-NTA column and allow to run through the column, collecting a small amount of sample for SDS-PAGE. Wash the Ni-NTA column with 3–5 × column volumes of wash buffer. Finally, elute the protein with 6–10 mL elution buffer.
7. Analyze by SDS-PAGE and dialyze into storage buffer. Proteins can be used immediately and then stored at 4°C for a few weeks. It is best to aliquot freshly purified proteins into volumes 200 μL or less and flash freeze and store at −80°C. Frozen protein is useable for up to a year, although some hydrolysis will occur. Proteins should be thawed on ice and then spun at top speed to pellet any aggregated proteins in the solution. Remove the supernatant and then remeasure protein concentration before using. (See Note 1 for challenges and considerations for protein purification.)

3.2 Transcription and Maintenance of Labeled RNA

1. Make sure the plasmid is linearized by PCR or restriction digest and that only one copy of the T7 promoter is present.
2. Assemble linearized DNA, nucleotides (ATP, UTP, CTP, GTP, and UTP cy3 or cy5), T7 polymerase and follow manufacturer instructions for RNA transcription using modified nucleotides: 1.5 μL each of ATP; UTP; CTP; GTP; and 0.1 μL UTP cy3 or cy5; 1.5 μL buffer; 10.9 μL of linear DNA, and 1.5 μL of T7 polymerase. This protocol uses NEB Hiscribe T7 in vitro transcription kit, but other kits can be used.
3. Incubate overnight at 37°C.

4. Bring the volume of reactions to 90 μL with nuclease-free H_20 and add 10 μL of Dnase reaction buffer plus 2 units of DNASE enzyme. Incubate at 37°C for 10 min. Stop reactions by adding 1 μL of 0.5 *M* EDTA at 65° C for 5 min.
5. Add 2.5 *M* LiCl solution (dissolved in nuclease-free water) to half the volume of the DNASE reaction and incubate at −20°C for 4–16 h.
6. Pellet RNA at top speed at 4°C for 10 min. Aspirate supernatant and wash pellet with ice-cold 80% ETOH 3 times.
7. Let pellet air-dry and resuspend in 200 μL TE buffer and store at −80°C.

3.3 Droplet Assembly and Imaging

1. Pretreat chambered cover glass with 30 mg/mL BSA (Sigma) for 30 min. Wash chambers 3 times with droplet buffer and let air-dry.
2. All proteins used in experiments contain 10:1 unlabeled to GFP-labeled versions. To lower salt concentration for protein-only phase separation, mix dialyzed proteins with no salt buffer (50 m*M* Tris [pH 8], 1 m*M* DTT) to obtain desired protein and salt concentrations. For RNA experiments, dilute proteins with droplet buffer (50 m*M* Tris [pH 8], 1 m*M* DTT, 150 m*M* KCl) to desired concentration and then add desired concentrations of unlabeled or labeled RNA.
3. Allow proteins and RNAs to incubate at room temperature for 1–4 h or until phase separation occurs and droplets form. This will vary depending on concentrations and specific components used as well as desired state of droplet, i.e., liquid or hardened (see Note 2).
4. Mount chamber on microscope and screening for droplets using phase contrast, epifluorescence, or laser light. When imaging, make sure the field contains a few saturated pixels, but that most of the image is unsaturated for proper quantification. Droplets can be imaged for quantification of size, protein and RNA recruitment and concentration, fusion kinetics and biophysical properties by performing microrheology or FRAP experiments (see methods of Brangwynne, Mitchison, & Hyman, 2011; Elbaum-Garfinkle et al., 2015; Langdon et al., 2018; Zhang et al., 2015).

3.4 Probing Intradroplet RNA Structure Using Selective 2′-Hydroxyl Acylation Analyzed by Primer Extension and Mutational Profiling (SHAPE-MaP)

1. Pretreat chambered cover glass with 30 mg/mL BSA (Sigma) for 30 min. Wash chambers 3 times with droplet buffer and let air-dry.

2. Dilute protein with droplet buffer (50 mM Tris [pH 8], 1 mM DTT, 150 mM KCl) to desired concentration and then add desired concentrations of unlabeled or labeled RNA in a total volume of 200 μL.
3. Allow proteins and RNAs to incubate at room temperature for 1–4 h or until phase separation occurs and droplets form. You will need two wells for each RNA probed with 1M7, a positive (1M7) and a negative (+DMSO).
4. To *empty* negative wells add 20 μL of DMSO and then add 200 μL of negative reactions from step 3 and mix. To *empty* positive wells, add 20 μL of 100 mM 1M7 dissolved in DMSO (final concentration 1 mM) and then add 200 μL of positive reactions from step 3 and mix. Incubate at room temperature for 10 min.
5. To denature the proteins and extract the modified RNA, heat the chambers on a heat block or in an oven to 95°C for 10–15 min.
6. Add 1.6 × volume of reaction of Agencourt RNAClean XP beads to the wells and incubate for 5 min. Move reaction to small PCR tubes and then place on a magnet and wait for liquid to separate from the beads. Remove liquid and wash beads 3 times with 80% ETOH. Let beads air-dry and elute RNA off beads in 30 μL of RNASE-free water.
7. Perform reverse transcription with specific primers (see Note): Add 10 μL of (+), (−) to separate PCR tubes. To each tube, add 1 μL of 2 μM reverse transcription primer. Incubate at 65°C for 5 min and then cool on ice (see Note 3).
8. Add 8 μL of 2.5 × MaP buffer to each tube. Mix well and incubate at 42°C for 2 min.
9. Add 1 μL of SuperScript II reverse transcriptase to each tube and mix well. Immediately incubate the reaction tubes at 42°C for 3 h.
10. Heat the reactions to 70°C for 15 min to inactivate SuperScript II reverse transcriptase. Place them at 4°C.
11. Purify cDNA from the (+) and (−) reactions using Agencourt AMPure XP beads, according to the manufacturer's instructions. Proceed to library preparation and sequencing.
12. Follow library preparation protocol (step 20A as per Smola et al., 2015). In brief,
    (a) Setup step 1 PCR per experimental condition in PCR plate. Use thermocycler to cycle through the program as described in Smola et al. (2015).
    (b) Purify step 1 PCR products from (+) and (−) reactions using PureLink PCR microspin columns. Elute into 10 μL H_2O.

(c) Setup step 2 PCR per experimental conditions in PCR plate. Use program as described in Smola et al. (2015).
(d) Remove PCR plate from thermocycler. Add 45 μL of Agencourt AMPure XP beads to each well. Mix thoroughly. Incubate at room temperature for 5 min.
(e) Plate PCR plate on magnetic stand for 2 min or until supernatant is clear. Remove and discard supernatant.
(f) Still on magnetic stand, add 200 μL of 80% ethanol to each well. Incubate for 30 s, then remove and discard ethanol.
(g) Repeat step 12f twice more. Remove any ethanol remaining by 10 μL pipet.
(h) Still on magnetic stand, let air-dry 15 min.
(i) Remove plate from magnetic stand. Add 17 μL H_2O to each well. Pipet up and down to mix well. Incubate at room temperature for 10 min.
(j) Place plate on magnetic stand for 2 min or until supernatant is clear.
(k) Remove 15 μL of *DNA-containing* supernatant from each well and place each in separate 1.7-mL microfuge tubes.

13. Measure the library concentration using a high-sensitivity DNA detection assay such as Qubit fluorometer using the Qubit™ dsDNA HS Assay Kit from Thermo Fischer Scientific.
14. Evaluate the library size distribution using an Agilent High Sensitivity DNA analysis kit to be run on an Agilent Bioanalyzer according to the manufacturer's instructions. Libraries should appear as a single, well-defined peak.
15. Calculate library molarity by calculating the molecular weight of dsDNA based on the input RNA sequence, and then add 81,000 g/mol to account for adapter sequences. Use this final molecular weight and the mass concentration from Step 13 to calculate the concentration (in n*M*) of the library. Depending on library preparations and starting sequence length the size range and mass concentration will vary. The final concentration of pooled libraries for sequencing should be 2 n*M*, with ideally each sample being close to that concentration before pooling.
16. Sequence libraries on an Illumina MiSeq or another sequencer according to the instrument instructions. Configure the sequencer to produce, FASTQ-formatted output.
17. Follow instructions in Smola et al. (2015) ShapeMapper analysis as described. In brief,

(a) Create a "RUN" folder. Ensure no spaces are present in the entire path to the directory.
(b) Copy/move uncompressed FASTQ files from previous step into RUN folder.
(c) Create a FASTA file containing each target sequence of interest. Use all capital letters for the nucleotide sequences.
(d) Copy EXAMPLE.cfg file provided with ShapeMapper into RUN folder and rename file.
(e) Edit configuration file using text editor as needed for your setup (e.g., paired-end, random priming, Nextera library prep). See Smola et al. (2015) for more details. Then, add sample names followed by alignment targets to the "[alignments]" section. Specify samples to be combined into reactivity profiles in "[profiles]" section.
(f) In command line (terminal), navigate to RUN directory. Run ShapeMapper by calling it like so: "ShapeMapper.py [config.cfg]." Upon completion, reactivity profiles will be written to "reactivity profiles" directory within the output folder.
(g) Use SHAPE reactivities to guide secondary structural modeling in RNA folding programs, such as SuperFold.

4. NOTES

1. Challenges and considerations for purification of phase-separated proteins:

The aforementioned protocol is a starting point for initial purification for phase-separating proteins. Due to their nature to precipitate, pull down unwanted DNA and RNA, one must consider alterations to this protocol to suit ones needs. For example, adding a nuclease into the lysate could aid in the removal of unwanted nucleic acids that could have deleterious downstream effects in experimental assays. In our hands, we have found that laboratory made or store bought Terrific Broth yields best induction and replication results of the bacteria producing the proteins. Further, the sonication step is crucial in assuring that the cells are lysed properly. In addition, you might find that your supernatant is sticky or very viscous due to the natural properties of these proteins, therefore, the filtering of the lysate before incubating it with the beads allows the proteins to pass through the column more efficiently and not aggregate the beads. Many tags can be used to purify these types of proteins; however, in our hands, a 6-histidine tag has worked best, but others have had success with MBP, HALO, and Streptavidin tags. Further, the solutions and salt concentrations are particular to our protein, but could serve as a good starting point, but other buffers such as HEPES and salts such as NaCl should also be considered in addition to different solution pH as well. Each of these phase separating proteins has their unique set of considerations, and this protocol is not an exhaustive list of them, but merely a starting point and general guideline.

2. Thoughts on starting points for RNA/protein droplet assay:

   To determine conditions for phase separation of protein and/or RNA in a simplified reconstitution of purified protein and RNAs, it is best to start with relatively low salt (20–75 m*M*) and high protein (25–100 μ*M*) concentrations in order to drive together the multivalent interactions between the phase-separating domains along the proteins. This can also be tried with RNA alone or by adding an electrostatic interactor into the mix such as the positively charged molecule spermine. RNA will typically lower the critical concentration for phase separation due to increased electrostatic and multivalent interactions between the RNA molecules and the proteins, thus that assay could be done in a more high, physiological salt such as 150 m*M* and lower protein and RNA concentrations in the 10 n*M*–10 μ*M* range. Depending on the goals of the assay, molecule constituents, concentrations, and buffer conditions, the time of phase separation can vary widely. The above considerations can be experimented with and fine-tuned in order to drive robust phase separation or to dissolve it. In addition, molecular crowders such as polyethylene glycol, ficoll, and others can also be added into the assay to change the phase separation process. Again, these are all considerations that should be taken into account, however, which considerations matter depends on the individual protein and RNA molecules in addition to the goals of the assay.

3. Modified RT-PCR primers for small RNAs:

   PCR is performed in two steps (see Smola et al. Tables 3 and 4). The first step uses RNA-specific primers (i.e., forward sense and reverse antisense), with an adjacent random 5-nt sequence for sequencing. The second step is performed using universal primers with a 6-nt "barcode" sequence for sample identification. These primers are not specific to the RNA of interest and can be reused.

5. OUTLOOK

This method combines biochemical and structural biology techniques to investigate RNA–protein interactions and their influence on RNA secondary structure within LLPS droplets. As biological phase transitions become more and more appreciated as a general mechanism of cell organization across many types of cells and organisms, we believe this method will be applicable to study LLPS in these diverse backgrounds, including pathological diseases. As RNA secondary structure is a way to control the composition of droplets, we hope this method will illuminate other situations in which RNA secondary structure may influence and be influenced by LLPS with RNA-binding proteins. Further, as SHAPE-MaP and other structural techniques that combine next-generational sequencing technologies are improved upon, these advances will be easily incorporated into this method.

REFERENCES

Banani, S. F., Lee, H. O., Hyman, A. A., & Rosen, M. K. (2017). Biomolecular condensates: Organizers of cellular biochemistry. *Nature Publishing Group*, *18*(5), 285–298. https://doi.org/10.1038/nrm.2017.7.

Banjade, S., & Rosen, M. K. (2014). Phase transitions of multivalent proteins can promote clustering of membrane receptors. *eLife*, *3*, 1–24. https://doi.org/10.7554/eLife.04123.

Brangwynne, C. P., Eckmann, C. R., Courson, D. S., Rybarska, A., Hoege, C., Gharakhani, J., et al. (2009). Germline P granules are liquid droplets that localize by controlled dissolution/condensation. *Science*, *324*(5935), 1729–1732. https://doi.org/10.1126/science.1172046.

Brangwynne, C. P., Mitchison, T. J., & Hyman, A. A. (2011). Active liquid-like behavior of nucleoli determines their size and shape in Xenopus laevis oocytes. *Proceedings of the National Academy of Sciences of the United States of America*, *108*(11), 4334–4339. https://doi.org/10.1073/pnas.1017150108.

Elbaum-Garfinkle, S., Kim, Y., Szczepaniak, K., Chen, C. C.-H., Eckmann, C. R., Myong, S., et al. (2015). The disordered P granule protein LAF-1 drives phase separation into droplets with tunable viscosity and dynamics. *Proceedings of the National Academy of Sciences of the United States of America*, *112*(23), 7189–7194. https://doi.org/10.1073/pnas.1504822112.

Hyman, A. A., Weber, C. A., & Jülicher, F. (2014). Liquid-liquid phase separation in biology. *Annual Review of Cell and Developmental Biology*, *30*(1), 39–58. https://doi.org/10.1146/annurev-cellbio-100913-013325.

Jain, N., Lin, H.-C., Morgan, C. E., Harris, M. E., & Tolbert, B. S. (2017). Rules of RNA specificity of hnRNP A1 revealed by global and quantitative analysis of its affinity distribution. *Proceedings of the National Academy of Sciences of the United States of America*, *114*(9), 2206–2211. https://doi.org/10.1073/pnas.1616371114.

Langdon, E. M., Qiu, Y., Ghanbari Niaki, A., McLaughlin, G. A., Weidmann, C. A., Gerbich, T. M., et al. (2018). mRNA structure determines specificity of a polyQ-driven phase separation. *Science*, *360*(6391), 922–927. https://doi.org/10.1126/science.aar7432.

Lee, C., Occhipinti, P., & Gladfelter, A. S. (2015). PolyQ-dependent RNA–protein assemblies control symmetry breaking. *The Journal of Cell Biology*, *208*(5), 533–544. https://doi.org/10.1083/jcb.201407105.

Lee, C., Zhang, H., Baker, A. E., Occhipinti, P., Borsuk, M. E., & Gladfelter, A. S. (2013). Protein aggregation behavior regulates cyclin transcript localization and cell-cycle control. *Developmental Cell*, *25*(6), 572–584. https://doi.org/10.1016/j.devcel.2013.05.007.

Li, P., Banjade, S., Cheng, H.-C., Kim, S., Chen, B., Guo, L., et al. (2012). Phase transitions in the assembly of multivalent signalling proteins. *Nature*, *483*(7389), 336–340. https://doi.org/10.1038/nature10879.

Lin, Y., Protter, D. S. W., Rosen, M. K., & Parker, R. (2015). Formation and maturation of phase-separated liquid droplets by RNA-binding proteins. *Molecular Cell*, *60*(2), 208–219. https://doi.org/10.1016/j.molcel.2015.08.018.

Molliex, A., Temirov, J., Lee, J., Coughlin, M., Kanagaraj, A. P., Kim, H. J., et al. (2015). Phase separation by low complexity domains promotes stress granule assembly and drives pathological fibrillization. *Cell*, *163*(1), 123–133. https://doi.org/10.1016/j.cell.2015.09.015.

Rog, O., Kohler, S., & Dernburg, A. F. (2017). The synaptonemal complex has liquid crystalline properties and spatially regulates meiotic recombination factors. *eLife*, *6*, 1–26. https://doi.org/10.7554/eLife.21455.001.

Saha, S., Weber, C. A., Nousch, M., Adame-Arana, O., Hoege, C., Hein, M. Y., et al. (2016). Polar positioning of phase-separated liquid compartments in cells regulated by an mRNA competition mechanism. *Cell*, *166*(6), 1572–1584.e16. https://doi.org/10.1016/j.cell.2016.08.006.

Shin, Y., & Brangwynne, C. P. (2017). Liquid phase condensation in cell physiology and disease. *Science (New York, NY)*, *357*(6357), eaaf4382–13. https://doi.org/10.1126/science.aaf4382.

Smith, J., Calidas, D., Schmidt, H., Lu, T., Rasoloson, D., & Seydoux, G. (2017). Spatial patterning of P granules by RNA-induced phase separation of the intrinsically-disordered protein MEG-3. *eLife*, *10*, 1–18. https://doi.org/10.7554/eLife.21337.001.

Smola, M. J., Rice, G. M., Busan, S., Siegfried, N. A., & Weeks, K. M. (2015). Selective 2′-hydroxyl acylation analyzed by primer extension and mutational profiling (SHAPE-MaP) for direct, versatile and accurate RNA structure analysis. *Nature Protocols*, *10*(11), 1643–1669. https://doi.org/10.1038/nprot.2015.103.

Strom, A. R., Emelyanov, A. V., Mir, M., Fyodorov, D. V., Darzacq, X., & Karpen, G. H. (2017). Phase separation drives heterochromatin domain formation. *Nature*, *547*(7662), 241–245. https://doi.org/10.1038/nature22989.

Taliaferro, J. M., Lambert, N. J., Sudmant, P. H., Dominguez, D., Merkin, J. J., Alexis, M. S., et al. (2016). RNA sequence context effects measured in vitro predict in vivo protein binding and regulation. *Molecular Cell*, *64*(2), 294–306. https://doi.org/10.1016/j.molcel.2016.08.035.

Tamayo, J. V., Teramoto, T., Chatterjee, S., Hall, T. M. T., & Gavis, E. R. (2017). The Drosophila hnRNP F/H homolog Glorund uses two distinct RNA-binding modes to diversify target recognition. *Cell Reports*, *19*(1), 150–161. https://doi.org/10.1016/j.celrep.2017.03.022.

Taylor, N., Elbaum-Garfinkle, S., Vaidya, N., Zhang, H., Stone, H. A., & Brangwynne, C. P. (2016). Biophysical characterization of organelle-based RNA/protein liquid phases using microfluidics. *Soft Matter*, *12*, 9142–9150. https://doi.org/10.1039/C6SM01087C.

Weber, S. C., & Brangwynne, C. P. (2015). Inverse size scaling of the nucleolus by a concentration-dependent phase transition. *Current Biology*, *25*(5), 641–646. https://doi.org/10.1016/j.cub.2015.01.012.

Zhang, H., Elbaum-Garfinkle, S., Langdon, E. M., Taylor, N., Occhipinti, P., Bridges, A. A., et al. (2015). RNA controls PolyQ protein phase transitions. *Molecular Cell*, *60*(2), 220–230. https://doi.org/10.1016/j.molcel.2015.09.017.

CHAPTER FIVE

The Use of ^{13}C Direct-Detect NMR to Characterize Flexible and Disordered Proteins

Erik C. Cook*, **Grace A. Usher**†, **Scott A. Showalter***,†,[1]
*Department of Chemistry, The Pennsylvania State University, University Park, PA, United States
†Department of Biochemistry and Molecular Biology, Center for Eukaryotic Gene Regulation, The Pennsylvania State University, University Park, PA, United States
[1]Corresponding author: e-mail address: sas76@psu.edu

Contents

Abstract

NMR spectroscopy remains the only experimental technique that provides (near) atomistic structural information for intrinsically disordered proteins (IDPs), but their sequence and structure characteristics still pose major challenges for high-resolution spectroscopy. Carbon-13 direct-detect NMR spectroscopy can overcome poor spectral dispersion and other difficulties associated with traditional ^{1}H-detected NMR of nonaggregating disordered proteins. This chapter presents spectroscopic protocols suitable for complete characterization of IDPs that rely exclusively on ^{13}C direct-detect experiments. The protocols described span initial characterization and chemical shift assignment; structure constraint through residual dipolar coupling and paramagnetic

Methods in Enzymology, Volume 611
ISSN 0076-6879
https://doi.org/10.1016/bs.mie.2018.08.025

relaxation enhancement measurements; and assessment of intramolecular dynamics through ^{15}N spin relaxation. The experiments described empower investigators to establish molecular mechanisms and structure–function relationships for IDPs and other proteins characterized by high internal flexibility.

1. INTRODUCTION

As this volume makes clear, intrinsically disordered proteins (IDPs) are the subject of intensive study, owing both to their critical role in physiological interactions such as in signal transduction or transcription, and to their pathological interactions in neurodegenerative and other diseases (Uversky, 2018). Despite clear interest in understanding the structure–function relationship of IDPs, their very nature—high conformational disorder and frequent strong bias in amino acid composition—poses a major barrier to their comprehensive investigation. However, tackling these challenges has provided novel insights into the structural, thermodynamic, and kinetic properties that adapt IDPs for their known functions (Gibbs & Showalter, 2015). Recently, we reviewed the broad and major breakthroughs that NMR has contributed to this field (Gibbs, Cook, & Showalter, 2017). While using NMR to study IDPs is established as a successful strategy, application of ^{13}C direct-detect experiments, which remain rare in biomolecular NMR, has emerged as an especially potent strategy that merits widespread adoption.

Early structural characterization of folded proteins by NMR relied almost exclusively on proton-detected experiments because the ^{1}H nucleus possesses a high gyromagnetic ratio. Experiments built on the $^{1}H,^{15}N$-HSQC detection platform are generally successful for folded systems, where resonances are well dispersed; in contrast, resonance overlap caused by poor spectral dispersion in the proton dimension is characteristic of IDPs and often precludes quantitative analysis. Motivated by the need to overcome the spectral dispersion problem, ^{13}C direct-detect experiments have been developed for the characterization of IDPs. The excellent spectral quality achieved through ^{13}C direct-detect spectroscopy has allowed us to interrogate IDP structure (Lawrence, Bonny, & Showalter, 2011), interactions with binding partners (Lawrence & Showalter, 2012), and posttranslational modification (Cordek et al., 2014). Others have recently adopted this strategy, leading to novel insights into transcription regulation (Martin et al., 2016) and liquid–liquid phase separation (Dao et al., 2018).

Given the success we and others have experienced with ^{13}C direct-detect NMR of IDPs, it is the objective of this chapter to empower NMR spectroscopists who wish to leverage this technology. To motivate our discussion, this chapter will present data acquired on the intrinsically disordered C-terminal domain of the pancreatic and duodenal homeobox protein I (Pdx1-C throughout). Pdx1 is a transcription factor that regulates the genes coding for insulin, glucagon, and the machinery necessary for endocrine pancreas-controlled glucose homeostasis (Andrali, Sampley, Vanderford, & Ozcan, 2008). The chapter introduces the $^{13}C,^{15}N$-CON as an ideal substitute for the $^{1}H,^{15}N$-HSQC detection platform, yielding multidimensional experiments for chemical shift determination, structure constraint, and the exploration of intramolecular dynamics. The collection of experimental protocols discussed will provide a robust foundation for comprehensive structure–function studies of IDPs and their interactions.

2. PREPARATION OF ^{15}N- AND ^{13}C-ENRICHED PROTEINS

Successful investigation of protein structure–function relationships through NMR spectroscopy depends on the reproducible generation of recombinant protein that is both stable and homogeneous at room temperature. By necessity, ^{13}C direct-detect experiments require the additional costs of ^{13}C enrichment for all samples, which elevates the need to develop an efficient strategy for optimized protein expression. Importantly, when utilizing "off-the-shelf" inverse configuration cryogenic probes, ^{13}C direct-detect NMR requires protein concentrations ≥ 1.0 mM for most applications. In contrast with traditional ^{1}H detection, these methods are tolerant to high salt, basic pH, and the presence of solubility-enhancing cosolutes (Bastidas, Gibbs, Sahu, & Showalter, 2015). We assume the reader is familiar with strategies for optimization of protein expression. While there are additional challenges associated with IDP expression and purification, appropriate solution conditions can be achieved for most nonaggregating disordered proteins.

2.1 Plasmid Construction and Protein Expression

Example spectra in this chapter were recorded on the 79 C-terminal amino acids from the human Pdx1 protein (Pdx1-C), which has a high proline abundance (18 of 79 residues) and that otherwise is characterized as a weak polyampholyte, with 0.22 fraction of charged residues and 0.01 net charge per residue (Das & Pappu, 2013). A gene encoding Pdx1-C with an

N-terminal tryptophan for ease of purification was purchased from GeneArt and subcloned into pET49b(+) (Novagen) using the *Xma*1 and *Bam*H1 restriction sites. This plasmid expresses a GST-6xHis-Pdx1-C fusion protein with a 3C protease site between the fusion tags and the Pdx1-C construct; cleavage leaves a nonnative N-terminal gly-pro-gly-trp, followed by native Pdx1-C, as shown in Fig. 1A. Pdx1-C was expressed in *fhuA2*$^{-}$ BL-21

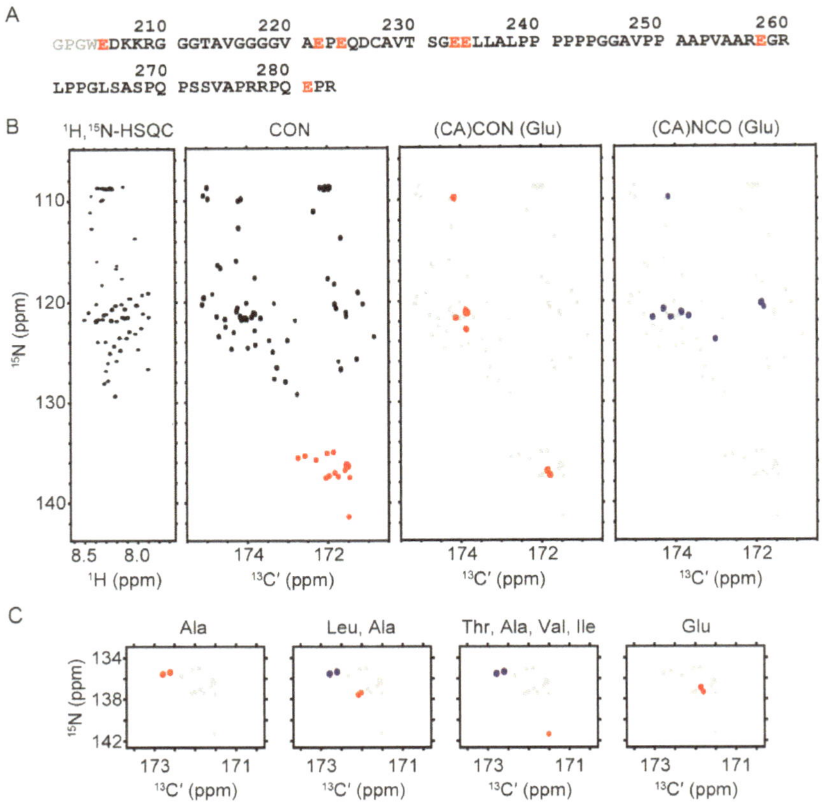

Fig. 1 Representative 2D spectra of Pdx1-C. (A) The amino acid sequence is shown with the cloning artifact in *gray* and glutamate residues in *red*. (B) *From the left*, the ^{1}H,^{15}N-HSQC; ^{13}C,^{15}N-CON, with Pro *red*; Glu-selective ^{13}C,^{15}N-(HACA)CON in *red*, with the nonselective spectrum in *gray*; and Glu-selective ^{13}C,^{15}N-(HACA)NCO in *blue*, with the nonselective spectrum in *gray*. (C) Focus on the Pro-resonances in representative CAS NMR ^{13}C,^{15}N-(HACA)CON spectra, with the nonselective spectrum in *gray*. *From the left*: Ala-selected resonances in *red*; Leu/Ala selection in *red*, with Ala-only selection superposed in *blue*; Thr, Ala, Val, Ile selection in *red*, with Ala selection superimposed in *blue*; Glu selection in *red*.

(DE3) *Escherichia coli* using quadruple density Luria broth (LB) starter cultures as described by Marley, Lu, and Bracken (2001), which improves yield over traditional M9-only expression.

2.2 Purification of Pdx1-C

1. Prepare 4 × 100 mL and 4 × 1 L of LB, each with 50 μg/mL kanamycin. Inoculate the 4 × 100 mL of LB with *E. coli* transformed with the Pdx1-C coding plasmid and incubate overnight at 37°C with agitation. The next day, use each 100 mL culture to inoculate one 1 L culture and grow until the OD_{600} reaches 0.6. Harvest the cells by centrifugation at 3300 × *g* and wash residual LB from the cells with 1 L of sterilized M9 salts, lacking glucose and ammonium chloride. Pellet again by centrifugation and resuspend the cell mass in 1 L of M9 supplemented with D-glucose-13C_6 and ^{15}N ammonium chloride. Measure the starting OD_{600} and continue incubation at 37°C until the OD_{600} has increased by ~0.1 U; at this time, add 500 μL of 1 *M* IPTG and incubate for an additional 3–4 h at 37°C.
2. Harvest the cells by centrifugation at 3300 × *g* and discard supernatant. All subsequent steps should be performed at 4°C. Resuspend the cells in lysis buffer (50 m*M* sodium phosphate, pH 8.0, 500 m*M* sodium chloride, 20 m*M* imidazole, and 5 m*M* β-mercaptoethanol), supplemented with 1 mg/mL lysozyme, 1 m*M* PMSF, and 1 × protease inhibitor cocktail (EMD Millipore). Incubate cells on ice for 30 min, lyse through sonication, and centrifuge the lysate at 14,000 × *g*.
3. Apply clarified supernatant to 5 mL of Ni^{2+}-loaded immobilized metal affinity chromatography sepharose resin (G Biosciences), preequilibrated with lysis buffer. Wash resin with 40 mL of lysis buffer, supplemented with 0.1% Triton X-100. Rinse resin with 20 mL lysis buffer and elute protein with 20 mL of lysis buffer, supplemented with 200 m*M* imidazole.
4. Add 3C protease to the eluent and dialyze overnight with stirring at 4°C against 1.6 L of 50 m*M* sodium phosphate, pH 8.0, 500 m*M* sodium chloride, and 5 m*M* β-mercaptoethanol in a 1000 MWCO dialysis tube.
5. Transfer the dialyzed solution to 5 mL of Ni^{2+}-loaded immobilized metal affinity chromatography sepharose resin, preequilibrated with the dialysate. Wash the resin with 15 mL of dialysate. Collect the flow through in both cases; these fractions should contain tag-free Pdx1-C.

6. Finish purification through size exclusion chromatography (SEC) using an S200 1.5 cm × 75 cm size exclusion column. Preequilibrate the column with SEC buffer (50 m*M* sodium phosphate, pH 6.5, 50 m*M* sodium chloride, and 5 m*M* dithiothreitol (DTT)). Buffer exchange Pdx1-C into SEC buffer and concentrate the protein solution to <500 μL prior to loading it onto the column. Set the flow rate to 1 mL/minute and collect 1 mL fractions.
7. Pool all fractions of pure Pdx1-C and concentrate using a 3000 MWCO centrifugal filter (MilliporeSigma) to a final volume of 500 μL. Add 1 × protease inhibitor cocktail, 0.01% (w/v) sodium azide, and 10% (v/v) D_2O, then transfer to an NMR tube.

3. RESONANCE ASSIGNMENT THROUGH ^{13}C DIRECT-DETECT SPECTRA

Chemical shift assignment is usually a means to an end, and is often the most substantial barrier to addressing biological function through NMR spectroscopy. In terms of applications to IDPs, we find that overcoming the assignment challenge is one of the greatest strengths of the ^{13}C direct-detect strategy presented in this chapter. Carbon-detected spectra generally offer superior resonance dispersion and are conducive to the assignment and study of proline-rich constructs. In the case of Pdx1-C, the ^{1}H,^{15}N-HSQC spectrum displayed in Fig. 1B demonstrates acceptable resolution, in the sense that most resonances are resolved (nearly) to baseline, but their occupancy of a very narrow chemical shift range is still evident. The contrast between the ^{1}H,^{15}N-HSQC and the ^{13}C,^{15}N-CON experiment (Bermel, Bertini, Felli, Kummerle, & Pierattelli, 2006), also shown in Fig. 1B, is dramatic; detection on ^{13}C′ imparts sharp linewidths and excellent chemical shift dispersion to the Pdx1-C spectrum. Pdx1-C features 18 proline residues in its sequence, which are absent in the ^{1}H,^{15}N-HSQC, but readily identified in the ^{13}C,^{15}N-CON by their characteristic ^{15}N chemical shift between 132 and 144 ppm (colored red in Fig. 1B for emphasis).

Given the benefits of ^{13}C direct-detect experiments, it is important to acknowledge their major limitation compared with conventional ^{1}H-detect NMR: the gyromagnetic ratio of ^{13}C is modest, which results in reduced sensitivity per scan, particularly when using an inverse probe configuration. Fortunately, experimental design can mitigate this disadvantage in most cases. While ^{13}C direct-detect spectroscopy was first developed as a

Table 1 ^{13}C Direct-Detect NMR Experiments for Chemical Shift Assignment

Experiment	Nuclei Observed	Reference
2D spectra		
CON, (HACA)CON	C_{i-1}', N_i	Bermel et al. (2008)
CAS (CA)CON	C_{i-1}', N_i where ^{13}C is from selected amino acid	Sahu, Bastidas, and Showalter (2014)
CAS (CA)NCO	C_{i-1}', N_i where ^{13}C is from selected amino acid C_i', N_i where ^{15}N is from the selected amino acid	Sahu et al. (2014)
3D spectra		
(HACA)N(CA)CON	C_{i-1}', N_i, N_{i-1}	Bastidas et al. (2015)
(HACA)N(CA)NCO	C_{i-1}', N_i, N_{i-1}, N_{i+1}	Bastidas et al. (2015)
CCCON	C_{i-1}', N_i, $C_{ali,\ i-1}$	Bermel et al. (2008)
H(CC)CON	C_{i-1}', N_i, $H_{ali,\ i-1}$	O'Hare, Benesi, and Showalter (2009)

"protonless" strategy (Bermel, Felli, Kümmerle, & Pierattelli, 2008), we favor utilizing proton-start experiments that enhance per-scan signal to noise. For example, the (HACA)-CON is initiated by magnetization transfer from the $^1H_\alpha$ nucleus, yielding improved per-scan signal intensity. Finally, the experiments presented here are amenable to nonuniform sampling strategies, which can yield dramatic and time-efficient improvements in spectral quality (Brutscher et al., 2015). Application of these principles led to complete chemical shift assignment of Pdx1-C (BMRB 19596), using only 2 weeks of instrument time on an 11.7 T Bruker Avance III NMR spectrometer, equipped with a TCI Cryoprobe. Table 1 shows an overview of the experiments required. The remainder of this section provides rationale and protocols for the specific experiments employed.

3.1 2D NMR

The ^{13}C,^{15}N-CON serves as the foundation for ^{13}C direct-detect chemical shift assignment, analogous to the ^{1}H,^{15}N-HSQC in conventional ^{1}H-detected spectroscopy. In addition, carbon-detect amino acid-specific (CAS NMR) variants of the ^{13}C,^{15}N-CON spectrum provide a powerful complement to 3D-based assignments (Bermel et al., 2012), as we have described for Pdx1-C (Sahu et al., 2014). The CAS NMR strategy utilizes pairs of amino acid-specific spectra, collected in (CA)CON and (CA)NCO format. The general approach is analogous to the proton-detected triple-resonance NMR assignment strategy: the (CA)CON is filtered such that the only resonances retained are for peptide planes in which the ^{13}C′ corresponds to the amino acid matching the selection filter; the (CA)NCO is bidirectional, building dipeptide linkages in which the selected amino acids are present either to the N-terminal or C-terminal side of the peptide plane. By convention, the resonances in CON-format spectra are labeled with the residue number of the amino acid that contributes the ^{15}N to the peptide plane; in this scheme, general references are made to the ^{15}N of residue i and the ^{13}C′ of residue $i-1$. This concept is demonstrated in Fig. 1B, which displays glutamate-selective (CA)CON and (CA)NCO spectra. In the (CA)CON spectrum, only one resonance appears in the ^{15}N chemical shift range associated with glycine ($\leq$114 ppm); thus G259 was unambiguously assigned. The set of resonances corresponding to peptide planes where the ^{15}N belongs to a glutamate are identified as the peaks appearing in the (CA)NCO, but not the (CA)CON. The utility of this technique is especially clear in the context of the resonances that arise from peptide bonds where proline contributes the ^{15}N atom. In Fig. 1C, we see how to combine the A-, TAVI-, LA-, and E-specific spectra to accelerate assignment of the proline resonances, which is advantageous for a proline-rich system like Pdx1-C.

3.1.1 Experimental Considerations for ^{13}C Direct-Detect 2D Spectra

1. Prepare an NMR sample, perhaps as described in Section 2.2, and collect preliminary ^{1}H,^{15}N-HSQC, ^{13}C,^{15}N-CON, and ^{13}C,^{15}N-(HACA)CON spectra. Collect both the carbon-start and the proton-start versions of the CON to determine which yields a better spectrum for the protein of interest. Typically, the proton-start spectrum is preferred, due to enhanced sensitivity, but we have experienced some situations in which internal dynamics or solution conditions (most notably high salt) make "protonless" spectra preferable. Suggested acquisition parameters are summarized in Table 2.
2. Homonuclear decoupling during acquisition is challenging, so "virtual decoupling" to remove ^{13}C′–^{13}C$_{\alpha}$ scalar coupling has been adopted as the standard method to simplify the presentation of ^{13}C direct-detect spectra. All of the pulse programs we have published and distribute

Table 2 Recommended Parameters for 2D NMR Experiments

	HSQC	CON	(HACA)CON	PRE	RDC	Relaxation
Detection	^{1}H	^{13}C	^{13}C	^{13}C	^{13}C	^{13}C
F2 TD (direct)	2048	1024	1024	1024	1024	1024
F1 TD (indirect)	256	256	256	512	512	512
Number of scans	≤8	16	8	32	32	32

record spectra in an in-phase/antiphase (IPAP) format that allows for script-based virtual decoupling. For example, executing the standard Bruker AU script "splitcomb" shifts and recombines the IPAP spectra for final presentation of the data.

3. Depending on sample conditions, CAS NMR may require up to 1 day for each (CA)CON, (CA)NCO pair. These spectra tend to be weaker than the nonselective $^{13}C,^{15}N$-CON, especially the (CA)NCO, so a minimum of 32 scans may be needed for optimal signal averaging.
4. The (CA)NCO often yields weak $i-1$ resonances, even when the self-identifying resonance is strong. This phenomenon is evident in Fig. 1, where some of the (CA)CON resonances do not appear in the glutamate-selective (CA)NCO. The absence of an $i-1$ resonance is not a cause for concern, so long as it was identified in the paired (CA) CON spectrum.
5. Many CAS NMR spectra are not selective for a single amino acid type, but rather for a subset of chemically similar amino acids. Ambiguity can be alleviated by collecting an exhaustive set of CAS NMR spectra. For example, the alanine-selective spectrum can be used to distinguish the leucine and alanine-associated resonances in the "LA" selective spectrum.

3.2 3D NMR

Most spectroscopists familiar with triple-resonance assignment strategies are accustomed to building backbone connectivity through the $^{13}C_{\alpha}$ and $^{13}C_{\beta}$ nuclei, which is also feasible with ^{13}C direct-detect spectra (Motackova et al., 2010). However, aliphatic carbon chemical shifts of IDPs generally do not deviate much from random coil values, which can produce assignment complications for repetitive sequences. Instead, we find great advantage in 3D spectra built from the (CA)CON and (CA)NCO that correlate the backbone amide ^{15}N chemical shifts of pairs or triplets of amino acids. The (HACA)N(CA)CON records the ^{15}N of residue *i* as an autocorrelation peak with a cross-correlation to the ^{15}N chemical shift for residue $i-1$. The (HACA)N(CA)NCO provides complementary information: the indirect ^{15}N dimension records cross-correlation peaks to the ^{15}N resonance of residues $i-1$ and $i+1$. Use of these experiments is illustrated in Fig. 2 for residues 255–259 from Pdx1-C, where the corresponding resonances in the 2D ^{13}C,^{15}N-CON are colored in red as a visual reference. The 3D CCCON (Bermel, Bertini, Felli, Piccioli, & Pierattelli, 2006) and H(CC)CON (O'Hare et al., 2009) can be used to complete aliphatic ^{13}C and ^{1}H chemical shift assignment, respectively, as demonstrated in Fig. 3 for the same Pdx1-C residues that were used to demonstrate backbone assignment.

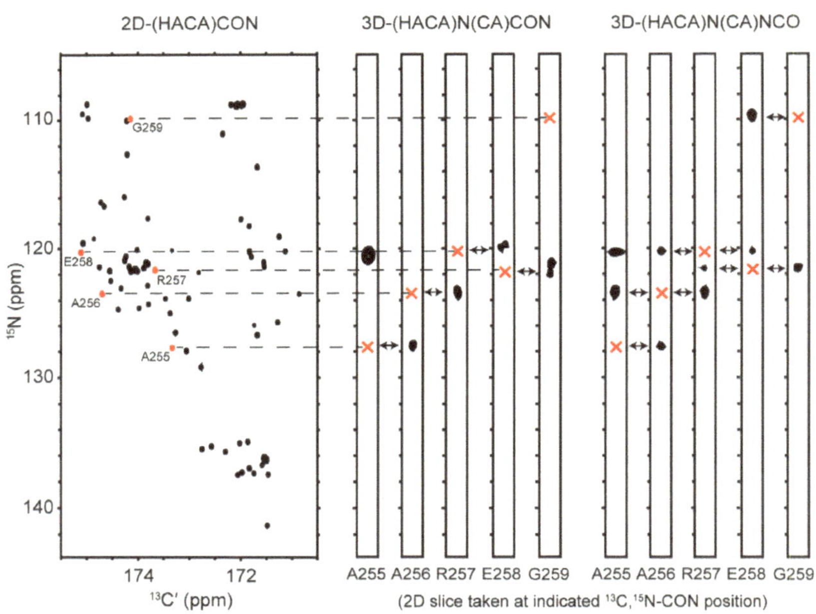

Fig. 2 Demonstrated use of the (HACA)N(CA)CON and (HACA)N(CA)NCO to build backbone sequence connectivity, using residues A255–G259 of Pdx1-C. The ^{13}C,^{15}N-(HACA)CON on the *left* has been colorized such that the five assigned residues are in *red*. For a representative segment of five residues, a *dashed line* is displayed that connects the labeled 2D ^{13}C,^{15}N-(HACA)CON resonance to the autocorrelation peak (displayed as a *red X*) in the 2D planes extracted from the 3D spectrum to highlight ^{15}N correlations in the third dimension. Cross-peak connectivity is indicated through *double-headed arrows* between the strips.

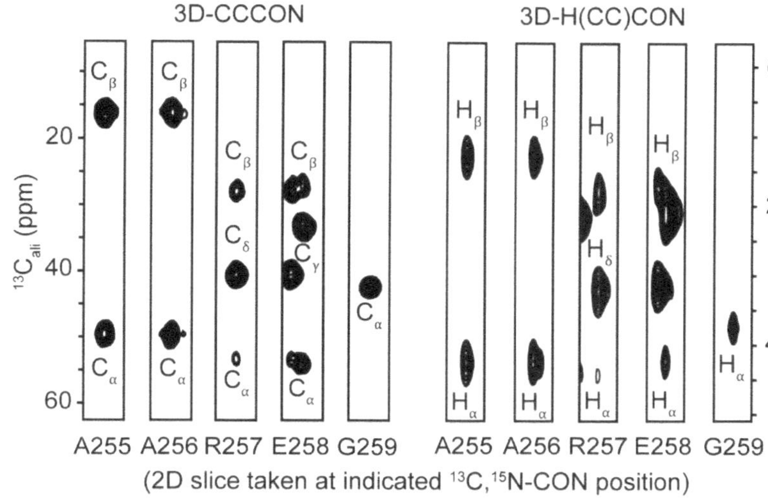

Fig. 3 The CCCON and H(CC)CON experiments identify the aliphatic carbon and proton chemical shifts, respectively. 2D projections show resonances associated with residues A255–G259.

3.2.1 Experimental Considerations for ^{13}C Direct-Detect 3D Spectra

1. There are both H_N-start (Bermel et al., 2012) and H_α-start (Bastidas et al., 2015) versions of the N(CA)CON and N(CA)NCO. In general, we favor the H_α-start versions, which offer better performance for proline-rich proteins.
2. The (HACA)N(CA)NCO is auxiliary, but can dramatically improve backbone coverage for difficult cases due to its ability to report on residues $i-1$ and $i+1$. This spectrum displays low per-scan signal to noise, so where time and sample longevity permit, additional signal averaging can be beneficial.
3. The CCCON and H(CC)CON offer facile assignment of all nuclei required for secondary structure determination. If collected in $^1H_{ali}$-start formats, both offer high sensitivity.

4. PARAMAGNETIC RELAXATION ENHANCEMENT

The dynamic nature of disordered protein ensembles presents a significant barrier to studying their structure–function relationships unless methods are employed that can report on low-population or short-lived states that may be critical for biological function. Paramagnetic relaxation enhancement (PRE) techniques offer an avenue to investigate transient intra- and intermolecular IDP interactions and, for this reason, have been widely adopted (Schneider et al., 2012). The PRE is similar in mechanism to the nuclear Overhauser effect (NOE), except the dipolar coupling exploited for the measurement exists between an unpaired electron and nearby nuclei. Due to the high gyromagnetic ratio of the electron, PRE can report on contacts that are short lived, or even on persistent colocalization out to the 20–30 Å length scale (Clore & Iwahara, 2009). Semiquantitative readout is achieved through recording paired paramagnetic and diamagnetic spectra, such as the 2D ^{13}C,^{15}N-CON spectra discussed in this chapter. The ratio of the paramagnetic and diamagnetic resonance intensity is proportional to the distance separating the spin probe and the nuclei that contribute the resonance peak.

Most IDPs do not contain native paramagnetic centers, so covalent modification with a spin probe is required. The most common method is to attach a nitroxide spin label, such as *S*-(1-oxyl-2,2,5,5-tetramethyl-2,5-dihydro-1*H*-pyrrol-3-yl)methyl methanesulfonothioate (MTSL) to a cysteine sidechain (Berliner, Grunwald, Hankovszky, & Hideg, 1982). Protein engineering to ensure that the system studied presents one cysteine for modification may be required; in any case, perform control experiments to verify that mutagenesis and MTSL conjugation have not altered the structure and biological activity of the IDP investigated.

In the case of Pdx1-C, the native cysteine at position 227 of the human sequence provides an excellent position for MTSL conjugation, leading to the PRE data shown in Fig. 4. Traditionally, PRE has been reported through the ^{1}H,^{15}N-HSQC, as displayed on the left, but it is clear that spectral crowding can impair quantitative peak height determination when IDPs are investigated using this spectrum. In contrast, the improved dispersion in the ^{13}C,^{15}N-CON spectrum provides excellent quantitative detection for IDPs. Notably, comparison of the ^{13}C-start and ^{1}H-start CON (middle and right panels, respectively) reveals several resonances that lose intensity only in the ^{1}H-start format (examples highlighted with blue boxes), which suggests that the excitation nucleus selected scales the sensitivity of CON-detected measurements to the relaxation effects of the paramagnetic center.

4.1 PRE Protocol

1. Purify the target protein using the method from Section 2.2 or equivalent, concentrate it to >500 μ*M*, and incubate for 30 min with 10 *M* equivalents of DTT to ensure near complete disulfide bond reduction.

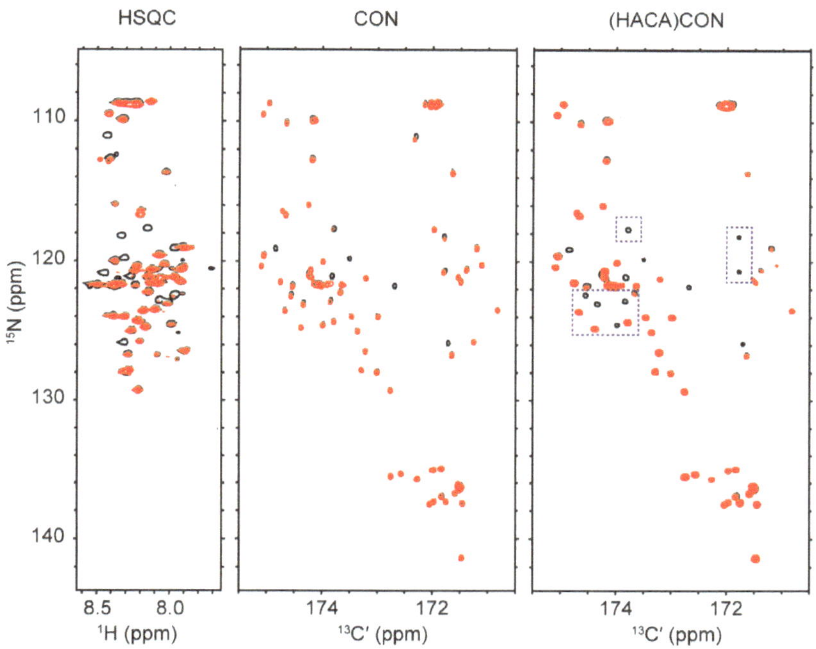

Fig. 4 Pdx1-C spectra for PRE measurement through ^{1}H,^{15}N-HSQC, ^{13}C,^{15}N-CON, and ^{13}C,^{15}N-(HACA)CON. Spectra acquired on the paramagnetic sample are shown in *red* as an overlay with the diamagnetic spectra in *black*. Under paramagnetic conditions, several resonances lose significant intensity in the ^{13}C,^{15}N-(HACA)CON, but not the ^{13}C,^{15}N-CON due to the difference in excitation nucleus. Three clusters of resonances that demonstrate this phenomenon are highlighted through enclosure in *dashed blue boxes*.

2. Remove the unreacted DTT by passing the protein over a PD-10 desalting column equilibrated with buffer that is free of reducing agent. Immediately combine the reduced protein with 20 *M* equivalents of MTSL dissolved in methanol. Use a blackout tube or aluminum foil to protect the sample, as MTSL is light sensitive. Allow the labeling reaction to proceed at 4°C for 1 h and then buffer exchange the MTSL-labeled protein into NMR buffer without reducing agent.
3. It is recommended that an aliquot of MTSL-labeled protein be reserved for analysis of the labeling efficiency through mass spectrometry or other means.
4. Collect a ^{13}C,^{15}N-CON spectrum of the paramagnetic sample using the experimental parameters suggested in Table 2. If sample lifetime permits, it is ideal to collect both the ^{13}C-start and the $^{1}H_{\alpha}$-start version of the CON, as they tend to report the PRE differently.

5. Quench the nitroxide radical by adding 5 M equivalents of sodium ascorbate to the sample. Repeat collection of the experiments recorded in Step 4 using the now diamagnetic sample. All spectroscopic parameters should be identical in order to ensure that the only changes in resonance intensity arise from differences in paramagnetism.
6. Calculate the PRE for each residue as the ratio of the peak intensities in the paramagnetic and diamagnetic forms (PRE $= I_{para}/I_{dia}$). Overlapping peaks should be omitted from analysis. As a best practice, perform measurements in duplicate and from independent biological replicates if it is possible to do so.

5. RESIDUAL DIPOLAR COUPLINGS

By virtue of their enhanced structural dynamics, most disordered proteins are not well suited for structural constraint by NOEs. For this reason, residual dipolar coupling (RDC) has emerged as a successful technique for generating IDP ensemble constraints. In an isotropic medium, the dipolar coupling between nuclear spins averages to zero; however, with slight solution anisotropy, averaging is incomplete and a "residual" dipolar coupling is preserved without loss of the sharp solution line shape. When dipolar coupling exists between nuclei that are covalently bonded, through up to two bonds, their internuclear distance is essentially fixed. Thus, the RDC serves as a proxy for internuclear vector orientation relative to a global coordinate system. The origin and applications of the RDC have been reviewed (Salmon & Blackledge, 2015), so the general theory will not be explained further here.

As a supplement to existing ^{1}H-detected RDC experiments, we have developed ^{13}C,^{15}N-CON variants that encode the NH, $C'C_{\alpha}$, $C'N$, and $H_{\alpha}C_{\alpha}$ RDC. In all cases, RDCs are recorded through a pair of spectra collected using the "in-phase (IP)/antiphase (AP)" technique in which the scalar coupling between the two nuclei under investigation is not refocused during the indirect chemical shift evolution period, resulting in the pair of spectra like those depicted on the left of Fig. 5. In the IP spectrum, both components of the doublet are phased equivalently, whereas the AP doublet features a 180° phase offset between its two components. Addition and subtraction of the IP and AP spectra leads to the spectra depicted on the right of Fig. 5, with the advantage that the scalar coupling constant can be determined from the peak offset without adding spectral crowding. Comparison of the splitting in the isotropic and aligned sample yields the RDC value for each spin pair through subtraction.

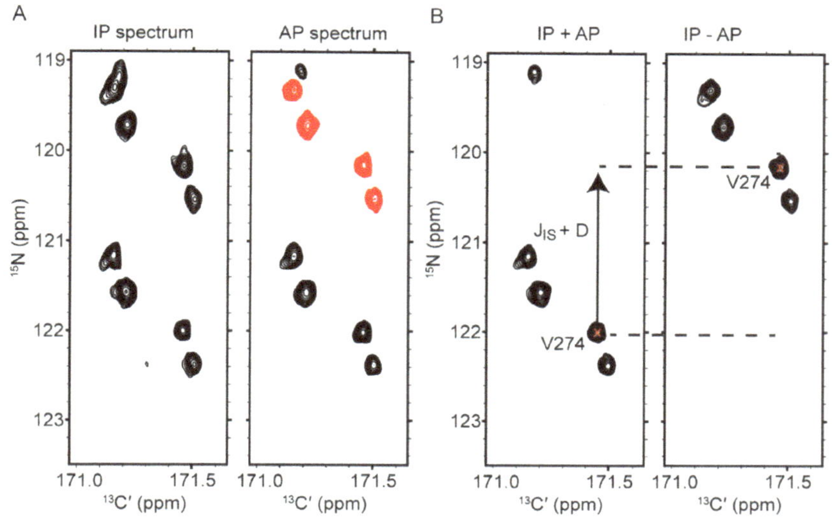

Fig. 5 RDC experiments are collected in an IPAP format as demonstrated for V274. (A) $^1J_{NH}$ is preserved to produce a doublet that is either in-phase or antiphase (negative phase is indicated through *red contours*). (B) Algebraic sum and difference spectra yield two spectral representations that preserve either the upfield or downfield component of the doublet, which facilitates quantitation of peak positions and determination of the scalar coupling constant.

5.1 RDC Protocol

1. Establish solution conditions that lead to good alignment without spectral distortion or perturbation of the IDP conformational ensemble. For general strategies, we refer the reader to established protocols (Chen & Tjandra, 2012). This protocol reflects studies of Pdx1-C, for which residual alignment was introduced with a compressed, neutral, 7% polyacrylamide gel (Sass, Musco, Stahl, Wingfield, & Grzesiek, 2000).
2. Collection of RDCs is achieved through variants of the ^{13}C,^{15}N-CON that preserve scalar coupling in the indirect dimension, such as the (HN)CON-IPAP, which preserves $^1J_{NH}$ in the ^{15}N dimension. Refer to Table 2 for suitable experimental parameters.
3. For alignment by gel compression, prepare the unaligned sample in a Shigemi-style tube. Collect the (HN)CON-IPAP and any other desired spectra needed to record, e.g., $^1J_{C'N}$, $^1J_{H\alpha C\alpha}$, or $^1J_{C'C\alpha}$.
4. Add the dehydrated gel to the Shigemi-style tube containing the sample. Set the plunger to the height needed for the desired compression ratio and secure it with parafilm. Gentle agitation of the tube is recommended for at least 20 min to prevent the gel from sticking to the wall of the tube

as it expands. It is best practice to allow overnight equilibration; if possible, gel hydration should occur in the bore of the magnet, such that equilibration of the aligned state with respect to the magnetic field is simultaneously established.

5. Collect the (HN)CON-IPAP spectrum, as well as any other desired spectra, on the aligned sample with the same experimental parameters as the unaligned sample.
6. The spectra acquired utilize the IPAP method in two different ways that must be sequentially resolved in processing. First, use the "splitcomb" AU script within the Bruker Topspin software, or equivalent, to achieve virtual decoupling in the direct dimension of the CON. After virtual decoupling, split the in-phase and antiphase spectra, using the "split" AU script or equivalent. The IP and AP spectra may now be Fourier transformed and processed as usual.
7. Algebraic addition and subtraction of the IP and AP spectra through the "add2d" command in Topspin, or equivalent, yields convenient subspectra from which the resonance frequency of the upfield and downfield components of the doublet pattern can be measured. Splitting in the unaligned sample yields ${}^1J_{NH}$, while splitting in the aligned spectrum yields ${}^1J_{NH} + {}^1D_{NH}$; subtraction of the two splitting values produces the final RDC measurement.

6. SPIN RELAXATION MEASUREMENT

Measurement of ^{15}N spin relaxation is a routine method used to assess protein backbone flexibility in both well-folded and intrinsically disordered systems. Adaptation of the ^{13}C,^{15}N-CON to report ^{15}N R_1 and R_2 relaxation rates has been reported (Lawrence & Showalter, 2012). While less common to employ, ^{13}C′ spin relaxation is also accessible in a ^{13}C direct-detect format (Bermel, Bertini, Felli, Peruzzini, & Pierattelli, 2010; Pasat, Zintsmaster, & Peng, 2008), as well as various cross-relaxation pathways (Bertini, Felli, Gonnelli, Kumar, & Pierattelli, 2011). In contrast with spin relaxation of cooperatively folded and globular systems, the ^{15}N R_2 of IDPs tends toward unexpectedly low values (i.e., long T_2 relaxation times) that are close in magnitude to the recorded R_1 values. In other words, the spin relaxation of IDPs often looks similar to that of small molecules, even for large systems. The molecular basis for this phenomenon is explained through polymer physics considerations (Pappu, Wang, Vitalis, & Crick, 2008), which lead to the conclusion that the thermal decorrelation length is short for IDPs compared to the overall chain length. Collection in a

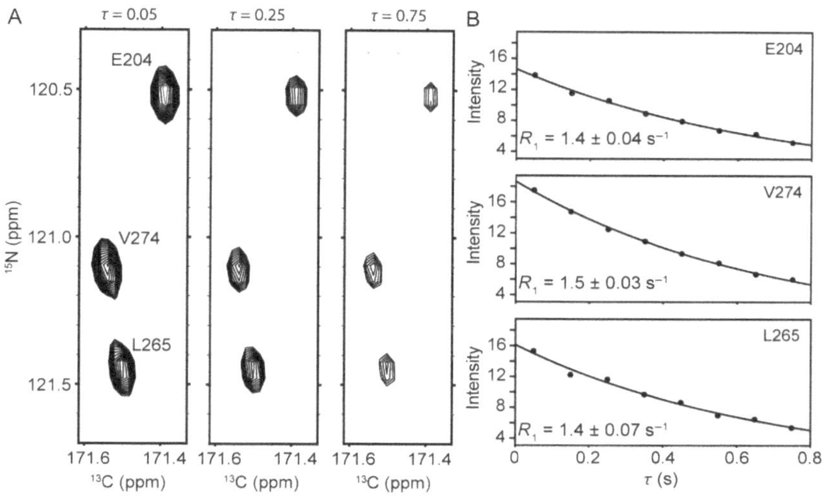

Fig. 6 ^{15}N spin relaxation can be recorded through intensity changes in ^{13}C,^{15}N-CON spectra. (A) The signal intensity for resonances E204, L265, and V274 are shown as a function of longitudinal relaxation delay time. (B) The intensities from a full set of ^{13}C,^{15}N-CON spectra are fit to a single exponential decay, yielding R_1, here, or R_2 for equivalent data.

^{13}C direct-detect format is directly analogous to standard ^{1}H-detected experiments. For example, residues E204, V274, and L265 from Pdx1-C display monoexponential relaxation, as demonstrated in Fig. 6.

6.1 Protocol for ^{15}N Relaxation Measurements

1. Construct the delay list for the R_1 and R_2 experiments. For R_1, a minimum delay of 50 ms with linear spacing to a final delay of ≥750 ms is recommended. Using Carr–Purcell–Meiboom–Gill (CPMG) for R_2 measurement constrains the relaxation delays to multiples of the CPMG supercycle (typically 16 ms) and delays ≥304 ms may be required.
2. Refer to Table 2 for suitable indirect dimension size and signal averaging.
3. It is preferable to collect each relaxation set as a "pseudo-3D," in which the data are interleaved, which will minimize the influence of systematic artifacts on the data. In this case, analysis requires splitting the data into serial 2D spectra by using the AU script "split-relax," or equivalent. All subsequent processing is standard.
4. For each resonance in the ^{13}C,^{15}N-CON, record the peak intensity as a function of relaxation delay time. Data should fit to a monoexponential decay model.

7. SUMMARY

This chapter has surveyed a purely ^{13}C direct-detect approach to biomolecular NMR that is designed to provide comprehensive and quantitative insight into the average structures and conformational dynamics of the solution ensembles adopted by IDPs and other highly flexible systems. The strategy described requires a probe with very high ^{13}C-sensitivity, which typically means a cryogenically cooled probe, and proteins that are soluble to high concentration ($\geq$1 m*M*). For most nonaggregating IDPs, the investigator should be able to achieve the needed concentration; the availability of suitable probes is also becoming more common. For ease of use, we have made Bruker-formatted pulse programs for all experiments described in this chapter available through the Penn State ScholarSphere through the stable URL: https://scholarsphere.psu.edu/collections/5712mv15x, or a keyword search on the ScholarSphere home page. Many of the described pulse sequences are already incorporated into contemporary distributions of the Bruker Topspin pulse program library. As a result of pulse program and hardware availability, and facilitated by the protocols described here, ^{13}C direct-detect NMR is poised to become a leading tool in the disordered protein community.

ACKNOWLEDGMENTS

We thank the staff of the Lloyd Jackman NMR Facility in the Penn State Chemistry Department for their consistent support of our research program. We thank the Penn State ScholarSphere team (http://scholarsphere.psu.edu) for establishing permanent and stable web hosting of our pulse programs and other aspects of the research contributing to this chapter. This work was funded in part by NSF award MCB-1515974 to S.A.S.

REFERENCES

Andrali, S. S., Sampley, M. L., Vanderford, N. L., & Ozcan, S. (2008). Glucose regulation of insulin gene expression in pancreatic beta-cells. *The Biochemical Journal*, *415*, 1–10.

Bastidas, M., Gibbs, E. B., Sahu, D., & Showalter, S. A. (2015). A primer for carbon-detected NMR applications to intrinsically disordered proteins in solution. *Concepts in Magnetic Resonance Part A*, *44*, 54–66.

Berliner, L. J., Grunwald, J., Hankovszky, H. O., & Hideg, K. (1982). A novel reversible thiol-specific spin label: Papain active site labeling and inhibition. *Analytical Biochemistry*, *119*, 450–455.

Bermel, W., Bertini, I., Chill, J., Felli, I. C., Haba, N., Kumar, M. V. V., et al. (2012). Exclusively heteronuclear (13) C-detected amino-acid-selective NMR experiments for the study of intrinsically disordered proteins (IDPs). *Chembiochem*, *13*, 2425–2432.

Bermel, W., Bertini, I., Felli, I. C., Gonnelli, L., Kozminski, W., Piai, A., et al. (2012). Speeding up sequence specific assignment of IDPs. *Journal of Biomolecular NMR*, *53*, 293–301.

Bermel, W., Bertini, I., Felli, I. C., Kummerle, R., & Pierattelli, R. (2006). Novel 13C direct detection experiments, including extension to the third dimension, to perform the complete assignment of proteins. *Journal of Magnetic Resonance*, *178*, 56–64.

Bermel, W., Bertini, I., Felli, I. C., Peruzzini, R., & Pierattelli, R. (2010). Exclusively heteronuclear NMR experiments to obtain structural and dynamic information on proteins. *Chemphyschem*, *11*, 689–695.

Bermel, W., Bertini, I., Felli, I. C., Piccioli, M., & Pierattelli, R. (2006). ^{13}C-detected protonless NMR spectroscopy of proteins in solution. *Progress in Nuclear Magnetic Resonance Spectroscopy*, *48*, 25–45.

Bermel, W., Felli, I. C., Kümmerle, R., & Pierattelli, R. (2008). 13C direct-detection biomolecular NMR. *Concepts in Magnetic Resonance Part A*, *32A*, 183–200.

Bertini, I., Felli, I. C., Gonnelli, L., Kumar, M. V. V., & Pierattelli, R. (2011). High-resolution characterization of intrinsic disorder in proteins: Expanding the suite of (13)C-detected NMR spectroscopy experiments to determine key observables. *Chembiochem*, *12*, 2347–2352.

Brutscher, B., Felli, I. C., Gil-Caballero, S., Hosek, T., Kummerle, R., Piai, A., et al. (2015). NMR methods for the study of instrinsically disordered proteins structure, dynamics, and interactions: General overview and practical guidelines. *Advances in Experimental Medicine and Biology*, *870*, 49–122.

Chen, K., & Tjandra, N. (2012). The use of residual dipolar coupling in studying proteins by NMR. *Topics in Current Chemistry*, *326*, 47–67.

Clore, G. M., & Iwahara, J. (2009). Theory, practice, and applications of paramagnetic relaxation enhancement for the characterization of transient low-population states of biological macromolecules and their complexes. *Chemical Reviews*, *109*, 4108–4139.

Cordek, D. G., Croom-Perez, T. J., Hwang, J., Hargittai, M. R., Subba-Reddy, C. V., Han, Q., et al. (2014). Expanding the proteome of an RNA virus by phosphorylation of an intrinsically disordered viral protein. *The Journal of Biological Chemistry*, *289*, 24397–24416.

Dao, T. P., Kolaitis, R. M., Kim, H. J., O'Donovan, K., Martyniak, B., Colicino, E., et al. (2018). Ubiquitin modulates liquid-liquid phase separation of UBQLN2 via disruption of multivalent interactions. *Molecular Cell*, *69*. 965–978.e6.

Das, R. K., & Pappu, R. V. (2013). Conformations of intrinsically disordered proteins are influenced by linear sequence distributions of oppositely charged residues. *Proceedings of the National Academy of Sciences of the United States of America*, *110*, 13392–13397.

Gibbs, E. B., Cook, E. C., & Showalter, S. A. (2017). Application of NMR to studies of intrinsically disordered proteins. *Archives of Biochemistry and Biophysics*, *628*, 57–70.

Gibbs, E. B., & Showalter, S. A. (2015). Quantitative biophysical characterization of intrinsically disordered proteins. *Biochemistry*, *54*, 1314–1326.

Lawrence, C. W., Bonny, A., & Showalter, S. A. (2011). The disordered C-terminus of the RNA polymerase II phosphatase FCP1 is partially helical in the unbound state. *Biochemical and Biophysical Research Communications*, *410*, 461–465.

Lawrence, C. W., & Showalter, S. A. (2012). Carbon-detected N-15 NMR spin relaxation of an intrinsically disordered protein: FCP1 dynamics unbound and in complex with RAP74. *Journal of Physical Chemistry Letters*, *3*, 1409–1413.

Marley, J., Lu, M., & Bracken, C. (2001). A method for efficient isotopic labeling of recombinant proteins. *Journal of Biomolecular NMR*, *20*, 71–75.

Martin, E. W., Holehouse, A. S., Grace, C. R., Hughes, A., Pappu, R. V., & Mittag, T. (2016). Sequence determinants of the conformational properties of an intrinsically disordered protein prior to and upon multisite phosphorylation. *Journal of the American Chemical Society*, *138*, 15323–15335.

Motackova, V., Novacek, J., Zawadzka-Kazimierczuk, A., Kazimierczuk, K., Zidek, L., Sanderova, H., et al. (2010). Strategy for complete NMR assignment of disordered proteins with highly repetitive sequences based on resolution-enhanced 5D experiments. *Journal of Biomolecular NMR*, *48*, 169–177.

O'Hare, B., Benesi, A. J., & Showalter, S. A. (2009). Incorporating ^{1}H-chemical shift determination into ^{13}C-direct detected spectroscopy of intrinsically disordered proteins in solution. *Journal of Magnetic Resonance*, *200*, 354–358.

Pappu, R. V., Wang, X., Vitalis, A., & Crick, S. L. (2008). A polymer physics perspective on driving forces and mechanisms for protein aggregation. *Archives of Biochemistry and Biophysics*, *469*, 132–141.

Pasat, G., Zintsmaster, J. S., & Peng, J. W. (2008). Direct 13C-detection for carbonyl relaxation studies of protein dynamics. *Journal of Magnetic Resonance*, *193*, 226–232.

Sahu, D., Bastidas, M., & Showalter, S. A. (2014). Generating NMR chemical shift assignments of intrinsically disordered proteins using carbon-detected NMR methods. *Analytical Biochemistry*, *449*, 17–25.

Salmon, L., & Blackledge, M. (2015). Investigating protein conformational energy landscapes and atomic resolution dynamics from NMR dipolar couplings: A review. *Reports on Progress in Physics*, *78*, 126601.

Sass, H. J., Musco, G., Stahl, S. J., Wingfield, P. T., & Grzesiek, S. (2000). Solution NMR of proteins within polyacrylamide gels: Diffusional properties and residual alignment by mechanical stress or embedding of oriented purple membranes. *Journal of Biomolecular NMR*, *18*, 303–309.

Schneider, R., Huang, J. R., Yao, M., Communie, G., Ozenne, V., Mollica, L., et al. (2012). Towards a robust description of intrinsic protein disorder using nuclear magnetic resonance spectroscopy. *Molecular BioSystems*, *8*, 58–68.

Uversky, V. N. (2018). Intrinsic disorder, protein-protein interactions, and disease. *Advances in Protein Chemistry and Structural Biology*, *110*, 85–121.

CHAPTER SIX

Using NMR Chemical Shifts to Determine Residue-Specific Secondary Structure Populations for Intrinsically Disordered Proteins

Wade M. Borcherds[1], Gary W. Daughdrill[1]

Department of Cell Biology, Microbiology, and Molecular Biology, University of South Florida, Tampa, FL, United States

[1]Corresponding authors: e-mail address: wborcher@mail.usf.edu; gdaughdrill@usf.edu

Contents

Abstract

Protein disorder is a pervasive phenomenon in biology and a natural consequence of polymer evolution that facilitates cell signaling by organizing sites for posttranslational modifications and protein–protein interactions into arrays of short linear motifs that can be rearranged by RNA splicing. Disordered proteins are missing the long-range nonpolar interactions that form tertiary structures, but they often contain regions with residual secondary structure that are stabilized by protein binding. NMR spectroscopy is uniquely suited to detect residual secondary structure in a disordered protein and it can provide atomic resolution data on the structure and dynamics of disordered protein interaction sites. Here we describe how backbone chemical shifts are used for assigning

Methods in Enzymology, Volume 611
ISSN 0076-6879
https://doi.org/10.1016/bs.mie.2018.09.011

residual secondary structure in disordered proteins and discuss some of the tools available for estimating secondary structure populations with a focus on disordered proteins containing different levels of alpha helical secondary structure which are stabilized by protein binding.

1. INTRODUCTION

Intrinsically disordered proteins (IDPs) or intrinsically disordered regions of proteins (IDRs) do not form stable tertiary structures, instead they are highly dynamic and sample a broad range of conformational space (Daughdrill, Pielak, Uversky, Cortese, & Dunker, 2005; Dunker, Brown, Lawson, Iakoucheva, & Obradovic, 2002; Dunker et al., 2001, 2008; Dyson & Wright, 2005; Uversky, 2002; van der Lee et al., 2014; Vendruscolo, 2007; Wright & Dyson, 1999). IDPs are abundant in nature with most eukaryotic proteins predicted to have at least one region of disorder longer than 40 residues (Dunker, Obradovic, Romero, Garner, & Brown, 2000; Ward, Sodhi, McGuffin, Buxton, & Jones, 2004). They are enriched in cell signaling and regulatory functions and sites for posttranslational modifications and protein–protein interactions are concentrated into groups of short linear motifs that can be reorganized by RNA splicing (Dunker, Cortese, Romero, Iakoucheva, & Uversky, 2005; Dunker et al., 2001, 2008; Dyson & Wright, 2005; Radivojac et al., 2007; Tompa, 2002, 2005; van der Lee et al., 2014; Vucetic et al., 2007; Xie et al., 2007a, 2007b). In many cases, IDPs and IDRs contain short linear motifs with some residual secondary structure that is stabilized by protein binding (Fuxreiter, Simon, Friedrich, & Tompa, 2004; Mohan et al., 2006; Oldfield et al., 2005; Si-Hyung et al., 2012). Identifying these protein binding sites using nuclear magnetic resonance (NMR) chemical shifts to estimate residual secondary structure has been a major focus of the lab for the last 10 years (Borcherds et al., 2014; Crabtree et al., 2017; Poosapati, Gregory, Borcherds, Chemes, & Daughdrill, 2018). During this time, several groups have observed that the presence of transient helical secondary structure in an IDR is usually predictive of a protein interaction site, providing some motivation to make chemical shift estimation of IDR secondary structure a routine part of drug target identification and validation (Fuxreiter et al., 2004; Kriwacki, Hengst, Tennant, Reed, & Wright, 1996; Mendoza-Espinosa, García-González, Moreno, Castillo, & Mas-Oliva, 2009; Oldfield et al., 2005; Shammas, Crabtree, Dahal, Wicky, & Clarke, 2016; Si-Hyung et al., 2012). The highly dynamic behavior of IDPs makes it difficult or impossible to crystallize them for X-ray crystallography. Even in proteins that do crystalize, if there is an IDR it is frequently represented by low electron density (Dunker et al., 1998). In contrast, NMR is a solution technique that provides atomic resolution dynamic and structural information for IDPs and IDRs (Dyson & Wright, 2004).

Chemical shifts are one of the principle observables from NMR spectroscopy. They are the NMR frequency of a given nucleus compared to a reference compound. Chemical shifts are mostly dependent on the external field produced by the NMR spectrometer, but are also influenced by the local electromagnetic environment of a given nucleus and therefore reports on this environment. They are measured in Hertz and reported in parts per million (PPM) to provide a value that is uniform across different magnetic field strengths. Chemical shifts are sensitive to protein secondary structure, a phenomenon first reported in the late 1960s, however, it was not until the early 1990s that there was a sufficient database of protein chemical shifts to establish a reliable correlation with secondary structure (Meadows, Markley, Cohen, & Jardetzky, 1967; Pastore & Saudek, 1990; Wishart, Sykes, & Richards, 1991). These newly established correlations allowed for the development of the Chemical Shift Index (CSI) which could be used to determine the secondary structure of ordered proteins (Wishart & Case, 2001; Wishart, Sykes, & Richards, 1992). The CSI is still in use today and continues to be updated with a CSI 3.0 version published as recently as 2015 (Hafsa, Arndt, & Wishart, 2015). Like other NMR observables, chemical shifts are a population-weighted average of the molecules in the sample and are therefore a powerful tool for examining transient secondary structures in IDPs and IDRs.

Chemical shifts of $^{13}C_{\alpha}$, $^{13}C'$, $^{1}H_{\alpha}$, $^{13}C_{\beta}$, ^{15}N, and $^{1}H_{N}$ nuclei are determined by the residue type, the Φ/Ψ torsion angles of the backbone, the side chain structure, ring current effects, hydrogen bonding, local electrical field, solvent, and temperature, as such they contain contributions from both the dynamic and structural properties of the polypeptide backbone (Berjanskii & Wishart, 2017; Wishart & Case, 2001). In particular, chemical shifts for $^{13}C_{\alpha}$, $^{13}C'$, and $^{1}H_{\alpha}$ nuclei are particularly sensitive to the Φ/Ψ torsion angles of the backbone. Chemical shifts provide precise estimates of secondary structure populations down to a few percent (Camilloni, De Simone, Vranken, & Vendruscolo, 2012; Tamiola & Mulder, 2012).

Using chemical shifts to estimate secondary structure populations in IDPs and IDRs have numerous practical advantages over other NMR measurements like residual dipolar couplings (RDCs). They are easier to measure and do not require special media to partially align the protein in the magnetic field (Chen & Tjandra, 2012). The chemical shift assignments needed to estimate secondary structure populations in IDPs and IDRs are routinely made during most protein NMR studies.

There are still some open challenges in using chemical shifts to determine secondary structure populations for IDPs. The biggest challenge is deciding which random coil chemical shift (RCCS) library to use as a reference, although there has been some convergence on this issue in the last decade (Kjaergaard & Poulsen, 2011; Nielsen & Mulder, 2018). RCCS libraries provide amino acid-specific chemical shift values that are intended to represent the structural properties of the denatured state. They are either measured using data from model peptides or large chemical shift databases (De Simone, Cavalli, Hsu, Vranken, & Vendruscolo, 2009; Kjaergaard & Poulsen, 2011; Nielsen & Mulder, 2018; Schwarzinger, Kroon, Foss, Wright, & Dyson, 2000; Tamiola, Acar, & Mulder, 2010; Wishart, Bigam, Holm, Hodges, & Sykes, 1995; Zhang, Neal, & Wishart, 2003). Both approaches have advantages and drawbacks, which we will illustrate using two IDRs that contain either low or high levels of transient helical secondary structure in important protein binding sites (Borcherds et al., 2014; Crabtree et al., 2017; Poosapati et al., 2018). We will provide examples showing the simplest interpretation of secondary chemical shifts as well as more elaborate methods that provide distributions of populations for alpha helix, beta-strand, polyproline II helix, and coil structures.

In addition to the choice of a random coil library, one must also pay attention to the referencing standard. Because chemical shifts are the resonant frequency of a given nucleus compared to a reference standard, they are relative not absolute measurements (Wishart & Case, 2001). Historically there were many chemical shift standards in use which could result in systematic inconsistencies in the chemical shifts when comparing between spectra that were referenced using a different standard (Wishart & Case, 2001). Errors as small as 0.3 PPM can have a significant impact on the identification of secondary structure for ordered proteins, errors that large are even more impactful for IDPs (Osapay & Case, 1994; Wishart et al., 1992; Wishart & Case, 2001). The IUPAC and IUBMB adopted a set of standards published in 1998 that recommends an internal standard DSS (2,2-dimethyl-2-silapentane-5-sulfonic acid) for referencing ^{1}H and ^{13}C and an external anhydrous liquid ammonia for ^{15}N referencing for biological molecules in aqueous solvents (Markley et al., 1998). Additionally it is possible to use the ^{1}H frequency of internal DSS to indirectly determine the other nuclei shifts, by using predetermined nucleus-specific frequency ratios, for example, the ratio from ^{1}H to ^{15}N is 1:0.101329118 (Bax & Subramanian, 1986; Markley et al., 1998; Wishart, Bigam, Yao, et al., 1995). If you are confident of the temperature regulation on your instrument the water peak can be used for referencing. However, the water peak is both pH and temperature sensitive, so it is recommended that the DSS standard can be used to establish the chemical shifts of the water peak under any specific experimental conditions. The Wishart lab has also developed a method to adjust the chemical shift referencing if necessary (Wang & Wishart, 2005).

2. EXPERIMENTAL CONSIDERATIONS

2.1 Design Considerations for IDR Expression Constructs

Although disordered proteins do not suffer from increasing transverse relaxation rates as the protein size increases, which limits the size of ordered proteins that can be studied with standard NMR techniques; there are practical limitations on the size of the IDR fragment we can easily study due to the resonance overlap problem. As mentioned earlier, chemical shifts are influenced by the local electromagnetic environment around a given nucleus. In ordered proteins, this environment will be heterogeneous leading to chemical shift values with either an upfield or downfield shift relative to the RCCSs. This leads to a property referred to as chemical shift dispersion, which is particularly strong for the backbone amide protons. Since the backbone amide protons of IDPs experience a more homogeneous structural environment their chemical shift dispersion is typically less than that of an ordered protein, which results in resonance overlap.

To limit the resonance overlap problem, we recommend designing the cDNA for your IDR fragment to contain 80–100 amino acids. This is usually long enough to incorporate the functionally relevant linear motif(s) that are sites for posttranslational modifications and protein–protein interactions. A disorder predictor like IUPred can be used to identify regions in an IDR sequence that may contain binding sites, the web portal can be found here http://iupred.enzim.hu/Help.php/ (Dosztányi, Csizmok, Tompa, & Simon, 2005; Zsuzsanna, 2018). Fig. 1 shows an IUPred plot of human MdmX. It contains regions that are both ordered (disorder tendency <0.5) and disordered (disorder tendency >0.5). MdmX is a homolog of the E3 ubiquitin ligase Mdm2 and a negative regulator of the p53 tumor

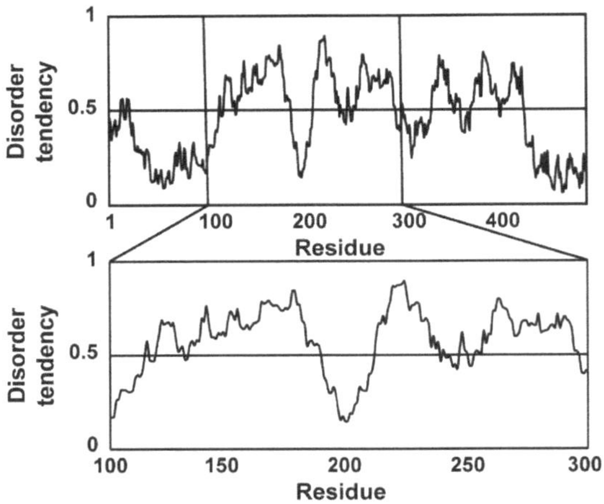

Fig. 1 Disorder prediction and construct design. IUPred plot showing the predicted disorder for the full sequence of MdmX, and an expanded view of residues 100–300.

suppressor (Wade, Li, & Wahl, 2013). It contains an ordered region from residues 23 to 111 that interacts with the p53 transactivation domain (TAD) and a long, disordered region from residues 111 to 300. This long, disordered region has a segment from residues 190 to 210 with disorder tendencies <0.5, which is characteristic of protein–protein interaction sites in IDRs. In the case of MdmX, this dip corresponds to an intramolecular binding motif that inhibits p53 binding. If you were interested in studying the transient secondary structure of the intramolecular binding motif, you could design a cDNA fragment that contains amino acids 112–210. There is another smaller dip in the IUPred plot from 220 to 270, so a fragment from residue 170 to 270 may also be a reasonable choice. We have investigated numerous fragments of MdmX that contain both ordered and disordered regions (Borcherds et al., 2017; Chen et al., 2015).

2.2 Resonance Overlap Problem

To demonstrate the resonance overlap problem, we show the ^{1}H–^{15}N heteronuclear single quantum coherence (HSQC) spectrum for a fragment of MdmX containing residues 23–210, which contain the p53 binding domain (residues 23–111) and the intramolecular binding motif (residues 190–210) (Fig. 2A). In this spectrum, chemical shift dispersion in the amide proton frequency dimension is sensitive to the heterogeneity of the local electromagnetic environment for a given amino acid created by the close proximity of other amino acids. The amide nitrogen frequency dimension is less

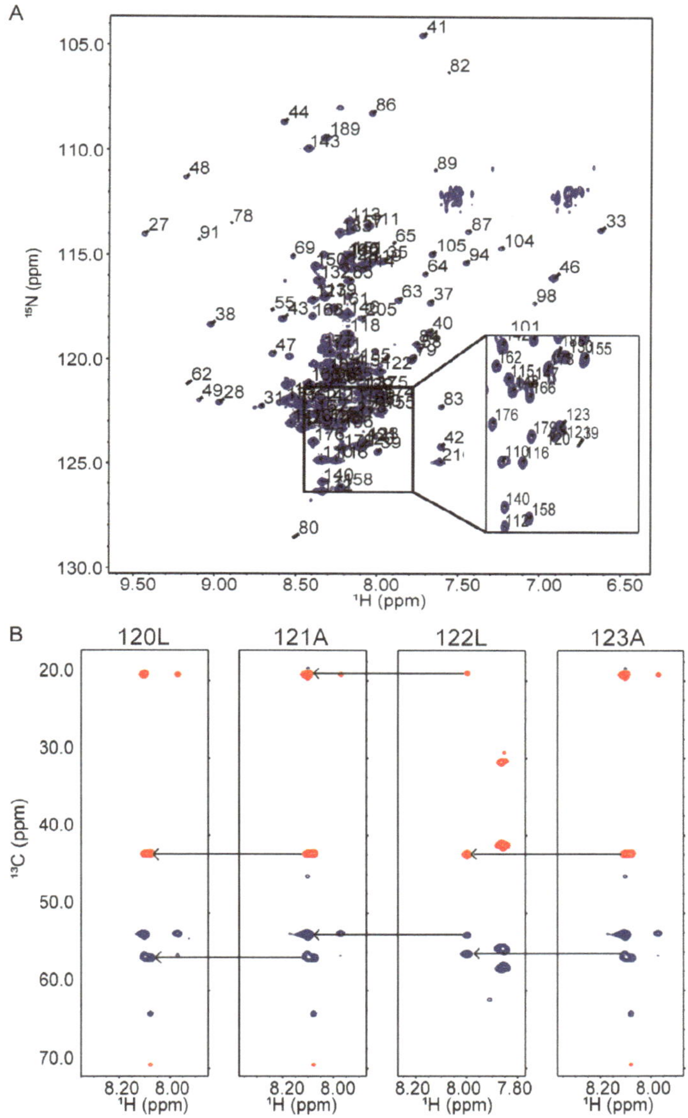

Fig. 2 Resonance overlap problem for IDPs. (A) ^{1}H–^{15}N HSQC of MdmX residues 23–210, a protein with both an ordered domain and disordered domain (Fig. 1). *Inset* shows crowding in the central region of the spectrum due resonance overlap from disordered residues. Panel (B) shows ^{1}H–^{13}C planes of correlated i and $i-1$ spin systems from the HNCACB spectrum of four residues and the connectivity of the resonances. *Blue resonances* correspond to alpha carbons (>45 PPM) and *red resonances* correspond to beta carbons (<45 PPM). *Each panel* shows the correlated spin systems for residue i and $i-1$. *Arrows* connect $i-1$ resonances from one spin system with the i resonances of the previous spin system.

sensitive to this effect and instead shows a very clear pattern of chemical shifts based on amino acid type (Wishart & Case, 2001). Most of the residues with disperse chemical shifts in the proton frequency dimension are in the ordered p53 binding domain (residues 23–111). Note the relatively high peak intensities and lack of chemical shift dispersion between the disordered peaks in the center of the amide proton frequency dimension, compared with the more dispersed but weaker ordered peaks. Resonance overlap makes resonance assignments for disordered fragments longer than 100 amino acids increasingly difficult, longer fragments may require using NMR pulse sequences with more than three frequency dimensions (see below), however, it depends greatly on a given construct's sequence complexity (Dyson & Wright, 2002). For instance, a standard NMR experiment used for making protein assignments is the HNCACB. This experiment correlates the amide nitrogen and proton of each residue with its own alpha and beta carbons, as well as the preceding residue's alpha and beta carbon. The example in Fig. 2B shows the HNCACB strip plot for MdmX residues 120–123, which show overlap in the HSQC spectra (see inset in Fig. 2A). These four residues have the sequence LALA, repetitive motifs like this are common in disordered proteins. The lower peak intensity observed for the ordered domain in Fig. 2A is due to the slower tumbling of the ordered domain and thus the increased transverse relaxation rate, this results in line broadening and the lower peak intensities (Kay, Torchia, & Bax, 1989; Pervushin, Riek, Wider, & Wüthrich, 1997).

2.3 NMR Experiments

Proteins for NMR should be at least 95% pure which can be checked by running a concentrated sample on SDS-PAGE. Following purification of a ^{13}C, ^{15}N isotope-labeled protein sample, a ^{1}H–^{15}N HSQC can be collected to determine whether the protein is ordered or disordered and provide some indication of aggregation, dimerization, or chemical exchange (Bodenhausen & Ruben, 1980; Klein et al., 2001; Sugiki et al., 2009). This is also a good opportunity to properly set the sweep widths of all future experiments as maximizing your experimental resolution is necessary for separating the overlapping peaks as much as possible. The most common experiments for making backbone resonance assignments are the HNCACB and HNCO. These experiments will provide chemical shifts of the amide proton, nitrogen, alpha carbon, beta carbon, and carbonyl carbon (Grzesiek & Bax, 1992b, 1992c; Kay, Ikura, Tschudin, & Bax, 1990;

Muhandiram & Kay, 1994). The CBCA(CO)NH and HNCA experiments are also useful for completing resonance assignments. If $^{1}H_{\alpha}$ and $^{1}H_{\beta}$ shifts are desired additional experiments like the HBHA(CO)NH and HCCH-TOCSY can be used (Bax, Clore, & Gronenborn, 1990; Farmer, Venters, Spicer, Wittekind, & Müller, 1992; Grzesiek & Bax, 1992a, 1992c, 1993; Kay et al., 1990; Olejniczak, Xu, & Fesik, 1992). The chemical shifts that most reliably report on alpha helical structure in order of decreasing reliability are $^{13}C_{\alpha}$, $^{13}C'$, $^{1}H_{\alpha}$, $^{13}C_{\beta}$, ^{15}N, and $^{1}H_{N}$, and the chemical shifts that report beta or extended structures most reliably in decreasing order are $^{1}H_{\alpha}$, $^{13}C_{\beta}$, $^{1}H_{N}$, ^{15}N, $^{13}C_{\alpha}$, and $^{13}C'$ (Tamiola & Mulder, 2012).

There are additional experiments that may be helpful for assigning particularly overlapped spectra, these include experiments of higher dimensionality that are possible due to the slower transverse relaxation rates of IDPs, as well as experiments that rely on the detection of more chemically dispersed nuclei than $^{1}H_{N}$ (Kjaergaard & Poulsen, 2012). There are pseudo 4D/3D experiments that cause differential shifting for i and $i-1$ peaks, like HN(CO)CA(CON)CA, and 7D experiments like the HNCO(CA) CBCANH that yields reduced signal overlap (Bagai, Ragsdale, & Zuiderweg, 2011; Hiller, Wasmer, Wider, & Wüthrich, 2007). There are also several experiments that are $^{1}H_{\alpha}$ detected rather than $^{1}H_{N}$, such as the HACABCANH and H(CA)CON experiments that are less sensitive to pH than the $^{1}H_{N}$ methods and possibly reduce signal overlap (Mäntylahti, Aitio, Hellman, & Permi, 2010; Motáčková et al., 2010). Finally there are a growing number of ^{13}C detected experiments that take advantage of the greater chemical shift dispersion of the carbon nuclei when compared with the proton nuclei, like the (H)CBCACO, or the CBCACON-IPAP and COCON-IPAP experiments that allow the connectivity to be made through three residues at a time from i to $i-1$, and through $i+1$ (Bermel et al., 2006, 2009; Hsu, Bertoncini, & Dobson, 2009).

3. SECONDARY CHEMICAL SHIFTS AND RCCS LIBRARIES

3.1 Random Coils and Secondary Chemical Shifts

The random coil shift for a polypeptide may be described as the state where the backbone Φ and Ψ torsion angles of one amino acid are independent of neighboring residues. As such the backbone samples conformational space broadly with no local or long-range interactions (Smith, Fiebig, Schwalbe, & Dobson, 1996). A random coil can be modeled using a random flight chain, where the global behavior of the polymer is defined by the number of monomers connected by bonds of fixed length oriented in random directions (Kuhn, 1934; Smith et al., 1996).

The secondary chemical shifts for the different nuclei of an amino acid in a polypeptide are the difference between its measured chemical shift and its RCCS (Wishart et al., 1992). Since they are referenced to a random coil standard, secondary chemical shifts can be used to identify and estimate transient secondary structure in disordered proteins. The sign of the shift for a given nucleus will provide an indication of the structural behavior of the polypeptide at that residue. Table 1 provides the sign of the secondary chemical shift for the given structure, for example, trends of positive secondary chemical shifts for the $^{13}C_{\alpha}$, and $^{13}C'$ indicate helical structure while positive $^{13}C_{\beta}$, ^{15}N, $^{1}H_{N}$, and $^{1}H_{\alpha}$ indicate beta structure.

Residue-specific RCCSs have been determined using either model peptides in denaturing conditions or estimated using chemical shift databases (De Simone et al., 2009; Kjaergaard & Poulsen, 2011; Nielsen & Mulder, 2018; Schwarzinger et al., 2000; Tamiola et al., 2010; Wishart, Bigam, Holm, et al., 1995; Zhang et al., 2003). Early studies used short peptides with the sequence GGXA, where X is one of the 20 common amino acids, which was thought to behave like a random coil (Bundi & Wuthrich, 1979; Richarz & Wuthrich, 1978). Over time it became clear that short peptides did have structural preferences and that neighboring residues affected the chemical shifts of residues at the X position. New libraries relying on different strategies have been developed to produce more accurate random coil values. This is especially important for disordered proteins that may have small secondary chemical shifts corresponding to small secondary structure populations, which can be easily overlooked if the RCCSs do not represent random coil behavior in a polypeptide. To analyze our chemical shift data, we use three RCCS libraries that were developed using different methodologies.

Table 1 Relationship Between Secondary Chemical Shift Sign and Secondary Structure for Each Nucleus

Structure	$^{13}C_{\alpha}$	$^{13}C_{\beta}$	^{13}CO	^{15}N	^{1}HN	$^{1}H_{\alpha}$
Helix	+	–	+	–	–	–
Beta	–	+	–	+	+	+

3.2 Calculating Secondary Chemical Shifts Using RCCS Libraries

3.2.1 IDRs Used for Chemical Shift Analysis

We present an analysis of the backbone chemical shifts from two model systems using three RCCS libraries and three methods for estimating secondary structure populations. One of our model systems is the N-terminal TAD of the tumor suppressor p53 and two mutants P27A and a mutant with all prolines mutated to alanines (all P to A) (Borcherds et al., 2014). p53 TAD is disordered in its free form with some helical propensity that is stabilized to form an amphipathic helix from residues 19 through 25 when bound by its negative regulator Mdm2 (Fig. 3A) (Lee et al., 2000; Vise, Baral, Latos, & Daughdrill, 2005; Wells et al., 2008). The P27A mutant increases the helical propensity of the Mdm2 binding site, and the all P to A mutant further increase the helical content (Borcherds et al., 2014). The second model system is the TAD of the myeloblastosis oncoprotein c-Myb which is also disordered and has a longer segment of partial helicity, from residues 291 through 315, that is stabilized when binding the KIX domain of the CREB binding protein (Fig. 3B) (Shammas, Travis, & Clarke, 2013, 2014; Zor, De Guzman, Dyson, & Wright, 2004). A range of mutations

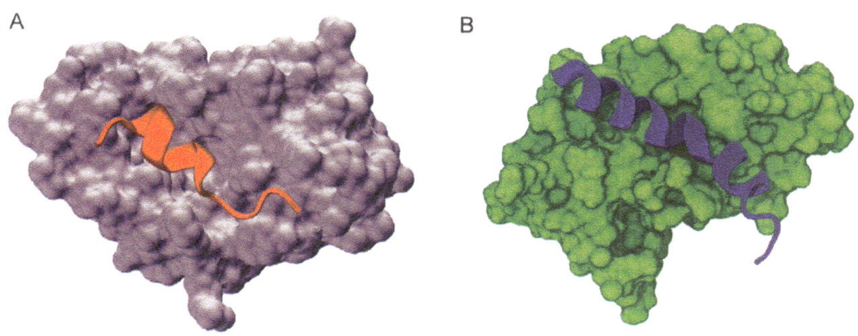

(A)
1 11 21
MEEPQSDPSV EPPLSQETFS DLWKLLPENN
31 41 51
VLSPLPSQAM DDLMLSPDDI EQWFTEDPGP
61 71
DEAPRMPEAA PRV

(B)
274 284 294
SPAAAAIQRH YNDEDPEKEK RIKELELLLM
304 314 324
STRNRLKGQQ VLPTQNHTAS YPGW

Fig. 3 Model systems used in this study. (A) Ribbon structure of p53 TAD peptide (*red*) forming a stable helix when bound to Mdm2 (*lavender*, PDB ID 1YCR). Sequence of p53 TAD residues 1–73 is shown below the structure. *Red bar* shows amino acids that fold upon binding to Mdm2. (B) Ribbon structure of c-Myb TAD peptide forming a stable helix when bound to KIX (PDB ID 1SB0). Sequence of c-Myb TAD residues 274–327 is shown below the structure. *Blue bar* shows amino acids that fold upon binding to KIX.

were designed to modulate the helicity while attempting to maintain the integrity of the c-Myb/KIX binding interface the mutants used for the random coil and structural method comparison are the L300G and L300P as they, respectively, have moderate and large effects on the helical content (Crabtree et al., 2017; Poosapati et al., 2018). We think these two models and the associated mutants provide examples of IDRs with a broad range of secondary structure content and demonstrate the effectiveness of the various RCCS libraries and the methods for estimating secondary structures. The data were all collected with at least a digital resolution for the $^{13}C_{\alpha}$ and $^{13}C'$ nuclei of 0.27 and 0.156 PPM, respectively.

3.2.2 Secondary Chemical Shifts Using the RCCS Library Developed by Wishart and Colleagues

The Wishart lab used chemically protected pentapeptides (GGXGG) where X can be any of the 20 common amino acids, to measure RCCSs (Wishart, Bigam, Holm, et al., 1995). The hexapeptide series GGXYGG is then used to calculate the neighbor correction, where X and Y can each be any of the 20 amino acids and the effect of X on Y and Y on X determines the neighbor corrections for each residue in the $i-1$, or $i+1$ position (Wishart, Bigam, Holm, et al., 1995). Other labs have used the same GGXGG peptide design to generate RCCSs for them under a range of conditions. This included measuring the RCCS of oxidized and reduced cysteines, phosphorylated serine, threonine, and tyrosine, measuring them under pressure, under a range of denaturing conditions, with the addition of cosolvents like 2,2,2-Trifluoroethanol (TFE) and, varying temperatures (Arnold, Kremer, Ludemann, & Kalbitzer, 2002; Bienkiewicz & Lumb, 1999; Merutka, Dyson, & Wright, 1995; Plaxco et al., 1997; Schwarzinger et al., 2000).

Fig. 4 shows the neighbor-corrected secondary chemical shifts for the $^{13}C_{\alpha}$, and $^{13}C'$ of p53 TAD calculated using the Wishart random coil library. These are the most sensitive shifts to alpha helical structures, a positive secondary chemical shift for either $^{13}C_{\alpha}$ or $^{13}C'$ spanning four or more residues is an indication of helix. Fig. 4A–C shows the WT (a), P27A mutant (b), and the all P to A mutant (c). It is hard to discern much helicity in the WT (a), however increased helicity for the 19 through 25 region is visible in the P27A mutant (b), and a general increase in overall helical content is visible in the all P to A mutant (c). Fig. 5A–C shows WT c-Myb (a), the L300G (b), and L300P (c) mutants. The WT c-Myb (Fig. 5A) has a greater degree of helical content than the WT p53 TAD (Fig. 4A) as can be seen by the broader region of positive secondary chemical shifts as well as the larger

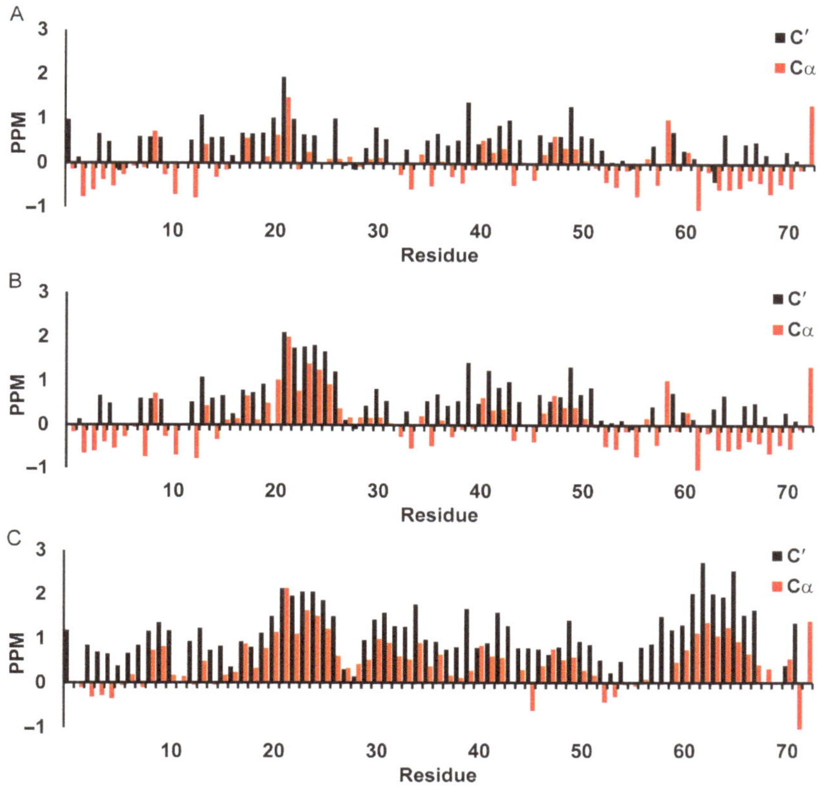

Fig. 4 Secondary chemical shifts for p53 TAD using Wishart library. *Red bars* show $^{13}C_\alpha$ secondary chemical shifts and *black bars* show $^{13}C'$ secondary chemical shifts in parts per million (PPM). (A) p53 TAD WT, (B) p53 TAD P27A, and (C) p53 TAD all P to A.

shifts. The positive secondary chemical shifts for L300G (Fig. 5B) span about the same number of residues as the WT secondary chemical shifts, however the secondary chemical shifts themselves are lower than the WT shifts, indicating a lower population of helical content. The L300P mutant (Fig. 5C) shows an even smaller positive shift than L300G, again indicating lower helical content. The negative $^{13}C_\alpha$ values seen mainly in WT p53 TAD (Fig. 4A) and P27A (Fig. 4B) may indicate beta sheet structure, however the $^{13}C'$ values do not corroborate this as they are either positive or near 0.

3.2.3 Secondary Chemical Shifts Using the RCCS Library Developed by Poulsen and Colleagues

Of the 20 common amino acids, glycine is unique in its lack of any side chain carbons, allowing it greater flexibility. Conformational studies demonstrate

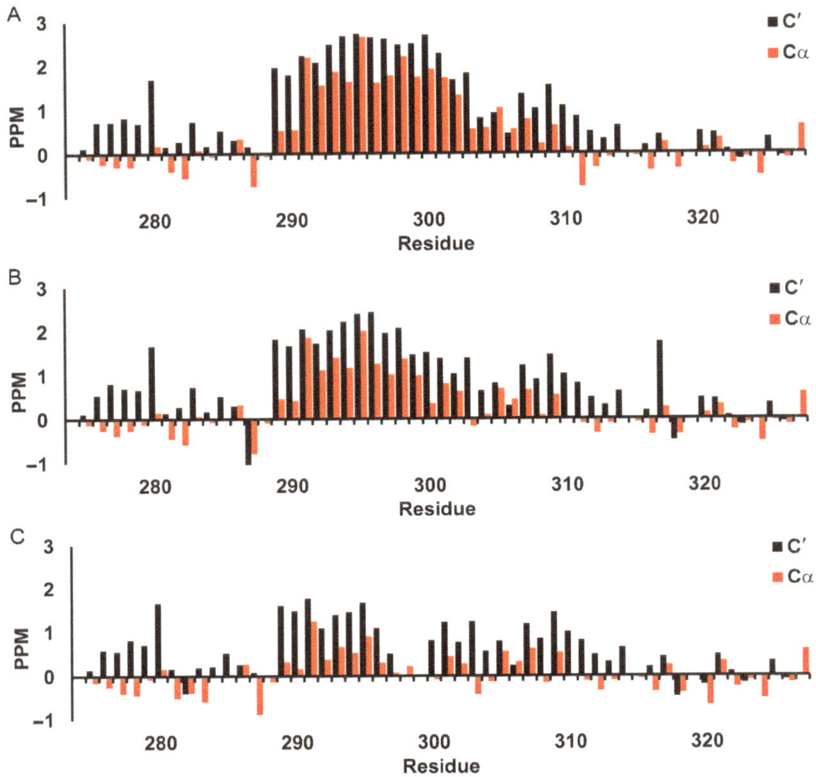

Fig. 5 Secondary chemical shifts for c-Myb TAD using Wishart library. *Red bars* show $^{13}C_{\alpha}$ secondary chemical shifts and *black bars* show $^{13}C'$ secondary chemical shifts in parts per million (PPM). (A) c-Myb TAD WT, (B) c-Myb TAD L300G, and (C) c-Myb TAD L300P.

that, like chemical shifts, the Ramachandran distribution of an amino acid varies depending on its neighboring residues. The same study also illustrates that the changes in the Ramachandran distribution caused by the neighboring residues were significantly different for glycine, when compared with the other residue types (Ting et al., 2010). Based on this new information, Poulsen et al. decided to create a new chemically protected pentapeptide, with the sequence QQXQQ, measuring the neighbor effects on each glutamine residue instead of glycine. They chose glutamine because it was found to have the Ramachandran distribution that was most representative of an average amino acid (Kjaergaard & Poulsen, 2011; Ting et al., 2010). The Poulsen lab has studied the effects of temperature on the RCCS. These corrections can be incorporated into the RCCS calculations (Kjaergaard, Brander, & Poulsen, 2011).

There is a web portal https://spin.niddk.nih.gov/bax/nmrserver/Poulsen_rc_CS/ that allows the user to provide the amino acid sequence and enter the temperature and pH for automatic correction of the random coil shifts, as the chemical shifts are sensitive to both (Kjaergaard et al., 2011). There is also an option to utilize the GGXGG values for glycine specifically, or to utilize perdeuterated shifts. They did not measure the chemical shifts of phosphorylated serine, threonine, and tyrosine, the oxidized chemical shifts for cysteine, or the chemical shifts in TFE.

The secondary chemical shifts of the $^{13}C_{\alpha}$ and $^{13}C'$ for the WT p53 TAD (a) and its two mutants P27A (b), and all P to A (c) calculated using Poulsen's RCCS library are shown in Fig. 6. Positive secondary chemical shift values are an indication of helical propensity. There is a small positive signal in the Mdm2 binding site of p53 TAD WT (Fig. 6A), with little positive or

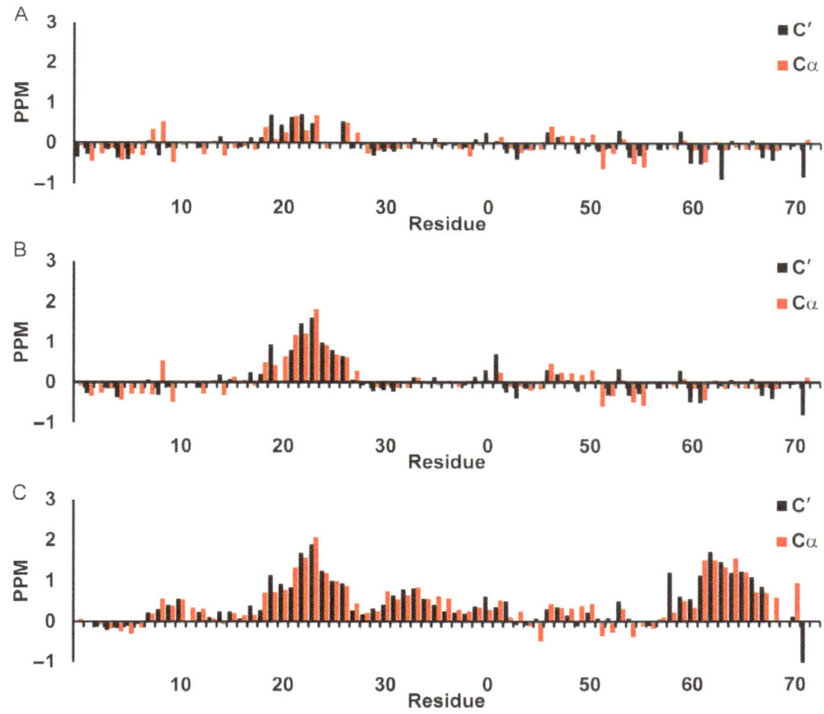

Fig. 6 Secondary chemical shifts for p53 TAD using Poulsen library. *Red bars* show $^{13}C_{\alpha}$ secondary chemical shifts and *black bars* show $^{13}C'$ secondary chemical shifts in parts per million (PPM). (A) p53 TAD WT, (B) p53 TAD P27A, and (C) p53 TAD all P to A.

negative shifts elsewhere to indicate much structural preference. The P27A mutant shows an increased positive shift with greater magnitude though confined to about the same number of residues (Fig. 6B). The all P to A mutant secondary chemical shifts seen in Fig. 6C in general indicate mostly helical tendencies spread throughout much of the polypeptide. Fig. 7 shows the $^{13}C'$ and $^{13}C_{\alpha}$ secondary chemical shifts for WT c-Myb (a) and its two mutants L300G (b) and L300P (c), using the Poulsen random coil library and corrections. The WT (Fig. 7C) shows the strongest positive shifts (a) with the L300G (b) and L300P (c) each seeing notable decreases in the helical signal.

The secondary chemical shifts generated using the Poulsen RCCS library (Figs. 6 and 7) show a better agreement in the $^{13}C_{\alpha}$ and $^{13}C'$ than the secondary chemical shifts generated using the Wishart RCCS library (Figs. 4 and 5). Additionally the nascent helices in both the Mdm2 binding

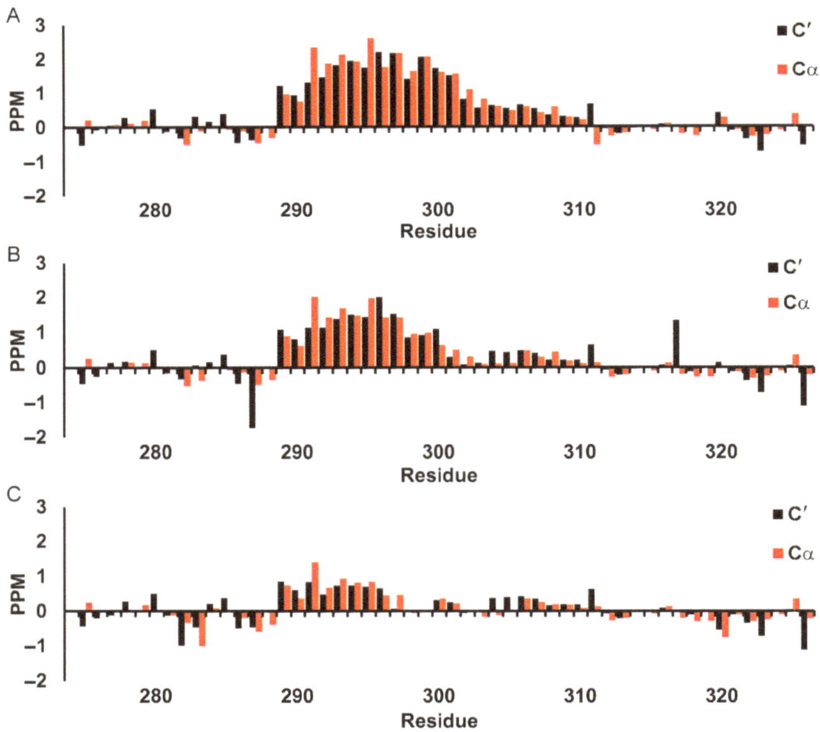

Fig. 7 Secondary chemical shifts for c-Myb TAD using Poulsen library. *Red bars* show $^{13}C_{\alpha}$ secondary chemical shifts and *black bars* show $^{13}C'$ secondary chemical shifts in parts per million (PPM). (A) c-Myb TAD WT, (B) c-Myb TAD L300G, and (C) c-Myb TAD L300P.

site of p53 TAD, and in the KIX binding site of c-Myb are more defined and readily apparent. The Poulsen referenced shifts also show fewer negative $^{13}C_{\alpha}$ shifts for all the constructs, the few negative shifts observed however are also in better agreements with the $^{13}C'$.

3.2.4 Secondary Chemical Shifts Using the RCCS Library Developed by Mulder and Colleagues

The Mulder lab has produced two neighbor-corrected RCCS libraries, the first one is termed the neighbor-corrected intrinsically disordered protein, ncIDP, library and the more recent prediction of temperature, neighbor and pH-corrected shifts for intrinsically disordered proteins (POTENCI) (Nielsen & Mulder, 2018; Tamiola et al., 2010). They took a different approach to generate these libraries compared with the two previously described. For the ncIDP they used 14 confirmed IDPs containing 6903 unique chemical shifts, reduced to 4439 with outliers removed and used a singular value decomposition methodology to elucidate the random coil value for each residue and the correction factor necessary for the nearest neighbor (Tamiola et al., 2010). With the newer POTENCI library, they drastically increased their training set of disordered proteins to 137, with 9810 confirmed disordered residues, with a total of 47,757 unique chemical shifts. The increase in database size as well as the incorporation of an algorithm to account for pH and ionic strength allows them to account for the protonation state of the side chains at different pH, ionic strength, and temperature, as well as expand their neighbor corrections out to the next nearest neighbor and incorporate second-order corrections (Nielsen & Mulder, 2018). Both the ncIDP and the POTENCI libraries are available at the web portal http://nmr.chem.rug.nl/wordpress/ where you provide the protein sequence and in the case of POTENCI the pH, temperature, and ionic strength are supplied as well as to allow for the correction for these parameters.

The secondary chemical shifts for the WT p53 TAD (a), its mutants P27A (b), and all P to A (c) calculated using the POTENCI RCCS are shown in Fig. 8. As before $^{13}C_{\alpha}$ and $^{13}C'$ have positive secondary chemical shifts indicating helical content. The nascent helix forming the Mdm2 binding site is clearly defined with gradual increases from WT (a) to P27A (b) and finally to all P to A (c) with the all P to A also having newly introduced nascent helical signals across much of the polypeptide. The WT c-Myb TAD (a) and its two mutants L300G (b) and L300P (c) secondary chemical shifts calculated using the POTENCI random coil library are shown in

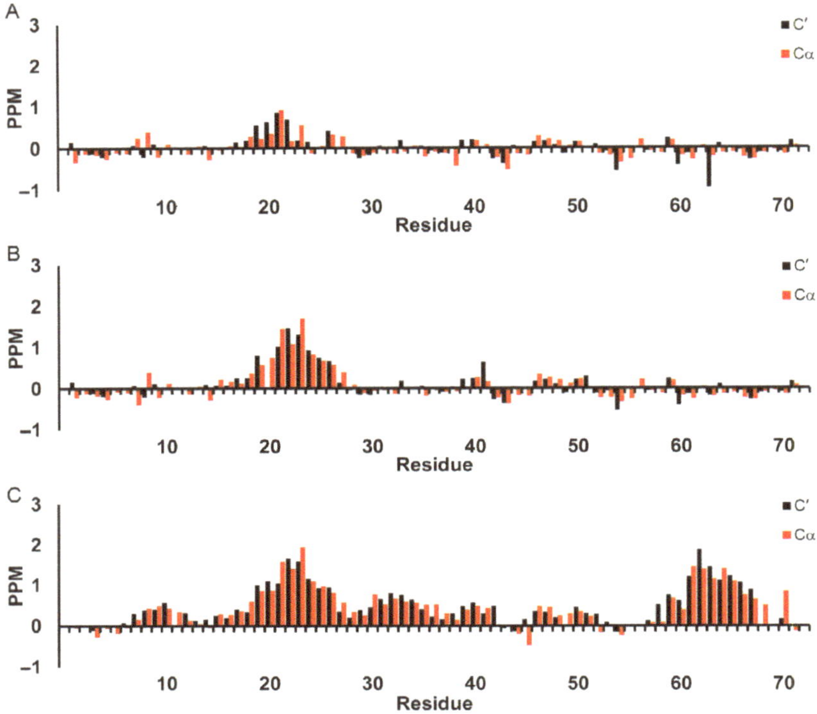

Fig. 8 Secondary chemical shifts for p53 TAD using POTENCI library. *Red bars* show $^{13}C_{\alpha}$ secondary chemical shifts and *black bars* show $^{13}C'$ secondary chemical shifts in parts per million (PPM). (A) p53 TAD WT, (B) p53 TAD P27A, and (C) p53 TAD all P to A.

Fig. 9. The KIX binding site again shows a strong helical signal in the WT (a) and is decreased in the L300G (b) and L300P (c). As can be seen in Figs. 8 and 9, the nascent helix is at least as well defined with the POTENCI as with the Poulsen Figs. 6 and 7. Like the Poulsen secondary chemical shifts, the $^{13}C_{\alpha}$ and $^{13}C'$ shifts are all in strong agreement, with POTENCI showing even smaller deviations from 0 PPM outside the defined helices of the Mdm2 KIX binding sites.

4. ESTIMATING SECONDARY STRUCTURE POPULATIONS OF IDRs

4.1 Direct Estimation of Secondary Structure Populations From Chemical Shifts

The structural propensities reported by secondary chemical shifts were noticed early in ordered proteins (Wishart et al., 1991). For example, in

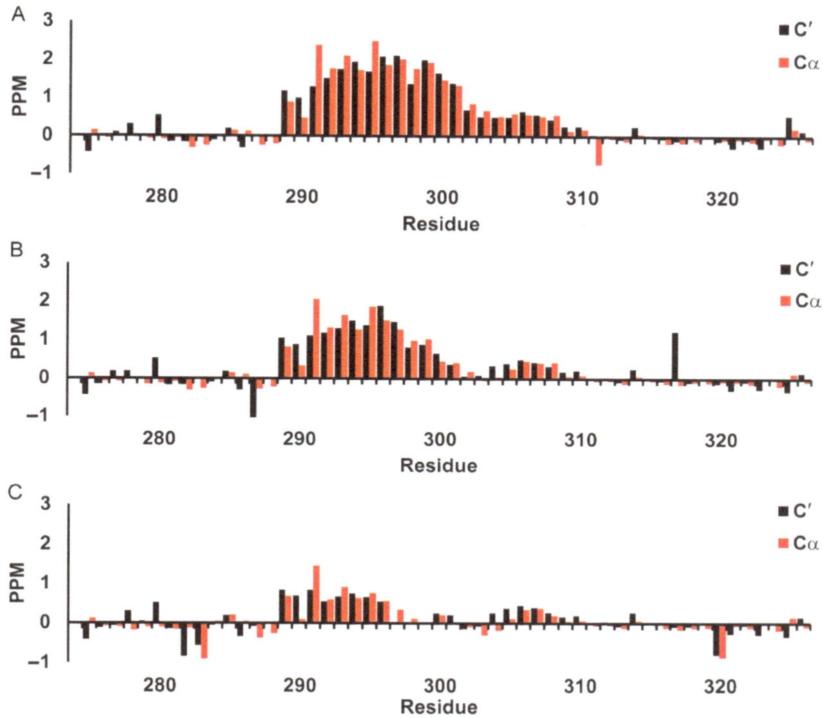

Fig. 9 Secondary chemical shifts for c-Myb TAD using POTENCI library. *Red bars* show $^{13}C_{\alpha}$ secondary chemical shifts and *black bars* show $^{13}C'$ secondary chemical shifts in parts per million (PPM). (A) c-Myb TAD WT, (B) c-Myb TAD L300G, and (C) c-Myb TAD L300P.

alpha helices the C_{α} shifts downfield by about 3.1 ± 1.0 PPM, whereas the C_{α} in beta sheets shifts upfield by about 1.5 ± 1.2 PPM (Spera & Bax, 1991). Because chemical shifts are a population-weighted average even small deviations from the RCCS values will correspond to small populations of helical or beta structures. For instance, the average helical shift for C_{α} can be used to estimate helical populations using the following relationship: $\sum_{i,j}(\Delta\delta C\alpha i,j)/3.1$, where $\Delta\delta C\alpha$ is the alpha carbon secondary chemical shifts for residues *i* through *j*. The sum of the secondary chemical shifts of the alpha carbon for residues *i* through *j* is divided by the average helical secondary chemical shift of 3.1 PPM, which yields the approximate helical propensity for that region (Dyson & Wright, 2002). However, not all residues will have the same secondary chemical shift when they assume a helical or beta structure and the standard deviation for the helical secondary chemical shift is 1 PPM which is much greater than the typical digital resolution of alpha carbon chemical shift measurements (≈0.2 PPM). A more precise method is to use a residue-specific database of helical chemical shifts, so that each residue type is being divided by its own residue type's average secondary chemical shift when in a helix. Wishart published a table of these chemical shift trends based on a database using 229,518 chemical shifts from ordered proteins, providing residue-specific average chemical shifts for helix, sheet, and coil for each of the backbone atoms (Neal, Nip, Zhang, & Wishart, 2003; Wishart, 2011; Wishart & Nip, 1998). For this method of calculation, the formula would be $\sum_i(\Delta\delta C\alpha i)/(H\delta i - RC\delta i)$ where $\Delta\delta C\alpha$ is the alpha carbon secondary chemical shift for residue i, $H\delta i$ is the helical chemical shift for the residue type i, and $RC\delta i$ is the RCCS of residue type i.

Fig. 10 shows the helical secondary structure populations for the WT p53TAD (a), P27A (b), and all P to A (c). The helicity is calculated by dividing the secondary chemical shifts by 3.1 PPM (black) or by dividing the secondary chemical shifts by the residue-specific alpha helical secondary

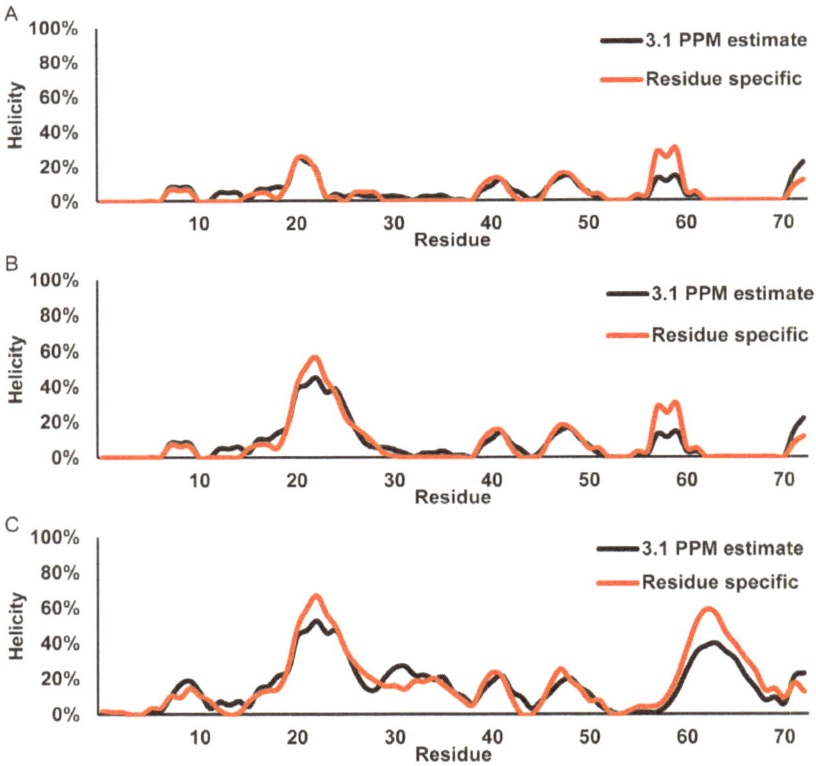

Fig. 10 Direct estimation of helical secondary structure populations for p53 TAD using $^{13}C_\alpha$ chemical shifts. *Red line* shows population estimate based on a residue-specific helical chemical shift library. *Black line* shows population estimate based on a 3.1 PPM helical secondary chemical shift for all residues. (A) p53 TAD WT, (B) p53 TAD P27A, and (C) p53 TAD all P to A.

chemical shifts as described earlier. Both the 3.1 PPM and the residue-specific plots detect helical content in the WT p53 TAD from residues 20 through 24, overlaying well with the stable helix formed when bound to Mdm2 with the 3.1 PPM method reporting about a helicity maximum of 19% and the residue-specific method reporting a maximum of 25%. Both of these methods also identify two small segments of helix from residues 40 to 43 and 47 to 52, very close to regions (residues 41–44 and 47–55) that are found to form stable helices when bound by RPA (Bochkareva et al., 2005). The spike of helicity at residue 59 for the residue-specific helicity is caused by an incorrect neighbor correction coming from the GP motif, the correction for a proline in the $i+1$ position is too high for a glycine residue. The P27A mutation (b) causes an increase in helicity as detected by both methods, but again the level is slightly different, with the 3.1 PPM method estimating a maximum helicity of 40% helicity, and the residue-specific method identifying a maximum 57% helicity in the Mdm2 binding site. A similar pattern of helical population is observed using the residue-specific method generally reporting an elevated helical content compared with the 3.1 PPM methodology is observed for the all P to A mutant (c) with the C-terminus displaying a significantly increased helical content compared with WT (a).

Fig. 11 shows the helical propensity calculated with the 3.1 PPM a specific methodology for c-Myb WT (a), its two mutants, L300G (b), and L300P (c). The helical region detected for the c-Myb WT (a) shows a weak helix starting around residue 281 and increasing to a maximum of 60% and 83% using the 3.1 PPM and residue-specific methods, respectively, before showing a dip at residue 304 and ending by residue 310. The L300G (b) shows a similar pattern, except with a maximum helical content of 47% and 63% for the 3.1 PPM and residue-specific content, the dip at 304 in WT now reaches 0% helicity by 303 as well. The helicity is then decreased further in the L300P mutant (c) with a maximum helicity of 25%, and 34% with the 3.1 PPM and residue-specific methodologies, respectively. Additionally, the helical plot for L300P reaches 0% by residue 299. The plots for Figs. 10 and 11 were created using a sliding average of three residues to smooth the curves.

4.2 Secondary Structure Population Estimates Using Chemical Shift Databases

4.2.1 Neighbor-Corrected Structural Propensity Calculator (NcSPC)

A potential problem with estimating secondary structure populations using chemical shift values from a single nucleus, as shown in Figs. 10 and 11, is that the information from the other chemical shifts are lost. In this section,

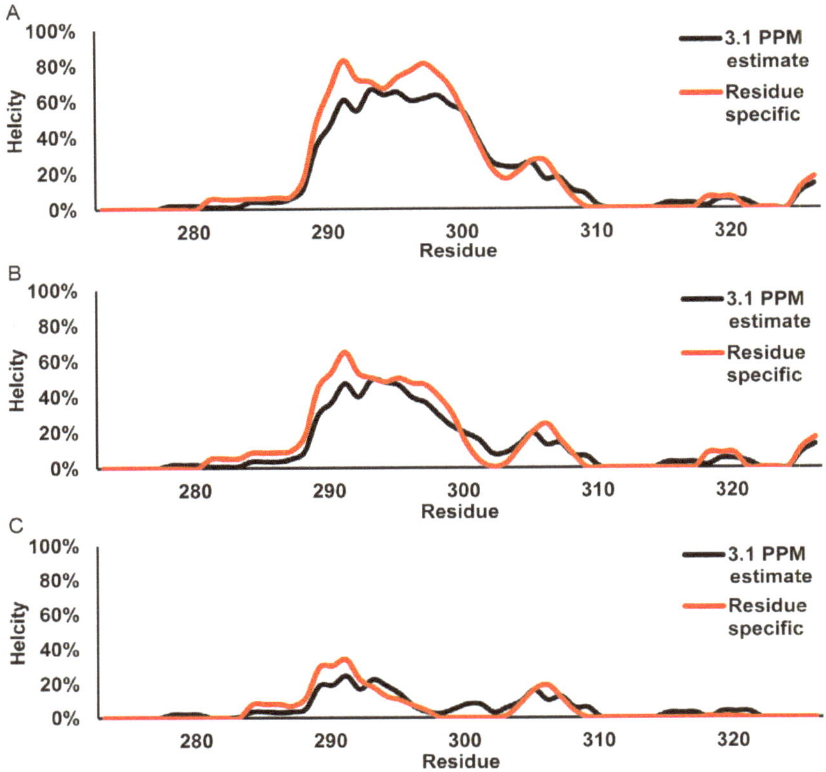

Fig. 11 Direct estimation of helical secondary structure populations for c-Myb TAD using $^{13}C_{\alpha}$ chemical shifts. *Red line* shows population estimate based on a residue-specific helical chemical shift library. *Black line* shows population estimate based on a 3.1 PPM helical secondary chemical shift for all residues. (A) c-Myb TAD WT, (B) c-Myb TAD L300G, and (C) c-Myb TAD L300P.

we will discuss methods that incorporate multiple chemical shifts to increase the accuracy of the population estimate and expand the secondary structure types. Since the chemical shifts of the nuclei in the polypeptide are not all equally sensitive to the structural changes, these methods rely on weighting algorithms that more heavily weight the data from the more sensitive nuclei. An early example of this approach was the secondary structure propensity (SSP) algorithm developed by Forman-Kay's lab (Marsh, Singh, Jia, & Forman-Kay, 2006). Mulder's lab based their neighbor-corrected structure propensity calculator (ncSPC) on SSP, adding an additional weighting procedure that accounts for the backbone conformational sensitivity of each amino acid (Tamiola & Mulder, 2012). An additional difference between

SSP and ncSPC is that ncSPC by default bases its calculations on the ncIDP random coil library rather than the older RefDB (Zhang et al., 2003). Both the SSP and the ncSPC generate a per residue score ranging from −1 to 1, with −1 representing complete beta structure, and +1 representing complete helical structure, with values in between representing the structural propensity as a fraction of 1. For example +0.25 would suggest about 25% helical propensity at that given position.

There is a web portal to access the ncSPC at: http://nmr.chem.rug.nl/wordpress/index.php/tools/ncspc/ where you can directly connect to the BMRB from which the program will directly pull previously uploaded chemical shifts. Chemical shifts can also be directly uploaded in TALOS format. There are tools available online to convert other file formats like NMRStar or Sparky. The portal lets you apply a manual reference offset correction if necessary, or apply an automatic ^{13}C rereferencing based on *Z*-scoring developed for SSP as well as account for deuteration (Marsh et al., 2006). The user will then have the option of choosing one of six possible random coil libraries, with the ncIDP being the default. As of this publication neither Poulsen's library nor the newer POTENCI library has been incorporated into ncSPC. The user can then select a series of analyses on their data with the default option being the ncSPC with a default moving window size of five residues.

The ncSPC plots seen in Fig. 12 are for the same p53 and c-Myb mutants described earlier. The ncSPC also detects a helical signal in the Mdm2 binding region, with a longer contiguous stretch of helical signal from residue 17 through 30, but a similar maximum as the residue-specific method described previously, at 26%. There is also a weak (11%) short helical signal at residues 48 through 51. The other weak helix around residues 41 through 44 is not detected here. Finally there is also a weak helical signal running from residues 57 through 61, however the random coil library ncIDP that the ncSPC uses is also prone to GP errors in the neighbor corrections of the random coil value, so that signal is likely erroneous. The ncSPC also reports on beta structural content which is largely absent here with a few maximums at residues 5, 32, and 56 showing 17%, 13%, and 18%, respectively. The P27A mutant (b) is similar but with an increased helical signal in the Mdm2 binding site with a new maximum of 50% higher than the 3.1 PPM method's 40%, but lower than the residue-specific value's 57%. The all P to A mutant (c) shows a slight increase in the Mdm2 binding site to 57% and an overall increase in helicity across much of the polypeptide with a notable increase in helical signal near the C-terminus of 58%,

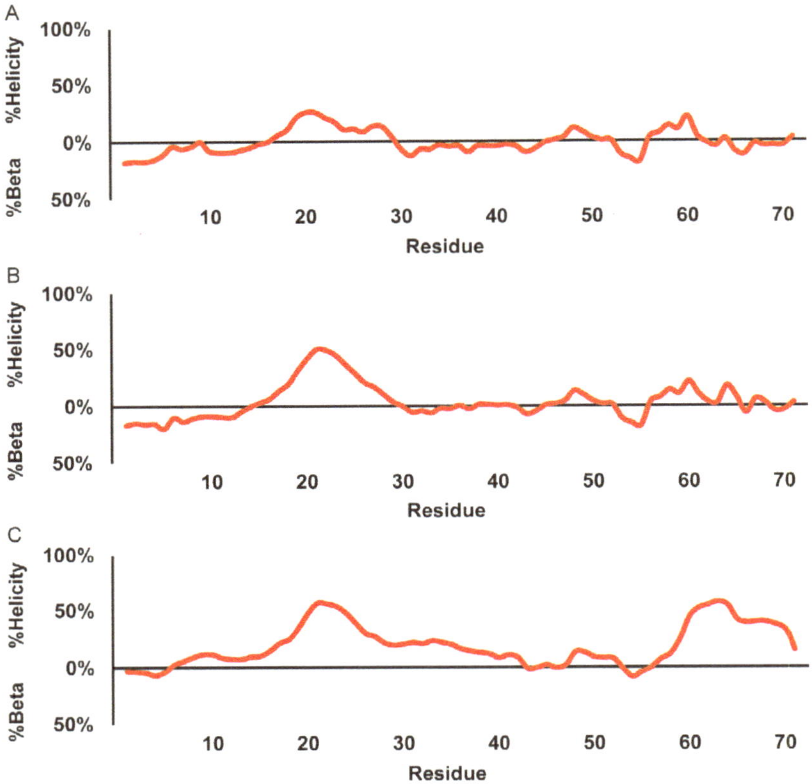

Fig. 12 Estimation of secondary structure populations for p53 TAD using NcSPC. The *red line* traces the secondary structure population estimate with values greater than zero indicating helical populations and values less than zero indicating beta populations. (A) p53 TAD WT, (B) p53 TAD P27A, and (C) p53 TAD all P to A.

compared to the 3.1 PPM and residue-specific methods' 41% and 59%. In Fig. 13, the c-Myb WT and its mutants do not show the weak helix running from 281 to 288 that the 3.1 PPM and residue-specific methods above showed, instead there is weak beta structure predicted here. The KIX binding site shows a 76% maximum with a gradual tapering rather than a dip at residue 304, before the helical signal gradually reaches 0%, and ncSPC starts reporting weak beta structure. The L300G (e) and L300P (f) mutants display similar weak beta signals, but now have a dip at residues 302 and 300, respectively. The maximum helicity is also decreased from the WT's 76%–70% for the L300G and 50% for the L300P, notably higher than either the 3.1 PPM or the residue-specific estimates.

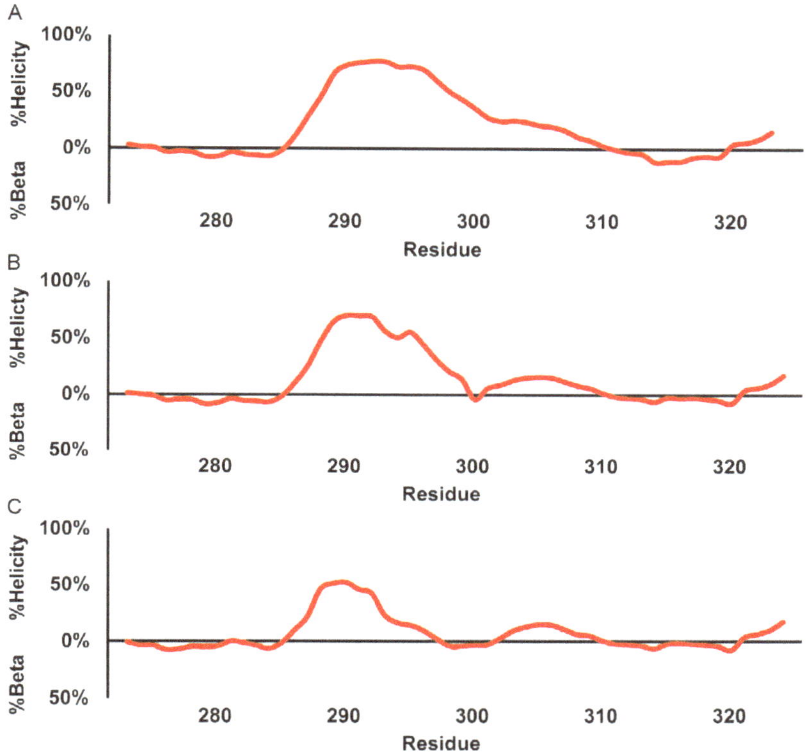

Fig. 13 Estimation of secondary structure populations for c-Myb TAD using NcSPC. The *red line* traces the secondary structure population estimate with values greater than zero indicating helical populations and values less than zero indicating beta populations. (A) c-Myb TAD WT, (B) c-Myb TAD L300G, and (C) c-Myb TAD L300P.

4.2.2 Calculating Secondary Structure Populations Using δ2D

The Vendruscolo lab has also produced a methodology to estimate secondary structure populations for disordered proteins called the δ2D method (Camilloni et al., 2012). They utilize a database of 1772 proteins with known chemical shifts and structures. The chemical shifts are divided into three libraries, one for all the alpha helical fragments longer than thee residues (with 55,470 residues), one for all the beta-strand fragments longer than 1 residue (36,871 residues), and for all polyprolyl II fragments longer than three residues (2419 residues) (Camilloni et al., 2012). Ninety percent of the database was used to calculate the average chemical shifts and correction factors for each residue type for each of the three structural elements, the remaining 10% was then used to test the method. The software uses an improved version of the CamCoil random coil library which is based on the same database, utilizing the chemical shifts in the loop regions of ordered proteins (De Simone et al., 2009). The δ2D software has been updated twice since it was first released, in 2013 with an update to version 1.2 that expanded the database and then again in 2015 again with a larger database and with the ability to take chemical shift correlations of neighboring residues into account. The δ2D method does not yet take into account the conformational bias of each residue type. The original iteration was calculated to be 86% accurate. The δ2D gives an output for the fractional secondary structure of alpha helix, beta sheet, polyproline helix, and coil.

There is a web portal to use δ2D that requires a login to be set up at http://www-mvsoftware.ch.cam.ac.uk/. The measured chemical shifts must be in Sparky format and can include the Ha, $^{13}C_{\alpha}$, $^{13}C_{\beta}$, $^{13}C'$, ^{15}N, $^{1}H_{N}$ shifts, the more complete the data the better the estimate will be. They do allow you to account for oxidized cysteines with an X instead of C, and the calculations can be done under either low pH conditions (less than or equal to pH 3) or "Neutral" (above pH 3).

δ2D provides population estimates for helical, beta, PPII, and coil. The estimates for WT p53 TAD (a), its mutants P27A (b), and all P to A (c) can be seen in Fig. 14. As with the previously described methods δ2D report a helical region for the Mdm2 binding site of p53, however it has a more gradual increase in the population and a more defined peak with maximums for WT, P27A, and all P to A of, 36%, 64%, and 73%, respectively. These are notably higher than any of the previous methodologies. δ2D also detects the two weaker helices from residues 40 to 44, and 47 to 51, however the signal is quite weak here with the WT p53 having 3.8%, and 3.9% helicity, P27A showing 5.8%, and 4.5%, and the all P to A mutant reporting 15.4%, and 6.4%. The beta populations for the WT seem to fluctuate in opposition to the helicity. The beta signal is the weakest, less than 1%, in the Mdm2 binding site, and then fluctuates to local maximums between the three helical regions, at residues 32 through 37, and 45 through 48, and again toward the C-terminus. The beta signal may correspond to the turn in between the two helices that form when bound by RPA. The beta population estimate is all but gone in the all P to A mutant. The PPII population estimate is nearly uniform across the sequence of WT and P27A with a value of approximately 28% except in the Mdm2 binding site. The PPII for all P to A is also pretty consistent, however it decreases wherever there is a helical peak, for example the Mdm2 binding site, near residues 35, 43, and the large helical region near the C-terminus.

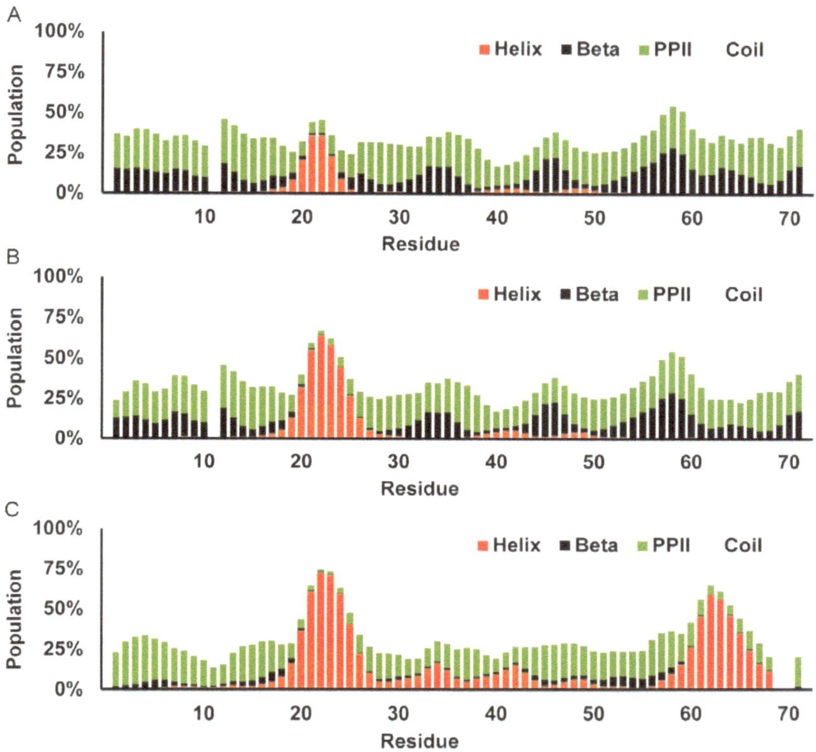

Fig. 14 Estimation of secondary structure populations for p53 TAD using δ2D. *Red bars* indicate helical population estimate, *black bars* indicate beta population estimates, *green bars* indicate PPII population estimates, and remaining *white* is random coil. (A) p53 TAD WT, (B) p53 TAD P27A, and (C) p53 TAD all P to A.

In the δ2D analysis on the WT c-Myb seen in Fig. 15A, one can see a clear strong helical population from residues 288 through 304 peaking at 92%, again higher than the previous methods, and a weak beta peak from residues 314 to 319 peaking at 15%. There are weak PPII signals near the N- and C-terminus of this construct with nearly 0% in the KIX binding site. The L300G mutation (Fig. 15B) decreases the helical signal to 73% while slightly increasing the beta and PPII populations in across the whole polypeptide, with the non KIX binding residues all showing approximately 29% PPII. The L300P further decreases the helicity of the KIX binding site and has the effect of increasing PPII, and beta populations are increased near the mutation (Fig. 15C). Again the PPII population outside the KIX binding site is near 28% for most of the polypeptide. This level of PPII population has

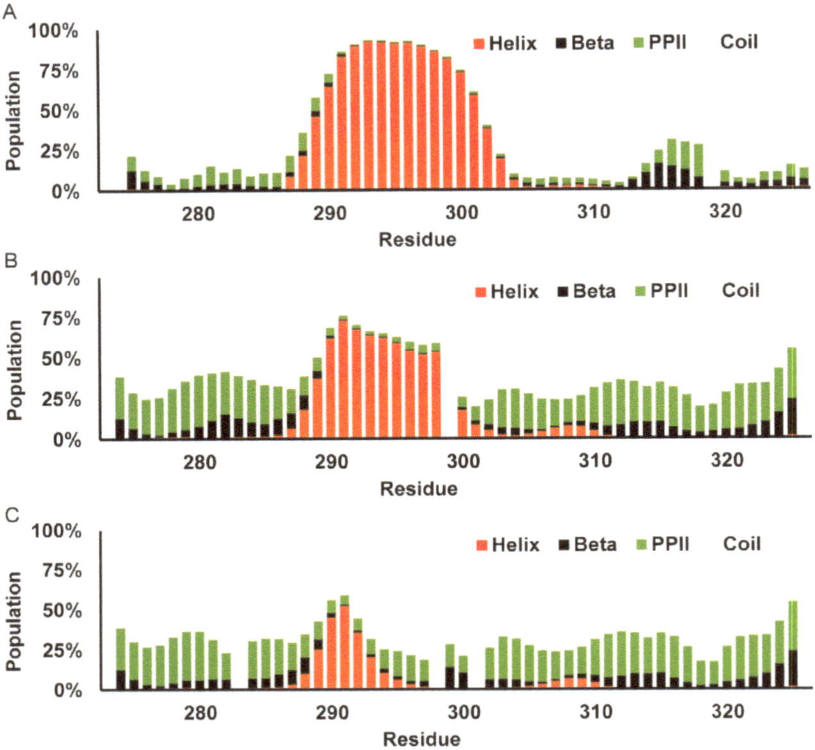

Fig. 15 Estimation of secondary structure populations for c-Myb TAD using δ2D. *Red bars* indicate helical population estimate, *black bars* indicate beta population estimates, *green bars* indicate PPII population estimates, and remaining *white* is random coil. (A) c-Myb TAD WT, (B) c-Myb TAD L300G, and (C) c-Myb TAD L300P.

been reported in other IDPs and in the denatured states of proteins before, and this partial PPII populated state may indeed be common for IDPs (Mantsyzov et al., 2014; Toal & Schweitzer-Stenner, 2014; Whittington, Chellgren, Hermann, & Creamer, 2005).

5. SUMMARY AND CONCLUSIONS

There has been significant progress refining RCCS libraries in the last 20 years. There is a clear convergence in the newer Poulsen and POTENCI methodologies over the older Wishart library. However, due to the extra conditions that the Wishart group and others have measured RCCS for the GGXGG pentapeptide, there are instances where those random coil values would be the most appropriate. Random coil libraries have improved to the point that secondary chemical shifts that do not fit the pattern of the adjacent residues' shifts are often an indication of an error in the assignment and should be checked.

We have compared the averaged helicity provided by each method described in Section 4 for a series of six c-Myb TAD constructs: WT, P289A, P289/316A, E292D, L300G, and L300P with circular dichroism (CD) data where the percent helicity was estimated using the molar residual ellipticity at 222 nm (Munoz & Serrano, 1994, 1995; Poosapati et al., 2018). As can be seen in Table 2, the c-Myb TAD fragments used for this comparison have a broad range of helical populations from 8% up to 28% based on CD measurements. According to Table 2, with the exception of the 3.1 PPM method, there is good agreement between the helicity estimates based on chemical shifts using the approaches described in this chapter and the global helicity estimates based on CD. Excluding the 3.1 PPM estimate, most of the average estimated helicity values were within 1% of the CD data, with only the L300P mutant showing any values differing by more than 3%. The population estimates for the L300P mutant are also the most variable between the different methods, with a spread of 4% average helicity where all the others were within 3% of one another (again excluding the 3.1 PPM estimate).

The methods described in this chapter allow a researcher to define regions of residual secondary structure in IDPs and IDRs. The presence of residual secondary structure in IDPs and IDRs is strongly correlated with protein–protein interaction sites, providing a direct link between IDP/IDR chemical shift analysis and function.

Table 2 Average Helical Content (%) of Full Polypeptide of c-Myb Constructs

	WT	P289A	P289,316A	E292D	L300G	L300P
CD	21	27	28	21	14	8
δ2D	21	27	28	20	14	6
NcSPC	21	26	28	18	15	9
Wishart/3.1 PPM	18	21	23	15	12	4
Wishart residue specific	21	28	30	20	14	5

ACKNOWLEDGMENT

G.W.D. is supported by the National Institutes of Health (2R01CA14124406-A1 and 1R01GM115556-01A1).

REFERENCES

Arnold, M. R., Kremer, W., Ludemann, H. D., & Kalbitzer, H. R. (2002). H-1-NMR parameters of common amino acid residues measured in aqueous solutions of the linear tetrapeptides Gly-Gly-X-Ala at pressures between 0.1 and 200 MPa. *Biophysical Chemistry*, *96*(2–3), 129–140. https://doi.org/10.1016/S0301-4622(02)00018-2.

Bagai, I., Ragsdale, S. W., & Zuiderweg, E. R. P. (2011). Pseudo-4D triple resonance experiments to resolve HN overlap in the backbone assignment of unfolded proteins. *Journal of Biomolecular NMR*, *49*(2), 69–74. https://doi.org/10.1007/s10858-010-9465-1.

Bax, A., Clore, G. M., & Gronenborn, A. M. (1990). 1H–1H correlation via isotropic mixing of 13C magnetization, a new three-dimensional approach for assigning 1H and 13C spectra of 13C-enriched proteins. *Journal of Magnetic Resonance (1969)*, *88*(2), 425–431. https://doi.org/10.1016/0022-2364(90)90202-K.

Bax, A., & Subramanian, S. (1986). Sensitivity-enhanced two-dimensional heteronuclear shift correlation NMR-spectroscopy. *Journal of Magnetic Resonance*, *67*(3), 565–569. https://doi.org/10.1016/0022-2364(86)90395-1.

Berjanskii, M. V., & Wishart, D. S. (2017). Unraveling the meaning of chemical shifts in protein NMR. *Biochimica et Biophysica Acta—Proteins and Proteomics*, *1865*(11), 1564–1576. https://doi.org/10.1016/j.bbapap.2017.07.005.

Bermel, W., Bertini, I., Csizmok, V., Felli, I. C., Pierattelli, R., & Tompa, P. (2009). H-start for exclusively heteronuclear NMR spectroscopy: The case of intrinsically disordered proteins. *Journal of Magnetic Resonance*, *198*(2), 275–281. https://doi.org/10.1016/j.jmr.2009.02.012.

Bermel, W., Bertini, I., Felli, I. C., Lee, Y.-M., Luchinat, C., & Pierattelli, R. (2006). Protonless NMR experiments for sequence-specific assignment of backbone nuclei in unfolded proteins. *Journal of the American Chemical Society*, *128*(12), 3918–3919. https://doi.org/10.1021/ja0582206.

Bienkiewicz, E. A., & Lumb, K. J. (1999). Random-coil chemical shifts of phosphorylated amino acids. *Journal of Biomolecular NMR*, *15*(3), 203–206. https://doi.org/10.1023/A:1008375029746.

Bochkareva, E., Kaustov, L., Ayed, A., Yi, G. S., Lu, Y., Pineda-Lucena, A., et al. (2005). Single-stranded DNA mimicry in the p53 transactivation domain interaction with replication protein A. *Proceedings of the National Academy of Sciences of the United States of America*, *102*(43), 15412–15417. https://doi.org/10.1073/pnas.0504614102.

Bodenhausen, G., & Ruben, D. J. (1980). Natural abundance nitrogen-15 NMR by enhanced heteronuclear spectroscopy. *Chemical Physics Letters*, *69*(1), 185–189. https://doi.org/10.1016/0009-2614(80)80041-8.

Borcherds, W., Becker, A., Chen, L., Chen, J., Chemes, L. B., & Daughdrill, G. W. (2017). Optimal affinity enhancement by a conserved flexible linker controls p53 mimicry in MdmX. *Biophysical Journal*, *112*(10), 2038–2042. https://doi.org/10.1016/j.bpj.2017.04.017.

Borcherds, W., Theillet, F.-X., Katzer, A., Finzel, A., Mishall, K. M., Powell, A. T., et al. (2014). Disorder and residual helicity alter p53-Mdm2 binding affinity and signaling in cells. *Nature Chemical Biology*, *10*, 1000. https://doi.org/10.1038/nchembio.1668.

Bundi, A., & Wuthrich, K. (1979). H-1-NMR parameters of the common amino-acid residues measured in aqueous-solutions of the linear tetrapeptides H-Gly-Gly-X-L-Ala-Oh. *Biopolymers*, *18*(2), 285–297. https://doi.org/10.1002/bip.1979.360180206.

Camilloni, C., De Simone, A., Vranken, W. F., & Vendruscolo, M. (2012). Determination of secondary structure populations in disordered states of proteins using nuclear magnetic resonance chemical shifts. *Biochemistry*, *51*(11), 2224–2231. https://doi.org/10.1021/bi3001825.

Chen, L., Borcherds, W., Wu, S., Becker, A., Schonbrunn, E., Daughdrill, G. W., et al. (2015). Autoinhibition of MDMX by intramolecular p53 mimicry. *Proceedings of the National Academy of Sciences*, *112*(15), 4624–4629. https://doi.org/10.1073/pnas.1420833112.

Chen, K., & Tjandra, N. (2012). The use of residual dipolar coupling in studying proteins by NMR. *NMR of Proteins and Small Biomolecules*, *326*, 47–67. https://doi.org/10.1007/128_2011_215.

Crabtree, M. D., Borcherds, W., Poosapati, A., Shammas, S. L., Daughdrill, G. W., & Clarke, J. (2017). Conserved helix-flanking prolines modulate intrinsically disordered protein: Target affinity by altering the lifetime of the bound complex. *Biochemistry*, *56*(18), 2379–2384. https://doi.org/10.1021/acs.biochem.7b00179.

Daughdrill, G. W., Pielak, G. J., Uversky, V. N., Cortese, M. S., & Dunker, A. K. (2005). Natively disordered proteins. In J. Buchner & T. Kiefhaber (Eds.), *Vol. 3. Protein folding handbook* (pp. 275–357). Darmstadt: Wiley-VCH.

De Simone, A., Cavalli, A., Hsu, S. T. D., Vranken, W., & Vendruscolo, M. (2009). Accurate random coil chemical shifts from an analysis of loop regions in native states of proteins. *Journal of the American Chemical Society*, *131*(45), 16332. https://doi.org/10.1021/ja904937a.

Dosztányi, Z., Csizmok, V., Tompa, P., & Simon, I. (2005). IUPred: Web server for the prediction of intrinsically unstructured regions of proteins based on estimated energy content. *Bioinformatics*, *21*(16), 3433–3434. https://doi.org/10.1093/bioinformatics/bti541.

Dunker, A. K., Brown, C. J., Lawson, J. D., Iakoucheva, L. M., & Obradovic, Z. (2002). Intrinsic disorder and protein function. *Biochemistry*, *41*(21), 6573–6582. https://doi.org/10.1021/bi012159+.

Dunker, A. K., Cortese, M. S., Romero, P., Iakoucheva, L. M., & Uversky, V. N. (2005). Flexible nets. The roles of intrinsic disorder in protein interaction networks. *FEBS Journal*, *272*(20), 5129–5148. https://doi.org/10.1111/j.1742-4658.2005.04948.x.

Dunker, A. K., Garner, E., Guilliot, S., Romero, P., Albrecht, K., Hart, J., et al. (1998). Protein disorder and the evolution of molecular recognition: Theory, predictions and observations. *Pacific Symposium on Biocomputing*, 473–484.

Dunker, A. K., Lawson, J. D., Brown, C. J., Williams, R. M., Romero, P., Oh, J. S., et al. (2001). Intrinsically disordered protein. *Journal of Molecular Graphics & Modelling*, *19*(1), 26–59.

Dunker, A. K., Obradovic, Z., Romero, P., Garner, E. C., & Brown, C. J. (2000). Intrinsic protein disorder in complete genomes. *Genome Informatics Workshop on Genome Informatics*, *11*, 161–171.

Dunker, A. K., Oldfield, C. J., Meng, J., Romero, P., Yang, J. Y., Chen, J. W., et al. (2008). The unfoldomics decade: An update on intrinsically disordered proteins. *BMC Genomics*, *9*(Suppl. 2), S1. https://doi.org/10.1186/1471-2164-9-S2-S1.

Dyson, H. J., & Wright, P. E. (2002). Insights into the structure and dynamics of unfolded proteins from nuclear magnetic resonance. *Unfolded Proteins*, *62*, 311–340.

Dyson, H. J., & Wright, P. E. (2004). Unfolded proteins and protein folding studied by NMR. *Chemical Reviews*, *104*(8), 3607–3622. https://doi.org/10.1021/cr030403s.

Dyson, H. J., & Wright, P. E. (2005). Intrinsically unstructured proteins and their functions. *Nature Reviews. Molecular Cell Biology*, *6*(3), 197–208.

Farmer, B. T., Venters, R. A., Spicer, L. D., Wittekind, M. G., & Müller, L. (1992). A refocused and optimized HNCA: Increased sensitivity and resolution in large macromolecules. *Journal of Biomolecular NMR*, *2*(2), 195–202. https://doi.org/10.1007/bf01875530.

Fuxreiter, M., Simon, I., Friedrich, P., & Tompa, P. (2004). Preformed structural elements feature in partner recognition by intrinsically unstructured proteins. *Journal of Molecular Biology*, *338*(5), 1015–1026. https://doi.org/10.1016/j.jmb.2004.03.017.

Grzesiek, S., & Bax, A. (1992a). Correlating backbone amide and side chain resonances in larger proteins by multiple relayed triple resonance NMR. *Journal of the American Chemical Society*, *114*(16), 6291–6293. https://doi.org/10.1021/ja00042a003.

Grzesiek, S., & Bax, A. (1992b). An efficient experiment for sequential backbone assignment of medium-sized isotopically enriched proteins. *Journal of Magnetic Resonance (1969)*, *99*(1), 201–207. https://doi.org/10.1016/0022-2364(92)90169-8.

Grzesiek, S., & Bax, A. (1992c). Improved 3D triple-resonance NMR techniques applied to a 31 kDa protein. *Journal of Magnetic Resonance (1969)*, *96*(2), 432–440. https://doi.org/10.1016/0022-2364(92)90099-S.

Grzesiek, S., & Bax, A. (1993). Amino acid type determination in the sequential assignment procedure of uniformly 13C/15N-enriched proteins. *Journal of Biomolecular NMR*, *3*(2), 185–204. https://doi.org/10.1007/bf00178261.

Hafsa, N. E., Arndt, D., & Wishart, D. S. (2015). CSI 3.0: A web server for identifying secondary and super-secondary structure in proteins using NMR chemical shifts. *Nucleic Acids Research*, *43*(W1), W370–W377. https://doi.org/10.1093/nar/gkv494.

Hiller, S., Wasmer, C., Wider, G., & Wüthrich, K. (2007). Sequence-specific resonance assignment of soluble nonglobular proteins by 7D APSY-NMR spectroscopy. *Journal of the American Chemical Society*, *129*(35), 10823–10828. https://doi.org/10.1021/ja072564+.

Hsu, S.-T. D., Bertoncini, C. W., & Dobson, C. M. (2009). Use of protonless NMR spectroscopy to alleviate the loss of information resulting from exchange-broadening. *Journal of the American Chemical Society*, *131*(21), 7222–7223. https://doi.org/10.1021/ja902307q.

Kay, L. E., Ikura, M., Tschudin, R., & Bax, A. (1990). Three-dimensional triple-resonance NMR spectroscopy of isotopically enriched proteins. *Journal of Magnetic Resonance (1969)*, *89*(3), 496–514. https://doi.org/10.1016/0022-2364(90)90333-5.

Kay, L. E., Torchia, D. A., & Bax, A. (1989). Backbone dynamics of proteins as studied by N-15 inverse detected heteronuclear Nmr-spectroscopy—Application to staphylococcal nuclease. *Biochemistry*, *28*(23), 8972–8979. https://doi.org/10.1021/bi00449a003.

Kjaergaard, M., Brander, S., & Poulsen, F. M. (2011). Random coil chemical shift for intrinsically disordered proteins: Effects of temperature and pH. *Journal of Biomolecular NMR*, *49*(2), 139–149. https://doi.org/10.1007/s10858-011-9472-x.

Kjaergaard, M., & Poulsen, F. M. (2011). Sequence correction of random coil chemical shifts: Correlation between neighbor correction factors and changes in the Ramachandran distribution. *Journal of Biomolecular NMR*, *50*(2), 157–165. https://doi.org/10.1007/s10858-011-9508-2.

Kjaergaard, M., & Poulsen, F. M. (2012). Disordered proteins studied by chemical shifts. *Progress in Nuclear Magnetic Resonance Spectroscopy*, *60*, 42–51. https://doi.org/10.1016/j.pnmrs.2011.10.001.

Klein, C., Planker, E., Diercks, T., Kessler, H., Kunkele, K. P., Lang, K., et al. (2001). NMR spectroscopy reveals the solution dimerization interface of p53 core domains bound to their consensus DNA. *Journal of Biological Chemistry*, *276*(52), 49020–49027. https://doi.org/10.1074/jbc.M107516200.

Kriwacki, R. W., Hengst, L., Tennant, L., Reed, S. I., & Wright, P. E. (1996). Structural studies of p21Waf1/Cip1/Sdi1 in the free and Cdk2-bound state: Conformational disorder mediates binding diversity. *Proceedings of the National Academy of Sciences of the United States of America*, *93*(21), 11504–11509.

Kuhn, W. (1934). Concerning the shape of thread shapes molecules in solution. *Kolloid-Zeitschrift*, *68*(1), 2–15. https://doi.org/10.1007/Bf01451681.

Lee, H., Mok, K. H., Muhandiram, R., Park, K. H., Suk, J. E., Kim, D. H., et al. (2000). Local structural elements in the mostly unstructured transcriptional activation domain of human p53. *Journal of Biological Chemistry*, *275*(38), 29426–29432. https://doi.org/10.1074/jbc.M003107200.

Mantsyzov, A. B., Maltsev, A. S., Ying, J., Shen, Y., Hummer, G., & Bax, A. (2014). A maximum entropy approach to the study of residue-specific backbone angle distributions in α-synuclein, an intrinsically disordered protein. *Protein Science*, *23*(9), 1275–1290. https://doi.org/10.1002/pro.2511.

Mäntylahti, S., Aitio, O., Hellman, M., & Permi, P. (2010). HA-detected experiments for the backbone assignment of intrinsically disordered proteins. *Journal of Biomolecular NMR*, *47*(3), 171–181. https://doi.org/10.1007/s10858-010-9421-0.

Markley, J. L., Bax, A., Arata, Y., Hilbers, C. W., Kaptein, R., Sykes, B. D., et al. (1998). Recommendations for the presentation of NMR structures of proteins and nucleic acids—IUPAC-IUBMB-IUPAB inter-union task group on the standardization of data bases of protein and nucleic acid structures determined by NMR spectroscopy. *Journal of Biomolecular NMR*, *12*(1), 1–23. https://doi.org/10.1023/A:1008290618449.

Marsh, J. A., Singh, V. K., Jia, Z. C., & Forman-Kay, J. D. (2006). Sensitivity of secondary structure propensities to sequence differences between alpha- and gamma-synuclein: Implications for fibrillation. *Protein Science*, *15*(12), 2795–2804. https://doi.org/10.1110/ps.062465306.

Meadows, D. H., Markley, J. L., Cohen, J. S., & Jardetzky, O. (1967). Nuclear magnetic resonance studies of structure and binding sites of enzymes. I. Histidine residues. *Proceedings of the National Academy of Sciences of the United States of America*, *58*(4), 1307. https://doi.org/10.1073/pnas.58.4.1307.

Mendoza-Espinosa, P., García-González, V., Moreno, A., Castillo, R., & Mas-Oliva, J. (2009). Disorder-to-order conformational transitions in protein structure and its relationship to disease. *Molecular and Cellular Biochemistry*, *330*(1), 105–120. https://doi.org/10.1007/s11010-009-0105-6.

Merutka, G., Dyson, H. J., & Wright, P. E. (1995). Random coil H-1 chemical-shifts obtained as a function of temperature and trifluoroethanol concentration for the peptide series Ggxgg. *Journal of Biomolecular NMR*, *5*(1), 14–24. https://doi.org/10.1007/Bf00227466.

Mohan, A., Oldfield, C. J., Radivojac, P., Vacic, V., Cortese, M. S., Dunker, A. K., et al. (2006). Analysis of molecular recognition features (MoRFs). *Journal of Molecular Biology*, *362*(5), 1043–1059. https://doi.org/10.1016/j.jmb.2006.07.087.

Motáčková, V., Nováček, J., Zawadzka-Kazimierczuk, A., Kazimierczuk, K., Žídek, L., Šanderová, H., et al. (2010). Strategy for complete NMR assignment of disordered proteins with highly repetitive sequences based on resolution-enhanced 5D experiments. *Journal of Biomolecular NMR*, *48*(3), 169–177. https://doi.org/10.1007/s10858-010-9447-3.

Muhandiram, D. R., & Kay, L. E. (1994). Gradient-enhanced triple-resonance three-dimensional NMR experiments with improved sensitivity. *Journal of Magnetic Resonance, Series B*, *103*(3), 203–216. https://doi.org/10.1006/jmrb.1994.1032.

Munoz, V., & Serrano, L. (1994). Elucidating the folding problem of helical peptides using empirical parameters. *Nature Structural Biology*, *1*(6), 399–409. https://doi.org/10.1038/nsb0694-399.

Munoz, V., & Serrano, L. (1995). Elucidating the folding problem of helical peptides using empirical parameters. 2. Helix macrodipole effects and rational modification of the helical content of natural peptides. *Journal of Molecular Biology*, *245*(3), 275–296. https://doi.org/10.1006/jmbi.1994.0023.

Neal, S., Nip, A. M., Zhang, H. Y., & Wishart, D. S. (2003). Rapid and accurate calculation of protein H-1, C-13 and N-15 chemical shifts. *Journal of Biomolecular NMR*, *26*(3), 215–240. https://doi.org/10.1023/A:1023812930288.

Nielsen, J. T., & Mulder, F. A. A. (2018). POTENCI: Prediction of temperature, neighbor and pH-corrected chemical shifts for intrinsically disordered proteins. *Journal of Biomolecular NMR*, *70*(3), 141–165. https://doi.org/10.1007/s10858-018-0166-5.

Oldfield, C. J., Cheng, Y. G., Cortese, M. S., Romero, P., Uversky, V. N., & Dunker, A. K. (2005). Coupled folding and binding with alpha-helix-forming molecular recognition elements. *Biochemistry*, *44*(37), 12454–12470. https://doi.org/10.1021/bi050736e.

Olejniczak, E. T., Xu, R. X., & Fesik, S. W. (1992). A 4D HCCH-TOCSY experiment for assigning the side chain 1H and13C resonances of proteins. *Journal of Biomolecular NMR*, *2*(6), 655–659. https://doi.org/10.1007/bf02192854.

Osapay, K., & Case, D. A. (1994). Analysis of proton chemical-shifts in regular secondary structure of proteins. *Journal of Biomolecular NMR*, *4*(2), 215–230.

Pastore, A., & Saudek, V. (1990). The relationship between chemical-shift and secondary structure in proteins. *Journal of Magnetic Resonance*, *90*(1), 165–176. https://doi.org/10.1016/0022-2364(90)90375-J.

Pervushin, K., Riek, R., Wider, G., & Wüthrich, K. (1997). Attenuated T2 relaxation by mutual cancellation of dipole–dipole coupling and chemical shift anisotropy indicates an avenue to NMR structures of very large biological macromolecules in solution. *Proceedings of the National Academy of Sciences*, *94*(23), 12366–12371. https://doi.org/10.1073/pnas.94.23.12366.

Plaxco, K. W., Morton, C. J., Grimshaw, S. B., Jones, J. A., Pitkeathly, M., Campbell, I. D., et al. (1997). The effects of guanidine hydrochloride on the 'random coil' conformations and NMR chemical shifts of the peptide series GGXGG. *Journal of Biomolecular NMR*, *10*(3), 221–230. https://doi.org/10.1023/A:1018340217891.

Poosapati, A., Gregory, E., Borcherds, W. M., Chemes, L. B., & Daughdrill, G. W. (2018). Uncoupling the folding and binding of an intrinsically disordered protein. *Journal of Molecular Biology*, *430*(16), 2389–2402. https://doi.org/10.1016/j.jmb.2018.05.045.

Radivojac, P., Iakoucheva, L. M., Oldfield, C. J., Obradovic, Z., Uversky, V. N., & Dunker, A. K. (2007). Intrinsic disorder and functional proteomics. *Biophysical Journal*, *92*(5), 1439–1456. https://doi.org/10.1529/biophysj.106.094045.

Richarz, R., & Wuthrich, K. (1978). C-13 NMR chemical-shifts of common amino-acid residues measured in aqueous-solutions of linear tetrapeptides H-Gly-Gly-X-L-Ala-oh. *Biopolymers*, *17*(9), 2133–2141. https://doi.org/10.1002/bip.1978.360170908.

Schwarzinger, S., Kroon, G. J. A., Foss, T. R., Wright, P. E., & Dyson, H. J. (2000). Random coil chemical shifts in acidic 8 M urea: Implementation of random coil shift data in NMR view. *Journal of Biomolecular NMR*, *18*(1), 43–48. https://doi.org/10.1023/A:1008386816521.

Shammas, S. L., Crabtree, M. D., Dahal, L., Wicky, B. I. M., & Clarke, J. (2016). Insights into coupled folding and binding mechanisms from kinetic studies. *Journal of Biological Chemistry*, *291*(13), 6689–6695. https://doi.org/10.1074/jbc.R115.692715.

Shammas, S. L., Travis, A. J., & Clarke, J. (2013). Remarkably fast coupled folding and binding of the intrinsically disordered transactivation domain of cMyb to CBP KIX. *Journal of Physical Chemistry B*, *117*(42), 13346–13356. https://doi.org/10.1021/jp404267e.

Shammas, S. L., Travis, A. J., & Clarke, J. (2014). Allostery within a transcription coactivator is predominantly mediated through dissociation rate constants. *Proceedings of the National Academy of Sciences of the United States of America*, *111*(33), 12055–12060. https://doi.org/10.1073/pnas.1405815111.

Si-Hyung, L., Do-Hyoung, K., Joan, J. H., Eun-Ji, C., Ji-Eun, L., Ye-Jin, C., et al. (2012). Understanding pre-structured motifs (PreSMos) in intrinsically unfolded proteins. *Current Protein & Peptide Science*, *13*(1), 34–54. https://doi.org/10.2174/138920312799277974.

Smith, L. J., Fiebig, K. M., Schwalbe, H., & Dobson, C. M. (1996). The concept of a random coil—residual structure in peptides and denatured proteins. *Folding & Design*, *1*(5), R95–R106. https://doi.org/10.1016/S1359-0278(96)00046-6.

Spera, S., & Bax, A. (1991). Empirical correlation between protein backbone conformation and C-alpha and C-beta C-13 nuclear-magnetic-resonance chemical-shifts. *Journal of the American Chemical Society*, *113*(14), 5490–5492. https://doi.org/10.1021/ja00014a071.

Sugiki, T., Yoshiura, C., Kofuku, Y., Ueda, T., Shimada, I., & Takahashi, H. (2009). High-throughput screening of optimal solution conditions for structural biological studies by fluorescence correlation spectroscopy. *Protein Science*, *18*(5), 1115–1120. https://doi.org/10.1002/pro.92.

Tamiola, K., Acar, B., & Mulder, F. A. (2010). Sequence-specific random coil chemical shifts of intrinsically disordered proteins. *Journal of the American Chemical Society*, *132*(51), 18000–18003. https://doi.org/10.1021/ja105656t.

Tamiola, K., & Mulder, F. A. A. (2012). Using NMR chemical shifts to calculate the propensity for structural order and disorder in proteins. *Biochemical Society Transactions*, *40*, 1014–1020. https://doi.org/10.1042/Bst20120171.

Ting, D., Wang, G. L., Shapovalov, M., Mitra, R., Jordan, M. I., & Dunbrack, R. L. (2010). Neighbor-dependent Ramachandran probability distributions of amino acids developed from a hierarchical Dirichlet process model. *PLoS Computational Biology*, *6*(4), 1–21. https://doi.org/10.1371/journal.pcbi.1000763.

Toal, S., & Schweitzer-Stenner, R. (2014). Local order in the unfolded state: Conformational biases and nearest neighbor interactions. *Biomolecules*, *4*(3), 725.

Tompa, P. (2002). Intrinsically unstructured proteins. *Trends in Biochemical Sciences*, *27*(10), 527–533.

Tompa, P. (2005). The interplay between structure and function in intrinsically unstructured proteins. *FEBS Letters*, *579*(15), 3346–3354. https://doi.org/10.1016/j.febslet.2005.03.072.

Uversky, V. N. (2002). What does it mean to be natively unfolded? *European Journal of Biochemistry*, *269*(1), 2–12.

van der Lee, R., Buljan, M., Lang, B., Weatheritt, R. J., Daughdrill, G. W., Dunker, A. K., et al. (2014). Classification of intrinsically disordered regions and proteins. *Chemical Reviews*, *114*(13), 6589–6631. https://doi.org/10.1021/cr400525m.

Vendruscolo, M. (2007). Determination of conformationally heterogeneous states of proteins. *Current Opinion in Structural Biology*, *17*(1), 15–20.

Vise, P. D., Baral, B., Latos, A. J., & Daughdrill, G. W. (2005). NMR chemical shift and relaxation measurements provide evidence for the coupled folding and binding of the p53 transactivation domain. *Nucleic Acids Research*, *33*(7), 2061–2077. https://doi.org/10.1093/nar/gki336.

Vucetic, S., Xie, H. B., Iakoucheva, L. M., Oldfield, C. J., Dunker, A. K., Obradovic, Z., et al. (2007). Functional anthology of intrinsic disorder. 2. Cellular components, domains, technical terms, developmental processes, and coding sequence diversities correlated with long disordered regions. *Journal of Proteome Research*, *6*(5), 1899–1916. https://doi.org/10.1021/pr060393m.

Wade, M., Li, Y.-C., & Wahl, G. M. (2013). MDM2, MDMX and p53 in oncogenesis and cancer therapy. *Nature Reviews Cancer*, *13*, 83. https://doi.org/10.1038/nrc3430.

Wang, Y. J., & Wishart, D. S. (2005). A simple method to adjust inconsistently referenced C-13 and N-15 chemical shift assignments of proteins. *Journal of Biomolecular NMR*, *31*(2), 143–148. https://doi.org/10.1007/s10858-004-7441-3.

Ward, J. J., Sodhi, J. S., McGuffin, L. J., Buxton, B. F., & Jones, D. T. (2004). Prediction and functional analysis of native disorder in proteins from the three kingdoms of life. *Journal of Molecular Biology*, *337*(3), 635–645. https://doi.org/10.1016/j.jmb.2004.02.002.

Wells, M., Tidow, H., Rutherford, T. J., Markwick, P., Jensen, M. R., Mylonas, E., et al. (2008). Structure of tumor suppressor p53 and its intrinsically disordered N-terminal transactivation domain. *Proceedings of the National Academy of Sciences of the United States of America*, *105*(15), 5762–5767. https://doi.org/10.1073/pnas.0801353105.

Whittington, S. J., Chellgren, B. W., Hermann, V. M., & Creamer, T. P. (2005). Urea promotes polyproline II helix formation: Implications for protein denatured states. *Biochemistry*, *44*(16), 6269–6275. https://doi.org/10.1021/bi050124u.

Wishart, D. S. (2011). Interpreting protein chemical shift data. *Progress in Nuclear Magnetic Resonance Spectroscopy*, *58*(1–2), 62–87. https://doi.org/10.1016/j.pnmrs.2010.07.004.

Wishart, D. S., Bigam, C. G., Holm, A., Hodges, R. S., & Sykes, B. D. (1995). H-1, C-13 and N-15 random coil NMR chemical-shifts of the common amino-acids. 1. Investigations of nearest-neighbor effects (Vol. 5, Pg 67, 1995). *Journal of Biomolecular NMR*, *5*(3), 332.

Wishart, D. S., Bigam, C. G., Yao, J., Abildgaard, F., Dyson, H. J., Oldfield, E., et al. (1995). H-1, C-13 and N-15 chemical-shift referencing in biomolecular NMR. *Journal of Biomolecular NMR*, *6*(2), 135–140.

Wishart, D. S., & Case, D. A. (2001). Use of chemical shifts in macromolecular structure determination. *Nuclear Magnetic Resonance of Biologica Macromolecules*, *338, Pt. A*, 3–34.

Wishart, D. S., & Nip, A. M. (1998). Protein chemical shift analysis: A practical guide. *Biochemistry and Cell Biology—Biochimie Et Biologie Cellulaire*, *76*(2–3), 153–163. https://doi.org/10.1139/bcb-76-2-3-153.

Wishart, D. S., Sykes, B. D., & Richards, F. M. (1991). Relationship between nuclear-magnetic-resonance chemical-shift and protein secondary structure. *Journal of Molecular Biology*, *222*(2), 311–333. https://doi.org/10.1016/0022-2836(91)90214-Q.

Wishart, D. S., Sykes, B. D., & Richards, F. M. (1992). The chemical-shift index—A fast and simple method for the assignment of protein secondary structure through NMR-spectroscopy. *Biochemistry*, *31*(6), 1647–1651. https://doi.org/10.1021/bi00121a010.

Wright, P. E., & Dyson, H. J. (1999). Intrinsically unstructured proteins: Re-assessing the protein structure–function paradigm. *Journal of Molecular Biology*, *293*(2), 321–331.

Xie, H. B., Vucetic, S., Iakoucheva, L. M., Oldfield, C. J., Dunker, A. K., Obradovic, Z., et al. (2007a). Functional anthology of intrinsic disorder. 3. Ligands, post-translational modifications, and diseases associated with intrinsically disordered proteins. *Journal of Proteome Research*, *6*, 1917–1932. https://doi.org/10.1021/pr060394e.

Xie, H. B., Vucetic, S., Iakoucheva, L. M., Oldfield, C. J., Dunker, A. K., Uversky, V. N., et al. (2007b). Functional anthology of intrinsic disorder. 1. Biological processes and functions of proteins with long disordered regions. *Journal of Proteome Research*, *6*, 1882–1898. https://doi.org/10.1021/pr060392u.

Zhang, H. Y., Neal, S., & Wishart, D. S. (2003). RefDB: A database of uniformly referenced protein chemical shifts. *Journal of Biomolecular NMR*, *25*(3), 173–195. https://doi.org/10.1023/A:1022836027055.

Zor, T., De Guzman, R. N., Dyson, H. J., & Wright, P. E. (2004). Solution structure of the KIX domain of CBP bound to the transactivation domain of c-Myb. *Journal of Molecular Biology*, *337*(3), 521–534. https://doi.org/10.1016/j.jmb.2004.01.038.

Zsuzsanna, D. (2018). Prediction of protein disorder based on IUPred. *Protein Science*, *27*(1), 331–340. https://doi.org/10.1002/pro.3334.

CHAPTER SEVEN

Experimental Characterization of Fuzzy Protein Assemblies: Interactions of Paramyxoviral N_{TAIL} Domains With Their Functional Partners

Francesca Troilo*,†, Christophe Bignon*, Stefano Gianni†, Monika Fuxreiter‡, Sonia Longhi*,1

*CNRS and Aix-Marseille Univ, Laboratoire Architecture et Fonction des Macromolecules Biologiques (AFMB), Marseille, France

†Istituto Pasteur—Fondazione Cenci Bolognetti, Dipartimento di Scienze Biochimiche 'A. Rossi Fanelli' and Istituto di Biologia e Patologia Molecolari del Consiglio Nazionale delle Ricerche, Sapienza Università di Roma, Rome, Italy

‡MTA-DE Laboratory of Protein Dynamics, Department of Biochemistry and Molecular Biology, University of Debrecen, Debrecen, Hungary

[1]Corresponding author: e-mail address: sonia.longhi@afmb.univ-mrs.fr

Contents

Methods in Enzymology, Volume 611
ISSN 0076-6879
https://doi.org/10.1016/bs.mie.2018.08.006

Abstract

In this chapter we detail various experimental approaches to characterize the fuzziness of complexes made of the C-terminal domain of the nucleoprotein (N_{TAIL}) from three representative paramyxoviruses and of the C-terminal X domain (XD) of the homologous phosphoprotein. We discuss the advantages, the limitations, as well as the caveats of the various methods. We describe experimental data showing that paramyxoviral N_{TAIL}–XD complexes are characterized by a considerable amount of conformational heterogeneity. We also detail recent data that revealed that N_{TAIL} is highly malleable, i.e., it displays a partner-mediated polymorphism. All the results suggest that N_{TAIL} plasticity and fuzziness play a role in the coordination and regulation of the N_{TAIL} interaction network so as to ensure efficient transcription and replication.

ABBREVIATIONS

k_{off}^{app}	apparent dissociation rate constant
k_{on}^{app}	apparent association rate constant
artN_{TAIL}	artificial N_{TAIL} sequence
CCS	collision cross section
CD	circular dichroism
CGFP	C-terminal moiety of GFP
CSD	charge-state distribution
Da	dalton
EM	electron microscopy
EPR	electron paramagnetic resonance
ESI-MS	electrospray ionization mass spectrometry
FPLC	fast protein liquid chromatography
GFP	green fluorescent protein
HeV	hendra virus
HPLC	high-pressure liquid chromatography
hsp70	70 kDa heat shock protein
HSQC	heterogeneous single quantum correlation
IDP	intrinsically disordered protein
IDR	intrinsically disordered region
IGR	intergenic region
IM	ion mobility
IPTG	isopropyl β-D-1-thiogalactopyranoside
ITC	isothermal titration calorimetry
K_D	equilibrium dissociation constant
k_{obs}	observed rate constant
L	large protein
MeV	measles virus
MM	molecular mass
MoRE	molecular recognition element
MTSL	methanethiosulfonate derivative
N	nucleoprotein
N_{CORE}	globular moiety of N

NGFP	N-terminal moiety of GFP
NiV	nipah virus
NMR	nuclear magnetic resonance
N_{TAIL}	C-terminal disordered region of N
OD	optical density
P	phosphoprotein
PCR	polymerase chain reaction
PCT	P C-terminal region
PMD	P multimerization domain
PNT	P N-terminal region
RDCs	residual dipolar couplings
RdRp	RNA-dependent RNA polymerase
R_g	radius of gyration
R_S	Stokes radius
SASA	solvent-accessible surface area
SAXS	small-angle X-ray scattering
SDSL	site-directed spin labeling
SDS-PAGE	sodium dodecyl sulfate polyacrylamide electrophoresis
SEC	size-exclusion chromatography
SEC-MALLS	SEC coupled to multiple angle laser light scattering
XD	X domain of P
τ_c	correlation time

1. INTRODUCTION

1.1 Fuzziness in Molecular Interactions

In contrast to the long-standing view of coupled folding to binding, protein complexes might retain considerable amount of structural and interaction multiplicity or exhibit a rapid conformational exchange even in the presence of a functional partner (Tompa & Fuxreiter, 2008). This phenomenon, termed as fuzziness, is quite common not only in intrinsically disordered (ID) proteins but also in structured proteins and has important biological outcomes (Fuxreiter, 2018; Miskei, Gregus, et al., 2017; Sharma, Raduly, Miskei, & Fuxreiter, 2015). Retaining a high degree of conformational flexibility upon assembly decreases the entropic penalty of binding as compared to a disorder-to-order transition, leading to enhanced affinity or specificity. Fuzziness confers many further functional advantages, including the plasticity of recognition modes both with the same or alternative partners (Fuxreiter & Tompa, 2012). Fuzzy regions can mediate both specific as well as nonspecific contacts, the latter having the potential to increase the local concentration nearby the partner. The long-range nature of fuzzy interactions manifests in allostery. As a result, amino acid substitutions in fuzzy regions located far away from the binding site and/or posttranslational modifications have the potential to impact affinity/specificity of interactions or may exert other regulatory effects (for recent reviews on fuzziness, see Duro, Miskei, & Fuxreiter, 2015; Fuxreiter, 2018; Miskei, Gregus, et al., 2017; Sharma et al., 2015).

Fuzzy complexes are defined by two criteria (Fuxreiter, 2012; Tompa & Fuxreiter, 2008). First, the structural heterogeneity in the bound state of the protein must be characterized by structural techniques. Second, the functional impacts of perturbations (mutations, truncations, deletions) in conformationally heterogeneous regions must be demonstrated. In other words, structural multiplicity or dynamical disorder directly contributes to the function of the complex. Here, we use paramyxovirus N_{TAIL} complexes as an example to overview biophysical and structural approaches enabling the characterization of the fuzziness of protein complexes.

1.2 The *Paramyxoviridae* Replicative Complex, the N and P Proteins, and the N_{TAIL}–XD Complex

The measles, Nipah, and Hendra viruses are members of the *Paramyxoviridae* family within the *Mononegavirales* order. Measles virus (MeV) is one of the most infectious viruses ever known (Lamb & Parks, 2013). The Nipah and Hendra (NiV and HeV) viruses are newly emerged BSL-4 pathogens (Eaton, Mackenzie, & Wang, 2007) that have been gathered within the *Henipavirus* genus. They are the only currently known zoonotic paramyxoviruses and are responsible for severe encephalitis in human beings, with case fatality rates exceeding 75% (Eaton, Broder, Middleton, & Wang, 2006; Wang et al., 2000).

Paramyxoviruses are enveloped viruses. Their envelope consists of a lipid bilayer derived from the plasma membrane of the host cell. Beneath the envelope, the viral matrix protein (M) bridges the viral envelop glycoproteins with the nucleocapsid. As in all *Mononegavirales* members, the genome of paramyxoviruses is encapsidated by the nucleoprotein (N) giving rise to a helical nucleocapsid (N-RNA) with a characteristic herringbone-like structure (Bhella, 2007; Desfosses, Goret, Farias Estrozi, Ruigrok, & Gutsche, 2011; Gutsche et al., 2015; Karlin, Longhi, & Canard, 2002; Longhi et al., 2003; Schoehn et al., 2004). Following fusion between the viral envelope and the host cell membrane, the nucleocapsid is released in the cytoplasm where primary transcription takes place. A peculiar feature of *Mononegavirales* members is that the nucleocapsid, and *not* naked RNA, is the template used by the RNA-dependent RNA polymerase (RdRp) during both transcription and replication. The RdRp is made of the large protein (L) and the phosphoprotein (P). During RNA synthesis, P tethers L onto the nucleocapsid template through the N–P interaction. The P protein is therefore an essential polymerase cofactor since it allows the L protein to be recruited onto the nucleocapsid template. The complex formed by the N, P, and L proteins constitutes the viral replicative unit (Fig. 1A). Once the viral ribonucleoprotein complexes are released in the cytoplasm of infected cells, transcription of viral genes starts. The polymerase enters the nucleocapsid at the 3′ end of the genome where are located the promoters for transcription and replication. Transcription of the genes occurs sequentially. At each intergenic region (IGR), the polymerase ends and reinitiates at the downstream gene. Following polyadenylation, which serves as gene end signal, the polymerase proceeds over three nucleotides without transcribing them and then restarts transcription upon recognition of a downstream gene start signal. The efficiency with which the polymerase reinitiates transcription decreases with increasing distance from the 3′ end thereby leading to a gradient in the transcripts, with the most distal genes being the least expressed (Lamb & Parks, 2013) (Fig. 1A).

Following primary transcription, and presumably due to an increase in the intracellular concentration of N, the polymerase switches to a replication mode, i.e., it ignores the gene junctions and synthesizes a full, complementary strand of the same length as that of the genome. This positive-stranded RNA (termed antigenome) is not used as a template for transcription and serves only as an intermediate in genome replication (see Blocquel, Bourhis, et al., 2012; Lamb & Parks, 2013 for reviews). During replication, the nascent RNA chain is concomitantly encapsidated by N.

Although in vitro L can synthesize short RNA transcripts using naked RNA as substrate in the absence of P (Jordan et al., 2018; Morin, Liang, Gardner, Ross, & Whelan, 2016; Morin, Rahmeh, & Whelan, 2012), in infected cells P is required to allow recognition of the nucleocapsid template. P also serves as a chaperon toward L, acting in conjunction with hsp90 (Bloyet, Welsch, et al., 2016; Katoh et al., 2017).

The N protein is the most abundant viral protein. It has a major structural role as it wraps the genomic RNA within a nuclease-resistant nucleocapsid. In paramyxoviruses, N proteins are more than 500 residues in length. They consist of a well-conserved globular N-terminal moiety, N_{CORE}, which contains all the regions necessary for self-assembly and RNA-binding, and of a C-terminal domain, N_{TAIL} (Fig. 1B) (for reviews, see Erales, Blocquel, et al., 2015; Habchi & Longhi, 2012, 2015; Habchi,

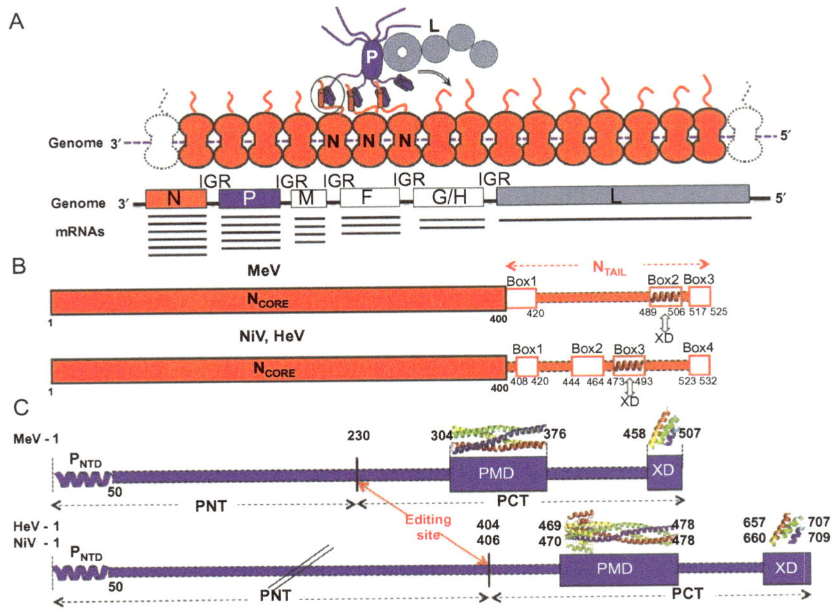

Fig. 1 (A) Schematic illustration of the *Paramyxoviridae* replicative complex and of their genome. The N protein (*red*) is drawn with a bilobal shape according to available structural data (Gutsche et al., 2015). P (*blue*) is represented as a tetramer to reflect the prevalence of this oligomeric state in paramyxoviral P proteins. The L protein (*light blue*) is represented as a donut decorated with a tail made of three globules according to available structural data (Liang et al., 2015). The intrinsically disordered protein regions of N and P are symbolized by lines and α-helixes by cylinders. The P–L complex forms the RNA-dependent RNA polymerase (RdRp) complex, which cartwheels onto the nucleocapsid complex via the X domain of P (XD). The negative-sense genomic RNA (*blue dashed line*) is presented in the 3′ to 5′ orientation. Below the genome is shown a schematic representation of the expression gradient of the mRNA as a result of inefficient transcription reinitiation by the polymerase during transcription (Lamb & Parks, 2013). (B) Modular organization of N showing that it consists of a folded domain, N_{CORE}, and a C-terminal disordered region, N_{TAIL}. The various boxes corresponding to putative or experimentally proven MoREs are shown. The α-MoRE involved in binding to XD is shown as a *red helix*, and its ability to interact with XD is symbolized by a *two-headed arrow*. (C) Modular organization of P. Domain organization of P showing two moieties, PNT and PCT. The P editing site (see text) is shown. Structured and disordered regions are represented as *large* and *narrow boxes*, respectively. *PCT*, C-terminal region of P; *PMD*, P multimerization domain; *PNT*, N-terminal region of P; *XD*, X domain consisting of a triple α-helical bundle. The α-MoRE at the N-terminal region of P (P_{NTD}), which is partly preconfigured in solution and shown to adopt a stable α-helical conformation upon binding to the monomeric form of N (Guryanov, Liljeroos, Kasaragod, Kajander, & Butcher, 2016; Yabukarski et al., 2014), is shown as a *blue helix*. The crystal structures of MeV PMD (PDB code: 4BHV) (Blocquel et al., 2014), NiV PMD (PDB code: 4N5B) (Bruhn-Johannsen et al., 2014), MeV XD (PDB code: 1OKS) (Johansson et al., 2003), and HeV XD (PDB code: 4HEO) (Communie, Habchi, et al., 2013) are shown. All structures were drawn using Pymol (DeLano, 2002).

Mamelli, & Longhi, 2012; Jensen et al., 2010; Longhi, 2015; Longhi, Bloyet, Gianni, & Gerlier, 2017; Longhi & Oglesbee, 2010). N_{TAIL} protrudes from N_{CORE} and is exposed at the surface of the viral nucleocapsid (Gutsche et al., 2015; Heggeness, Scheid, & Choppin, 1980, 1981; Karlin, Longhi, & Canard, 2002; Ringkjøbing Jensen et al., 2011) thereby allowing recruitment of P (Bankamp et al., 1996; Houben, Marion, Tarbouriech, Ruigrok, & Blanchard, 2007; Jensen et al., 2010; Kingston, Hamel, Gay, Dahlquist, & Matthews, 2004; Kingston, Walter, & Gay, 2004; Liston, Batal, DiFlumeri, & Briedis, 1997; Longhi et al., 2003).

The amino acid sequence of N_{TAIL} domains in *Paramyxoviridae* members is highly variable and features a compositional bias (i.e., it is enriched in polar and charged residues and depleted in hydrophobic residues) (Bourhis et al., 2004; Habchi, Mamelli, Darbon, & Longhi, 2010; Karlin, Ferron, Canard, & Longhi, 2003; Longhi et al., 2003). MeV N_{TAIL} is hypersensitive to proteolysis (Karlin, Longhi, & Canard, 2002) and cannot be visualized in (cryo)-electron microscopy (EM) reconstructions of nucleocapsids (Bhella, Ralph, & Yeo, 2004; Desfosses et al., 2011; Longhi et al., 2003). In a seminal study carried out in the early 2000s, these features were regarded as possible hints of intrinsic structural disorder, and the disordered nature of MeV N_{TAIL} was indeed experimentally proven (Bourhis et al., 2004; Longhi et al., 2003). This seminal study seeded subsequent studies focused on other paramyxoviral N proteins (e.g., Sendai virus and henipaviruses) (Habchi et al., 2010; Houben et al., 2007) that unveiled that the presence of a long disordered domain within paramyxoviral N proteins is a conserved feature.

EM studies failed to provide direct structural data on the disordered N_{TAIL} domain because this region was either unresolved (Desfosses et al., 2011) or removed by limited proteolysis (Gutsche et al., 2015). Thanks to the combination of nuclear magnetic resonance (NMR) and small-angle scattering, a model of the RNA-bound form of full-length MeV N was obtained (Ringkjøbing Jensen et al., 2011). In this model, N_{TAIL} points toward the inner channel of the nucleocapsid and then escapes back to the solvent through the confined interstitial space between successive turns of the nucleocapsid (Ringkjøbing Jensen et al., 2011). This model provides a mechanistic basis to explain the increased rigidity of MeV nucleocapsids in which the flexible N_{TAIL} region has been proteolytically removed (Gutsche et al., 2015; Longhi et al., 2003; Schoehn et al., 2004). Subsequent studies carried out on *Henipavirus* nucleocapsid-like particles unveiled a very similar structural organization and showed that NiV and HeV N_{TAIL} domains too remain disordered when appended to the N_{CORE} domain (Baronti et al., 2015; Communie, Habchi, et al., 2013).

Being at least partly exposed at the surface of the nucleocapsid, N_{TAIL} can bind to various partners including the P protein.

The P protein is encoded by the P gene. Paramyxoviral P genes encode multiple proteins through either overlapping reading frames (C protein) or messenger editing (V and W proteins). The latter is a process whereby one or more nontemplated G are inserted cotranscriptionally at a specific site. This insertion causes a shift in the open reading frame leading to a new protein product. Depending on the number of nucleotides inserted at the edition site, the V or W proteins are the end products. The V and W proteins share therefore with P their N-terminal domain and possess each a unique C-terminal domain. The P N-terminal (PNT) and the P C-terminal (PCT) regions, as well as the C-terminal region of V and W, are therefore bona fide domains. Bioinformatics analyses showed that paramyxoviral P proteins share a common structural organization with PNT predicted to be intrinsically disordered and PCT consisting of alternating structured and disordered regions (Habchi et al., 2010; Karlin et al., 2003) (Fig. 1C). The PNT domain of henipaviruses is spectacularly large, consisting of 404 (HeV) or 406 (NiV) residues, and accounts for the extra length of *Henipavirus* P proteins. The disordered state of MeV, NiV, and HeV PNT was experimentally confirmed using a wide range of biochemical and biophysical approaches (Habchi et al., 2010; Karlin, Longhi, Receveur, & Canard, 2002). Interestingly, we have recently shown that *Henipavirus* PNT domains retain their disordered nature also in the context of the V protein (Salladini, Delauzun, & Longhi, 2017).

Within PCT, there are two structured regions: the P multimerization domain (PMD) and the C-terminal X domain (XD), which are responsible, respectively, for P oligomerization and for binding to N_{TAIL} (Bourhis et al., 2004; Houben et al., 2007; Johansson et al., 2003; Kingston, Hamel, et al., 2004; Kingston, Walter, et al., 2004) (Fig. 1C). While PMD is a coiled-coil (Blocquel et al., 2014; Bruhn-Johannsen et al., 2014; Communie, Crepin, et al., 2013), the structure of XD consists of a bundle of three antiparallel α-helices (Blanchard et al., 2004; Communie, Habchi, et al., 2013; Gely et al., 2010; Johansson et al., 2003; Kingston, Hamel, et al., 2004) (Fig. 1C). Its NiV counterpart is predicted to adopt a similar folding (Habchi et al., 2011). Although MeX XD is folded, electron spray ionization mass spectrometry (ESI-MS) studies unveiled that it is structurally heterogeneous, populating at least two alternative conformations under native conditions (D'Urzo et al., 2015). This conclusion is further strengthened by a recent study where the structure of a folding intermediate of MeV XD, which accounts for as much as 30% of the protein population under native conditions, was identified (Bonetti et al., 2016).

N_{TAIL} and XD interact with each other forming a 1:1 stoichiometric complex with an equilibrium dissociation constant (K_D) in the μM range (Dosnon et al., 2015; Habchi et al., 2011). Binding to XD triggers α-helical folding of a short N_{TAIL} region (Box2, aa 486–504 of MeV N, and Box3, aa 473–493 of *Henipavirus* N), referred to as a molecular recognition element or α-MoRE (Fig. 1B) (Bourhis et al., 2004; Habchi et al., 2011; Johansson et al., 2003; Kingston, Hamel, et al., 2004). MoREs are short, transiently populated secondary structures within ID regions that are biased toward their bound state (Fuxreiter, Simon, Friedrich, & Tompa, 2004).

Both random and site-directed mutagenesis allowed identifying the MeV N_{TAIL} residues most critical for binding to XD (Bloyet, Brunel, et al., 2016; Gruet et al., 2013). Further investigation of single-site N_{TAIL} variants bearing substitutions within the α-MoRE revealed that the affinity of the N_{TAIL}–XD pair is directly related to the efficiency with which the viral polymerase reinitiates transcription at each IGR (Bloyet, Brunel, et al., 2016).

The Cα chemical shifts of both MeV and *Henipavirus* N_{TAIL} and the mobility of paramagnetic spin labels grafted at various N_{TAIL} positions indicate that the α-MoRE is partly preconfigured as an α-helix in the absence of XD (Baronti et al., 2015; Belle et al., 2008; Communie, Habchi, et al., 2013; Gely et al., 2010; Martinho et al., 2013; Morin et al., 2006). The Box3 region of NiV N_{TAIL} exhibits a higher degree of preconfiguration (Baronti et al., 2015). It also populates longer-lived interconverting α-helical segments than in its HeV N_{TAIL} counterpart, as detected by site-directed spin-labeling (SDSL) electron paramagnetic resonance (EPR) measurements (Martinho et al., 2013). A subsequent study that combined residual dipolar couplings (RDCs) measurements and ensemble optimization methods (Bernado et al., 2005) showed that the α-MoRE of MeV N_{TAIL} exists in a conformational equilibrium between a completely unfolded form and four partially helical conformers (Ringkjøbing Jensen et al., 2011). In the closely related Sendai virus, the α-MoRE has a similar conformational behavior, although it samples only three helical conformers (Jensen et al., 2010, 2008).

Partial preconfiguration of the XD-binding region of N_{TAIL} is therefore a conserved feature shared by various members of the *Paramyxoviridae* family, implying a functional significance. The partial preconfiguration of MoREs facilitates the folding-upon-binding process: the residual structure restrains the conformational space sampled by the ID protein, thereby reducing the number of interconverting conformers in solution and rendering the structural transition to the (partially) folded conformation energetically less demanding (Fuxreiter et al., 2004). Therefore, although disorder-to-order transitions are entropically disfavored, ID proteins can finely tune their affinity toward partners by varying the extent of preconfiguration of their partially structured elements (Fuxreiter et al., 2004).

In spite of this preconfiguration, molecular dynamic simulations and kinetics studies showed that MeV N_{TAIL} folds according to a folding after binding mechanism (Dosnon et al., 2015; Wang et al., 2013). Subsequent mutational studies coupled to Φ-value analysis led to a detailed structural description of the folding and binding events occurring in the recognition between N_{TAIL} and XD (Bonetti et al., 2017). By Φ-value analysis, residue-specific structural information is obtained by comparing the kinetics of the reaction (folding and/or binding) of the *wild-type* protein with that of a series of conservative single-site variants, yielding the so-called Φ-value. The latter represents an index of native-like structure of the mutated residue in the transition state (Fersht, Matouschek, & Serrano, 1992). In the case of MeV N_{TAIL}, we measured the effect of single-amino acid substitutions in N_{TAIL} on the reaction mechanism. Analysis of the experimental data allowed the identification of key residues involved in the initial recognition between the two molecules and enabled the general features of the folding pathway of N_{TAIL} to be unraveled. In addition, analysis of the changes in stability for all the variants highlighted the inherent poor folding efficiency of N_{TAIL}. We have proposed that this feature might be a consequence of the weakly funneled nature of the energy landscape of ID proteins in their unbound state that might bias for a highly heterogeneous bound state (Bonetti et al., 2017).

A folding upon binding mechanism is not a unique feature of the MeV N_{TAIL}–XD interaction. Indeed, quantitative analysis of peak intensities in heterogeneous single quantum correlation (HSQC) spectra of HeV N_{TAIL} at various XD titration points showed a differential broadening consistent with a mechanism where XD would first bind to a short, central helix within the α-MoRE, and then this helix would be gradually extended via helical folding of the adjacent residues (Communie, Habchi, et al., 2013). A folding upon binding mechanism is also shared by the Sendai virus N_{TAIL}–XD pair, where a detailed atomic description of the molecular recognition trajectory could be obtained from relaxation dispersion studies (Schneider et al., 2015).

The crystal structure of a MeV chimeric construct in which XD is covalently attached to the α-MoRE of N_{TAIL} (aa 486–504) was solved at 1.8 Å (Kingston, Hamel, et al., 2004). The structure consists of a pseudo-four helix complex in which the α-MoRE of N_{TAIL} adopts a parallel orientation with respect to XD and is embedded in a large hydrophobic cleft delimited by XD helices α2 and α3 (Kingston, Hamel, et al., 2004). The structure of the MeV XD/α-MoRE complex was used as a template to model the structure of the homologous *Henipavirus* complexes (Habchi et al., 2011). These models were successively validated by SDSL EPR spectroscopy studies (Martinho et al., 2013).

In spite of the lack of direct structural data on *Henipavirus* N_{TAIL}–XD complexes, NMR studies provided insights into the structure of the HeV N_{TAIL}–XD complex and allowed further model refinement using constraints based on chemical shifts (Communie, Habchi, et al., 2013). Analysis of chemical shift perturbations in reciprocal titration studies, combined with the availability of the crystal structure of XD, allowed the residues involved in the interaction to be identified (Communie, Habchi, et al., 2013). These studies revealed that albeit the binding interface is prevalently hydrophobic, the binding pocket of XD is surrounded by acidic residues that may establish electrostatic interactions with basic residues of Box3. Subsequent isothermal titration calorimetry (ITC) studies carried out at different pH values did in fact confirm the role of electrostatics in complex formation, a conclusion further strengthened by mutational studies that targeted charged residues of both N_{TAIL} and XD (Erales, Beltrandi, Roche, Maté, & Longhi, 2015). Accordingly, we proposed that the formation of the HeV N_{TAIL}–XD complex may rely on the so-called "electrostatic steering mechanism" (Schreiber, Haran, & Zhou, 2009). This model also implies a folding after binding mechanism, in agreement with NMR titration data (Communie, Habchi, et al., 2013). Neither chemical shifts nor electrostatic interactions enable drawing definite conclusions on the orientation of the MoRE at the XD surface. Through a combination of mutational and small-angle X-ray scattering (SAXS) studies, the α-MoRE was shown to adopt a parallel orientation at the surface of HeV XD (Erales, Beltrandi, et al., 2015), as in the case of the MeV complex (Kingston, Hamel, et al., 2004). The conserved parallel orientation of the MoRE at the XD surface in both MeV and HeV might reflect the need for a precise relative orientation of the whole P protein with respect to the N_{TAIL} region protruding from the nucleocapsid. In its turn, this might be related to optimal positioning of the polymerase onto the nucleocapsid template and might impart directionality to the polymerase movement along the nucleocapsid.

In both MeV and henipaviruses, following binding to XD, the majority of N_{TAIL} exhibits a conformational exchange and does not establish stable contacts with XD (Baronti et al., 2015; Belle et al., 2008; Belle, Rouger, Costanzo, Longhi, & Fournel, 2010; Blocquel, Habchi, Gruet, Blangy, & Longhi, 2012; Bourhis et al., 2005; Communie, Habchi, et al., 2013; Gely et al., 2010; Habchi et al., 2011; Martinho et al., 2013; Morin et al., 2006). These N_{TAIL}–XD complexes are therefore illustrative examples of fuzziness (Fuxreiter, 2012, 2018; Tompa & Fuxreiter, 2008).

Below, we summarize the available experimental data supporting the fuzziness of paramyxoviral N_{TAIL}–XD complexes and describe experiments that allowed unraveling its functional impact. We focus onto the experimental design, and for each of the various experimental approaches detailed below we emphasize the advantages and limitations and provide a few caveats.

2. EXPERIMENTAL APPROACHES TO PROBE FUZZINESS IN PARAMYXOVIRAL N_{TAIL}–XD COMPLEXES

As detailed later, various experimental approaches can be used to document fuzziness in IDP complexes. They broadly vary in terms of the molecular information provided, of protein requirements, and of ease of implementation. Fig. 2 illustrates a general suggested procedure for the experimental assessment of fuzziness with increasing level of complexity.

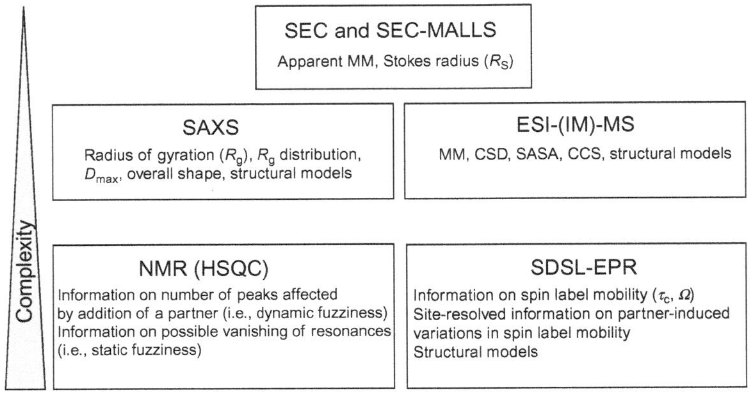

Fig. 2 General suggested procedure for the experimental assessment of fuzziness in IDP complexes. The molecular information provided by each approach is indicated, as it is the degree of complexity of the various approaches.

2.1 SEC and SEC-MALLS

One of the easiest approaches to document lack of significant compaction upon complex formation is size-exclusion chromatography (SEC). The latter allows separating proteins according to their hydrodynamic volume. In practice, upon injection of a mixture containing the two interacting proteins at a concentration that has to be well above the K_D so as to ensure complex formation, the Stokes radius (R_S) of the complex can be determined from the elution volume using a calibration curve obtained using globular proteins of known molecular mass (MM, in daltons). The R_S (in Å) of the molecular mass standards is calculated according to Uversky (1993):

$$\log\left(R_S^{Obs}\right) = 0.369^* \left(\log \text{MM}\right) - 0.254 \quad (1)$$

From this calibration curve one can thus infer the R_S of the complex ($R_S^{Obs\text{-}Complex}$). This experimentally observed value is then compared to the value expected for a fully folded complex ($R_S^{FF\text{-}Complex}$) made of protein A and protein B. The latter is calculated according to Uversky (2002):

$$\log\left(R_S^{FF\text{-}Complex}\right) = 0.357^* \left[\log\left(\text{MM}^A + \text{MM}^B\right)\right] - 0.204 \quad (2)$$

where MM^A and MM^B are the molecular masses (in Daltons) of protein A and protein B. It is recommended to check that the elution profile is not ionic strength- and buffer-dependent. It also goes without saying that one has to check by sodium dodecyl sulfate polyacrylamide electrophoresis (SDS-PAGE) the effective presence of the two interacting proteins in the fractions of the peak supposed to correspond to the complex.

Alternatively, the Stokes radius can be determined by SEC coupled to multiple angle laser light scattering (SEC-MALLS). In this case, the Stokes radius and the molecular mass are directly measured and not inferred from a calibration curve. While SEC relies on fast protein liquid chromatography (FPLC), SEC-MALLS requires the more sophisticated and hence more expensive high-pressure liquid chromatography (HPLC) system.

In both cases, the major technical limitation resides in the ability to document complex formation. Indeed, upon injection of 500 μL of a mixture containing both MeV N_{TAIL} and XD at protein concentrations up to 5 mg/mL each onto a SEC Superdex 75 HR 10/30 column (GE, Healthcare) no peak corresponding to the complex could ever be detected. This is likely due to complex dissociation during elution in the column induced by dilution. This technical limitation can be sometimes overcome using SEC-MALLS where both the volume of the injected sample (typically, 30 μL of a protein solution at 0.3–1.5 m*M*) and that of the column are reduced. Typically, silica-based columns (Shodex) have a smaller volume (15 mL) compared to Superdex columns routinely used in SEC. Using this approach we could indeed successfully document NiV N_{TAIL}–XD complex formation (Habchi et al., 2011): upon addition of stoichiometric amounts of XD, the N_{TAIL} peak shifted toward a lower elution volume indicating formation of the complex, as also judged from the estimated mass (25,610 ± 80 Da), a value consistent with a 1:1 complex (expected mass = ~22 kDa). Upon adding a twofold molar excess of XD, no further significant shift in the elution volume of N_{TAIL} was observed, indicating that saturation was achieved for a 1:1 complex. The measured R_S of the complex is 35.4 ± 3.1 Å, a value much larger than that expected (22.3 Å) for a fully folded complex with a molecular mass equal to the sum of the masses of N_{TAIL} and XD (~22 kDa) (Uversky, 2002). This large value of the R_S reflects a complex that is not fully compact and that rather retains considerable conformational entropy (Habchi et al., 2011). Notably, no shift in the elution profile of HeV N_{TAIL} was observed upon addition of as much as a fourfold molar excess of XD, indicating lack of complex formation and probably reflecting a less tight complex. That the NiV N_{TAIL}–XD complex is tighter than its HeV counterpart is also supported by SDSL EPR and NMR spectroscopy data that unveiled the contribution of an additional region (i.e., Box2, aa 444–464) to binding to XD (Baronti et al., 2015; Martinho et al., 2013) (see below).

2.2 Small-Angle X-Ray Scattering

SAXS is a powerful method to obtain low-resolution (20–10 Å) structural information on flexible molecules in solutions. The scattering curve, as recorded by irradiating a protein solution with X-rays and using a detector located at hundreds of cm from the sample (typically from 1 to 3 m), depends on the size and the shape of the protein. The higher the distance of the detector, the lower the scattering angle (2θ), and hence the lower the resolution. Although SAXS studies can also be performed using laboratory equipment, the use of synchrotron radiation dramatically speeds up the process of data collection and yields data of better quality. Typically 30–50 μL of protein samples are used in a concentration range of 1–10 mg/mL. It is recommended to collect data at various protein concentrations so as to rule out any possible attractive and/or repulsive interparticle interactions. It is also recommended to use *strictly* the same buffer in which the protein is prepared when recording background scattering.

From the scattering curve one can derive the radius of gyration, R_g, the R_g distribution (for IDPs), the maximal intramolecular distance, and the molecular mass. The R_g and the R_S correspond to two distinct parameters. The R_g is the average root-mean-square distance to the center of density in the molecule weighted by the scattering length density, whereas the R_S is the radius of a spherical particle of the same molecular mass as that of the particle under study. From the scattering data, Guinier plots (where $\ln(I)$ is plotted as a function of q^2) can be built. The slope of the plots in the Guinier region ($qR_g < 1.0$) gives the value of the Rg, while the intercept of the straight line gives the $I(0)$, which is proportional to the molecular mass of the scatterer.

Not only SAXS is very well suited to study IDPs (Bernado & Svergun, 2010, 2012) but can also be used to study complexes made of IDPs bound to their partner(s). We successfully used SAXS to investigate the MeV N_{TAIL}–XD complex (Bourhis et al., 2005).

We first ascertain that XD possesses the same structure in solution as in the crystal. SAXS experiments performed on XD showed that it has a globular shape with an R_g of 12.1 $\pm$0.8 Å and a maximum diameter D_{max} of 41 $\pm$ 1 Å. The scattering profile back calculated from the crystal structure (PDB code: 1OKS) using the program CRYSOL (Svergun, Baraberato, & Koch, 1995) is close to the experimental one ($\chi^2 = 1.4$) and yields a theoretical R_g of 12.3 Å, thereby indicating that the overall conformation of XD in solution is similar to that observed in the crystal.

The N_{TAIL}–XD complex was prepared by mixing N_{TAIL} with a twofold molar excess of XD at a final protein concentration of 10 mg/mL. SAXS data were also collected on N_{TAIL} at various protein concentrations ranging from 1.8 to 10 mg/mL. Background scattering was measured before or after each protein sample using the buffer solution and then subtracted from the protein scattering patterns. The experimental scattering curves at the various protein concentrations were merged to yield a single merged curve.

The scattering profile of the N_{TAIL}–XD complex was obtained as follows. To avoid any possible bias on the absolute intensities due to the concentration, the experimental merged scattering curves obtained as described earlier were normalized by their theoretical I(0). The scattering pattern of XD was then subtracted from the scattering pattern of the mixture containing N_{TAIL} and XD at a 1:2 M ratio. Since the K_D between N_{TAIL} and XD is in the μM range, the concentration of uncomplexed N_{TAIL} in solution can be assumed to be negligible at the concentration used in the SAXS experiments. The scattering pattern of the complex obtained as described earlier was then used for further calculations. Guinier analysis of the curve yielded an R_g of 32.7 ± 0.7 Å. The molecular mass calculated from the forward scattering intensity $I(0)$ is 23.5 ± 2 kDa, in agreement with the value expected for a 1:1 stoichiometric complex (21.3 kDa), thus confirming complex formation. The high value of R_g indicated that the overall structure of the N_{TAIL}–XD complex is not compact. The distance distribution function inferred from the scattering curve of the N_{TAIL}–XD complex is asymmetrical with a maximum at 20 Å, a shoulder at about 30 Å, and a maximum value at 146 Å, which is typical of an elongated object (see Bourhis et al., 2005). The bump most probably corresponds to the intramolecular distance within the globular portion of the complex (see below), while the tail indicates that N_{TAIL} possesses regions with an extended conformation.

The overall envelope of the complex was calculated ab initio from its scattering profile using the program GASBOR (Svergun, Petoukhov, & Koch, 2001). Several independent runs yielded different shapes with recurrent features: they were all elongated, with a globular cluster of always the same size at one extremity and an elongated protuberance with different curvatures and cross sections. The quality of the fit to the experimental data was similar in all cases, with χ^2 values ranging from 1.3 to 1.6. The globular part was assumed to correspond to XD packed against the folded region of N_{TAIL} (i.e., the α-MoRE), while the appendage corresponds to unfolded regions of the latter. The crystal structure of the chimera made of XD and of the 486–505 region of N_{TAIL} (PDB code: 1T6O) was inserted in the shape using the program SUPCOMB (Kozin & Svergun, 2001) (Fig. 3). In order to get insights into the conformation that the regions of N_{TAIL} preceding and following the α-MoRE (aa 401–485 and 506–525, respectively) adopt in the complex with XD, the program package CREDO was used. CREDO is an extension of GASBOR and calculates the position of missing loops in crystal structures. The results obtained with several independent runs using the crystal structure of the chimera as template, yielded multiple models all consisting of at one side a peptide chain in an extended conformation and at the other side a chain packed against the bulky portion of the XD/α-MoRE complex. The extended conformation, protruding from the globular cluster of the complex and pointing to the solvent, corresponds to the long N-terminal region of N_{TAIL}. In contrast, the C-terminal region of N_{TAIL} (aa 506–525) always packs against the globular moiety of the envelope but at varying distances and positions. The back-calculated scattering curves from these models were found all to well (and equally) fit the experimental data (χ^2 of 0.8–1.0).

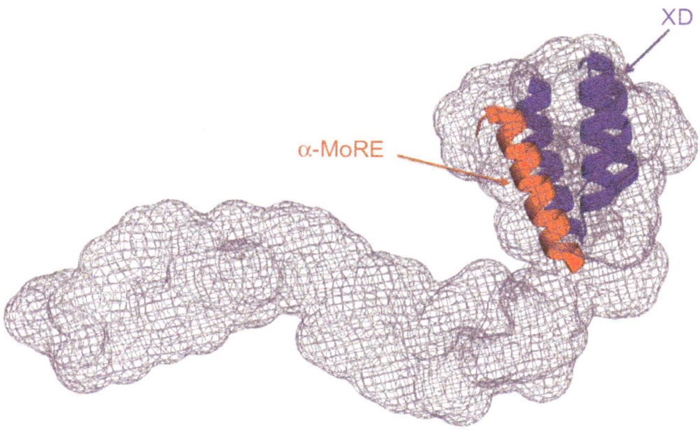

Fig. 3 SAXS model of the MeV N_{TAIL}–XD complex. The structure of the chimeric construct made of XD and of the α-MoRE is shown within the globular part of the envelope obtained by ab initio shape reconstruction. *Data are taken from Bourhis, J. M., Receveur-Bréchot, V., Oglesbee, M., Zhang, X., Buccellato, M., Darbon, H., et al. (2005). The intrinsically disordered C-terminal domain of the Measles virus nucleoprotein interacts with the C-terminal domain of the phosphoprotein via two distinct sites and remains predominantly unfolded.* Protein Science, 14, *1975–1992.*

In conclusion SAXS studies clearly unveiled the presence of a long flexible appendage accommodating the N-terminal region of N_{TAIL}.

2.3 Nuclear Magnetic Resonance

NMR spectroscopy is among the most (if not the most) powerful techniques to investigate IDPs and the structural transitions they undergo upon binding to their partner(s) (Felli & Pierattelli, 2015). Thanks to recent improvements in NMR instrumentation, pulse sequence design, and sample preparation, new NMR tools have become available for the characterization of IDPs at atomic resolution. NMR spectroscopy provides a unique tool to access dynamics information at atomic resolution, from the picosecond timescale to slow exchange processes on the second (or even slower) timescale. As such, NMR is ideally well suited to study IDPs that are often characterized by heterogeneous dynamic properties along the polypeptide chain. Peak positions in NMR spectra depend on the resonance frequencies of the observed nuclei that in turn depend on their chemical environment. Resonance positions in NMR spectroscopy are expressed as "chemical shifts" relative to a standard and are expressed in parts *per* million (ppm). In IDPs, the lack of a stable structure results in averaging of a large part of the contributions to the chemical shift coming from the local chemical environment, thereby leading to the typical low chemical shift dispersion observed in the NMR spectra (Fig. 4). The chemical shifts of backbone

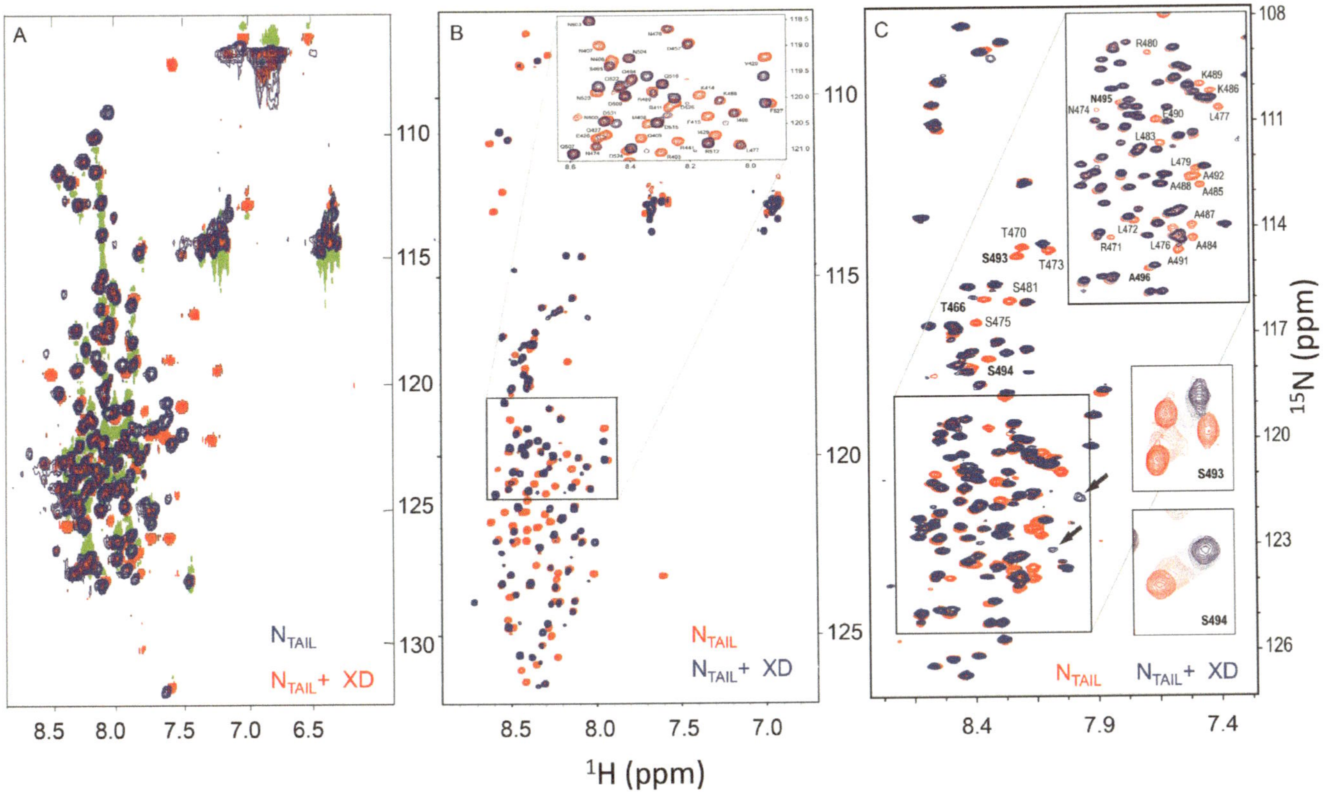

Fig. 4 See legend on opposite page.

nuclei are sensitive to the local backbone geometry, and thus provide useful information on the nature and level of secondary structural elements within the protein (Spera & Bax, 1991). The chemical shifts make NMR an atomic resolution technique in that once the resonances are assigned to their respective nuclei their response to perturbations (such as addition of partners/ligands) can be monitored.

HSQC spectra are very useful to highlight structural transitions in IDPs. The approach relies on the use of a ^{15}N uniformly labeled IDP and on the monitoring of possible chemical shift changes in the backbone amide and proton resonances upon addition of an unlabeled partner. In this way, one can estimate the number of affected residues without the need of a prior full assignment of the spectrum.

^{1}H–^{15}N HSQC studies of ^{15}N uniformly labeled MeV N_{TAIL} XD either alone or in the presence of a twofold molar excess of XD (Bourhis et al., 2005) allowed a quantitative estimation of the number of residues involved in the interaction with XD (Fig. 4A). Eleven peaks undergo an upfield shift, which is indicative of a random coil to α-helix transition, and five additional ones undergo a less dramatic displacement reflecting only a change in the chemical environment without concomitant gain of regular secondary structure. The change in the chemical environment of these latter residues can result either by direct interaction with the partner or by local magnetic perturbations of residues spatially close to the newly formed α-helix. Although this approach allows drawing conclusions about the number of residues affected by the partner and the nature of the structural transition,

Fig. 4 2D-HSQC NMR spectra of N_{TAIL} domains. MeV (A), HeV (B), and NiV (C) N_{TAIL}. (A) HSQC spectrum of a 0.5-m*M* solution of MeV ^{15}N-N_{TAIL} alone (*blue*) or in the presence of a twofold molar excess of XD (*red*) in 10 m*M* sodium phosphate at pH 7. The *green spots* correspond to peaks of negative intensity. The spectra were recorded at 283 K on a 500-MHz DRX Bruker spectrometer. ppm quotes for resonance shifts in parts *per* million of the spectrophotometer frequency (Bourhis et al., 2005). (B) Superposition of the ^{1}H–^{15}N HSQC spectra of HeV N_{TAIL} alone at 40 μ*M* (*red*) and in the presence of a 4.4-fold molar excess of XD (*blue*) (Communie, Habchi, et al., 2013). The spectra were obtained at a ^{1}H frequency of 800 MHz and 293 K. The expanded regions show the spectral assignment of a number of resonances in the spectrum of the isolated N_{TAIL} domain. (C) Superimposition of the ^{1}H–^{15}N HSQC spectra of isolated N_{TAIL} at 180 μ*M* (*red*) and in the presence of a fourfold molar excesses of XD (*blue*) (Baronti et al., 2015). The spectra were obtained at 500 MHz and 285.5 K. The expanded region shows the spectral assignment of a number of resonances in the spectrum of the isolated N_{TAIL} domain. Two additional peaks that appear in the presence of saturating amounts of XD are indicated by an *arrow*.

it does not provide information on the involved residues. Identifying the latter requires assignment of the spectrum.

Subsequently, the HSQC spectrum of MeV N_{TAIL} could be assigned using a double-labeled ($^{13}C/^{15}N$) N_{TAIL} sample and a series of BEST-type triple-resonance experiments (Gely et al., 2010). NMR titration experiments with XD were then performed. In these experiments, ^{1}H–^{15}N HSQC spectra were collected on a 0.3-m*M* sample of ^{15}N-labeled N_{TAIL} in the presence of different concentrations of unlabeled XD. At low XD concentrations (0.037, 0.075, and 0.150 m*M*), most resonances in the HSQC experience no chemical shift changes (residues 401–474), some resonances undergo fast exchange and experience chemical shift changes with minimal line broadening (residues 475–482 and residues 507–525), and some resonances undergo intermediate exchange and experience significant line broadening (residues 483–506). For most resonances in the latter category, line broadening was so severe that prevented signal detection. At high XD concentrations (0.3 and 0.6 m*M*), resonances for residues 492–501, and 503–506 reappeared at new positions, suggesting the formation of a stable complex. The most pronounced amide ^{1}H and ^{15}N chemical shift changes upon binding XD concern N_{TAIL} residues 483–506. In addition, several residues in the Box3 region have chemical shift differences in the range of 20–30 Hz (Gely et al., 2010). From the chemical shifts of the free and bound form of N_{TAIL}, the secondary chemical shifts ($\Delta\delta$) were computed and allowed estimating the amount of helical structure along the polypeptide chain. $\Delta\delta$ values greater than one indicate a fractional population of helical structure in the range of 50%. This is the case of residues 491–499 in free NTAIL. The 13Ca and 13CO resonances of these residues undergo positive shifts in their $\Delta\delta$ values in the presence of XD, indicating that the transient helical structure in free NTAIL is stabilized in the NTAIL–XD complex.

Altogether, NMR studies revealed that the majority of N_{TAIL} residues are not affected by binding to XD (i.e., aa 401–474) and therefore remain fuzzy. Among affected residues, two types of interaction are discernible: one, mediated by residues belonging to Box2, involves a significant gain of α-helicity, while the other, attributable to Box3 residues, is not accompanied by a significant structural transition.

Similar NMR experiments were successively carried out on HeV and NiV N_{TAIL} (Habchi et al., 2011). Their spectra could be assigned using standard triple-resonance experiments (Baronti et al., 2015; Communie, Habchi, et al., 2013). Experimental chemical shifts and nuclear relaxation data unveiled the presence of a transiently populated α-helix encompassing residues 470–490 in HeV N_{TAIL} and residues 470–495 in NiV N_{TAIL}, a region largely overlapping with Box3 (Fig. 1B) and previously shown to be the XD-binding region (Blocquel, Habchi, et al., 2012; Habchi et al., 2011). Like in the case of MeV N_{TAIL}, titration studies with unlabeled XD revealed that most resonances in the HSQC experienced no chemical shift changes (Fig. 4B and C). For both HeV and NiV N_{TAIL} samples, the largest chemical shift perturbations were observed within the 470–490 region (Baronti et al., 2015; Communie, Habchi, et al., 2013). In the case of HeV, the addition of an increasing amount of unlabeled XD to ^{15}N-labeled N_{TAIL} results in severe line broadening of the NMR resonances for residues 470–490 even at substoichiometric amounts of XD. In the case of NiV, peaks corresponding to this region disappear at the first titration point. Resonance vanishing without concomitant appearance of new peaks in the α-helical region is consistent with an intermediate exchange regime occurring on the micro- to millisecond time scale (i.e., the exchange rate between free and bound N_{TAIL} is comparable to the chemical shift difference between the free and the bound form) and reflects a highly dynamic binding of Box3 at the surface of XD.

In the case of NiV, 2D ^{13}C–^{15}N CON experiments revealed the additional involvement in the interaction of residues 452–456 that undergo slow to intermediate exchange upon addition of XD.

Collectively, titration studies revealed that, in contrast with MeV, for both NiV and HeV N_{TAIL} domains, the resonances of the α-MoRE vanish upon addition of the homologous XD, a finding consistent with a highly dynamic complex in which the α-MoRE undergoes α-helical fraying at the surface of XD (Baronti et al., 2015; Communie, Habchi, et al., 2013; Habchi et al., 2011). The NiV N_{TAIL}–XD complex is slightly tighter than that of HeV (Baronti et al., 2015; Blocquel, Habchi, et al., 2012; Communie, Habchi, et al., 2013; Habchi et al., 2011), a property that is reflected in our ability to document by SEC complex formation in the case of NiV but not in the case of HeV (Habchi et al., 2011). In further support of a tighter complex, SDSL EPR experiments showed that in the case of NiV, an additional N_{TAIL} region (i.e., Box2) is involved in the interaction with XD (Martinho et al., 2013) (see below).

Therefore, the *Henipavirus* N_{TAIL}–XD complexes are even fuzzier than the MeV one, as the fuzziness also pertains the binding region itself. This likely provides an additional way of tuning the strength of the interaction, rendering it more prone to be competed out by other partners.

2.4 Site-Directed Spin-Labeling Electron Paramagnetic Resonance (SDSL-EPR) Spectroscopy

EPR spectroscopy is a technique that specifically detects unpaired electrons. It is based on the observation of the energy absorbed by a paramagnetic system in a magnetic field. SDSL combined with EPR is a powerful technique for detecting structural changes in proteins that are devoid of paramagnetic centers. The basic strategy of SDSL involves the insertion of a paramagnetic label at a selected site of a protein and its observation by EPR spectroscopy (for reviews, see Belle et al., 2010; Hubbell, Gross, Langen, & Lietzow, 1998; Longhi, Belle, Fournel, Guigliarelli, & Carrière, 2011). This is generally achieved through cysteine-substitution mutagenesis, followed by covalent modification of the sulfhydryl group with a selective nitroxide reagent, such as the methanethiosulfonate derivative (MTSL). For a detailed procedure for the generation of variants (from site-directed mutagenesis to spin-labeling and EPR spectra recording) the interested reader may refer to Habchi, Martinho, et al. (2012).

The EPR spectrum of the labeled protein is then recorded, and variations in spin-label mobility under various conditions can be monitored. The main advantages of this technique reside in the fact that (i) spectra can be recorded at room temperature, (ii) paramagnetic probes introduce a minimal perturbation of the system, and (iii) they probe the very local environment of the spin label. Any change in this environment resulting from either a physical or chemical event can then be identified from the change in the EPR signature. Indeed, the mobility of the spin label is reflected in the spectral shape of the spin-labeled protein. When paramagnetic labels are allowed to tumble rapidly in an isotropic way, magnetic interactions are completely averaged and the EPR spectrum displays three narrow lines. By contrast, with slower motions the magnetic anisotropy is no longer totally averaged, thus giving rise to spectral variations. In the so-called *fast-motional regime* (i.e., correlation time τ_c up to 10^{-9} s), the spectrum is characterized by a differential broadening of lines although line positions remain constant. For $\tau_r > 10^{-9}$ s, the averaging of the magnetic interactions becomes progressively less efficient, leading to shape distortions of the EPR spectrum.

Spin labels grafted within IDPs typically display a fast-motional regime. To derive quantitative information on the spin-label mobility, the most suited parameter is the ratio between the peak-to-peak amplitude of the low-field and central lines, $h(+1)/h(0)$ ratio (Morin et al., 2006). This ratio decreases with decreasing spin-label mobility (Morin et al., 2006).

SDSL EPR spectroscopy has emerged in the last decade as a powerful method to investigate structural transitions in IDPs (for reviews, see Casey et al., 2014; Drescher, 2012).

We have successfully used it to investigate the structural transitions that N_{TAIL} domains undergo in the presence of the homologous XD (Belle et al., 2008; Martinho et al., 2013; Morin et al., 2006). As detailed later, the approach allowed us to precisely map the N_{TAIL} region undergoing α-helical folding.

In the case of MeV N_{TAIL}, we targeted for spin-labeling 14 sites, 12 of which are scattered in the 488–525 region, while 2 (positions 407 and 460) are located outside the reported region of interaction with XD (Fig. 5A). Fourteen single-site cysteine variants were designed and purified from the soluble fraction of *Escherichia coli* and labeled as described in Belle et al. (2008). Labeling yields ranged from 50% to 80%. The concentration of labeled proteins was evaluated by double integration of the EPR signal recorded under nonsaturating conditions and comparison with that given by a 3-carboxy-proxyl standard sample. The labeling yields were estimated by calculating the ratio between the concentration of labeled proteins and the total protein concentration.

Like for all experimental approaches that rely on site-directed mutagenesis and/or on the covalent modification of the protein, it is recommended to check that the substitution and/or the chemical modification does not affect the overall secondary structure content and the global folding (or lack of) of the protein. In agreement, the possibility that the cysteine substitution and the introduction of the spin label might affect the overall structure and the folding propensities of the N_{TAIL} variants was checked and ruled out by far-UV circular dichroism (CD) (Belle et al., 2008).

The EPR spectra of all the spin-labeled N_{TAIL} variants exhibit a relatively narrow, single-component shape and are indicative of a high radical mobility, as judged on the basis of ($h(+1)/h(0)$ ratio) (Fig. 5A, left panel). In spite of the overall high mobility, the spin labels grafted within Box2 exhibit a slightly reduced mobility, a behavior likely reflecting the partial preconfiguration of the α-MoRE. EPR spectra were also recorded in the presence of a molar excess of XD (both either in the absence or in the presence of 30% sucrose) (Fig. 5B, right panel). For each EPR spectrum, with the only exception of variant S491C that constitutes a special case of highly restricted motional regime, we measured the $h(+1)/h(0)$. For position 491, which is the only one whose side chain is located at the XD-binding interface, the

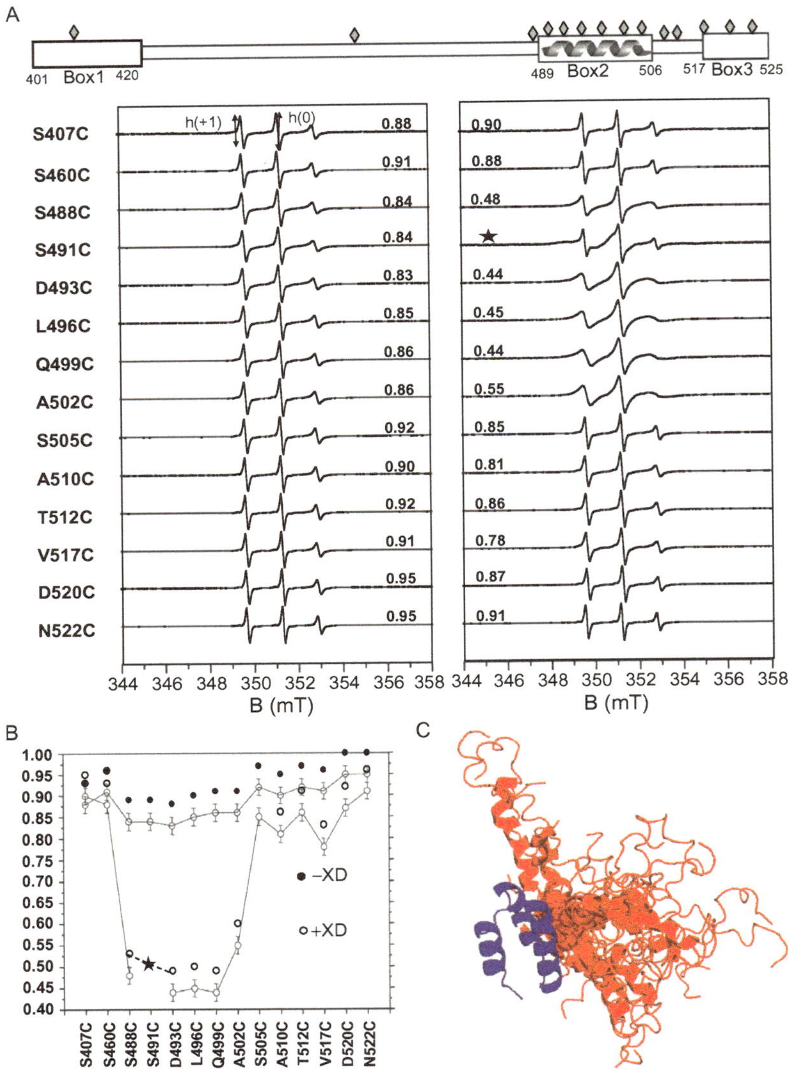

Fig. 5 (A) Scheme of N_{TAIL} and of the positions targeted for cysteine substitution and spin labeling (*diamonds*) and amplitude normalized room temperature EPR spectra of the spin-labeled MeV N_{TAIL} proteins (100 μ*M*) in the absence (*left panel*) or presence of a molar excess of XD (*right panel*). EPR spectra were recorded at 296 K on an ESP 300E Bruker spectrometer equipped with an ELEXSYS Super High Sensitivity resonator operating at 9.9 GHz. Samples were injected into 20-μL quartz capillaries. The microwave power was 10 mW, and the magnetic field modulation frequency and amplitude were 100 kHz and 0.1 mT, respectively. (B) $h(+1)/h(0)$ ratios of the spin-labeled N_{TAIL} proteins either alone or in the presence of saturating amounts of XD as a function of spin-label

(Continued)

mobility of the spin label was so dramatically reduced that the $h(+1)/h(0)$ parameter is no longer appropriate (for further details, see Belle et al., 2008). The effects of XD on the mobility of the spin label range from null (positions 407 and 460) to either dramatic (positions 488–502) or significant though less pronounced (positions 505–522). EPR experiments in the presence of 30% sucrose (a condition where the contribution of protein rotation to the EPR spectral line shape is reduced) allowed us to precisely map to residues 488–502, the N_{TAIL} region undergoing α-helical folding (Belle et al., 2008), in agreement with NMR data (Bourhis et al., 2005; Gely et al., 2010).

The possibility that lack of impact of XD on the spin-label mobility of spin labels grafted at positions 407 and 460 reflects the inability of these spin-labeled variants to interact with XD was checked and ruled out by means of equilibrium displacement experiments. These results therefore clearly indicate the binding to XD has no impact on positions 407 and 460, consistent with a scenario where they do not establish contacts with XD and retain their disordered nature.

In a subsequent study, we used the information on spin-label mobility gathered in the aforementioned study as constraints to obtain an atomistic ensemble description of XD in complex with the N_{TAIL} region encompassing residues 488–525 (Kavalenka et al., 2010).

By comparing measured and modeled temperature-dependent restrictions of the side-chain conformational spaces of 12 spin-labeled N_{TAIL} variants, we showed that the 490–500 region of N_{TAIL} is prestructured in the absence of the partner and could quantitatively estimate the extent of the α-helical sampling of the free form. By combining motional pattern analysis and experimental SDSL EPR data we could infer global structural

Fig. 5—Cont'd position. The $h(+1)/h(0)$ ratio of the spin-labeled S491C variant was not indicated, as it is not a reliable indicator of the mobility of this spin label (see text). (C) Population of the 20 best-fit structures of the MeV complex made of XD (*blue*) and the 488–525 region of N_{TAIL} (*red*) as obtained after structural modeling at 281 K. *Data of panels (A) and (B) were taken from Belle, V., Rouger, S., Costanzo, S., Liquiere, E., Strancar, J., Guigliarelli, B., et al. (2008). Mapping alpha-helical induced folding within the intrinsically disordered C-terminal domain of the measles virus nucleoprotein by site-directed spin-labeling EPR spectroscopy.* Proteins: Structure, Function and Bioinformatics, 73*(4), 973–988. https://doi.org/10.1002/prot.22125. Data of panel (C) were taken from Kavalenka, A., Urbancic, I., Belle, V., Rouger, S., Costanzo, S., Kure, S., et al. (2010). Conformational analysis of the partially disordered Measles virus NTAIL-XD complex by SDSL EPR spectroscopy.* Biophysical Journal, 98*(6), 1055–1064.*

information and could propose an ensemble model of MeV XD bound to the 488–525 region of N_{TAIL} (Fig. 5C). The obtained population of structures of the complex shows that while the Box2 region adopts a similar conformation in all the structures, the C-terminal region of N_{TAIL}, including Box3, samples a much wider conformational space. It therefore conserves a significant degree of freedom even in the bound form. The latter finding indicates that the fuzziness of N_{TAIL} is not restricted to the region preceding the α-MoRE and rather pertains also the C-terminal region. Therefore, in the MeV N_{TAIL}–XD complex N_{TAIL} is bound to XD through its α-MoRE, the latter being flanked by two fuzzy appendages, of which the N-terminal one is much longer.

SDSL EPR spectroscopy studies were also carried out to investigate the partner-induced folding of *Henipavirus* N_{TAIL} domains (Martinho et al., 2013). In this case six positions were targeted for spin labeling, one within Box1, one within Box2, three within Box3, and one within Box4 (see Fig. 1 in Martinho et al., 2013). Spectra of spin-labeled variants either alone or in the presence of a molar excess of XD were recorded in the presence of 30% sucrose. Spectra were then simulated using the EPR-SIM-C software program (Strancar et al., 2005). Spectral simulation allows extracting quantitative information on spin-label mobility from EPR spectra (Budil, Lee, Saxena, & Freed, 1996). In particular, it gives access to the rotational correlation time τ_c, a dynamic parameter, as well as to Ω, a normalized geometric parameter that represents the free rotational space, which varies from 0 (totally restricted movement) to 1 (totally unrestricted movement). The simulation also allows an EPR spectrum to be decomposed into possibly different components and to estimate their relative contribution, thereby allowing possible structural heterogeneity (i.e., polymorphism) to be appreciated. While in the case of HeV, XD has no effect on the mobility of the spin label grafted within Box2, in the case of NiV the spectrum observed in the presence of XD can be decomposed in three spectral components, two of which reflect a reduced mobility and one corresponds to a fraction of unbound N_{TAIL}. In the case of N_{TAIL} variants bearing a spin label grafted within Box3, the conformational heterogeneity of the bound form is even higher, being particularly pronounced in the case of NiV. This polymorphism reflects the occurrence of multiple conformations of the side chain onto which is grafted the spin label. Notably, for both HeV and NiV, no XD impact was detected on the mobility of spin labels grafted within Box1 and Box4, arguing for lack of direct interaction of these sites with

XD and supporting their disordered nature in the bound form. Therefore, as in the case of MeV, these studies unveiled that the N- and C-terminal regions of N_{TAIL} remain fuzzy in the complex.

2.5 Electrospray Ionization Mass Spectrometry

This techniques relies on the generation of gas-phase ions through non-denaturing electrospray ionization (ESI) and on analysis of the charge-state distribution (CSD) by mass spectrometry (MS). Protein CSDs can deliver important structural information, thanks to the fact that protein compactness in the original solution has a strong influence on the extent of protein ionization under electrospray conditions. The higher the structural compactness, the lower the average net charge that will be observed for any given protein (Kaltashov, Bobst, & Abzalimov, 2013). The average charge state of each component yields an estimate of solvent-accessible surface area (SASA) for the structure in solution, at the moment of transfer to the gas phase.

MS is a powerful method for the analysis of complex mixtures, as it allows detection not only of distinct masses but also of different conformers endowed with variable degrees of compactness in the molecular ensemble (Kaltashov et al., 2013). In addition, it also allows poorly populated states to be detected. Because of this exquisite sensitivity and of its ability to directly assess species distributions, without averaging over the molecular population, ESI-MS is a valuable tool for IDP analysis (Beveridge, Chappuis, Macphee, & Barran, 2013). Ion mobility (IM) spectrometry is a separation technique that is increasingly used in combination with mass spectrometry (IM-MS) to measure gas-phase conformation of desolvated molecules (Stuchfield & Barran, 2018). The mobility of an ion is a measure of its velocity in a neutral buffer gas under the influence of a weak electric field. The number of collisions with the gas, together with the charge and shape of the ion, lead to an elution profile as an arrival-time distribution, which can be used to determine the rotationally averaged collision cross section (CCS) of the selected ion. The CCS serves as an indicator of the shape of the molecule in the gas phase (Stuchfield & Barran, 2018). ESI-IM-MS therefore allows separating ions based on their compactness, thereby unveiling possible conformational heterogeneity for a given CSD. One technical limitation of ESI-MS is that samples have to be dissolved in volatile buffers (such as ammonium acetate or formate) without salts, a condition that is not suitable for all proteins and can lead to precipitation. One of the advantages of this approach is that it requires relatively small amounts of protein (i.e., in the μg range).

We have used ESI-MS combined with IM techniques to investigate the MeV N_{TAIL}–XD complex (D'Urzo et al., 2015). Prior to infusion, equimolar amounts of the two proteins were mixed and incubated for 10 min at room temperature without agitation. The ESI-MS spectrum, as obtained upon infusing a mixture at 20 μ*M* of each protein, features peaks corresponding to the free proteins along with peaks specific to the N_{TAIL}–XD complex (Fig. 6A). These results indicate that the complex, which relies on hydrophobic interactions, is preserved, at least partially, during the electrospray process and gas-phase ion separation. The measured mass of the complex perfectly matches the value expected for a 1:1 complex. The complex disappears upon applying denaturing solvent conditions or high declustering potentials (D'Urzo et al., 2015).

The CSD of the complex-specific peaks is quite broad and bimodal, indicating the coexistence of at least two distinct conformational states. The high-charge component of the complex (18 +) was assumed to correspond to an "open" conformation, in which the disordered arms of N_{TAIL}, upstream and downstream to the α-MoRE, fluctuate maintaining high solvent accessibility. A contrario, the low-charge component (11 +), probably represents a compact or "closed" conformation of the complex, in which the N_{TAIL} arms are packed on the surface of the partner, as suggested by the low extent of ionization, which approaches the expected value for folded globular proteins of the same mass (9.46+) (Testa, Brocca, & Grandori, 2011). The average SASA values of the "open" and "closed" states, as calculated from the empiric relation with the average charge state (Testa et al., 2011), are 16,100 and 9500 Å^2, respectively. Notably, the closed state escaped detection in previous studies, thus raising the question as to whether this conformation reflects a real conformational heterogeneity in solution, captured only by the exquisite sensitivity of native MS, or an altered conformational ensemble induced by the electrospray conditions. The fact that the CSDs of IDPs, as detected by nano-ESI-MS, and their response to solvent conditions are protein specific (Testa et al., 2011) advocates in favor of the former hypothesis.

In order to get further insights onto the compact state of the complex, we carried out IM-MS experiments using a Waters Synapt G2 HDMS instrument (D'Urzo et al., 2015). The IM-MS spectrum obtained under the same conditions described earlier is shown in Fig. 6B. The CSDs are similar to those observed with the QSTAR Elite instrument, although with higher

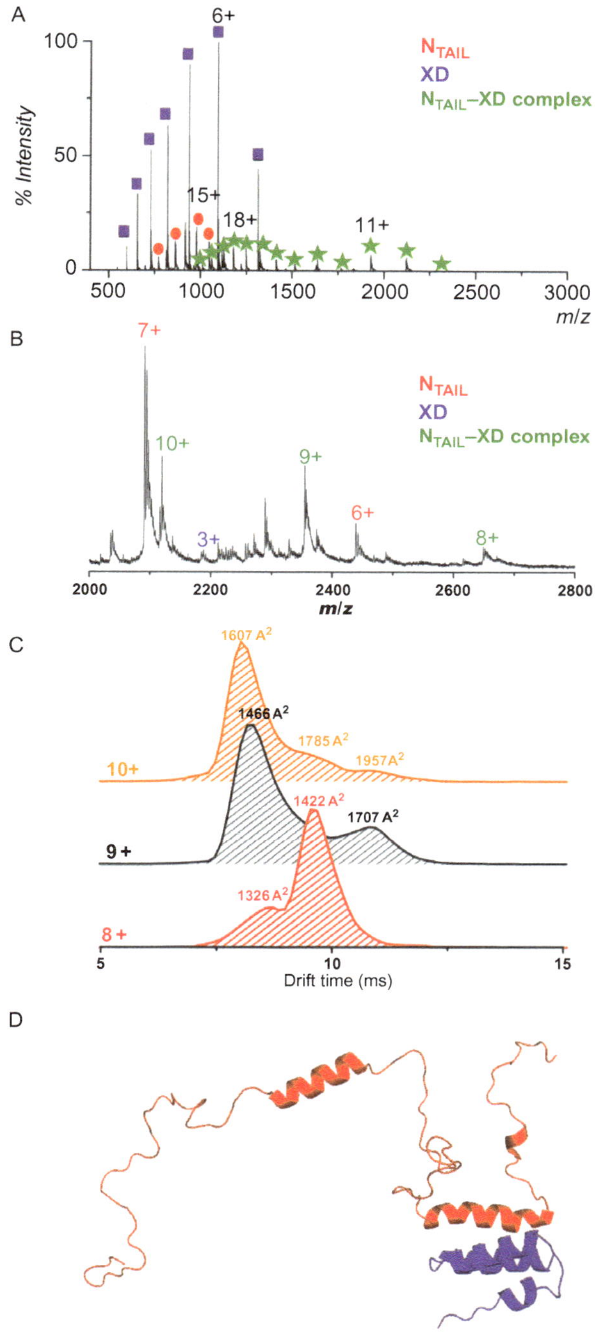

Fig. 6 See legend on next page.

relative amounts of the compact form. The additional separation of ions afforded by IM allowed detection of distinct structural species even at the same charge state (Fig. 6C). Two slightly different conformations are found to populate the compact state of the N_{TAIL}–XD complex (charge states 8+ to 10+). Average CCS of 1326 and 1422 Å^2 can be derived for the two main peaks detectable for the 8+ charge state (Fig. 6C). Thus, the compact state of the complex displays further structural heterogeneity: although its overall ionization suggests a collapsed structure, its arrival-time distribution reveals distinct peaks rather than the single peak typically observed for folded proteins. The presence of multiple conformers, although with different CCS values, is also observed for the 9+ and 10+ ions (Fig. 6C). The predominant conformer displays progressively larger CCS values (1466, 1607 Å^2) with increasing charge state (Fig. 6C). This trend is consistent with the lowest charge state (8+) corresponding to the most compact species in the original ensemble and to the species with lowest interference of Coulomb repulsions upon transfer to the gas phase. Altogether, these results suggest that both the open and closed states of the complex consist of conformational ensembles.

Next we generated structural models of the N_{TAIL}–XD complex by combining homology modeling and PROFASI (D'Urzo et al., 2015). The latter is an all-atom Monte Carlo-based simulation tool (Jonsson, Mohanty, & Irback, 2012) that allows exploring the conformational ensemble of the complex in implicit solvent. The quality of the conformational sampling was assessed based on the agreement between calculated and experimental chemical shifts data of N_{TAIL} bound to XD (Gely et al., 2010). A quite good agreement was also found between the calculated CD spectra

Fig. 6 (A) Nano-ESI-MS spectrum of an equimolar mixture of MeV N_{TAIL} and XD at 20 μ*M* each in 10 m*M* ammonium acetate, pH 6.5. The spectrum was recorded using an AB Sciex QSTAR Elite instrument. (B) Excerpt of the higher *m/z* region of the nano-ESI-IM-MS spectrum of a MeV N_{TAIL} and XD mixture (same conditions as in panel (A)) recorded using a Waters Synapt G2 HDMS instrument. (C) Arrival times of the low charge (8–10+) states of the N_{TAIL}–XD complex. The 8–10+ charge states of complex show at least two different conformations, with corresponding CCSs values that vary over a wide range. (D) Cartoon representation of a representative structural model of the N_{TAIL}–XD complex selected from the conformational ensemble based on best agreement with the chemical shifts from NMR and the calculated SASA from ESI-MS. XD is shown in *blue* and N_{TAIL} in *red*. *Data were taken from D'Urzo, A., Konijnenberg, A., Rossetti, G., Habchi, J., Li, J., Carloni, P., et al. (2015). Molecular basis for structural heterogeneity of an intrinsically disordered protein bound to a partner by combined ESI-IM-MS and modeling.* Journal of the American Society for Mass Spectrometry, 26*(3), 472–481. https://doi.org/10.1007/s13361-014-1048-z.*

of the conformational ensemble and the experimental ones (Shu et al., 2012). Finally, the SASA mean value calculated from the models is $17{,}920 \pm 690\,Å^2$, a value in good agreement with that of the open state of the complex detected by ESI-MS. Fig. 6D shows one representative model selected from the conformational ensemble based on best agreement with the experimental SASA. Analysis of the contacts established between the fuzzy region with XD in this model and other models equally well fitting the experimental SASAs, unveiled their prevalently hydrophobic nature (D'Urzo et al., 2015).

While previous atomic models only concerned the α-MoRE region of N_{TAIL}, this study represents the first attempt at modeling the entire N_{TAIL}–XD complex at atomic resolution. By including the disordered arms in the molecular modeling, it was possible to extend interaction analysis to the dynamic regions of the complex, thereby setting the basis for targeting specific conformations of the complex by small-molecule inhibitors.

In conclusion, IM-MS data indicate that the compact state of the complex is characterized by high structural heterogeneity in the gas phase. The broad arrival-time distributions observed for the low charge-state component could be a distinct feature of collapsed states of IDPs, as opposed to globular structures. Additional studies are, however, required to rule out possible artifacts induced by the electrospray process.

3. ASSESSING THE IMPACT OF THE LONG, N-TERMINAL N_{TAIL} FUZZY APPENDAGE ON BINDING

3.1 Deletion Studies Coupled to Protein Complementation Assays

In Section 2, we have described various methods allowing assessment of the presence of residual disorder in paramyxoviral N_{TAIL}–XD complexes. These studies allowed us to divide MeV N_{TAIL} (aa 401–525) into three regions: a fuzzy N-terminal region (aa 401–485), the α-MoRE (aa 486–504), and a fuzzy C-terminal region (aa 505–525).

Here, we describe how we have assessed the impact of the N-terminal fuzzy region of N_{TAIL} on the interaction of the latter with two known binding partners, namely, XD and the major inducible heat shock protein 70 (hsp70). To that end, we have serially truncated the N-terminal fuzzy region of N_{TAIL} from residue 401 to residue 481 by 10 residues intervals (see Fig. 7).

Among the various possible approaches to quantitatively assess protein–protein interactions (i.e., ITC, microscale thermophoresis, surface plasmon

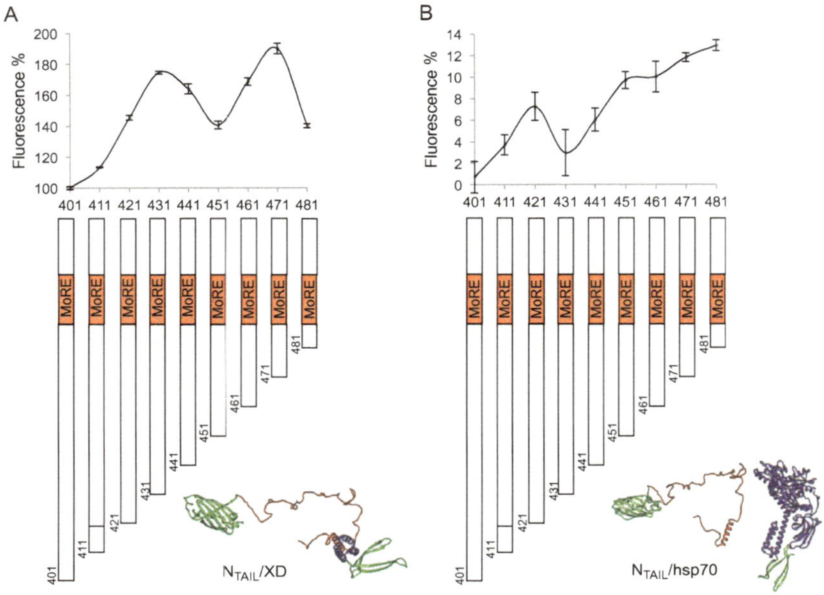

Fig. 7 Split-GFP complementation assay using MeV N_{TAIL} truncation variants and MeV XD (A) or hsp70 (B). *E. coli* T7-pRos (i.e., T7 cells (New England Biolabs) bearing the pLysS plasmid from Rosetta(DE3) pLysS cell (Novagen)) were cotransformed with plasmid pNGG bearing the coding sequence of either full-length N_{TAIL} (401) or its truncation variants (411–481), and plasmid pMRBAD-link-CGFP bearing the coding sequence of either XD (A) or hsp70 (B), and then plated on ampicillin- and kanamycin-containing agar plates. The next day, several colonies were scraped off the plate and used to seed 4 mL of Luria Bertani medium containing ampicillin, kanamycin, and chloramphenicol in 24-wells deep-wells. Chloramphenicol is used to maintain the pRos plasmid. After one night at 37°C under 200 rpm shaking, 400 μL of preculture were used to seed each of 3 × 4 mL of terrific broth containing the same antibiotics. Cells were grown for an additional 2.5 h at 37°C. IPTG (0.5 m*M*) and arabinose (2%) were added to each 4 mL culture, and the deep-well was incubated over night at 17°C under shaking. The next day, deep-wells were spun for 5′ at 4000 g in a swinging rotor and the supernatant was discarded. Cell pellets were resuspended in 1 mL of PBS, and 10 μL of each cell suspension were diluted in 100 μL PBS in each well of a 96-wells clear bottom black plate. The optical density at 600 nm (i.e., the number of cells) (OD_{600}) and GFP fluorescence (Fluo) of each well were measured using a TECAN GENios Plus spectrofluorimeter. The fluorescence value of each well was divided by the OD_{600} of the same well (Fluo/OD_{600}), and the mean value of each triplicate and standard deviation were calculated. Results are expressed in percentage (with 100% being the value provided by full-length N_{TAIL}). Below the *X*-axis is represented a scheme of each variant. At the *bottom right* is a cartoon of each interacting pair. *Data were taken from Gruet, A., Dosnon, M., Blocquel, D., Brunel, J., Gerlier, D., Das, R. K., et al. (2016). Fuzzy regions in an intrinsically disordered protein impair protein-protein interactions.* FEBS Journal, 283, *576–594. https://doi.org/10.1111/febs.13631.*

resonance, and biointerferometry), we have chosen to evaluate the impact of these truncations on the binding with XD and with hsp70 using a protein complementation assay based on the use of the green fluorescent protein (GFP) (Gruet et al., 2013; Wilson, Magliery, & Regan, 2004). GFP is a β-barrel made of 11 β-strands mainly linked by loops. By cutting the loop linking β-strands 7 and 8, GFP can be split into two uneven halves made of the first 7 (NGFP) and the last 4 (CGFP) β-strands, respectively. Separately, NGFP or CGFP are unable to fluoresce. After splitting GFP into NGFP and CGFP, two proteins X and Y, known to interact with each other, can be fused to the C-terminal end of NGFP and to the N-terminal end of CGFP, respectively. When NGFP-X and Y-CGFP fusion proteins are coexpressed in *E. coli*, X and Y bind to each other within the cell, which also brings their respective fusion partners NGFP and CGFP in close proximity. This allows NGFP and CGFP moieties to reconstitute the full-length GFP, which is fluorescent in contrast with its isolated NGFP and CGFP constituents. Thus, the interaction between two proteins X and Y (in both their *wild-type* or mutated form) can be visualized by the fluorescence of bacteria in which NGFP-X and Y-CGFP are coexpressed. Since the fluorescence is proportional to the affinity between X and Y, the interaction between different combinations of NGFP-X and Y-CGFP can be compared by simply measuring the fluorescence of the bacteria coexpressing NGFP-X and Y-CGFP. Of course, one has to rule out the possibility that observed fluorescence variations may reflect differences in expression between the variants under study. It is therefore mandatory to check expression by SDS-PAGE analysis.

In our case, X was N_{TAIL} or its truncation variants (Fig. 7), and Y was either XD (Fig. 7A) or hsp70 (Fig. 7B). For practical reasons, we have modified the pET11a-link-NGFP plasmid originally created by Magliery et al. (2005) by replacing the cloning site made of restriction sites with a Gateway® recombination cassette (Invitrogen). This allows an easier subcloning of the different X coding sequences under study. A detailed description of this new vector (pNGG) can be found in Gruet, Longhi, & Bignon (2012). For the expression of Y-CGFP, the original plasmid (pMRBAD-link-CGFP) was used, in which XD and hsp70 coding sequences were inserted by restriction cloning. The two plasmids can be comaintained in the same bacterial cell because they have compatible origins of replications (ColE1 for pNGG and p15A for pMRBAD-link-CGFP), and because they confer a resistance to different antibiotics (ampicillin for pNGG and kanamycin for pMRBAD-link-CGFP). In these constructs, expression of NGFP-X is under the control of a T7 promoter inducible by isopropyl β-D-1-thiogalactopyranoside (IPTG), and expression of Y-CGFP is under the control of the arabinose promoter inducible by arabinose.

In practice, the N_{TAIL} coding sequence or that of its truncation variants (Fig. 7) was subcloned in pNGG and coexpressed in the *E. coli* strain T7-pRos with either the coding sequence of XD (Fig. 7A) or hsp70 (Fig. 7B) that had been subcloned in pMRBAD-link-CGFP (Gruet et al., 2016). Expression was induced by addition of 0.5 m*M* IPTG and 2% arabinose, and the cell fluorescence was measured after an overnight incubation at 17°C in a shaking incubator. In order to compensate possible differences in cell growth between cultures, the fluorescence is divided by the number of cells in the considered culture (i.e., by dividing the fluorescence by the optical density (OD) at 600 nm (OD_{600})). Each experimental point was performed in triplicate. Results are reported in Fig. 7. As one can see, the fluorescence increases with the truncation in a nonmonotonic manner, with this being observed with both XD (Fig. 7A) and hsp70 (Fig. 7B). However, the overall fluorescence was higher for XD than for hsp70 (see the different Y-axis scales in Fig. 7), a finding in line with the known higher affinity of N_{TAIL} for XD (3 μ*M*) (Dosnon et al., 2015) than for hsp70 (70 μ*M*) (Couturier et al., 2010). In conclusion, the fuzzy N-terminal region of N_{TAIL} (aa 401–485) has a negative effect on the binding of N_{TAIL} with two binding partners that differ in both size and affinity. Interestingly, we obtained identical results when N_{TAIL} and XD from NiV and HeV were used or when another protein complementation assay based on split-luciferase (Cassonnet et al., 2011) and not on split-GFP was used. Thus, the negative effect of the fuzzy N-terminal region of N_{TAIL} (aa 401–485) on the binding seems to be a feature shared by at least three paramyxoviruses and does not depend on the assay used (Gruet et al., 2016).

We next investigated the possible reasons of this negative effect. As we have seen in Sections 2.2–2.4, the N-terminal region of N_{TAIL} (aa 401–485) is fuzzy, i.e., it remains disordered after binding. The first possible reason for the observed negative effect of this region on binding could be its sole disordered nature. If this were the case, then swapping the *wild-type* sequence of the fuzzy appendage of N_{TAIL} with another unrelated sequence is expected to lead to the same dampening effect provided that it is fuzzy. To test this hypothesis, we replaced the *wild-type* fuzzy N-terminal region of N_{TAIL} (aa 401–480) with another nonnatural sequence (compare Fig. 8A and B). Compared to its *wild-type* counterpart, this artificial sequence (i) has the same number of residues, (ii) is predicted to be slightly more disordered (Fig. 8C), (iii) shares only 6% identity. This artificial sequence was fused with the remaining part (481–525) of the sequence of *wild-type* N_{TAIL}

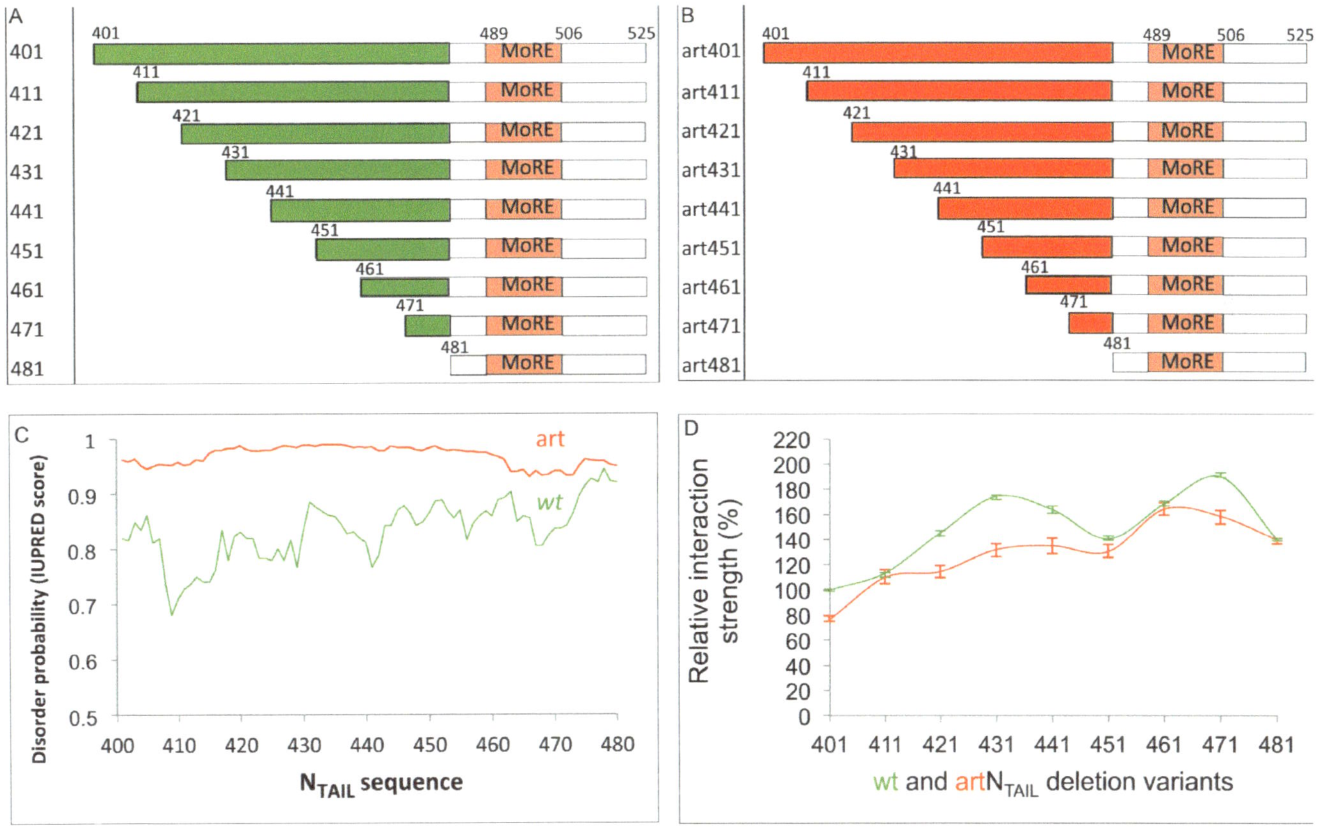

Fig. 8 See legend on opposite page.

to reconstitute an artificial full-length N_{TAIL} (aa 401–525) (artN_{TAIL}, Fig. 8B). We then generated the same truncation variants as those we had generated from the *wild-type* sequence (Fig. 8B) and compared their effect on the binding to XD. As shown in Fig. 8D, wtN_{TAIL} and artN_{TAIL} truncation variants feature similar binding patterns, with the binding strength increasing with the truncation. However, two differences can be noted. First, the artificial variants curve is more linear than that of the *wild-type* variants. Second, for the same truncation each artificial variant systematically provides a lower binding than its *wild-type* counterpart. One possible explanation for this behavior could be the higher disorder probability of the artificial sequence compared to that of the *wild-type* sequence (Fig. 8C). In conclusion, the negative effect of the fuzzy N-terminal region of N_{TAIL} (aa 401–485) on the binding to XD is not related to its specific sequence but to its fuzzy nature.

3.2 Kinetics Studies

We further investigated the molecular mechanisms by which the fuzzy appendage of MeV N_{TAIL} influences the interaction with XD by characterizing binding kinetics. The binding between MeV N_{TAIL} and XD is an extremely rapid reaction that occurs in the μs time scale and, therefore, escapes classical stopped-flow mixing methods (dead time of about 1–2 ms). Therefore, to study the kinetics of interaction between N_{TAIL} and XD we used the temperature jump method, a kinetic technique that allows measuring ultrarapid reactions. The temperature jump methodology is a relaxation method whereby the equilibrium of a solution containing a mixture of the interacting proteins is perturbed by a rapid increase in temperature. After trigger, the relaxation to the new equilibrium is then observed. Rapid heating is due to the electrical discharge in the solution of high voltage of tens of kV, which is capable of inducing an increase in temperature of typically 5–9°C within few microseconds.

Fig. 8 Split-GFP complementation assay using *wild-type* and artificial MeV N_{TAIL} truncation variants and MeV XD. *Wild-type* (wt) (A) and artificial (art) (B) MeV N_{TAIL} truncation variants. In (A) and (B), numbers above the schemes indicate the first and last residue of the variant. (C) IUPred disorder prediction of wt (*green line*) and art (*red line*) N_{TAIL} region encompassing residues 400–480. (D) Binding values obtained by split-GFP complementation assays using wt (*green line*) and art (*red line*) MeV N_{TAIL} truncation variants. See legend of Fig. 7 for experimental details. *Data were taken from Gruet, A., Dosnon, M., Blocquel, D., Brunel, J., Gerlier, D., Das, R. K., et al. (2016). Fuzzy regions in an intrinsically disordered protein impair protein-protein interactions.* FEBS Journal, 283, *576–594. https://doi.org/10.1111/febs.13631.*

In the case of N_{TAIL} and XD we measured the change in the tryptophan fluorescence upon binding in one of the two proteins. In particular, a pseudo-*wt* tryptophan variant of XD, Y480W, was produced. The tryptophan substitution was tested and it was demonstrated that it does not affect the thermodynamic stability and structure of XD nor the stability of the complex with N_{TAIL} (Dosnon et al., 2015).

The experiments were performed under pseudo-first order conditions, with one of the two interacting proteins in large excess. In these conditions the reaction rate will depend solely on the concentration of the reactant present at low concentration, which simplifies quantitative analysis of the data. In practice a constant concentration of pseudo-*wt* Y480W XD (2 μ*M*) was mixed with different N_{TAIL} concentrations usually ranging from 2 to 50 μ*M*. The buffer used was 10 m*M* sodium phosphate pH 7.0, 150 m*M* NaCl. The presence of salt is required to ensure conductivity. Each sample was degassed and slowly pumped through the 0.5 × 2 mm quartz flow cell of a Hi-Tech PTJ-64 capacitor-discharge Temperature-jump apparatus (Hi-Tech, Salisbury, UK). In the cell, a rapid discharge of 12 kV leads to a temperature jump of 9°C (from 11°C to 20°C). The change in tryptophan fluorescence, caused by the perturbation of the equilibrium and the consequent relaxation to the initial temperature, is followed as a function of the time. Usually 10–20 traces were averaged for each sample. The excitation wavelength was 296 nm, and the fluorescence emission of the tryptophan was measured using a 320-nm cutoff glass filter. The fluorescence change of *N*-acetyl-tryptophanamide was used in control measurements.

The experiment was performed using the full-length N_{TAIL} and a synthetic peptide designed to mimic just the MoRE region of N_{TAIL} (residues 485–506), in order to observe whether removal of the fuzzy region influences the rate of folding of N_{TAIL}. The full-length N_{TAIL} and the tryptophan variant of XD were expressed and purified as described in Dosnon et al. (2015), the peptide mimicking the MoRE region was purchased from JPT (Berlin, Germany).

In all cases, the reaction followed a single exponential decay of the fluorescence and all the traces were satisfactorily fitted by the following single exponential equation:

$$F(t) = \Delta F \cdot e^{(k_{obs} \cdot t)} + F_{\infty} \tag{3}$$

where ΔF is the amplitude of the trace, k_{obs} is the observed rate constant of the reaction, t is the time of acquisition, and F_{∞} is the final value of the fluorescence. We collected the observed rate constants (k_{obs}) for each N_{TAIL} or

MoRE concentrations, and we plotted them as a function of N_{TAIL} or MoRE concentrations.

In the case of full-length N_{TAIL}, a hyperbolic dependence of the k_{obs} on ligand concentration was observed and ascribed to a change in the rate-limiting step, with the folding of N_{TAIL} becoming rate-limiting at high reactant concentrations (Dosnon et al., 2015). Conversely, when a truncated synthetic peptide mimicking the isolated MoRE was used, a linear dependence was observed (Fig. 9). This behavior suggests that the fuzzy appendage of N_{TAIL} has a direct effect on the folding of the MoRE, possibly lowering its rate constant of folding (Gruet et al., 2016). The precise underlying molecular mechanisms remain, however, to be established.

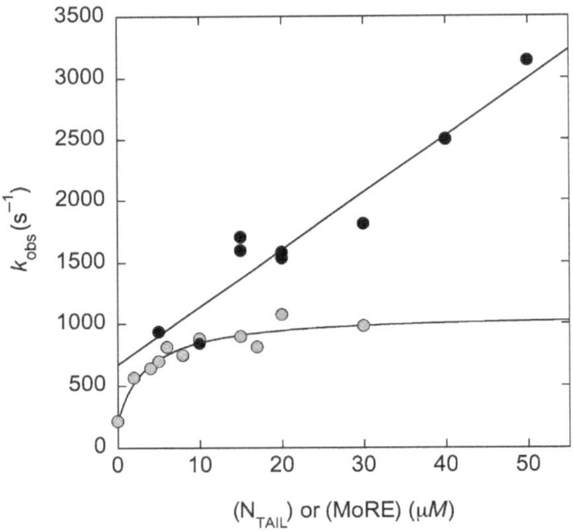

Fig. 9 Pseudo-first order kinetics of the binding between Y480W XD (at a constant concentration of 2 μ*M*) with excess concentrations of either wtN_{TAIL} (*gray circles*) or a peptide mimicking the MoRE (*black circles*) in 10 m*M* sodium phosphate buffer and 150 m*M* NaCl at pH 7.0. Under all conditions, there was an at least fivefold difference in concentration between the two proteins to ensure pseudo-first order conditions. Experiments were carried out using a PTJ-64 capacitor-discharge T-jump apparatus (Hi-Tech, Salisbury, UK). The temperature was rapidly changed with a jump size of 9°C, from 11°C to 20°C. For both N_{TAIL} and MoRE peptide, the observed rate constant increases with increasing concentration of reactants. The data obtained with the MoRE were fitted to a linear function, while those obtained with N_{TAIL} were fitted to a hyperbolic function. The binding reaction with N_{TAIL}, but not that with the MoRE, appears to be limited, at high reactant concentrations, by a monomolecular step of $\sim 10^3\,s^{-1}$. *Data were taken from Gruet, A., Dosnon, M., Blocquel, D., Brunel, J., Gerlier, D., Das, R. K., et al. (2016). Fuzzy regions in an intrinsically disordered protein impair protein-protein interactions.* FEBS Journal, 283, *576–594. https://doi.org/10.1111/febs.13631.*

4. ASSESSING PARTNER-MEDIATED POLYMORPHISM OF MEASLES VIRUS N_{TAIL}

4.1 Alanine-Scanning and Rational Mutagenesis Coupled to Protein Complementation Assay

4.1.1 Alanine-Scanning

In Section 3.1, we have seen that the fuzzy N-terminal region of MeV N_{TAIL} (401–485) has the same negative effect on the binding to two partners of N_{TAIL} in spite of their profound differences. Indeed, XD is a small viral protein (Johansson et al., 2003) that binds the MoRE with a relatively high affinity (3 μ*M*) (Dosnon et al., 2015), whereas hsp70 is a large cellular protein that binds the MoRE with a much lower affinity than XD (70 μ*M*) (Couturier et al., 2010). Albeit the major hsp70-binding site had been already mapped to the MoRE (aa 486–504) (Couturier et al., 2010; Zhang et al., 2005), no structural information was available for the N_{TAIL}–hsp70 complex. In particular, it remained to be established whether the MoRE undergoes α-helical folding also upon binding to hsp70 and whether the interaction relies on the same N_{TAIL} residues. The relevance of investigating the molecular mechanisms governing the N_{TAIL}–hsp70 interaction is related to its well-documented functional impact on viral transcription and replication (Zhang et al., 2005, 2002) and on the innate immune response (Kim et al., 2013).

We therefore have undertaken a study aimed at analyzing the contribution of each residue of the MoRE to binding to both XD and hsp70. To that end, we have used the alanine-scanning technique. Alanine scanning consists of systematically and individually replacing each residue of a considered sequence with an alanine residue (or a glycine residue if the *wild-type* residue is already an alanine), and then assessing the effect of these substitutions on the properties of the protein. We have thus alanine scanned each residue of the MoRE from residue 486 to residue 504, and tested the effect of these substitutions on the binding of each individual variant using the split-GFP complementation assay described in Section 3.1. Note that we used N_{TAIL} truncation variant 471 (aa 471–525) as backbone to derive single-site variants because it binds XD better than full-length N_{TAIL} (Fig. 7A). Results are reported in Fig. 10. In the case of XD binding (Fig. 10, upper panel), most of the single alanine variants exhibit a decreased binding compared to that of the *wild-type* sequence. In a few cases, and in perfect agreement with the known 3D structure of the MeV XD/α-MoRE complex

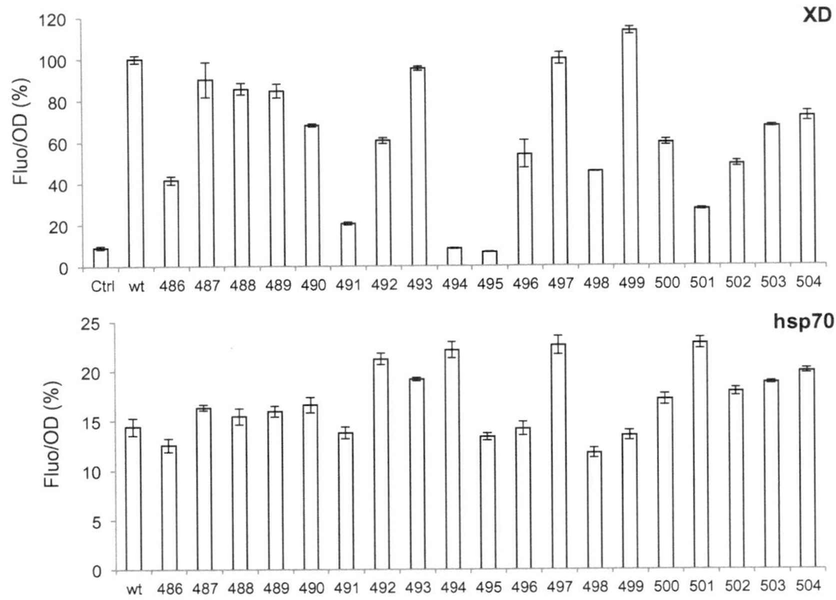

Fig. 10 Alanine-scanning mutagenesis of MeV N_{TAIL} MoRE. The MoRE residues (aa 486–504) of MeV N_{TAIL} truncation variant 471 (see Fig. 7) were individually mutated into an alanine residue (or a glycine residue when the *wild-type* residue was already an alanine). The binding ability of each single variant was then compared to that of the *wild-type* sequence by split-GFP complementation assay using either XD or hsp70. *Ctrl*, negative control (fluorescence background as obtained using empty pNGG vector without N_{TAIL}); *wt*, *wild-type* truncation variant 471. See legend of Fig. 7 for experimental details. *Data were taken from Bignon, C., Troilo, F., Gianni, S., & Longhi, S. (2017). Partner-mediated polymorphism of an intrinsically disordered protein.* Journal of Molecular Biology 430(*16*), *2493–2507.*

(Kingston, Hamel, et al., 2004), the single alanine substitution partially or totally abrogated binding (residues 491, 494, 495, 501). These residues can therefore be considered as critical for XD binding. A quite different profile was obtained with hsp70 (Fig. 10, lower panel). First, many substitutions resulted in an increased binding compared to the wild-type sequence. Second, no single residue proved to be mandatory for binding hsp70. In conclusion, although XD and hsp70 exhibit the same sensitivity to the negative effect of the fuzzy N-terminal region of N_{TAIL} (aa 401–485) (Fig. 7, Section 3.1), they bind the MoRE of N_{TAIL} using different mechanisms.

These results also suggest that the MoRE of N_{TAIL} likely adopts a different conformation at the surface of XD. The latter hypothesis was further investigated using rational mutagenesis as described in the next paragraph.

4.1.2 Rational Mutagenesis

We know that the MoRE of N_{TAIL} folds as an α-helix upon XD binding (see Sections 2.3 and 2.4), but nothing is known about the conformation that it adopts upon binding to hsp70. Based on the results discussed in Section 4.1.1, it is plausible that the MoRE does not fold into an α-helix once bound to hsp70.

To test this hypothesis, we conceived and generated two MoRE variants with opposite folding capabilities. In the first one, called AlaMoRE, all the residues that proved noncritical for XD binding on the basis of the alanine scanning were replaced with an alanine. In the second one, called GlyMoRE, all the residues that proved noncritical for XD binding were replaced with a glycine. Because alanine is known to favor α-helix formation, whereas glycine is known to have the opposite effect (Borcherds et al., 2014; Otieno & Kriwacki, 2012; Rogers, Wong, & Clarke, 2014; Scott, Alonso, Sato, Fersht, & Daggett, 2007), AlaMoRE and GlyMoRE were expected to be more and less α-helical than wtMoRE, respectively. This assumption, reinforced by disorder prediction (Fig. 11A) and modeling (Fig. 11B), was experimentally confirmed by CD analysis (Fig. 11C).

WtMoRE, AlaMoRE, and GlyMoRE were then tested for their ability to bind XD or hsp70 by split-GFP complementation assay as described in Section 3.1. This was performed using truncation variant 471 as backbone because of its higher binding ability compared to that of full-length N_{TAIL} (see Section 4.1.1). Results, reported in Fig. 11D, unambiguously indicated that increasing the α-helicity of the MoRE (AlaMoRE) slightly increases its binding to XD compared to that of wtMoRE, whereas the lack of α-helicity (GlyMoRE) results in a complete loss of binding in spite of the presence of the residues revealed to be critical for XD binding by the alanine scanning.

The lower XD-binding ability of GlyMoRE compared to that of wtMoRE and AlaMoRE was also confirmed by kinetics experiments. In those experiments, the binding of XD to wtMoRE, AlaMoRE, and GlyMoRE was measured by incubating a constant concentration ($5\,\mu M$) of XD Y480W with varying concentrations of either wtMoRE or AlaMoRE or GlyMoRE (typically ranging from 2 to $50\,\mu M$). In analogy to what we described previously, in all cases (un)binding was induced by a rapid discharge of 12 kV on a quartz cell, corresponding to a rapid increase in temperature of 9°C. The fluorescence traces conformed all to a single exponential transition (see supplementary fig. S2 in Bignon, Troilo, Gianni, & Longhi, 2017). Analysis of kinetic data (Fig. 11E) revealed that while the behavior of AlaMoRE is close to that of wtMoRE (wtMoRE:

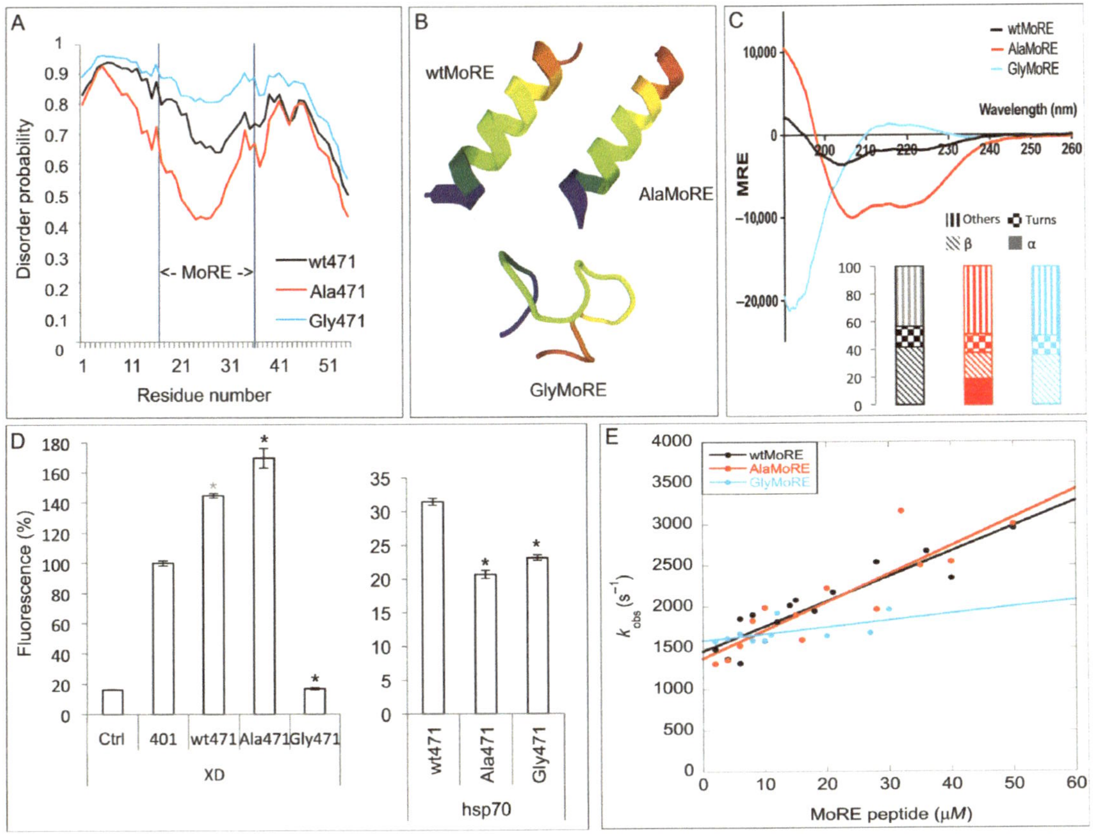

Fig. 11 See legend on opposite page..

$k_{off}^{app} = 1460 \pm 80\ s^{-1}$, $k_{on}^{app} = 30 \pm 4\ s^{-1}\ \mu M^{-1}$; AlaMoRE: $k_{off}^{app} = 1400 \pm 100\ s^{-1}$, $k_{on}^{app} = 34 \pm 5\ s^{-1}\ \mu M^{-1}$), there is a detectable destabilization of the complex in the case of GlyMoRE ($k_{off}^{app} = 1610 \pm 60\ s^{-1}$, $k_{on}^{app} = 5 \pm 2\ s^{-1}\ \mu M^{-1}$) as judged from the lower association rate constant, which is represented by the slope of the observed rate constant as a function of reactant concentration. These results indicate that while an increase in helicity of the MoRE compared to the wt sequence does not seemingly contribute to a stabilization of the complex in vitro, a destabilization of the secondary structure of the MoRE is paralleled by a weaker binding to XD. A possible reason for the discrepancy between split-GFP and kinetics data, as far as results with AlaMoRE are concerned, may lie in the differences in the experimental setup between the two approaches. In particular, it is conceivable that the differences in the concentrations used in split-GFP reassembly assays and in kinetic studies may be responsible for the observed discrepancies.

By contrast with what was observed with XD, AlaMoRE and GlyMoRE behaved similarly when assessed for their binding to hsp70: they both exhibited a partial decreased binding compared to that of wtMoRE. These results definitely indicate that XD and hsp70 do not rely on the same structural requirements to bind to the α-MoRE of N_{TAIL}. More specifically, increasing the α-helicity of the α-MoRE increases XD binding but decreases hsp70 binding, suggesting that hsp70 binding does not induce α-helical folding of the α-MoRE. This conclusion is strengthened by the relative insensitiveness of hsp70 to a MoRE displaying a reduced α-helicity.

Fig. 11 Binding ability of N_{TAIL} MoRE variants with different degrees of α-helicities. (A) IUPred disorder prediction (Dosztanyi, Csizmok, Tompa, & Simon, 2005). (B) Structural models as obtained using PEP-FOLD3 (Lamiable et al., 2016). (C) Far-UV CD spectra of MoRE peptides. The *inset* shows the secondary structure content as derived using BESTSEL (Micsonai et al., 2015). (D) Split-GFP complementation assay. On the *X*-axis are shown the NGFP fusions. The coexpressed CGFP fusion (XD or hsp70) used is indicated below the *X*-axis. *401*, full-length wild-type N_{TAIL}; *Ala471*, 471 truncated variant with AlaMoRE; *Ctrl*, negative control using empty pNGG vector; *Gly471*, 471 truncated variant with GlyMoRE; *wt471*, 471 truncated variant with a wtMoRE. The *black asterisk* indicates a statistically significant difference (*T* Student's test, with $P < 0.01$) with respect to wt471. The *gray asterisk* indicates a statistically significant difference (*T* Student's test, with $P < 0.01$) with respect to 401. See legend of Fig. 6 for experimental details. (E) Pseudo-first-order kinetics of binding of MoRE peptides to XD Y480W variant. See legend of Fig. 7 for experimental details. *Data were taken from Bignon, C., Troilo, F., Gianni, S., & Longhi, S. (2017). Partner-mediated polymorphism of an intrinsically disordered protein.* Journal of Molecular Biology 430(*16*), *2493–2507.*

In conclusion, in addition to use a different set of N_{TAIL} residues (see Section 4.1.1), XD and hsp70 do not induce and/or require the same folding within the MoRE of N_{TAIL} and are therefore likely to interact with the latter through completely different mechanisms. The discovery that the conformation sampled by the free form of the MoRE of N_{TAIL} does not necessarily commit N_{TAIL} to adopt an α-helical conformation in the bound form provides an additional example of partner-mediated polymorphism and of the relative insensitiveness of the bound structure to the prerecognition state. These results are in line with recent findings that revealed that the closely related HeV N_{TAIL} folds as antiparallel β-sheet at the air/water interface (Bénarouche et al., 2017) thereby further underscoring the extreme polymorphism of this IDP.

5. CONCLUSIONS

Conformational heterogeneity in protein complexes is quite common leading to multiplicity of interaction modes (Fuxreiter, 2018; Miskei, Antal, & Fuxreiter, 2017). The experimental characterization of fuzzy assemblies is, however, challenging owing to the inherent multiplicity of the species, which entails resolution, spectral broadening, and averaging problems.

In this chapter, we have reviewed a variety of techniques enabling the characterization of different aspects of fuzziness. Conformational heterogeneity in the bound state could be detected by various techniques such as NMR, EPR, and fluorescence spectroscopy in case of highly dynamical complexes, whereas X-ray crystallography can provide information on stable, polymorphic complexes. The shape distribution of the ensemble could be obtained by SEC, SEC-MALLS, and SAXS. The high-resolution ensemble can be described by molecular simulation methods that make use of experimental data as constraints (for a recent example based on SAXS data, see Cordeiro et al., 2017). Notably, computational methods can also provide insights into the mechanisms of molecular recognition. Some issues, such as the averaging over collective motions with different timescales, have still to be solved. Single-molecule approaches are ideally suited to study fuzzy complexes as they circumvent the inherent averaging problem from which all the other approaches suffer (Borgia et al., 2018). The combination of different techniques, such as NMR and SAXS for instance (Sterckx et al., 2014), is of course an additional step forward as it allows limiting data overfitting. Future developments will focus on achieving detection of poorly populated states that might play crucial roles in function.

All these techniques together provided a comprehensive view of paramyxoviral N_{TAIL} interactions and the role of fuzziness in N_{TAIL} function.

Among all the interactions established by N_{TAIL}, the interaction with XD is critical as it allows the L–P complex to be recruited onto the nucleocapsid, triggering transcription and replication (Longhi, 2007, 2009, 2011; Longhi et al., 2017). The so-called cartwheeling mechanism proposed for *Paramyxoviridae* posits that the polymerase complex cartwheels from one N monomer to another within the nucleocapsid in order to allow transcription and replication to take place (Kolakofsky, Le Mercier, Iseni, & Garcin, 2004) (Fig. 12A). Hence, the N_{TAIL}–XD interaction needs to be dynamically established and broken to ensure RNA synthesis.

The affinity of the N_{TAIL}–XD interaction was shown to be a critical parameter governing both viral messenger accumulation rates (Brunel et al., 2014) and the efficiency with which the polymerase complex reinitiates transcription at each IGR (Bloyet, Brunel, et al., 2016; Bloyet, Roche, Gerlier, & Longhi, 2017). The N_{TAIL}–XD interaction thus tightly controls viral polymerase progression along the nucleocapsid. Its interaction strength has consequently to be kept into a precise window to ensure

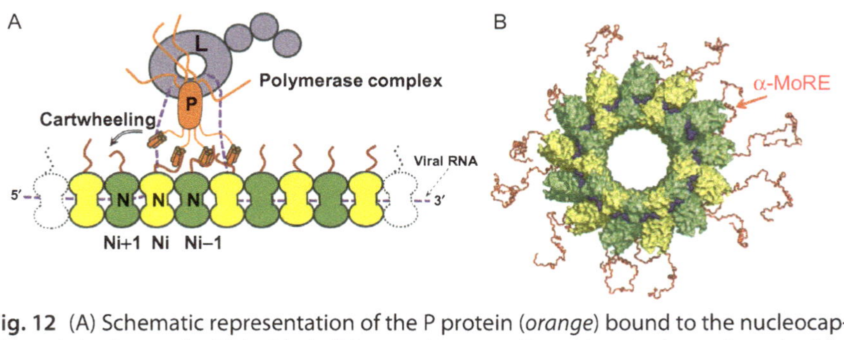

Fig. 12 (A) Schematic representation of the P protein (*orange*) bound to the nucleocapsid and the L protein (*light blue*). IDRs are shown as *lines*. N_{TAIL} is shown in *red*, while successive N_{CORE} monomers are colored in *yellow* and *green*. The negative-sense genomic RNA (*blue dashed line*) is presented in the 5′–3′ orientation. Because of this, the polymerase complex is shown to cartwheel onto the nucleocapsid template from *right* to *left*. (B) Structural model of a single turn of the nucleocapsid made of 13N monomers. N_{TAIL}, N_{CORE}, and the RNA are colored as in panel (A). For the sake of clarity, only one α-MoRE is highlighted by a *red arrow*. *Data were taken from Ringkjøbing Jensen, M., Communie, G., Ribeiro, E. D., Jr., Martinez, N., Desfosses, A., Salmon, L., et al. (2011). Intrinsic disorder in Measles virus nucleocapsids.* Proceedings of the National Academy of Sciences of the United States of America, 108*(24), 9839–9844.*

efficient transcription and replication and to keep a balanced relative amount of the various viral proteins. The fuzziness of paramyxoviral N_{TAIL}–XD complexes provides an exquisite means to modulate the binding affinity. Dynamic binding of the binding motif and the resulting polymorphism at the surface of the partner, as observed in *Henipavirus* N_{TAIL}–XD complexes, brings an additional means for regulating the interaction strength. By tuning the extent of preconfiguration of the binding motifs and/or the length of neighboring fuzzy appendages the virus can achieve an optimal binding strength.

The role of the fuzzy region preceding the α-MoRE in dampening the N_{TAIL}–XD interaction (Gruet et al., 2016) sheds light on why paramyxoviral N_{TAIL} domains require such a long (i.e., >80 residues) disordered arm (Fig. 12B). While such a long arm might in principle be thought as affording spatial flexibility to the MoRE thereby facilitating recruitment of the L–P complex, it was recently shown that relocation of the MoRE on the structured moiety of N (i.e., N_{CORE}) is tolerated by MeV (Cox, Krumm, Thakkar, Sohn, & Plemper, 2017). In light of the observed impact of shortening the fuzzy appendage on XD binding, it is rather tempting to speculate that in the course of evolution, the length of this region has been under selective pressure so as to ensure a "balanced" affinity toward XD. In other words, we propose that the arm has been elongated until an optimal dampening level of the α-MoRE–XD interaction was achieved. Recent data that documented enhanced transcription and concurrent reduced viral growth in a mutated MeV in which the central region of N_{TAIL} preceding the MoRE was removed (Thakkar et al., 2018), afford additional support in favor of a critical role of this region in virus replication and pathogenesis.

Fuzziness also plays an important role in tuning the interaction strength toward additional partners. The fuzzy Box3 region of MeV N_{TAIL} was shown to serve as a low-affinity binding site for hsp70 (Couturier et al., 2010; Zhang et al., 2005), a cellular protein known to stimulate viral transcription and replication (Carsillo, Zhang, Vasconcelos, Niewiesk, & Oglesbee, 2006; Oglesbee, 2007; Zhang et al., 2002), with the primary binding site having being mapped to Box2 (Carsillo, Traylor, Choi, Niewiesk, & Oglesbee, 2006; Zhang et al., 2002). Since hsp70 competitively inhibits XD binding to N_{TAIL} (Zhang et al., 2005), hsp70 could enhance transcription and genome replication by reducing the stability of N_{TAIL}–XD complexes and hence promoting P–L cartwheeling on the nucleocapsid (Bourhis et al., 2005; Zhang et al., 2005). The enhanced interaction strength with XD displayed by MeV N_{TAIL} variants devoid of Box3 may result not only from an entropic effect (like for the N-terminal fuzzy region) but also from abrogation of a destabilizing interaction between Box3 and DnaK, the *E. coli* counterpart of hsp70 (Gruet et al., 2013).

Finally, fuzzy regions flanking MoREs can also be allosterically coupled to the binding region. In line with this, a random mutagenesis study of MeV N_{TAIL} led to the identification of five regulatory regions (i.e., eBoxes) that are located in the N-terminal fuzzy region of N_{TAIL} and act as natural dampeners of the interaction (Gruet et al., 2013). The underlying molecular mechanisms remain to be elucidated.

Taken together all these results highlight that conformational heterogeneity or fuzziness in N_{TAIL} complexes, either as static or dynamic, importantly contributes to the key functions of this domain in the replicative complex of *Paramyxoviridae* members and has been optimized by evolution. Although characterization of fuzziness is complex, it enables a more holistic view of viral protein functions and mechanisms, than the one that can be achieved through the study of shorter, more structured, constructs.

ACKNOWLEDGMENTS

S.L. wishes to thank former members of her lab who contributed to these studies: J.M. Bourhis (IBS, Grenoble, FR), J. Habchi (Dept. of Chemistry, University of Cambridge, UK), D. Blocquel (The Scripps Institute, La Jolla, CA, USA), A. Gruet (Memorial Sloan Kettering Cancer Center, New York, USA), and M. Dosnon. The authors also thank D. Gerlier (CIRI, INSERM U758, Lyon, FR), R. Pierattelli and Isabella Felli (CERM, Florence, IT), M. Blackledge and M. Ringkjøbing-Jensen (IBS, Grenoble, FR), Gary Daughdrill (University of South Florida, Tampa, FL, USA), V. Receveur-Bréchot (BIP, Marseille, FR), V. Belle and B. Guigliarelli (BIP, Marseille, FR), J. Strancar (EPR Center, Institute Josez Stephan, Ljubljana, SI), R. Grandori (University of Milano Bicocca, IT), Paolo Carloni (German Research School for Simulation Sciences and Computational Biomedicine, Jülich, D), R. Das (Dept. of Biomedical Engineering and Center for Biological Systems Engineering, Washington University in St. Louis, St. Louis, MO, USA). The studies herein reviewed were carried out with the financial support of the Agence Nationale de la Recherche, specific programs "Physico-Chimie du Vivant", ANR-08-PCVI-0020-01, and "ASTRID," ANR-11-ASTR-003-01 to S.L. They were also supported by the Italian Ministero dell'Istruzione dell'Università e della Ricerca (Progetto di Interesse "Invecchiamento" to S.G.) and by Sapienza University of Rome (C26A155S48 to S.G.). They also benefited from support from the CNRS, the Direction Générale de l'Armement (DGA), and the Fondation pour la Recherche Médicale (FRM). F.T. is a recipient of a PhD fellowship from the Italo-French University. This work was also financially supported by the Hungarian Academy of Sciences and the GINOP-2.3.2-15-2016-00044 project (M.F.). The funders had no role in study design, data collection and analysis, decision to publish, or preparation of the manuscript.

REFERENCES

Bankamp, B., Horikami, S. M., Thompson, P. D., Huber, M., Billeter, M., & Moyer, S. A. (1996). Domains of the Measles virus N protein required for binding to P protein and self-assembly. *Virology*, *216*(1), 272–277.

Baronti, L., Erales, J., Habchi, J., Felli, I. C., Pierattelli, R., & Longhi, S. (2015). Dynamics of the intrinsically disordered C-terminal domain of the Nipah virus nucleoprotein and interaction with the X domain of the phosphoprotein as unveiled by NMR spectroscopy. *Chem BioChem*, *16*(2), 268–276. https://doi.org/10.1002/cbic.201402534.

Belle, V., Rouger, S., Costanzo, S., Liquiere, E., Strancar, J., Guigliarelli, B., et al. (2008). Mapping alpha-helical induced folding within the intrinsically disordered C-terminal domain of the Measles virus nucleoprotein by site-directed spin-labeling EPR spectroscopy. *Proteins: Structure, Function and Bioinformatics*, *73*(4), 973–988. https://doi.org/10.1002/prot.22125.

Belle, V., Rouger, S., Costanzo, S., Longhi, S., & Fournel, A. (2010). Site-directed spin labeling EPR spectroscopy. In V. N. Uversky & S. Longhi (Eds.), *Instrumental analysis of intrinsically disordered proteins: Assessing structure and conformation*. John Wiley and Sons: Hoboken, NJ.

Bénarouche, A., Habchi, J., Cagna, A., Maniti, O., Girard-Egrot, A., Cavalier, J., et al. (2017). Interfacial properties of NTAIL, an intrinsically disordered protein. *Biophysical Journal*, *113*(12), 2723–2735.

Bernado, P., Blanchard, L., Timmins, P., Marion, D., Ruigrok, R. W., & Blackledge, M. (2005). A structural model for unfolded proteins from residual dipolar couplings and small-angle X-ray scattering. *Proceedings of the National Academy of Sciences of the United States of America*, *102*(47), 17002–17007.

Bernado, P., & Svergun, D. I. (2010). Structural insights into intrinsically disordered proteins by small-angle X-ray scattering. In V. N. Uversky & S. Longhi (Eds.), *Instrumental analysis of intrinsically disordered proteins: Assessing structure and conformation* (pp. 451–476). John Wiley and Sons: Hoboken, NJ.

Bernado, P., & Svergun, D. I. (2012). Structural analysis of intrinsically disordered proteins by small-angle X-ray scattering. *Molecular BioSystems*, *8*(1), 151–167. https://doi.org/10.1039/c1mb05275f.

Beveridge, R., Chappuis, Q., Macphee, C., & Barran, P. (2013). Mass spectrometry methods for intrinsically disordered proteins. *Analyst*, *138*(1), 32–42. https://doi.org/10.1039/c2an35665a.

Bhella, D. (2007). Measles virus nucleocapsid structure, conformational flexibility and the rule of six. In S. Longhi (Ed.), *Measles virus nucleoprotein*. Hauppauge, NY: Nova Publishers Inc.

Bhella, D., Ralph, A., & Yeo, R. P. (2004). Conformational flexibility in recombinant Measles virus nucleocapsids visualised by cryo-negative stain electron microscopy and real-space helical reconstruction. *Journal of Molecular Biology*, *340*(2), 319–331.

Bignon, C., Troilo, F., Gianni, S., & Longhi, S. (2017). Partner-mediated polymorphism of an intrinsically disordered protein. *Journal of Molecular Biology*, *430*(16), 2493–2507.

Blanchard, L., Tarbouriech, N., Blackledge, M., Timmins, P., Burmeister, W. P., Ruigrok, R. W., et al. (2004). Structure and dynamics of the nucleocapsid-binding domain of the Sendai virus phosphoprotein in solution. *Virology*, *319*(2), 201–211.

Blocquel, D., Bourhis, J. M., Eléouët, J. F., Gerlier, D., Habchi, J., Jamin, M., et al. (2012). Transcription et réplication des Mononégavirales: une machine moléculaire originale. *Virologie*, *16*(4), 225–257.

Blocquel, D., Habchi, J., Durand, E., Sevajol, M., Ferron, F., Erales, J., et al. (2014). Coiled-coil deformations in crystal structures: The Measles virus phosphoprotein multimerization domain as an illustrative example. *Acta Crystallographica. Section D, Biological Crystallography*, *70*(6), 1589–1603.

Blocquel, D., Habchi, J., Gruet, A., Blangy, S., & Longhi, S. (2012). Compaction and binding properties of the intrinsically disordered C-terminal domain of henipavirus nucleoprotein as unveiled by deletion studies. *Molecular BioSystems, 8*(1), 392–410. https://doi.org/10.1039/c1mb05401e.

Bloyet, L., Brunel, J., Dosnon, M., Hamon, V., Erales, J., Gruet, A., et al. (2016). Modulation of re-initiation of Measles virus transcription at intergenic regions by PXD to NTAIL binding strength. *PLoS Pathogens, 12*(12), e1006058.

Bloyet, L. M., Roche, P., Gerlier, D., & Longhi, S. (2017). How is transcription reinitiation governed in Measles virus? *Medical Science (Paris), 33*(10), 843–845. https://doi.org/10.1051/medsci/20173310010.

Bloyet, L. M., Welsch, J., Enchery, F., Mathieu, C., de Breyne, S., Horvat, B., et al. (2016). HSP90 chaperoning in addition to phosphoprotein required for folding but not for supporting enzymatic activities of Measles and Nipah virus L polymerases. *Journal of Virology, 90*(15), 6642–6656. https://doi.org/10.1128/JVI.00602-16.

Bonetti, D., Camilloni, C., Visconti, L., Longhi, S., Brunori, M., Vendruscolo, M., et al. (2016). Identification and structural characterization of an intermediate in the folding of the Measles virus X domain. *Journal of Biological Chemistry, 291*(20), 10886–10892. https://doi.org/10.1074/jbc.M116.721126.

Bonetti, D., Troilo, F., Toto, A., Brunori, M., Longhi, S., & Gianni, S. (2017). Analyzing the folding and binding steps of an intrinsically disordered protein by protein engineering. *Biochemistry, 56*(29), 3780–3786.

Borcherds, W., Theillet, F. X., Katzer, A., Finzel, A., Mishall, K. M., Powell, A. T., et al. (2014). Disorder and residual helicity alter p53-Mdm2 binding affinity and signaling in cells. *Nature Chemical Biology, 10*(12), 1000–1002. https://doi.org/10.1038/nchembio.1668.

Borgia, A., Borgia, M. B., Bugge, K., Kissling, V. M., Heidarsson, P. O., Fernandes, C. B., et al. (2018). Extreme disorder in an ultrahigh-affinity protein complex. *Nature, 555*(7694), 61–66. https://doi.org/10.1038/nature25762.

Bourhis, J., Johansson, K., Receveur-Bréchot, V., Oldfield, C. J., Dunker, A. K., Canard, B., et al. (2004). The C-terminal domain of Measles virus nucleoprotein belongs to the class of intrinsically disordered proteins that fold upon binding to their physiological partner. *Virus Research, 99*, 157–167.

Bourhis, J. M., Receveur-Bréchot, V., Oglesbee, M., Zhang, X., Buccellato, M., Darbon, H., et al. (2005). The intrinsically disordered C-terminal domain of the Measles virus nucleoprotein interacts with the C-terminal domain of the phosphoprotein via two distinct sites and remains predominantly unfolded. *Protein Science, 14*, 1975–1992.

Bruhn-Johannsen, J. F., Barnett, K., Bibby, J., Thomas, J., Keegan, R., Rigden, D., et al. (2014). Crystal structure of the Nipah virus phosphoprotein tetramerization domain. *Journal of Virology, 88*(1), 758–762. https://doi.org/10.1128/JVI.02294-13.

Brunel, J., Chopy, D., Dosnon, M., Bloyet, L. M., Devaux, P., Urzua, E., et al. (2014). Sequence of events in Measles virus replication: Role of phosphoprotein-nucleocapsid interactions. *Journal of Virology, 88*(18), 10851–10863. https://doi.org/10.1128/JVI.00664-14.

Budil, D. E., Lee, S., Saxena, S., & Freed, J. H. (1996). Non linear least-squares analysis of slow motion EPR spectra in one and two dimensions using a modified Levenberg-Marquardt algorithm. *Journal of Magnetic Resonance Series A, 120*(2), 155–189.

Carsillo, T., Traylor, Z., Choi, C., Niewiesk, S., & Oglesbee, M. (2006). Hsp72, a host determinant of Measles virus neurovirulence. *Journal of Virology, 80*(22), 11031–11039. https://doi.org/10.1128/JVI.01438-06.

Carsillo, T., Zhang, X., Vasconcelos, D., Niewiesk, S., & Oglesbee, M. (2006). A single codon in the nucleocapsid protein C terminus contributes to in vitro and in vivo fitness of Edmonston Measles virus. *Journal of Virology, 80*(6), 2904–2912. https://doi.org/10.1128/JVI.80.6.2904-2912.2006.

Casey, T. M., Liu, Z., Esquiaqui, J. M., Pirman, N. L., Milshteyn, E., & Fanucci, G. E. (2014). Continuous wave W- and D-band EPR spectroscopy offer "sweet-spots" for characterizing conformational changes and dynamics in intrinsically disordered proteins. *Biochemical and Biophysical Research Communications*, *450*(1), 723–728. https://doi.org/10.1016/j.bbrc.2014.06.045.

Cassonnet, P., Rolloy, C., Neveu, G., Vidalain, P. O., Chantier, T., Pellet, J., et al. (2011). Benchmarking a luciferase complementation assay for detecting protein complexes. *Nature Methods*, *8*(12), 990–992. https://doi.org/10.1038/nmeth.1773.

Communie, G., Crepin, T., Maurin, D., Jensen, M. R., Blackledge, M., & Ruigrok, R. W. (2013). Structure of the tetramerization domain of Measles virus phosphoprotein. *Journal of Virology*, *87*(12), 7166–7169. https://doi.org/10.1128/JVI.00487-13.

Communie, G., Habchi, J., Yabukarski, F., Blocquel, D., Schneider, R., Tarbouriech, N., et al. (2013). Atomic resolution description of the interaction between the nucleoprotein and phosphoprotein of Hendra virus. *PLoS Pathogens*, *9*(9), e1003631.

Cordeiro, T. N., Chen, P. C., De Biasio, A., Sibille, N., Blanco, F. J., Hub, J. S., et al. (2017). Disentangling polydispersity in the PCNA-p15PAF complex, a disordered, transient and multivalent macromolecular assembly. *Nucleic Acids Research*, *45*(3), 1501–1515. https://doi.org/10.1093/nar/gkw1183.

Couturier, M., Buccellato, M., Costanzo, S., Bourhis, J. M., Shu, Y., Nicaise, M., et al. (2010). High affinity binding between Hsp70 and the C-terminal domain of the Measles virus nucleoprotein requires an Hsp40 co-chaperone. *Journal of Molecular Recognition*, *23*(3), 301–315.

Cox, R. M., Krumm, S. A., Thakkar, V. D., Sohn, M., & Plemper, R. K. (2017). The structurally disordered paramyxovirus nucleocapsid protein tail domain is a regulator of the mRNA transcription gradient. *Science Advances*, *3*(2), e1602350. https://doi.org/10.1126/sciadv.1602350.

DeLano, W. L. (2002). The PyMOL molecular graphics system. *Proteins: Structure, Function and Bioinformatics*, *30*, 442–454.

Desfosses, A., Goret, G., Farias Estrozi, L., Ruigrok, R. W., & Gutsche, I. (2011). Nucleoprotein-RNA orientation in the measles virus nucleocapsid by three-dimensional electron microscopy. *Journal of Virology*, *85*(3), 1391–1395. https://doi.org/10.1128/JVI.01459-10.

Dosnon, M., Bonetti, D., Morrone, A., Erales, J., di Silvio, E., Longhi, S., et al. (2015). Demonstration of a folding after binding mechanism in the recognition between the measles virus NTAIL and X domains. *ACS Chemical Biology*, *10*(3), 795–802.

Dosztanyi, Z., Csizmok, V., Tompa, P., & Simon, I. (2005). IUPred: Web server for the prediction of intrinsically unstructured regions of proteins based on estimated energy content. *Bioinformatics*, *21*(16), 3433–3434.

Drescher, M. (2012). EPR in protein science: Intrinsically disordered proteins. *Topics in Current Chemistry*, *321*, 91–119. https://doi.org/10.1007/128_2011_235.

Duro, N., Miskei, M., & Fuxreiter, M. (2015). Fuzziness endows viral motif-mimicry. *Molecular BioSystems*, *11*(10), 2821–2829. https://doi.org/10.1039/c5mb00301f.

D'Urzo, A., Konijnenberg, A., Rossetti, G., Habchi, J., Li, J., Carloni, P., et al. (2015). Molecular basis for structural heterogeneity of an intrinsically disordered protein bound to a partner by combined ESI-IM-MS and modeling. *Journal of the American Society for Mass Spectrometry*, *26*(3), 472–481. https://doi.org/10.1007/s13361-014-1048-z.

Eaton, B. T., Broder, C. C., Middleton, D., & Wang, L. F. (2006). Hendra and Nipah viruses: Different and dangerous. *Nature Reviews: Microbiology*, *4*(1), 23–35.

Eaton, B. T., Mackenzie, J. S., & Wang, L. F. (2007). Henipaviruses. In B. N. Fields, D. M. Knipe, & P. M. Howley (Eds.), *Fields virology* (5th ed., pp. 1587–1600). Philadelphia: Lippincott-Raven.

Erales, J., Beltrandi, M., Roche, J., Maté, M., & Longhi, S. (2015). Insights into the Hendra virus NTAIL-XD complex: Evidence for a parallel organization of the helical MoRE at the XD surface stabilized by a combination of hydrophobic and polar interactions. *Biochimica and Biopysical Acta*, *1854*(8), 1038–1053.

Erales, J., Blocquel, D., Habchi, J., Beltrandi, M., Gruet, A., Dosnon, M., et al. (2015). Order and disorder in the replicative complex of paramyxoviruses. *Advances in Experimental Medicine and Biology*, *870*, 351–381.

Felli, I. C., & Pierattelli, R. (2015). *Intrinsically disordered proteins studied by NMR spectroscopy*. Printforce, The Netherlands: Springer.

Fersht, A. R., Matouschek, A., & Serrano, L. (1992). The folding of an enzyme. I. Theory of protein engineering analysis of stability and pathway of protein folding. *Journal of Molecular Biology*, *224*(3), 771–782. https://doi.org/10.1016/0022-2836(92)90561-W.

Fuxreiter, M. (2012). Fuzziness: Linking regulation to protein dynamics. *Molecular BioSystems*, *8*(1), 168–177. https://doi.org/10.1039/c1mb05234a.

Fuxreiter, M. (2018). Fuzziness in protein interactions—A historical perspective. *Journal of Molecular Biology*, *430*(16), 2278–2287. https://doi.org/10.1016/j.jmb.2018.02.015.

Fuxreiter, M., Simon, I., Friedrich, P., & Tompa, P. (2004). Preformed structural elements feature in partner recognition by intrinsically unstructured proteins. *Journal of Molecular Biology*, *338*(5), 1015–1026.

Fuxreiter, M., & Tompa, P. (2012). Fuzzy complexes: A more stochastic view of protein function. *Advances in Experimental Medicine and Biology*, *725*, 1–14. https://doi.org/10.1007/978-1-4614-0659-4_1.

Gely, S., Lowry, D. F., Bernard, C., Ringkjobing-Jensen, M., Blackledge, M., Costanzo, S., et al. (2010). Solution structure of the C-terminal X domain of the Measles virus phosphoprotein and interaction with the intrinsically disordered C-terminal domain of the nucleoprotein. *Journal of Molecular Recognition*, *23*, 435–447.

Gruet, A., Dosnon, M., Blocquel, D., Brunel, J., Gerlier, D., Das, R. K., et al. (2016). Fuzzy regions in an intrinsically disordered protein impair protein-protein interactions. *FEBS Journal*, *283*, 576–594. https://doi.org/10.1111/febs.13631.

Gruet, A., Dosnon, M., Vassena, A., Lombard, V., Gerlier, D., Bignon, C., et al. (2013). Dissecting partner recognition by an intrinsically disordered protein using descriptive random mutagenesis. *Journal of Molecular Biology*, *425*(18), 3495–3509. https://doi.org/10.1016/j.jmb.2013.06.025.

Gruet, A., Longhi, S., & Bignon, C. (2012). One-step generation of error-prone PCR libraries using Gateway® technology. *Microbial Cell Factories*, *11*(1), 14. https://doi.org/10.1186/1475-2859-11-14.

Guryanov, S. G., Liljeroos, L., Kasaragod, P., Kajander, T., & Butcher, S. J. (2016). Crystal structure of the Measles virus nucleoprotein core in complex with an N-terminal region of phosphoprotein. *Journal of Virology*, *90*(6), 2849–2857. https://doi.org/10.1128/JVI.02865-15.

Gutsche, I., Desfosses, A., Effantin, G., Ling, W. L., Haupt, M., Ruigrok, R. W., et al. (2015). Near-atomic cryo-EM structure of the helical Measles virus nucleocapsid. *Science*, *348*, 704–707. https://doi.org/10.1126/science.aaa5137.

Habchi, J., Blangy, S., Mamelli, L., Ringkjobing Jensen, M., Blackledge, M., Darbon, H., et al. (2011). Characterization of the interactions between the nucleoprotein and the phosphoprotein of Henipaviruses. *Journal of Biological Chemistry*, *286*(15), 13583–13602.

Habchi, J., & Longhi, S. (2012). Structural disorder within paramyxovirus nucleoproteins and phosphoproteins. *Molecular BioSystems*, *8*(1), 69–81. https://doi.org/10.1039/c1mb05204g.

Habchi, J., & Longhi, S. (2015). Structural disorder within paramyxoviral nucleoproteins and phosphoproteins in their free and bound forms: From predictions to experimental assessment. *International Journal of Molecular Sciences*, *16*(7), 15688–15726. https://doi.org/10.3390/ijms160715688.

Habchi, J., Mamelli, L., Darbon, H., & Longhi, S. (2010). Structural disorder within Henipavirus nucleoprotein and phosphoprotein: From predictions to experimental assessment. *PLoS One*, *5*(7), e11684. https://doi.org/10.1371/journal.pone.0011684.

Habchi, J., Mamelli, L., & Longhi, S. (2012). Structural disorder within the nucleoprotein and phosphoprotein from measles, nipah and hendra viruses. In V. N. Uversky & S. Longhi (Eds.), *Flexible viruses: Structural disorder in viral proteins* (pp. 47–94). Hoboken, NJ: John Wiley and Sons.

Habchi, J., Martinho, M., Gruet, A., Guigliarelli, B., Longhi, S., & Belle, V. (2012). Monitoring structural transitions in IDPs by site-directed spin labeling EPR spectroscopy. *Methods in Molecular Biology*, *895*, 361–386. https://doi.org/10.1007/978-1-61779-927-3_21.

Heggeness, M. H., Scheid, A., & Choppin, P. W. (1980). Conformation of the helical nucleocapsids of paramyxoviruses and vesicular stomatitis virus: Reversible coiling and uncoiling induced by changes in salt concentration. *Proceedings of the National Academy of Sciences of the United States of America*, 77(5), 2631–2635.

Heggeness, M. H., Scheid, A., & Choppin, P. W. (1981). The relationship of conformational changes in the Sendai virus nucleocapsid to proteolytic cleavage of the NP polypeptide. *Virology*, *114*(2), 555–562.

Houben, K., Marion, D., Tarbouriech, N., Ruigrok, R. W., & Blanchard, L. (2007). Interaction of the C-terminal domains of sendai virus N and P proteins: Comparison of polymerase-nucleocapsid interactions within the paramyxovirus family. *Journal of Virology*, *81*(13), 6807–6816.

Hubbell, W. L., Gross, A., Langen, R., & Lietzow, M. A. (1998). Recent advances in site-directed spin labeling of proteins. *Current Opinion in Structural Biology*, *8*(5), 649–656.

Jensen, M. R., Bernado, P., Houben, K., Blanchard, L., Marion, D., Ruigrok, R. W., et al. (2010). Structural disorder within sendai virus nucleoprotein and phosphoprotein: Insight into the structural basis of molecular recognition. *Protein and Peptide Letters*, *17*(8), 952–960. https://doi.org/10.2174/092986610791498885.

Jensen, M. R., Houben, K., Lescop, E., Blanchard, L., Ruigrok, R. W., & Blackledge, M. (2008). Quantitative conformational analysis of partially folded proteins from residual dipolar couplings: Application to the molecular recognition element of Sendai virus nucleoprotein. *Journal of the American Chemical Society*, *130*(25), 8055–8061. https://doi.org/10.1021/ja801332d.

Johansson, K., Bourhis, J. M., Campanacci, V., Cambillau, C., Canard, B., & Longhi, S. (2003). Crystal structure of the Measles virus phosphoprotein domain responsible for the induced folding of the C-terminal domain of the nucleoprotein. *Journal of Biological Chemistry*, *278*(45), 44567–44573.

Jonsson, S. A., Mohanty, S., & Irback, A. (2012). Distinct phases of free alpha-synuclein—A Monte Carlo study. *Proteins*, *80*(9), 2169–2177. https://doi.org/10.1002/prot.24107.

Jordan, P. C., Liu, C., Raynaud, P., Lo, M. K., Spiropoulou, C. F., Symons, J. A., et al. (2018). Initiation, extension, and termination of RNA synthesis by a paramyxovirus polymerase. *PLoS Pathogens*, *14*(2), e1006889. https://doi.org/10.1371/journal.ppat.1006889.

Kaltashov, I. A., Bobst, C. E., & Abzalimov, R. R. (2013). Mass spectrometry-based methods to study protein architecture and dynamics. *Protein Science*, *22*(5), 530–544. https://doi.org/10.1002/pro.2238.

Karlin, D., Ferron, F., Canard, B., & Longhi, S. (2003). Structural disorder and modular organization in paramyxovirinae N and P. *Journal of General Virology*, *84*(Pt. 12), 3239–3252.

Karlin, D., Longhi, S., & Canard, B. (2002). Substitution of two residues in the Measles virus nucleoprotein results in an impaired self-association. *Virology*, *302*(2), 420–432.
Karlin, D., Longhi, S., Receveur, V., & Canard, B. (2002). The N-terminal domain of the phosphoprotein of morbilliviruses belongs to the natively unfolded class of proteins. *Virology*, *296*(2), 251–262.
Katoh, H., Kubota, T., Nakatsu, Y., Tahara, M., Kidokoro, M., & Takeda, M. (2017). Heat shock protein 90 ensures efficient mumps virus replication by assisting with viral polymerase complex formation. *Journal of Virology*, *91*(6), e02220–16. https://doi.org/10.1128/JVI.02220-16.
Kavalenka, A., Urbancic, I., Belle, V., Rouger, S., Costanzo, S., Kure, S., et al. (2010). Conformational analysis of the partially disordered Measles virus NTAIL-XD complex by SDSL EPR spectroscopy. *Biophysical Journal*, *98*(6), 1055–1064.
Kim, M. Y., Shu, Y., Carsillo, T., Zhang, J., Yu, L., Peterson, C., et al. (2013). Hsp70 and a novel axis of type I interferon-dependent antiviral immunity in the Measles virus-infected brain. *Journal of Virology*, *87*(2), 998–1009. https://doi.org/10.1128/JVI.02710-12.
Kingston, R. L., Hamel, D. J., Gay, L. S., Dahlquist, F. W., & Matthews, B. W. (2004). Structural basis for the attachment of a paramyxoviral polymerase to its template. *Proceedings of the National Academy of Sciences of the United States of America*, *101*(22), 8301–8306.
Kingston, R. L., Walter, A. B., & Gay, L. S. (2004). Characterization of nucleocapsid binding by the Measles and the mumps virus phosphoprotein. *Journal of Virology*, *78*(16), 8630–8640.
Kolakofsky, D., Le Mercier, P., Iseni, F., & Garcin, D. (2004). Viral DNA polymerase scanning and the gymnastics of Sendai virus RNA synthesis. *Virology*, *318*(2), 463–473.
Kozin, M. B., & Svergun, D. I. (2001). Automated matching of high- and low-resolution structural models. *Journal of Applied Crystallography*, *34*, 33–41.
Lamb, R. A., & Parks, G. D. (2013). Paramyxoviridae. In D. M. Knipe & P. M. Howley (Eds.), *Fields virology* (pp. 957–995). Philadelphia: Lippincott Williams & Wilkins.
Lamiable, A., Thevenet, P., Rey, J., Vavrusa, M., Derreumaux, P., & Tuffery, P. (2016). PEP-FOLD3: Faster de novo structure prediction for linear peptides in solution and in complex. *Nucleic Acids Research*, *44*(W1), W449–W454. https://doi.org/10.1093/nar/gkw329.
Liang, B., Li, Z., Jenni, S., Rahmeh, A. A., Morin, B. M., Grant, T., et al. (2015). Structure of the L protein of vesicular stomatitis virus from electron cryomicroscopy. *Cell*, *162*(2), 314–327. https://doi.org/10.1016/j.cell.2015.06.018.
Liston, P., Batal, R., DiFlumeri, C., & Briedis, D. J. (1997). Protein interaction domains of the Measles virus nucleocapsid protein (NP). *Archives of Virology*, *142*(2), 305–321.
Longhi, S. (2007). *Measles virus nucleoprotein*. Hauppauge, NY: Nova Publishers Inc.
Longhi, S. (2009). Nucleocapsid structure and function. *Current Topics in Microbiology and Immunology*, *329*, 103–128.
Longhi, S. (2011). Structural disorder within the measles virus nucleoprotein and phosphoprotein: Functional implications for transcription and replication. In M. Luo (Ed.), *Negative strand RNA virus* (pp. 95–125). Singapore: World Scientific Publishing.
Longhi, S. (2015). Structural disorder within paramyxoviral nucleoproteins. *FEBS Letters*, *589*(19 Pt A), 2649–2659. https://doi.org/10.1016/j.febslet.2015.05.055.
Longhi, S., Belle, V., Fournel, A., Guigliarelli, B., & Carrière, F. (2011). Probing structural transitions in both structured and disordered proteins by site-directed spin-labeling EPR spectroscopy. *Journal of Peptide Science*, *17*(5), 315–328.
Longhi, S., Bloyet, L. M., Gianni, S., & Gerlier, D. (2017). How order and disorder within paramyxoviral nucleoproteins and phosphoproteins orchestrate the molecular interplay of transcription and replication. *Cellular and Molecular Life Sciences*, *74*(17), 3091–3118. https://doi.org/10.1007/s00018-017-2556-3.

Longhi, S., & Oglesbee, M. (2010). Structural disorder within the Measles virus nucleoprotein and phosphoprotein. *Protein and Peptide Letters*, *17*(8), 961–978.
Longhi, S., Receveur-Brechot, V., Karlin, D., Johansson, K., Darbon, H., Bhella, D., et al. (2003). The C-terminal domain of the Measles virus nucleoprotein is intrinsically disordered and folds upon binding to the C-terminal moiety of the phosphoprotein. *Journal of Biological Chemistry*, *278*(20), 18638–18648.
Magliery, T. J., Wilson, C. G., Pan, W., Mishler, D., Ghosh, I., Hamilton, A. D., et al. (2005). Detecting protein-protein interactions with a green fluorescent protein fragment reassembly trap: Scope and mechanism. *Journal of the American Chemical Society*, *127*(1), 146–157.
Martinho, M., Habchi, J., El Habre, Z., Nesme, L., Guigliarelli, B., Belle, V., et al. (2013). Assessing induced folding within the intrinsically disordered C-terminal domain of the Henipavirus nucleoproteins by site directed spin labeling EPR spectroscopy. *Journal of Biomolecular Structure and Dynamics*, *31*(5), 453–471.
Micsonai, A., Wien, F., Kernya, L., Lee, Y. H., Goto, Y., Refregiers, M., et al. (2015). Accurate secondary structure prediction and fold recognition for circular dichroism spectroscopy. *Proceedings of the National Academy of Sciences of the United States of America*, *112*(24), E3095–E3103. https://doi.org/10.1073/pnas.1500851112.
Miskei, M., Antal, C., & Fuxreiter, M. (2017). FuzDB: Database of fuzzy complexes, a tool to develop stochastic structure-function relationships for protein complexes and higher-order assemblies. *Nucleic Acids Research*, *45*(D1), D228–D235. https://doi.org/10.1093/nar/gkw1019.
Miskei, M., Gregus, A., Sharma, R., Duro, N., Zsolyomi, F., & Fuxreiter, M. (2017). Fuzziness enables context dependence of protein interactions. *FEBS Letters*, *591*(17), 2682–2695. https://doi.org/10.1002/1873-3468.12762.
Morin, B., Bourhis, J. M., Belle, V., Woudstra, M., Carrière, F., Guigliarelli, B., et al. (2006). Assessing induced folding of an intrinsically disordered protein by site-directed spin-labeling EPR spectroscopy. *The Journal of Physical Chemistry. B*, *110*(41), 20596–20608.
Morin, B., Liang, B., Gardner, E., Ross, R. A., & Whelan, S. P. (2016). An in vitro RNA synthesis assay for rabies virus defines critical ribonucleoprotein interactions for polymerase activity. *Journal of Virology*, *91*(1), e01508-16. https://doi.org/10.1128/JVI.01508-16.
Morin, B., Rahmeh, A. A., & Whelan, S. P. (2012). Mechanism of RNA synthesis initiation by the vesicular stomatitis virus polymerase. *EMBO Journal*, *31*(5), 1320–1329. https://doi.org/10.1038/emboj.2011.483.
Oglesbee, M. (2007). Nucleocapsid protein interactions with the major inducible heat shock protein. In S. Longhi (Ed.), *Measles virus nucleoprotein* (pp. 53–98). Hauppauge, NY: Nova Publishers Inc.
Otieno, S., & Kriwacki, R. (2012). Probing the role of nascent helicity in p27 function as a cell cycle regulator. *PLoS One*, 7(10), e47177. https://doi.org/10.1371/journal.pone.0047177.
Ringkjøbing Jensen, M., Communie, G., Ribeiro, E. D., Jr., Martinez, N., Desfosses, A., Salmon, L., et al. (2011). Intrinsic disorder in Measles virus nucleocapsids. *Proceedings of the National Academy of Sciences of the United States of America*, *108*(24), 9839–9844.
Rogers, J. M., Wong, C. T., & Clarke, J. (2014). Coupled folding and binding of the disordered protein PUMA does not require particular residual structure. *Journal of the American Chemical Society*, *136*(14), 5197–5200. https://doi.org/10.1021/ja4125065.
Salladini, E., Delauzun, V., & Longhi, S. (2017). The Henipavirus V protein is a prevalently unfolded protein with a zinc-finger domain involved in binding to DDB1. *Molecular BioSystems*, *13*(11), 2254–2267. https://doi.org/10.1039/c7mb00488e.
Schneider, R., Maurin, D., Communie, G., Kragelj, J., Hansen, D. F., Ruigrok, R. W., et al. (2015). Visualizing the molecular recognition trajectory of an intrinsically disordered

protein using multinuclear relaxation dispersion NMR. *Journal of the American Chemical Society, 137*(3), 1220–1229. https://doi.org/10.1021/ja511066q.
Schoehn, G., Mavrakis, M., Albertini, A., Wade, R., Hoenger, A., & Ruigrok, R. W. (2004). The 12 A structure of trypsin-treated Measles virus N-RNA. *Journal of Molecular Biology, 339*(2), 301–312.
Schreiber, G., Haran, G., & Zhou, H. X. (2009). Fundamental aspects of protein-protein association kinetics. *Chemical Reviews, 109*(3), 839–860. https://doi.org/10.1021/cr800373w.
Scott, K. A., Alonso, D. O., Sato, S., Fersht, A. R., & Daggett, V. (2007). Conformational entropy of alanine versus glycine in protein denatured states. *Proceedings of the National Academy of Sciences of the United States of America, 104*(8), 2661–2666. https://doi.org/10.1073/pnas.0611182104.
Sharma, R., Raduly, Z., Miskei, M., & Fuxreiter, M. (2015). Fuzzy complexes: Specific binding without complete folding. *FEBS Letters, 589*(19 Pt. A), 2533–2542. https://doi.org/10.1016/j.febslet.2015.07.022.
Shu, Y., Habchi, J., Costanzo, S., Padilla, A., Brunel, J., Gerlier, D., et al. (2012). Plasticity in structural and functional interactions between the phosphoprotein and nucleoprotein of Measles virus. *Journal of Biological Chemistry, 287*, 11951–11967.
Spera, S., & Bax, A. (1991). Empirical correlation between protein backbone conformation and Ca and Cb 13C nuclear magnetic resonance chemical shifts. *Journal of the American Chemical Society, 113*, 5490–5492.
Sterckx, Y. G., Volkov, A. N., Vranken, W. F., Kragelj, J., Jensen, M. R., Buts, L., et al. (2014). Small-angle X-ray scattering- and nuclear magnetic resonance-derived conformational ensemble of the highly flexible antitoxin PaaA2. *Structure, 22*(6), 854–865. https://doi.org/10.1016/j.str.2014.03.012.
Strancar, J., Koklic, T., Arsov, Z., Filipic, B., Stopar, D., & Hemminga, M. A. (2005). Spin label EPR-based characterization of biosystem complexity. *Journal of Chemical Information and Modeling, 45*(2), 394–406. https://doi.org/10.1021/ci049748h.
Stuchfield, D., & Barran, P. (2018). Unique insights to intrinsically disordered proteins provided by ion mobility mass spectrometry. *Current Opinion in Chemical Biology, 42*, 177–185. https://doi.org/10.1016/j.cbpa.2018.01.007.
Svergun, D. I., Baraberato, C., & Koch, M. H. (1995). CRYSOL—A program to evaluate X-ray solution scattering of biological macromolecules from atomic coordinates. *Journal of Applied Crystallography, 28*, 768–773.
Svergun, D. I., Petoukhov, M. V., & Koch, M. H. (2001). Determination of domain structure of proteins from X-ray solution scattering. *Biophysical Journal, 80*(6), 2946–2953.
Testa, L., Brocca, S., & Grandori, R. (2011). Charge-surface correlation in electrospray ionization of folded and unfolded proteins. *Analytical Chemistry, 83*(17), 6459–6463. https://doi.org/10.1021/ac201740z.
Thakkar, V. D., Cox, R. M., Sawatsky, B., da Fontoura Budaszewski, R., Sourimant, J., Wabbel, K., et al. (2018). The unstructured paramyxovirus nucleocapsid protein tail domain modulates viral pathogenesis through regulation of transcriptase activity. *Journal of Virology, 92*(8). https://doi.org/10.1128/JVI.02064-17.
Tompa, P., & Fuxreiter, M. (2008). Fuzzy complexes: Polymorphism and structural disorder in protein-protein interactions. *Trends in Biochemical Sciences, 33*(1), 2–8.
Uversky, V. N. (1993). Use of fast protein size-exclusion liquid chromatography to study the unfolding of proteins which denature through the molten globule. *Biochemistry, 32*(48), 13288–13298.
Uversky, V. N. (2002). What does it mean to be natively unfolded? *European Journal of Biochemistry, 269*(1), 2–12.
Wang, Y., Chu, X., Longhi, S., Roche, P., Han, W., Wang, E., et al. (2013). Multiscaled exploration of coupled folding and binding of an intrinsically disordered molecular

recognition element in Measles virus nucleoprotein. *Proceedings of the National Academy of Sciences of the United States of America*, *110*(40), E3743–E3752. https://doi.org/10.1073/pnas.1308381110.

Wang, L. F., Yu, M., Hansson, E., Pritchard, L. I., Shiell, B., Michalski, W. P., et al. (2000). The exceptionally large genome of Hendra virus: Support for creation of a new genus within the family Paramyxoviridae. *Journal of Virology*, *74*(21), 9972–9979.

Wilson, C. G., Magliery, T. J., & Regan, L. (2004). Detecting protein-protein interactions with GFP-fragment reassembly. *Nature Methods*, *1*(3), 255–262. https://doi.org/10.1038/nmeth1204-255.

Yabukarski, F., Lawrence, P., Tarbouriech, N., Bourhis, J. M., Delaforge, E., Jensen, M. R., et al. (2014). Structure of Nipah virus unassembled nucleoprotein in complex with its viral chaperone. *Nature Structural & Molecular Biology*, *21*(9), 754–759. https://doi.org/10.1038/nsmb.2868.

Zhang, X., Bourhis, J. M., Longhi, S., Carsillo, T., Buccellato, M., Morin, B., et al. (2005). Hsp72 recognizes a P binding motif in the Measles virus N protein C-terminus. *Virology*, *337*(1), 162–174.

Zhang, X., Glendening, C., Linke, H., Parks, C. L., Brooks, C., Udem, S. A., et al. (2002). Identification and characterization of a regulatory domain on the carboxyl terminus of the Measles virus nucleocapsid protein. *Journal of Virology*, *76*(17), 8737–8746.

CHAPTER EIGHT

Characterization of Dynamic IDP Complexes by NMR Spectroscopy

Andreas Prestel, Katrine Bugge, Lasse Staby, Ruth Hendus-Altenburger, Birthe B. Kragelund[1]

Structural Biology and NMR Laboratory, Department of Biology, University of Copenhagen, Copenhagen, Denmark

[1]Corresponding author: e-mail address: bbk@bio.ku.dk

Contents

Methods in Enzymology, Volume 611
ISSN 0076-6879
https://doi.org/10.1016/bs.mie.2018.08.026

Abstract

NMR spectroscopy has proven to be a key method for studying intrinsically disordered proteins (IDPs). Nonetheless, traditional NMR methods developed for solving structures of ordered protein complexes are insufficient for the full characterization of dynamic IDP complexes, where the energy landscape is broader and more rugged. Furthermore, due to their high sensitivity to environmental changes, NMR studies of IDP complexes must be conducted with extra care and the observed NMR parameters thoroughly evaluated to enable disentanglement of binding events from ensemble distribution changes. In this chapter, written for the non-NMR expert, we start out by outlining sample preparation for IDP complexes, guide through the recording and evaluation of diagnostic $^{1}H,^{15}N$-HSQC spectra, and delineate more sophisticated NMR strategies to follow for the particular type of complex. The most relevant experiments are then described in terms of aims, needs, pitfalls, analysis, and expected outcomes, with references to recent examples.

1. INTRODUCTION

Intrinsically disordered proteins (IDPs) or regions (IDRs) (here collectively termed IDPs) are proteins that are functional while existing in a broad ensemble of near isoenergetic conformations. Despite their lack of tertiary structure, IDPs are involved in communication with other molecules by forming all kinds of associations ranging from binary, discrete complexes to large multicomponent assemblies. IDP complexes form with affinities similar to those of globular folded proteins, and are on average only 2.5 kcal mol^{-1} less stable due to a conformational entropy loss (Teilum, Olsen, & Kragelund, 2015). Similar to globular proteins their complexes serve structural, functional, and regulatory roles (Uversky, 2018; Wright & Dyson, 2015), but due to their dynamic nature, they expand the types of possible complexes (Miskei, Antal, & Fuxreiter, 2016; Mittag, Kay, & Forman-Kay, 2010; Motlagh, Wrabl, Li, & Hilser, 2014). The fast dynamics characteristic of IDPs may persist in their complexes and the degree of disorder can vary greatly. In one end of the scale, IDPs may completely fold upon binding to form globular-like complexes with little disorder (Rogers, Steward, & Clarke, 2013; Wright & Dyson, 2009), while at the other end disorder may persist and result in complexes where both binding partners stay disordered (Borgia et al., 2018) (Fig. 1). Furthermore, some IDP regions have limited sequence diversity which under certain conditions can lead to large-scale associations involving an undefined number of molecules forming liquid–liquid phase-separated states referred to as condensates (Fung, Birol, & Rhoades, 2018) (see also

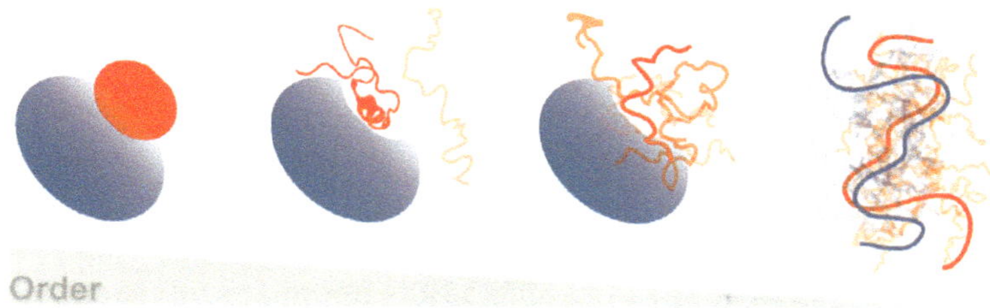

Fig. 1 Order–disorder spectrum of protein complexes. IDPs may form complexes with varying degrees of disorder and stoichiometries, from almost completely ordered complexes to complexes staying almost fully disordered. Only binary complexes are illustrated.

chapter "Visualization and quantitation of phase-separated droplet formation by human HP1α" by Keenen et al.

Solution-state NMR spectroscopy has proven to be a valuable tool in the studies of IDPs, as this method allows data acquisition in the absence of structure formation. The dynamical properties of IDPs can even be advantageous in NMR studies, as this may result in narrow line widths. Where X-ray crystallography may solve structures of IDP complexes in cases where folding occurs (Qi et al., 2017) or when the binding region can be contained within a small peptide (Gulbis, Kelman, Hurwitz, O'Donnell, & Kuriyan, 1996), NMR spectroscopy allows studying complexes retaining disorder. Sometimes NMR data are integrated with computation (Delaforge et al., 2018), or X-ray crystallography (Gógl et al., 2016; Leyrat et al., 2011). Yet, the dynamics inherent to IDPs pose both conceptual and technical challenges to the study of their complexes by NMR spectroscopy. From the technical side, line broadening from dynamics on unfavorable timescales, lack of intermolecular nuclear Overhauser effects (NOEs), peak overlap from disordered regions, and weak affinities challenge information retrieval. Conceptually, for designing the study and interpreting the data it must be taken into account that the complex may not exist in a single state.

This chapter has been written for the non-NMR expert and focuses on the use of NMR spectroscopy to study IDP complexes where disorder is at least partly retained. Thus, we will not go into details in cases where complete folding upon binding is achieved, as relevant and well-written reviews and textbooks are available (see Bonvin, Boelens, & Kaptein, 2005; Gronenborn & Clore, 1995; Hass & Ubbink, 2014; Schieborr et al., 2005; Thompson, Beck, & Campbell, 2015). Likewise, NMR methods for the study of free IDPs have been well described (Brutscher et al., 2015; Jensen et al., 2009; Jensen, Zweckstetter, Huang, & Blackledge, 2014; Konrat, 2014). The reader will be taken from the sample preparation of the complex through the recording of diagnostic 2D ^{1}H,^{15}N-heteronuclear single quantum coherence (HSQC) spectra that may reveal important information on the system under investigation and help define further possible NMR strategies for the specific type of complex. NMR experiments of particular relevance for IDP complexes are then suggested, including aims, needs, pitfalls, analysis, and expected outcomes, as well as references to relevant examples.

2. TRIMMING IDPs TO OPTIMIZE COMPLEXES FOR NMR STUDIES

The first step in the successful characterization of an IDP complex by NMR spectroscopy is to consider if the protein sequence needs to be subdivided to simplify analysis. Sometimes the binding region is part of a longer disordered chain and sometimes it is limited to a smaller region. In this case, several factors need consideration, including the length of the region, the presence of transient structures or folded domains, and whether the interaction sites are known or can be predicted. Due to the narrow NMR peak dispersion of IDPs, the length of the constructs should be limited to less than ~150 residues to avoid severe peak overlap. Long IDPs can be divided into two—or more—fragments with 5–20 residue overlaps (divide and conquer approach) as has been done for , e.g., Tau (Lippens et al., 2006) and the intracellular domains of cytokine receptors (Haxholm et al., 2015). The fragments should be designed to prevent interference with binding or interruption of transient structures, and the spectra of the shorter variants should be compared to the full IDP to assure its representativeness by overlay of ^{1}H,^{15}N-HSQC spectra (see later). If no experimental data on interaction sites and secondary structure elements are available to guide the design, bioinformatics can be used to predict short linear motifs (SLiMs) (Davey, Cyert, & Moses, 2015; Van Roey et al., 2014) and structured/disordered regions (Dinkel et al., 2014; Dosztányi, Csizmok, Tompa, & Simon, 2005; Gu & Hilser, 2008). At least 10–20 residues should be included at each end of a predicted interaction site. However, if the exact binding site is unknown or additional contacts are anticipated, design of expression constructs needs to be done more carefully. In such cases, several constructs of varying lengths should be made for screening purposes, which later may be shortened for advanced experiments based on the screening results.

There are other design features to be considered. Use of tags such as hexahistidine or FLAG may interfere with the ensemble distribution and should be used with care or better be removed before data acquisition. The same caution is advised for mutational analyses and use of phosphorylation mimics, which may stabilize or destabilize transient structures and/or otherwise change the structural ensemble. The region of interest should preferably start and end with charged or polar residues to ensure solubility. Finally, due to the low content of aromatic residues in IDPs, concentration determination using absorbance at 280 nm can be impossible and as alternatives, albeit less precise, one can use absorbance at 205 nm (Anthis & Clore, 2013) or 214 nm (Kuipers & Gruppen, 2007), or BCA analysis. For precise quantitative determinations, amino acid analyses are recommended.

3. SELECTING AND OPTIMIZING NMR SAMPLE CONDITIONS FOR IDP COMPLEXES

Optimization of NMR sample conditions is critical for the successful acquisition of NMR data. For all proteins, sample condition optimization aims at striking the right balance between capturing the essential characteristics of the native environment that supports functionality and achieving sufficient NMR data quality so that the intensity and number of observable peaks in the NMR spectra are maximized. With the higher average exposure of residues to the surrounding solvent of IDPs compared to globular proteins, the structural features and interactions of IDPs are likely to be more sensitive to solvent characteristics. Consequently, in titration series of IDP complexes, small changes in pH may induce chemical shift perturbations (CSPs) that are difficult to distinguish from those arising from interactions with the binding partner. Additionally, certain conditions may infer conformational dynamics on a timescale that leads to extensive line broadening in the NMR spectra, resulting in a decreased signal-to-noise ratio and/or missing peaks.

The most basic parameters to consider during optimization include buffer composition, pH, ionic strength, and temperature. Buffers are often used at 10–50 m*M* concentrations, and nonprotonated buffers are preferred to prevent overlap of NMR signals from the protein. Phosphate buffers are commonly used as an economical nonprotonated buffer with a small temperature coefficient, as large temperature coefficients change the pH with temperature. In terms of pH, a physiologically meaningful value, typically close to pH 7, is often chosen, although the quality of the NMR data may increase by the slower amide exchange rate at lower pH (slowest rate around pH 3.8). However, at physiological pH, residues in proximity to His are highly pH sensitive, and a pH titration is advisable. Furthermore, to avoid a zero net charge that enhances aggregation, the pH should preferably differ by at least 1 pH unit from the pI of each binding partner as well as the complex.

The ionic strength of the solution is particularly critical for proteins with a high number of charged residues, which is a characteristic trait of IDPs. Under physiological conditions, electrostatic interactions are partially screened by the presence of ions and it is essential to correctly mimic this to acquire biologically meaningful data. The ionic strength may have opposing effects on different IDPs; at low ionic strength, IDPs with a high net charge may experience considerable expansion due to charge repulsion (Mao, Crick, Vitalis, Chicoine, & Pappu, 2010; Muller-Spath et al., 2010), while IDPs with a similar number of positively and negatively charged residues may collapse due to attraction along the chain (Muller-Spath et al., 2010). Importantly, the ionic strength may impact the association and dissociation rates of complexes (Wicky, Shammas, & Clarke, 2017). An extreme example is the charge-driven IDP complex between ProTα and Histone H1 for which a mere doubling of the ionic strength from 165 to 340 m*M* resulted in a six orders of magnitude affinity reduction (Borgia et al., 2018). However, different ions also have an impact on the association kinetics going beyond a sheer electrostatic effect, causing changes in the residual structure (Wicky et al., 2017), and there are also possibilities for chain compaction/expansion and specific ion-binding sites. Besides the considerations on physiological relevance, the ionic strength also affects the NMR data quality, as a high ionic strength (>200 m*M*) demands longer pulses, thus increasing sample heating and reducing signal to noise (especially using cryogenic probes).

Several other components are, or may be, necessary to add to the NMR sample, such as an internal chemical shift reference, typically 4,4-dimethyl-4-silapentane-1-sulfonic acid (DSS), for standardization of the resonance frequency to chemical shift values in ppm. The NMR sample must also include 5%–10% (v/v) D_2O to lock the instrument frequency and 4 m*M* NaN_3 can be added to prevent microbial growth. Lastly, several IDPs have free cysteines and need long-lasting reducing agents for NMR measurements. During purification, β-mercaptoethanol or dithiothreitol (DTT) can be added, while for stability reasons tris(2-carboxyethyl)phosphine hydrochloride (TCEP) is preferred for NMR samples, typically at final concentrations of 1–10 m*M* (Getz, Xiao, Chakrabarty, Cooke, & Selvin, 1999).

NMR data acquisition of IDPs is often conducted at 5–10°C to slow down amide exchange. However, be aware that temperature may affect the population of transient structures, such as a loss of polyproline structure with increasing temperatures (Kjaergaard et al., 2010), as well as chain expansion at lower temperatures (Wuttke et al., 2014).

4. THE ^{1}H,^{15}N-HSQC SPECTRUM—THE PROTEIN FINGERPRINT

Evaluation of different NMR sample conditions is often done with ^{1}H,^{15}N-HSQC spectra, which give rise to a correlation peak for each covalently bonded ^{1}H–^{15}N group (Fig. 2(1)). In the ideal situation, the ^{1}H,^{15}N-HSQC spectrum will contain one peak for each backbone amide (except those of prolines) and peaks for each side-chain amide or indole groups (and, under certain conditions, guanidino groups). Therefore, this spectrum is considered the protein "fingerprint." It will provide the first insights into the protein properties and the feasibility (or complexity) of the coming NMR studies on the sample. A reasonably good ^{1}H,^{15}N-HSQC spectrum can often be acquired in less than an hour, but if specific residue information is needed, the resonances have to be assigned. This is a time-consuming task for which strategies are available (see López-Méndez & Güntert, 2006; Sahu, Bastidas, & Showalter, 2014).

The required concentration of the isotope-labeled component for acquiring a ^{1}H,^{15}N-HSQC spectrum is typically 50–300 μ*M* for IDPs, depending on their properties. However, high-quality ^{1}H,^{15}N-HSQC spectra of an IDP in a complex have been obtained with concentrations down to 10 μ*M* (Borgia et al., 2018), while acquisition of 3D backbone assignment spectra usually requires concentrations above 100 μ*M*. The relatively high concentrations needed for NMR can influence the oligomerization state of IDPs (Danielsson et al., 2008), and it is therefore advisable to perform a concentration series with each complex component in isolation, to investigate potential self-association.

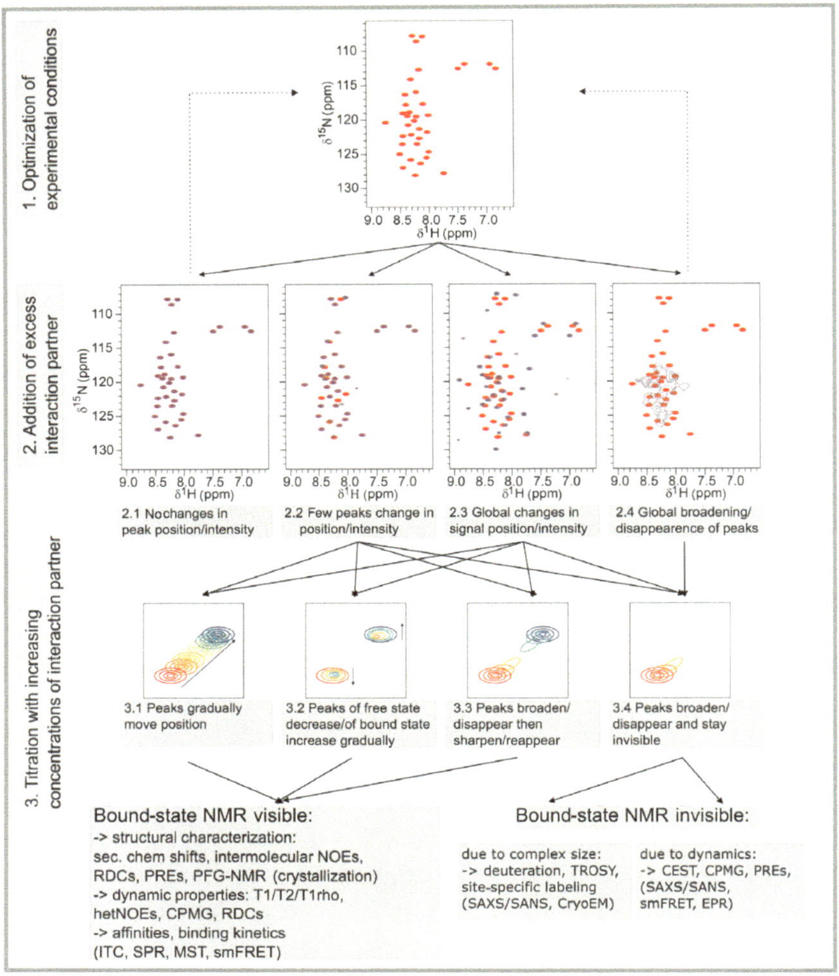

Fig. 2 Suggested workflow for the characterization of an IDP complex by NMR spectroscopy. (1) Schematic ^{1}H,^{15}N-HSQC spectrum of an ^{15}N-labeled IDP after optimization of the experimental conditions. (2) Comparison of schematic ^{1}H,^{15}N-HSQC spectra of the IDP in its free state (*red*) and in complex (*blue*) with excess of unlabeled interaction partner for four different binding modes. (3) Possible scenarios of signal changes upon titration of the ^{15}N-labeled IDP with increasing concentrations of the interaction partner (rainbow coloring: from *red* = no interaction partner to *blue* = saturated with interaction partner).

In cases of low proton dispersion and severe overlap in the ^{1}H,^{15}N-HSQC spectrum, or where the sequence has a high content of prolines, direct ^{13}C and ^{15}N detection methods are great alternatives, as ^{13}C and ^{15}N nuclei show better chemical shift dispersion than ^{1}H (Felli & Pierattelli, 2014; Gibbs & Kriwacki, 2018).

5. INTERPRETATION OF BINDING-INDUCED CHANGES IN THE $^{1}H,^{15}N$-HSQC SPECTRUM

A major strength of NMR spectroscopy is that it allows mapping of residue-specific effects of complex formation through relatively fast and simple experiments. Usually, this is achieved by the acquisition of $^{1}H,^{15}N$-HSQC spectra in the absence and presence of an interaction partner, simply requiring sufficient amounts of isotope-labeled and unlabeled forms of both partners. To prepare a sample of an IDP in the presence of the interaction partner, some considerations are needed: First, it is advisable to make sure the ^{15}N-labeled IDP and the unlabeled interaction partner are brought to the exact same buffer conditions by codialysis before mixing. Second, to avoid complications arising from exchange processes one should make sure that the ^{15}N-labeled IDP is saturated in the bound state, so the unlabeled interaction partner often needs to be added in excess. For a medium-affinity complex with a K_D of 25 μM, saturation will only be 60% at protein concentrations of 100 μM of each binding partner. Thus, if information on the stoichiometry and affinity are available, concentration and saturation should be considered. In some cases it may, however, be advantageous to record NMR data at substoichiometric amounts (see Sections 7.3 and 7.6).

In the resulting $^{1}H,^{15}N$-HSQC spectra, each peak is defined by its position (the resonance frequency or chemical shift), intensity (the concentration of each population), and width (inversely correlated to the transverse relaxation time), and perturbations of these parameters provide different information on the changes occurring upon complex formation. The position of a peak, i.e., the chemical shift in δ [ppm], is dependent on the local chemical environment, and CSPs often arise from binding of the unlabeled interaction partner, induction of secondary structure, or other ensemble changes. Importantly, from the CSPs alone it is not possible to determine the cause of the perturbations.

Once the complex with a partner in excess is ready for NMR acquisition, there are several possible outcomes; four illustrative cases are depicted schematically in Fig. 2 and described later.

5.1 No Changes in Signal Position/Intensity (Fig. 2(2.1))

If there are no or only subtle changes in the spectrum after addition of the interaction partner, there is either no interaction or the experimental conditions are not allowing for an efficient interaction and the buffer composition might have to be optimized. While a low pH is often ideal for analysis and assignment of the free IDP, it might prevent complex formation. If the interaction has been identified in, e.g., cellular studies, there is a risk that a crucial cofactor or a posttranslational modification is lacking in the in vitro NMR setup.

5.2 Few Signals Change Position and/or Intensity (Fig. 2(2.2))

If only a small fraction of the signals is affected by the addition of the interaction partner, the IDP probably interacts via a SLiM and the rest of the protein remains disordered in the complex. In cases where assignments are available one can often directly identify the region of interaction by mapping the chemical shift and intensity changes to the amino acid sequence of the IDP.

5.3 Global Changes in Signal Position/Intensity (Fig. 2(2.3))

Another possible mode of interaction is a large-scale folding of the IDP upon binding to its interaction partner. This often leads to global changes in the appearance of the $^{1}H,^{15}N$-HSQC spectra, typically resulting in a larger dispersion of the signals in the ^{1}H dimension, which is characteristic of residues adopting ordered structure. With these types of complexes it is in most cases feasible to apply the full arsenal of techniques available for characterization of interactions between two structured proteins that have been described and reviewed elsewhere (Aguirre, Cala, & Krimm, 2015; Hass & Ubbink, 2014).

5.4 Global Broadening or Disappearance of Signals (Fig. 2(2.4))

In NMR spectroscopy, signals can be broadened severely or disappear completely due to the enhanced relaxation in large macromolecular complexes or due to chemical exchange processes (see Section 6.4). It is therefore possible that the overall spectral quality after addition of the ligand is drastically decreased and if the interaction partner is a large globular molecule (>50 kDa), the disappearance of signals from directly interacting residues is indeed an expected outcome. Even when the interaction partner is small, there are circumstances where the spectral quality can decrease severely. This is often caused by chemical exchange between states on timescales similar to the difference in resonance frequencies of the states or because of oligomerization. Aggregation of the protein complex or formation of liquid/liquid phase separation can also be encountered and lead to low-quality NMR spectra that are difficult to interpret. It is then advisable to rescreen the experimental setup (see Section 4) to find conditions where chemical exchange/oligomerization/aggregation/phase separation can be avoided—if this is indeed not the focus of the study.

6. TITRATION WITH AN INTERACTION PARTNER

The $^{1}H,^{15}N$-HSQC spectrum of the fully bound state of the IDP can give information about the mode of interaction, but to obtain insight into the dynamical properties of the complex a titration with varying concentrations of interaction partner can be performed. To maximize the level of information, it is advisable to record several concentration ratios until no further changes in the NMR spectrum are detectable. The key to an interpretable $^{1}H,^{15}N$-HSQC titration series lies in careful sample preparation as described earlier, ensuring that the observable perturbations only stem from complex formation. Furthermore, for analyzing changes in peak intensities it is a prerequisite that the $^{1}H,^{15}N$-HSQC spectra have been acquired with the same NMR parameters and the exact same concentration of the isotope-labeled protein. A straightforward way to achieve the latter is to make a stock of the isotope-labeled protein, which is pipetted in the same volume to obtain different samples. Serial titrations can be designed in a way that requires only samples of the titration end points, which are subsequently and consecutively mixed to keep the concentration of isotope-labeled protein constant while varying the nonlabeled ligand concentration. For example, if sample A contains 200 μM ^{15}N–P1 and 0 μM ^{14}N–P2 (1:0) and sample C contains 200 μM ^{15}N–P1 and 400 μM ^{14}N–P2 (1:2), a sample B containing 200 μM ^{15}N–P1 and 200 μM ^{14}N–P2 (1:1) can be constructed by mixing 50% of sample A with 50% of sample C.

There are several aspects that decide the course of a titration: The size of the complex, the number and population of sampled bound conformations, and the exchange rate between these. A schematic depiction of some of the possible outcomes from this experiment is given in Fig. 2. A detailed theoretical description of the processes leading to these different scenarios has been provided elsewhere (Waudby, Ramos, Cabrita, & Christodoulou, 2016) and practical considerations have also been discussed (Teilum, Kunze, Erlendsson, & Kragelund, 2017). In the following the possible outcomes of the titrations will be explained along with ways to analyze the data.

6.1 Signals Move Gradually With Increasing Ligand Concentration (Fig. 2(3.1))

A gradual movement of peaks with increasing ligand concentration is characteristic of complexes with a fast exchange rate k_{ex} ($k_{ex} = k_{on}[L] + k_{off}$) between free and bound states compared to the difference in their resonance frequency $\Delta\omega$ in Hz ($k_{ex} \gg \Delta\omega$). This can be calculated from the chemical shift difference in ppm ($\Delta\delta$) and the operating frequency of the spectrometer ($\omega_{carrier}$) as $\Delta\omega = \Delta\delta^* \omega_{carrier}/10^6$. The peak position at each titration point is a population-weighted average of the respective chemical shifts of free and bound states. The CSPs ($\Delta\,\delta$) of ^{1}H, ^{15}N or their combination is then calculated as the distance given by the weighted Euclidian distance

$$\Delta\delta = \sqrt{(\Delta\delta\mathrm{H})^2 + (0.154*\Delta\delta\mathrm{N})^2} \quad (1)$$

where the scaling factor 0.154 has been experimentally determined (Mulder, Schipper, Bott, & Boelens, 1999). The CSPs may be used to evaluate the perturbation of each residue as well as plotting a binding curve allowing for determination of the dissociation constant, K_D (Teilum et al., 2017). Be aware that the magnitude of CSPs is not necessarily proportional to the significance of the effects they represent.

6.2 Signals of the Free State Gradually Disappear While Bound-State Signals Increase in Intensity (Fig. 2(3.2))

If the exchange rate is slow compared to the difference in resonance frequency ($k_{ex} \ll \Delta\omega$), the free and the bound states can be observed individually. In this case the signal positions do not change throughout the titration, but peaks of the free state will gradually decrease in intensity, while signals of the bound state get more intense with increasing ligand concentration. It is often necessary to perform a resonance assignment of the free as well as the bound state to map CSPs as relying on "minimum chemical shift procedures"—that is choosing the nearest peak after perturbation—can be highly inaccurate. The changes in peak intensities provided as the ratio between bound and free states (I_{bound}/I_{free}) in the slow exchange regime may in principle be used to quantify the binding. However, this is ill-advised as the peak intensity is sensitive to changes in dynamics (e.g., affected by system size) and movement into the intermediate exchange regime (see Teilum et al., 2017).

6.3 Signals Broaden/Disappear and Then Sharpen/Reappear With Increasing Ligand Concentrations (Fig. 2(3.3))

At exchange rates on the intermediate timescale ($k_{ex} \approx \Delta\omega$), peaks can be severely broadened or disappear beyond detection upon addition of substoichiometric amounts of ligand but reappear when saturation of the bound state is approached. In these cases it is difficult to extract information on the K_D of the complex, but it is sometimes possible to escape this exchange regime by altering the total protein concentration and/or the temperature (Teilum et al., 2017).

6.4 Signals Broaden/Disappear and Stay Invisible Even at High Ligand Concentrations (Fig. 2(3.4))

In the previous cases the fully saturated bound state gave rise to a single set of sharp signals, but this is, as mentioned, not always the case. If the size of the complex increases, its relaxation processes are accelerated, leading to severe peak broadening. In this case, deuteration of the proteins, site-specific labeling, and application of the transverse relaxation-optimized spectroscopy methodology might be necessary (Pervushin, Riek, Wider, & Wüthrich, 1997; Riek, Pervushin, & Wüthrich, 2000; Rosenzweig & Kay, 2014). Especially for IDPs, examples have been found where multiple conformations are sampled in the complex (Charlier et al., 2017). So even at saturation, exchange processes take place that can lead to peak broadening depending on the timescale of the exchange and the number and diversity of states sampled. In cases where the bound state is not directly accessible due to loss of resonances, there are possibilities to access information on the bound state under nonsaturated conditions (see later).

7. NMR METHODS TO STUDY DYNAMIC IDP COMPLEXES—AIMS, NEEDS, AND PITFALLS

In this section, different NMR experiments relevant to the characterization of dynamic IDP complexes are presented along with the expected outcomes, pitfalls, and examples, some of which are shown in Fig. 3.

7.1 Secondary Chemical Shift Changes

IDPs can adopt transiently populated secondary structures, which are abundant in interaction sites and can contribute to complex formation (Kragelj, Ozenne, Blackledge, & Jensen, 2013; Wright & Dyson, 2015). Thus, changes in transient secondary structures upon complex formation can be observed, contributing to changes in the ensemble upon binding (Mollica et al., 2016) (see also chapter "Spectroscopic characterization of structure–function relationships in the intrinsically disordered protein complexin" by Snead and Eliezer). The most direct NMR method to identify secondary structure is based on the secondary chemical shift (SCS) analysis, utilizing that chemical shifts of specific backbone nuclei correlate strongly with local backbone structure. In a SCS analysis, the position and populations of secondary structures can be derived by correlating the chemical shifts of the C^{α} and C' to random coil chemical shifts

$$\mathrm{SCS} = \Delta\delta = \delta(\mathrm{observed}) - \delta(\mathrm{random\ coil}) \tag{2}$$

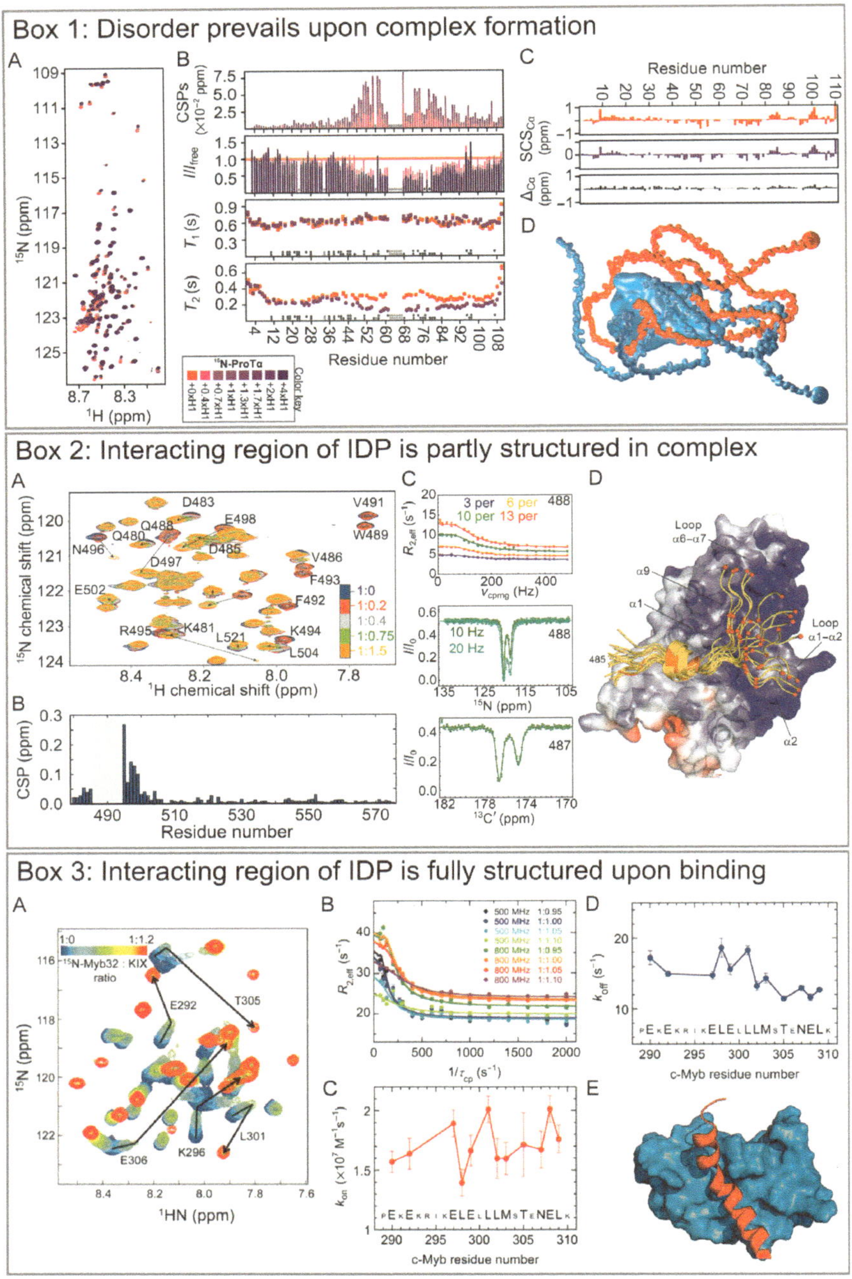

Fig. 3 Examples of IDP complexes studied by NMR. **Box 1:** Characterization of the interaction between the acidic and disordered prothymosin-α (ProTα) and the mostly disordered and highly positively charged linker Histone H1 (H1) (Borgia et al., 2018). (A) ^{1}H,^{15}N-HSQC spectrum of ^{15}N-ProTα titrated with zero- to fourfold molar addition

of H1 (*colors* correspond to key), revealing fast exchange on the NMR timescale. (B) Two *top plots*: CSPs and peak intensity ratios for assigned residues of ProTα induced by zero- to fourfold molar addition of H1 (*bar colors* correspond to key), showing perturbation of the NMR data in more than half of the ProTα chain. Third plot from the top: T_1 ^{15}N relaxation times of free (*red*) and H1-bound (*purple*) ^{15}N-ProTα, with $\langle T_1 \rangle = 610$ ms (free) and 636 ms (complex). *Bottom plot*: T_2 ^{15}N relaxation times of free (*red*) and H1-bound (*purple*) ^{15}N-ProTα, with $\langle T_2 \rangle = 302$ ms (free) and 217 ms (complex). *Light gray stars* highlight prolines and unassigned residues; *dark gray stars* highlight overlap and/or insufficient data quality. (C) C^{α} SCSs of ProTα free (*red*), in complex with H1 (*purple*), and their differences (*black*), showing the absence of induction of secondary structure upon complex formation. In both states, three transiently populated α-helices are present in ProTα: residues Ser9–Glu19 (~10% populated), Ala82–Thr86 (~13% populated), and Val99–Lys102 (~18% populated). (D) A structure from the ensemble of the ProTα-H1 complex, with ProTα in *red* and H1 in *blue* (Borgia et al., 2018). **Box 2:** Interaction between the disordered C-terminal region of Artemis ($Art^{480-575}$) and the DNA-binding domain (DBD) of Ligase IV (Charlier et al., 2017). (A) $^1H,^{15}N$-HSQC spectra of ^{15}N-labeled $Art^{480-575}$ upon titration with unlabeled DBD. (B) CSPs show that the signals of the C-terminal part are barely affected by addition of DBD, while the resonances of residues 485–499 comprising the interaction motif change position and are severely broadened, indicating conformational exchange in the bound state. (C) CPMG relaxation dispersion (*top panel*), and CEST experiments on amide ^{15}N (*middle panel*) and $^{13}C'$ (*lower panel*) performed with substoichiometric DBD present (3%–13%) providing information on the conformation of the bound state. (D) HADDOCK model generated from the CPMG, CEST, and CSP data on the folded DBD domain (not shown here): Residues 485–496 were identified to fold upon interaction forming a short α-helix, while the residues 497–501 remained flexible and engaged in electrostatic interactions between negatively charged residues of Artemis (D497, E498, and D501; *red dots*) and an extended positive surface patch on the DBD of Ligase IV, consisting of three helices (α1, α2, α9) and two loops (α1–α2, α6–α7). **Box 3:** Interaction between the disordered C-terminal region of c-Myb (c-$Myb_{284-315}$) and the KIX domain of CREB-binding protein (CREB-binding $protein_{586-672}$) (Arai, Sugase, Dyson, & Wright, 2015). (A) $^1H,^{15}N$-HSQC spectra of ^{15}N-labeled c-Myb titrated with unlabeled KIX domain. Ratios according to key. Resonances move in a nonlinear manner and within different exchange regimes (see Section 6), suggesting multiple binding sites with different affinities. (B) Representative CPMG relaxation dispersion curves for residue L301 of ^{15}N-labeled c-Myb upon titration scanned around equimolar amounts of unlabeled KIX domain at two field strengths. The concentration-dependent curvature of the dispersion curves suggests exchange between bound and unbound states. (C and D) Association rates (k_{on}) and dissociation rates (k_{off}) for residues in c-Myb showing substantial dispersions. Rates are determined from fitting of the dispersion curves for each residue independently to a two-state model. (E) NMR solution structure of the complex between c-$Myb_{291-315}$ and the KIX domain of CREB (PDB: 2SB0). ***Box 2:*** *Adapted and reprinted with permission from Charlier, C., Bouvignies, G., Pelupessy, P., Walrant, A., Marquant, R., Kozlov, M., et al. (2017). Structure and dynamics of an intrinsically disordered protein region that partially folds upon binding by chemical-exchange NMR.* Journal of the American Chemical Society, 139, *12219–12227. Copyright 2017 American Chemical Society.* ***Box 3(A)–(D):*** *Adapted and reprinted with permission from Arai, M., Sugase, K., Dyson, H. J., & Wright, P. E. (2015). Conformational propensities of intrinsically disordered proteins influence the mechanism of binding and folding.* Proceedings of the National Academy of Sciences of the United States of America, 112, *9614–9619.*

For SCSs derived from C^α and C' shifts, consecutive positive SCSs indicate helicity (≧4 residues), negative values β-strand or extended structure (≧3 residues), and values close to 0 indicate that the ensemble distributes randomly within the allowed regions in the Ramachandran plot (Wishart & Sykes, 1994; Wishart, Sykes, & Richards, 1992). An example of SCSs is provided in Fig. 3, Box 1C. The random coil chemical shifts can be the theoretical chemical shifts of the protein in the absence of structure, which varies for the individual residues depending on their sequential neighbors (Kjaergaard & Poulsen, 2011; Wishart, Bigam, Holm, Hodges, & Sykes, 1995) and the experimental conditions (Kjaergaard, Brander, & Poulsen, 2011). Several datasets are available as reference chemical shifts (Kjaergaard et al., 2011; Kjaergaard & Poulsen, 2011; Marsh, Singh, Jia, & Forman-Kay, 2006; Nielsen & Mulder, 2018; Tamiola, Acar, & Mulder, 2010); however, the transient nature of secondary structures in IDPs results in small SCSs and makes the analysis prone to bias. In addition, the effects of chemical modifications, which may affect populations of secondary structures (Hendus-Altenburger et al., 2017; Sibille et al., 2011), are not yet included in any reference library. One way around this is to use the chemical shifts of the same protein in urea as an internal intrinsic reference (Haxholm et al., 2015; Modig et al., 2007), eliminating library bias under the assumption that the ensemble in urea is a good random coil approximation.

To quantify the population of transient structures in an IDP, the SCSs of fully formed secondary structure elements are needed. However, this is not trivial, partly because the chemical shift difference between the random coil state and the structured state varies significantly between residue types (Kjaergaard & Poulsen, 2012; Zhang, Neal, & Wishart, 2003) and because tertiary contacts and sample conditions also influence the chemical shifts. One measure of helicity has been provided from an isolated helix for which the C^α shifts did not change compared to those of the intact protein. This gave a value of 3.1 as an approximation of the SCS for a fully formed α-helix (Spera & Bax, 1991; Weinstock, Narayanan, Baum, & Levy, 2008).

7.2 Relaxation Properties and Chain Dynamics

Different nuclear spin relaxation rates (R_x) or relaxation times ($T_x = 1/R_x$) are valuable reporters on local as well as global molecular motions. The quantum mechanical processes leading to NMR relaxation are beyond the scope of this chapter but are discussed elsewhere (see Cavanagh, Fairbrother, Palmer, Rance, & Skelton, 2007; Kay, Torchia, & Bax, 1989;

Kleckner & Foster, 2011; Palmer, 2016). Often, the longitudinal relaxation rates (R_1) characterizing the process by which magnetization returns to thermodynamic equilibrium is determined along with the transverse relaxation rates (R_2) describing the loss of spin coherence and the heteronuclear Overhauser effect (hetNOE) resulting from cross-relaxation between dipolar-coupled spins. While other nuclei can be used, the ^{15}N relaxation rates are most commonly investigated. To limit other relaxation pathways, relaxation rates must be acquired on NMR samples containing only uniformly ^{15}N-labeled protein. Due to the slow relaxation rates of IDPs, care should be taken to ensure that equilibrium is reached between consecutive scans by using sufficiently long recycle delays. This is especially important for the hetNOE experiments where relaxation delays of up to 20 s may be needed to ensure complete relaxation of water magnetization (Renner, Schleicher, Moroder, & Holak, 2002).

To determine R_1 and R_2 values, a series of $^1H,^{15}N$-HSQC experiments with varying relaxation delays are recorded. These delays are typically in the range of 20–1500 ms for R_1 and 20–500 ms for R_2, depending on the properties of the complex under study. The change in signal amplitude for each amide fitted to a single exponential yielding the relaxation parameter (R_1 or R_2)

$$I = I_0 e^{-R_x t} \tag{3}$$

HetNOEs are obtained by acquisition of two $^1H,^{15}N$-HSQC spectra with and without proton presaturation (Kay et al., 1989). The hetNOEs are then calculated for each amide from the peak amplitude ratio according to

$$\mathrm{NOE} = \frac{I_{\mathrm{sat}}}{I_{\mathrm{ref}}} \tag{4}$$

where I_{sat} and I_{ref} are the peak intensities in the saturated and reference spectra, respectively. Typically, hetNOEs are negative for highly disordered chains (Pierce et al., 2016) but can be positive in regions with transient secondary structures.

After measuring the relaxation rates of the IDP in free and complexed states, ΔR_2, ΔR_1, $\Delta(R_2/R_1)$, and ΔhetNOE are typically plotted on a per residue basis to pinpoint regions with altered dynamics (Fig. 3, Box 1B) (Bah et al., 2015; Borgia et al., 2018; Milles et al., 2015). Interpreting these changes is complex as they depend in different ways on internal and overall dynamics, but typically R_2s and hetNOEs can be used to pinpoint potential binding sites because binding often leads to decreased molecular motions for residues involved in the interaction. This is especially true when secondary structure is induced (Hendus-Altenburger et al., 2016) since R_2s and het-NOEs correlate inversely with the molecular tumbling rate. Chemical exchange between bound and free states or different bound states also contributes to R_2 on the ms timescale (Hansen, Vallurupalli, & Kay, 2008; Kleckner & Foster, 2011), which is exploited in relaxation dispersion-type experiments (see later).

For ordered proteins, the model-free formalism is often used to extract more tangible meaning of the relaxation parameters through the order parameter S^2 (Lipari & Szabo, 1982). However, this type of analysis requires a number of assumptions only valid for globular proteins such as a uniform overall correlation time, thus making it inapplicable for IDPs (Dyson, 2011). An extended protocol may, however, be used in some cases as exemplified in Gill, Byrd, and Palmer (2016).

7.3 Relaxation Dispersion Experiments

Many proteins interconvert between different conformational states with important functional implications. This also applies to IDPs in complexes where multiple bound states may coexist. Despite their importance, these "excited states" are often lowly populated and therefore difficult to investigate. Relaxation dispersion experiments such as the Carr–Purcell–Meiboom–Gill (CPMG) experiment (Hansen et al., 2008; Kovrigin, Kempf, Grey, & Loria, 2006) can be used to probe conformational transitions that occur on the ms timescale with k_{ex} values ranging from 100 to $3000\,s^{-1}$. For faster exchange processes one can alternatively measure rotating frame relaxation dispersion, $T_{1\rho}$ (sometimes referred to as the inverse value $R_{1\rho}$), which can probe transitions with k_{ex} ranging from 10,000 to $50{,}000\,s^{-1}$. Depending on the model used in the analysis, relaxation dispersion data may provide exchange rates, populations, and the difference in chemical shifts between ground and excited states, and thus information on thermodynamics and kinetics. Binding can induce changes in R_{eff} (Ban, Iconaru, Ramanathan, Zuo, & Kriwacki, 2017), and on and off rates for the formation of IDP complexes can be probed by titrating sub-stoichiometric concentrations of an unlabeled partner into an ^{15}N-labeled IDP while measuring relaxation dispersion (Fig. 3, Box 3B–D). From such data, it is possible to gain insights into the binding mechanism, revealing whether the interaction occurs via intermediates (Arai et al., 2015; Sugase, Dyson, & Wright, 2007).

Similar to other nuclear spin relaxation rates, backbone amide ^{15}N relaxation dispersions are probed using a series of $^{1}H,^{15}N$-HSQC-like experiments with different CPMG pulse frequencies or spin-lock frequencies for CPMG and $T_{1\rho}$ experiments, respectively. Peak amplitudes are then extracted and used to calculate the effective transverse relaxation rate ($R_{2,\text{eff}}$) or $T_{1\rho}$ which is plotted against the CPMG pulse frequency or, for $T_{1\rho}$, the spin-lock frequency. In cases where chemical exchange occurs on the μs–ms timescale within the exchange range covered by the experiment, $R_{2,\text{eff}}$ or $T_{1\rho}$ will vary with the frequency (Fig. 3, Box 3B), whereas it will be constant if no exchange occurs. Many programs such as relax and GLOVE have been developed for the analysis of relaxation dispersion data and are freely available for download (Morin et al., 2014; Sugase, Konuma, Lansing, & Wright, 2013). These programs include a wide set of two- and three-state models for different timescales that can be tested. A careful statistical analysis is needed to ensure that the model represents the best possible fit to the data. Furthermore, due to parameter correlations it is recommended to record dispersion data at different static magnetic fields (Kovrigin et al., 2006). Other nuclei than ^{15}N may be probed and isotope-labeling schemes for specific focus on side-chain dynamics exist (Lundström, Vallurupalli, Hansen, & Kay, 2009).

7.4 Paramagnetic Relaxation Enhancements

Transient long-range interactions in an IDP complex can be probed by paramagnetic relaxation enhancements (PREs). This is achieved by chemically attaching a paramagnetic spinlabel at a defined position, typically via a thiol group of naturally occurring or introduced Cys residues. The spinlabel, e.g., a nitroxide moiety like TEMPO or MTSL, lanthanide tags, or caged paramagnetic ions (Clore & Iwahara, 2009; Göbl et al., 2016; Otting, 2010; Schilder et al., 2016; Teilum, Kragelund, & Poulsen, 2002; Yang, Wang, Pan, & Su, 2016), possesses unpaired electrons, resulting in magnetic dipolar interactions that increase the transverse relaxation rates R_2 of adjacent nuclei (Otting, 2010). This increase depends on the inverse sixth power of the average distance between the unpaired electron and the observed nucleus. Due to the large magnetic moment of unpaired electrons, distances up to 20–25 Å typically can be probed, depending on the label.

A potential pitfall for PRE measurements is the selection of the labeling site, which should not interfere with the interaction or change the ensemble. This can be tested by comparison of the $^{1}H,^{15}N$-HSQC spectrum of the wild type to that of the variant with the spinlabel in both the paramagnetic and diamagnetic states, the latter of which can be introduced via reduction by, e.g., ascorbic acid.

After spinlabeling, signals from residues surrounding the labeling site typically disappear or get considerably broadened, whereas effects on signals at sequentially distant sites on the same protein or on the interaction partner suggest long-range interactions. For analyzing dimerization or interactions with a partner in a complex, the label is attached to the ^{14}N-labeled protein, while the intensities (or relaxation rates) are recorded on the ^{15}N-labeled partner protein in the mixture of the two (Danielsson et al., 2008). The PRE data may then be used in MD simulations to calculate distances and conformational ensembles (Battiste & Wagner, 2000; Lindorff-Larsen et al., 2004). More information on the PRE method (Clore & Iwahara, 2009; Otting, 2010) and examples (Clore, 2013; Danielsson et al., 2008) can be found in the literature.

7.5 Diffusion Coefficients

Diffusion experiments can be used to obtain information about the overall size of a complex in terms of the radius of hydration, R_h, which for IDPs is especially useful as it can assess the compactness of free and bound states. Since diffusion constants are population averaged, diffusion measurements can also be used to determine the K_D by measuring diffusion constants at different ligand concentrations (Danielsson, Jarvet, Damberg, & Gräslund, 2004). This is highly relevant in cases where CSP-based methods are impeded by line broadening beyond detection (Fig. 2(3.4)). In order to measure diffusion on free and bound states of the IDP, isotope labeling is combined with the use of filtered pulse schemes to avoid signals from the unlabeled binding partner (Tillett, Horsfield, Lian, & Norwood, 1999).

Certain pitfalls need consideration when setting up the experiment. Diffusion coefficients depend heavily on solvent viscosity and are therefore sensitive toward solvent conditions and temperature. For example, the addition of D_2O, a common additive in NMR samples, needs to be corrected for due to the viscosity differences between H_2O and D_2O (Danielsson, Jarvet, Damberg, & Gräslund, 2002). It is also necessary to calibrate the temperature in the NMR probe to ensure agreement between set and measured temperatures. Finally, it is important to calibrate the gradient strength, which varies for different spectrometers. These pitfalls can be avoided by including an internal diffusion reference such as 1,4-dioxane with a known hydrodynamic radius of 2.12 Å in the sample (Wilkins et al., 1999).

The typical outcome of a pulsed-field gradient diffusion experiment is a series of spectra recorded at varying gradient strengths. Signal amplitudes are then extracted, and the translational diffusion coefficients (D) obtained by fitting the signal decay to the Stejskal–Tanner equation (Eq. 5)

$$I = I_0 e^{-g^2\gamma^2\delta^2\left(\Delta - \frac{\delta}{3}\right)D} \tag{5}$$

where g is the gradient strength, γ is the gyromagnetic ratio, δ is the length of the gradient, and Δ is the diffusion time. If an internal reference is included, the hydrodynamic radius R_h can be calculated from the simple relation

$$R_h(\text{protein}) = \frac{R_h(\text{reference})}{D(\text{reference})} \cdot D(\text{protein}) \tag{6}$$

Otherwise R_h can be calculated from the Stokes–Einstein relation (Eq. 7)

$$R_h = \frac{k_B T}{6\pi\eta D} \tag{7}$$

where k_B is the Boltzmann constant, T the temperature, and η the solvent viscosity.

Diffusion constants can be plotted against chemical shifts, commonly referred to as a diffusion-ordered spectroscopy (DOSY) spectrum (Ribeiro et al., 2010; Stilbs, 1981), which can in principle reveal different states of an IDP complex. However, this requires that the exchange between the states is slow (see Section 6) and that the diffusion rates of the different states are sufficiently different.

7.6 Saturation Transfer Experiments

Saturation transfer difference NMR (STD-NMR) is another useful method to quantitatively analyze protein interactions by NMR. It requires that the complex is in fast exchange on the NMR timescale, excess of the smaller partner (typically 100-fold), and that the complex consists of a larger protein and a smaller molecule (e.g., peptides derived from a longer IDP). Selective pulses are used to saturate proton resonances of the larger protein only, and the magnetization is then transferred to protons of the smaller molecule that are proximal to the saturated protons of the larger protein (Furukawa, Konuma, Yanaka, & Sugase, 2016; Viegas, Manso, Nobrega, & Cabrita, 2011). Practically, in an STD experiment, two spectra are recorded on the complex of interest, which are then subtracted: one in which the bigger protein is selectively saturated (on-resonance spectrum) and the other recorded without saturation (off-resonance spectrum). When the on-resonance signal intensities are subtracted from those of the off-resonance spectrum, only the signals from the smaller protein that received saturation transfer from the bigger protein through the NOE effect will remain (Viegas et al., 2011). For saturation transfer to occur, distances need to be 5 Å or less, with more intense signals at closer proximity (Erlendsson et al., 2014; Pfaff et al., 2013; Viegas et al., 2011).

Chemical exchange saturation transfer (CEST) and dark-state exchange saturation transfer (DEST) are useful techniques for studying sparsely populated and/or transiently formed complexes. The principle of both experiments is to study a state of the protein that is NMR invisible due to low population and/or dynamics through its slow exchange with a major, visible state (for further details, see Fawzi, Ying, Torchia, & Clore, 2012; Vallurupalli, Bouvignies, & Kay, 2012; Vallurupalli, Sekhar, Yuwen, & Kay, 2017; Yuwen, Kay, & Bouvignies, 2018). While CEST utilizes the difference in chemical shift between the exchanging states, DEST distinguishes between states based on differences in transverse relaxation rates (R_2) (Fawzi, Ying, Ghirlando, Torchia, & Clore, 2011; Fawzi et al., 2012), and is particularly useful when a bound state becomes NMR invisible due to association with a large entity. Analysis of CEST and DEST profiles allows extraction of the minor-state population, lifetime, chemical shifts, and line widths, which has been exploited in several studies of complexes involving IDPs (Fig. 3, Box 2C) (Charlier et al., 2017; Delaforge et al., 2018; Fawzi et al., 2011).

7.7 Residual Dipolar Couplings

Residual dipolar couplings (RDCs) provide a multitude of information ranging from long-range orientational constraints to dynamic properties on the μs–ms timescale. For IDPs that are poorly suited for characterization based on NOEs, RDCs can be used to generate ensemble descriptions (Jensen et al., 2008, 2009; Milles et al., 2015) and its combination with PREs can be particularly useful (Salmon et al., 2010). Since the angular constraints extracted from RDCs are not distance dependent, relative orientations of domains and proteins in macromolecular complexes can be determined by RDCs (Berlin, O'Leary, & Fushman, 2010). To record RDCs the molecule or complex of interest needs to be present in an anisotropic environment often achieved using filamentous phages, stretched polyacrylamide gels, or liquid crystalline phases (Jain, 2009). The preparation of these alignment media can be troublesome, and it is essential that the molecules under investigation do not engage in strong interactions with the alignment media. If possible, multiple alignment media should be used, enabling a detailed analysis of the dynamic properties of the molecule (Tolman, 2002). The RDCs are obtained from a comparison of selectively non-decoupled multidimensional NMR spectra in the absence (only scalar couplings observable) and presence (scalar and dipolar couplings observable) of the alignment medium. Subtracting one from the other will provide the size as well as the sign of the RDC. While the RDCs between amide ^{1}H,^{15}N are most widely used, there is a multitude of experiments available for different pairs of coupled nuclei (Prestegard, Al-Hashimi, & Tolman, 2000).

8. CONCLUSIONS AND OUTLOOK

As outlined in this chapter, NMR spectroscopy offers a wide range of experiments allowing residue-specific characterization of interactions, distances, and dynamics in IDP complexes. As recent work has highlighted, the characteristic dynamics of IDPs may be conserved in their complexes (Bah et al., 2015; Borgia et al., 2018; Charlier et al., 2017; Delaforge et al., 2018; Milles et al., 2015; Sparks, Temel, Rout, & Cowburn, 2018), requiring increased focus on methodological developments that will serve to capture the characteristics of dynamic interactions. Furthermore, interaction determinants between proteins are not always related to specific residues, but may rather involve sets of residues, leading to mean field-type complexes relying on overall electrostatics (Borgia et al., 2018). Similarly, IDPs governed by low-complexity, repetitive sequences are capable of liquid/liquid-phase separation, leading to condensate formation. Their studies by NMR spectroscopy are still sparse, but CSPs, intensity changes, PREs, NOEs, diffusion, and dynamics have all been used to gain insight into their liquid-state behavior (Ambadipudi, Biernat, Riedel, Mandelkow, & Zweckstetter, 2017; Lin, Song, Forman-Kay, & Chan, 2017; Ryan et al., 2018).

Although NMR spectroscopy is powerful on its own, the efforts toward understanding the details of multivalent, fuzzy, or even fully disordered IDP complexes introduce significant technical challenges and ambiguities, making an integrative structural biology approach necessary. Here, NMR spectroscopy can be combined with complimentary methods such as SAXS, simulations, and single-molecule spectroscopies (Borgia et al., 2018; Milles et al., 2015), and a variety of techniques that can either support or complement the information obtained from NMR spectroscopy exist. In Table 1, strategies to answer a specific question in the NMR study of a

Table 1 Methods for Studying IDP Complexes

Question	NMR Methods	Complementary Methods
What is the affinity of my complex?	CSPs (low affinity) Line-shape analysis (high affinity) Pulsed-field gradient NMR (Danielsson et al., 2008)	Isothermal titration calorimetry (ITC) (O'Shea et al., 2017) Fluorescence methods Single-molecule Förster resonance energy transfer (smFRET) (Borgia et al., 2018) MicroScale Thermophoresis (MST) (Mysling et al., 2016)
What are the kinetic properties of my complex?	Line-shape analysis (Waudby et al., 2016) Carr–Purcell–Meiboom–Gill (CPMG) (Sugase et al., 2007)	Stopped-flow techniques (Rogers et al., 2013) Surface plasmon resonance (SPR) (Lacy et al., 2004)
What is the binding interface of my complex?	CSPs (Hendus-Altenburger et al., 2016) Line-shape analysis Longitudinal and transverse relaxation (R_1/R_2) (Hendus-Altenburger et al., 2016) Saturation transfer difference (STD) (Pfaff et al., 2013) Hydrogen–deuterium exchange (HDX) Nuclear Overhauser effect (NOE) Chemical exchange saturation transfer (CEST) Dark-state exchange saturation transfer (DEST) (Fawzi et al., 2011)	Truncation studies (O'Shea et al., 2017) Mutagenesis (Bugge et al., 2018) Crystallography (De Biasio et al., 2012) Hydrogen–deuterium exchange mass spectroscopy (HDX MS) (Keppel, Howard, & Weis, 2011) Chemical cross-linking MS (Pfaff et al., 2013)

How dynamic is my complex?	Longitudinal and transverse relaxation (R_1/R_2) (Borgia et al., 2018) Heteronuclear NOEs (HetNOE) (Mittag et al., 2008) Paramagnetic relaxation enhancement (PRE) Residual dipolar couplings (RDCs) Chemical exchange saturation transfer (CEST) (Charlier et al., 2017) Dark-state exchange saturation transfer (DEST) (Fawzi et al., 2011) Hydrogen–deuterium exchange	Electron paramagnetic resonance (EPR) Single-molecule methods (Borgia et al., 2018) Hydrogen–deuterium exchange-Mass spectroscopy (HDX MS)
Does my protein fold upon binding?	Amide proton chemical shift dispersion Nuclear Overhauser effect (NOE) Secondary chemical shifts (O'Shea et al., 2017)	Circular dichroism (CD) spectroscopy Crystallography (Gulbis et al., 1996) Electron paramagnetic resonance (EPR)
Does it interconvert between different bound conformations? What is the structural ensemble?	Nuclear Overhauser effect (lack of) (Landrieu et al., 2015) Paramagnetic relaxation enhancement (PRE) (Brzovic et al., 2011) Carr–Purcell–Meiboom–Gill (CPMG) (Delaforge et al., 2018) $R_1\rho$ (Delaforge et al., 2018) Chemical exchange saturation transfer (CEST) (Delaforge et al., 2018)	Single-molecule methods (Borgia et al., 2018) Electron paramagnetic resonance (EPR) Computation—coarse grained (Borgia et al., 2018) Small-Angle X-ray Scattering (SAXS)/Small-Angle Neutron Scattering (SANS) (Arbesú et al., 2017; Sparks et al., 2018)

Continued

Table 1 Methods for Studying IDP Complexes—cont'd

Question	NMR Methods	Complementary Methods
	Dark-state exchange saturation transfer (DEST) (Fawzi et al., 2011)	
How extended/compact is my complex?	Pulsed-field gradient NMR (Borgia et al., 2018) Paramagnetic relaxation enhancement (PRE) (Dedmon, Lindorff-Larsen, Christodoulou, Vendruscolo, & Dobson, 2005)	SAXS/SANS Size-exclusion chromatography (SEC) Dynamic light scattering (DLS)
What is the structure of my complex?	Nuclear Overhauser effect (Brzovic et al., 2011) HADDOCK—use of NMR data (Landrieu et al., 2015) Computation—use of NMR data Secondary chemical shifts (Hendus-Altenburger et al., 2016)	Crystallography (De Biasio et al., 2015) HADDOCK—use of other data (Bugge et al., 2018) Cryo-electron microscopy

disordered IDP complex are provided, along with suggestions for complementary methods and, whenever possible, examples of the application to an IDP complex.

Finally, we hope to have conveyed in this chapter that the multifaceted use of NMR spectroscopy makes it at the same time a crucial, collaborative, and explorative technique for the studies of dynamic IDP complexes.

ACKNOWLEDGMENTS

We are grateful to all the questions raised by the students at *SBiNLab* struggling with IDP complexes and for their comments. This work was supported by grants from the Danish Research Councils (#4181-00344) and the Novo Nordisk Foundation (to B.B.K.).

REFERENCES

Aguirre, C., Cala, O., & Krimm, I. (2015). Overview of probing protein-ligand interactions using NMR. In *Current protocols in protein science* (pp. 17.18.1–17.18.24). Hoboken, NJ, USA: John Wiley & Sons, Inc.

Ambadipudi, S., Biernat, J., Riedel, D., Mandelkow, E., & Zweckstetter, M. (2017). Liquid-liquid phase separation of the microtubule-binding repeats of the Alzheimer-related protein Tau. *Nature Communications*, *8*, 275.

Anthis, N. J., & Clore, G. M. (2013). Sequence-specific determination of protein and peptide concentrations by absorbance at 205 nm. *Protein Science*, *22*, 851–858.

Arai, M., Sugase, K., Dyson, H. J., & Wright, P. E. (2015). Conformational propensities of intrinsically disordered proteins influence the mechanism of binding and folding. *Proceedings of the National Academy of Sciences of the United States of America*, *112*, 9614–9619.

Arbesú, M., Maffei, M., Cordeiro, T. N., Teixeira, J. M. C., Pérez, Y., Bernadó, P., et al. (2017). The unique domain forms a fuzzy intramolecular complex in Src family kinases. *Structure*, *25*, 630–640.e4.

Bah, A., Vernon, R. M., Siddiqui, Z., Krzeminski, M., Muhandiram, R., Zhao, C., et al. (2015). Folding of an intrinsically disordered protein by phosphorylation as a regulatory switch. *Nature*, *519*, 106–109.

Ban, D., Iconaru, L. I., Ramanathan, A., Zuo, J., & Kriwacki, R. W. (2017). A small molecule causes a population shift in the conformational landscape of an intrinsically disordered protein. *Journal of the American Chemical Society*, *139*, 13692–13700.

Battiste, J. L., & Wagner, G. (2000). Utilization of site-directed spin labeling and high-resolution heteronuclear nuclear magnetic resonance for global fold determination of large proteins with limited nuclear overhauser effect data. *Biochemistry*, *39*, 5355–5365.

Berlin, K., O'Leary, D. P., & Fushman, D. (2010). Structural assembly of molecular complexes based on residual dipolar couplings. *Journal of the American Chemical Society*, *132*, 8961–8972.

Bonvin, A. M. J. J., Boelens, R., & Kaptein, R. (2005). NMR analysis of protein interactions. *Current Opinion in Chemical Biology*, *9*, 501–508.

Borgia, A., Borgia, M. B., Bugge, K., Kissling, V. M., Heidarsson, P. O., Fernandes, C. B., et al. (2018). Extreme disorder in an ultrahigh-affinity protein complex. *Nature*, *555*, 61–66.

Brutscher, B., Felli, I. C., Gil-Caballero, S., Hošek, T., Kümmerle, R., Piai, A., et al. (2015). NMR methods for the study of instrinsically disordered proteins structure, dynamics, and interactions: General overview and practical guidelines. *Advances in Experimental Medicine and Biology*, *870*, 49–122.

Brzovic, P. S., Heikaus, C. C., Kisselev, L., Vernon, R., Herbig, E., Pacheco, D., et al. (2011). The acidic transcription activator Gcn4 binds the mediator subunit Gal11/Med15 using a simple protein interface forming a fuzzy complex. *Molecular Cell*, *44*, 942–953.

Bugge, K., Staby, L., Kemplen, K. R., O'Shea, C., Bendsen, S. K., Jensen, M. K., et al. (2018). Structure of radical-induced cell death1 hub domain reveals a common αα-scaffold for disorder in transcriptional networks. *Structure*, *26*, 734–746.e7.

Cavanagh, J., Fairbrother, W., Palmer, A., III, Rance, M., & Skelton, N. (2007). *Protein NMR spectroscopy. Principles and practice.* Academic Press.

Charlier, C., Bouvignies, G., Pelupessy, P., Walrant, A., Marquant, R., Kozlov, M., et al. (2017). Structure and dynamics of an intrinsically disordered protein region that partially folds upon binding by chemical-exchange NMR. *Journal of the American Chemical Society*, *139*, 12219–12227.

Clore, G. M. (2013). Generating accurate contact maps of transient long-range interactions in intrinsically disordered proteins by paramagnetic relaxation enhancement. *Biophysical Journal*, *104*, 1635–1636.

Clore, G. M., & Iwahara, J. (2009). Theory, practice, and applications of paramagnetic relaxation enhancement for the characterization of transient low-population states of biological macromolecules and their complexes. *Chemical Reviews*, *109*, 4108–4139.

Danielsson, J., Jarvet, J., Damberg, P., & Gräslund, A. (2002). Translational diffusion measured by PFG-NMR on full length and fragments of the Alzheimer Aβ(1-40) peptide. Determination of hydrodynamic radii of random coil peptides of varying length. *Magnetic Resonance in Chemistry*, *40*, S89–S97.

Danielsson, J., Jarvet, J., Damberg, P., & Gräslund, A. (2004). Two-site binding of beta-cyclodextrin to the Alzheimer Abeta(1-40) peptide measured with combined PFG-NMR diffusion and induced chemical shifts. *Biochemistry*, *43*, 6261–6269.

Danielsson, J., Liljedahl, L., Bárány-Wallje, E., Sønderby, P., Kristensen, L. H., Martinez-Yamout, M. A., et al. (2008). The intrinsically disordered RNR inhibitor Sml1 is a dynamic dimer. *Biochemistry*, *47*, 13428–13437.

Davey, N. E., Cyert, M. S., & Moses, A. M. (2015). Short linear motifs—Ex nihilo evolution of protein regulation. *Cell Communication and Signaling: CCS*, *13*, 43.

De Biasio, A., Campos-Olivas, R., Sánchez, R., López-Alonso, J. P., Pantoja-Uceda, D., Merino, N., et al. (2012). Proliferating cell nuclear antigen (PCNA) interactions in solution studied by NMR. *PLoS One*, *7*, e48390.

De Biasio, A., de Opakua, A. I., Mortuza, G. B., Molina, R., Cordeiro, T. N., Castillo, F., et al. (2015). Structure of p15(PAF)-PCNA complex and implications for clamp sliding during DNA replication and repair. *Nature Communications*, *6*, 6439.

Dedmon, M. M., Lindorff-Larsen, K., Christodoulou, J., Vendruscolo, M., & Dobson, C. M. (2005). Mapping long-range interactions in alpha-synuclein using spin-label NMR and ensemble molecular dynamics simulations. *Journal of the American Chemical Society*, *127*, 476–477.

Delaforge, E., Kragelj, J., Tengo, L., Palencia, A., Milles, S., Bouvignies, G., et al. (2018). Deciphering the dynamic interaction profile of an intrinsically disordered protein by NMR exchange spectroscopy. *Journal of the American Chemical Society*, *140*, 1148–1158.

Dinkel, H., Van Roey, K., Michael, S., Davey, N. E., Weatheritt, R. J., Born, D., et al. (2014). The eukaryotic linear motif resource ELM: 10 years and counting. *Nucleic Acids Research*, *42*, D259–D266.

Dosztányi, Z., Csizmok, V., Tompa, P., & Simon, I. (2005). IUPred: Web server for the prediction of intrinsically unstructured regions of proteins based on estimated energy content. *Bioinformatics*, *21*, 3433–3434.

Dyson, H. J. (2011). Expanding the proteome: Disordered and alternatively folded proteins. *Quarterly Reviews of Biophysics*, *44*, 467–518.

Erlendsson, S., Rathje, M., Heidarsson, P. O., Poulsen, F. M., Madsen, K. L., Teilum, K., et al. (2014). Protein interacting with C-kinase 1 (PICK1) binding promiscuity relies on unconventional PSD-95/discs-large/ZO-1 homology (PDZ) binding modes for non-class II PDZ ligands. *The Journal of Biological Chemistry*, *289*, 25327–25340.

Fawzi, N. L., Ying, J., Ghirlando, R., Torchia, D. A., & Clore, G. M. (2011). Atomic-resolution dynamics on the surface of amyloid-β protofibrils probed by solution NMR. *Nature*, *480*, 268–272.

Fawzi, N. L., Ying, J., Torchia, D. A., & Clore, G. M. (2012). Probing exchange kinetics and atomic resolution dynamics in high-molecular-weight complexes using dark-state exchange saturation transfer NMR spectroscopy. *Nature Protocols*, 7, 1523–1533.

Felli, I. C., & Pierattelli, R. (2014). Novel methods based on 13C detection to study intrinsically disordered proteins. *Journal of Magnetic Resonance*, *241*, 115–125.

Fung, H. Y. J., Birol, M., & Rhoades, E. (2018). IDPs in macromolecular complexes: The roles of multivalent interactions in diverse assemblies. *Current Opinion in Structural Biology*, *49*, 36–43.

Furukawa, A., Konuma, T., Yanaka, S., & Sugase, K. (2016). Quantitative analysis of protein-ligand interactions by NMR. *Progress in Nuclear Magnetic Resonance Spectroscopy*, *96*, 47–57.

Getz, E. B., Xiao, M., Chakrabarty, T., Cooke, R., & Selvin, P. R. (1999). A comparison between the sulfhydryl reductants tris(2-carboxyethyl)phosphine and dithiothreitol for use in protein biochemistry. *Analytical Biochemistry*, *273*, 73–80.

Gibbs, E. B., & Kriwacki, R. W. (2018). Direct detection of carbon and nitrogen nuclei for high-resolution analysis of intrinsically disordered proteins using NMR spectroscopy. *Methods*, *138–139*, 39–46.

Gill, M. L., Byrd, R. A., & Palmer, A. G., III. (2016). Dynamics of GCN4 facilitate DNA interaction: A model-free analysis of an intrinsically disordered region. *Physical Chemistry Chemical Physics*, *18*, 5839–5849.

Göbl, C., Resch, M., Strickland, M., Hartlmüller, C., Viertler, M., Tjandra, N., et al. (2016). Increasing the chemical-shift dispersion of unstructured proteins with a covalent lanthanide shift reagent. *Angewandte Chemie International Edition*, *55*, 14847–14851.

Gógl, G., Alexa, A., Kiss, B., Katona, G., Kovács, M., Bodor, A., et al. (2016). Structural basis of ribosomal S6 kinase 1 (RSK1) inhibition by S100B protein: Modulation of the extracellular signal-regulated kinase (ERK) signaling cascade in a calcium-dependent way. *The Journal of Biological Chemistry*, *291*, 11–27.

Gronenborn, A. M., & Clore, G. M. (1995). Structures of protein complexes by multidimensional heteronuclear magnetic resonance spectroscopy. *Critical Reviews in Biochemistry and Molecular Biology*, *30*, 351–385.

Gu, J., & Hilser, V. J. (2008). Predicting the energetics of conformational fluctuations in proteins from sequence: A strategy for profiling the proteome. *Structure*, *16*, 1627–1637.

Gulbis, J. M., Kelman, Z., Hurwitz, J., O'Donnell, M., & Kuriyan, J. (1996). Structure of the C-terminal region of p21(WAF1/CIP1) complexed with human PCNA. *Cell*, *87*, 297–306.

Hansen, D. F., Vallurupalli, P., & Kay, L. E. (2008). Using relaxation dispersion NMR spectroscopy to determine structures of excited, invisible protein states. *Journal of Biomolecular NMR*, *41*, 113–120.

Hass, M. A., & Ubbink, M. (2014). Structure determination of protein–protein complexes with long-range anisotropic paramagnetic NMR restraints. *Current Opinion in Structural Biology*, *24*, 45–53.

Haxholm, G. W., Nikolajsen, L. F., Olsen, J. G., Fredsted, J., Larsen, F. H., Goffin, V., et al. (2015). Intrinsically disordered cytoplasmic domains of two cytokine receptors mediate conserved interactions with membranes. *The Biochemical Journal*, *468*, 495–506.

Hendus-Altenburger, R., Lambrughi, M., Terkelsen, T., Pedersen, S. F., Papaleo, E., Lindorff-Larsen, K., et al. (2017). A phosphorylation-motif for tuneable helix stabilisation in intrinsically disordered proteins—Lessons from the sodium proton exchanger 1 (NHE1). *Cellular Signalling*, *37*, 40–51.

Hendus-Altenburger, R., Pedraz-Cuesta, E., Olesen, C. W., Papaleo, E., Schnell, J. A., Hopper, J. T. S., et al. (2016). The human Na(+)/H(+) exchanger 1 is a membrane scaffold protein for extracellular signal-regulated kinase 2. *BMC Biology*, *14*, 31.

Jain, N. U. (2009). Use of residual dipolar couplings in structural analysis of protein-ligand complexes by solution NMR spectroscopy. *Methods in Molecular Biology*, *544*, 231–252.

Jensen, M. R., Houben, K., Lescop, E., Blanchard, L., Ruigrok, R. W. H., & Blackledge, M. (2008). Quantitative conformational analysis of partially folded proteins from residual dipolar couplings: Application to the molecular recognition element of Sendai virus nucleoprotein. *Journal of the American Chemical Society*, *130*, 8055–8061.

Jensen, M. R., Markwick, P. R. L., Meier, S., Griesinger, C., Zweckstetter, M., Grzesiek, S., et al. (2009). Quantitative determination of the conformational properties of partially folded and intrinsically disordered proteins using NMR dipolar couplings. *Structure*, *17*, 1169–1185.

Jensen, M. R., Zweckstetter, M., Huang, J.-R., & Blackledge, M. (2014). Exploring free-energy landscapes of intrinsically disordered proteins at atomic resolution using NMR spectroscopy. *Chemical Reviews*, *114*, 6632–6660.

Kay, L. E., Torchia, D. A., & Bax, A. (1989). Backbone dynamics of proteins as studied by 15N inverse detected heteronuclear NMR spectroscopy: Application to staphylococcal nuclease. *Biochemistry*, *28*, 8972–8979.

Keppel, T. R., Howard, B. A., & Weis, D. D. (2011). Mapping unstructured regions and synergistic folding in intrinsically disordered proteins with amide H/D exchange mass spectrometry. *Biochemistry*, *50*, 8722–8732.

Kjaergaard, M., Brander, S., & Poulsen, F. M. (2011). Random coil chemical shift for intrinsically disordered proteins: Effects of temperature and pH. *Journal of Biomolecular NMR*, *49*, 139–149.

Kjaergaard, M., Nørholm, A.-B., Hendus-Altenburger, R., Pedersen, S. F., Poulsen, F. M., & Kragelund, B. B. (2010). Temperature-dependent structural changes in intrinsically disordered proteins: Formation of alpha-helices or loss of polyproline II? *Protein Science*, *19*, 1555–1564.

Kjaergaard, M., & Poulsen, F. M. (2011). Sequence correction of random coil chemical shifts: Correlation between neighbor correction factors and changes in the Ramachandran distribution. *Journal of Biomolecular NMR*, *50*, 157–165.

Kjaergaard, M., & Poulsen, F. M. (2012). Disordered proteins studied by chemical shifts. *Progress in Nuclear Magnetic Resonance Spectroscopy*, *60*, 42–51.

Kleckner, I. R., & Foster, M. P. (2011). An introduction to NMR-based approaches for measuring protein dynamics. *Biochimica et Biophysica Acta*, *1814*, 942–968.

Konrat, R. (2014). NMR contributions to structural dynamics studies of intrinsically disordered proteins. *Journal of Magnetic Resonance*, *241*, 74–85.

Kovrigin, E. L., Kempf, J. G., Grey, M. J., & Loria, J. P. (2006). Faithful estimation of dynamics parameters from CPMG relaxation dispersion measurements. *Journal of Magnetic Resonance*, *180*, 93–104.

Kragelj, J., Ozenne, V., Blackledge, M., & Jensen, M. R. (2013). Conformational propensities of intrinsically disordered proteins from NMR chemical shifts. *Chemphyschem*, *14*, 3034–3045.

Kuipers, B. J. H., & Gruppen, H. (2007). Prediction of molar extinction coefficients of proteins and peptides using UV absorption of the constituent amino acids at 214 nm to enable quantitative reverse phase high-performance liquid chromatography-mass spectrometry analysis. *Journal of Agricultural and Food Chemistry*, *55*, 5445–5451.

Lacy, E. R., Filippov, I., Lewis, W. S., Otieno, S., Xiao, L., Weiss, S., et al. (2004). p27 binds cyclin-CDK complexes through a sequential mechanism involving binding-induced protein folding. *Nature Structural & Molecular Biology*, *11*, 358–364.

Landrieu, I., Verger, A., Baert, J.-L., Rucktooa, P., Cantrelle, F.-X., Dewitte, F., et al. (2015). Characterization of ERM transactivation domain binding to the ACID/PTOV domain of the Mediator subunit MED25. *Nucleic Acids Research*, *43*, 7110–7121.

Leyrat, C., Yabukarski, F., Tarbouriech, N., Ribeiro, E. A., Jensen, M. R., Blackledge, M., et al. (2011). Structure of the vesicular stomatitis virus N0-P complex. *PLoS Pathogens*, *7*, e1002248.

Lin, Y. H., Song, J., Forman-Kay, J. D., & Chan, H. S. (2017). Random-phase-approximation theory for sequence-dependent, biologically functional liquid-liquid phase separation of intrinsically disordered proteins. *Journal of Molecular Liquids*, *228*, 176–193.

Lindorff-Larsen, K., Kristjansdottir, S., Teilum, K., Fieber, W., Dobson, C. M., Poulsen, F. M., et al. (2004). Determination of an ensemble of structures representing the denatured state of the bovine acyl-coenzyme A binding protein. *Journal of the American Chemical Society*, *126*, 3291–3299.

Lipari, G., & Szabo, A. (1982). Model-free approach to the interpretation of nuclear magnetic resonance relaxation in macromolecules. 1. Theory and range of validity. *Journal of the American Chemical Society*, *104*, 4546–4559.

Lippens, G., Sillen, A., Smet, C., Wieruszeski, J.-M., Leroy, A., Buée, L., et al. (2006). Studying the natively unfolded neuronal Tau protein by solution NMR spectroscopy. *Protein and Peptide Letters*, *13*, 235–246.

López-Méndez, B., & Güntert, P. (2006). Automated protein structure determination from NMR spectra. *Journal of the American Chemical Society*, *128*, 13112–13122.

Lundström, P., Vallurupalli, P., Hansen, D. F., & Kay, L. E. (2009). Isotope labeling methods for studies of excited protein states by relaxation dispersion NMR spectroscopy. *Nature Protocols*, *4*, 1641–1648.

Mao, A. H., Crick, S. L., Vitalis, A., Chicoine, C. L., & Pappu, R. V. (2010). Net charge per residue modulates conformational ensembles of intrinsically disordered proteins. *Proceedings of the National Academy of Sciences of the United States of America*, *107*, 8183–8188.

Marsh, J. A., Singh, V. K., Jia, Z., & Forman-Kay, J. D. (2006). Sensitivity of secondary structure propensities to sequence differences between α- and γ-synuclein: Implications for fibrillation. *Protein Science*, *15*, 2795–2804.

Milles, S., Mercadante, D., Aramburu, I. V., Jensen, M. R., Banterle, N., Koehler, C., et al. (2015). Plasticity of an ultrafast interaction between nucleoporins and nuclear transport receptors. *Cell*, *163*, 734–745.

Miskei, M., Antal, C., & Fuxreiter, M. (2016). FuzDB: Database of fuzzy complexes, a tool to develop stochastic structure-function relationships for protein complexes and higher-order assemblies. *Nucleic Acids Research*, *45*, D228–D235.

Mittag, T., Kay, L. E., & Forman-Kay, J. D. (2010). Protein dynamics and conformational disorder in molecular recognition. *Journal of Molecular Recognition*, *23*, 105–116.

Mittag, T., Orlicky, S., Choy, W.-Y., Tang, X., Lin, H., Sicheri, F., et al. (2008). Dynamic equilibrium engagement of a polyvalent ligand with a single-site receptor. *Proceedings of the National Academy of Sciences of the United States of America*, *105*, 17772–17777.

Modig, K., Jürgensen, V. W., Lindorff-Larsen, K., Fieber, W., Bohr, H. G., & Poulsen, F. M. (2007). Detection of initiation sites in protein folding of the four helix bundle ACBP by chemical shift analysis. *FEBS Letters*, *581*, 4965–4971.

Mollica, L., Bessa, L. M., Hanoulle, X., Jensen, M. R., Blackledge, M., & Schneider, R. (2016). Binding mechanisms of intrinsically disordered proteins: Theory, simulation, and experiment. *Frontiers in Molecular Biosciences*, *3*, 52.

Morin, S., Linnet, T. E., Lescanne, M., Schanda, P., Thompson, G. S., Tollinger, M., et al. (2014). Relax: The analysis of biomolecular kinetics and thermodynamics using NMR relaxation dispersion data. *Bioinformatics*, *30*, 2219–2220.

Motlagh, H. N., Wrabl, J. O., Li, J., & Hilser, V. J. (2014). The ensemble nature of allostery. *Nature*, *508*, 331–339.

Mulder, F. A., Schipper, D., Bott, R., & Boelens, R. (1999). Altered flexibility in the substrate-binding site of related native and engineered high-alkaline Bacillus subtilisins. *Journal of Molecular Biology*, *292*, 111–123.

Muller-Spath, S., Soranno, A., Hirschfeld, V., Hofmann, H., Ruegger, S., Reymond, L., et al. (2010). Charge interactions can dominate the dimensions of intrinsically disordered proteins. *Proceedings of the National Academy of Sciences of the United States of America*, *107*, 14609–14614.

Mysling, S., Kristensen, K. K., Larsson, M., Beigneux, A. P., Gårdsvoll, H., Fong, L. G., et al. (2016). The acidic domain of the endothelial membrane protein GPIHBP1 stabilizes lipoprotein lipase activity by preventing unfolding of its catalytic domain. *eLife*, *5*, e12095.

Nielsen, J. T., & Mulder, F. A. A. (2018). POTENCI: Prediction of temperature, neighbor and pH-corrected chemical shifts for intrinsically disordered proteins. *Journal of Biomolecular NMR*, *70*, 141–165.

O'Shea, C., Staby, L., Bendsen, S. K., Tidemand, F. G., Redsted, A., Willemoës, M., et al. (2017). Structures and short linear motif of disordered transcription factor regions provide clues to the interactome of the cellular hub protein radical-induced cell death 1. *The Journal of Biological Chemistry*, *292*, 512–527.

Otting, G. (2010). Protein NMR using paramagnetic ions. *Annual Review of Biophysics*, *39*, 387–405.

Palmer, A. G. (2016). A dynamic look backward and forward. *Journal of Magnetic Resonance*, *266*, 73–80.

Pervushin, K., Riek, R., Wider, G., & Wüthrich, K. (1997). Attenuated T2 relaxation by mutual cancellation of dipole-dipole coupling and chemical shift anisotropy indicates an avenue to NMR structures of very large biological macromolecules in solution. *Proceedings of the National Academy of Sciences of the United States of America*, *94*, 12366–12371.

Pfaff, J., Hennig, J., Herzog, F., Aebersold, R., Sattler, M., Niessing, D., et al. (2013). Structural features of Argonaute-GW182 protein interactions. *Proceedings of the National Academy of Sciences of the United States of America*, *110*, E3770–E3779.

Pierce, W. K., Grace, C. R., Lee, J., Nourse, A., Marzahn, M. R., Watson, E. R., et al. (2016). Multiple weak linear motifs enhance recruitment and processivity in SPOP-mediated substrate ubiquitination. *Journal of Molecular Biology*, *428*, 1256–1271.

Prestegard, J. H., Al-Hashimi, H. M., & Tolman, J. R. (2000). NMR structures of biomolecules using field oriented media and residual dipolar couplings. *Quarterly Reviews of Biophysics*, *33*, 371–424.

Qi, S., Li, Z., Schulze-Gahmen, U., Stjepanovic, G., Zhou, Q., & Hurley, J. H. (2017). Structural basis for ELL2 and AFF4 activation of HIV-1 proviral transcription. *Nature Communications*, *8*, 14076.

Renner, C., Schleicher, M., Moroder, L., & Holak, T. A. (2002). Practical aspects of the 2D 15N-[1h]-NOE experiment. *Journal of Biomolecular NMR*, *23*, 23–33.

Ribeiro, J. P., Palczewska, M., André, S., Cañada, F. J., Gabius, H.-J., Jiménez-Barbero, J., et al. (2010). Diffusion nuclear magnetic resonance spectroscopy detects substoichiometric concentrations of small molecules in protein samples. *Analytical Biochemistry*, *396*, 117–123.

Riek, R., Pervushin, K., & Wüthrich, K. (2000). TROSY and CRINEPT: NMR with large molecular and supramolecular structures in solution. *Trends in Biochemical Sciences*, *25*, 462–468.

Rogers, J. M., Steward, A., & Clarke, J. (2013). Folding and binding of an intrinsically disordered protein: Fast, but not "diffusion-limited". *Journal of the American Chemical Society*, *135*, 1415–1422.

Rosenzweig, R., & Kay, L. E. (2014). Bringing dynamic molecular machines into focus by methyl-TROSY NMR. *Annual Review of Biochemistry*, *83*, 291–315.
Ryan, V. H., Dignon, G. L., Zerze, G. H., Chabata, C. V., Silva, R., Conicella, A. E., et al. (2018). Mechanistic view of hnRNPA2 low-complexity domain structure, interactions, and phase separation altered by mutation and arginine methylation. *Molecular Cell*, *69*, 465–479.e7.
Sahu, D., Bastidas, M., & Showalter, S. A. (2014). Generating NMR chemical shift assignments of intrinsically disordered proteins using carbon-detected NMR methods. *Analytical Biochemistry*, *449*, 17–25.
Salmon, L., Nodet, G., Ozenne, V., Yin, G., Jensen, M. R., Zweckstetter, M., et al. (2010). NMR characterization of long-range order in intrinsically disordered proteins. *Journal of the American Chemical Society*, *132*, 8407–8418.
Schieborr, U., Vogtherr, M., Elshorst, B., Betz, M., Grimme, S., Pescatore, B., et al. (2005). How much NMR data is required to determine a protein-ligand complex structure? *Chembiochem*, *6*, 1891–1898.
Schilder, J., Liu, W.-M., Kumar, P., Overhand, M., Huber, M., & Ubbink, M. (2016). Protein docking using an ensemble of spin labels optimized by intra-molecular paramagnetic relaxation enhancement. *Physical Chemistry Chemical Physics*, *18*, 5729–5742.
Sibille, N., Huvent, I., Fauquant, C., Verdegem, D., Amniai, L., Leroy, A., et al. (2011). Structural characterization by nuclear magnetic resonance of the impact of phosphorylation in the proline-rich region of the disordered Tau protein. *Proteins*, *80*, 454–462.
Sparks, S., Temel, D. B., Rout, M. P., & Cowburn, D. (2018). Deciphering the "Fuzzy" interaction of FG nucleoporins and transport factors using small-angle neutron scattering. *Structure*, *26*, 477–484.e4.
Spera, S., & Bax, A. (1991). Empirical correlation between protein backbone conformation and C.alpha. and C.beta. 13C nuclear magnetic resonance chemical shifts. *Journal of the American Chemical Society*, *113*, 5490–5492.
Stilbs, P. (1981). Molecular self-diffusion coefficients in Fourier transform nuclear magnetic resonance spectrometric analysis of complex mixtures. *Analytical Chemistry*, *53*, 2135–2137.
Sugase, K., Dyson, H. J., & Wright, P. E. (2007). Mechanism of coupled folding and binding of an intrinsically disordered protein. *Nature*, *447*, 1021–1025.
Sugase, K., Konuma, T., Lansing, J. C., & Wright, P. E. (2013). Fast and accurate fitting of relaxation dispersion data using the flexible software package GLOVE. *Journal of Biomolecular NMR*, *56*, 275–283.
Tamiola, K., Acar, B., & Mulder, F. a. a. (2010). Sequence-specific random coil chemical shifts of intrinsically disordered proteins. *Journal of the American Chemical Society*, *132*, 18000–18003.
Teilum, K., Kragelund, B. B., & Poulsen, F. M. (2002). Transient structure formation in unfolded acyl-coenzyme A-binding protein observed by site-directed spin labelling. *Journal of Molecular Biology*, *324*, 349–357.
Teilum, K., Kunze, M. B. A., Erlendsson, S., & Kragelund, B. B. (2017). (S)Pinning down protein interactions by NMR. *Protein Science*, *26*, 436–451.
Teilum, K., Olsen, J. G., & Kragelund, B. B. (2015). Globular and disordered-the non-identical twins in protein-protein interactions. *Frontiers in Molecular Biosciences*, *2*, 1–6.
Thompson, P. M., Beck, M. R., & Campbell, S. L. (2015). Protein-protein interaction analysis by nuclear magnetic resonance spectroscopy. *Methods in Molecular Biology*, *1278*, 267–279.
Tillett, M. L., Horsfield, M. A., Lian, L. Y., & Norwood, T. J. (1999). Protein-ligand interactions measured by 15N-filtered diffusion experiments. *Journal of Biomolecular NMR*, *13*, 223–232.
Tolman, J. R. (2002). A novel approach to the retrieval of structural and dynamic information from residual dipolar couplings using several oriented media in biomolecular NMR spectroscopy. *Journal of the American Chemical Society*, *124*, 12020–12030.

Uversky, V. N. (2018). Intrinsic disorder, protein-protein interactions, and disease. *Advances in Protein Chemistry and Structural Biology*, *110*, 85–121.

Vallurupalli, P., Bouvignies, G., & Kay, L. E. (2012). Studying "invisible" excited protein states in slow exchange with a major state conformation. *Journal of the American Chemical Society*, *134*, 8148–8161.

Vallurupalli, P., Sekhar, A., Yuwen, T., & Kay, L. E. (2017). Probing conformational dynamics in biomolecules via chemical exchange saturation transfer: A primer. *Journal of Biomolecular NMR*, *67*, 243–271.

Van Roey, K., Uyar, B., Weatheritt, R. J., Dinkel, H., Seiler, M., Budd, A., et al. (2014). Short linear motifs: Ubiquitous and functionally diverse protein interaction modules directing cell regulation. *Chemical Reviews*, *114*, 6733–6778.

Viegas, A., Manso, J., Nobrega, F. L., & Cabrita, E. J. (2011). Saturation-transfer difference (STD) NMR: A simple and fast method for ligand screening and characterization of protein binding. *Journal of Chemical Education*, *88*, 990–994.

Waudby, C. A., Ramos, A., Cabrita, L. D., & Christodoulou, J. (2016). Two-dimensional NMR lineshape analysis. *Scientific Reports*, *6*, 24826.

Weinstock, D. S., Narayanan, C., Baum, J., & Levy, R. M. (2008). Correlation between 13Calpha chemical shifts and helix content of peptide ensembles. *Protein Science*, *17*, 950–954.

Wicky, B. I. M., Shammas, S. L., & Clarke, J. (2017). Affinity of IDPs to their targets is modulated by ion-specific changes in kinetics and residual structure. *Proceedings of the National Academy of Sciences of the United States of America*, *114*, 9882–9887.

Wilkins, D. K., Grimshaw, S. B., Receveur, V., Dobson, C. M., Jones, J. A., & Smith, L. J. (1999). Hydrodynamic radii of native and denatured proteins measured by pulse field gradient NMR techniques. *Biochemistry*, *38*, 16424–16431.

Wishart, D. S., Bigam, C. G., Holm, A., Hodges, R. S., & Sykes, B. D. (1995). 1H, 13C and 15N random coil NMR chemical shifts of the common amino acids. I. Investigations of nearest-neighbor effects. *Journal of Biomolecular NMR*, *5*, 332.

Wishart, D. S., & Sykes, B. D. (1994). The 13C chemical-shift index: A simple method for the identification of protein secondary structure using 13C chemical-shift data. *Journal of Biomolecular NMR*, *4*, 171–180.

Wishart, D. S., Sykes, B. D., & Richards, F. M. (1992). The chemical shift index: A fast and simple method for the assignment of protein secondary structure through NMR spectroscopy. *Biochemistry*, *31*, 1647–1651.

Wright, P. E., & Dyson, H. J. (2009). Linking folding and binding. *Current Opinion in Structural Biology*, *19*, 31–38.

Wright, P. E., & Dyson, H. J. (2015). Intrinsically disordered proteins in cellular signalling and regulation. *Nature Reviews. Molecular Cell Biology*, *16*, 18–29.

Wuttke, R., Hofmann, H., Nettels, D., Borgia, M. B., Mittal, J., Best, R. B., et al. (2014). Temperature-dependent solvation modulates the dimensions of disordered proteins. *Proceedings of the National Academy of Sciences of the United States of America*, *111*, 5213–5218.

Yang, F., Wang, X., Pan, B.-B., & Su, X.-C. (2016). Single-armed phenylsulfonated pyridine derivative of DOTA is rigid and stable paramagnetic tag in protein analysis. *Chemical Communications (Cambridge, England)*, *52*, 11535–11538.

Yuwen, T., Kay, L. E., & Bouvignies, G. (2018). Dramatic decrease in CEST measurement times using multi-site excitation. *Chemphyschem*, *19*, 1707–1710.

Zhang, H., Neal, S., & Wishart, D. S. (2003). RefDB: A database of uniformly referenced protein chemical shifts. *Journal of Biomolecular NMR*, *25*, 173–195.

CHAPTER NINE

Spectroscopic Characterization of Structure–Function Relationships in the Intrinsically Disordered Protein Complexin

David Snead, David Eliezer[1]
Department of Biochemistry, Weill Cornell Medicine, New York, NY, United States
[1]Corresponding author: e-mail address: dae2005@med.cornell.edu

Contents

Methods in Enzymology, Volume 611
ISSN 0076-6879
https://doi.org/10.1016/bs.mie.2018.08.005

Abstract

Complexins play a critical role in the regulation of neurotransmission by regulating SNARE-mediated exocytosis of synaptic vesicles. Complexins can exert either a facilitatory or an inhibitory effect on neurotransmitter release, depending on the context, and different complexin domains contribute differently to these opposing roles. Structural characterization of the central helix domain of complexin bound to the assembled SNARE bundle provided key insights into the functional mechanism of this domain of complexin, which is critical for both complexin activities, but many questions remain, particularly regarding the roles and mechanisms of other complexin domains. Recent progress has clarified the structural properties of these additional domains, and has led to various proposals regarding how they contribute to complexin function. This chapter describes spectroscopic approaches used in our laboratory and others, primarily involving circular dichroism and solution-state NMR spectroscopy, to characterize structure within complexins when isolated or when bound to interaction partners. The ability to characterize complexin structure enables structure/function studies employing in vitro or in vivo assays of complexin function. More generally, these types of approaches can be used to study the binding of other intrinsically disordered proteins or protein regions to membrane surfaces or for that matter to other large physiological binding partners.

1. INTRODUCTION

Complexins are a family of highly conserved presynaptic proteins that regulate SNARE (for SNAP receptor proteins, where SNAPs are soluble NSF attachment proteins, and NSF is the *N*-ethylmaleimide sensitive factor) complex assembly and thus SNARE-mediated synaptic vesicle exocytosis (Brose, 2008a, 2008b; Mohrmann, Dhara, & Bruns, 2015; Rizo & Xu, 2015; Trimbuch & Rosenmund, 2016). Knockout of the two predominant mammalian neuronal isoforms (complexin-1 and complexin-2) is lethal in mice (Brose, 2008a; Reim et al., 2001), and single knockout complexin-1 or complexin-2 mice have varied neurological deficits (Drew, Kyd, & Morton, 2007; Gibson, Reim, Brose, Morton, & Jones, 2005; Glynn, Bortnick, & Morton, 2003; Glynn, Drew, Reim, Brose, & Morton, 2005; Glynn, Sizemore, & Morton, 2007; Huang et al., 2000; Reim et al., 2001; Takahashi et al., 1999; Yamauchi et al., 2005). Complexin expression levels are altered in a number of neurological and psychiatric disorders, and though these changes are generally not thought to be causative, they may contribute to disease symptomatology (Brose, 2008a). Recently, autosomal recessive variants in human complexin-1 were identified through whole exome sequencing in cases of severe infantile myoclonic epilepsy and intellectual disability (Redler et al., 2017) and in cases of malignant migrating epilepsy and cortical atrophy (Karaca et al., 2015), suggesting that mutations in complexin might be causative for some human diseases.

There is general consensus based on a wide range of observations both in vivo and in vitro that complexin has dual roles in synaptic vesicle exocytosis: inhibition of spontaneous synaptic vesicle exocytosis and facilitation of stimulus-evoked synchronous exocytosis (Brose, 2008b; Buhl et al., 2013; Cho et al., 2014; Giraudo, Eng, Melia, & Rothman, 2006; Hobson, Liu, Watanabe, & Jorgensen, 2011; Huntwork & Littleton, 2007; Kaeser-Woo, Yang, & Sudhof, 2012; Lai et al., 2014; Malsam et al., 2012, 2009; Martin, Hu, Fenz, Fernandez, & Dittman, 2011; Maximov, Tang, Yang, Pang, & Sudhof, 2009; Schaub, Lu, Doneske, Shin, & McNew, 2006; Wragg et al., 2013; Xue et al., 2009; Yang, Cao, & Sudhof, 2013; Yoon et al., 2008). The detailed mechanisms by which complexin executes these functions, however, remain incompletely understood. Initial models suggested complexin binds to and clamps assembling SNARE complexes in an intermediate, inhibited, but on-pathway and fusion competent state; during evoked release, calcium-bound synaptotagmin could then displace complexin to allow rapid completion of SNARE assembly and synaptic vesicle/plasma membrane fusion (Brose, 2008b; Giraudo et al., 2006; Schaub et al., 2006; Sudhof, 2013; Tang et al., 2006). In this model, complexin inhibits complete SNARE assembly but sets up primed vesicles such that rapid assembly occurs upon complexin displacement. Recent evidence suggests, however, that complexin inhibitory and facilitatory functions are in fact discrete and separable: mutations have been identified that selectively impair one of the two functions (Buhl et al., 2013; Cho et al., 2014), and the concentration range at which complexin facilitates fusion appears lower than that at which complexin inhibits fusion (Lai et al., 2014).

Complexin is a small, soluble protein that consists of one stable alpha helix flanked by two largely intrinsically disordered regions (Pabst et al., 2000). Complexin is considered to contain four distinct domains with discrete functions (Brose, 2008b; Cho, Song, & Littleton, 2010; Hobson et al., 2011; Huntwork & Littleton, 2007; Kaeser-Woo et al., 2012; Martin et al., 2011; Xue et al., 2009, 2008) (Fig. 1). An essential central helical domain directly binds to the SNARE four helix bundle and is necessary for both inhibitory and facilitatory functions (Bracher, Kadlec, Betz, & Weissenhorn, 2002; Chen et al., 2002; Cho et al., 2010; Choi, Zhao, Zhang, Lai, & Brunger, 2016; Lai et al., 2014; Martin et al., 2011; Maximov et al., 2009; Pabst et al., 2000; Xue et al., 2009, 2007;

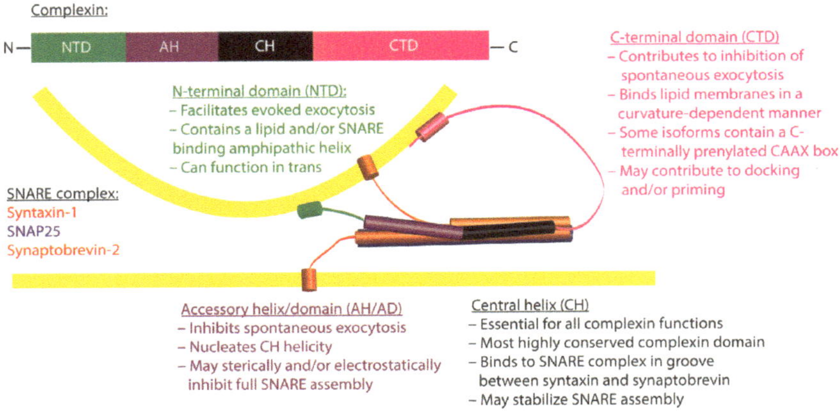

Fig. 1 Structure and function of complexin. Complexin is a small, ~15 kDa protein containing four distinct domains with discrete functions in synaptic vesicle exocytosis. A central helix domain (*black*) binds the SNARE complex in a groove between syntaxin (*red*) and synaptobrevin (*orange*) and is essential for all complexin function. A stable accessory helix (*purple*) contributes to inhibition of exocytosis. A disordered N-terminal domain (*green*) helps facilitate vesicle release, while a disordered C-terminal domain (*pink*) is critical for inhibition of exocytosis. The *curved vesicle* and *flat plasma membranes* are shown in *yellow*.

Yin, Kim, & Shin, 2016; Zhou et al., 2017). An adjacent accessory helix/accessory domain forms a continuous helix with the central helix and contributes to inhibition of exocytosis (Bykhovskaia, Jagota, Gonzalez, Vasin, & Littleton, 2013; Giraudo et al., 2009; Krishnakumar et al., 2011; Kummel et al., 2011; Li et al., 2011; Lu, Song, & Shin, 2010; Prinslow, Brautigam, & Rizo, 2017; Radoff et al., 2014; Trimbuch et al., 2014; Vasin, Volfson, Littleton, & Bykhovskaia, 2016; Xue et al., 2007); a number of mechanisms have been proposed for accessory domain function, including ones based on binding to and steric blockade of an assembling *trans*-SNARE complex (Bykhovskaia et al., 2013; Krishnakumar et al., 2011; Kummel et al., 2011; Li et al., 2011; Lu et al., 2010; Trimbuch et al., 2014), on electrostatic repulsion (Trimbuch et al., 2014), and on helix nucleation and propagation (Radoff et al., 2014). An intrinsically disordered C-terminal domain contributes critically to inhibition of exocytosis (Cho et al., 2010; Kaeser-Woo et al., 2012; Malsam et al., 2009; Martin et al., 2011; Seiler, Malsam, Krause, & Sollner, 2009; Snead, Wragg, Dittman, & Eliezer, 2014; Wragg et al., 2013; Xue et al., 2009), and likely mediates synaptic vesicle localization through direct complexin/lipid membrane interactions (Seiler et al., 2009; Snead et al., 2014; Wragg et al., 2013).

Finally, a disordered N-terminal domain facilitates exocytosis, perhaps through an interaction with the SNARE complex and/or with lipid membranes (Lai et al., 2016; Xue et al., 2010, 2007).

Solution-state nuclear magnetic resonance (NMR) spectroscopy provides a powerful tool to characterize intrinsically disordered protein structure and dynamics, and to probe the functionally significant structural and dynamical changes that can occur in disordered regions upon interacting with macromolecular binding partners such as other proteins, DNA, carbohydrates, and lipid membranes (Eliezer, 2007, 2009; Gibbs, Cook, & Showalter, 2017; Kurzbach, Kontaxis, Coudevylle, & Konrat, 2015). Indeed, NMR analyses yielded structural descriptions of the free state of complexin (Pabst et al., 2000; Snead et al., 2014) and have been used extensively to characterize binding of complexin to the SNARE complex and to the lipid membranes (Chen et al., 2002; Pabst et al., 2000; Snead et al., 2017, 2014; Wragg et al., 2017, 2013; Xue et al., 2010). Importantly, detailed structure–function relationships for complexin can be determined by identifying mutations in complexin that impair its function in some in vivo or in vitro assay, and then characterizing the effect of those mutations on complexin structure, dynamics, or interactions with proteins or lipids. These types of approaches have bolstered our mechanistic understanding of each individual complexin domain as noted above.

In this chapter, we describe the approach taken in our lab to clone, express, and purify wild-type and mutant worm and mouse complexin-1 protein for structural analyses by circular dichroism (CD) and solution-state NMR spectroscopy. Because high quality NMR spectra require isotopically labeled protein and relatively high concentrations, we convey relevant methods to purify complexin at high yields and with multiple different isotopic labeling strategies. As our work has focused on the spectroscopic characterization of complexin/membrane interactions and their functional sequelae, we then discuss liposome and micelle preparation. We finish by describing the key NMR-spectroscopic techniques necessary to assign NMR backbone resonances even in crowded IDP spectra, and the methods available to study protein/lipid membrane binding by NMR. After applying these techniques to wild-type complexin, it is possible to rationally design mutations in the complexin sequence expected to perturb specific properties of complexin such as membrane binding or secondary structure. The structural effects of such mutations can be assessed by spectroscopic techniques in vitro, and their functional impact can be determined using a well-established behavioral assay of complexin function in *Caenorhabditis elegans*. Though these latter techniques are beyond the scope of this chapter, complementary in vitro and in vivo approaches provide a powerful combination for characterizing complexin structure–function relationships.

2. SAMPLE PREPARATION FOR IN VITRO SPECTROSCOPIC CHARACTERIZATION OF COMPLEXIN–LIPOSOME INTERACTIONS USING SOLUTION-STATE PROTEIN NMR SPECTROSCOPY AND CD SPECTROSCOPY

As noted above, solution-state protein NMR spectroscopy is a powerful and well-established technique to study intrinsically disordered protein structure and dynamics, as well as their interactions with varied binding partners, including other proteins and lipid membranes. Careful sample preparation is by far the best way to ensure high quality NMR data. It is possible to optimize spectroscopic parameters to compensate for common issues that arise with poorly behaved protein samples such as protein aggregation or low solubility, and techniques are available to circumvent spectral limitations such as crowded spectra (with, for example, a highly degenerate polypeptide sequence), or limited signal (common, for example, with larger molecular weight proteins that are typically difficult to study with a size-limited technique such as solution NMR). However, if the protein construct and purification protocol can be optimized to yield large amounts of reasonably pure protein that can be dissolved to at least 200–500 μ*M*, the vast gains in NMR spectral quality and corresponding reductions in required NMR experimental time are usually worth the upfront investment. Thus, we start by describing our approach to produce complexin protein, including general methods for cloning, mutagenesis, protein expression with multiple isotopic labeling schemes, purification, and sample preparation.

In general, we assume the user has access to basic lab equipment such as PCR tubes, microfuge tubes, conical centrifuge/culture tubes, protein and DNA gel electrophoresis apparatus, pipets and tips, etc.

2.1 Expression Vector Cloning

We have expressed and purified hexahistidine (his_6)-tagged complexin protein from the standard pET28a expression vector, as well as hexahistidine-SUMO-tagged protein from the pET SUMO vector (SUMO is a small, ubiquitin-like protein). Expression of complexin as a SUMO-tagged protein is preferable for a few reasons. Protein solubility can be enhanced upon fusion to the SUMO protein, and this increased solubility can improve protein behavior and enhance initial purification yield prior to cleavage. His_6-SUMO-tagged complexin is cleaved during the purification protocol to release untagged complexin; interestingly, the SUMO protease recognizes SUMO structure rather than a simple peptide sequence (Mossessova & Lima, 2000). Because his_6-SUMO can then be separated from complexin by a second nickel-affinity step, we found that cleaved protein purified through the his_6-SUMO affinity protocol was of adequate purity for use without any further column purification. Finally, and most importantly, cleavage of the his_6-SUMO tag yields complexin protein with no residual tag. Thus, because his_6-SUMO allowed us to purify large quantities of pure complexin protein with its full native sequence and no additional tag, we did not need to consider other tagging and purification strategies such as GST or MBP tags.

Standard kits are commercially available to clone a sequence of interest into pET SUMO, and in general we cloned complexin into this vector according to the manufacturer's instructions using the general workflow described below.

2.1.1 Equipment

- Thermocycler
- Standard agarose DNA gel set-up

2.1.2 Buffers and Reagents

- PCR amplification primers
- DNA template for PCR product
- Taq polymerase
- Gel extraction/purification kit (e.g., QIAquick)
- Champion pET SUMO Protein expression system cloning kit
- Plasmid Miniprep Kit (e.g., QIAGEN QIAprep spin kit)

2.1.3 Procedure

1. Manually design primers to PCR amplify a desired protein-coding sequence of interest (in our case, full-length wild-type worm complexin lacking its N-terminal methionine residue, or mouse complexin containing its full sequence). The first amino acid should not be proline, lysine, valine, or leucine, as these decrease cleavage efficiency by the SUMO protease. We had success with primer melting temperatures

around 80°C as very roughly estimated by the relation $T=(4*\#GC)+(2*\#AT)$, where T is the melting temperature, #GC is the total number of guanine and cytidine nucleotides, and #AT is the number of adenine and thymine nucleotides. Primers should contain a guanine or cytidine nucleotide at either end, and extremes of GC content percentage should be avoided.

2. PCR amplify the desired sequence to be cloned into the SUMO vector. We typically used the cycling parameters: (1) 95°C, 3 min; (2) 95°C, 30 s; (3) 60°C, 30 s; (4) 72°C, 2 min; (5) repeat steps 2–4 50×; (6) 72°C, 10 min; (7) hold at 4°C. Primer annealing temperature can be adjusted from 55°C to 60°C to optimize PCR amplification, and extension time can be adjusted depending on desired product length. A single, abundant product band is ideal.
3. Run the PCR reaction on a 1% agarose gel and extract the appropriate PCR band using a gel extraction/purification kit per manufacturer's instructions. In general, our attempts to SUMO-clone without a gel purification step were unsuccessful.
4. Ligate the purified PCR product into the pET SUMO vector per manufacturer's instructions; the pET SUMO kit comes with the vector prelinearized, and the Taq polymerase used for the above PCR reaction provides adenine-overhangs that allow immediate ligation.
5. Use the ligation reaction to transform *Escherichia coli* OneShot Mach1™-T1(R) competent cells using the transformation protocol per manufacturer's instructions and plate the cells on LB/kanamycin plates.
6. Pick 10–20 individual, transformed colonies and inoculate each into 10 mL LB medium with 50 μg/mL kanamycin. If cloning efficiency was low, inoculate as many colonies as possible. Incubate with shaking overnight (~16 h).
7. Isolate plasmid DNA using a Plasmid Miniprep Kit of your choice (e.g., QIAGEN QIAprep spin). Overnight culture volumes of 10 mL increased our DNA yield, but buffer volumes for subsequent miniprep steps should be increased according to manufacturer's instructions.
8. Sequence verify the SUMO-cloned DNA sequence. The T7 promoter can be used to sequence using commonly available primers

2.1.4 Notes

1. Calculators to more precisely estimate primer melting temperature are available online, and we have also had success using these to design primers with a salt adjusted melting temperature around 60–62 degrees

(many are available, though we prefer Northwestern's OligoCalc: Oligonucleotide properties calculator, http://biotools.nubic.northwestern.edu/OligoCalc.html).

2. We typically used glass beads to plate transformed cells, though any preferred method is appropriate.
3. LB/kanamycin plates containing transformed colonies can be kept at 4°C for up to 2 weeks.

2.2 Expression Vector Mutagenesis

To generate complexin mutant variants, we use the above-described vector coding for full-length, wild-type complexin—SUMO fusion as a starting template for site-directed mutagenesis. Single amino acid changes can be generated using the QuikChange II site-directed mutagenesis kit per manufacturer's instructions. Multiamino acid changes can be performed one at a time, though kits are also available to generate multiple mutations simultaneously (e.g., QuikChange multisite-directed mutagenesis kit). Complexin truncations can be generated simply by using site-directed mutagenesis to introduce an early stop codon at the desired location.

2.2.1 Equipment

- Thermocycler

2.2.2 Buffers and Reagents

- Mutagenesis primers
- DNA template to be mutated
- Site-directed mutagenesis kit (e.g., QuikChange)

2.2.3 Procedure

1. Design mutagenesis primers using the online program PrimerX (http://www.bioinformatics.org/primerx/). This program allows the user to prespecify the specific mutagenesis kit to be used, and to specify *E. coli* as the desired organism for codon optimization (codon optimization improves ultimate protein expression and purification yield).
2. PCR amplification of mutated plasmid, dpn I digestion, and transformation are all performed according to manufacturer's instructions.
3. Isolate plasmid DNA and sequence verify according to steps 5–8 in Section 2.1.3, but using XL1-Blue supercompetent cells. Kanamycin selection is appropriate for the pET-SUMO vector, though in general antibiotic should be adjusted as necessary for a given plasmid.

2.3 Expression of Unlabeled Complexin

NMR spectroscopy requires samples containing isotopically labeled protein; unlabeled protein is easier to express in larger quantities and is appropriate for most other biochemical and biophysical approaches. We primarily express and purify unlabeled complexin for analysis by CD spectroscopy. The protocol below is for a single 1 L culture, though we often grow up multiple 1 L cultures simultaneously as this is no more onerous, and protein expression pellets can be stored for months at −20°C. A typical growth and subsequent purification of 1 L unlabeled complexin yields enough protein for 2–3 full liposome titrations by CD spectroscopy. Though not specifically noted below, in general it is good practice to work in as sterile conditions as possible. This can be achieved by simple techniques such as working in the vicinity of a Bunsen-burner flame, and by keeping plates and flasks covered as much as possible.

2.3.1 Equipment

- Incubator
- LB/kanamycin agar plates
- Standard 14-mL culture tubes
- 250-mL Erlenmeyer flasks with bevels for aeration (bevels are not absolutely necessary but enhance growth of *E. coli*)
- 2500-mL Erlenmeyer flasks with bevels for aeration (bevels are not absolutely necessary but enhance growth of *E. coli*)
- UV/Vis spectrometer

2.3.2 Buffers and Reagents

- BL21(DE3) competent cells (or other protein expression competent *E. coli* cell line)
- LB media, autoclaved, with 50 μg/mL kanamycin (or other antibiotic as appropriate)
- IPTG (isopropyl-β-thiogalactoside, 0.8–1 *M* stock)

2.3.3 Procedure

1. Transform BL21(DE3) competent cells: (1) Mix 0.5–1 μL of purified plasmid with 10 μL competent cells. (2) Leave on ice for 5 min. (3) Heat shock for 45 s at 42°C. (4) Add 100 μL LB media and incubate with shaking for 45–60 min at 37°C. Plate onto LB/kanamycin agar plates and leave overnight (~16 h).

2. Pick a single transformed colony and inoculate into a starter culture containing 4 mL LB with 50 μg/mL kanamycin. Incubate the starter culture with shaking at 37°C for ~ 5 h.
3. Prepare 1 L LB media containing 50 μg/mL kanamycin.
4. Use 2 mL starter culture to inoculate 100 mL LB/50 μg/mL kanamycin (this can be taken from the above prepared 1 L LB/kanamycin). Let grow overnight at 37°C, ~16–20 h.
5. Add the 100 mL overnight culture to the remaining 900 mL LB/kanamycin to yield a final volume of 1 L.
6. Incubate with shaking at 37°C to grow to an OD_{600} (optical density at 600 nm) of ~0.6. This typically took 30–60 min.
7. Add IPTG to a final concentration of 0.8–1 m*M* to induce protein expression
8. Induce expression for 3–4 h. Check the OD_{600} every hour to ensure cell viability, and check every half hour after 3 h. Stop growth after 3 h if the OD_{600} stalls or decreases.
9. Centrifuge cells (6500 rpm, 15 min, 4°C).
10. Pellet can be stored for months at −20°C.

2.3.4 Notes

1. We typically used glass beads to plate transformed cells, though any preferred method is appropriate.
2. Plates containing transformed colonies can be kept at 4°C for up to 2 weeks.

2.4 Expression of ^{15}N- and ^{15}N^{13}C-Labeled Complexin

We generally follow the protocol of Marley, Lu, and Bracken (2001) to express isotopically labeled protein. Using this technique, 4 L of LB media are pooled into minimal medium containing isotopically labeled nitrogen and/or carbon sources, and grown to high density under aeration. Note that leaky expression can lead to the production of partially labeled or unlabeled protein. It is therefore advisable to confirm the efficiency of isotopic labeling using mass spectrometry of the final purified proteins. ^{15}N-labeled protein is appropriate for standard heteronuclear single quantum coherence (HSQC) experiments and is typically expressed in greater quantities than ^{15}N^{13}C-labeled protein, which is required for triple-resonance experiments (such as those used to obtain backbone resonance assignments). A typical growth and purification of ^{15}N-complexin as described below yields enough protein for one full NMR-based liposome binding titration. A typical growth and purification of $^{15}N^{13}C$-complexin as described below yields enough protein for one high quality, high concentration NMR sample for triple-resonance experiments.

2.4.1 Equipment

- Incubator
- LB/kanamycin agar plates
- Standard 14-mL culture tubes
- 250-mL Erlenmeyer flasks with bevels for aeration (bevels are not absolutely necessary but enhance growth of *E. coli*)
- 2500-mL Erlenmeyer flasks with bevels for aeration (bevels are not absolutely necessary but enhance growth of *E. coli*)
- Water bath and aeration set-up
- UV/Vis spectrometer

2.4.2 Buffers and Reagents

- BL21(DE3) competent cells (or other protein expression competent *E. coli* cell line)
- LB media, autoclaved, with 50 μg/mL kanamycin (or other antibiotic as appropriate)
- IPTG (1 *M* stock)
- ^{15}N-ammonium chloride, D-glucose-U-$^{13}C_6$
- 5 × M9 stock, autoclaved: 2.5 g/L NaCl, 34 g/L Na_2HPO_4, 15 g/L KH_2PO_4
- $MgSO_4$ stock, autoclaved(1 *M*)
- $CaCl_2$ stock, autoclaved (1 *M*)
- BME vitamins 100 × solution
- Wash medium: 1 × M9 medium, 50 μg/mL kanamycin, 1 m*M* $MgSO_4$, 100 μ*M* $CaCl_2$ (*note*: calcium chloride may precipitate upon addition but can be dissolved with stirring)
- ^{15}N expression medium: 1 × M9 medium, 50 μg/mL kanamycin, 1 × BME vitamins, 1 m*M* $MgSO_4$, 100 μ*M* $CaCl_2$ (*note*: calcium chloride may precipitate upon addition but can be dissolved with stirring), 1 g ^{15}N-labeled ammonium chloride, 4 g dextrose
- $^{15}N^{13}C$ expression medium: 1 × M9 medium, 50 μg/mL kanamycin, 1 × BME vitamins, 1 m*M* $MgSO_4$, 100 μ*M* $CaCl_2$ (*note*: calcium chloride may precipitate upon addition but can be redissolved with stirring), 1 g ^{15}N-labeled ammonium chloride, 2 g D-glucose-U-$^{13}C_6$.

2.4.3 Procedure

1. Transform cells as in step 1 of Section 2.3.3.
2. Grow cells to an OD_{600} ~0.6 as described in Section 2.3.3 steps 2–6, with one crucial exception: protocol should be scaled up to 4L of rich media.
3. Pellet 4L cells (6500rpm, 15min, room temperature).
4. Resuspend all 4L pellets in 1L wash medium. We typically use a sterile pipet tip to manually scrape/agitate the pellet for resuspension.
5. Pellet washed cells (6500rpm, 15min, room temperature)
6. Resuspend pellet in expression medium (^{15}N or $^{15}N^{13}C$), transfer to 2500-mL Erlenmeyer culture flask, place in 37°C water bath, and aerate.
7. Allow cells to equilibrate at 37°C with aeration for 30min.
8. Add IPTG to a final concentration of 0.8–1m*M* to induce protein expression
9. Induce expression for 3–4h. Check the OD_{600} every hour to ensure cell viability, and check every half hour after 3h. Stop growth after 3h if the OD_{600} stalls or decreases.
10. Centrifuge cells (6500rpm, 15min, 4°C).
11. Pellet can be stored for months at −20°C.

2.5 Expression of Deuterated Complexins

Use of fractionally or perdeuterated proteins can significantly improve spectral quality, particularly for large proteins or protein complexes or, in our case, for dodecylphosphocholine (DPC)-micelle bound complexin (Snead et al., 2017). By decreasing proton density, deuteration achieves two effects: (1) it minimizes spin diffusion and, by eliminating this source of proton relaxation, increases NMR signal to noise and (2) it reduces the dipolar interaction between ^{13}C or ^{15}N and a directly bound proton, decreasing the relaxation of these heteronuclei (Sattler & Fesik, 1996). Deuteration is not necessary for simple HSQC spectra or for triple-resonance experiments with free complexin. However, perdeuterated complexin combined with TROSY-based triple-resonance experiments was essential for us to fully assign a particularly difficult stretch within the complexin accessory and central helices (Radoff et al., 2014; Snead et al., 2014). Furthermore, after unsuccessful attempts to obtain triple-resonance spectra with protonated complexin and protonated micelles, it was possible to rapidly obtain very high quality data for perdeuterated complexin bound to deuterated DPC micelles. Though we considered a variety of micellar systems, deuterated DPC is relatively inexpensive and is a well-known membrane mimetic that contains a phosphocholine headgroup and is therefore reasonably comparable to bona fide phospholipids. A typical growth and purification of 1 L of deuterated complexin as described below yields enough protein for one high quality, high concentration NMR sample for triple-resonance experiments.

2.5.1 Equipment

- Incubator
- LB/kanamycin agar plates
- Standard 14-mL culture tubes
- 250-mL Erlenmeyer flasks with bevels for aeration (bevels are not absolutely necessary but enhance growth of *E. coli*)
- 2500-mL Erlenmeyer flasks with bevels for aeration (bevels are not absolutely necessary but enhance growth of *E. coli*)
- UV/Vis spectrometer

2.5.2 Buffers and Reagents

We describe fractional deuteration and perdeuteration media below for $^{15}N^{13}C$-labeled protein, though deuterated ^{15}N, deuterated ^{13}C, and "unlabeled" deuterated protein could be made by substituting the appropriate labeled isotope source with its unlabeled equivalent.

- BL21(DE3) competent cells (or other protein expression competent *E. coli* cell line)
- IPTG (1 *M* stock)
- ^{15}N-ammonium chloride, D-glucose-U-$^{13}C_6$, D-glucose-U-$^{13}C_6$-1,2,3,4,5,6,6-d7
- BME vitamins 100 × solution
- Deuterium oxide (D_2O)
- Fractional deuteration medium: 1 L D_2O plus 6.8 g Na_2HPO_4, 3 g KH_2PO_4, 0.5 g NaCl, 100 μL $CaCl_2$ (1 *M*), 1 mL $MgSO_4$ (1 *M*), 1 g ^{15}N-ammonium chloride, 2 g of D-glucose-U-$^{13}C_6$, 1 × BME vitamin (10 mL 100 ×), 50 μg/mL kanamycin
- Perdeuteration medium: 1 L D_2O plus 6.8 g Na_2HPO_4, 3 g KH_2PO_4, 0.5 g NaCl, 100 μL $CaCl_2$ (1 *M*), 1 mL $MgSO_4$ (1 *M*), 1 g ^{15}N-ammonium chloride, 2 g of D-glucose-U-$^{13}C_6$-1,2,3,4,5,6,6-d7 (deuterated glucose), 1 × BME vitamin (10 mL 100 ×), 50 μg/mL kanamycin

2.5.3 Procedure

1. Transform cells as in step 1 of Section 2.3.3.
2. Pick a single transformed colony and inoculate into a starter culture containing 4 mL LB with 50 μg/mL kanamycin. Incubate the starter culture with shaking at 37°C for ~5 h.
3. Prepare 1 L of the appropriate deuteration medium, containing 50 μg/mL kanamycin.
4. Use 2 mL starter culture to inoculate 100 mL deuteration medium (this can be taken from the medium prepared in step 3). Let grow overnight at 37°C, ~16–20 h.
5. Add the 100 mL overnight culture to the remaining 900 mL deuteration medium.
6. Incubate with shaking at 37°C to grow to an OD_{600} of ~0.6. Growth in deuterated media takes longer than growth in H_2O-based media (a few hours)
7. Add IPTG to a final concentration of 0.8–1 m*M* to induce protein expression
8. Induce expression for 3–4 h. Check the OD_{600} every hour to ensure cell viability, and check every half hour after 3 h. Stop growth after 3 h if the OD_{600} stalls or decreases.
9. Centrifuge cells (6500 rpm, 15 min, 4°C)
10. Pellet can be stored for months at −20°C.

2.5.4 Notes

1. On rare occasions, the 100 mL overnight culture fails to grow. If this happens repeatedly, minimize D_2O waste by making just 100 mL of deuteration medium for the overnight culture and then making 900 mL of deuteration medium the next day if there is adequate growth in the starter culture.

2.6 Selective Labeling Strategies: Single Amino Acid Labeling

A stretch of residues within the worm complexin accessory and central helices yields highly overlapping, difficult to assign resonances, in the proton, nitrogen, and carbon dimensions. Though we ultimately circumvented this problem by using TROSY-based experiments with perdeuterated protein, we also used a selective labeling strategy to label only lysine residues. This strategy can prove useful in a number of contexts. We describe a protocol for uniformly ^{13}C-labeled, ^{15}N-lysine-labeled protein, though the basic approach can be adjusted for a variety of labeling strategies (e.g., using 4 g unlabeled dextrose in place of 2 g D-glucose-U-$^{13}C_6$ to make protein that is only ^{15}N-lysine labeled, or, e.g., by using 1 g ^{15}N-ammonium chloride and 80 mg ^{13}C-labeled valine to make uniformly ^{15}N-labled, ^{13}C-valine labeled protein). Note that isotopically labeled amino acids can be metabolized, potentially leading to scrambling of the label. Amino acids that are at the terminus of biosynthetic pathways work best.

2.6.1 Equipment

- Incubator
- LB/kanamycin agar plates
- Standard 14-mL culture tubes
- 250-mL Erlenmeyer flasks with bevels for aeration (bevels are not absolutely necessary but enhance growth of *E. coli*)
- 2500-mL Erlenmeyer flasks with bevels for aeration (bevels are not absolutely necessary but enhance growth of *E. coli*)
- Water bath and aeration set-up
- UV/Vis spectrometer

2.6.2 Buffers and Reagents

- BL21(DE3) competent cells (or other protein expression competent *E. coli* cell line)
- LB media, autoclaved, with 50 μg/mL kanamycin (or other antibiotic as appropriate)
- IPTG (what concentration)
- ^{15}N-ammonium chloride, D-glucose-U-$^{13}C_6$, alpha-^{15}N-lysine
- 5× M9 stock, autoclaved: 2.5 g/L NaCl, 34 g/L Na_2HPO_4, 15 g/L KH_2PO_4
- BME vitamins 100× solution
- Wash medium: 1× M9 medium, 50 μg/mL kanamycin, 1 m*M* $MgSO_4$, 100 μ*M* $CaCl_2$(*note*: calcium chloride may precipitate upon addition but can be dissolved with stirring)
- Growth medium: 1× M9 medium, 50 μg/mL kanamycin, 1× BME vitamins, 1 m*M* $MgSO_4$ (1 *M*), 100 μ*M* $CaCl_2$ (1 *M*), 5 g unlabeled (^{14}N) ammonium chloride, 2 g D-glucose-U-$^{13}C_6$

2.6.3 Procedure

1. Transform cells as in step 1 of Section 2.3.3.
2. Grow cells to an OD_{600} ~0.6 as described in Section 2.3.3 steps 2–6, though scaled up to 4 L.

3. Pellet 4 L cells (6500 rpm, 15 min, room temperature).
4. Resuspend all 4 L pellets in 1 L wash medium. We typically use a sterile pipet tip to manually scrape/agitate the pellet for resuspension.
5. Pellet washed cells (6500 rpm, 15 min, room temperature).
6. Resuspend pellet in expression medium, transfer to 2500-mL Erlenmeyer culture flask, place in 37°C water bath, and aerate.
7. Allow cells to equilibrate at 37°C with aeration for 30 min.
8. Add IPTG to a final concentration of 0.8–1 m*M* to induce protein expression. Add 80 mg alpha-^{15}N-lysine powder to culture at this point as well, and mix to dissolve
9. Induce expression for 3–4 h. Check the OD_{600} every hour to ensure cell viability, and check every half hour after 3 h. Stop growth after 3 h if the OD_{600} stalls or decreases.
10. Centrifuge cells (6500 rpm, 15 min, 4°C).
11. Pellet can be stored for months at −20°C.

2.7 Affinity Purification of His-SUMO-CPX

We affinity purify hexahistidine-SUMO-tagged complexin on a nickel-NTA agarose gravity column by manually pouring clarified cell lysates over the column and manually collecting elution fractions. It is also possible, however, to adopt nickel-NTA columns for FPLC use. We initially dialyzed freshly purified protein directly into a given experimental working buffer. However, we found that both lyophilized and freshly purified complexin protein behave similarly and yielded similar HSQC spectra and comparable membrane binding behavior (Snead et al., 2014).

2.7.1 Equipment

- Ultracentrifuge (e.g., Beckman Coulter Optima XL-100 K)
- Ultracentrifuge rotor (e.g., 50.2 Ti)
- Ultracentrifuge tubes (e.g., polycarbonate 26.3 mL)
- Sonifier (e.g., Branson)
- Chromatography column filled with 30–40 mL nickel-NTA agarose resin
- Transfer pipets
- Lyophilizer
- Standard SDS-page protein gel set-up
- SDS-page gels (15% agarose)

2.7.2 Buffers and Reagents

We keep stock solutions of Tris (1 *M*), NaCl (1 *M*), and EDTA (0.5 *M*) at room temperature for months. Dithiothreitol (DTT) is stored dessicated at 4°C. PMSF is dissolved in ethanol and stored at −20°C (note: PMSF is hazardous and protective eyewear must be used). Imidazole (2 *M*) is stored in the dark (wrapped in aluminum foil) either at room temperature or at 4°C.

- Lysis Buffer (100 mL): 20 m*M* Tris pH 8, 350 m*M* NaCl, 20 m*M* Imidazole, 1 m*M* EDTA, 2 m*M* DTT, 1.7 m*M* 2-mercaptoethanol, 1 m*M* PMSF
- Wash Buffer (200 mL): 20 m*M* Tris pH 8, 350 m*M* NaCl, 20 m*M* Imidazole, 2 m*M* DTT, 1.7 m*M* 2-mercaptoethanol
- Elution Buffer #1 (100 mL): 20 m*M* Tris pH 8, 350 m*M* NaCl, 250 m*M* Imidazole, 2 m*M* DTT, 1.7 m*M* 2-mercaptoethanol
- Elution Buffer #2 (200 mL, same as #1 but used on the second purification day): 20 m*M* Tris pH 8, 350 m*M* NaCl, 250 m*M* Imidazole, 2 m*M* DTT, 1.7 m*M* 2-mercaptoethanol
- Dialysis Buffer #1 (2 L, larger quantity for dialysis): 20 m*M* Tris pH 8, 150 m*M* NaCl, 1 m*M* DTT
- Dialysis Buffer #2 (same as above, smaller volume, 200 mL): 20 m*M* Tris pH 8, 150 m*M* NaCl, 1 m*M* DTT
- Appropriate electrophoresis buffers, including running buffer (15 g Tris base, 72 g glycine, 5 g SDS in 5 L deionized water), Coomassie gel stain (1 g Coomassie Blue R-250, 450 mL methanol, 250 mL water, 100 mL glacial acetic acid), Coomassie gel destain (100 mL methanol, 100 mL glacial acetic acid, 800 mL water), 5 × sample reducing/loading buffer (1.2 mL 1 *M* Tris pH 6.8, 5 mL 100% glycerol, 4 mL 10% SDS, 1 mL beta mercaptoethanol (BME), 2 mL 1% bromophenol blue, 6.8 mL deionized water).
- Dialysis membrane (e.g., Spectrum Laboratories, 1 or 3.5 kDa)
- Liquid nitrogen

2.7.3 Procedure

1. Make the lysis buffer, wash buffer, elution buffer #1, and dialysis buffer #1, and store at 4°C until needed.
2. Resuspend bacterial cell pellet (in 50-mL falcon tube) in 40 mL lysis buffer by agitating with sterile pipet or spatula and/or vortexing. Take 40 μL sample for verification by SDS-Page gel electrophoresis
3. Using a tip sonifier, lyse cells by sonication on ice. We typically use two cycles of 4 min each at power level output 6, 50% duty cycle. Let cells cool for a few minutes between sonication cycles.

4. Clarify supernatant by centrifugation at 40,000 rpm for 45 min in an ultracentrifuge. It is important to ensure ultracentrifuge samples are adequately balanced—weigh the tube/sample prior to centrifugation adjust balanced samples to equal mass using remaining lysis buffer. Transfer supernatant to a clean tube, and take a 40 μL sample for gel electrophoresis.
5. Run 150 mL wash buffer over the column to equilibrate nickel-NTA resin
6. Run clarified supernatant over the column to allow hexa-histidine-tagged SUMO complexin to bind to the nickel-NTA-agarose beads. Collect all 40–50 mL flow through in a single tube. Set aside 40 μL for gel electrophoresis.
7. Run remaining 50 mL wash buffer over the column to wash. Collect flow through in 5 × 10 mL fractions, and take 40 μL of each fraction for gel electrophoresis.
8. Run 100 mL elution buffer #1 over the column to elute protein and to rid the column of any residual bound protein. We collect roughly 5 mL fractions. Take 40 μL from the first 10–20 fractions for gel electrophoresis (one can collect samples from, e.g., each of the first five, then from every other or every third fraction after that). If a concentrated working stock of protein is desired directly after purification, it is important to precisely assess which fractions contain complexin protein via gel electrophoresis; since we typically lyophilize complexin, however, we usually just take the first 50–70 mL that elutes from the column (all complexin protein reliably elutes by this volume).
9. Verify purification success by running the above collected fractions on SDS-page protein gel (we typically use 15% gels). It is important to determine which elution fractions contain substantial amounts of purified protein the first few times a protein is purified. However, as noted above, we often would simply collect the first 50–70 mL that elute, without verification by gel for further purification steps, below.
10. Dialyze purified hexahistidine-SUMO-complexin into dialysis buffer overnight with stirring at 4°C (usually around 50 mL sample into 2 L dialysis buffer).
11. The following day, after dialysis, take a 40 μL sample of protein precleavage for gel electrophoresis.
12. Make the elution buffer #2 and dialysis buffer #2, and store at 4°C until needed.

13. Add SUMO-protease to approximately 1 μ*M* and incubate at 4°C for 2–3 h. Take 40 μL sample postcleavage for gel electrophoresis.
14. Run 150 mL dialysis buffer over the nickel-NTA column to equilibrate.
15. Run the entire volume of cleaved protein over the column. Collect the flow through. This contains purified, cleaved complexin! Take 40 μL sample for gel electrophoresis.
16. Run 50 mL dialysis buffer over the column to elute remaining cleaved, nonnickel-NTA-bound complexin. Collect in 10 mL fractions and take 40 μL of each for gel electrophoresis. These will also contain some purified, cleaved complexin.
17. Run 200 mL elution buffer over the column to elute bound hexahistidine-SUMO protein. This can be collected but will contain SUMO protein rather than the desired, complexin protein. We typically collected the first 40 mL and took a 40 μL sample for gel electrophoresis.
18. Run 200 mL deionized water over the column to equilibrate for storage at 4°C. For long-term storage equilibrate in 20% isopropanol.
19. Run the above collected samples on SDS-page gel to verify which fractions contain substantial amounts of purified, cleaved protein. As above, this is particularly important when first performing a purification.
20. Dialyze purified complexin into deionized water overnight with stirring at 4°C (usually around 100 mL protein into 4 L water).
21. Flash freeze dialyzed, purified protein in liquid nitrogen.
22. Lyophilize protein and store dessicated at −20°C.

2.7.4 Notes

1. Because the complexin purification proved to be highly reproducible and because we lyophilize the protein for long-term storage, we typically pooled the protein load flow through with wash fractions 1–3 (usually around 60 mL protein load, 30 mL wash). However, if the protein is to be dialyzed directly into a working buffer, more care should be taken to selectively pool protein-containing fractions and so to minimize overall solution volume.

2.8 Liposome Formation by Extrusion

2.8.1 Equipment

- Pyrex tubes
- Evaporating unit (e.g., Pierce Reacti-Vap), heating/stirring module (e.g., Pierce Reacti-Therm)

– Nitrogen gas
– Vacuum centrifuge
– Liquid nitrogen
– Mini Extruder (e.g., Avanti)

2.8.2 Buffers and Reagents

– Lipids (e.g., from Avanti Polar Lipids) in chloroform stock solutions
– PBS or other standard aqueous buffer solution (we typically aim to have at least as much salt in our buffer, usually as NaCl, as the final lipid concentration). We match our liposome buffer to that of our purified protein solution, which can vary depending on the desired application (e.g., CD spectroscopy vs NMR, as discussed in Section 2.11).

2.8.3 Procedure

1. In a pyrex tube, mix lipids dissolved in chloroform in the appropriate amounts to achieve the desired molar ratio and concentration for a given desired final volume
   a. Sample calculation to prepare, e.g., a 1 mL stock of 30 m*M* liposomes consisting of 70% POPC, 30% POPS. A 70/30 mix of POPC/POPS at 30 m*M* total consists of 21 m*M* POPC, 9 m*M* POPS. At a final volume of 1 mL, this requires 21 μmol POPC and 9 μmol POPS. The corresponding mass required can be calculated from the molecular mass of POPC and POPS; finally, the amount of lipid/chloroform to add to the initial mixture is calculated given the stock chloroform lipid concentration (usually 10 or 25 mg/mL).
2. Evaporate the chloroform solvent from the lipids by applying a stream of nitrogen gas using the evaporating unit (tubes will sit in the heating/stirring module). For starting volumes of 1–2 mL solvent this usually takes 15–30 min.
3. Centrifuge under vacuum for at least 1 h. This yields a film of lipids at the bottom of the pyrex tube.
4. Redissolve the lipid film in buffer to achieve a final desired concentration by pipetting up and down, with periodic vortexing. Visually verify complete dissolution of the lipid film.
5. Freeze the aqueous liquid solution in liquid nitrogen, and then thaw in warm water (37–42°C). Repeat the freeze/thaw cycle 10 times.

6. Set up the lipid extruder. For each extrusion step, sandwich a membrane of a given pore size between 2 × 2 filters (four total, two on each side) within the extruder, after wetting the filters in deionized water. Close the extruder. Insert a 1-mL syringe in one side of the extruder. Test for extruder integrity by taking up 0.5–1 mL deionized water in the second syringe, and then flowing it through the closed extruder. Discard the test water from both syringes and replace one of them into the extruder. To extrude (steps 7–8), take up the aqueous lipid solution into a syringe and insert into the extruder. We typically made vesicles with a diameter of 120–130 nm. As described below in steps 7–8, to achieve this we first extrude through a 400 nm pore membrane, and then through a 100 nm pore membrane.
7. Extrude using a membrane with a 400 nm pore size. Extrude back and forth through the extruder 21 times. Store the 1 × extruded lipid solution in the pyrex tube momentarily.
8. Discard the used filters and membrane, and set up the extruder using a membrane with a 100 nm pore size. Load the 400 nm liposome solution prepared in step 7, using the same protocol described in step 6. Extrude back and forth through the extruder 21 times. This yields the final large unilamellar vesicle stock solution.
9. Store liposomes at 4°C, and use within 1 week. It is best to displace air in the storage container with an inert gas such as argon or nitrogen to limit lipid oxidation.
10. If desired, calculate the final lipid concentration using an established colorimetric phospholipid quantification assay that has been described elsewhere (Ploier & Menon, 2016). We typically achieve a final yield of 60%–70%.
11. Characterize vesicle size and quality using electron microscopy or dynamic light scattering. This does not need to be done every time but should be performed periodically to verify adequate liposome preparation.

2.8.4 Notes

1. In step 4, if it is difficult to dissolve the lipids, it is also possible to physically scrape them from the tube with the pipet tip while pipetting up and down, taking care to ensure that all lipids get from the pipet tip back into solution.
2. Extrusion can be performed at room temperature, though it is also possible to extrude within a heat block at elevated temperatures (e.g., 37°C).

2.9 Liposome Formation by Sonication

2.9.1 Equipment

- Pyrex tubes
- Evaporating unit (e.g., Pierce Reacti-Vap), heating/stirring module (e.g., Pierce Reacti-Therm)
- Nitrogen gas
- Vacuum centrifuge
- Bath sonicator
- Ultracentrifuge (e.g., Sorvall RC M120 EX Ultracentrifuge)
- Ultracentrifuge rotor (e.g., S120AT2)
- Ultracentrifuge tubes (e.g., 1.0 mL PC Thick-Walled Tubes)

2.9.2 Buffers and Reagents

- Lipids (e.g., from Avanti Polar Lipids) in chloroform stock solutions
- PBS or other standard aqueous buffer solution (we typically aim to have at least as much salt in our buffer, usually as NaCl, as the final lipid concentration).

2.9.3 Procedure

1. Prepare lipid solution in aqueous buffer following the same steps 1–4 noted in Section 2.8.3
2. Insert the sample into a bath sonicator, with the top of the sample solution just below the surface of the water. Sonicate until the solution becomes transparent. For high quality sonicators this usually takes three cycles of 5 min of sonication each. In between cycles, centrifuge the solution to spin down any liquid that has landed on the sides of the tube as a result of agitation by sonication. This also gives the sample a chance to cool down, as sonication heats the solution somewhat, though typically incubations on ice are not necessary.
3. Clarify SUVs by centrifugation in an ultracentrifuge, 60,000 rpm, 2 h, 4° C. The supernatant contains the desired small unilamellar vesicles.
4. Store liposomes at 4°C, and use within 1 week. It is best to displace air in the storage container with an inert gas such as argon or nitrogen to limit lipid oxidation.
5. Characterize liposome yield/concentration and size as per steps 10 and 11 in Section 2.8.3.

2.9.4 Notes

1. Depending on the bath sonicator, if the solution does not become transparent in a timely fashion, it can help to look closely at the sonicated bath water to identify regions with a particularly high density of sound waves (where the water appears most agitated).

2.10 DPC Micelle Preparation

Micelles are commonly used as membrane mimetics in the study of protein–membrane interactions, and while not comprised of bona fide lipid bilayers, are significantly smaller that liposomes; thus, in contrast with protein/liposome interactions, protein/micelle interactions can be directly visualized by solution-state NMR spectroscopy.

2.10.1 Equipment

- Pyrex tubes
- Evaporating unit (e.g., Pierce Reacti-Vap), heating/stirring module (e.g., Pierce Reacti-Therm)
- Nitrogen gas
- Vacuum centrifuge

2.10.2 Buffers and Reagents

- DPC detergent (e.g., Avanti)—we typically purchase DPC as a powder
- PBS or other standard aqueous buffer solution (we typically aim to have at least as much salt in our buffer, usually as NaCl, as the final detergent concentration).

2.10.3 Procedure

1. Dissolve DPC powder in chloroform to up to 50 mg/mL
2. Measure out the appropriate amount of DPC in chloroform for a desired final volume and concentration, and dry according to steps 2–3 in Section 2.8.3.
3. Resuspend the detergent film/detergent pellet in aqueous buffer to a desired stock concentration. Stock solutions of up to 120 m*M* were generally well behaved, but were typically diluted in a timely fashion to a 50–60 m*M* working concentration in the NMR samples used for triple-resonance experiments, or to a 30 m*M* stock concentration for CD experiments. Store at room temperature and use within a few days.

2.11 NMR and CD Sample Preparation: General Considerations

Basic sample preparation for NMR or CD spectroscopic analysis of complexin and complexin/membrane binding involves preparation of protein and liposome stock solutions for subsequent mixture in desired protein:lipid ratios. Protein stocks are made by dissolving purified, lyophilized complexin protein to high concentration. We always filter our protein stock solution through a 100 kDa filter membrane after dissolution prior to use. Though this can decrease protein concentration, it can lead to better behaved samples and higher quality data; because we dissolve complexin to such high initial concentrations, the improvement in sample behavior and data quality is worth some protein loss. Liposome stocks are generated as in Sections 2.8–2.10, and, if desired, the lipid concentration can be accurately determined using the protocol outlined in Ploier and Menon (2016). Of note, we prefer a buffer pH of ~6.1, as the slightly reduced pH improves NMR spectral quality by reducing amide proton exchange with solvent water, a process than can decrease NMR signal intensity.

2.11.1 Equipment

- Eppendorf tubes
- NMR tubes and caps
- Glass pipet and bulb
- CD sample plates, with pathlength 0.02 cm (e.g., Aviv, Starna Cells Inc., or Hellma Analytics)

2.11.2 Buffers and Reagents

We use different buffers for NMR and CD. Namely, dithiothreitol (DTT) is omitted from the CD-buffer, which improves CD spectral quality. Piperazine-*N*,*N′*-bis(2-ethanesulfonic acid) (PIPES) is used as the NMR buffer because it is compatible with the colorimetric assay used to assess precise lipid concentrations (Ploier & Menon, 2016)—because this assay measures the lipid phosphate phosphorous atom, it cannot be performed in phosphate buffer. It is critical that the NMR buffer contain D_2O (usually 10% by volume), which provides the NMR spectrometer lock signal. This can be included in the stock NMR buffer per the recipe below, though in practice we usually prepared NMR buffer in 100% protonated solvent and add in D_2O when actually preparing the final NMR sample.

- purified, cleaved, and lyophilized protein, prepared above
- SUV or LUV stock, prepared above

- NMR buffer: 50 m*M* PIPES pH 6.1, 100 m*M* NaCl, 0.5 m*M* EDTA, 2 m*M* DTT, 10% D_2O
- CD buffer: 20 m*M* phosphate buffer pH 6.1, 100 m*M* NaCl, 0.5 m*M* EDTA

2.11.3 Procedure

1. Determine the concentrations and volumes of reagents required for a given experiment. For a full liposome titration by CD spectroscopy, we generally use 1/3 of the yield from expression and purification of unlabeled complexin in 1 L of bacterial culture; we dissolve the lyophilized protein in 1–2 mL and use around 20–30 μL per sample. For a full liposome titration by NMR spectroscopy, we typically dissolve the entire yield from a 4 L → 1 L ^{15}N-complexin expression and purification in 6–7 mL and use 200–300 μL per sample. For triple-resonance experiments, we dissolve the entire yield from a 4 L → 1 L isotopically labeled complexin expression and purification in 1 mL buffer.
2. Dissolve lyophilized protein in the desired volume, using the rough guidelines in step 1. If desired, a specific mass of lyophilized protein can be weighed for dissolution, though this becomes more difficult for complexin variants that lyophilize as a thin film or as a somewhat viscous, oil-like material instead of as fluffy powder.
3. Measure the pH of the sample and adjust with HCl or NaOH to get to pH 6.1 (start with small volumes of dilute acid or base so as to not over-adjust the pH).
4. Filter the protein stock solution through a 100 kDa membrane.
5. If desired, measure the liposome stock solution and/or final sample lipid concentration using published colorimetric phosphate assays (Ploier & Menon, 2016).
6. If desired, measure stock complexin protein concentration. Unfortunately, complexin lacks any native tryptophan residues and so protein concentration determination can be difficult. Rough estimates for the expected extinction coefficient can be obtained from online calculators (e.g., expasy.org ProtParam) for concentration determination using OD_{280} (absorbance at 280 nm). Alternatively, Bradford reagent can be used. We usually, however, estimate complexin concentration by running SDS-Page protein gels for dilutions of BSA standards at known concentrations in parallel with dilutions of complexin. Standard curves can be generated for integrated BSA gel band intensities, and complexin protein concentration is determined through comparison of integrated

complexin band intensities with this standard curve. Gel image analysis is performed in ImageJ.

7. Prepare NMR or CD samples by mixing protein and lipid stocks in the desired concentrations. We typically make samples fresh, as close to running the experiment as possible.
8. Load CD sample into demountable cuvette for CD analysis. Alternately, load NMR sample into NMR tube. In either case, it is important to avoid bubbles, as these can interfere with the optical path (CD) or with shimming (NMR).

3. SPECTROSCOPIC CHARACTERIZATION OF COMPLEXIN AND COMPLEXIN–LIPID INTERACTIONS

With a high quality, carefully prepared sample in hand, solution-state NMR spectroscopy provides a powerful method to assess protein structure, dynamics, and intermolecular interactions. Solution-state NMR and other spectroscopic techniques are particularly valuable for small, intrinsically disordered proteins like complexin that are not amenable to structural analysis by crystallography or cryoEM. NMR in particular provides the key advantage of delivering site-resolved information for every residue in the polypeptide sequence in the same experiment. For example, HSQC spectra can be obtained in less than 2 h each and can be performed with and without liposomes to derive site-resolved data on membrane binding by complexin (Snead et al., 2017, 2014). It is thus possible to complete entire titrations in 1–2 days, particularly given that data analysis for these experiments is straightforward. Standard sets of triple-resonance experiments can be used to obtain backbone resonance assignments and alpha- and beta-carbon chemical shifts derived from these experiments provide some information on secondary structure content and location as well; with a concentrated sample, these can be completed in 1–2 weeks (or less using more modern reduced dimensionality and nonuniform sampling approaches). With high quality spectra an experienced user can complete a backbone resonance assignment for a small protein like complexin (143 amino acids for worm complexin-1) in a few days. Given that samples can be prepared from lyophilized complexin powder, it is possible to prepare larger volume (600–700 μL), highly concentrated (>300 μ*M*) samples that give adequate data in standard NMR tubes; however, Shigemi tubes are available to improve spectral quality for low volume samples. As noted above, one purification usually yielded enough protein for one full HSQC-based liposome titration, or for one highly concentrated sample for triple-resonance experiments (where we typically dissolve the entire amount of lyophilized protein in one sample). We use NMRPipe and NMRView for data analysis, though other programs are available.

3.1 Characterization of Secondary Structure of Complexin by CD Spectroscopy

One major shortcoming of solution-state NMR spectroscopy is the inability to directly observe proteins that are bound to large slowly tumbling species such as liposomes. Far-ultraviolet CD spectroscopy, on the other hand, can be used to analyze secondary structure for both free and liposome- or micelle-bound proteins. Although the information obtained reflects global secondary structure and is not site-specific, CD has proven to be an important complementary technique for the assessment of complexin–lipid interactions (Snead et al., 2017, 2014). CD spectroscopy specifically measures the ellipticity of circularly polarized light passing through a sample (Greenfield & Fasman, 1969; Sreerama & Woody, 2004). Different elements of secondary structure display unique CD spectral signatures—i.e., distinct spectral lineshapes, with ellipticity minima and maxima at stereotypical wavelengths. Spectra for random coil or unstructured proteins contain a characteristic minimum around 195 nm but are otherwise relatively featureless (Venyaminov, Baikalov, Shen, Wu, & Yang, 1993). Helical secondary structure leads to spectra with characteristic minima at 208 and 222 nm, and a maximum at ~193 nm (Holzwarth & Doty, 1965). The beta-sheet signature consists of a minimum at 218 and a maximum at ~195 nm (Greenfield & Fasman, 1969). In theory, CD spectra can be quantitatively decomposed into fractional helix, sheet, and coil contributions (Lees, Smith, Wien, Miles, & Wallace, 2004; Whitmore & Wallace, 2008). However, such analysis requires extremely high quality data extending down to short wavelengths (190 nm or below)—this is often difficult to achieve practically. Instead, signature wavelengths are often monitored as a reliable proxy of secondary structure formation or loss upon some perturbation to the sample such as thermal denaturation (Xiao & Raleigh, 2010), addition of a binding partner (e.g., another protein (Johansson et al., 2003; Longhi et al., 2003), detergent micelles (Dikiy & Eliezer, 2014; Ferreon & Deniz, 2007), or liposomes (Davidson, Jonas, Clayton, & George, 1998)), chemical modification (Paleologou et al., 2010; Fauvet et al., 2012; Maltsev, Ying, & Bax, 2012; Dikiy & Eliezer, 2014), protein aggregation (Fares et al., 2014;

Khalaf et al., 2014), etc. Indeed, many disordered amphipathic sequences like those found in complexin adopt helical structure upon binding to liposomes (Snead et al., 2017, 2014). It is thus possible to titrate lipid vesicles into the protein and monitor binding-induced helix formation based on the stereotypical increase in negative ellipticity at 222 nm.

3.1.1 Equipment

- CD spectrometer (e.g., Aviv)
- Demountable CD cuvette, with short (e.g., 0.02 cm) pathlength (e.g., Aviv, Starna Cells Inc., or Hellma Analytics)
- Nitrogen gas

3.1.2 Buffers and Reagents

- CD samples from protein and/or lipid stocks, prepared as per Section 2.11.

3.1.3 Procedure

We try to use a liposome stock at 25–30 m*M* lipids, and a (prefiltration) protein stock at around 600 μ*M*. A typical titration is run at roughly 100 μ*M* final protein concentration, ranging from 0 to 15 m*M* final lipid concentrations. Because we were rarely limited by sample quantity, we typically would just dissolve the entire protein yield resulting from purification of a 1 L growth into around 1–2 mL buffer (1/3 of the yield from a 1 L growth should suffice if protein is being conserved). We typically prepare 80 μL CD samples using 0–40 μL liposome stock per 80 μL sample, and 10–30 μL protein per 80 μL sample. It is difficult to accurately determine the concentration of complexin. To ensure that different experiments from day to day were performed at roughly the same concentration, we empirically vary the amount of complexin stock added into the CD sample to achieve a common, comparable free-state spectrum (e.g., with the same raw ellipticity at 222 nm). This allows us to roughly match protein concentration from experiment to experiment despite the difficulty in measuring it directly.

1. With the CD spectrometer off, connect an (ideally full) tank of high purity nitrogen gas to the CD spectrometer and flow nitrogen gas over the spectrometer optics for at least 30 min to purge the chamber of oxygen gas. A full nitrogen tank is typically sufficient for an entire day of experiments.

2. Turn on the CD spectrometer (turn on the light source, ignite when ready, then turn on the electronics and computer).
3. Open the appropriate software and adjust settings as necessary. We typically acquire spectra from 250 to 200 nm, with 3–4 scans, and an averaging time of 1.7 s. If dynode voltage is not limiting at low wavelengths, it is possible to acquire below 200 nm, though we typically cannot safely achieve this.
4. Prepare stock protein and lipid solutions as in Section 2.11. Measure absolute protein and lipid concentrations if desired. General sample considerations were discussed above.
5. Fill the demountable CD cuvette with 60–70 μL buffer, load into the CD spectrometer, and acquire. After data acquisition, wash the cuvette plates thoroughly with ethanol and deionized water to clean.
6. Load the CD cuvette with 60–70 μL protein diluted to the desired concentration, and acquire.
7. Adjust the protein dilution to match the observed free-state spectrum with that generally seen in previous experiments.
8. Once an optimal protein concentration is found, prepare samples containing liposomes at a given concentration plus or minus protein. Run these samples. When performing a "lipid into protein" titration, we make multiple different samples at the same protein concentration with varied liposome concentrations.
9. Repeat step 8 for all desired liposome concentrations.
10. If CD spectra are acquired with 3–4 scans, these must be averaged for each condition. This can usually be performed in the CD software.
11. For each liposome concentration, subtract the averaged liposome-only spectrum from the averaged liposome plus protein spectrum.
12. Analyze data as desired.

3.1.4 Notes

1. It is crucial to have a matched liposome "blank" spectrum to subtract from the protein sample at each lipid concentration.

3.2 HSQC Experiments

HSQC experiments are a core technique in protein NMR spectroscopy and correlate two chemically bonded NMR-active nuclei, usually ^{1}H and ^{15}N (i.e., amide groups) or ^{1}H and ^{13}C. A wealth of data can be gleaned from ^{1}H/^{15}N HSQC spectra, though it is necessary to first assign NMR resonances/peaks to their corresponding residues/nuclei using the experiments discussed here. Resonance assignments have been successfully obtained for worm (Snead et al., 2014; Wragg et al., 2013) and mouse (Pabst et al., 2000) complexins in the free state, and for worm complexin bound to DPC micelles (Snead et al., 2017). HSQC experiments have been used to derive site-specific information about the interaction of complexin with, for example, the SNARE complex (Pabst et al., 2000; Xue et al., 2010), and with liposomes (Snead et al., 2014).

For the multidimensional NMR experiments discussed in this section, magnetization transfer occurs through scalar couplings (through space transfer via the nuclear Overhauser effect is discussed in Section 3.6). In the most basic HSQC pulse sequence, an INEPT sequence first transfers longitudinal proton magnetization (I_z) into an antiphase heteronuclear single quantum (SQ) coherence ($2I_zS_y$), hence the name of the experiment. INEPT sequences provide an efficient way to transfer magnetization from a more sensitive nucleus (e.g., ^{1}H) to a less sensitive one (e.g., ^{15}N or ^{13}C). This SQ coherence then evolves during an initial "t_1" period; during this evolution period heteronuclear chemical shifts modulate signal intensity ultimately resulting in their appearance in the final spectrum. A second INEPT sequence transfers the frequency-modulated heteronuclear SQ coherence back to ^{1}H for direct detection. Multiple scans are then performed with differing t_1 evolution periods; the resulting data consists of a two-dimensional time domain signal (directly detected ^{1}H in one dimension, and indirectly detected ^{15}N or ^{13}C in the other), that can be Fourier transformed into the frequency domain to yield the HSQC spectrum (Cavanagh, Fairbrother, Palmer, & Skelton, 2007).

^{1}H/^{15}N HSQC spectra are considered to be a "fingerprint" spectrum of a protein, as they typically contain a single peak corresponding to the amide group of each residue of the protein, excepting proline residues, which lack an N—H group. These peaks can be specifically assigned to a given residue using triple-resonance experiments as discussed below. HSQC peak position is defined by both proton and nitrogen chemical shifts, which are sensitive to factors that influence the amide group electrochemical environment. Thus, residue-specific changes in peak position observed upon perturbation of the sample can reflect structural changes, binding interactions, or chemical modifications (e.g., phosphorylation, protonation/deprotonation, methionine oxidation, etc.). In the slow exchange limit, sample heterogeneity from, e.g., multiple conformations or states can lead to multiple HSQC peaks for a single site, from whose intensities or volumes it may be possible to estimate the relative populations of these distinct states. Here, conversion between the two states is slow relatively to the NMR timescale, which is determined by the chemical shift difference between the two interconverting resonances. In the fast exchange limit, a single averaged peak is observed, as interconversion between distinct states is rapid relative to the NMR timescale. At intermediate timescales, exchange broadening can occur and can lead to peak broadening or disappearance (Cavanagh et al., 2007). HSQC experiments are often used to monitor titrations of some (unlabeled) binding partner into a given (^{15}N labeled) protein, where the partner can be, e.g., liposomes, detergents, another protein, a ligand, or metal ions. Changes in the HSQC spectra such as chemical shift perturbations or intensity decreases then reveal which specific region of the protein interacts with a given binding partner (or responds to such an interaction indirectly). Chemical shift perturbations are typically defined as

$$\Delta\delta_{avg} = \sqrt{\left(\frac{1}{2}\right)\left(\Delta\delta_{HN}^2 + \left(\frac{\Delta\delta_N}{5}\right)^2\right)}$$

where $\Delta\delta_{HN}$ and $\Delta\delta_N$ are the amide proton and amide nitrogen chemical shift differences, respectively. Nitrogen shifts are scaled by a factor of 5 because of the intrinsically larger chemical shift values of amide nitrogens. Regions with larger chemical shift perturbations are in many cases those either directly interacting with or somehow otherwise affected by the experimental perturbation. In other cases, peak intensities can change; intensity decreases can arise from exchange broadening (when the relevant environmental changes occur on the intermediate timescale), or from an increase in tumbling time upon binding of the protein to a larger, object (e.g., liposomes). In the latter case, in the slow exchange limit a visible free-state peak is in equilibrium with an invisible, bound peak. As the bound state population increases, the intensity of the observed free-state peak decreases accordingly (Bodner, Dobson, & Bax, 2009; Bussell & Eliezer, 2004; Cavanagh et al., 2007).

3.2.1 Equipment

- High field NMR spectrometer, ideally equipped with a CryoProbe
- NMR tubes (standard or Shigemi) and caps
- Spinner

3.2.2 Buffers and Reagents

- NMR sample

3.2.3 Pulse Sequences

– $^{1}H/^{15}N$ or $^{1}H/^{13}C$ HSQC

3.2.4 Procedure

Before using any NMR spectrometer, users should ensure they are adequately trained in the relevant safety issues.

We describe our general workflow without regard to spectrometer brand, aiming to convey the basic principles underlying practical aspects of setting up these NMR experiments. Users should coordinate with NMR core facility staff to learn further command and pulse sequence details. Specific commands can vary between spectrometer brand and software, but the basic aspects of locking, shimming, and tuning/matching can be found in spectrometer manuals. HSQC pulse sequences are commercially available and can be loaded relatively easily according to manufacturer's parameters. Initially this should be done with assistance from an expert user.

1. Using an NMR depth gauge, insert the NMR sample tube into the spinner to the appropriate depth. Incorrect sample position can lead to nonoptimal shimming or, in the worst case, can break the sample tube and damage the NMR probe.
2. Insert the sample-loaded spinner into the NMR spectrometer. To do this, first eject any sample that was left in the spectrometer from a previous user. Listen for the sound of airflow, indicating activation of the air cushion that will support the sample and gently load it into the spectrometer. Never insert a sample without first verifying the air cushion is on, as it will fall into the spectrometer, break, and damage the probe. Gently place the spinner on the air cushion to verify it is supported. Use the computer software to insert the sample.
3. Lock the spectrometer, indicating H_2O/D_2O as the sample solvent (we include 10% D_2O in all of our samples). The lock procedure varies depending on the spectrometer.
4. Shim the spectrometer. Shim coils provide adjustable magnetic fields to correct for inhomogeneity of the main, static magnetic field. An inhomogeneous magnetic field causes different molecules within the sample to sense a different external magnetic field and so to resonate at slightly different frequencies. By optimizing field homogeneity, shimming improves signal intensity and linewidth and so is necessary for acquisition of high quality data. Many modern spectrometers have the ability to shim automatically, but it is also possible to manually shim, or to

manually touch up the shims after autoshimming. Commands will vary depending on the spectrometer.

5. Tune the spectrometer. Different samples may have different dielectric constants, and so tuning and matching are required to optimize energy transfer from the probe to the sample. Depending on the experiment, it is necessary to tune and match the probe proton, nitrogen, and/or carbon channels. Some probes will have the ability to tune and match automatically, while others require this to be performed manually. Commands will vary depending on the spectrometer.
6. Calibrate the proton 90 degree pulse width using the proton signal from water. We usually do this by first calibrating the 360 degree pulse width and then dividing by four (systematically varying the proton pulse with from 0 to 360 degrees yields positive signal from 0 to 180 degree, with a maximum at 90 degree and zero signal at 180 degree, and negative signal from 180 to 360, with a minimum at 270 degree, and again zero signal at 360 degree). The 90 degree pulse transfers longitudinal proton magnetization to transverse magnetization, and precession in the transverse plane results in the detected signal. Thus, a calibrated 90 degree pulse is necessary to optimize NMR signal—over or under rotation of longitudinal magnetization causes some magnetization to remain along the *z*-axis rather than the transverse plane and thereby decreases the observed signal. It is also possible to calibrate the nitrogen or carbon 90 degree pulse width for each sample, though these pulses exhibit much less variation between samples, as long as they have reasonably similar dielectric constants. HSQC experiments usually provide sufficient sensitivity such that small errors in the nitrogen or carbon pulse widths do not drastically impact spectral quality.
7. Load the HSQC pulse sequence. Adjust the proton 90 degree pulse with to that calculated in step 6.
8. Adjust HSQC parameters as desired. We typically use a proton spectral width of 12–20 ppm centered at 4.7 ppm (the water resonance) with 1024 complex data points. Total acquisition time in the direct dimension should be kept under ~150 ms if high power decoupling is used to prevent probe damage. The TROSY version of an HSQC, where there is no decoupling during direct detection allows for longer acquisition times. A nitrogen spectral width of 24 ppm centered on 117.5 or 118 ppm is typical for a disordered protein such as complexin, with as many complex data points as reasonable for a desired experiment length (usually 256 complex data points). Carbon spectral widths

and carrier frequency settings will depend on the region of the spectrum that is desired. The number of repeat scans performed and integrated at each nitrogen/carbon evolution delay must be adjusted to achieve the desired signal to noise. Notably, the signal to noise only scales as the square root of the number of scans. With high quality, high concentration samples run on a 600 MHz spectrometer equipped with a CryoProbe, 2 h of data acquisition yields high quality $^{1}H/^{15}N$ spectra.

9. Run the experiment.

10. When data collection has finished, save the data and export to a computer for analysis.

11. We process our data using NMRPipe (Delaglio et al., 1995). Within NMRPipe, the raw NMR data are first converted to an appropriate PIPE file; the "bruker" or "varian" macro/command opens an interactive window to facilitate this process. During this conversion the user inputs the actual sample temperature, spectral settings (e.g., spectral width and central frequency), and data acquisition mode. If run from within a folder containing the appropriate spectrometer files, the "read parameters" command automatically loads many of the appropriate parameters, but careful manual adjustment/verification is required. This process automatically references the dataset based on the position of the water resonance and the sample temperature, assuming that the transmitter frequency is set on the water signal.

12. The commands in step 11 generate a fid.com file that can be run to generate a test.fid file for analysis in the program NMRDraw (command "nmrDraw"). NMRDraw facilitates phasing of the spectrum in the proton and nitrogen dimensions and contains a number of sample processing macros. For HSQC spectra, we adjust the basic 2D processing macro as:

```
nmrPipe -in test.fid \
| nmrPipe -fn SOL \
| nmrPipe -fn SP -off 0.5 -end 1.00 -pow 1 -c 1.0 \
| nmrPipe -fn ZF -auto \
| nmrPipe -fn FT -auto \
| nmrPipe -fn PS -p0 0.0 -p1 0.0 -di -verb \
| nmrPipe -fn TP \
| nmrPipe -fn SP -off 0.5 -end 1.00 -pow 1 -c 0.5
| nmrPipe -fn PS -p0 0.0 -p1 0.0 -di -verb \
| nmrPipe -fn TP \
| pipe2xyz -nv -out SAMPLENAME.nv -verb -ov
```

where –p0 0.0 and –p1 0.0 should be adjusted (from 0.0) to the actual calculated phase corrections for proton (top line) and nitrogen (bottom line). "SAMPLENAME.nv" should be adjusted as per the desired sample name, but with the .nv file type retained. Note that optimal window functions such as the Kaiser function can provide superior results.

This script, in order, loads the test.fid file, applies a solvent filter to remove residual signal from solvent/water, and then starting in the proton dimension applies an adjustable sine window function, automatically zero-fills the data set, Fourier transforms the proton dimension, phases the proton dimension, transposes the spectrum, applies an adjustable sine window function to the indirect dimension, zero fills the data, Fourier transforms the indirect dimension, phases it, and then transposes the spectrum back before converting the file to an NMRView compatible .nv file.

13. Once the final processed file is saved as a .nv file, we analyze our data using NMRView (Johnson & Blevins, 1994). Very briefly, NMRView allows the user to pick and annotate individual NMR peaks in a given spectrum, and it gives relevant parameters such as chemical shift and peak intensity. Peaklists can be generated and saved. After generating a peaklist for free complexin, for example, it can be copied, assigned to a different HSQC spectrum (e.g., for a sample of complexin with liposomes), and used to analyze, e.g., peak intensities for this different sample condition.

3.2.5 Notes

1. Modern improvements to these, as well as other experiments, such as nonuniform sampling and BEST/SOFAST pulse sequences can significantly reduce experimental acquisition time.

3.3 Backbone Resonance Assignment Using Standard Triple-Resonance Experiments

Peaks in the $^{15}N/^{1}H$ HSQC spectrum must be specifically assigned to a given residue for the above types of HSQC-based experiments to be maximally useful. Further, HSQC experiments do not reveal much information regarding local secondary structure. This type of information is instead derived from triple-resonance experiments, where spectra of $^{15}N^{13}C$ double-labeled protein typically correlate the amide proton, nitrogen, and one or more carbons of the peptide backbone and/or side chain (side chain assignment is also feasible, but has not been performed for complexin and is outside the scope of this chapter). A standard suite of triple-resonance experiments has been developed and optimized for protein backbone resonance assignment and is widely available. These rely on magnetization transfer through scalar couplings, with chemical shift evolution in two indirect dimensions, nitrogen and carbon, followed by direct detection on proton; this generates three-dimensional spectra containing peaks whose chemical shifts correlate an amide proton and nitrogen with a nearby carbon—intra- and or inter-residue alpha carbons, beta carbons, or carbonyl carbons (Bax & Ikura, 1991; Cavanagh et al., 2007; Clubb, Thanabal, & Wagner, 1992; Kay, Ikura, Tschudin, & Bax, 1990). Note that more recently, experiments using direct detection of ^{13}C or even ^{15}N magnetization have been developed for protein backbone resonance assignment, especially in the case of highly disordered proteins like complexin. These experiments typically rely on CryoProbes where the ^{13}C and/or ^{15}N channel preamplifiers are also cooled.

Importantly these three-dimensional experiments come in complementary pairs: one experiment correlates the amide group of a given residue, i, and a carbon of the preceding residue, $i-1$, while a second experiment correlates that same amide group for residue i, with the same type of carbon within the same residue i (or with both the i and $i-1$ carbons, leading to two peaks at a given proton and nitrogen chemical shift). For example, an HNCO correlates the nitrogen and proton in residue i with the carbonyl carbon (CO) of the $i-1$ residue; an HNCACO correlates the same proton and nitrogen of residue i with the carbonyl carbon of both residue i and residue $i-1$. CBCA(CO)NH and HNCACB spectra represent a similar pair but correlate amide groups with both alpha and beta carbons (for residue $i-1$ with CBCA(CO)NH, and for both residue i and $i-1$ for the HNCACB). Lastly, amide group/alpha carbon correlations can be obtained from HN(CO)CA (correlates with $i-1$ alpha carbon) and HNCA (correlates with both i and $i-1$ alpha carbons) spectra (Cavanagh et al., 2007). For HNCACO, HNCA, and HNCACB spectra, the peaks resulting from intraresidue correlations tend to be stronger (i.e., of greater intensity) than interresidue peaks; this feature is often useful during assignment. Armed with chemical shifts of both intra (i) and interresidue ($i-1$) carbons for a given amide peak, it is usually possible to scan through the spectrum to find the amide peak corresponding to the $i-1$ residue—i.e., that peak whose intra (i) carbon chemical shifts match the interresidue ($i-1$) shifts of the initial residue. In this way, it is possible to essentially "walk through" the protein sequence and connect each peak with its $i-1$ peak, which can then be connected to its $i-1$ peak, and so on (Fig. 2). Importantly, distinct amino

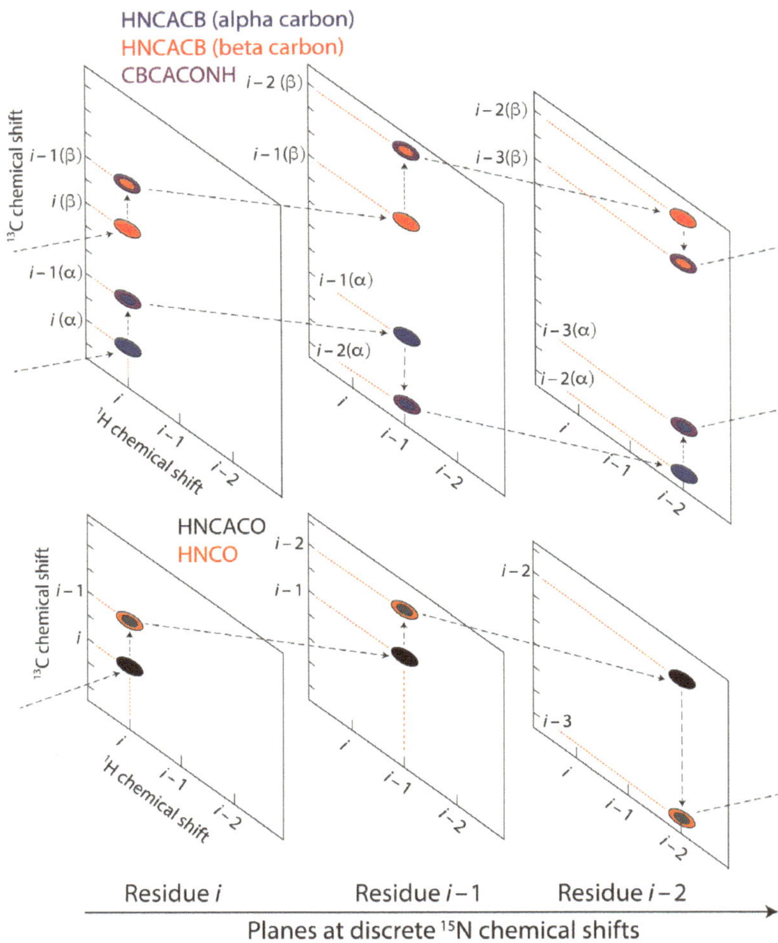

Fig. 2 Sequential backbone resonance assignment using standard triple-resonance experiments. Triple-resonance pulse sequences typically correlate three different nuclei (typically proton, nitrogen, and carbon) via through-bond scalar couplings. Different pulse sequences correlate the backbone amide group with distinct carbon atoms. The HNCO experiment (*orange*), for example, correlates the carbonyl carbon of the preceding residue ($i-1$) with the amide group at residue i. An HNCACO experiment (*black*) correlates both the carbonyl carbon of the same residue, i, and of the preceding residue, $i-1$, to the amide group of residue i, though the intraresidue correlation (i) is typically more efficient, yielding a higher peak intensity. The HNCO and HNCACO thus provide a matched pair of experiments. Similarly, the CBCACONH experiment (*purple*) labels the alpha and beta carbons of the preceding, $i-1$ residue, while an HNCACB experiment labels the intraresidue (i) and interresidue ($i-1$) alpha (*blue*) and beta (*red*) carbons. The CBCACONH and HNCACB thus provide another complementary pair. Starting from a given amide group for residue i, correlated with carbon chemical shifts for both intraresidue (i) and interresidue ($i-1$) carbons, one can then search to find an amide group with appropriate matching carbon values corresponding to the $i-1$ residue. This $i-1$ amide group can then be matched to the $i-2$ residue, and so one can sequentially assign a polypeptide sequence.

acids yield somewhat stereotypical alpha- and beta-carbon chemical shifts, which further aids assignment when the protein sequence is known (Wishart, Bigam, Holm, Hodges, & Sykes, 1995).

Of note, CO-, alpha-, and beta-carbon chemical shifts reflect local secondary structure (Spera & Bax, 1991; Wishart, Sykes, & Richards, 1991). Chemical shifts have been tabulated for each amino acid for the expected value if that amino acid populates a fully random coil conformation (we often use random coil values obtained from linear hexapeptides in 1 *M* urea at pH 5.0 and 25°C (Wishart et al., 1995), or more recently tabulated values optimized for disordered proteins (Kjaergaard, Brander, & Poulsen, 2011; Kjaergaard & Poulsen, 2011)). Comparison of observed alpha- and beta-carbon chemical shifts to these tabulated random coil values yields important structural information. We primarily depend on a parameter known as a "secondary shift," which is calculated as

$$\text{Secondary shift} = (C\alpha_{obs} - C\alpha_{rc}) - (C\beta_{obs} - C\beta_{rc})$$

where $C\alpha_{obs}$ and $C\alpha_{rc}$ are the observed and tabulated random coil alpha-carbon chemical shifts, respectively, and $C\beta_{obs}$ and $C\beta_{rc}$ are the observed and tabulated random coil beta-carbon chemical shifts, respectively. Notably, this secondary shift is independent of ^{13}C referencing errors, which would affect alpha and beta carbons equally and so cancel out upon subtraction (Kunze et al., 2012). For a given residue, positive values for this secondary shift indicate a tendency toward helical structure, negative values correspond to more extended or beta-strand structure, and values around zero imply random coil structures (Wishart et al., 1991). Carbonyl secondary shifts also reflect local secondary structure, with positive and negative values corresponding to helical and extended structure, respectively (Zhang, Neal, & Wishart, 2003).

The spectra of many proteins can be assigned through the basic set of triple-resonance experiments noted above. However, while intrinsically disordered proteins yield spectra with intense, narrow bands, they also feature poor chemical shift dispersion and many highly overlapping peaks. Multiple degenerate peaks can make resonance assignment quite difficult in these cases. A variety of additional three- and four-dimensional NMR experiments can be performed to facilitate analysis when the standard dataset above is insufficient. For example, a four-dimensional HN(COCA)NH correlates a given amide group with the preceding ($i-1$) amide group; the four spectral dimensions are thus H ($i-1$), N ($i-1$), N (i), and H (i). Higher dimensionality experiments, often run using a reduced dimensionality sampling approach, can be very useful for large disordered proteins such as tau (Harbison, Bhattacharya, & Eliezer, 2012; Narayanan et al., 2010). Larger proteins or large protein complexes can also present problems. For example, the complexin/DPC complex is difficult owing to low signal intensity and broad peak linewidths (Snead et al., 2017).

For both the complexin/DPC micelle complex (Snead et al., 2017), and for difficult-to-assign, crowded regions of the complexin free-state spectrum (Radoff et al., 2014; Snead et al., 2014), despite different experimental challenges, we were able to obtain an adequate dataset resonance assignment using perdeuterated protein (and, for the DPC micelle bound state, using deuterated DPC) combined with TROSY (transverse relaxation-optimized spectroscopy) versions of the standard triple-resonance experiments. Both perdeuteration and TROSY-based pulse sequences drastically decrease linewidths and so decrease spectral overlap. Very briefly, TROSY experiments depend on the fact that different coherences (different component signals of coupled multiplets for a particular nucleus) that are generated during a given pulse sequence have different relaxation rate constants. Larger transverse relaxation rate constants lead to faster signal decay and thus broader signals, as the NMR lineshape arises from the Fourier transform of the exponential decay of the free induction signal; in other words, if the signal decays too quickly, there is not enough information available in the recorded signal to precisely determine exact frequencies. In a standard (i.e., non-TROSY) experiment, total magnetization of all multiplet components is measured as they are decoupled into a single peak; TROSY experiments instead select a particular multiplet peak (or subset of peaks) with the smallest transverse relaxation rate constants and thus with the narrowest lineshape (Cavanagh et al., 2007). Though TROSY decreases the inherent experiment sensitivity, for a protein like complexin, where high protein concentrations are easily achievable, it is often worth trading pulse sequence sensitivity for enhanced resolution.

3.3.1 Equipment

- High field NMR spectrometer equipped with a CryoProbe
- NMR tubes (regular or Shigemi) and caps
- Spinner

3.3.2 Buffers and Reagents

- NMR sample

3.3.3 Pulse Sequences

We found the following groups of triple-resonance experiments to be the most useful, though a number of others are available. TROSY-versions of these experiments are also available and have proven helpful for complexin, as noted above.

- HNCACB/CBCA(CO)NH/HNCA (note that for perdeuterated proteins, CBCA(CO)NH must be replaced with an HN(CO)CACB experiment)
- HN(CA)CO/HNCO
- HN(COCA)NH

3.3.4 Procedure

1. Insert the sample, lock, shim, and tune according to steps 1–5 in Section 3.2.4.
2. Calibrate the proton 90 degree pulse width, as in step 6 of Section 3.2.4. Triple-resonance experiments have substantially reduced sensitivity relative to the two-dimensional HSQC experiment, and so it becomes more important, here, to also carefully adjust the nitrogen and carbon 90 degree pulse widths. Pulse sequences are available to facilitate this calibration process.
3. Load the appropriate triple-resonance experiments and adjust parameters as necessary. A detailed discussion of this process is beyond the scope of this chapter, and users should coordinate with their local NMR core facility staff or with an expert NMR user.
4. Run experiment, save the data files, and transfer to computer for analysis.
5. We process and analyze three-dimensional data using NMRPipe and NMRView using the same workflow as above in steps 11–13 of Section 3.2.4. Below is a typical NMRPipe script for three-dimensional spectra:

```
xyz2pipe -in fid/test%03d.fid -x -verb
| nmrPipe -fn SOL \
| nmrPipe -fn SP -off 0.5 -end 0.98 -pow 2 -c 0.5 \
| nmrPipe -fn ZF -auto \
| nmrPipe -fn FT -auto \
| nmrPipe -fn PS -p0 0.0 -p1 0.0 -di \
| nmrPipe -fn EXT -left -sw \
| nmrPipe -fn TP \
| nmrPipe -fn SP -off 0.5 -end 0.98 -pow 1 -c 1.0
| nmrPipe -fn PS -p0 0.0 -p1 0.0 -di \
| pipe2xyz -out ft/test%03d.ft2 -y
```

```
xyz2pipe -in ft/test%03d.ft2 -z -verb
| nmrPipe -fn SP -off 0.5 -end 0.98 -pow 2 -c 0.5 \
| nmrPipe -fn ZF -auto \
| nmrPipe -fn FT -auto \
| nmrPipe -fn PS -p0 0.0 -p1 0.0 -di \
| pipe2xyz -out ft/test%03d.ft3 -z
xyz2pipe -in ft/test%03d.ft3 -x -verb
| nmrPipe -fn ZTP \
| nmrPipe -fn TP \
| nmrPipe -fn ZTP \
| pipe2xyz -nv -out SAMPLENAME.nv -verb -ov
```

As above, phasing values should be adjusted as determined in nmrDraw. Depending on the pulse sequence used, FT (Fourier transform) command modifiers such as –alt or –inv may be required in place of –auto.

3.4 Derivation and Quantitative Analysis of Complexin Membrane Binding Curves

On a size scale relevant for solution-state NMR, liposomes are large, slowly tumbling objects. Thus, for proteins that bind to the liposome surface, the large, slowly tumbling protein/liposome complex cannot be directly observed by solution-state NMR spectroscopy. It is possible, however, to titrate liposomes into protein and monitor the disappearance of NMR peaks as a measure of membrane binding by the protein (Bodner et al., 2009; Bussell & Eliezer, 2003, 2004). Importantly, these experiments allow specific identification of those residues that interact with liposomes, as their corresponding peaks decrease in intensity upon binding to the membrane (Fig. 3). In the slow exchange limit, the observed intensity ratio (intensity in the presence of liposomes divided by the intensity for the free state), subtracted from one, directly reflects the population of membrane bound states. Noninteracting residues, to the extent that their motions are decoupled from those of bound residues, should in theory retain a high degree of mobility and a high signal intensity. Equilibrium partitioning of complexin into a lipid bound state can thus be probed in a site-specific manner, and the population of bound vs free protein can be calculated from the ratio of bound-state to free-state signal intensities (the intensity ratio) (Bodner et al., 2009; Bussell & Eliezer, 2004). These experiments can be performed with a variety of liposome sizes and lipid compositions. Intensity ratios calculated at multiple lipid concentrations from lipid into protein titrations allow derivation of binding curves for any given

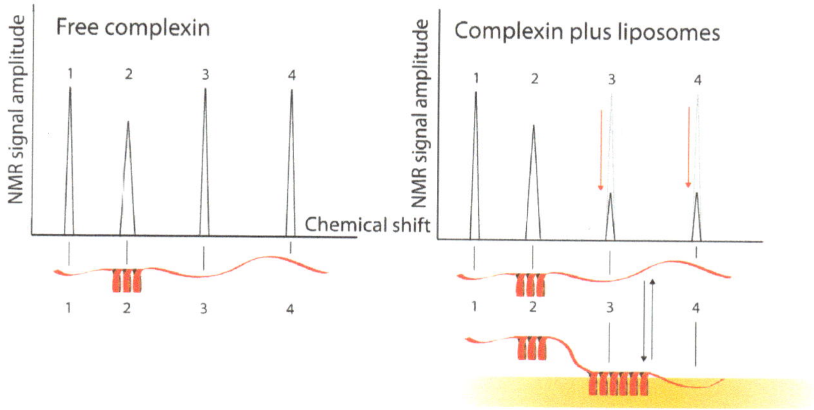

Fig. 3 HSQC intensity ratios provide a quantitative measure of membrane binding by disordered proteins. In the free state (*left*), residues within disordered regions exhibit sharp, narrow resonances (peaks 1, 3, and 4), while secondary and tertiary structure can lead to weaker, broader peaks (peak 2). In the presence of liposomes (*right*), when membrane binding is in the slow exchange limit relative to the NMR timescale, an NMR-invisible bound state is in equilibrium with an NMR-observable free state. This leads to a drop in the intensity of the observed free-state peak, without peak broadening, for residues that bind to the membrane (peaks 3 and 4). The remaining peak intensity results from the fraction of protein for which the corresponding residues remain unbound. In the limit that unbound residues are motionally decoupled from any bound regions of the protein, the ratio of the peak intensity in the presence of liposomes to that in the absence of liposomes quantitatively reflects the fraction of protein for which these residues are unbound. For disordered proteins, residues that do not bind liposomes typically retain enough mobility that their peaks are essentially unaffected (peaks 1 and 2).

specific residue, or for stretches of polypeptide (averaging the intensity ratio/fraction bound across multiple residues).

We interpret our binding curves using a method established by Pfefferkorn and Lee (2010). Briefly, binding curves are fit to the equation

$$F_B = \frac{1}{2P_t}\left(B_{\max}L + P_t + \frac{1}{K_a^{\text{app}}} - \sqrt{\left(B_{\max}L + P_t + \frac{1}{K_a^{\text{app}}}\right)^2 - 4P_t B_{\max}L}\right)$$

where P_t is the total protein concentration, L is the total lipid concentration, K_a^{app} is the apparent association constant, and $B_{\max}$ is the number of binding sites per lipid molecule. Of note, this equation contains two floating parameters, K_a^{app}, the apparent association constant, and $B_{\max}$, the number of binding sites per lipid molecule (the number of lipids per binding site can be calculated as $1/B_{\max}$). This equation is simply the solution to the quadratic equation that results from the equation describing an equilibrium of a bimolecular reaction (Bussell & Eliezer, 2004; Pfefferkorn & Lee, 2010). It is important to note, however, that binding sites on a membrane surface are not independent molecular entities, and therefore the reaction is not accurately described as bimolecular. Instead, the reaction would be better described as a partitioning of the protein onto the membrane surface (Ducas & Rhoades, 2012). Nevertheless, the $B_{\max}$ and K_a^{app} paramenters that result from fitting this equation can be useful for a qualitative understanding of the binding interaction.

3.4.1 Equipment

- Equipment as noted above necessary to obtain and analyze HSQC spectra
- Computer with curve fitting software

3.4.2 Buffers and Reagents

- NMR samples as prepared in Section 2.11.

3.4.3 Procedure

1. Acquire an HSQC spectrum of free protein in the absence of liposomes as per Section 3.2. We typically prepare 600 μL samples, with 50 μL D_2O, 250 μL complexin protein stock, and 0–300 μL liposomes (corresponding to 0–15 m*M* final lipid concentration). For a binding analysis by NMR to be maximally useful, the HSQC spectrum should be assigned using the experiments discussed in Section 3.3. Take note of relevant NMR parameters such as receiver gain, number of scans, spectral widths, and numbers of complex data points—these must be kept constant for all HSQCs.
2. Prepare the next NMR sample, containing the same concentration of complexin as the free-state sample, and the desired lipid concentration.
3. Acquire an HSQC spectrum as described in step 1, for the sample prepared in step 2. The sample must be relocked, reshimmed, and retuned, but otherwise HSQC spectral parameters (receiver gain, number of scans, spectral widths, and numbers of complex data points) must be kept constant.
4. Acquire HSQC spectra for multiple lipid concentrations spanning the desired range.
5. Process and analyze the data as noted in Section 3.2. We perform our analyses in NMRViewJ, which facilitates HSQC peak picking and annotation. Peaklists can be saved and transferred from the free-state

HSQC to each subsequent lipid/complexin HSQC. Briefly, we first manually generate and save a peaklist for the free-state spectrum; peak intensities can be found under peak > show peak table. This peaklist can be copied, renamed, and opened during analysis of subsequent HSQC spectra. We manually touch up peak positions, and then associate the copied peaklist with its relevant spectrum (peak > integrate > associate dataset), before calculating intensities for the new HSQC (peak > integrate > get intensities).

6. Once peak intensities have been calculated for all relevant peaks in all HSQC spectra, intensity ratios can be calculated as intensity in the lipid bound state divided by intensity in the free state.
7. Binding curves can now be generated for any given residue by plotting fraction bound (1 minus the intensity ratio) as a function of lipid concentration. We typically averaged the fraction bound across multiple residues within a region of interest, typically a region that is expected or observed to bind cooperatively.
8. Fit curves to the equation discussed above, typically using a nonlinear least squares algorithm.

3.4.4 Notes

1. Because titration analyses require acquisition of many HSQC spectra, it is beneficial to optimize HSQC settings such that data of adequate quality can be obtained as quickly as possible. For a disordered protein like complexin that intrinsically yields strong, sharp signals, and that can be prepared at relatively high concentrations in low pH buffer, we usually ran 1–4 h HSQC spectra.

3.5 Relaxation Experiments and Membrane Binding On-Rate Calculation

Insight into protein dynamics can be obtained from relaxation of the NMR signals. Relaxation parameters and their measurement are discussed in more detail below. Importantly, however, on rates for complexin binding to liposomes can be estimated through comparison of ^{15}N transverse relaxation rates obtained in the absence of liposomes vs in the presence of large or small unilamellar vesicles (Bodner, Maltsev, Dobson, & Bax, 2010; Snead et al., 2014). Transient binding of a population of protein residues to liposomes leads to very rapid relaxation of their nuclear magnetization, and effectively eliminates their contribution to the NMR signal (Fig. 4). As a result, liposome binding can contribute to the decay of observable magnetization in

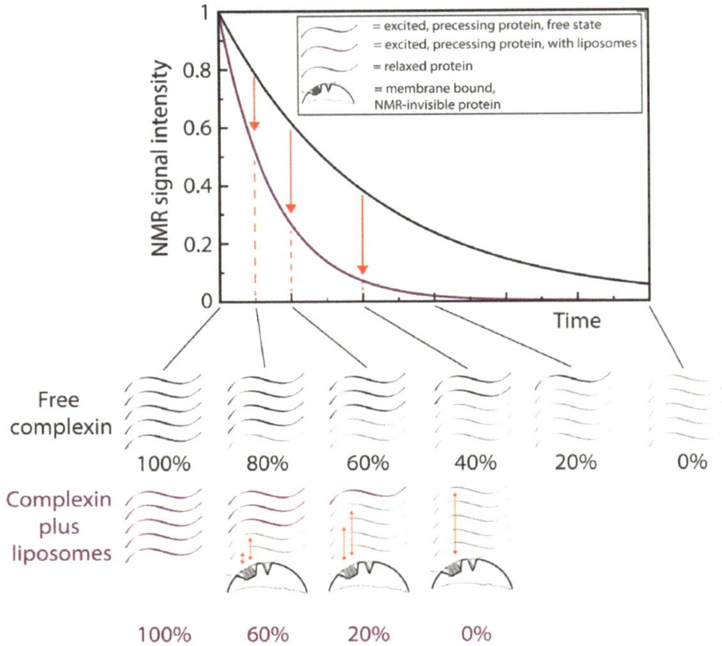

Fig. 4 Membrane binding contributes to T_2 relaxation, allowing calculation of binding on rates. In the free state (*black*), complexin NMR resonances exhibit transverse relaxation with some intrinsic relaxation rate, T_2 (illustrated by the transition of individual molecules from *black* to *gray*, leading to the *black decay curve*). In the presence of liposomes (*purple*), transient binding of individual molecules to the membrane surface results in highly efficient (effectively instantaneous) relaxation in the bound state, eliminating their contribution to the observable NMR signal and resulting in a more rapid decay (*purple decay curve*). Thus, the observed T_2 relaxation is enhanced, with $T_2^{obs} = T_2^{intrinsic} + k_{on}$.

NMR experiments, provided that it occurs on a time scale similar to that dictated by the intrinsic R_2 relaxation rate of the free state. Under these conditions, the R_2 value observed with vesicles approximately equals the sum of the intrinsic, free state R_2, and the vesicle binding on rate (Bodner et al., 2010). Normalization of on rates derived in this way for lipid concentration (or for membrane binding site concentration) yields an approximation of a second order association rate constant (Snead et al., 2014).

Longitudinal relaxation (known as "spin–lattice" relaxation, with relaxation time T_1 and relaxation rate R_1) and transverse relaxation (known as "spin–spin" relaxation, with relaxation time T_2 and relaxation rate R_2) reflect local polypeptide mobility/flexibility. Longitudinal relaxation represents decay of excited nuclei back to the ground state (i.e., relaxation back toward alignment with the externally applied magnetic field along the longitudinal axis). Transverse relaxation arises from a dephasing of excited state magnetization precessing in the transverse plane: subtle inhomogeneities across an observed ensemble cause the local environment surrounding a given nucleus in distinct molecules to differ slightly and so lead to slightly different precession rates. Eventually, this leads to signal decay as the nuclear magnetization's phase coherence is lost (Cavanagh et al., 2007). Nitrogen ^{15}N R_2 relaxation rates are a particularly useful and easily measured indicator of protein motions on the picosecond to nanosecond timescale or of chemical exchange processes on the millisecond timescale.

Practically, ^{15}N R_2 relaxation rates are derived from a series of HSQC-like experiments with a variable time delay following preparation of transverse ^{15}N magnetization through a refocused INEPT pulse sequence. Cross-relaxation from $^{1}H—^{15}N$ dipole/dipole coupling and ^{15}N chemical shift anisotropy are suppressed by a series of 180 degree proton refocusing pulses, and chemical shift evolution is prevented through CPMG (Carr–Purcell–Meiboom–Gill) refocusing pulse trains (Cavanagh et al., 2007; Palmer, Skelton, Chazin, Wright, & Rance, 1992). During this period, nitrogen undergoes transverse relaxation, after which an HSQC-like sequence provides proton nitrogen correlations. The resulting signal decreases with increasing relaxation time delay; the signal intensity (or volume) for each amide cross-peak is thus plotted as a function of the time delay and fit to a single exponential decay function using an internal NMRViewJ fitting module (Johnson & Blevins, 1994):

$$I = I_0 e^{-R_2 t}$$

where I is the signal intensity at time t, I_0 is the intensity at time 0, R_2 is the transverse relaxation rate, and t is the relaxation time. Fits yield residue-specific transverse relaxation rates, which in turn reflect local polypeptide flexibility and mobility.

3.5.1 Equipment

- High field NMR spectrometer equipped with a CryoProbe
- NMR tubes (regular or Shigemi) and caps
- Spinner

3.5.2 Buffers and Reagents

- NMR sample

3.5.3 Pulse Sequences

– ^{15}N T2 CPMG

3.5.4 Procedure

1. Load the NMR sample, and then lock, shim, and tune according to steps 1–5 in Section 3.2.4.
2. Calibrate proton and nitrogen 90 degree pulse widths as per step 2 of Section 3.3.4.
3. Load the appropriate relaxation experiment pulse sequence and adjust parameters as necessary. A detailed discussion of this process is beyond the scope of this chapter, and users should coordinate with their local NMR core facility staff on an expert NMR user. Relaxation experiments can damage the spectrometer probe if not set up appropriately, as they entail trains of high-power pulses. We used 8–13 relaxation delays from approximately 15–340 ms, though longer delays are required for free complexin vs liposome bound complexin (intrinsically disordered proteins like complexin have particularly long T_2 relaxation times, and so lengthy relaxation delays are required for optimal fits to the signal decay). We typically use a 900 μs delay between CPMG pi-pulses.
4. Run experiment, save the data files, and transfer to computer for analysis.
5. The pulse sequence used to measure T_2 relaxation essentially generates a series of HSQC spectra at different relaxation delays. Each of these can be processed as per the standard HSQC processing noted above.
6. We analyze our T_2 relaxation data using NMRView, which contains an internal fitting module (Johnson & Blevins, 1994).
7. Create a time file summarizing the number of relaxation delays and their lengths, with the format below (in this fictional example from a first relaxation delay of 20 ms to a final relaxation delay of 300 ms). Save as a .txt file.

   1 20
   2 40
   3 60
   4 80
   5 100
   6 120
   7 140
   8 160
   9 180
   10 200

11 220
12 240
13 260
14 280
15 300

8. For the first HSQC with the shortest time delay, create and annotate a peaklist with each peak of interest picked and labeled. It is crucial to have each HSQC saved according to some standard nomenclature, e.g., HSQC_1, HSQC_2, …, HSQC_15. The peaklist should have the name of the HSQC it is linked to, e.g., HSQC_1.
9. With all HSQCs and the peaklist for HSQC_1 loaded and open in NMRView, go to analysis > rate analysis to open the internal fitting module. Input "preface for matrix numbers" as the standardized nomenclature from step 8, in this example "HSQC_". Select relaxation type T_2. Otherwise leave the default settings (e.g., suffix for matrix names as .nv and peak intensity mode as jitter). Select the peaklist to be used (HSQC_1). Load the time file created in step 7. Hit "measure all" to perform the analysis. This fits the data for each peak selected to the exponential decay function noted above. Under "fitdata," the intensity at time 0 is noted as parameter *a*, while the relaxation rate corresponds to the max value for parameter *b*.

3.6 NOESY Experiments

Through bond magnetization transfer can occur in NMR experiments through scalar coupling interactions. Magnetization transfer may also be achieved through-space, and one through-space transfer mechanism is known as the nuclear Overhauser effect (NOE). This process is a result of dipolar couplings between nuclei; it is thus proportional to r^{-6}, where r is the distance between two nuclei (Cavanagh et al., 2007). NOE transfer is only efficient for nuclei within a short distance of each other, usually <5 Å; NOE transfer is therefore enhanced in regions of relatively rigid and collapsed secondary structure, such as alpha-helix, beta-sheet, or turn regions, compared to more disordered protein regions. NOE-type experiments can facilitate resonance assignment, particularly for the assignment of structured regions, which will display stronger $i,i\pm1$ NH-NH NOE cross-peaks (Wuthrich, 1986). The $^{1}H/^{15}N$ HSQC-NOESY-HSQC experiment is particularly useful (Zhang, Forman-Kay, Shortle, & Kay, 1997). In this experiment, an initial nitrogen evolution period is followed by magnetization transfer to proton (essentially completing the first HSQC), an NOE mixing period where NOE transfer may occur between nearby protons, and a second HSQC where both nitrogen and proton are detected (indirectly and directly, respectively); this yields a three-dimensional spectrum whose axes are ^{15}N, ^{15}N, and ^{1}H, with NOE cross-peaks (Cavanagh et al., 2007). NOEs are critical for most structure determinations using solution NMR spectroscopy, though complete structure calculations require side chain restraints in addition to backbone type experiments. Different types of secondary structure elements exhibit unique patterns of NH-NH NOE connectivity along the peptide backbone. Alpha-helical regions display strong $i,i+1$ NH-NH cross-peaks as well as detectable $i,i+2$, $i,i+3$, and $i,i+4$ NH-NH NOE cross-peaks. Beta-sheets, on the other hand, are characterized by weaker $i,i+1$ NH-NH NOE cross-peaks. Disordered regions tend to exhibit weaker NOE cross-peaks (Wuthrich, 1986). For complexin, sequential $i,i+1$ NH-NH NOESY cross-peak intensities were measured and used to obtain information about the free and DPC-micelle bound structures of the protein (Snead et al., 2017, 2014).

The structural characterization of protein/vesicle interactions by solution-state NMR spectroscopy is intrinsically limited by the fact that the slowly tumbling vesicle bound state is "NMR-invisible"—increased T_2 relaxation enhances signal decay and effectively broadens the NMR peaks beyond detection. For this reason, experiments are often performed with membrane mimetics, such as with DPC as described above (Snead et al., 2017). Nonetheless, a few solution-state NMR techniques have been developed to probe the membrane bound "dark state." For example, some information can be recovered from transferred NOE experiments, which depend on protein exchange onto and off of the vesicle surface during the NOE mixing period (Bodner et al., 2009, 2010). In other words, a given molecule binds to the membrane, undergoes NOE transfer in the bound state, unbinds, and then is directly detected (Fig. 5). NOE transfer occurring in the membrane bound state is thus "read out" in the free state. Helical structure confers enhanced NOE transfer, but these experiments suffer from the fact that the slowly tumbling proton-rich liposomes also enhance multistep spin diffusion, which ultimately limits the structural information that can be obtained from such NOEs. Transferred NOEs have been calculated for complexin using HSQC-NOESY-TROSY(HSQC) pulse sequences, a TROSY version of the HSQC-NOESY-HSQC experiment discussed above (Snead et al., 2014).

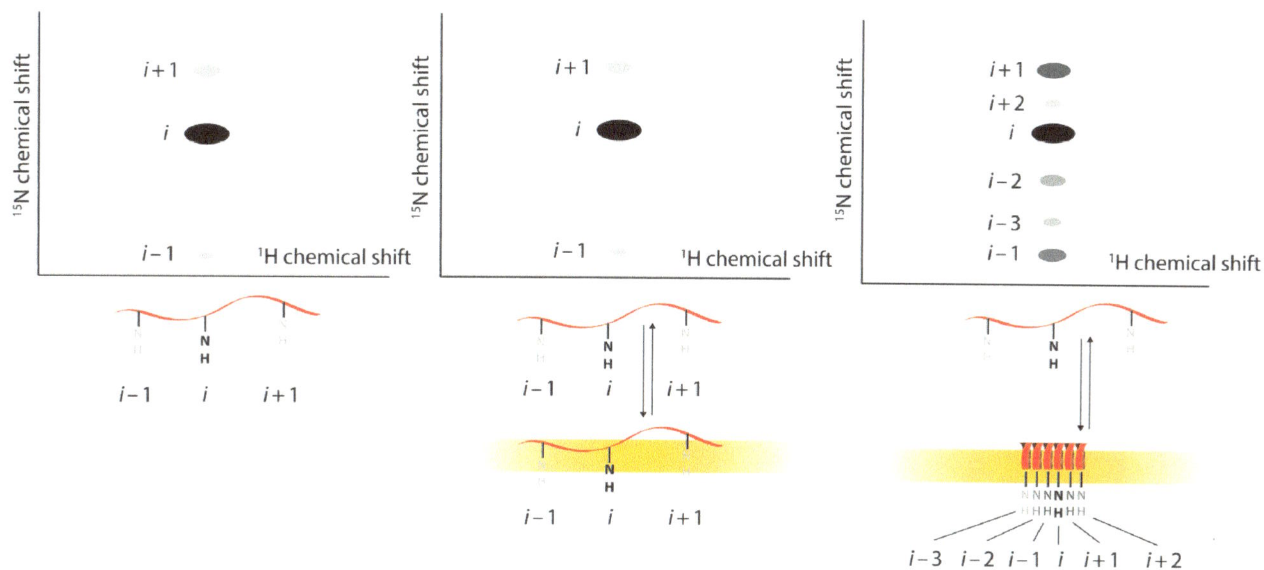

Fig. 5 Transferred NOEs recover structural information about the membrane bound dark state. Characterization of membrane-bound structure by solution-state NMR is limited by the fact that the large, slowly tumbling liposome bound state is "NMR-invisible," as rapid transverse relaxation in the bound state causes rapid signal decay. Transferred NOE experiments provide one method to recover information about the dark state, as NOE transfer in the membrane bound state is "read out" in the free state. During, e.g., an HSQC-NOESY-HSQC type experiment, some subset of protein in the ensemble will bind and release the vesicle during the NOE transfer period. NOE transfer that occurs in this transiently bound state generates NOESY cross-peaks in the observable free state spectrum. For example, helix formation by an intrinsically disordered region upon membrane binding (*right panel*) leads to decreased interresidue distances between sequential backbone amide groups, which can enhance magnetization transfer and thereby generate NOESY cross-peaks that are not observed in the absence of membrane binding (*left panel*) or in the absence of helix formation (*middle panel*). One important caveat to note, however, is that cross-peaks can also result from efficient multistep spin diffusion through the protein or the local proton dense lipid environment.

Other techniques to assess membrane bound protein structure include spin state selective methyl group relaxation dispersion experiments and dark state exchange saturation transfer (DEST) experiments, which probe the 50 μs—10 ms and 10 ms—100 ms timescales, respectively (Fawzi, Ying, Ghirlando, Torchia, & Clore, 2011). These have yet to be applied to complexin and so are beyond the scope of this chapter. Lastly, solid state NMR allows direct visualization of the membrane-bound proteins through the use of either magic angle spinning or sample (membrane) alignment, but suffers from a number of drawbacks. Sample preparation for solid state NMR, particularly with membrane-bound proteins, can be difficult and requires extensive optimization, and solid state NMR has not yet been used successfully to characterize complexin in its membrane bound state.

3.6.1 Equipment

- High field NMR spectrometer equipped with a CryoProbe
- NMR tubes (regular or Shigemi) and caps
- Spinner

3.6.2 Buffers and Reagents

- NMR sample

3.6.3 Pulse Sequences

- HSQC-NOESY-HSQC
- HSQC-NOESY-HSQC(TROSY)

3.6.4 Procedure

1. Load the NMR sample, and then lock, shim, and tune according to steps 1–5 in Section 3.2.4.
2. Calibrate proton and nitrogen 90 degree pulse widths as per step 2 of Section 3.3.4.
3. Load the appropriate NOESY pulse sequence and adjust parameters as necessary. A detailed discussion of this process is beyond the scope of this chapter, and users should coordinate with their local NMR core facility staff or an expert NMR user. We generally use NOESY mixing times of 100–200 ms.
4. Run experiment, save the data files, and transfer to computer for analysis.
5. We process and analyze three-dimensional data using NMRPipe and NMRView using the same workflow as above in steps 11–13 of Section 3.2.4 and step 5 of Section 3.3.4. As above, phasing values should be adjusted as determined in nmrDraw. A typical NMRPipe script for

NOESY processing essentially resembles those for triple-resonance experiments noted above. Depending on the pulse sequence used, Fourier transform modifiers such as –alt or –inv may be required in place of –auto. Phasing values must also be varied in these scripts from experiment to experiment.

4. SUMMARY AND CONCLUSIONS

Here, we have provided a comprehensive description of the complementary spectroscopic methods that have been applied to characterize complexin structure at site-resolved detail in the free-, micelle-, and membrane-associated states. Complexin binds to liposomes through two discrete motifs: one remains disordered in the bound state, while the other undergoes a membrane curvature dependent disorder-to-order transition upon binding. These same general methods are applicable to a wide range of peripherally membrane-associated (not transmembrane) intrinsically disordered proteins. While the techniques discussed can be applied regardless of membrane-bound structure, they have proven particularly useful for proteins that, like complexin, fold into amphipathic helical structure upon membrane binding. Indeed, much of our work was inspired by similar types of experiments performed with alpha synuclein, a protein associated with Parkinson's disease (Dikiy & Eliezer, 2012; Snead & Eliezer, 2014) that has been extensively studied as a canonical membrane-associated amphipathic helix (Bussell & Eliezer, 2003, 2004; Bussell, Ramlall, & Eliezer, 2005; Dikiy & Eliezer, 2014; Dikiy et al., 2016; Sung & Eliezer, 2006). A similar approach has also been taken for the larger disordered protein tau, which is associated with Alzheimer's disease and other neurodegenerative disorders, examining its binding either to membranes/micelles (Ait-Bouziad et al., 2017; Barre & Eliezer, 2006, 2013; Eliezer et al., 2005) or to microtubules (Mukrasch et al., 2007; Sillen et al., 2007). Importantly, the NMR-based techniques discussed above are further generalizable to disordered proteins that bind to any large, slowly tumbling species with binding kinetics that are in slow exchange relative to the NMR timescale. While we have focused on protein–membrane interactions, NMR analyses similar to those presented above have also revealed important and detailed information regarding monomer binding to the ends of amyloid fibrils, for example (Fawzi et al., 2011). Moving forward, solution-state NMR could be of use for any system where a highly disordered protein or protein region binds to a large, slowly tumbling, NMR-invisible species.

ACKNOWLEDGMENTS

This work was supported by NIH Grants AG019391 and GM117518.

REFERENCES

Ait-Bouziad, N., Lv, G., Mahul-Mellier, A. L., Xiao, S., Zorludemir, G., Eliezer, D., et al. (2017). Discovery and characterization of stable and toxic Tau/phospholipid oligomeric complexes. *Nature Communications*, *8*, 1678.

Barre, P., & Eliezer, D. (2006). Folding of the repeat domain of tau upon binding to lipid surfaces. *Journal of Molecular Biology*, *362*, 312–326.

Barre, P., & Eliezer, D. (2013). Structural transitions in tau k18 on micelle binding suggest a hierarchy in the efficacy of individual microtubule-binding repeats in filament nucleation. *Protein Science*, *22*, 1037–1048.

Bax, A., & Ikura, M. (1991). An efficient 3D NMR technique for correlating the proton and 15N backbone amide resonances with the alpha-carbon of the preceding residue in uniformly 15N/13C enriched proteins. *Journal of Biomolecular NMR*, *1*, 99–104.

Bodner, C. R., Dobson, C. M., & Bax, A. (2009). Multiple tight phospholipid-binding modes of alpha-synuclein revealed by solution NMR spectroscopy. *Journal of Molecular Biology*, *390*, 775–790.

Bodner, C. R., Maltsev, A. S., Dobson, C. M., & Bax, A. (2010). Differential phospholipid binding of alpha-synuclein variants implicated in Parkinson's disease revealed by solution NMR spectroscopy. *Biochemistry*, *49*, 862–871.

Bracher, A., Kadlec, J., Betz, H., & Weissenhorn, W. (2002). X-ray structure of a neuronal complexin-SNARE complex from squid. *The Journal of Biological Chemistry*, *277*, 26517–26523.

Brose, N. (2008a). Altered complexin expression in psychiatric and neurological disorders: Cause or consequence? *Molecules and Cells*, *25*, 7–19.

Brose, N. (2008b). For better or for worse: Complexins regulate SNARE function and vesicle fusion. *Traffic*, *9*, 1403–1413.

Buhl, L. K., Jorquera, R. A., Akbergenova, Y., Huntwork-Rodriguez, S., Volfson, D., & Littleton, J. T. (2013). Differential regulation of evoked and spontaneous neurotransmitter release by C-terminal modifications of complexin. *Molecular and Cellular Neurosciences*, *52*, 161–172.

Bussell, R., Jr., & Eliezer, D. (2003). A structural and functional role for 11-mer repeats in alpha-synuclein and other exchangeable lipid binding proteins. *Journal of Molecular Biology*, *329*, 763–778.

Bussell, R., Jr., & Eliezer, D. (2004). Effects of Parkinson's disease-linked mutations on the structure of lipid-associated alpha-synuclein. *Biochemistry*, *43*, 4810–4818.

Bussell, R., Jr., Ramlall, T. F., & Eliezer, D. (2005). Helix periodicity, topology, and dynamics of membrane-associated alpha-synuclein. *Protein Science*, *14*, 862–872.

Bykhovskaia, M., Jagota, A., Gonzalez, A., Vasin, A., & Littleton, J. T. (2013). Interaction of the complexin accessory helix with the C-terminus of the SNARE complex: Molecular-dynamics model of the fusion clamp. *Biophysical Journal*, *105*, 679–690.

Cavanagh, J., Fairbrother, W. J., Palmer, A. G., III, & Skelton, N. J. (2007). *Protein NMR spectroscopy: Principles and practice*. Burlington, MA: Elsevier Academic Press.

Chen, X., Tomchick, D. R., Kovrigin, E., Arac, D., Machius, M., Sudhof, T. C., et al. (2002). Three-dimensional structure of the complexin/SNARE complex. *Neuron*, *33*, 397–409.

Cho, R. W., Kummel, D., Li, F., Baguley, S. W., Coleman, J., Rothman, J. E., et al. (2014). Genetic analysis of the Complexin trans-clamping model for cross-linking SNARE complexes in vivo. *Proceedings of the National Academy of Sciences of the United States of America*, *111*, 10317–10322.

Cho, R. W., Song, Y., & Littleton, J. T. (2010). Comparative analysis of Drosophila and mammalian complexins as fusion clamps and facilitators of neurotransmitter release. *Molecular and Cellular Neurosciences*, *45*, 389–397.

Choi, U. B., Zhao, M., Zhang, Y., Lai, Y., & Brunger, A. T. (2016). Complexin induces a conformational change at the membrane-proximal C-terminal end of the SNARE complex. *eLife*, *5*, e16886.

Clubb, R. T., Thanabal, V., & Wagner, G. (1992). A constant-time 3-dimensional triple-resonance pulse scheme to correlate intraresidue H-1(N), N-15, and C-13(') chemical-shifts in N-15-C-13-labeled proteins. *Journal of Magnetic Resonance*, *97*, 213–217.

Davidson, W. S., Jonas, A., Clayton, D. F., & George, J. M. (1998). Stabilization of alpha-synuclein secondary structure upon binding to synthetic membranes. *The Journal of Biological Chemistry*, *273*, 9443–9449.

Delaglio, F., Grzesiek, S., Vuister, G. W., Zhu, G., Pfeifer, J., & Bax, A. (1995). NMRPipe: A multidimensional spectral processing system based on UNIX pipes. *Journal of Biomolecular NMR*, *6*, 277–293.

Dikiy, I., & Eliezer, D. (2012). Folding and misfolding of alpha-synuclein on membranes. *Biochimica et Biophysica Acta*, *1818*, 1013–1018.

Dikiy, I., & Eliezer, D. (2014). N-terminal acetylation stabilizes N-terminal helicity in lipid- and micelle-bound alpha-synuclein and increases its affinity for physiological membranes. *The Journal of Biological Chemistry*, *289*, 3652–3665.

Dikiy, I., Fauvet, B., Jovicic, A., Mahul-Mellier, A. L., Desobry, C., El-Turk, F., et al. (2016). Semisynthetic and in vitro phosphorylation of alpha-synuclein at Y39 promotes functional partly helical membrane-bound states resembling those induced by PD mutations. *ACS Chemical Biology*, *11*, 2428–2437.

Drew, C. J., Kyd, R. J., & Morton, A. J. (2007). Complexin 1 knockout mice exhibit marked deficits in social behaviours but appear to be cognitively normal. *Human Molecular Genetics*, *16*, 2288–2305.

Ducas, V. C., & Rhoades, E. (2012). Quantifying interactions of beta-synuclein and gamma-synuclein with model membranes. *Journal of Molecular Biology*, *423*, 528–539.

Eliezer, D. (2007). Characterizing residual structure in disordered protein States using nuclear magnetic resonance. *Methods in Molecular Biology*, *350*, 49–67.

Eliezer, D. (2009). Biophysical characterization of intrinsically disordered proteins. *Current Opinion in Structural Biology*, *19*, 23–30.

Eliezer, D., Barre, P., Kobaslija, M., Chan, D., Li, X., & Heend, L. (2005). Residual structure in the repeat domain of tau: Echoes of microtubule binding and paired helical filament formation. *Biochemistry*, *44*, 1026–1036.

Fares, M. B., Ait-Bouziad, N., Dikiy, I., Mbefo, M. K., Jovicic, A., Kiely, A., et al. (2014). The novel Parkinson's disease linked mutation G51D attenuates in vitro aggregation and membrane binding of alpha-synuclein, and enhances its secretion and nuclear localization in cells. *Human Molecular Genetics*, *23*, 4491–4509.

Fauvet, B., Fares, M. B., Samuel, F., Dikiy, I., Tandon, A., Eliezer, D., et al. (2012). Characterization of semisynthetic and naturally Nalpha-acetylated alpha-synuclein in vitro and in intact cells: Implications for aggregation and cellular properties of alpha-synuclein. *The Journal of Biological Chemistry*, *287*, 28243–28262.

Fawzi, N. L., Ying, J., Ghirlando, R., Torchia, D. A., & Clore, G. M. (2011). Atomic-resolution dynamics on the surface of amyloid-beta protofibrils probed by solution NMR. *Nature*, *480*, 268–272.

Ferreon, A. C., & Deniz, A. A. (2007). Alpha-synuclein multistate folding thermodynamics: Implications for protein misfolding and aggregation. *Biochemistry*, *46*, 4499–4509.

Gibbs, E. B., Cook, E. C., & Showalter, S. A. (2017). Application of NMR to studies of intrinsically disordered proteins. *Archives of Biochemistry and Biophysics*, *628*, 57–70.

Gibson, H. E., Reim, K., Brose, N., Morton, A. J., & Jones, S. (2005). A similar impairment in CA3 mossy fibre LTP in the R6/2 mouse model of Huntington's disease and in the complexin II knockout mouse. *The European Journal of Neuroscience*, *22*, 1701–1712.

Giraudo, C. G., Eng, W. S., Melia, T. J., & Rothman, J. E. (2006). A clamping mechanism involved in SNARE-dependent exocytosis. *Science*, *313*, 676–680.

Giraudo, C. G., Garcia-Diaz, A., Eng, W. S., Chen, Y., Hendrickson, W. A., Melia, T. J., et al. (2009). Alternative zippering as an on-off switch for SNARE-mediated fusion. *Science*, *323*, 512–516.

Glynn, D., Bortnick, R. A., & Morton, A. J. (2003). Complexin II is essential for normal neurological function in mice. *Human Molecular Genetics*, *12*, 2431–2448.

Glynn, D., Drew, C. J., Reim, K., Brose, N., & Morton, A. J. (2005). Profound ataxia in complexin I knockout mice masks a complex phenotype that includes exploratory and habituation deficits. *Human Molecular Genetics*, *14*, 2369–2385.

Glynn, D., Sizemore, R. J., & Morton, A. J. (2007). Early motor development is abnormal in complexin 1 knockout mice. *Neurobiology of Disease*, *25*, 483–495.

Greenfield, N., & Fasman, G. D. (1969). Computed circular dichroism spectra for the evaluation of protein conformation. *Biochemistry*, *8*, 4108–4116.

Harbison, N. W., Bhattacharya, S., & Eliezer, D. (2012). Assigning backbone NMR resonances for full length tau isoforms: Efficient compromise between manual assignments and reduced dimensionality. *PLoS One*, *7*, e34679.

Hobson, R. J., Liu, Q., Watanabe, S., & Jorgensen, E. M. (2011). Complexin maintains vesicles in the primed state in *C. elegans*. *Current Biology*, *21*, 106–113.

Holzwarth, G., & Doty, P. (1965). The ultraviolet circular dichroism of polypeptides. *Journal of the American Chemical Society*, *87*, 218–228.

Huang, G. Z., Ujihara, H., Takahashi, S., Kaba, H., Yagi, T., & Inoue, S. (2000). Involvement of complexin II in synaptic plasticity in the CA1 region of the hippocampus: The use of complexin II-lacking mice. *Japanese Journal of Pharmacology*, *84*, 179–187.

Huntwork, S., & Littleton, J. T. (2007). A complexin fusion clamp regulates spontaneous neurotransmitter release and synaptic growth. *Nature Neuroscience*, *10*, 1235–1237.

Johansson, K., Bourhis, J. M., Campanacci, V., Cambillau, C., Canard, B., & Longhi, S. (2003). Crystal structure of the measles virus phosphoprotein domain responsible for the induced folding of the C-terminal domain of the nucleoprotein. *The Journal of Biological Chemistry*, *278*, 44567–44573.

Johnson, B. A., & Blevins, R. A. (1994). NMRView: A computer program for the visualization and analysis of NMR data. *Journal of Biomolecular NMR*, *4*, 603–614.

Kaeser-Woo, Y. J., Yang, X., & Sudhof, T. C. (2012). C-terminal complexin sequence is selectively required for clamping and priming but not for Ca2+ triggering of synaptic exocytosis. *The Journal of Neuroscience*, *32*, 2877–2885.

Karaca, E., Harel, T., Pehlivan, D., Jhangiani, S. N., Gambin, T., Coban Akdemir, Z., et al. (2015). Genes that affect brain structure and function identified by rare variant analyses of Mendelian neurologic disease. *Neuron*, *88*, 499–513.

Kay, L. E., Ikura, M., Tschudin, R., & Bax, A. (1990). 3-Dimensional triple-resonance NMR-spectroscopy of isotopically enriched proteins. *Journal of Magnetic Resonance*, *89*, 496–514.

Khalaf, O., Fauvet, B., Oueslati, A., Dikiy, I., Mahul-Mellier, A. L., Ruggeri, F. S., et al. (2014). The H50Q mutation enhances alpha-synuclein aggregation, secretion, and toxicity. *The Journal of Biological Chemistry*, *289*, 21856–21876.

Kjaergaard, M., Brander, S., & Poulsen, F. M. (2011). Random coil chemical shift for intrinsically disordered proteins: Effects of temperature and pH. *Journal of Biomolecular NMR*, *49*, 139–149.

Kjaergaard, M., & Poulsen, F. M. (2011). Sequence correction of random coil chemical shifts: Correlation between neighbor correction factors and changes in the Ramachandran distribution. *Journal of Biomolecular NMR*, *50*, 157–165.

Krishnakumar, S. S., Radoff, D. T., Kummel, D., Giraudo, C. G., Li, F., Khandan, L., et al. (2011). A conformational switch in complexin is required for synaptotagmin to trigger synaptic fusion. *Nature Structural & Molecular Biology*, *18*, 934–940.

Kummel, D., Krishnakumar, S. S., Radoff, D. T., Li, F., Giraudo, C. G., Pincet, F., et al. (2011). Complexin cross-links prefusion SNAREs into a zigzag array. *Nature Structural & Molecular Biology*, *18*, 927–933.

Kunze, G., Barre, P., Scheidt, H. A., Thomas, L., Eliezer, D., & Huster, D. (2012). Binding of the three-repeat domain of tau to phospholipid membranes induces an aggregated-like state of the protein. *Biochimica et Biophysica Acta*, *1818*, 2302–2313.

Kurzbach, D., Kontaxis, G., Coudevylle, N., & Konrat, R. (2015). NMR spectroscopic studies of the conformational ensembles of intrinsically disordered proteins. *Advances in Experimental Medicine and Biology*, *870*, 149–185.

Lai, Y., Choi, U. B., Zhang, Y., Zhao, M., Pfuetzner, R. A., Wang, A. L., et al. (2016). N-terminal domain of complexin independently activates calcium-triggered fusion. *Proceedings of the National Academy of Sciences of the United States of America*, *113*, E4698–E4707.

Lai, Y., Diao, J., Cipriano, D. J., Zhang, Y., Pfuetzner, R. A., Padolina, M. S., et al. (2014). Complexin inhibits spontaneous release and synchronizes $Ca2^{+}$-triggered synaptic vesicle fusion by distinct mechanisms. *eLife*, *3*, e03756.

Lees, J. G., Smith, B. R., Wien, F., Miles, A. J., & Wallace, B. A. (2004). CDtool-an integrated software package for circular dichroism spectroscopic data processing, analysis, and archiving. *Analytical Biochemistry*, *332*, 285–289.

Li, F., Pincet, F., Perez, E., Giraudo, C. G., Tareste, D., & Rothman, J. E. (2011). Complexin activates and clamps SNAREpins by a common mechanism involving an intermediate energetic state. *Nature Structural & Molecular Biology*, *18*, 941–946.

Longhi, S., Receveur-Brechot, V., Karlin, D., Johansson, K., Darbon, H., Bhella, D., et al. (2003). The C-terminal domain of the measles virus nucleoprotein is intrinsically disordered and folds upon binding to the C-terminal moiety of the phosphoprotein. *The Journal of Biological Chemistry*, *278*, 18638–18648.

Lu, B., Song, S., & Shin, Y. K. (2010). Accessory alpha-helix of complexin I can displace VAMP2 locally in the complexin-SNARE quaternary complex. *Journal of Molecular Biology*, *396*, 602–609.

Malsam, J., Parisotto, D., Bharat, T. A., Scheutzow, A., Krause, J. M., Briggs, J. A., et al. (2012). Complexin arrests a pool of docked vesicles for fast $Ca2^{+}$-dependent release. *The EMBO Journal*, *31*, 3270–3281.

Malsam, J., Seiler, F., Schollmeier, Y., Rusu, P., Krause, J. M., & Sollner, T. H. (2009). The carboxy-terminal domain of complexin I stimulates liposome fusion. *Proceedings of the National Academy of Sciences of the United States of America*, *106*, 2001–2006.

Maltsev, A. S., Ying, J., & Bax, A. (2012). Impact of N-terminal acetylation of alpha-synuclein on its random coil and lipid binding properties. *Biochemistry*, *51*, 5004–5013.

Marley, J., Lu, M., & Bracken, C. (2001). A method for efficient isotopic labeling of recombinant proteins. *Journal of Biomolecular NMR*, *20*, 71–75.

Martin, J. A., Hu, Z., Fenz, K. M., Fernandez, J., & Dittman, J. S. (2011). Complexin has opposite effects on two modes of synaptic vesicle fusion. *Current Biology*, *21*, 97–105.

Maximov, A., Tang, J., Yang, X., Pang, Z. P., & Sudhof, T. C. (2009). Complexin controls the force transfer from SNARE complexes to membranes in fusion. *Science*, *323*, 516–521.

Mohrmann, R., Dhara, M., & Bruns, D. (2015). Complexins: Small but capable. *Cellular and Molecular Life Sciences*, *72*, 4221–4235.

Mossessova, E., & Lima, C. D. (2000). Ulp1-SUMO crystal structure and genetic analysis reveal conserved interactions and a regulatory element essential for cell growth in yeast. *Molecular Cell*, *5*, 865–876.

Mukrasch, M. D., von Bergen, M., Biernat, J., Fischer, D., Griesinger, C., Mandelkow, E., et al. (2007). The "jaws" of the tau-microtubule interaction. *The Journal of Biological Chemistry*, *282*, 12230–12239.

Narayanan, R. L., Durr, U. H., Bibow, S., Biernat, J., Mandelkow, E., & Zweckstetter, M. (2010). Automatic assignment of the intrinsically disordered protein Tau with 441-residues. *Journal of the American Chemical Society*, *132*, 11906–11907.

Pabst, S., Hazzard, J. W., Antonin, W., Sudhof, T. C., Jahn, R., Rizo, J., et al. (2000). Selective interaction of complexin with the neuronal SNARE complex. Determination of the binding regions. *The Journal of Biological Chemistry*, *275*, 19808–19818.

Paleologou, K. E., Oueslati, A., Shakked, G., Rospigliosi, C. C., Kim, H. Y., Lamberto, G. R., et al. (2010). Phosphorylation at S87 is enhanced in synucleinopathies, inhibits alpha-synuclein oligomerization, and influences synuclein-membrane interactions. *The Journal of Neuroscience*, *30*, 3184–3198.

Palmer, A. G., Skelton, N. J., Chazin, W. J., Wright, P. E., & Rance, M. (1992). Suppression of the effects of cross-correlation between dipolar and anisotropic chemical-shift relaxation mechanisms in the measurement of spin spin relaxation rates. *Molecular Physics*, *75*, 699–711.

Pfefferkorn, C. M., & Lee, J. C. (2010). Tryptophan probes at the alpha-synuclein and membrane interface. *The Journal of Physical Chemistry. B*, *114*, 4615–4622.

Ploier, B., & Menon, A. K. (2016). A fluorescence-based assay of phospholipid scramblase activity. *Journal of Visualized Experiments*, *115*, e54635.

Prinslow, E. A., Brautigam, C. A., & Rizo, J. (2017). Reconciling isothermal titration calorimetry analyses of interactions between complexin and truncated SNARE complexes. *eLife*, *6*, e30286.

Radoff, D. T., Dong, Y., Snead, D., Bai, J., Eliezer, D., & Dittman, J. S. (2014). The accessory helix of complexin functions by stabilizing central helix secondary structure. *eLife*, *3*, e04553.

Redler, S., Strom, T. M., Wieland, T., Cremer, K., Engels, H., Distelmaier, F., et al. (2017). Variants in CPLX1 in two families with autosomal-recessive severe infantile myoclonic epilepsy and ID. *European Journal of Human Genetics*, *25*, 889–893.

Reim, K., Mansour, M., Varoqueaux, F., McMahon, H. T., Sudhof, T. C., Brose, N., et al. (2001). Complexins regulate a late step in $Ca2^+$-dependent neurotransmitter release. *Cell*, *104*, 71–81.

Rizo, J., & Xu, J. (2015). The synaptic vesicle release machinery. *Annual Review of Biophysics*, *44*, 339–367.

Sattler, M., & Fesik, S. W. (1996). Use of deuterium labeling in NMR: Overcoming a sizeable problem. *Structure*, *4*, 1245–1249.

Schaub, J. R., Lu, X., Doneske, B., Shin, Y. K., & McNew, J. A. (2006). Hemifusion arrest by complexin is relieved by $Ca2^+$-synaptotagmin I. *Nature Structural & Molecular Biology*, *13*, 748–750.

Seiler, F., Malsam, J., Krause, J. M., & Sollner, T. H. (2009). A role of complexin-lipid interactions in membrane fusion. *FEBS Letters*, *583*, 2343–2348.

Sillen, A., Barbier, P., Landrieu, I., Lefebvre, S., Wieruszeski, J. M., Leroy, A., et al. (2007). NMR investigation of the interaction between the neuronal protein tau and the microtubules. *Biochemistry*, *46*, 3055–3064.

Snead, D., & Eliezer, D. (2014). Alpha-synuclein function and dysfunction on cellular membranes. *Experimental Neurobiology*, *23*, 292–313.

Snead, D., Lai, A. L., Wragg, R. T., Parisotto, D. A., Ramlall, T. F., Dittman, J. S., et al. (2017). Unique structural features of membrane-bound C-terminal domain motifs modulate complexin inhibitory function. *Frontiers in Molecular Neuroscience*, *10*, 154.

Snead, D., Wragg, R. T., Dittman, J. S., & Eliezer, D. (2014). Membrane curvature sensing by the C-terminal domain of complexin. *Nature Communications*, *5*, 4955.

Spera, S., & Bax, A. (1991). Empirical correlation between protein backbone conformation and C-alpha and C-beta C-13 nuclear magnetic resonance chemical shifts. *Journal of the American Chemical Society*, *113*, 5490–5492.

Sreerama, N., & Woody, R. W. (2004). Computation and analysis of protein circular dichroism spectra. *Methods in Enzymology*, *383*, 318–351.

Sudhof, T. C. (2013). Neurotransmitter release: The last millisecond in the life of a synaptic vesicle. *Neuron*, *80*, 675–690.

Sung, Y. H., & Eliezer, D. (2006). Secondary structure and dynamics of micelle bound beta- and gamma-synuclein. *Protein Science*, *15*, 1162–1174.

Takahashi, S., Ujihara, H., Huang, G. Z., Yagyu, K. I., Sanbo, M., Kaba, H., et al. (1999). Reduced hippocampal LTP in mice lacking a presynaptic protein: Complexin II. *The European Journal of Neuroscience*, *11*, 2359–2366.

Tang, J., Maximov, A., Shin, O. H., Dai, H., Rizo, J., & Sudhof, T. C. (2006). A complexin/synaptotagmin 1 switch controls fast synaptic vesicle exocytosis. *Cell*, *126*, 1175–1187.

Trimbuch, T., & Rosenmund, C. (2016). Should I stop or should I go? The role of complexin in neurotransmitter release. *Nature Reviews. Neuroscience*, *17*, 118–125.

Trimbuch, T., Xu, J., Flaherty, D., Tomchick, D. R., Rizo, J., & Rosenmund, C. (2014). Re-examining how complexin inhibits neurotransmitter release. *eLife*, *3*, e02391.

Vasin, A., Volfson, D., Littleton, J. T., & Bykhovskaia, M. (2016). Interaction of the Complexin accessory Helix with synaptobrevin regulates spontaneous fusion. *Biophysical Journal*, *111*, 1954–1964.

Venyaminov, S., Baikalov, I. A., Shen, Z. M., Wu, C. S., & Yang, J. T. (1993). Circular dichroic analysis of denatured proteins: Inclusion of denatured proteins in the reference set. *Analytical Biochemistry*, *214*, 17–24.

Whitmore, L., & Wallace, B. A. (2008). Protein secondary structure analyses from circular dichroism spectroscopy: Methods and reference databases. *Biopolymers*, *89*, 392–400.

Wishart, D. S., Bigam, C. G., Holm, A., Hodges, R. S., & Sykes, B. D. (1995). ^{1}H, ^{13}C and ^{15}N random coil NMR chemical shifts of the common amino acids. I. Investigations of nearest-neighbor effects. *Journal of Biomolecular NMR*, *5*, 67–81.

Wishart, D. S., Sykes, B. D., & Richards, F. M. (1991). Relationship between nuclear magnetic resonance chemical shift and protein secondary structure. *Journal of Molecular Biology*, *222*, 311–333.

Wragg, R. T., Parisotto, D. A., Li, Z., Terakawa, M. S., Snead, D., Basu, I., et al. (2017). Evolutionary divergence of the C-terminal domain of Complexin accounts for functional disparities between vertebrate and invertebrate Complexins. *Frontiers in Molecular Neuroscience*, *10*, 146.

Wragg, R. T., Snead, D., Dong, Y., Ramlall, T. F., Menon, I., Bai, J., et al. (2013). Synaptic vesicles position complexin to block spontaneous fusion. *Neuron*, *77*, 323–334.

Wuthrich, K. (1986). *NMR of proteins and nucleic acids*. New York: J. Wiley & Sons.

Xiao, S., & Raleigh, D. P. (2010). A critical assessment of putative gatekeeper interactions in the villin headpiece helical subdomain. *Journal of Molecular Biology*, *401*, 274–285.

Xue, M., Craig, T. K., Xu, J., Chao, H. T., Rizo, J., & Rosenmund, C. (2010). Binding of the complexin N terminus to the SNARE complex potentiates synaptic-vesicle fusogenicity. *Nature Structural & Molecular Biology*, *17*, 568–575.

Xue, M., Lin, Y. Q., Pan, H., Reim, K., Deng, H., Bellen, H. J., et al. (2009). Tilting the balance between facilitatory and inhibitory functions of mammalian and Drosophila Complexins orchestrates synaptic vesicle exocytosis. *Neuron*, *64*, 367–380.

Xue, M., Reim, K., Chen, X., Chao, H. T., Deng, H., Rizo, J., et al. (2007). Distinct domains of complexin I differentially regulate neurotransmitter release. *Nature Structural & Molecular Biology*, *14*, 949–958.

Xue, M., Stradomska, A., Chen, H., Brose, N., Zhang, W., Rosenmund, C., et al. (2008). Complexins facilitate neurotransmitter release at excitatory and inhibitory synapses in mammalian central nervous system. *Proceedings of the National Academy of Sciences of the United States of America*, *105*, 7875–7880.

Yamauchi, Y., Qin, L. H., Nishihara, M., Sawada, K., Kato, K., & Inoue, S. (2005). Vulnerability of synaptic plasticity in the complexin II knockout mouse to maternal deprivation stress. *Brain Research*, *1056*, 59–67.

Yang, X., Cao, P., & Sudhof, T. C. (2013). Deconstructing complexin function in activating and clamping Ca2+−triggered exocytosis by comparing knockout and knockdown phenotypes. *Proceedings of the National Academy of Sciences of the United States of America*, *110*, 20777–20782.

Yin, L., Kim, J., & Shin, Y. K. (2016). Complexin splits the membrane-proximal region of a single SNAREpin. *The Biochemical Journal*, *473*, 2219–2224.

Yoon, T. Y., Lu, X., Diao, J., Lee, S. M., Ha, T., & Shin, Y. K. (2008). Complexin and Ca2 + stimulate SNARE-mediated membrane fusion. *Nature Structural & Molecular Biology*, *15*, 707–713.

Zhang, O., Forman-Kay, J. D., Shortle, D., & Kay, L. E. (1997). Triple-resonance NOESY-based experiments with improved spectral resolution: Applications to structural characterization of unfolded, partially folded and folded proteins. *Journal of Biomolecular NMR*, *9*, 181–200.

Zhang, H., Neal, S., & Wishart, D. S. (2003). RefDB: A database of uniformly referenced protein chemical shifts. *Journal of Biomolecular NMR*, *25*, 173–195.

Zhou, Q., Zhou, P., Wang, A. L., Wu, D., Zhao, M., Sudhof, T. C., et al. (2017). The primed SNARE-complexin-synaptotagmin complex for neuronal exocytosis. *Nature*, *548*, 420–425.

CHAPTER TEN

Accurate Transfer Efficiencies, Distance Distributions, and Ensembles of Unfolded and Intrinsically Disordered Proteins From Single-Molecule FRET

Erik D. Holmstrom*[,1], Andrea Holla*, Wenwei Zheng[†,1], Daniel Nettels*, Robert B. Best[‡,1], Benjamin Schuler*[,§,1]

*Department of Biochemistry, University of Zurich, Zurich, Switzerland
[†]College of Integrative Sciences and Arts, Arizona State University, Mesa, AZ, United States
[‡]Laboratory of Chemical Physics, National Institute of Diabetes and Digestive and Kidney Diseases, National Institutes of Health, Bethesda, MD, United States
[§]Department of Physics, University of Zurich, Zurich, Switzerland
[1]Corresponding authors: e-mail addresses: e.holmstrom@bioc.uzh.ch; wenweizheng@asu.edu; robertbe@helix.nih.gov; schuler@bioc.uzh.ch

Contents

Abstract

Intrinsically disordered proteins (IDPs) sample structurally diverse ensembles. Characterizing the underlying distributions of conformations is a key step toward understanding

Methods in Enzymology, Volume 611
ISSN 0076-6879
https://doi.org/10.1016/bs.mie.2018.09.030

the structural and functional properties of IDPs. One increasingly popular method for obtaining quantitative information on intramolecular distances and distributions is single-molecule Förster resonance energy transfer (FRET). Here we describe two essential elements of the quantitative analysis of single-molecule FRET data of IDPs: the sample-specific calibration of the single-molecule instrument that is required for determining accurate transfer efficiencies, and the use of state-of-the-art methods for inferring accurate distance distributions from these transfer efficiencies. First, we illustrate how to quantify the correction factors for instrument calibration with alternating donor and acceptor excitation measurements of labeled samples spanning a wide range of transfer efficiencies. Second, we show how to infer distance distributions based on suitably parameterized simple polymer models, and how to obtain conformational ensembles from Bayesian reweighting of molecular simulations or from parameter optimization in simplified coarse-grained models.

1. INTRODUCTION

For more than a century (Fischer, 1902), one of the fundamental concepts of molecular biology, and in particular enzymology, has been that the functions of proteins are closely coupled to their folded, three-dimensional structures. However, it is now clear that proteins can be functional without stable tertiary or even secondary structure (Dyson & Wright, 2005; Forman-Kay & Mittag, 2013; Tompa, 2005). Such intrinsically disordered proteins (IDPs) are involved in many essential biological processes, particularly in higher eukaryotes (Dunker et al., 2001; Oates et al., 2013; van der Lee et al., 2014). Nevertheless, understanding their functional mechanisms requires a rigorous and quantitative analysis of the structurally diverse ensembles they populate.

There are many powerful methods available for characterizing the broad conformational distributions and dynamics of IDPs (Gibbs & Showalter, 2015; Uversky & Keith Dunker, 2012), such as NMR (Konrat, 2014) and small-angle X-ray scattering (SAXS) (Kikhney & Svergun, 2015). NMR provides a wealth of detailed structural information, especially regarding short-range interactions, secondary structure content, and conformational dynamics over a broad range of timescales. SAXS can report on the overall dimensions and shape of a biomolecule. It is becoming increasingly apparent that these methods can be ideally complemented by single-molecule Förster resonance energy transfer (FRET), which has been used successfully as a "spectroscopic ruler" to probe the dimensions and dynamics of IDPs (Ferreon, Moran, Gambin, & Deniz, 2010; Schuler, Soranno, Hofmann, & Nettels, 2016). Key strengths of such single-molecule experiments are the ability to quantify specific long-range intra- and intermolecular distances, to distinguish static and dynamic heterogeneity, to resolve coexisting subpopulations, and to probe conformational dynamics ranging from rapid conformational fluctuations on the nanosecond timescale all the way to the formation of higher-order assemblies on the timescale of days and weeks (Schuler & Hofmann, 2013). Finally, the small volumes and low concentrations used in single-molecule FRET require only minute amounts of sample, provide access to concentrations down to the picomolar range, and can easily be used in a broad spectrum of solutions conditions, even within live cells (König et al., 2015; Sustarsic & Kapanidis, 2015). Together, these advantages have made single-molecule FRET a versatile tool for biophysical studies of conformationally heterogeneous biological molecules like IDPs.

In many applications, *qualitative* distance information provided by single-molecule FRET is sufficient (e.g., in many kinetic analyses), but FRET can also be used to obtain *quantitative* distance information, not only for structured biomolecules (Hellenkamp, Wortmann, Kandzia, Zacharias, & Hugel, 2017; Kalinin et al., 2012; Muschielok et al., 2008), but also for IDPs (Gomes & Gradinaru, 2017; Schuler et al., 2016), as demonstrated by comparison with other methods (Aznauryan et al., 2016; Borgia et al., 2016; Fuertes et al., 2017). However, this task requires two key steps: First, an accurate transfer efficiency must be obtained from the experimental data acquired on a calibrated instrument. Second, the distance distribution within the IDP must be inferred from this transfer efficiency based on a reasonable model. The first step poses essentially the same challenges for IDPs as for structured biomolecules (Hellenkamp et al., 2018); additionally, a detailed analysis of photon statistics can be used to identify the presence of broad distance distributions (Gopich & Szabo, 2012; Schuler et al., 2016). The second step is even more demanding, since information about a broad distribution of distances must be inferred, corresponding to a highly underdetermined inverse problem. However, advances in the use of analytical polymer models and molecular simulations can now be employed to infer increasingly accurate distance distributions of unfolded and intrinsically disordered proteins from single-molecule FRET experiments, ideally in combination with data from complementary methods (Borgia et al., 2016; Fuertes et al., 2017; Zheng et al., 2018). In this chapter, we outline the steps required for performing accurate transfer efficiency measurements of fluorescently labeled IDPs and for inferring the underlying distance distributions.

2. FÖRSTER RESONANCE ENERGY TRANSFER

FRET is a photophysical phenomenon involving the nonradiative dipole–dipole coupling between chromophores (Förster, 1948) that is often exploited to measure distances on biomolecular length scales. An electronically excited donor chromophore (D^*) transfers energy to a nearby acceptor chromophore in the ground state (A), resulting in deexcitation of the donor (D) and excitation of the acceptor (A^*). The rate coefficient for energy transfer, k_{FRET}, from D^* to A depends on the donor fluorescence lifetime, τ_D, and the sixth power of the Förster radius, R_0, divided by the distance, r, between the two fluorophores:

$$k_{\mathrm{FRET}} = \frac{1}{\tau_D}\left(\frac{R_0}{r}\right)^6. \tag{1}$$

The FRET efficiency, ε, at a given r is the probability that a $D^* \rightarrow D$ transition will occur via energy transfer (k_{FRET}) rather than other nonradiative (k_{nrad}) or radiative (k_{rad}) decay processes, and thus $\varepsilon = ½$ when $r = R_0$:

$$\varepsilon = \frac{k_{\mathrm{FRET}}}{k_{\mathrm{FRET}} + k_{\mathrm{rad}} + k_{\mathrm{nrad}}} = \frac{R_0^6}{R_0^6 + r^6}. \tag{2}$$

The value of R_0 can be calculated from the refractive index of the medium, the donor quantum yield, the relative orientation of the two fluorophores, and the overlap integral of the donor emission and acceptor absorption spectra (Van Der Meer, Coker, & Chen, 1994). If R_0 is known, the efficiency of energy transfer between two nearby fluorophores can be used to quantify the distance between them, which is why FRET has often been referred to as a "spectroscopic ruler" (Stryer, 1978; Stryer & Haugland, 1967). An important aspect of single-molecule FRET is that typical values of R_0 are between 5 and 7 nm, which makes FRET ideally suited for probing biological macromolecules.

Although rate coefficients are used to define ε, they are challenging to determine precisely for single fluorophores because of the low photon emission rates. A common alternative is to obtain ε from ratiometric measurements of the single-molecule transfer efficiency as $E = N_A/(N_A + N_D)$, using the number of donor (N_D) and acceptor (N_A) photons (Deniz et al., 2001). Accordingly, the average of many such measurements from individual fluorescence bursts or time bins corresponds directly to the FRET efficiency:

$$\varepsilon = \langle E \rangle = \left\langle \frac{N_A}{N_A + N_D} \right\rangle = \frac{R_0^6}{R_0^6 + r^6}. \tag{3}$$

However, this approach requires that N_D and N_A are corrected for factors such as differences in quantum yields and detection efficiencies for the donor and acceptor fluorophores, which is one of the crucial challenges associated with obtaining accurate transfer efficiencies.

3. EXPERIMENTAL CONSIDERATIONS

Single-molecule experiments require individual molecules to be spatially separated from one another. In practice, this can be achieved in one of two ways: either by immobilizing molecules on a substrate at low surface densities or by studying freely diffusing molecules at extremely low concentrations. In the latter case, individual fluorescently labeled biomolecules randomly diffuse through the confocal observation volume and give rise to bursts of photons that are readily distinguishable from the background photon detection rates (Deniz et al., 1999) (Fig. 1). This chapter focuses on how to accurately determine transfer efficiencies of fluorescently labeled biomolecules in confocal free-diffusion experiments, because they are spectroscopically most versatile. However, the principles described here are also applicable to experiments on surface-immobilized molecules (Hellenkamp et al., 2018).

3.1 Sample Design and Preparation

The success of any single-molecule FRET experiment is highly dependent on the design and quality of the sample. Essential design criteria are the spectral properties of the fluorophores and their Förster radius, which determines the accessible distance range, and the position of the dyes within the protein or nucleic acid. Over the last two decades, a preference has emerged for specific classes of fluorophores, primarily because of their commercial availability with versatile coupling chemistries, photostability, quantum yield, and compatibility with commonly available laser lines (Gust et al., 2014; Ha & Tinnefeld, 2012). Those two classes are rhodamine-based fluorophores (e.g., Alexa 488 as a donor and Alexa 594 as an acceptor), which are more commonly used for proteins, and cyanine-based fluorophores (e.g., Cy3 as a donor and Cy5 as an acceptor), which are more frequently used for nucleic acids. Additional considerations involve viable coupling chemistries, solvent exposure of the labeling sites, proximity to potential quenchers (e.g., other aromatic groups), and electrostatic interactions. Finally, the preparation and purification should minimize the amount of donor-only and acceptor-only contaminants within the FRET-labeled sample. For IDPs, this goal is most stringently achieved with high-resolution reversed-phase or ion exchange chromatography. Many of these issues have recently been discussed in detail elsewhere (Zosel, Holla, & Schuler, 2018).

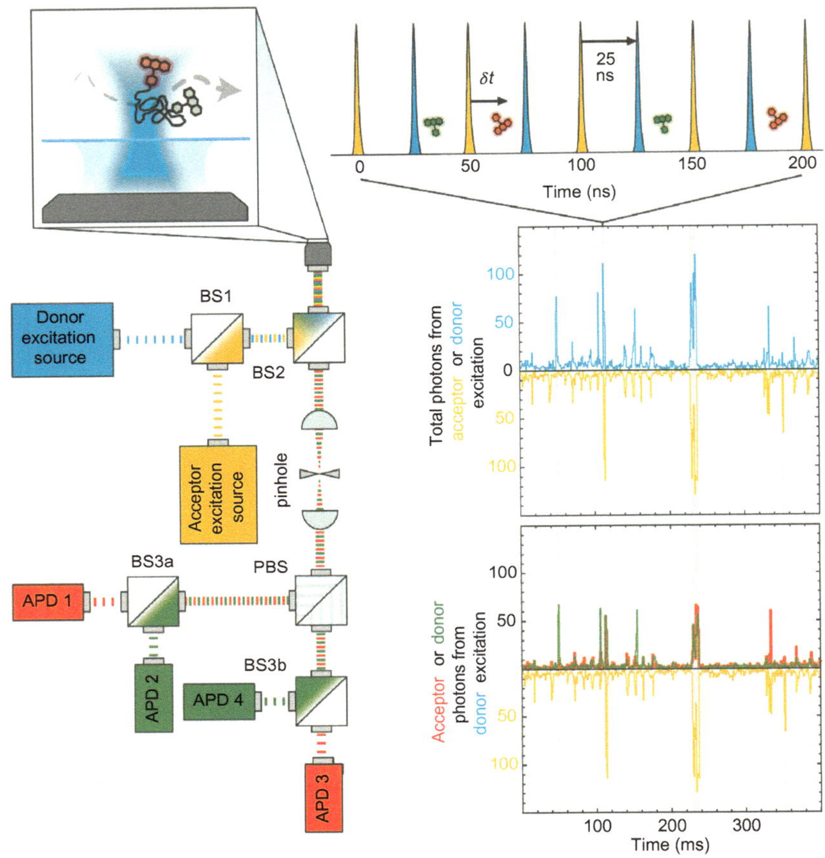

Fig. 1 Confocal single-molecule fluorescence spectroscopy of freely diffusing molecules using pulsed interleaved excitation (PIE) and four-channel detection. Two interleaved pulsed lasers (*blue* and *yellow*) for donor and acceptor excitation, respectively, are coaxially aligned using a dichroic mirror (BS1) and then directed into the back aperture of a high-numerical-aperture microscope objective. The spatial selection of the femtoliter confocal observation volume enables single-molecule detection at subnanomolar concentrations. When a single molecule diffuses through this volume, it produces a short (~1 ms) burst of donor (*green*) and acceptor (*red*) photons. The signal upon direct excitation of the acceptor is shown in *yellow*. Fluorescence photons are then collected by the same objective in an epifluorescence configuration, spatially separated from the excitation light using a second dichroic mirror (BS2), focused through a pinhole to reject out-of-focus fluorescence, split by polarization (PBS), and directed toward avalanche photodiodes (APDs 1–4) via a pair of dichroic mirrors (BS3a and BS3b) chosen to spectrally separate donor and acceptor fluorescence. The resulting four detection channels correspond to parallel and perpendicular polarized fluorescence from the donor and acceptor fluorophores.

3.2 Excitation Scheme

A second important aspect for accurately determining transfer efficiencies in single-molecule FRET experiments is the excitation scheme (Fig. 1). In principle, it is possible to use a single laser to directly excite the donor fluorophore and then determine E from N_D and N_A for each burst. However, in practice, corrections for differences in quantum yields, detection efficiencies, spectral cross-talk, and direct excitation of the acceptor are required to accurately determine the transfer efficiency. A simple approach for obtaining these correction factors involves measurements of high concentrations of uncoupled fluorophores (Schuler, 2007), but it requires that the correction factors do not change upon coupling to the biomolecule, which is not always the case (Haenni, Zosel, Reymond, Nettels, & Schuler, 2013; Kretschy, Sack, & Somoza, 2016; Sanborn, Connolly, Gurunathan, & Levitus, 2007; Zosel, Haenni, Soranno, Nettels, & Schuler, 2017). This limitation can be circumvented by using an additional laser to alternatingly excite donor and acceptor fluorophores (Kapanidis et al., 2004; Müller, Zaychikov, Bräuchle, & Lamb, 2005), which makes it possible to determine all correction factors directly from the labeled samples (Hellenkamp et al., 2018; Kudryavtsev et al., 2012; Lee et al., 2005). For this reason, most studies that aim to accurately measure single-molecule FRET efficiencies utilize some form of alternating excitation.

The basic idea behind this method is to alternate between donor and acceptor excitation and use time gating to analyze the photons from the two excitation sources separately. In practice, this can either be achieved using rapidly alternating continuous-wave lasers (referred to as alternating laser excitation, ALEX) (Kapanidis et al., 2005, 2004) or via interleaved pulsed lasers (referred to as pulsed interleaved excitation, PIE) (Kudryavtsev et al., 2012; Müller et al., 2005). Although the continuous-wave lasers used for ALEX often yield higher photon detection rates, the fluorescence lifetime information afforded by PIE is very useful for identifying undesired photophysical effects (e.g., quenching) or the presence of rapidly sampled distance distributions (see below). Therefore, this chapter will focus on PIE (Fig. 1). However, apart from the lifetime information, the other aspects of PIE and ALEX are essentially identical with respect to instrument calibration.

3.3 Detection Scheme

Another important aspect of any single-molecule FRET experiment is the detection system used for recording donor and acceptor fluorescence.

Spectral separation of photons is easily achieved using dichroic mirrors, resulting in two detection channels: one for donor photons and another for acceptor photons. Additionally, it is useful to separate photons by polarization, resulting in four detection channels: donor parallel, donor perpendicular, acceptor parallel, and acceptor perpendicular (relative to the polarization of the excitation light). This four-channel approach provides access to information about the fluorescence anisotropy of the donor and acceptor, and, much like the lifetime data afforded by PIE, serves as an invaluable tool for identifying potential complications (especially hindered fluorophore rotation) that invalidate the basic assumptions necessary for quantitative transfer efficiency measurements (Sisamakis, Valeri, Kalinin, Rothwell, & Seidel, 2010). The detection scheme can be extended with additional spectral channels, e.g., to accommodate multicolor FRET, but here we focus on the commonly used and also commercially available (Wahl, Koberling, Patting, Rahn, & Erdmann, 2004) four-channel, two-color, configuration (Fig. 1).

4. ACCURATE FRET EFFICIENCIES

The value of E for each fluorescence burst in a free-diffusion measurement can be determined from the numbers of donor and acceptor photons after donor excitation. However, obtaining accurate values of E is complicated by several effects: direct excitation of the acceptor fluorophore by the donor excitation source; leakage of donor emission into the acceptor detection channel; different quantum yields of donor and acceptor; different detection efficiencies for donor and acceptor photons; and background. These effects depend on many experimental factors, including the photophysical properties of the fluorophores; the excitation wavelengths and radiant fluxes; the combination of filters and dichroic mirrors in the detection system; the sensitivity of each detector; and the alignment of the instrument. Thus, even after correcting for background, we only obtain an apparent mean transfer efficiency, $\langle \widehat{E} \rangle = \left\langle \widehat{N}_A^d / \left(\widehat{N}_A^d + \widehat{N}_D^d \right) \right\rangle$, from the detected numbers of donor, $\widehat{N}_D^d$, and acceptor, $\widehat{N}_A^d$, photons in each burst. For the actual mean transfer efficiency, $\langle E \rangle = \langle N_A^d / (N_A^d + N_D^d) \rangle$, which is the desired quantity related to the distance between donor and acceptor via Eq. (3), we require the properly corrected values, N_D^d and N_A^d. This section provides an overview of the analytical methods and workflow required to generate experimental correction factors from subpopulations of samples containing fluorescently labeled molecules (Hellenkamp et al., 2018; Kudryavtsev et al., 2012; Lee et al., 2005), which enable us to calculate N_D^d and N_A^d from $\widehat{N}_D^d$, and $\widehat{N}_A^d$.

4.1 Data Analysis: Apparent Fluorescence Stoichiometry Ratio and Apparent FRET Efficiency

One of the principle advantages of single-molecule experiments is their ability to separate individual subpopulations, provided that their photophysical properties are sufficiently different and their interconversion kinetics are slow relative to the burst duration. This is particularly obvious in PIE experiments that make use of both donor and acceptor excitation (Fig. 2). Such FRET measurements will typically be comprised of at least three distinct subpopulations: molecules that contain both an active donor and an active acceptor (FRET-active), molecules with only active donor fluorophores (donor-only), and molecules with only active acceptor fluorophores (acceptor-only), where the latter two subpopulations almost inevitable arise from photobleaching and imperfectly labeled molecules.

To separate these subpopulations, we utilize the total numbers of photons in a burst after donor excitation, N_{tot}^d, and acceptor excitation, N_{tot}^a, to define a parameter called the fluorescence stoichiometry ratio, $S = N_{\text{tot}}^d / \left(N_{\text{tot}}^d + N_{\text{tot}}^a\right)$. Bursts from FRET-active molecules are expected to have a stoichiometry ratio of $S = ½$, whereas donor-only and acceptor-only molecules should produce bursts of photons with $S = 1$, and $S = 0$, respectively. However, due to the previously mentioned complications associated with FRET measurements, the detected numbers of photons after donor excitation, $\widehat{N}_{\text{tot}}^d$, and acceptor excitation, $\widehat{N}_{\text{tot}}^a$, usually do not equal the values of N_{tot}^d and N_{tot}^a. As a result, the mean apparent fluorescence stoichiometry ratio, $\widehat{S} = \left\langle \widehat{N}_{\text{tot}}^d / \left(\widehat{N}_{\text{tot}}^d + \widehat{N}_{\text{tot}}^a\right) \right\rangle$, differs slightly from $S = \left\langle N_{\text{tot}}^d / \left(N_{\text{tot}}^d + N_{\text{tot}}^a\right) \right\rangle$. Nevertheless, these three subpopulations can easily be identified via a histogram of $\widehat{S}$ and used to determine the correction factors needed for proper instrument calibration (Fig. 2).

4.1.1 Correction for Cross-Talk and Acceptor Direct Excitation From Donor-Only and Acceptor-Only Subpopulations

Donor-only molecules (i.e., $\widehat{S} \approx 1$) only emit donor photons and should thus ideally have an apparent mean transfer efficiency of $\left\langle \widehat{E}_{\text{donor-only}} \right\rangle = 0$. However, due to spectral cross-talk, some donor photons leak into the

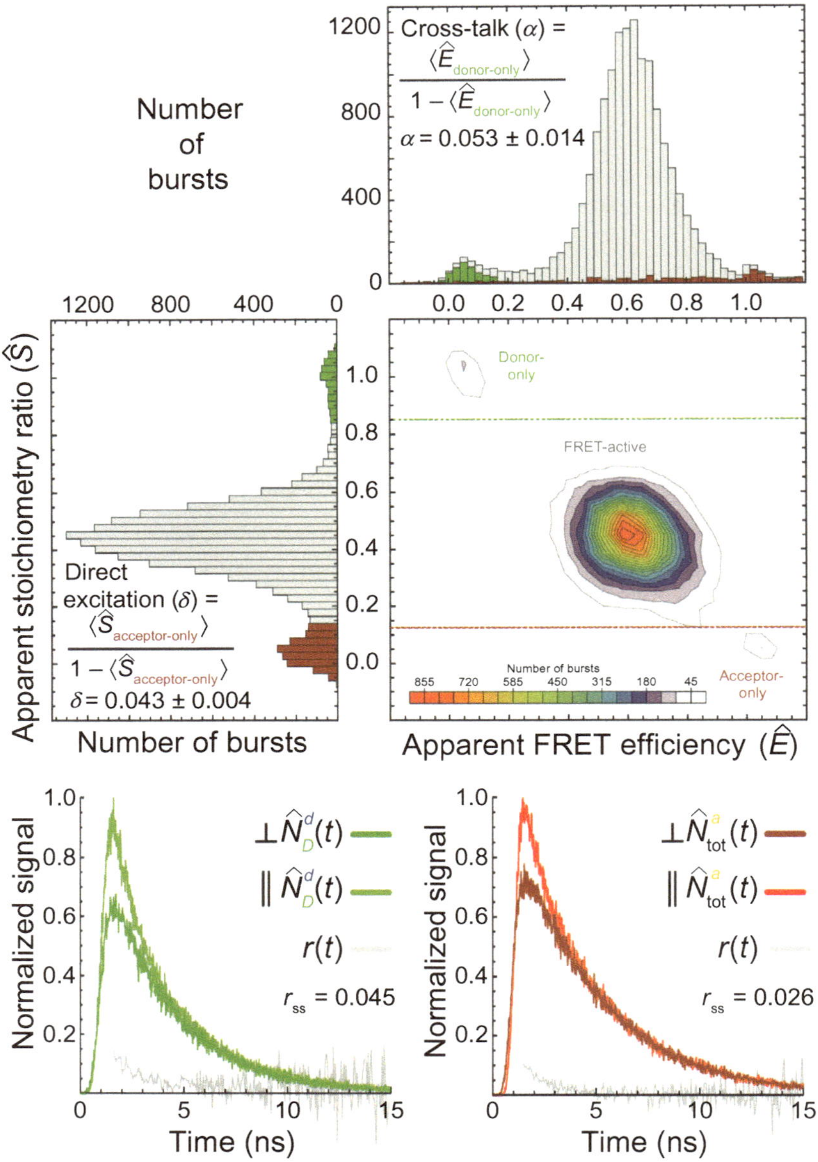

Fig. 2 Determining correction factors for acceptor direct excitation (δ) and donor cross-talk (α). Using the apparent fluorescence stoichiometry ratio, $\hat{S}$, it is possible to identify bursts from molecules with either only active donor (i.e., $\hat{S} \approx 1$) or only active acceptor (i.e., $\hat{S} \approx 0$) fluorophores in the 2D $\hat{S}$ vs $\hat{E}$ histogram. The apparent mean transfer efficiency, $\langle \hat{E} \rangle$, of the donor-only subpopulation is used to determine the correction factor

acceptor detection channel, and as a result, $\langle \widehat{E}_{\text{donor-only}} \rangle > 0$ (Fig. 2). This nonzero value is used to define the cross-talk correction factor, $\alpha = \langle \widehat{E}_{\text{donor-only}} \rangle / (1 - \langle \widehat{E}_{\text{donor-only}} \rangle)$. Similarly, acceptor-only molecules (i.e., $\widehat{S} \approx 0$) do not contain a donor fluorophore and therefore should not be excited by the donor excitation laser. However, due to residual direct excitation of the acceptor fluorophore by the donor excitation laser, $\langle \widehat{S}_{\text{acceptor-only}} \rangle > 0$ (Fig. 2). Again, this nonzero value is used to define the direct excitation correction factor, $\delta = \langle \widehat{S}_{\text{acceptor-only}} \rangle / (1 - \langle \widehat{S}_{\text{acceptor-only}} \rangle)$. Then, α and δ are used to correct the detected number of acceptor photons after donor excitation for both cross-talk and direct excitation, $N_A^d = \widehat{N}_A^d - \left(\alpha \cdot \widehat{N}_D^d\right) - \left(\delta \cdot \widehat{N}_{\text{tot}}^a\right)$. Cross-talk of acceptor emission into the donor channel is usually negligible.

4.1.2 Correction for Excitation and Detection Efficiencies From the FRET-Active Subpopulation

The values of N_A^d are then used to redefine the apparent transfer efficiency and apparent fluorescence stoichiometry ratio of each burst:

$$\widehat{E} = \frac{N_A^d}{N_A^d + \widehat{N}_D^d} \text{ and } \widehat{S} = \frac{N_A^d + \widehat{N}_D^d}{N_A^d + \widehat{N}_D^d + \widehat{N}_{\text{tot}}^a}. \tag{4}$$

Regardless of its value of $\langle \widehat{E} \rangle$, every FRET-active subpopulation should have an apparent mean stoichiometry ratio of $\langle \widehat{S} \rangle = ½$. However, because of the different excitation and detection efficiencies for the two fluorophores,

---

for cross-talk, $\alpha = \langle \hat{E}_{\text{donor-only}} \rangle / (1 - \langle \hat{E}_{\text{donor-only}} \rangle)$. The apparent mean fluorescence stoichiometry ratio, $\hat{S}$, associated with the acceptor-only subpopulation is used to determine the correction factor for direct excitation, $\delta = \langle \hat{S}_{\text{acceptor-only}} \rangle / (1 - \langle \hat{S}_{\text{acceptor-only}} \rangle)$ (due to background correction, $\hat{E}$ values for the acceptor-only bursts extend beyond the range shown). With pulsed interleaved excitation (PIE), time-correlated single-photon counting, and four-channel detection, it is also possible to obtain time-resolved fluorescence intensity and anisotropy decay plots from the parallel and perpendicular emission of the donor-only subpopulation after donor excitation (i.e., $\| \widehat{N}_D^d$ and $\perp \widehat{N}_D^d$) and the parallel and perpendicular emission of the acceptor-only subpopulations after acceptor excitation (i.e, $\| \widehat{N}_{\text{tot}}^a$ and $\perp \widehat{N}_{\text{tot}}^a$). This information is then used to determine the fluorescence lifetimes and the time-resolved and the steady-state anisotropies (r_{ss}) of the donor and acceptor fluorophores.

this is generally not the case. Two additional correction factors account for these experimental imperfections. The relative excitation efficiency, β, describes how efficiently the two fluorophores are excited by their respective excitation lasers, $\beta = (P^a \cdot \varepsilon_A^a)/(P^d \cdot \varepsilon_D^d)$, where P^a and P^d represent the relative powers of the acceptor and donor excitation lasers, and ε_A^a and ε_D^d are the extinction coefficients of the acceptor and donor fluorophores at their respective excitation wavelengths. The second correction factor, γ, accounts for the relative quantum yields and detection efficiencies for donor and acceptor emission, with $\gamma = (\phi_A \cdot \eta_A)/(\phi_D \cdot \eta_D)$, where ϕ_A and ϕ_D are the quantum yields of donor and acceptor, and η_A and η_D are the detection efficiencies for donor and acceptor photons, respectively. The values of β and γ can be determined from the dependence of $\langle \widehat{S} \rangle$ on $\langle \widehat{E} \rangle$ using at least two subpopulations with different donor–acceptor distances (Lee et al., 2005):

$$\langle \widehat{S} \rangle = \frac{1}{1 + (\beta \cdot \gamma) + \langle \widehat{E} \rangle (\beta - (\beta \cdot \gamma))} \tag{5}$$

This relation shows that if $\gamma = 1$, $\langle \widehat{S} \rangle$ is independent of $\langle \widehat{E} \rangle$. If, additionally, the two fluorophores are excited with identical efficiency (i.e., $\beta = 1$), then $\langle \widehat{S} \rangle = ½$. The factor γ is used to correct the number of donor photons detected after donor excitation for the different detection efficiencies of donor and acceptor photons (i.e., $N_D^d = \gamma \cdot \widehat{N}_D^d$), and the correction factor β is used to correct the total number of photons detected after acceptor excitation for the different excitation efficiencies of the two fluorophores (i.e., $N_{\text{tot}}^a = \widehat{N}_{\text{tot}}^a / \beta$). These two correction factors are then used to calculate the transfer efficiency, $E = N_A^d / \left(N_A^d + N_D^d\right)$, and stoichiometry ratio, $S = \left(N_A^d + N_D^d\right) / \left(N_A^d + N_D^d + N_{\text{tot}}^a\right)$, for each burst. The values of $\langle S \rangle$ and $\langle E \rangle$ for a given subpopulation can then be determined via a 2D-Gaussian fit to a 2D S vs E histogram of the FRET-active bursts. Because the correction factors γ and β are determined from the dependence of $\langle \widehat{S} \rangle$ on $\langle \widehat{E} \rangle$, we need to measure multiple subpopulations with different values of $\langle \widehat{E} \rangle$. This can be achieved in a variety of ways; the most straightforward is by working with a single sample with two or more well-separated subpopulations (e.g., native/denatured, bound/unbound, *cis*/*trans*, phosphorylated/dephosphorylated). However, it is often difficult to cleanly determine $\langle \widehat{S} \rangle$ and $\langle \widehat{E} \rangle$ when there are more than a few subpopulations in a single sample, which in turn limits the ability to determine β and γ. A more robust, albeit more time-consuming approach is to measure multiple independent samples (e.g., different biomolecules or different experimental conditions) labeled with the same fluorophores. Regardless of the approach, it is important to ensure that the differences in $\langle\widehat{E}\rangle$ are solely caused by different donor–acceptor distances and not by differences in rotational flexibility, quenching of the fluorophores, changes in refractive index, or other effects that would lead to a change in R_0. It is thus important to quantify such contributions, e.g., for rotational motion via fluorescence anisotropies, for dynamic quenching via changes in fluorescence lifetimes, or for static quenching via nanosecond fluorescence correlation spectroscopy (Haenni et al., 2013; Zosel et al., 2017).

4.2 Calibration Samples and Measurements

To demonstrate this calibration approach, we performed single-molecule FRET measurements of different IDPs and polyproline peptides labeled with Alexa 488 and Alexa 594 via maleimide chemistry (Fig. 3A). None of the samples showed detectable signs of hindered fluorophore rotation or quenching and yielded correction factors for acceptor direct excitation and donor cross-talk of $\alpha = 0.042 \pm 0.014$ and $\delta = 0.043 \pm 0.004$, respectively. These values were used to generate 2D histograms of $\widehat{S}$ vs $\widehat{E}$ for each sample using Eq. (4), with mean values determined from 2D-Gaussian fits. The resulting values of $\langle\widehat{S}\rangle$ and $\langle\widehat{E}\rangle$ were then analyzed with Eq. (5), yielding $\beta = 1.16 \pm 0.03$ and $\gamma = 1.27 \pm 0.02$ (Fig. 3A), which in turn were used to determine $\langle S\rangle$ and $\langle E\rangle$ for each of the eight samples (Fig. 3B). Once the correction factors are established for a given FRET pair and instrument configuration, they can be used for any future samples labeled with the same dyes, provided that the photophysical properties of the fluorophores do not differ from the reference samples. As recently demonstrated in a multilaboratory benchmark study, this methodology typically results in transfer efficiencies with experimental uncertainties between ± 0.02 and ± 0.05 (Hellenkamp et al., 2018).

One of the advantages of determining $\langle S\rangle$ and $\langle E\rangle$ values for a larger sample set is that it is possible to identify cases where the photophysical properties (e.g., quantum yields) of the dyes deviate from the calibration set based on deviations from $\langle S\rangle = ½$. To demonstrate this behavior, we use the N-terminal domain of HIV-1 integrase (IN), an IDP with a tryptophan residue at position 23 that is known to quench Alexa 488 (Zosel et al., 2017).

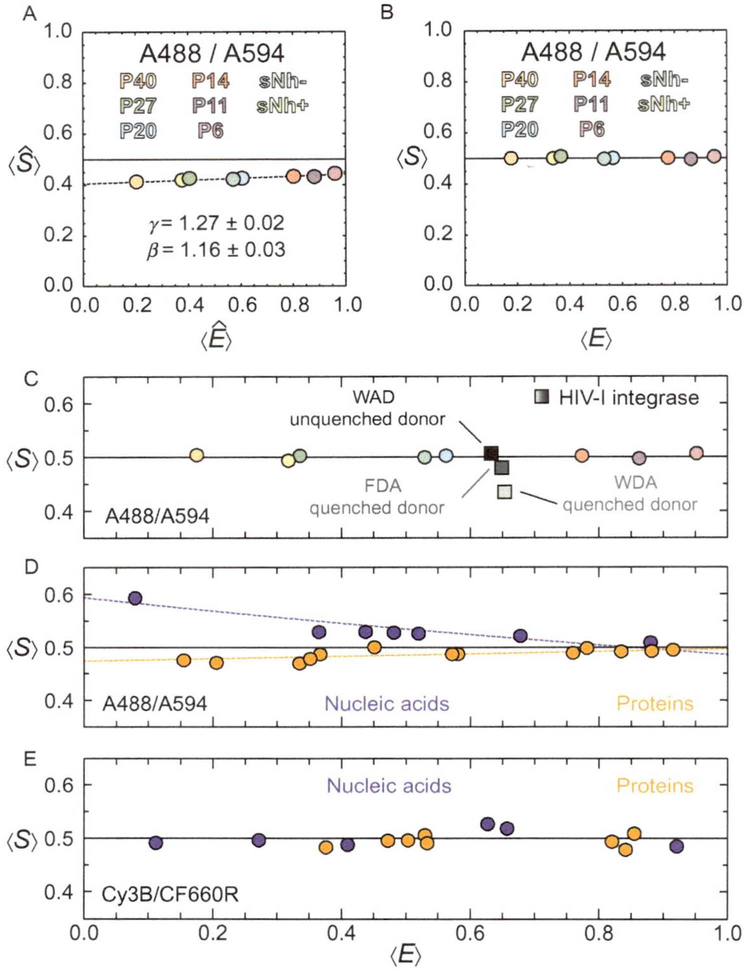

Fig. 3 Quantifying correction factors for relative excitation (β) and detection (γ) efficiencies. For each of eight different samples, a 2D histogram of $\hat{S}$ vs $\hat{E}$ is generated using Eq. (4) (Fig. 2), and the values of $\langle\hat{S}\rangle$ and $\langle\hat{E}\rangle$ for the FRET-active subpopulation of each sample determined using 2D-Gaussian fits. (A) These data are plotted and fit to Eq. (5) to determine the correction factors β and γ. (B) All four correction factors (i.e., α, δ, β, and γ) are then used to determine $\langle S\rangle$ and $\langle E\rangle$ for the eight samples (see Table 1). Note that $\langle S\rangle = 1/2$ for all samples, independent of $\langle E\rangle$. (C) Species with photophysical irregularities can be identified via a deviation from $\langle S\rangle = 1/2$, for example, tryptophan-induced quenching of Alexa 488 in HIV-1 integrase (*gray squares*; see text for details). (D) Example of a set of correction factors derived from a collection of biomolecules labeled with Alexa 488/594 that results in $\langle S\rangle$ values that systematically deviate from 1/2, depending on the type of biomolecule. This deviation is quantified by fitting $\langle S\rangle$ vs $\langle E\rangle$ for proteins and nucleic acids separately with Eq. (5) (*colored dashed lines*). This analysis indicates that the different local chemical environments of the fluorophores give rise to variations in their photophysical properties, resulting in different correction factors. (E) A more broadly applicable set of correction factors is obtained for another diverse set of proteins and nucleic acids labeled with Cy3B/CF660R, indicating that the photophysical properties of this FRET pair are less dependent on the local environment than those of the Alexa 488/594 FRET pair.

Table 1 Fluorescence Parameters for IDP/Polyproline Samples

	Sample	δ	τ_D	α	τ_A	r_{ss}(donor)	r_{ss}(acceptor)	$\langle E \rangle$	$\langle S \rangle$
	Free dyes	0.035	3.92	0.04	3.9	a	a	—	—
Reference data set	○ P40	0.045	4.00	0.05	3.9	0.02	0.02	0.18	0.504
	○ P27	0.040	3.99	0.04	3.9	0.01	0.02	0.34	0.503
	○ P20	0.040	3.95	0.05	3.9	0.01	0.01	0.56	0.502
	○ P14	0.050	3.79	0.02	3.9	0.02	0.01	0.77	0.502
	○ P11	0.038	3.98	0.05	3.9	0.02	0.02	0.86	0.497
	○ P6	0.045	3.95	0.02	3.8	0.02	0.02	0.95	0.506
	○ sNh$^-$	0.043	3.98	0.05	4.3	0.05	0.03	0.53	0.501
	○ sNh$^+$	0.040	3.97	0.05	4.0	0.04	0.01	0.32	0.494
	Average	0.043	3.95	0.04	3.9	0.02	0.02	—	0.501
	Standard deviation	0.004	0.07	0.01	0.2	0.01	0.01	—	0.004
Integrase data set	□ IN-WDA	0.042	*3.37*	0.06	4.1	0.05	0.02	0.65	*0.435*
	■ IN-FDA	*0.057*	*3.66*	0.05	4.1	0.03	0.03	0.65	*0.480*
	■ IN-WAD	0.044	3.81	0.05	4.3	0.03	0.04	0.63	0.506

Orange italic values in the integrase data set are more than three standard deviations away from the mean of the reference data set.

[a]Steady-state anisotropies of the free dyes are ~0.02, virtually indistinguishable from zero within the experimental error of confocal measurements. For simplicity, we thus assumed their anisotropies to be zero for the purpose of correcting for differential collection efficiencies of the parallel and perpendicular detection channels.

Photophysical parameters: δ is the correction factor for acceptor direct excitation; τ_D is the donor-only fluorescence lifetime (from tail fits); α is the correction factor for spectral cross-talk; τ_A is the acceptor-only fluorescence lifetime (from tail fits); r_{ss}(donor) is the donor steady-state anisotropy of the donor-only subpopulation; and r_{ss}(acceptor) is the stead-state anisotropy of the acceptor-only subpopulation upon acceptor direct excitation. Note that δ increases upon coupling to proteins, an indication that the use of free dyes would result in nonrepresentative correction factors. After the correction factors are applied, the mean fluorescence stoichiometry ratio of the reference data set is $\langle S \rangle = 0.501 \pm 0.004$. The decreased donor lifetimes of the HIV-1 integrase sample result in a substantially decreased fluorescence stoichiometry ratio. Replacing the tryptophan near the donor fluorophore (IN-WDA) with a phenylalanine (IN-FDA) leads to an increase in donor lifetime and fluorescence stoichiometry ratio. A similar effect is apparent when the labeling positions of the donor and acceptor are swapped (IN-WAD). Color code as in Fig. 4.

The fluorescence stoichiometry ratio in the IN sample where Alexa 488 is close to the tryptophan residue (IN-WDA) deviates detectably from $\langle S \rangle = ½$ (Fig. 3C), concomitant with a reduced donor fluorescence lifetime (Table 1). Replacing the tryptophan residue with phenylalanine (IN-FDA) shifts $\langle S \rangle$ closer to ½. Also, the donor lifetime determined from the donor-only population of IN-FDA is closer to the corresponding values of the calibration set, whose members lack any aromatic residues. A similar shift occurs when swapping the positions of the donor and acceptor (IN-WAD). This example illustrates that quenched samples can be identified based on the fluorescence stoichiometry ratio without having to directly monitor the fluorescence lifetime; the effects of quenching can then be taken into consideration when calculating $\langle S \rangle$ and $\langle E \rangle$. The slight shift of IN-WAD to lower E, however, which is likely due to static quenching of the acceptor by the tryptophan (Haenni et al., 2013), is not obvious from this analysis and requires alternative methods for detection, such as nanosecond fluorescence correlation spectroscopy (Doose, Neuweiler, & Sauer, 2005; Haenni et al., 2013; Zosel et al., 2017).

The photophysical parameters, and thus the correction factors, associated with the fluorophores can vary depending on the molecules they are coupled to. Therefore, we measured a diverse collection of FRET-labeled samples to determine how robust the correction factors are. This collection of molecules (Fig. 3D and E) is comprised of different types of biomolecules (folded proteins, IDPs, as well as single- and double-stranded nucleic acids) labeled with two different FRET pairs (Cy3B/CF600R and Alexa 488/594) using different coupling chemistries (maleimide or *N*-succinimidyl ester). The molecules labeled with Alexa 488/594 exhibit significantly more scatter in $\langle S \rangle$ than the reference samples shown in Fig. 3C. Closer inspection reveals slightly but systematically different behavior for the nucleic acid and protein samples. Furthermore, the fluorescence stoichiometry ratios generated from this set are not independent of the transfer efficiency. These differences can be quantified by analyzing the protein and nucleic acid data points separately in the $\langle S \rangle$ vs $\langle E \rangle$ plot (Fig. 3D) using Eq. (5), which results in $\gamma' = 0.65 \pm 0.04$ for nucleic acids and $\gamma' = 1.10 \pm 0.04$ for proteins. The significant deviations from the expected value of $\gamma' = 1$ indicate that different correction factors should be used for the protein and nucleic acid samples with this dye pair. For instance, the error in the mean transfer efficiency of a protein sample at $\langle \hat{E} \rangle = 0.5$ analyzed using the correction factors from the nucleic acid samples would be $\Delta \langle E \rangle \approx 0.14$. However, this effect is highly fluorophore-dependent: The data set in Fig. 3E for the Cy3B/CF600R FRET pair, e.g.,yields correction factors ($\alpha = 0.038 \pm 0.004$, $\delta = 0.113 \pm 0.007$, $\beta = 0.97 \pm 0.03$, $\gamma = 0.60 \pm 0.04$ in this case) that are largely independent of the biomolecules the dyes are attached to.

The above examples (Fig. 3) illustrate how large sets of samples can provide robust correction factors necessary for determining accurate transfer efficiencies. These measurements can also reveal variability in the photophysical properties of fluorophores coupled to biomolecules. Both features make this approach ideally suited for quantifying distances in biomolecules, including IDPs.

4.3 Evidence for Distance Distributions

Thus far, we have focused on obtaining accurate transfer efficiencies. The next step is to relate these values to the distances within the molecule. The key complication for IDPs is that they usually sample broad distance distributions, $P(r)$, on timescales much shorter than the microsecond interphoton times of typical single-molecule FRET experiments (Schuler, 2018; Schuler et al., 2016). If these conformational fluctuations occur on a timescale much longer than τ_D, then

$$\langle E \rangle = \langle \varepsilon \rangle = \int P(r)\varepsilon(r)\, dr. \tag{6}$$

Note that the mean values of ε and E are equal if the conformational dynamics occur between these two limiting timescales, but the distribution of E in a FRET efficiency histogram is determined primarily by shot noise arising from the 100 or so photons within each burst and not by the underlying $P(r)$ that we would like to characterize.

However, in addition to $\langle \varepsilon \rangle$, it is possible to extract the variance, $\sigma^2 = \langle \varepsilon^2 \rangle - \langle \varepsilon \rangle^2$, of the distribution of transfer efficiencies from single-molecule data acquired using PIE and time-correlated single-photon counting (Gopich & Szabo, 2012; Kalinin, Valeri, Antonik, Felekyan, & Seidel, 2010). The reason is that the relevant observation time in this case is the fluorescence lifetime, which is on the order of a few nanoseconds and thus much faster than the relaxation time of the interdye distance (typically in the range of tens to hundreds of nanoseconds for IDPs in the length range accessible to single-molecule FRET) (Schuler, 2018; Schuler et al., 2016). To obtain σ^2, we estimate the donor fluorescence lifetime of each FRET-active burst, τ_{DA}, from the mean of the excitation–emission delay time, Δt_{DA}, of donor photons from the corresponding bursts, i.e., $\tau_{DA} = \langle \Delta t_{DA} \rangle$. For a single fixed distance, $\langle \tau_{DA} \rangle / \tau_D = 1 - \varepsilon$, which follows directly from Eq. (2). However, since IDPs rapidly sample broad distance distributions, the value of τ_{DA} is biased toward longer times, because expanded conformations, for which the donor lifetime is longer, emit more donor photons than compact conformations. In a plot of τ_{DA}/τ_D vs E (Fig. 4), this bias can be visualized as a displacement from the diagonal associated with a single fixed distance, also referred to as the "static FRET line" (Kalinin et al., 2010). This displacement is then used to quantify σ^2 (Chung, Louis, & Gopich, 2016; Gopich & Szabo, 2012):

$$\frac{\langle \tau_{DA} \rangle}{\tau_D} = 1 - \langle \varepsilon \rangle + \frac{\sigma^2}{1 - \langle \varepsilon \rangle} \tag{7}$$

Correspondingly, bursts arising from molecules that have a broad but *static* transfer efficiency distribution (i.e., slow interconversion during the 1-ms burst duration), exemplified by the polyproline peptide shown in Fig. 4

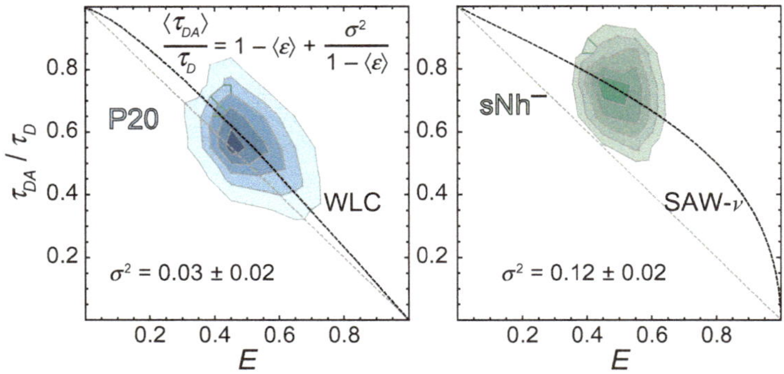

Fig. 4 Assessing distributions of interdye distances with fluorescence lifetime information. With pulsed interleaved excitation, it is possible to determine the relative donor lifetime in the presence of the acceptor, τ_{DA}/τ_D. This parameter provides information about the variance of the underlying distribution of transfer efficiencies. (*Left*) In the case of a static distribution of distances in a 20-mer polyproline peptide (Best et al., 2007; Schuler, Lipman, Steinbach, Kumke, & Eaton, 2005), the values of $\langle\tau_{DA}\rangle/\tau_D$ cluster close to the diagonal, which corresponds to a single fixed distance (static FRET line), $\langle\tau_{DA}\rangle/\tau_D = 1 - \varepsilon$. (*Right*) However, for a broad and rapidly sampled distribution, for example, the intrinsically disordered peptide sNh^-, values of τ_{DA}/τ_D cluster above the diagonal. This vertical displacement provides a measure of the variance of the underlying distribution of transfer efficiencies, σ^2. In this way, lifetime vs transfer efficiency plots can in principle be used to assess the quality of polymer models (i.e., self-avoiding walk (Zheng et al., 2018), or worm-like chain (O'Brien, Morrison, Brooks, & Thirumalai, 2009)) commonly used to describe FRET-labeled biomolecules. The error of σ^2 was estimated assuming an uncertainty of ~0.1 ns for both τ_D and τ_{DA}.

(Best et al., 2007; Schuler et al., 2005), will cluster close to the diagonal, whereas IDPs that dynamically sample a broad transfer efficiency distribution on the timescale of the interphoton time or faster will cluster further above the diagonal. The lifetime information in these experiments can thus provide evidence for rapid conformational dynamics within a subpopulation. The experimental values of $\langle\varepsilon\rangle$ and σ^2 afford a model-independent assessment of $P(r)$ and can in principle be used to parameterize the underlying distance distribution.

5. INFERRING DISTRIBUTIONS OF DISTANCES AND CONFORMATIONS

Thus far, we have discussed how the mean efficiency, $\langle\varepsilon\rangle$, and its variance, σ^2, are related to an unspecified distribution, $P(r)$, of the distance r between the chromophores. For folded proteins, the distance r usually fluctuates relatively little about its mean, so that a given $\langle E \rangle$ is commonly mapped directly to a mean distance via Eq. (3) (even though more quantitative approaches accounting for the flexibility of dye linkers are available; Kalinin et al., 2012; Muschielok et al., 2008). Disordered and unfolded proteins, however, populate a very broad range of r. Therefore, it is necessary to characterize the distribution $P(r)$ to obtain properties such as its mean and higher moments. In practice though, how can one reconstruct $P(r)$ using the limited experimental information available? A reasonable choice, when it is safe to assume that the protein is disordered, is to take $P(r)$ from a homopolymer model characterized by one or a small number of adjustable parameters that can be determined uniquely by fitting to the experimental data, so that the problem is well posed. This is what is most commonly done for intrinsically disordered and unfolded proteins (Schuler et al., 2016), but some care is needed in order to obtain quantitatively accurate results (O'Brien et al., 2009), as we outline in Section 5.1. In particular, it may be necessary to allow the form of $P(r)$ to vary to accommodate changes in solution conditions (Zheng et al., 2018). In addition to inferring end-to-end distances, we would often like to know the radius of gyration $R_g = \left(1/2\left\langle r_{ij}^2(m)\right\rangle_{i,j,m}\right)^{1/2}$ (here defined in terms of the average distances $r_{ij}(m)$ over all atom pairs i, j, and also over all conformations m). This facilitates comparison with SAXS, which measures R_g almost directly; R_g is also a fundamental property of interest of a disordered chain. However, FRET measures only $P(r)$ for a single distance, and there is no fixed relationship between R_g^2 and $\langle r^2 \rangle$—indeed it is known that the ratio of these quantities also varies with solution conditions (Borgia et al., 2016; Fuertes et al., 2017; Schäfer, 1999), which must be considered when interconverting them.

While analytical polymer models are by far the simplest to use, they can only be applied to single, fully disordered chains. Many proteins containing intrinsically disordered regions also contain folded domains (Oldfield & Dunker, 2014; van der Lee et al., 2014); others assemble into higher-order complexes with other biomolecules (Wright & Dyson, 2015). Even among isolated disordered proteins, extreme variations of charge or hydrophobic patterning can cause deviations from the distributions that would be obtained from analytical homopolymer models (Das & Pappu, 2013; Fuertes et al., 2017). These more complex scenarios, which usually cannot be treated analytically and call for the use of molecular models, are the subject of Sections 5.2 and 5.3. The chemical diversity inherent in disordered proteins means that it is also necessary to run molecular dynamics (or Monte Carlo) simulations of the molecular model in order to sample representative configurations of the system, which makes this procedure much more time-consuming.

A further complication is that, unlike analytical polymer models, predictive simulation models typically have many parameters—how, then, should one adapt the model if it does not perfectly reproduce experiment? Bayesian statistics (or similarly, the maximum entropy principle) provide a solution in which the ensemble of structures obtained by simulation is minimally perturbed in order to match experiment. This is described in Section 5.2. An alternative scheme is to use a minimalist coarse-grained model, characterized by only a handful of free parameters. One may then fit the parameters of such a model directly to the data, as is described in Section 5.3. This "Occam's Razor" approach avoids overfitting by using only a limited number of parameters in the model. In contrast, the Bayesian procedure in Section 5.2 has many more parameters than can possibly be determined by the available data; it therefore requires a "regularization" procedure to avoid overfitting.

5.1 Polymer Model Description of Intrinsically Disordered Proteins

Given a functional form for a distance distribution, $P(r;a)$, characterized by parameter(s) a, one can straightforwardly determine a by numerically solving the integral equation (Eq. 6) above. A variety of models have been employed for this purpose, including the Gaussian chain, worm-like chain, and self-avoiding walk (SAW) (Schuler et al., 2016). The distribution which has been most commonly used in the past is the Gaussian chain (GC). This is an idealized polymer in which the displacements between monomer units are governed by Gaussian statistics (a type of random walk), and there are no interactions between the monomer units (i.e., a "phantom chain"). The Gaussian chain $P(r)$ is given by

$$P(r;R) = \left(\frac{3}{2\pi}\right)^{3/2} \frac{4\pi}{R} \exp\left[-\frac{3}{2}\left(\frac{r}{R}\right)^2\right] \tag{8}$$

where the single parameter R is the root mean square end-to-end distance. One could then infer the radius of gyration by using the exact relation for a Gaussian chain, $R_g = R/\sqrt{6}$. This model is often a quite acceptable first approximation, in particular, since unfolded proteins in water are frequently close to the so-called theta state (Hofmann et al., 2012), in which protein–protein and protein–solvent interactions are balanced—the situation in which a Gaussian chain works best (Borgia et al., 2016; Zheng et al., 2018). However, if the protein interacts more favorably with the solvent than with itself (for example, in high concentrations of chemical denaturant, or in a chain with high net charge), a Gaussian chain tends to overestimate both the mean-square distance, as well as the inferred radius of gyration (Borgia et al., 2016; Fuertes et al., 2017; O'Brien et al., 2009; Song, Gomes, Gradinaru, & Chan, 2015). This overestimation arises from the neglect of excluded volume in a Gaussian chain, an effect that contributed partly to an apparent discrepancy between inferences of R_g from FRET and from SAXS experiments (Yoo et al., 2012). This discrepancy has recently been resolved by improving the methods for analyzing both FRET and SAXS (Borgia et al., 2016; Fuertes et al., 2017; Riback et al., 2017; Zheng & Best, 2018; Zheng et al., 2018).

The end-to-end distance distribution when protein–solvent interactions are very favorable is better approximated by a SAW, a model in which the residues in the protein have only short-range repulsive interactions between them. An approximation to the end-to-end distance distribution of the SAW is given by

$$P(r; R, \nu) = \frac{4\pi A}{R}\left(\frac{r}{R}\right)^{2+(\gamma-1)/\nu} \exp\left[-\alpha\left(\frac{r}{R}\right)^{1/(1-\nu)}\right] \tag{9}$$

where R is the root mean square distance, $\gamma \approx 1.1615$ is a critical exponent (Le Guillou & Zinn-Justin, 1977), and $\nu \approx 0.6$ is the Flory exponent for a SAW. The constants A and α are determined from the requirements $\int_0^\infty P(r)\,dr = 1$ and $\int_0^\infty P(r)\,r^2\,dr = R^2$. For a SAW, $R_g = R/\sqrt{6.26}$ for long chains. Examples of $P(r)$ for a Gaussian chain and a SAW corresponding to three different FRET efficiencies are given in Fig. 5. As is clear from the figure, it is possible to infer somewhat different average end-to-end distances from the same FRET efficiency when using different models; the change of the estimated average distance with FRET efficiency is also larger for the Gaussian chain. How can one minimize the model dependence of the inferred properties? Naively, one might assume that always using the same model to fit the FRET efficiency should give self-consistent results. Unfortunately, the most appropriate model depends on the effective solvent quality, which can vary with denaturant concentration, ionic strength,

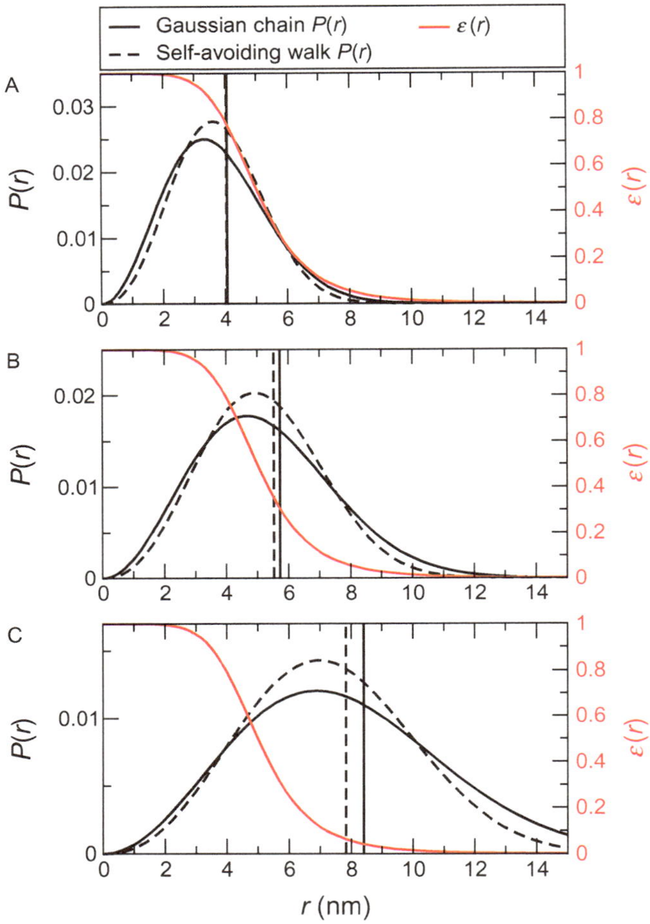

Fig. 5 Illustration of the ambiguity in inferring a distance distribution from limited FRET data. Distributions with mean efficiencies of $\langle \varepsilon \rangle = 0.75$, 0.5, and 0.25 are shown in A, B, and C, respectively. In each case, the *P*(*r*) for a Gaussian chain and a self-avoiding walk corresponding to the same FRET efficiency are shown. The variances of the corresponding transfer efficiency distributions, σ^2, for the Gaussian chain and self-avoiding walk, respectively, are 0.08 and 0.07 in (A), 0.13 and 0.11 in (B), and 0.10 and 0.09 in (C).

temperature, or amino acid sequence. In principle, it should be possible to use the additional information provided by σ^2 from fluorescence lifetime data (Fig. 4) to choose the most appropriate distribution. However, in practice, this variance is often very similar for the different models and does not have much discriminating power once the experimental error is considered. For example, in Fig. 5B at $\langle \varepsilon \rangle = 0.5$, σ^2 is 0.13 for a GC and 0.11 for a SAW (see figure legend), whereas the intrinsically disordered peptide sNh$^-$ in Fig. 4, which resides near $\langle \varepsilon \rangle = 0.5$, has a variance of $\sigma^2 = 0.12 \pm 0.02$—i.e., the difference between the two models is within experimental error. Therefore, some additional information is needed to constrain the form of $P(r)$. Furthermore, the Gaussian chain and SAW represent two simplified limiting scenarios, while in reality a continuum of intermediate distributions is expected (for instance due to changes in solvent quality between theta solvent and good solvent for finite chains).

To overcome these difficulties, we recently proposed a semiempirical approach, referred to as SAW-ν, in which the distance distribution is described via Eq. (9), but the exponent ν is treated as an adjustable parameter, rather than being fixed to 0.6 (Zheng et al., 2018). Consequently, two parameters, ν and R, must be determined, rather than one (R) for the GC or SAW. We have found that these parameters could not be reliably determined from the combination of $\langle \varepsilon \rangle$ and σ^2 readily available from FRET experiments (Zheng et al., 2018), because these parameters are highly correlated and σ^2 does not discriminate sufficiently between distributions, as discussed earlier. Instead, we use an approximate scaling law for the end-to-end distance in proteins to relate R to ν,

$$R = bN^{\nu} \tag{10}$$

in which the prefactor b is approximately 0.55 nm for proteins with well-mixed sequences (Hofmann et al., 2012) (for some low-complexity sequences or stiff homopolymers such as polyproline (Fig. 4A), a different prefactor may be required). This additional relation makes it possible to solve for the single free parameter characterizing the distribution, i.e., ν. In order to convert R^2 to R_g^2, we employ the approximate relation (Witten & Schäfer, 1978)

$$\frac{R^2}{R_g^2} = \frac{2(\gamma + 2\nu)(\gamma + 2\nu + 1)}{\gamma(\gamma + 1)}. \tag{11}$$

The SAW-ν model is thus able to interpolate between different forms of $P(r)$, characterized by the scaling exponent ν. The quality of the fit can be tested against molecular simulation models, in which varying the temperature modulates the effective solvent quality, represented by ν. In Fig. 6,

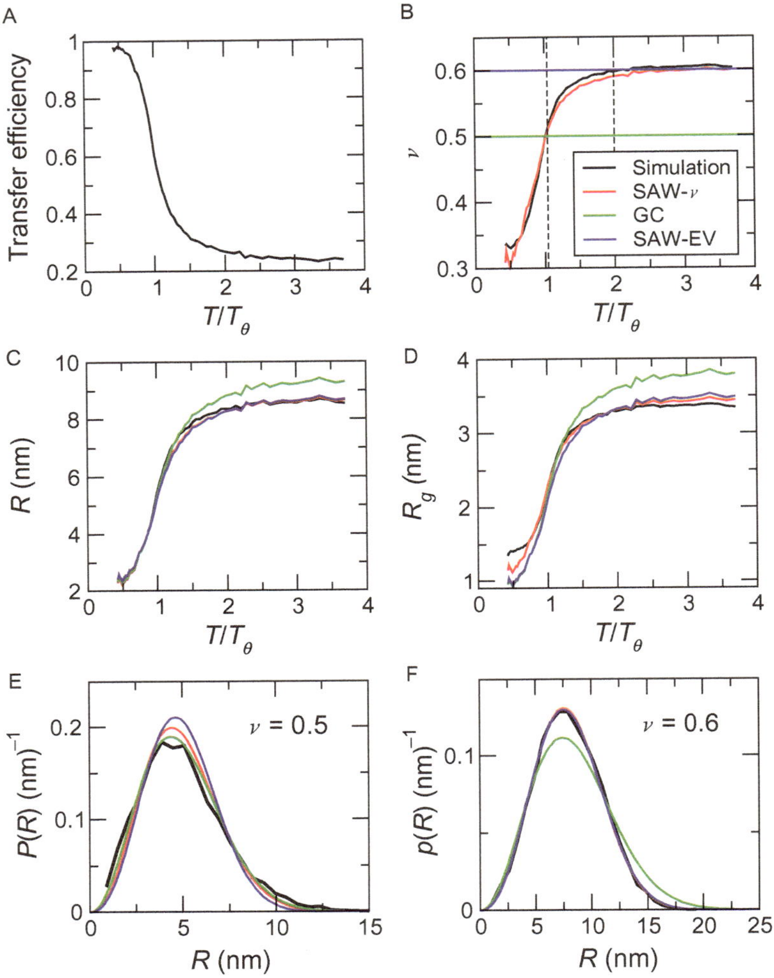

Fig. 6 Ability of polymer models to recover ensemble properties from simulation data. (A) Mean transfer efficiency computed from simulations of a simple homopolymer (Val_{100}), as a function of simulation temperature. (B) Scaling exponent ν computed directly from internal distance scaling (Borgia et al., 2016) in the simulation, and from the SAW-ν model (Zheng et al., 2018). Fixed ν implicit in Gaussian chain and SAW-EV models shown for reference (*horizontal lines*). *Broken vertical lines* indicate temperatures corresponding to $\nu \approx 0.5$ and $\nu \approx 0.6$ in (E) and (F). (C) End-to-end distance, *R*, and (D) radius of gyration, R_g, as a function of temperature, as calculated directly from the simulation, and as inferred from each polymer model (see legend in (B)). (E and F) *p*(*R*) at $\nu \approx 0.5$ and $\nu \approx 0.6$, respectively, as calculated directly from the simulation and as determined from fitting polymer models to the FRET efficiencies in (A).

we apply the model to synthetic values of ε calculated from a simulation of a homopeptide (poly-Val) chain of length 100, at different temperatures (Fig. 6A). We can then use different models to attempt to reconstruct $P(r)$ from the knowledge of the efficiencies only. Since we have data at multiple temperatures, we can see how well each model performs as a function of temperature, with each temperature corresponding to a different effective solvent quality.

We see that the SAW-ν can recover the scaling exponent ν computed directly from the protein coordinates in the simulations (Fig. 6B), while the Gaussian chain or a SAW with fixed $\nu \approx 0.6$ (SAW-EV) of course cannot capture this variation. Both SAW-ν and SAW-EV do a reasonable job of recovering the average end-to-end distance and radius of gyration (Fig. 6C and D). As noted earlier, the Gaussian chain tends to overestimate these quantities from the mean transfer efficiency when the chain is expanded. Looking in more detail at the distributions of end-to-end distance for simulations near the theta state ($\nu = 0.5$, Fig. 6E) and near the good solvent limit ($\nu = 0.6$, Fig. 6F), the Gaussian chain captures $P(R)$ where $\nu = 0.5$, but is a poor approximation to the true distribution at $\nu = 0.6$. On the other hand, the SAW-EV model captures $P(R)$ where $\nu = 0.6$, but performs worst in capturing $p(R)$ where $\nu = 0.5$. These results simply reflect the relative strengths of each model discussed earlier. The SAW-ν model is able to fit both of these extreme scenarios similarly well. In summary, in the absence of additional information about the IDP of interest, the SAW-ν model is a good choice for approximating the underlying distance distributions.

Finally, to relate the experimentally observed average interdye distance to the distance between the two labeled residues, we need to account for the length of the dyes and linkers in this framework of analytical polymer models. The most common approach is to simply treat the dyes and linkers as part of the polymer chain and consider them equivalent to a suitably chosen number of amino acid residues. With the scaling exponent of the specific model used for the analysis (i.e., $\nu = 0.5$ for GC, 0.6 for SAW-EV, or the respective value obtained with SAW-ν), the residue–residue distance is then estimated by rescaling the interdye distance via Eq. (10) and $N = N_{aa} + L$, where N is the sequence length of the interdye segment, comprising both the number of peptide bonds, N_{aa}, and the contribution from both dye linkers, L. The appropriate value of L can be estimated, e.g., based on atomistic simulations (but the force field may need to be adjusted to prevent unrealistic dye sticking; Best, Hofmann, Nettels, & Schuler, 2015).

An experimental alternative is to infer it from a global analysis of a set of measurements of the same unfolded protein labeled at different sites, so that N_{aa} is known and different in every variant, but L is the same for all and can be treated as a fit parameter. For the Alexa 488/594 maleimide dye pair, e.g., such an analysis yielded $L = 9 \pm 2$, corresponding to four or five residues per dye and linker (Aznauryan et al., 2016).

5.2 Combining Experiment With Simulations Using Bayesian Inference

An alternative to analytical polymer models for generating a distance distribution, $P(r)$, is the use of molecular simulations. The advantage of such simulations is that, in addition to providing $P(r)$, they also yield an ensemble of conformations, which we will denote $\{x_i\}$, that can be further analyzed. The protocol for running the simulations themselves is beyond the scope of this chapter to describe, but really any type of molecular representation can be used (e.g., from one bead per residue to all atom), as long as the distance r in question can be computed from each conformation x_i. For the purposes of this section, we are thinking of simulations in which all atoms of the protein are explicitly represented. Such an ensemble allows additional details beyond $P(r)$ to be determined, such as local secondary structure and the types of contacts that are formed most frequently. In an ideal world, a highly accurate, predictive simulation model would be used and the FRET data would merely be a check on the results. However, that level of accuracy is rarely achieved in practice, and the mean efficiency computed from the simulations will be somewhat different from experiment (Best, Zheng, & Mittal, 2014). If the simulation model is a reasonably good approximation, and only a small shift in the distribution of configurations is needed, one might imagine reweighting the observed conformations x_i, each by a corresponding weight w_i, to give a reweighted efficiency

$$\langle \varepsilon \rangle_{\mathrm{rw}}(\{w_i\}) = \frac{\sum_i w_i \varepsilon(x_i)}{\sum_i w_i} \tag{12}$$

In this expression, $\varepsilon(x_i)$ is the FRET efficiency computed for conformation x_i. This can be optimized to match experiment by minimizing the χ^2 function, defined as

$$\chi^2(\{w_i\}) = \sum_{k=1}^{N_{\rm obs}} \frac{\left(\langle \varepsilon \rangle_{\rm rw}(k; \{w_i\}) - \langle \varepsilon \rangle_{\rm expt}(k)\right)^2}{\delta\varepsilon^2(k)}. \tag{13}$$

Here, we describe the most general situation, in which several ($N_{\rm obs}$) FRET efficiency observations are available, with the mean efficiency and experimental uncertainty of observation k (arising mostly from systematic calibration errors) given by $\langle \varepsilon \rangle_{\rm expt}(k)$ and $\delta\varepsilon(k)$, respectively. Multiple observed $\langle \varepsilon \rangle_{\rm expt}(k)$ may come from different labeling positions of the protein (Borgia et al., 2016; Hoffmann et al., 2007), or from three-color FRET (Gambin & Deniz, 2010), but the procedure is still applicable even if only one observation is available. In the language of Bayesian statistics, minimizing χ^2 corresponds to maximizing the log-likelihood of the observed efficiencies, with a prior distribution given by the molecular simulation.

An obvious problem, considering the large and diverse ensemble of structures sampled in a typical simulation of a disordered protein, is that the solution $\{w_i\}$ minimizing χ^2 is highly underdetermined by the data (Hummer & Koefinger, 2015). This difficulty may be overcome by requiring that the weights deviate minimally from the original (uniform) weights. One way of achieving this is to introduce an additional term to the optimization function, penalizing sets of weights which differ from being uniform, and yielding the modified target function (Hummer & Koefinger, 2015).

$$G(\{w_i\}) = \chi^2(\{w_i\}) - KS(\{w_i\}) \tag{14}$$

where the deviation from uniform weights is measured by the Shannon entropy of the weight set, $S = \sum_i w_i \ln w_i$. The factor K controls the relative importance of the penalty (or regularization) term. It is chosen to be as large as possible, thus keeping the weights as uniform as possible, without causing a large increase in χ^2. The weights are chosen using a Monte Carlo (MC) simulation, in which attempts to vary weights are accepted according to a Metropolis criterion (Borgia et al., 2016; Hummer & Koefinger, 2015). These simulations are run until χ^2 reaches a stable value. We note here that a similar formalism (Leung et al., 2016) was developed starting from the principal of maximum entropy (Jaynes, 1957) and has also been applied to refinement of disordered protein ensembles (Fuertes et al., 2017; Leung et al., 2016). Outside of the context of single-molecule FRET, a number of methods have been developed for ensemble refinement of proteins against experiment, as recently summarized in an excellent review (Bonomi, Heller, Camilloni, & Vendruscolo, 2017).

In Fig. 7, we show a practical example of this procedure, applied to FRET data from the IDP ACTR as a function of denaturant concentration (Borgia et al., 2016). This data set includes FRET efficiencies for three different pairs of labeling positions of the protein under each set of conditions, which are combined in the ensemble fit (shown in Fig. 7A). In Fig. 7B, we show a plot of χ^2 vs S at each denaturant concentration. Different points on each curve correspond to different weight factors K. These curves show a plateau region at low χ^2 and low entropy, corresponding to smaller K. As K is increased, the entropy is increased, corresponding to more uniform sets of $\{w_i\}$, but this initially has little effect on χ^2, because there are many sets of $\{w_i\}$ which can achieve this low χ^2. If K is increased too far, keeping the weights uniform becomes more important than reducing χ^2, and there is a sharp increase in χ^2. We chose the value of K immediately before this increase, as it corresponds to the highest entropy solution that is still able to fit the data (Borgia et al., 2016; Hummer & Koefinger, 2015; Mantsyzov et al., 2014). In Fig. 7A, we compare the experimentally observed efficiencies with the fits from the model with this optimal choice of K, and in Fig. 7C and D we show the average end-to-end distance, R, and R_g.

The penalty for nonuniform weights helps to avoid the solution being dominated by a few large weights w_i with the remainder being small. Nonetheless, a reweighting procedure of this sort always works best when there is good "overlap" between the initial ensemble of structures generated by the simulation and the final solution (i.e., weights are all approximately uniform) (Hummer & Koefinger, 2015). For example, if the initial distribution of conformations from simulation consisted almost entirely of compact structures, the only way to match experimental data reflecting an expanded distribution of conformations would be to have nonzero weights only for those few (if any) structures that are more expanded. Clearly this is undesirable when the underlying ensemble is expected to be a smooth distribution of diverse structures, not a few outliers, and this would be reflected in a lower entropy for the weights. One intuitive way of assessing this is to plot the distributions of various properties (e.g., R, R_g) from the initial, unweighted ensemble and also from the reweighted ensemble—it should not look like the reweighting is picking out only the tails of the distribution, and the

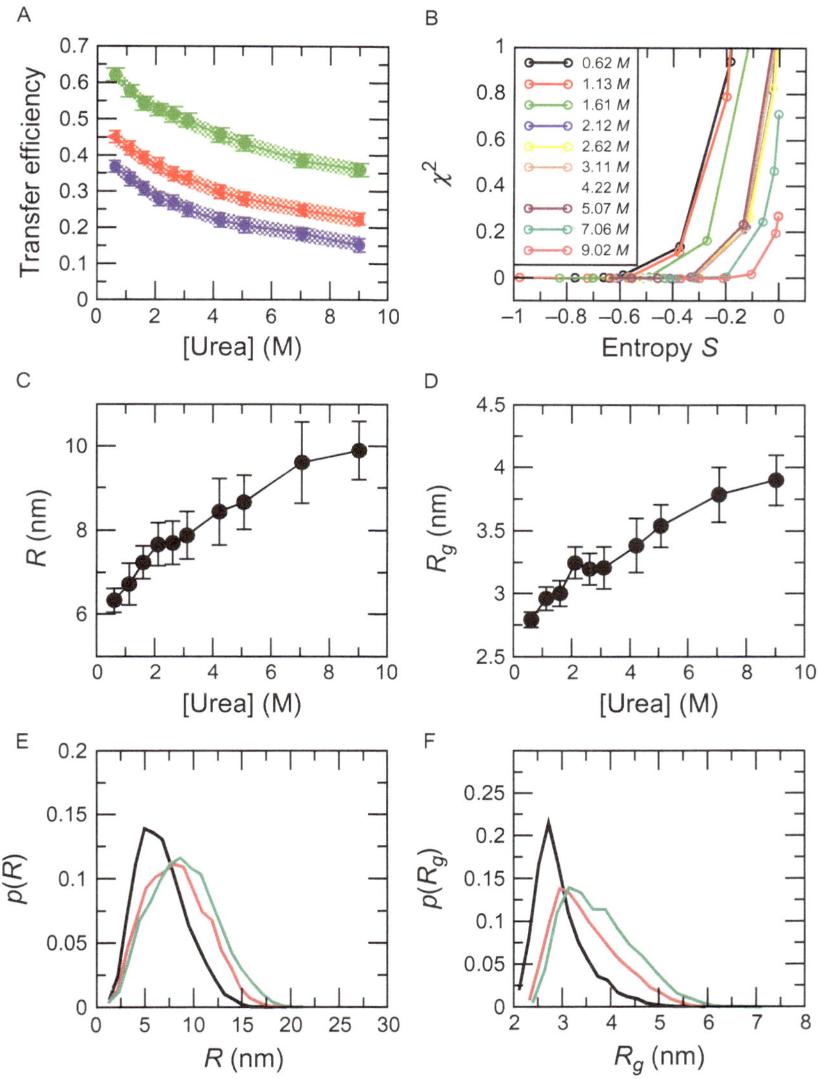

Fig. 7 Bayesian ensemble reweighting of simulations of unfolded R17. (A) FRET efficiency data vs denaturant concentration. *Shaded regions* show experimental confidence intervals (standard error), *symbols and error bars* show the values calculated from the reweighted ensemble. (B) Variation of χ^2 and S as the control parameter K is varied. Each *curve* corresponds to a given denaturant concentration (see legend) and each point to a particular value of K. Root mean square (C) distance, R, and (D) radius of gyration, R_g, are recovered as a function of denaturant. In (E) and (F) we show, respectively, examples of distribution functions for R and R_g (color code is: *black*, 1.13 M urea; *magenta*, 5.07 M urea; *cyan*, 9.02 M urea).

reweighted distributions should (generally) look smooth, as illustrated for R and R_g in the example of ACTR introduced above (Fig. 7E and F). Another objective procedure is to compute the partition function for the weights, i.e., $Z = \left(\sum_{i=1}^{N_s} w_i\right) / \max\left[\{w_i\}\right]$, where N_S is the number of structures in the simulation ensemble. With this definition, the value of Z measures the number of structures contributing to the average. A good rule of thumb is that Z/N_S should be at least 0.2. In practice, there is no guarantee that a standard force field will have such a good overlap; therefore, it is often helpful to generate a range of ensembles, for example, at different simulation temperatures, and pick the temperature which yields the highest entropy fit or best overlap with the experimental data as the starting point for reweighting (Best et al., 2018; Borgia et al., 2016).

An attractive feature of ensemble reweighting is that it is applicable to other types of experimental data besides FRET, as long as they can be computed from each structure in the ensemble. For example, we have recently applied it to joint refinement of unfolded state ensembles against FRET and SAXS data (Best et al., 2018; Borgia et al., 2016). The main drawback is clearly that the procedure requires running a simulation of the system of interest, often several simulations under different conditions, in order to find one that overlaps well with the experimental data. In addition, some care is required in the reweighting procedure to avoid overfitting to a few structures. Therefore, if the protein in question can be described as a disordered chain, using a method such as SAW-ν to fit the data directly is a much easier alternative, but it cannot capture aspects such as secondary structure formation or other residue-specific interactions. Another advantage of detailed molecular simulations is that the fluorophores can be included explicitly if a suitable force field parameterization is available (Best et al., 2015; Zheng et al., 2016). Alternatively, rotamer libraries of fluorophores can be used to add the dyes after the simulation, albeit at the expense of information about their dynamics (Fuertes et al., 2017; Grotz et al., 2018).

5.3 Optimizing Simulation Parameters for Simplified Models

An alternative philosophy to the Bayesian reweighting scheme described earlier is a minimalist model. Rather than taking the view that atomistic simulations are a reasonable approximation needing only small improvements, one includes in the model only those features that are required to reproduce the experimental data, together with basic assumptions about protein structure that can be expected to be valid, such as covalent bonding between residues, and the size of a given residue type (Kim & Hummer, 2008; Tozzini, 2005). Such simplified models can potentially highlight which physical properties are most important for determining the observed conformational ensemble (i.e., properties that must be included in the model). The resulting model is naturally going to be quite coarse-grained, with little structural detail. An advantage, however, is that simulations of such models are much cheaper and faster to perform, and the energy landscape of the system is of lower dimensionality and smoother than those associated with all-atom simulations. Consequently, it is much easier to sample the system sufficiently to ensure that the results are independent of initial conditions (i.e., "converged" results), which is required for this method (it is desirable for Bayesian reweighting, but not absolutely essential in practice).

A typical minimalist model would represent each residue by a bead centered on its alpha carbon, with an energy function or force field, V_{ff}, of the form:

$$V_{ff} = V_{\text{bonded}} + \sum_{i<j} \frac{q_i q_j}{4\pi\epsilon_0 D r_{ij}} \exp\left[-\frac{r_{ij}}{\lambda_D}\right] + \sum_{i<j} 4e_{ij}\left(\left(\frac{s_{ij}}{r_{ij}}\right)^{12} - \left(\frac{s_{ij}}{r_{ij}}\right)^{6}\right). \tag{15}$$

The first part of the energy, V_{bonded}, includes typical force field terms describing covalent bonding of the chain of residues, i.e., harmonic bond terms (or constraints), harmonic angle terms, and dihedral angle terms. These terms are described in more detail elsewhere (Borgia et al., 2018), and being transferable between different proteins, they would not generally be subject to optimization to fit experiment. The second term describes electrostatics in terms of a Debye–Hückel screening potential, with q_i being the charge on atom i, ϵ_0 the permittivity of free space, D the dielectric constant and λ_D the screening length. Being a physically derived term, this is also not subject to optimization and is defined by the experimental ionic strength. The last term is the contact potential describing short-range interactions between the beads, here given in terms of a Lennard–Jones potential with contact distances s_{ij} between beads i and j and optimal interaction energies e_{ij}; r_{ij} is the distance between the beads. This is where the flexibility lies for fitting to experimental data. Although among the combinations of s_{ij} and e_{ij} there is in principle a large number of parameters, this can be greatly reduced.

First, the contact distances s_{ij} can be estimated from the known average volumes of different residue types in crystal structures (Kim & Hummer, 2008). The energies e_{ij} can be based on a standard contact potential for amino acids, such as the Miyazawa–Jernigan potential (Miyazawa & Jernigan, 1996). The simplest type of optimization would involve a global shifting and/or scaling of e_{ij} in the simulation in order to match the experimental transfer efficiency data. The optimal parameters can be obtained by a parameter search, most simply using a binary search in parameter space. Alternatively, a more sophisticated gradient-based parameter optimization could be used, as has recently been proposed for general force field refinement (Wang, Martinez, & Pande, 2014). The simplicity of the model allows multiple simulations with different parameters during optimization. A coarse-grained representation of the fluorophores and linkers can also be included in the model to assess their effect on the observed transfer efficiencies (Borgia et al., 2018).

Having given this very general outline of the form of the coarse-grained models used, we will illustrate the concept with the example of the binding of two highly charged proteins, prothymosin α (ProTα) and histone H1 (H1) (Borgia et al., 2018). These are both IDPs, except for the presence of a small folded domain in H1, and they gain no structure upon complex formation. The model used is similar to that given in Eq. (15), except for an extra term for the folded H1 domain, which is described in more detail by Borgia et al. (2018); its purpose in this example is only to keep the domain folded throughout the simulation. The contact potential adopted is extremely simple, being just a single, common e_{ij} for all protein residue pairs, optimized against the single-molecule FRET data; more sophisticated potentials did not show better agreement with experiment. FRET efficiencies were measured for 28 inter- and intramolecular labeling pairs. Remarkably, it was possible to fit virtually all of the experimental FRET efficiencies, and in particular their pattern along the protein sequences (Fig. 8). From the resulting ensemble, we can compute the average distances between residue pairs, as well as distance distributions between any pair of residues of interest, some examples of which are shown also in Fig. 8. Lastly, one can go beyond distributions to characterize the ensemble of structures, a few of which are shown in Fig. 8. In this case, the simplicity of the energy function revealed that electrostatic interactions were the key factor in distinguishing the involvement of different regions of the protein sequence in the structure of the complex.

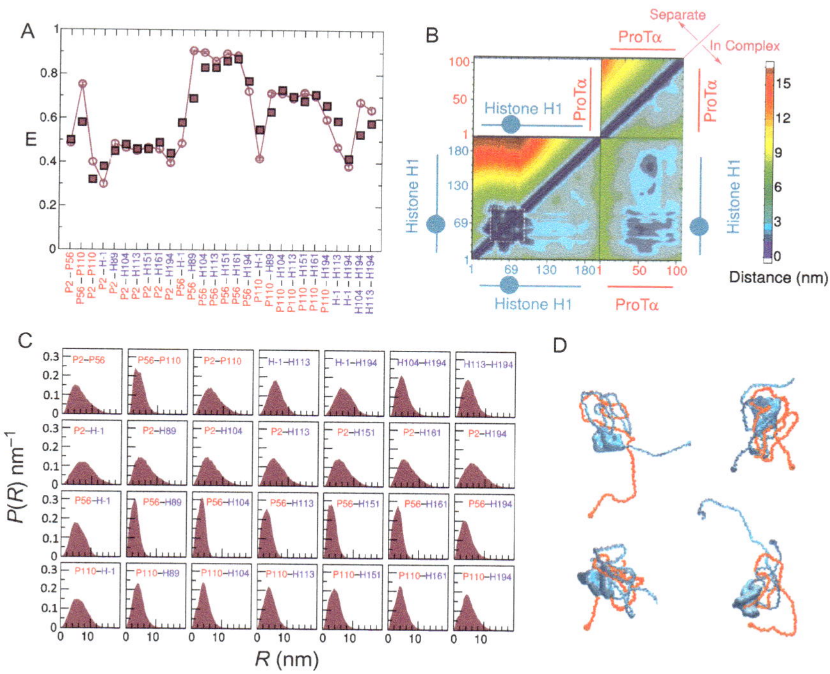

Fig. 8 Fitting a coarse-grained model to experimental FRET efficiencies. (A) Experimental (*solid symbols*) transfer efficiency data for inter- and intramolecular labeling pairs in the complex between ProTα and H1 (*red*: ProTα residue; *blue*: H1 residue). *Empty symbols* show results from the fitted model. (B) Average inter- and intramolecular pair distances between the two proteins. *Blue circle* in schematic of H1 represents the folded domain. (C) Pair distance distributions for labeled residue pairs from simulations with optimized energy function. (D) Example structures of the highly disordered ProTα–H1 complex.

6. CONCLUSIONS

Quantifying distances and distance distributions in IDPs with single-molecule FRET requires two key steps, which we have described here: (1) obtaining accurate transfer efficiencies, which rely on careful instrument calibration and (2) extracting information about the structurally diverse ensemble of conformations based on a suitable model. Once accurate transfer efficiencies are available (ideally from multiple labeling positions), the simplest approach for inferring intramolecular distance distributions is the use of analytical polymer models, such as the SAW-ν model (Eq. 9), which often provides a good approximation. For a more detailed description that considers specific interactions, residual structure, or involves the complex of an IDP with a binding partner, the methods of choice are reweighting of molecular simulations or fitting of coarse-grained models to experimental data. These approaches further enable the direct integration with additional experimental results that afford complementary information, e.g., from NMR or scattering experiments. The resulting distance distributions can also be combined with nanosecond fluorescence correlation spectroscopy for quantifying intra- and interchain dynamics (Schuler, 2018). Single-molecule FRET thus increasingly contributes to our understanding of the structural, dynamic, and functional properties of IDPs.

ACKNOWLEDGMENTS

This work was supported by the Swiss National Science Foundation (to B.S.), the European Molecular Biology Organization (EMBO ALTF-471-2015, to E.D.H.), and the Intramural Research Program of the National Institute of Diabetes and Digestive and Kidney Diseases of the National Institutes of Health.

REFERENCES

Aznauryan, M., Delgado, L., Soranno, A., Nettels, D., Huang, J. R., Labhardt, A. M., et al. (2016). Comprehensive structural and dynamical view of an unfolded protein from the combination of single-molecule FRET, NMR, and SAXS. *Proceedings of the National Academy of Sciences of the United States of America, 113*(37), E5389–E5398. https://doi.org/10.1073/pnas.1607193113.

Best, R. B., Hofmann, H., Nettels, D., & Schuler, B. (2015). Quantitative interpretation of FRET experiments via molecular simulation: Force field and validation. *Biophysical Journal, 108*(11), 2721–2731. https://doi.org/10.1016/j.bpj.2015.04.038.

Best, R., Merchant, K., Gopich, I. V., Schuler, B., Bax, A., & Eaton, W. A. (2007). Effect of flexibility and cis residues in single molecule FRET studies of polyproline. *Proceedings of the National Academy of Sciences of the United States of America, 104*(48), 18964–18969.

Best, R. B., Zheng, W., Borgia, A., Buholzer, K., Borgia, M. B., Hofmann, H., et al. (2018). Comment on "innovative scattering analysis shows that hydrophobic disordered proteins are expanded in water" *Science, 361*, eaar7101.

Best, R. B., Zheng, W., & Mittal, J. (2014). Balanced protein-water interactions improve properties of disordered proteins and non-specific protein association. *Journal of Chemical Theory and Computation, 10*, 5113–5124.

Bonomi, M., Heller, G. T., Camilloni, C., & Vendruscolo, M. (2017). Principles of protein structural ensemble determination. *Current Opinion in Structural Biology, 42*, 106–116. https://doi.org/10.1016/j.sbi.2016.12.004.

Borgia, A., Borgia, M. B., Bugge, K., Kissling, V. M., Heidarsson, P. O., Fernandes, C. B., et al. (2018). Extreme disorder in an ultrahigh-affinity protein complex. *Nature, 555*(7694), 61–66. https://doi.org/10.1038/nature25762.

Borgia, A., Zheng, W., Buholzer, K., Borgia, M. B., Schuler, A., Hofmann, H., et al. (2016). Consistent view of polypeptide chain expansion in chemical denaturants from multiple experimental methods. *Journal of the American Chemical Society, 138*(36), 11714–11726. https://doi.org/10.1021/jacs.6b05917.

Chung, H. S., Louis, J. M., & Gopich, I. V. (2016). Analysis of fluorescence lifetime and energy transfer efficiency in single-molecule photon trajectories of fast-folding proteins. *The Journal of Physical Chemistry B, 120*(4), 680–699. https://doi.org/10.1021/acs.jpcb.5b11351.

Das, R. K., & Pappu, R. V. (2013). Conformations of intrinsically disordered proteins are influenced by linear sequence distributions of oppositely charged residues. *Proceedings of the National Academy of Sciences of the United States of America, 110*(33), 13392–13397. https://doi.org/10.1073/pnas.1304749110.

Deniz, A. A., Dahan, M., Grunwell, J. R., Ha, T., Faulhaber, A. E., Chemla, D. S., et al. (1999). Single-pair fluorescence resonance energy transfer on freely diffusing molecules: Observation of Forster distance dependence and subpopulations. *Proceedings of the National Academy of Sciences of the United States of America, 96*(7), 3670–3675.

Deniz, A. A., Laurence, T. A., Dahan, M., Chemla, D. S., Schultz, P. G., & Weiss, S. (2001). Ratiometric single-molecule studies of freely diffusing biomolecules. *Annual Review of Physical Chemistry, 52*, 233–253.

Doose, S., Neuweiler, H., & Sauer, M. (2005). A close look at fluorescence quenching of organic dyes by tryptophan. *Chemphyschem, 6*(11), 2277–2285.

Dunker, A. K., Lawson, J. D., Brown, C. J., Williams, R. M., Romero, P., Oh, J. S., et al. (2001). Intrinsically disordered protein. *Journal of Molecular Graphics & Modelling, 19*(1), 26–59.

Dyson, H. J., & Wright, P. E. (2005). Intrinsically unstructured proteins and their functions. *Nature Reviews Molecular Cell Biology, 6*(3), 197–208. https://doi.org/10.1038/nrm1589.

Ferreon, A. C., Moran, C. R., Gambin, Y., & Deniz, A. A. (2010). Single-molecule fluorescence studies of intrinsically disordered proteins. *Methods in Enzymology, 472*, 179–204. https://doi.org/10.1016/S0076-6879(10)72010-3.

Fischer, E. (1902). *Nobel lecture: Syntheses in the purine and sugar group*. Retrieved May 29, 2018, from https://www.nobelprize.org/nobel_prizes/chemistry/laureates/1902/fischer-lecture.pdf.

Forman-Kay, J. D., & Mittag, T. (2013). From sequence and forces to structure, function, and evolution of intrinsically disordered proteins. *Structure, 21*(9), 1492–1499. https://doi.org/10.1016/j.str.2013.08.001.

Förster, T. (1948). Zwischenmolekulare energiewanderung Und Fluoreszenz. *Annalen der Physik, 2*(1–2), 55–75.

Fuertes, G., Banterlea, N., Ruff, K. M., Chowdhury, A., Mercadante, D., Koehler, C., et al. (2017). Decoupling of size and shape fluctuations in heteropolymeric sequences reconciles discrepancies in SAXS vs. FRET measurements. *Proceedings of the National Academy of Sciences of the United States of America, 114*(31), E6342–E6351. https://doi.org/10.1073/pnas.1704692114.

Gambin, Y., & Deniz, A. A. (2010). Multicolor single-molecule FRET to explore protein folding and binding. *Molecular BioSystems, 6*(9), 1540–1547. https://doi.org/10.1039/c003024d.

Gibbs, E. B., & Showalter, S. A. (2015). Quantitative biophysical characterization of intrinsically disordered proteins. *Biochemistry, 54*(6), 1314–1326. https://doi.org/10.1021/bi501460a.

Gomes, G. N., & Gradinaru, C. C. (2017). Insights into the conformations and dynamics of intrinsically disordered proteins using single-molecule fluorescence. *Biochimica et Biophysica Acta, 1865*(11 Pt. B), 1696–1706. https://doi.org/10.1016/j.bbapap.2017.06.008.

Gopich, I. V., & Szabo, A. (2012). Theory of the energy transfer efficiency and fluorescence lifetime distribution in single-molecule FRET. *Proceedings of the National Academy of Sciences of the United States of America, 109*(20), 7747–7752. https://doi.org/10.1073/pnas.1205120109.

Grotz, K. K., Nüesch, M., Holmstrom, E., Heinz, M., Stelzl, L. S., Schuler, B., et al. (2018). Dispersion correction alleviates dye stacking of single-stranded DNA and RNA in simulations of single-molecule fluorescence experiments. *The Journal of Physical Chemistry B*. in press, https://doi.org/10.1021/acs.jpcb.8b07537.

Gust, A., Zander, A., Gietl, A., Holzmeister, P., Schulz, S., Lalkens, B., et al. (2014). A starting point for fluorescence-based single-molecule measurements in biomolecular research. *Molecules*, *19*(10), 15824–15865. https://doi.org/10.3390/molecules191015824.

Ha, T., & Tinnefeld, P. (2012). Photophysics of fluorescent probes for single-molecule biophysics and super-resolution imaging. *Annual Review of Physical Chemistry*, *63*, 595–617. https://doi.org/10.1146/annurev-physchem-032210-103340.

Haenni, D., Zosel, F., Reymond, L., Nettels, D., & Schuler, B. (2013). Intramolecular distances and dynamics from the combined photon statistics of single-molecule FRET and photoinduced electron transfer. *The Journal of Physical Chemistry B*, *117*(42), 13015–13028. https://doi.org/10.1021/jp402352s.

Hellenkamp, B., Schmid, S., Doroshenko, O., Opanasyuk, O., Kühnemuth, R., Adariani, S. R., et al. (2018). Precision and accuracy of single-molecule FRET measurements—A worldwide benchmark study. *Nature Methods*, *15*(15), 669–676.

Hellenkamp, B., Wortmann, P., Kandzia, F., Zacharias, M., & Hugel, T. (2017). Multidomain structure and correlated dynamics determined by self-consistent FRET networks. *Nature Methods*, *14*(2), 174–180. https://doi.org/10.1038/nmeth.4081.

Hoffmann, A., Kane, A., Nettels, D., Hertzog, D. E., Baumgärtel, P., Lengefeld, J., et al. (2007). Mapping protein collapse with single-molecule fluorescence and kinetic synchrotron radiation circular dichroism spectroscopy. *Proceedings of the National Academy of Sciences of the United States of America*, *104*(1), 105–110.

Hofmann, H., Soranno, A., Borgia, A., Gast, K., Nettels, D., & Schuler, B. (2012). Polymer scaling laws of unfolded and intrinsically disordered proteins quantified with single-molecule spectroscopy. *Proceedings of the National Academy of Sciences of the United States of America*, *109*(40), 16155–16160. https://doi.org/10.1073/pnas.1207719109.

Hummer, G., & Koefinger, J. (2015). Bayesian ensemble refinement by replica simulations and reweighting. *The Journal of Chemical Physics*, *143*(24). https://doi.org/10.1063/1.4937786. Artn 243150.

Jaynes, E. T. (1957). Information theory and statistical mechanics. *Physics Review*, *106*, 620–630.

Kalinin, S., Peulen, T., Sindbert, S., Rothwell, P. J., Berger, S., Restle, T., et al. (2012). A toolkit and benchmark study for FRET-restrained high-precision structural modeling. *Nature Methods*, *9*(12), 1218–1225. https://doi.org/10.1038/nmeth.2222.

Kalinin, S., Valeri, A., Antonik, M., Felekyan, S., & Seidel, C. A. (2010). Detection of structural dynamics by FRET: A photon distribution and fluorescence lifetime analysis of systems with multiple states. *The Journal of Physical Chemistry. B*, *114*(23), 7983–7995. https://doi.org/10.1021/jp102156t.

Kapanidis, A. N., Laurence, T. A., Lee, N. K., Margeat, E., Kong, X., & Weiss, S. (2005). Alternating-laser excitation of single molecules. *Accounts of Chemical Research*, *38*(7), 523–533. https://doi.org/10.1021/ar0401348.

Kapanidis, A. N., Lee, N. K., Laurence, T. A., Doose, S., Margeat, E., & Weiss, S. (2004). Fluorescence-aided molecule sorting: Analysis of structure and interactions by alternating-laser excitation of single molecules. *Proceedings of the National Academy of Sciences of the United States of America*, *101*(24), 8936–8941. https://doi.org/10.1073/pnas.0401690101.

Kikhney, A. G., & Svergun, D. I. (2015). A practical guide to small angle X-ray scattering (SAXS) of flexible and intrinsically disordered proteins. *FEBS Letters*, *589*(19 Pt. A), 2570–2577. https://doi.org/10.1016/j.febslet.2015.08.027.

Kim, Y. C., & Hummer, G. (2008). Coarse-grained models for simulations of multiprotein complexes: Application to ubiquitin binding. *Journal of Molecular Biology*, *375*, 1416–1433.
König, I., Zarrine-Afsar, A., Aznauryan, M., Soranno, A., Wunderlich, B., Dingfelder, F., et al. (2015). Single-molecule spectroscopy of protein conformational dynamics in live eukaryotic cells. *Nature Methods*, *12*(8), 773–779. https://doi.org/10.1038/nmeth.3475.
Konrat, R. (2014). NMR contributions to structural dynamics studies of intrinsically disordered proteins. *Journal of Magnetic Resonance*, *241*, 74–85. https://doi.org/10.1016/j.jmr.2013.11.011.
Kretschy, N., Sack, M., & Somoza, M. M. (2016). Sequence-dependent fluorescence of Cy3- and Cy5-labeled double-stranded DNA. *Bioconjugate Chemistry*, *27*(3), 840–848. https://doi.org/10.1021/acs.bioconjchem.6b00053.
Kudryavtsev, V., Sikor, M., Kalinin, S., Mokranjac, D., Seidel, C. A., & Lamb, D. C. (2012). Combining MFD and PIE for accurate single-pair Forster resonance energy transfer measurements. *Chemphyschem*, *13*(4), 1060–1078. https://doi.org/10.1002/cphc.201100822.
Le Guillou, J. C., & Zinn-Justin, J. (1977). Critical exponents for N-vector model in 3 dimensions from field-theory. *Physical Review Letters*, *39*(2), 95–98. https://doi.org/10.1103/Physrevlett.39.95.
Lee, N. K., Kapanidis, A. N., Wang, Y., Michalet, X., Mukhopadhyay, J., Ebright, R. H., et al. (2005). Accurate FRET measurements within single diffusing biomolecules using alternating-laser excitation. *Biophysical Journal*, *88*(4), 2939–2953. https://doi.org/10.1529/biophysj.104.054114.
Leung, H. T. A., Bignucolo, O., Aregger, R., Dames, S. A., Mazur, A., Bernèche, S., et al. (2016). A rigorous and efficient method to reweight very large conformational ensembles using average experimental data and to determine their relative information content. *Journal of Chemical Theory and Computation*, *12*, 383–394.
Mantsyzov, A. B., Maltsev, A. S., Ying, J., Shen, Y., Hummer, G., & Bax, A. (2014). A maximum entropy approach to the study of residue-specific backbone angle distributions in alpha-synuclein, an intrinsically disordered protein. *Protein Science*, *23*, 1275–1290.
Miyazawa, S., & Jernigan, R. L. (1996). Residue-residue potentials with a favourable contact pair term and an unfavourable high packing density term, for simulation and threading. *Journal of Molecular Biology*, *256*, 623–644.
Müller, B. K., Zaychikov, E., Bräuchle, C., & Lamb, D. C. (2005). Pulsed interleaved excitation. *Biophysical Journal*, *89*(5), 3508–3522. https://doi.org/10.1529/biophysj.105.064766.
Muschielok, A., Andrecka, J., Jawhari, A., Bruckner, F., Cramer, P., & Michaelis, J. (2008). A nano-positioning system for macromolecular structural analysis. *Nature Methods*, *5*(11), 965–971. https://doi.org/10.1038/nmeth.1259.
Oates, M. E., Romero, P., Ishida, T., Ghalwash, M., Mizianty, M. J., Xue, B., et al. (2013). D(2)P(2): Database of disordered protein predictions. *Nucleic Acids Research*, *41*(Database issue), D508–D516. https://doi.org/10.1093/nar/gks1226.
O'Brien, E. P., Morrison, G., Brooks, B. R., & Thirumalai, D. (2009). How accurate are polymer models in the analysis of Forster resonance energy transfer experiments on proteins? *The Journal of Chemical Physics*, *130*(12), 124903. https://doi.org/10.1063/1.3082151.
Oldfield, C. J., & Dunker, A. K. (2014). Intrinsically disordered proteins and intrinsically disordered protein regions. *Annual Review of Biochemistry*, *83*, 553–584.
Riback, J. A., Bowman, M. A., Zmyslowski, A. M., Knoverek, C. R., Jumper, J. M., Hinshaw, J. R., et al. (2017). Innovative scattering analysis shows that hydrophobic disordered proteins are expanded in water. *Science*, *358*(6360), 238–241. https://doi.org/10.1126/science.aan5774.

Sanborn, M. E., Connolly, B. K., Gurunathan, K., & Levitus, M. (2007). Fluorescence properties and photophysics of the sulfoindocyanine Cy3 linked covalently to DNA. *The Journal of Physical Chemistry. B*, *111*(37), 11064–11074. https://doi.org/10.1021/jp072912u.

Schäfer, L. (1999). *Excluded volume effects in polymer solutions as explained by the renormalization group*. Berlin: Springer.

Schuler, B. (2007). Application of single molecule Forster resonance energy transfer to protein folding. *Methods in Molecular Biology*, *350*, 115–138.

Schuler, B. (2018). Perspective: Chain dynamics of unfolded and intrinsically disordered proteins from nanosecond fluorescence correlation spectroscopy combined with single-molecule FRET. *The Journal of Chemical Physics*, *149*(1), 010901.

Schuler, B., & Hofmann, H. (2013). Single-molecule spectroscopy of protein folding dynamics—Expanding scope and timescales. *Current Opinion in Structural Biology*, *23*(1), 36–47. https://doi.org/10.1016/j.sbi.2012.10.008.

Schuler, B., Lipman, E. A., Steinbach, P. J., Kumke, M., & Eaton, W. A. (2005). Polyproline and the "spectroscopic ruler" revisited with single molecule fluorescence. *Proceedings of the National Academy of Sciences of the United States of America*, *102*, 2754–2759.

Schuler, B., Soranno, A., Hofmann, H., & Nettels, D. (2016). Single-molecule FRET spectroscopy and the polymer physics of unfolded and intrinsically disordered proteins. *Annual Review of Biophysics*, *45*, 207–231. https://doi.org/10.1146/annurev-biophys-062215-010915.

Sisamakis, E., Valeri, A., Kalinin, S., Rothwell, P. J., & Seidel, C. A. M. (2010). Accurate single-molecule FRET studies using multiparameter fluorescence detection. *Methods in Enzymology*, *475*, 455–514.

Song, J., Gomes, G.-N., Gradinaru, C., & Chan, H. S. (2015). An adequate account of excluded volume is necessary to infer compactness and asphericity of disordered proteins by Förster resonance energy transfer. *The Journal of Physical Chemistry B*, *119*, 15191–15202.

Stryer, L. (1978). Fluorescence energy transfer as a spectroscopic ruler. *Annual Review of Biochemistry*, *47*, 819–846. https://doi.org/10.1146/annurev.bi.47.070178.004131.

Stryer, L., & Haugland, R. P. (1967). Energy transfer: A spectroscopic ruler. *Proceedings of the National Academy of Sciences of the United States of America*, *58*(2), 719–726.

Sustarsic, M., & Kapanidis, A. N. (2015). Taking the ruler to the jungle: Single-molecule FRET for understanding biomolecular structure and dynamics in live cells. *Current Opinion in Structural Biology*, *34*, 52–59. https://doi.org/10.1016/j.sbi.2015.07.001.

Tompa, P. (2005). The interplay between structure and function in intrinsically unstructured proteins. *FEBS Letters*, *579*(15), 3346–3354. https://doi.org/10.1016/j.febslet.2005.03.072.

Tozzini, V. (2005). Coarse-grained models for proteins. *Current Opinion in Structural Biology*, *15*, 144–150.

Uversky, V. N. A., & Keith Dunker, A. (2012). Intrinsically disordered protein analysis. *Vol. 896*. New York: Springer.

van der Lee, R., Buljan, M., Lang, B., Weatheritt, R. J., Daughdrill, G. W., Dunker, A. K., et al. (2014). Classification of intrinsically disordered regions and proteins. *Chemical Reviews*, *114*(13), 6589–6631. https://doi.org/10.1021/cr400525m.

Van Der Meer, B. W., Coker, G., III, & Chen, S. Y. S. (1994). *Resonance energy transfer: Theory and data*. New York: VCH Publishers, Inc.

Wahl, M., Koberling, F., Patting, M., Rahn, H., & Erdmann, R. (2004). Time-resolved confocal fluorescence imaging and spectrocopy system with single molecule sensitivity and sub-micrometer resolution. *Current Pharmaceutical Biotechnology*, *5*(3), 299–308.

Wang, L.-P., Martinez, T. J., & Pande, V. S. (2014). Building force fields: An automatic, systematic and reproducible approach. *Journal of Chemical Theory and Computation*, *5*, 1885–1891.

Witten, T. A., & Schäfer, L. (1978). Two critical ratios in polymer solutions. *Journal of Physics A*, *11*(9), 1843–1854.
Wright, P. E., & Dyson, H. J. (2015). Intrinsically disordered proteins in cellular signalling and regulation. *Nature Reviews Molecular Cell Biology*, *16*, 18–29.
Yoo, T. Y., Meisberger, S. P., Hinshaw, J., Pollack, L., Haran, G., Sosnick, T. R., et al. (2012). Small-angle X-ray scattering and single-molecule FRET spectroscopy produce highly divergent views of the low-denaturant unfolded state. *Journal of Molecular Biology*, *418*, 226–236.
Zheng, W., & Best, R. B. (2018). An extended Guinier analysis for intrinsically disordered proteins. *Journal of Molecular Biology*, *430*, 2540–2553.
Zheng, W., Borgia, A., Buholzer, K., Grishaev, A., Schuler, B., & Best, R. B. (2016). Probing the action of chemical denaturant on an intrinsically disordered protein by simulation and experiment. *Journal of the American Chemical Society*, *138*(36), 11702–11713. https://doi.org/10.1021/jacs.6b05443.
Zheng, W., Zerze, G. H., Borgia, A., Mittal, J., Schuler, B., & Best, R. B. (2018). Inferring properties of disordered chains from FRET transfer efficiencies. *The Journal of Chemical Physics*, *148*(12), 123329. https://doi.org/10.1063/1.5006954.
Zosel, F., Haenni, D., Soranno, A., Nettels, D., & Schuler, B. (2017). Combining short- and long-range fluorescence reporters with simulations to explore the intramolecular dynamics of an intrinsically disordered protein. *The Journal of Chemical Physics*, *147*(15), 152708. https://doi.org/10.1063/1.4992800.
Zosel, F., Holla, A., & Schuler, B. (2018). Labeling of proteins for single-molecule fluorescence spectroscopy. *Methods in Molecular Biology*. in press.

CHAPTER ELEVEN

Probing Differential Binding Mechanisms of Phenylalanine-Glycine-Rich Nucleoporins by Single-Molecule FRET

Piau Siong Tan*, Edward A. Lemke*,†,‡,1

*Structural and Computational Biology Unit & Cell Biology and Biophysics Unit, European Molecular Biology Laboratory (EMBL), Heidelberg, Germany
†Biocenter, Departments of Biology and Chemistry, Pharmacy and Geosciences, Johannes Gutenberg-University Mainz, Mainz, Germany
‡Institute of Molecular Biology (IMB), Mainz, Germany
[1]Corresponding author: e-mail address: edlemke@uni-mainz.de

Contents

Abstract

Phenylalanine-glycine-rich nucleoporins (FG-Nups) are intrinsically disordered proteins, constituting the selective barrier of the nuclear pore complex. They are highly dynamic under physiological conditions and studying their interaction with nuclear transport receptors (NTRs) is key to understanding the molecular mechanism of nucleocytoplasmic transport. Distinct conformational features of FG-Nups interacting with diverse NTRs can be detected by multiparameter single-molecule fluorescence energy transfer (smFRET), which is a powerful technique for studying the dynamics

Methods in Enzymology, Volume 611
ISSN 0076-6879
https://doi.org/10.1016/bs.mie.2018.08.034

and interactions of biomolecules in solution. Here we provide a detailed protocol utilizing smFRET to reveal differential binding mechanisms of FG-Nups to NTRs, with a focus on practical considerations on sample preparation of unglycosylated and glycosylated FG-Nups, site-specific dual-labeling, smFRET measurements, and data analysis.

ABBREVIATIONS

AcF *p*-acetylphenylalanine
Amp ampicillin
CBD chitin binding domain
Cm chloramphenicol
Cys cysteine
ddH$_2$O double-distilled water
DTT dithiothreitol
E. coli *Escherichia coli*
EDTA ethylenediaminetetraacetic acid
FCS fluorescence correlation spectroscopy
FG-Nup phenylalanine-glycine nucleoporin
FRET fluorescence resonance energy transfer
GdHCl guanidine hydrochloride
IDP intrinsically disordered protein
MgAc magnesium acetate
MgCl$_2$ magnesium chloride
NaAc sodium acetate (anhydrous)
NaCl sodium chloride
ncAAs noncanonical amino acids
NTR nuclear transport receptor
OD$_{600}$ optical density measured at a wavelength of 600 nm
O-GlcNAc O-linked-*N*-acetylglycosamine
OGT O-linked *N*-acetylglucosaminyl transferase
PBS phosphate-buffered saline
PMSF phenylmethylsulfonyl fluoride
POI protein of interest
rpm revolutions per minute
S stoichiometry
sm single molecule
smFRET single-molecule fluorescence resonance energy transfer
TCEP tris(2-carboxyethyl)phosphine
UDP-GlcNAc uridine 5′-diphospho-*N*-acetylglucosamine sodium salt

1. INTRODUCTION

1.1 Nuclear Pore Complex

The nuclear pore complex (NPC), which resides in the nuclear envelope, mediates the molecular exchange between the nucleus and the cytoplasm.

The metazoan NPC is around 125 MDa large and formed by 30 different nucleoporins (Nups) where roughly one-third of them are largely intrinsically disordered proteins (IDPs) rich in phenylalanine-glycine motifs, termed FG-Nups (Denning, Patel, Uversky, Fink, & Rexach, 2003; Hurt, 1988; Ori et al., 2013; Wente, Rout, & Blobel, 1992). FG-Nups have high conformational flexibility and multiple binding sites (FG motifs) which allows them to interact with transport receptors (Aramburu & Lemke, 2017; Gorlich & Kutay, 1999; Lim et al., 2006; Tetenbaum-Novatt & Rout, 2010; Tran, King, & Corbett, 2014). In metazoan they have been shown to be targets of O-linked β-*N*-acetylglucosamine (O-GlcNAc) modification, an omnipresent posttranslational modification in metazoan NPCs, which may play a role in regulating nucleocytoplasmic transport (Favreau, Worman, Wozniak, Frappier, & Courvalin, 1996; Labokha et al., 2013; Zhu et al., 2016). FG-Nups form the permeability barrier of the NPC and selectively control molecular transport, e.g., the barrier is permeable for small molecules below ~40 kDa that can cross via passive diffusion and enables the translocation of large macromolecules when the transport is mediated by nuclear transport receptors (NTRs) (Cook, Bono, Jinek, & Conti, 2007; Gorlich & Kutay, 1999; Timney et al., 2016). While structured Nups contributing to the eightfold symmetric ring of the NPC have been extensively studied by X-ray crystallography, cryo-electron tomography, and biochemical methods (Brohawn & Schwartz, 2009; Chug, Trakhanov, Hulsmann, Pleiner, & Gorlich, 2015; Eibauer et al., 2015; Kim et al., 2018; Kosinski et al., 2016; Lin et al., 2016; von Appen et al., 2015), conformational features of FG-Nups and changes imposed by their interactions with NTRs during nucleocytoplasmic transport remain unclear.

1.2 Coupled Reconfiguration-Binding vs Archetypal-Fuzzy Complex Formation in the NPC

Recently, we showed that at least two differential binding mechanisms between FG-Nups and NTRs coexist (Fig. 1A) (Milles et al., 2015; Tan et al., 2018). First, in vitro, NTRs like Importinβ were found to interact with FG-Nups (e.g., FG-Nup153) via the formation of an archetypal-fuzzy complex, in which the conformational ensemble of the unbound FG-Nup153 is binding prone and ready to bind Importinβ without substantial conformational change on the length scale probed by single-molecule fluorescent measurements. Such a binding mechanism can explain how fast, yet specific transport of molecules across the NPC is possible, in agreement with complementary results and other techniques from us and other labs

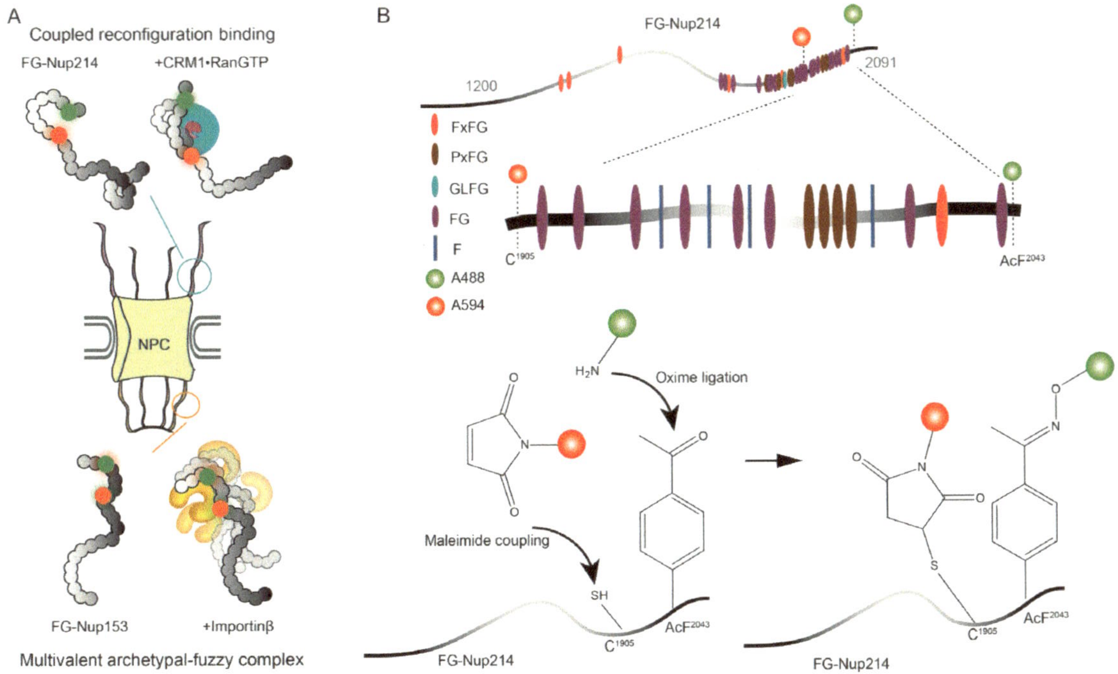

Fig. 1 Differential binding modes of FG-Nup•NTR interactions and site-specific labeling of FG-Nups. (A) Disordered FG-Nup214 undergoes coupled reconfiguration-binding with the NTR CRM1•RanGTP complex, in contrast to archetypal-fuzzy complex formation of FG-Nup153•Importinβ (Tan et al., 2018). (B) *Top*: scheme of FG-Nup214 constructs and labeling sites. The different FG motifs are color-coded as distributed across the FG-Nup214 sequence. F residues are only shown in the zoom-in region. *Bottom*: scheme of labeling FG-Nup214 with *p*-AcF via oxime ligation with Alexa 488-hydroxylamine dye and coupling of Alexa 594-maleimide dye to the Cys residue of FG-Nup214. *Panels (A) and (B) are adapted/modified from Tan, P. S., Aramburu, I. V., Mercadante, D., Tyagi, S., Chowdhury, A., Spitz, D., et al. (2018). Two differential binding mechanisms of FG-nucleoporins and nuclear transport receptors.* Cell Reports, *22(13), 3660–3671. https://doi.org/10.1016/j.celrep.2018.03.022.*

(Hayama et al., 2018; Hough et al., 2015; Milles et al., 2015; Raveh et al., 2016; Sparks, Temel, Rout, & Cowburn, 2018; Tan et al., 2018). Second, we found that FG-Nup214, which localizes to the cytoplasmic side of the NPC, undergoes a clear and distinct conformational change when binding the CRM1•RanGTP export complex. This indicated that the complex formation of FG-Nup214•CRM1•RanGTP follows rather a coupled reconfiguration-binding mechanism which can regulate spatial localization of export complexes that help cargo and NTR undocking processes (Tan et al., 2018).

Much of the conclusions drawn in these studies were driven by multiple experimental tools such as molecular dynamics simulations, stopped-flow kinetics, nuclear magnetic resonance, thermodynamic measurements, scattering, and (single-molecule) fluorescence measurements. Nevertheless, a key indication of these differential binding mechanisms can be easily derived from single-molecule fluorescence resonance energy transfer (smFRET) experiments. Here we compare side by side the FG-Nup153•Importinβ complex (archetypal-fuzzy complex formation) with the FG-Nup214•CRM1•RanGTP complex (coupled reconfiguration-binding mechanism) to showcase how smFRET enables to distinguish characteristic fingerprints of these binding mechanisms. We provide a detailed step-by-step protocol of sample preparation, including FG-Nup expression, purification, and labeling steps. Then we discuss smFRET, including experimental setup, measurement, and data analysis, and how evidence for coupled reconfiguration-binding mechanism can be distinguished from archetypal-fuzzy complexes formation. We also report on the preparation of glycosylated FG-Nups and their study using smFRET.

2. SITE-SPECIFIC DUAL-LABELING OF FG-NUPS/GLYCOSYLATED FG-NUPS

FRET is the nonradiative transfer from a fluorophore-labeled donor molecule in the singlet excited state to another fluorophore-labeled acceptor molecule via a dipole–dipole coupling mechanism. The prerequisite to perform quantitative smFRET experiments as a structural biology tool is to site-specifically introduce a FRET dye pair into the protein. This permits to probe conformational features in the region sandwiched between the two dyes. The FRET efficiency (E_{FRET}) reports the distance between dye pair (r), described by the equation

$$E_{FRET} = \frac{1}{1 + \left(\frac{r}{R_0}\right)^6}$$

where R_0 being the Förster radius of this dye pair at the distance at which $E_{FRET} = 0.5$. Due to the power 6 dependence, probing intra-dye distance changes is most sensitive around R_0 (Roy, Hohng, & Ha, 2008).

Here we designed an FG-Nup214 mutant for smFRET measurements to probe the part of Nup214 involved in CRM1 binding as seen from the crystal structure of FG-Nup214•CRM1•RanGTP•SNUP1 (Port et al., 2015). We site-specifically labeled the FG-Nup214 mutant at cysteine (Cys, aa 1905) with Alexa 594-maleimide (acceptor dye) and at the incorporated noncanonical amino acids (ncAAs) *p*-acetylphenylalanine (AcF, aa 2043) with Alexa 488-hydroxylamine (donor dye) via maleimide–cysteine coupling and oxime ligation reaction, respectively (Fig. 1B). In this way, specific labeling of the reactive group can be achieved (Brustad, Lemke, Schultz, & Deniz, 2008; Lemke, 2011) (see Note 1 for FG-Nup153).

2.1 Materials

- Codon-optimized FG-Nup gene (Mr. Gene, Regensburg, Germany)
- pEVOL-*p*-AcF plasmid (Addgene)
- *Escherichia coli* (*E. coli*) BL21(DE3) AI cells, arabinose-inducible promoter araBAD upstream of T7 RNA pol (Invitrogen)
- LB agar plates with 50 μg/L ampicillin (Amp) and 33 μg/L chloramphenicol (Cm)
- Luria–Bertani (LB) growth medium
- Terrific Broth (TB) growth medium
- *p*-acetylphenylalanine (*p*-AcF)
- 20% L-arabinose, filter-sterilized through a 0.22 μm Millipore® Stericup
- Lysis buffer (2 *M* Urea, 4 × phosphate-buffered saline (PBS), 5 m*M* imidazole, 1 m*M* phenylmethylsulfonyl fluoride (PMSF), and 0.2 m*M* tris (2-carboxyethyl)phosphine (TCEP))
- Storage buffer (4 *M* GdHCl, 1 × PBS, pH 7.4)
- Ni-NTA superflow resin (Thermo Fisher Scientific: 88223)
- Chitin resin (NEB)
- TEV cleavage buffer (50 m*M* Tris, 150 m*M* sodium chloride (NaCl), 1 m*M* PMSF, 0.2 m*M* TCEP, pH 7.5)
- Glycosylation buffer (50 m*M* Tris, 200 m*M* NaCl, 20 m*M* magnesium chloride ($MgCl_2$), 0.2 m*M* TCEP, and 1% Tween20, pH 7.5)
- TEV protease

- O-linked *N*-acetylglucosaminyl transferase (OGT)
- Uridine 5′-diphospho-*N*-acetylglucosamine sodium salt (UDP-GlcNAc) (Sigma: U4375)
- β-mercaptoethanol (Sigma-Aldrich: M6250)
- Polypropylene chromatography column with filter (Qiagen)
- Oxime labeling buffer (4 *M* GdHCl, 0.05 *M* anhydrous sodium acetate (NaAc), 0.15 *M* NaCl, pH 4)
- Maleimide labeling buffer (4 *M* GdHCl, 1 × PBS, 0.1 m*M* ethylenediaminetetraacetic acid (EDTA), pH 7)
- EDTA-free protease inhibitor cocktail (Roche: 11873580001)
- Alexa Fluor® 488-hydroxylamine dye (Thermo Fisher Scientific)
- Alexa Fluor® 594-maleimide dye (Thermo Fisher Scientific)
- Dithiothreitol (DTT) (Biomol: 04010)
- Anhydrous acetonitrile (Thermo Fisher Scientific)
- S200 chromatography column (GE Healthcare)
- NuPAGE denaturing SDS-PAGE gel (Thermo Fisher Scientific)
- SDS-PAGE MOPS running buffer
- SDS-PAGE loading dye
- SDS-PAGE molecular weight markers
- 3 kDa MWCO Amicon® ultra centrifugal filter units (Merck)
- Sonicator or microfluidizer
- UV–visible spectrophotometer

2.2 Expression and Purification of FG-Nup214 Containing ncAAs

1. Perform site-directed mutagenesis of codon-optimized FG-Nup214 disordered region (see Note 2) to introduce Amber stop codon at the site of the interest. Mutate the codon for amino acid 2043 in the FG-Nup214 sequence to TAG and 1905 FG-Nup214 to Cys. The gene of interest containing a single Cys and a TAG codon at the desired positions is then cloned into a pBAD-intein-chitin binding domain (CBD)-12His plasmid, generating pBAD-Nup214^{1905C,2043TAG}-intein-CBD-12His.
2. Cotransform *E. coli* BL21 (DE3) AI cells with pEVOL-*p*-AcF plasmid, which encodes for the *p*-AcF Amber suppressor tRNA/tRNA synthetase pair and Cm resistance with plasmids pBAD-Nup214^{1905C,2043TAG}-intein-CBD-12His, which encodes for protein of interest (POI) and Amp resistance (see Note 3). Spread transformed colonies on LB agar plates with 50 μg/L Amp and 33 μg/L Cm. Let the cells grow overnight in a 37°C incubator.

3. Pick an individual colony and grow overnight in LB medium (containing Amp and Cm) at 37°C.
4. Add 20 mL (1:100) overnight culture and 2 mL (1:1000) Amp/Cm in 2 L TB medium. Let the cell grow at 37°C, 180 revolutions per minute (rpm) in the incubator.
5. Add 1 m*M* (final) *p*-AcF when the optical density measured at a wavelength of 600 nm (OD_{600}) = 0.4.
6. Induce the recombinant expression of the POI with 0.02% arabinose at $0.7 < OD_{600} < 1.2$. Let cells grow at 37°C for 6–8 h.
7. Harvest cells after expression time at 4°C, 4000 rpm for 20 min.
8. Dispose of supernatant in S1 waste, resuspend cell pellet with one pellet volume of lysis buffer (see Note 4).
9. Sonicate the sample 3 × 30 s on ice. Let the cells rest on ice for at least 30 s between each cycle, or microfluidize sample 3 times for smaller volumes (4–8 L expression/less than 200 mL volume of lysate) or 4–7 times for larger volumes (12–24 L expression/more than 200 mL volume of lysate). Spin down the lysate at 4°C, 15,000 rpm for 1 h.
10. Wash Ni-beads with double-distilled water (ddH_2O) (1 mL beads per 1 L expression) and equilibrate in lysis buffer. Take the clear lysate (supernatant) and load it on the Ni-beads. Incubate 2 h at 4°C on the shaker or roller.
11. Collect flow-through with polypropylene column in the column rack in the following step (see Note 5). For in vitro glycosylation of FG-Nup, jump to step 12.
    a. Wash each column with ddH_2O and once with lysis buffer.
    b. Wash with 20 mL/mL Ni-bead with lysis buffer.
    c. Wash with 20 mL/mL Ni-bead with lysis buffer of 15 m*M* imidazole.
    d. Wash with 20 mL/mL Ni-bead with lysis buffer of 20 m*M* imidazole.
    e. Wash with 10 mL/mL Ni-bead with lysis buffer of 25 m*M* imidazole.
    f. Wash with 10 mL/mL Ni-bead with lysis buffer of 35 m*M* imidazole.
    g. Elute protein with 10 mL elution buffer (lysis buffer with 400 m*M* imidazole) per 2 mL Ni-beads.
12. In vitro glycosylation of FG-Nup, adapted from previously published protocol for in vitro glycosylation of Nup98 (Labokha et al., 2013).

   a. Wash the Ni-bead with 1.5 times volume of Ni-beads of glycosylation buffer.
   b. Add 20 time excess Ni-bead volume of glycosylation buffer containing 5 μ*M* OGT (see Note 6) and 1 m*M* UDP-GlcNAc.
   c. Stir the Ni-beads at room temperature for at least 16 h.
   d. Wash Ni-beads with glycosylation buffer and follow step 11.
13. Add 100 m*M* β-mercaptoethanol and 1 × EDTA-free protease inhibitor cocktail to the eluted protein. Incubate overnight at room temperature on a shaker (see Note 7).
14. Dialyze the sample in a suitable dialysis bag and put it in 2 L of lysis buffer (without imidazole) for 4 h (on a magnetic stirrer in the cold room). Repeat dialysis in fresh 2 L lysis buffer (without imidazole) for another 2 h (see Note 8).
15. Wash 0.5 mL Ni-beads/L of expression with ddH_2O and then with lysis buffer.
16. Load buffer exchanged protein on the Ni-beads and incubate 2 h at 4°C on a shaker.
17. Collect flow-through with polypropylene column in the column rack. Concentrate the protein and run size exclusion chromatography.
18. Check purification success by SDS-PAGE. Concentrate fractions of interest and buffer exchange to storage buffer.

2.3 Fluorescent Labeling of FG-Nup214

2.3.1 Oxime Ligation of FG-Nup214 With Alexa 488-Hydroxylamine Dye

1. Exchange the protein buffer with oxime labeling buffer, filtering the buffer before use.
2. Dissolve 1 mg of Alexa 488-hydroxylamine dye in pH 4 oxime labeling buffer for 1 m*M* final concentration. Store frozen in the dark at −80°C and thaw prior to use.
3. Label protein with fivefold molar excess of Alexa 488-hydroxylamine dye.
4. Allow the reaction to proceed for 12 h at room temperature in the dark (without stirring).

2.3.2 Coupling of FG-Nup214 With Alexa 594-Maleimide Dye

1. Exchange buffer to maleimide labeling buffer with 10 m*M* DTT at pH 7, filter the buffer before use.
2. Wash sample (5 times) with maleimide labeling buffer with 0.2 m*M* TCEP in 3k Centricon® device (see Note 9).

3. Solubilize 1 mg of Alexa 594-maleimide dye in anhydrous acetonitrile. Aliquot the solution in 25 nmol fractions into dry microcentrifuge tubes. Lyophilize the aliquots and store at −80°C.
4. Label protein with twofold molar excess of Alexa 594-maleimide dye.
5. Allow the reaction to proceed for 2 h at room temperature in the dark.
6. Quench the labeling reaction with maleimide labeling buffer containing 10 m*M* DTT.
7. Run size exclusion chromatography (Superdex 200). Check labeling success by SDS-PAGE and UV–visible spectrophotometry. Concentrate fractions of interest and buffer exchange in storage buffer.

3. SINGLE-MOLECULE FRET STUDIES OF FG-Nup AND NTR INTERACTIONS

3.1 Single-Molecule Fluorescence Resonance Energy Transfer

In contrast to ensemble FRET measurement, which provides the average E_{FRET} value of a high number of molecules, smFRET is able to resolve the FRET signal of individual molecules. Hence, it can provide direct measurements of conformational distributions and stochastic dynamics even in the context of complex structural landscapes and mixtures of subpopulations (Ferreon, Gambin, Lemke, & Deniz, 2009; Mukhopadhyay, Krishnan, Lemke, Lindquist, & Deniz, 2007; Muller-Spath et al., 2010). Our smFRET measurement is performed on a custom-built confocal detection-based microscope as shown in the schematic diagram (Fig. 2A). A pulsed laser diode (LDH 485, Picoquant, Germany), pulsed at 26.67 MHz, is used for donor excitation; a supercontinuum white light laser source (SuperK Extreme, NKT photonics, Denmark) is used for acceptor excitation. We use pulsed interleaved excitation (PIE) scheme (Muller, Zaychikov, Brauchle, & Lamb, 2005) to alternatingly excite donor and acceptor fluorophores to retrieve the stoichiometry (*S*) information. Both lasers are cleaned up with excitation filters (Brightline FF01-482/18 and Brightline FF01-572/15, Semrock) and passed through the polarizer (GL-10A, Thorlabs) before entering the objective lens (60×, water immersion, Nikon) through the central dichroic mirror (ZT488/561 Chroma). The diffusing doubly labeled proteins are excited within a confocal volume (Fig. 2B). The excited fluorescent light is reflected into the emission pathway, through the tube lens (*f* = 200 mm, Nikon) via the central dichroic mirror. The emission light is then focused to a 100 μm pinhole to block

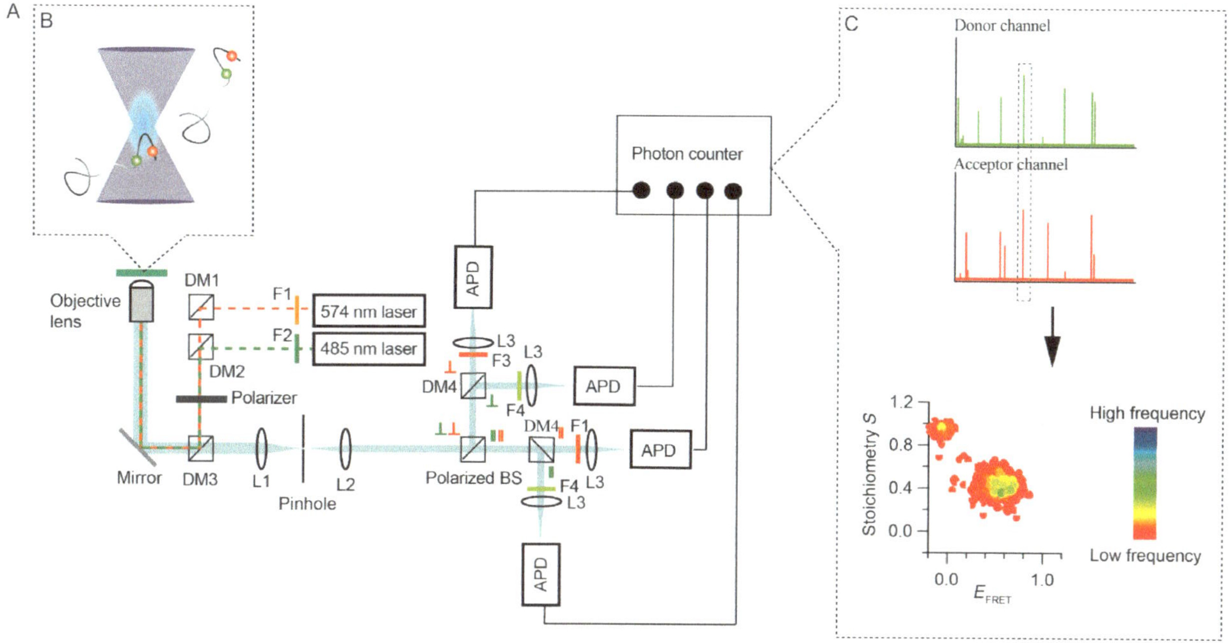

Fig. 2 Scheme of sm multiparameter setup. (A) The PIE pathway is highlighted as *green* (donor excitation) and *orange* (acceptor excitation) *dotted lines*. A confocal volume is used to excite singly freely diffusing FRET-labeled proteins as shown in (B). The emitted photons (highlighted in *thick*, *light blue*) are collected by the objective lens, passed through a pinhole, split based on their emission wavelengths and polarization states (parallel: ∥, perpendicular: ⊥), and eventually detected by APDs (see details in the text). (C) Separate bursts of photons, each corresponding to a single-molecule event, are collected by avalanche photon-counting modules, and analyzed using custom-written software. For each burst, E_{FRET} and S are calculated and the results are placed in the E_{FRET} vs S 2D histogram. DM1: dichroic mirror (R488-Di01, Semrock); DM2: dichroic mirror (FF605-Di02, Semrock); DM3: dichroic mirror (ZT488/561, Chroma); DM4: dichroic mirror (ZT561rdc-UF2, Chroma); L1: lenses (f=200mm, Nikon); L2: lenses (f=160mm, Thorlabs); L3: lenses (f=50mm, Thorlabs); BS: beam splitter (Thorlabs); APD: avalanche photodiode; F1: laser clean-up filter (FF01-572/15, Semrock); F2: laser clean-up filter (FF01-482/18, Semrock); F3: emission filter (FF01-525/50, Semrock); F4: emission filter (FF01-612/69, Semrock).

the out-of-focus signal and directed onto a lens (*f*=160 mm, Thorlabs) for beam collimation. After that, the fluorescence emission of the donor and the acceptor dyes is split into parallel and perpendicular polarization directions via a polarized beam splitter (BS), followed by being separated into donor and acceptor color channels via the dichroic mirror (zt561rdc-UF2, Chroma), and detected by avalanche photodiodes (APDs) (Picoquant, Berlin, Germany). Photon signals are counted using a Hydraharp400 (Picoquant, Berlin, Germany). Acquired data are subject to multiparameter fluorescence analysis and processed burstwise for fluorescence intensities, donor lifetime, and donor anisotropies (Eggeling et al., 2001; Sisamakis, Valeri, Kalinin, Rothwell, & Seidel, 2010). In this chapter, we will focus on the analysis of the fluorescence intensity to calculate E_{FRET} and S values.

3.2 Experimental Measurement and Analysis

All sm experiments are performed at protein concentration of 50 p*M* (to ensure the detection of single molecule) (see Note 10). CRM1, Importinβ, and RanGTP concentrations are 1 μ*M* (see Note 6). The sample is placed in 8-well chamber slides with a final volume of 200 μL. Traces are recorded for 30 min and each fluorescent molecule is identified by the start and the end of a fluorescent burst (Fig. 2C) (see Note 11). A burst search algorithm (Eggeling et al., 2001) is used and detected photons are collected into 1 ms bins. After burst recognition, a threshold of 80 photons is set for every detected burst (see Note 12). All analyses are performed using a custom-written code in IgorPro (Wavemetrics, Lake Oswego, OR).

E_{FRET} and S values are then calculated burstwise using the following equations

$$E_{FRET} = \frac{I_A}{\gamma I_D + I_A}; S = \frac{I_A + I_D}{I_D + I_A + I_A^{dir}}$$

where I_D and I_A are the donor and acceptor intensity, respectively. I_A^{dir} is the intensity from directly excited acceptor molecules by acceptor laser, and γ is a correction factor dependent on quantum yields and detection efficiencies of donor and acceptor. The γ value can be estimated from leakage and direct excitation-corrected E_{FRET} and stoichiometries following a basic procedure developed by Lee et al. (2005) and described in further detail by Fuertes et al. (2017) with the preassumption that the quantum yields of both donor and acceptor do not vary across samples.

For every single-molecule burst E_{FRET} and S are calculated and plotted into 2D histogram. The histogram is color-coded for number of events. The laser power of donor and acceptor laser is adjusted such that a FRET-labeled molecule (1:1 donor to acceptor dye-labeled sample) lies at $S\sim0.5$. In such 2D frequency histogram of E_{FRET} vs S (Fig. 2C) at least two peaks emerge. A $S=1$ and $E_{FRET}=0$ molecules can be found that are labeled with donor only (or where the acceptor dye bleached or was photophysically inactive for other reasons (Kong, Nir, Hamadani, & Weiss, 2007; Sisamakis et al., 2010)), while a population shifting toward zero along the x axis (E_{FRET} value) and staying constant at $S\sim0.5$ indicates an increase of the end-to-end distance between the dyes.

In the absence of CRM1•RanGTP complex doubly labeled FG-Nup214 molecules showed a single FRET population with an E_{FRET} value of 0.6 (Fig. 3A, top panel). If the CRM1•RanGTP complex is added (1 μM), we observed a single population with significant shift in the E_{FRET} values of FG-Nup214 corresponding to a bound ($E_{FRET}=0.1$ and $S\sim0.5$) state. This suggests a much more expanded and docked conformation than in the unbound state. FG-Nup214 appears to undergo a drastic conformational change upon binding the CRM1•RanGTP complex, which is distinct from the core signature of archetypal-fuzzy complex formation of FG-Nup153•Importinβ which did not show any substantial change in the E_{FRET} peak positions upon binding (Fig. 3B, top panel).

When there is a conformational change detected in smFRET, it usually indicates that binding occurred. However, when there is no substantial conformational change detected, it can be that no binding occurred, which is very common due to the inherently low concentrations used in smFRET experiments. Hence, one needs additional experimental validation to confirm that under single-molecule concentrations the complex is indeed formed, before the absence of detecting a conformational change can be taken as an indication for formation of an archetypal-fuzzy complex. Among other tools, fluorescence correlation spectroscopy (FCS), stopped-flow kinetics, or fluorescence anisotropy measurement as used by us previously can be recruited for such purposes, and we refer the reader to the literature to learn more about those (Milles & Lemke, 2014; Milles et al., 2015; Tan et al., 2018). In particular FCS analysis can also be performed on the same single-molecule data set as obtained for smFRET and can be directly used to confirm binding interactions. The acceptor burst signal excited by the donor laser (corresponding to FRET signal) is extracted from smFRET measurement and autocorrelated to obtain information about the translational

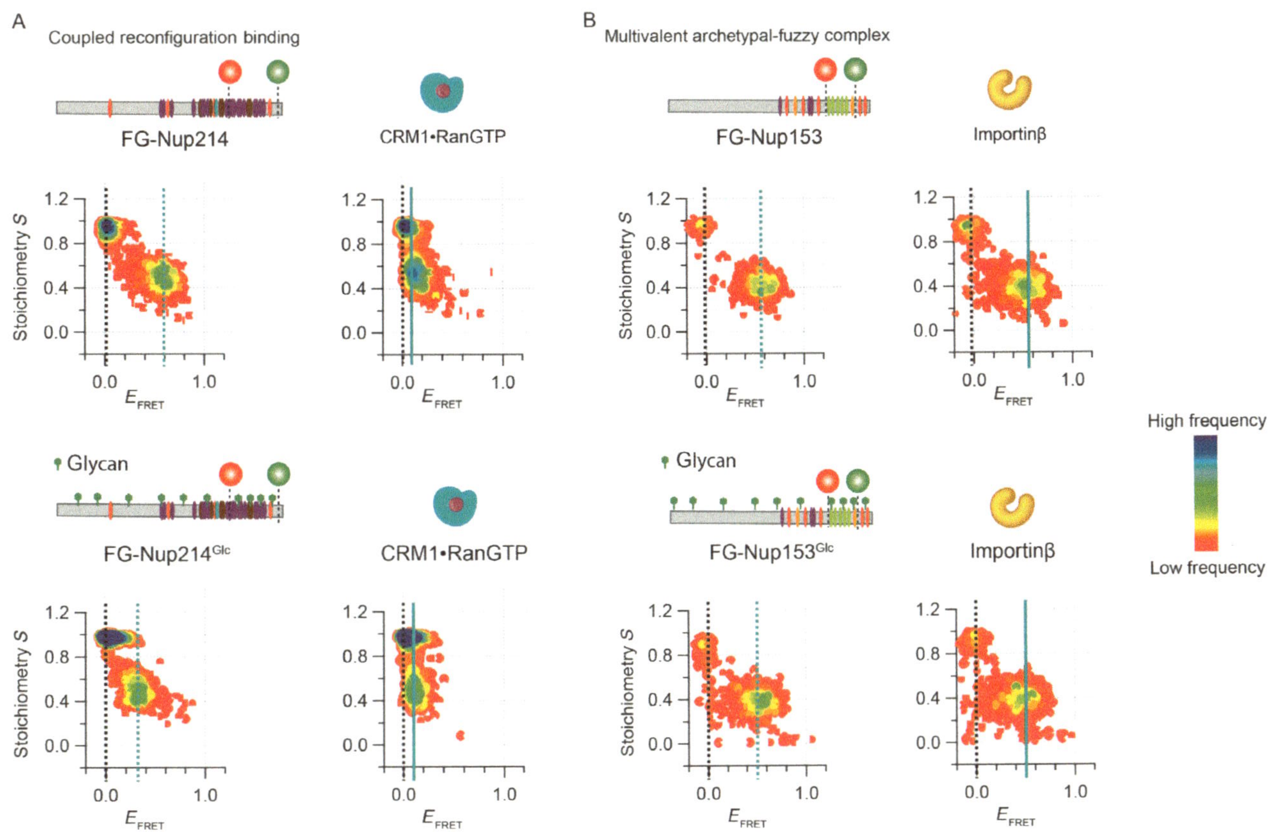

Fig. 3 See legend on opposite page.

motion of the FG-Nups. When the translational diffusion of FG-Nup is slower in the presence of NTR, it indicates that FG-Nup•NTR complexes formed (Milles et al., 2015; Sisamakis et al., 2010).

We repeated the smFRET experiments using glycosylated variants of FG-Nups (FG-Nup214^{Glc} and FG-Nup153^{Glc}). As shown in Fig. 3A (bottom panel), a mildly lower E_{FRET} was measured in comparison to unglycosylated FG-Nup214 in the unbound form, indicating that the glycosylated Nup is more extended. In the presence of CRM1•RanGTP, we detected a single population with an E_{FRET} nearly identical to the unglycosylated bound state. Likewise, the FG-Nup153^{Glc} behaves similar in binding to Importinβ (Fig. 3B, bottom panel). The smFRET data strongly suggest that the glycosylated and unglycosylated Nups behave remarkably similar in their respective binding mechanisms, highlighting that glycosylation only mildly tunes the binding mechanism between FG-Nup and NTR. Taken together, we summarize that in the NPC at least two fundamentally different binding mechanisms coexist between FG-Nups and NTRs (Tan et al., 2018). The conformational features of these binding mechanisms can be distinguished well by smFRET.

4. SUMMARY

In this chapter, we report a workflow combining smFRET and site-specific dual-labeling method to study the binding mechanism of

Fig. 3 Two differential binding mechanisms of FG-Nups•NTRs detected by smFRET. (A) E_{FRET} vs S histograms of 50 pM FG-Nup214 (*bottom panel*: FG-Nup214^{Glc}) in the absence and presence of CRM1•RanGTP (from *left to right*, at 1 μM for both CRM1 and RanGTP concentration). Donor-only labeled population ($E_{FRET}=0$) is shown in *black dotted line*, indicating the imperfect labeling efficiency or photobleaching of acceptor dye. In the presence of CRM1•RanGTP, one distinct conformation with very low E_{FRET} state (*blue solid line*) is seen suggesting a much more expanded conformation than in the unbound state (*blue dotted line*) for both FG-Nup214 and FG-Nup214^{Glc}. (B) E_{FRET} vs S histograms of 50 pM FG-Nup153 (*bottom* panel: FG-Nup153^{Glc}) in the absence (*blue dotted line*) and presence of Importinβ (*blue solid line*), indicating no conformational change in FG-Nup153 (and FG-Nup153^{Glc}) as detectable by smFRET, in the presence of Importinβ. In both cases, FG-Nup214^{Glc} and FG-Nup153^{Glc} show a mildly lower E_{FRET} compared to unglycosylated case in the unbound form, indicating that the glycosylated FG-Nup is more extended. However, the two differential binding mechanisms are conserved. *Panels (A) and (B) are adapted from Tan, P. S., Aramburu, I. V., Mercadante, D., Tyagi, S., Chowdhury, A., Spitz, D., et al. (2018). Two differential binding mechanisms of FG-nucleoporins and nuclear transport receptors.* Cell Reports, 22*(13), 3660–3671. https://doi.org/10.1016/j.celrep.2018.03.022.*

intrinsically disordered FG-Nups. SmFRET allows to differentiate the conformational features of FG-Nup upon interacting with NTRs at the single-molecule level, while the labeling technique enables us to site-specifically probe specific regions of interest. The ability of the presented smFRET technique to resolve structural and dynamic heterogeneity is particularly well suited for the investigation of IDP structure and function. It provides not only detailed information about the folding and ligand/protein binding of IDPs, but could also be used to detect folding intermediates and pathways which are very challenging to extract from ensemble-average data. The presented technique should enable the design and construction of novel tools for biological studies.

5. NOTES

Note 1. The codon-optimized FG-Nup153 disordered region (amino acids 875–1475 of the full-length Nup153; numbering with respect to the full-length protein as in UniProt: P49790) is cloned into a pTXB3-6His-intein-CBD vector, generating pTXB3-6His-Nup153-intein-CBD. The His-tag is cleaved via TEV cleavage and the intein-CBD is cleaved via intein cleavage. We site-specifically labeled FG-Nup153 mutant at the cysteine (Cys, aa 1312) with Alexa 594-maleimide (acceptor dye) and at the incorporated noncanonical amino acids (ncAAs) *p*-acetylphenylalanine (AcF, aa 1391) with Alexa 488-hydroxylamine (donor dye) via maleimide–cysteine coupling and oxime ligation reaction, respectively.

Note 2. The codon-optimized FG-Nup214 disordered region: aa 1392–2090 is numbered with respect to the full-length protein (Uniprot: P35658).

Note 3. We use two plasmids for ncAAs incorporation in *E. coli*, one for the expression of the gene of interest, containing the Amber (TAG) stop codon mutation, and a second that encodes the orthogonal tRNA and aminoacyl-tRNA synthetase pair. When the plasmid containing the Amber codon (UAG) is being translated in the ribosome, the Amber codon is recognized by the orthogonal tRNA charged with the ncAA and subsequently the ncAA is added to the growing polypeptide chain. In this chapter, FG-Nup214$^{1905C,\ 2043TAG}$ is used as an example and the protocol can be applied to other FG-Nup variants unless otherwise stated.

Note 4. Usually 10 mL of lysis buffer/L expression at pH 8.5.

Note 5. For FG-Nup153, collect flow-through with polypropylene column in the column rack in the following steps. Additional chitin beads purification steps are required for higher protein purity.

a. Wash each column with ddH_2O and once with lysis buffer.

b. Wash with 10 mL/mL Ni-bead with lysis buffer of 10 m*M* imidazole.

c. Wash with 10 mL/mL Ni-bead with lysis buffer of 15 m*M* imidazole.

d. Elute protein with 10 mL elution buffer (lysis buffer with 400 m*M* imidazole) per 2 mL Ni-beads.

e. Pour Ni-bead elution on chitin beads (2–5 mL per liter expression).

f. Close column(s) and incubate elution on chitin beads for at least 8 h to overnight at 4°C on shaker.

g. Discard the flow-through and wash the chitin beads with lysis buffer and equilibrate on TEV cleavage buffer.

h. Add 0.5 mg TEV protease per liter expression and supplement 100 m*M* BME in the TEV cleavage buffer.

i. Perform the cleavage for at least 6 h at room temperature under mild rotation.

j. Collect the flow-through and continue with the step 14.

Note 6. OGT, CRM1, Importin β, and RanGTP are expressed in *E. coli* BL21 (DE3) AI cell and purified using standard nickel affinity and size exclusion chromatography purification procedures as described in Tan et al. (2018).

Note 7. This step is performed to achieve intein cleavage in order to remove the intein-CBD-12His tag of the protein.

Note 8. This step is to eliminate β-mercaptoethanol and imidazole in the sample.

Note 9. This step is to remove the excess of DTT which would quench the maleimide dye.

Note 10. The doubly labeled protein is diluted to 1 μ*M* from stock solution to prevent protein from sticking to the wall of Eppendorf tube. The protein is further diluted to a final concentration of 50 p*M* in solution in 1× PBS pH 7.4 containing 2 m*M* DTT and 2 m*M* MgAc.

Note 11. In the cases where very few bursts were detected per measurement, signal from different replicates can be accumulated.

Note 12. In order to remove aggregated clusters from analysis, all bursts longer than 8 ms and brighter than 600 photons were discarded.

ACKNOWLEDGMENTS

We thank Dr. Iker Valle Aramburu for critical proofreading of this manuscript. We are grateful for the fruitful discussions from Lemke group. P.S.T. is supported by a fellowship from the EMBL Interdisciplinary Postdoc (EIPOD) program under Marie Curie Actions COFUND program (grant number 291772). E.A.L. acknowledges funding from the ERC grant SMPF v2.0 and the Gutenberg-Research College.

REFERENCES

Aramburu, I. V., & Lemke, E. A. (2017). Floppy but not sloppy: Interaction mechanism of FG-nucleoporins and nuclear transport receptors. *Seminars in Cell and Developmental Biology*, *68*, 34–41. PMID: 28669824. https://doi.org/10.1016/j.semcdb.2017.06.026.

Brohawn, S. G., & Schwartz, T. U. (2009). Molecular architecture of the Nup84-Nup145C-Sec13 edge element in the nuclear pore complex lattice. *Nature Structural & Molecular Biology*, *16*(11), 1173–1178. https://doi.org/10.1038/nsmb.1713.

Brustad, E. M., Lemke, E. A., Schultz, P. G., & Deniz, A. A. (2008). A general and efficient method for the site-specific dual-labeling of proteins for single molecule fluorescence resonance energy transfer. *Journal of the American Chemical Society*, *130*(52), 17664–17665. https://doi.org/10.1021/ja807430h.

Chug, H., Trakhanov, S., Hulsmann, B. B., Pleiner, T., & Gorlich, D. (2015). Crystal structure of the metazoan Nup62.Nup58.Nup54 nucleoporin complex. *Science*, *350*(6256), 106–110. https://doi.org/10.1126/science.aac7420.

Cook, A., Bono, F., Jinek, M., & Conti, E. (2007). Structural biology of nucleocytoplasmic transport. *Annual Review of Biochemistry*, *76*, 647–671. https://doi.org/10.1146/annurev.biochem.76.052705.161529.

Denning, D. P., Patel, S. S., Uversky, V., Fink, A. L., & Rexach, M. (2003). Disorder in the nuclear pore complex: The FG repeat regions of nucleoporins are natively unfolded. *Proceedings of the National Academy of Sciences of the United States of America*, *100*(5), 2450–2455. https://doi.org/10.1073/pnas.0437902100.

Eggeling, C., Berger, S., Brand, L., Fries, J. R., Schaffer, J., Volkmer, A., et al. (2001). Data registration and selective single-molecule analysis using multi-parameter fluorescence detection. *Journal of Biotechnology*, *86*(3), 163–180.

Eibauer, M., Pellanda, M., Turgay, Y., Dubrovsky, A., Wild, A., & Medalia, O. (2015). Structure and gating of the nuclear pore complex. *Nature Communications, 6*. 7532. https://doi.org/10.1038/ncomms8532.

Favreau, C., Worman, H. J., Wozniak, R. W., Frappier, T., & Courvalin, J. C. (1996). Cell cycle-dependent phosphorylation of nucleoporins and nuclear pore membrane protein Gp210. *Biochemistry*, *35*(24), 8035–8044. https://doi.org/10.1021/bi9600660.

Ferreon, A. C., Gambin, Y., Lemke, E. A., & Deniz, A. A. (2009). Interplay of alpha-synuclein binding and conformational switching probed by single-molecule fluorescence. *Proceedings of the National Academy of Sciences of the United States of America*, *106*(14), 5645–5650. https://doi.org/10.1073/pnas.0809232106.

Fuertes, G., Banterlea, N., Ruff, K. M., Chowdhury, A., Mercadante, D., Koehler, C., et al. (2017). Decoupling of size and shape fluctuations in heteropolymeric sequences reconciles discrepancies in SAXS vs. FRET measurements. *Proceedings of the National*

Academy of Sciences of the United States of America, *114*(31), E6342–E6351. https://doi.org/10.1073/pnas.1704692114.
Gorlich, D., & Kutay, U. (1999). Transport between the cell nucleus and the cytoplasm. *Annual Review of Cell and Developmental Biology*, *15*, 607–660. https://doi.org/10.1146/annurev.cellbio.15.1.607.
Hayama, R., Sparks, S., Hecht, L. M., Dutta, K., Karp, J. M., Cabana, C. M., et al. (2018). Thermodynamic characterization of the multivalent interactions underlying rapid and selective translocation through the nuclear pore complex. *The Journal of Biological Chemistry*, *293*(12), 4555–4563. https://doi.org/10.1074/jbc.AC117.001649.
Hough, L. E., Dutta, K., Sparks, S., Temel, D. B., Kamal, A., Tetenbaum-Novatt, J., et al. (2015). The molecular mechanism of nuclear transport revealed by atomic-scale measurements. *eLife4*, e10027. https://doi.org/10.7554/eLife.10027.
Hurt, E. C. (1988). A novel nucleoskeletal-like protein located at the nuclear periphery is required for the life-cycle of Saccharomyces-cerevisiae. *EMBO Journal*, 7(13), 4323–4334.
Kim, S. J., Fernandez-Martinez, J., Nudelman, I., Shi, Y., Zhang, W. Z., Raveh, B., et al. (2018). Integrative structure and functional anatomy of a nuclear pore complex. *Nature*, *555*(7697), 475–482. https://doi.org/10.1038/nature26003.
Kong, X. X., Nir, E., Hamadani, K., & Weiss, S. (2007). Photobleaching pathways in single-molecule FRET experiments. *Journal of the American Chemical Society*, *129*(15), 4643–4654. https://doi.org/10.1021/Ja068002s.
Kosinski, J., Mosalaganti, S., von Appen, A., Teimer, R., DiGuilio, A. L., Wan, W., et al. (2016). Molecular architecture of the inner ring scaffold of the human nuclear pore complex. *Science*, *352*(6283), 363–365. https://doi.org/10.1126/science.aaf0643.
Labokha, A. A., Gradmann, S., Frey, S., Hulsmann, B. B., Urlaub, H., Baldus, M., et al. (2013). Systematic analysis of barrier-forming FG hydrogels from Xenopus nuclear pore complexes. *EMBO Journal*, *32*(2), 204–218. https://doi.org/10.1038/emboj.2012.302.
Lee, N. K., Kapanidis, A. N., Wang, Y., Michalet, X., Mukhopadhyay, J., Ebright, R. H., et al. (2005). Accurate FRET measurements within single diffusing biomolecules using alternating-laser excitation. *Biophysical Journal*, *88*(4), 2939–2953. https://doi.org/10.1529/biophysj.104.054114.
Lemke, E. A. (2011). Site-specific labeling of proteins for single-molecule FRET measurements using genetically encoded ketone functionalities. *Methods in Molecular Biology*, *751*, 3–15. https://doi.org/10.1007/978-1-61779-151-2_1.
Lim, R. Y. H., Huang, N. P., Koser, J., Deng, J., Lau, K. H. A., Schwarz-Herion, K., et al. (2006). Flexible phenylalanine-glycine nucleoporins as entropic barriers to nucleocytoplasmic transport. *Proceedings of the National Academy of Sciences of the United States of America*, *103*(25), 9512–9517. https://doi.org/10.1073/pnas.0603521103.
Lin, D. H., Stuwe, T., Schilbach, S., Rundlet, E. J., Perriches, T., Mobbs, G., et al. (2016). Architecture of the symmetric core of the nuclear pore. *Science*, *352*(6283), aaf1015. https://doi.org/10.1126/science.aaf1015.
Milles, S., & Lemke, E. A. (2014). Mapping multivalency and differential affinities within large intrinsically disordered protein complexes with segmental motion analysis. *Angewandte Chemie (International Ed. in English)*, *53*(28), 7364–7367. https://doi.org/10.1002/anie.201403694.
Milles, S., Mercadante, D., Aramburu, I. V., Jensen, M. R., Banterle, N., Koehler, C., et al. (2015). Plasticity of an ultrafast interaction between nucleoporins and nuclear transport receptors. *Cell*, *163*(3), 734–745. https://doi.org/10.1016/j.cell.2015.09.047.
Mukhopadhyay, S., Krishnan, R., Lemke, E. A., Lindquist, S., & Deniz, A. A. (2007). A natively unfolded yeast prion monomer adopts an ensemble of collapsed and rapidly

fluctuating structures. *Proceedings of the National Academy of Sciences of the United States of America*, *104*(8), 2649–2654. https://doi.org/10.1073/pnas.0611503104.
Muller, B. K., Zaychikov, E., Brauchle, C., & Lamb, D. C. (2005). Pulsed interleaved excitation. *Biophysical Journal*, *89*(5), 3508–3522. https://doi.org/10.1529/biophysj.105.064766.
Muller-Spath, S., Soranno, A., Hirschfeld, V., Hofmann, H., Ruegger, S., Reymond, L., et al. (2010). Charge interactions can dominate the dimensions of intrinsically disordered proteins. *Proceedings of the National Academy of Sciences of the United States of America*, *107*(33), 14609–14614. https://doi.org/10.1073/pnas.1001743107.
Ori, A., Banterle, N., Iskar, M., Andres-Pons, A., Escher, C., Bui, H. K., et al. (2013). Cell type-specific nuclear pores: A case in point for context-dependent stoichiometry of molecular machines. *Molecular Systems Biology*, *9*, 648. https://doi.org/10.1038/msb.2013.4.
Port, S. A., Monecke, T., Dickmanns, A., Spillner, C., Hofele, R., Urlaub, H., et al. (2015). Structural and functional characterization of CRM1-Nup214 interactions reveals multiple FG-binding sites involved in nuclear export. *Cell Reports*, *13*(4), 690–702. https://doi.org/10.1016/j.celrep.2015.09.042.
Raveh, B., Karp, J. M., Sparks, S., Dutta, K., Rout, M. P., Sali, A., et al. (2016). Slide-and-exchange mechanism for rapid and selective transport through the nuclear pore complex. *Proceedings of the National Academy of Sciences of the United States of America*, *113*(18), E2489–E2497. https://doi.org/10.1073/pnas.1522663113.
Roy, R., Hohng, S., & Ha, T. (2008). A practical guide to single-molecule FRET. *Nature Methods*, *5*(6), 507–516. https://doi.org/10.1038/Nmeth.1208.
Sisamakis, E., Valeri, A., Kalinin, S., Rothwell, P. J., & Seidel, C. A. (2010). Accurate single-molecule FRET studies using multiparameter fluorescence detection. *Methods in Enzymology*, *475*, 455–514. https://doi.org/10.1016/S0076-6879(10)75018-7.
Sparks, S., Temel, D. B., Rout, M. P., & Cowburn, D. (2018). Deciphering the "fuzzy" interaction of FG nucleoporins and transport factors using small-angle neutron scattering. *Structure*, *26*(3), 477–484.e4. https://doi.org/10.1016/j.str.2018.01.010.
Tan, P. S., Aramburu, I. V., Mercadante, D., Tyagi, S., Chowdhury, A., Spitz, D., et al. (2018). Two differential binding mechanisms of FG-nucleoporins and nuclear transport receptors. *Cell Reports*, *22*(13), 3660–3671. https://doi.org/10.1016/j.celrep.2018.03.022.
Tetenbaum-Novatt, J., & Rout, M. P. (2010). The mechanism of nucleocytoplasmic transport through the nuclear pore complex. *Cold Spring Harbor Symposia on Quantitative Biology*, *75*, 567–584. https://doi.org/10.1101/sqb.2010.75.033.
Timney, B. L., Raveh, B., Mironska, R., Trivedi, J. M., Kim, S. J., Russel, D., et al. (2016). Simple rules for passive diffusion through the nuclear pore complex. *Journal of Cell Biology*, *215*(1), 57–76. https://doi.org/10.1083/jcb.201601004.
Tran, E. J., King, M. C., & Corbett, A. H. (2014). Macromolecular transport between the nucleus and the cytoplasm: Advances in mechanism and emerging links to disease. *Biochimica et Biophysica Acta-Molecular Cell Research*, *1843*(11), 2784–2795. https://doi.org/10.1016/j.bbamcr.2014.08.003.
von Appen, A., Kosinski, J., Sparks, L., Ori, A., DiGuilio, A. L., Vollmer, B., et al. (2015). In situ structural analysis of the human nuclear pore complex. *Nature*, *526*(7571), 140–143. https://doi.org/10.1038/nature15381.
Wente, S. R., Rout, M. P., & Blobel, G. (1992). A new family of yeast nuclear-pore complex proteins. *Journal of Cell Biology*, *119*(4), 705–723. https://doi.org/10.1083/jcb.119.4.705.
Zhu, Y. P., Liu, T. W., Madden, Z., Yuzwa, S. A., Murray, K., Cecioni, S., et al. (2016). Post-translational O-GlcNAcylation is essential for nuclear pore integrity and maintenance of the pore selectivity filter. *Journal of Molecular Cell Biology*, *8*(1), 2–16. https://doi.org/10.1093/jmcb/mjv033.

CHAPTER TWELVE

Fluorescence Depolarization Kinetics to Study the Conformational Preference, Structural Plasticity, Binding, and Assembly of Intrinsically Disordered Proteins

Anupa Majumdar*,†, Samrat Mukhopadhyay*,†,1

*Centre for Protein Science, Design and Engineering, Department of Biological Sciences, Indian Institute of Science Education and Research (IISER) Mohali, Punjab, India
†Department of Chemical Sciences, Indian Institute of Science Education and Research (IISER) Mohali, Punjab, India
[1]Corresponding author: e-mail address: mukhopadhyay@iisermohali.ac.in

Contents

Abstract

Fluorescence depolarization kinetics measured by the time-resolved fluorescence anisotropy decay serves as a sensitive and powerful methodology to study the conformational dynamics of macromolecules. This methodology allows us to delineate the

Methods in Enzymology, Volume 611
ISSN 0076-6879
https://doi.org/10.1016/bs.mie.2018.09.031

different modes of biomolecular motional dynamics including the local, segmental, and global rotational dynamics on the timescale ranging from picoseconds to nanoseconds. In this chapter, we describe the principles and applications of this methodology to obtain unique molecular insights into the intrinsically disordered proteins (IDPs). Fluorescence depolarization kinetics, when performed in a site-specific manner, can offer a reliable tool to monitor the intrinsic backbone torsional dynamics of expanded IDPs and is capable of discerning the conformational preference of IDPs. Additionally, the time-resolved fluorescence anisotropy measurements allow us to investigate the mechanism of binding and assembly of a wide range of IDPs that are involved in crucial function and disease.

1. INTRODUCTION

Proteins are the molecular workhorses that fold up into unique three-dimensional shapes to perform their biological functions (Alberts et al., 2014; Redfern, Dessailly, & Orengo, 2008). However, current investigations have revealed the presence of a large number of proteins and polypeptide segments possessing high net charge and low hydrophobicity (Dunker et al., 2001; Mao, Crick, Vitalis, Chicoine, & Pappu, 2010; Marsh & Forman-Kay, 2010; Uversky, Santambrogio, Brocca, & Grandori, 2012). They lack a well-defined structure under physiological conditions, and therefore, the traditional sequence–structure–function paradigm cannot account for such natively unstructured proteins (Tompa, 2005; Wright & Dyson, 1999). They belong to a distinct class of proteins commonly known as intrinsically disordered proteins (IDPs) (Dunker et al., 2013, 2001; Habchi, Tompa, Longhi, & Uversky, 2014; Uversky, 2013a, 2013b, 2014). IDPs do not possess any unique three-dimensional tertiary structure but rather exist as an ensemble of dynamic and heterogeneous conformers. They perform numerous biological functions (Dyson & Wright, 2005) and constitute a significantly large subset of the eukaryotic proteome. The characteristic structural flexibility of the IDPs is encoded in their amino acid sequences (Das, Ruff, & Pappu, 2015; Uversky, 2013a). IDPs are predominantly enriched in polar residues and disorder-promoting residues and are generally devoid of order-promoting hydrophobic residues (Uversky, 2013b). The combined effect of low mean hydrophobicity and high net charge in IDPs predisposes these proteins toward having intrinsically expanded coil-like structures under physiological conditions. While low hydrophobicity provides less driving force for protein compaction, charge–charge repulsion also contributes toward disorder (van der Lee et al., 2014). The conformational preference of an IDP to adopt either an extended coil-like structure or collapsed globule is modulated by the net charge per residue of the protein as well as the immediate solvent microenvironment. Interestingly, protein disorder is also modulated by the presence of low-complexity domains within the amino acid sequences of IDPs having long stretches of low overall diversity of amino acids (Lin, Currie, & Rosen, 2017; Uversky, 2017a, 2017b). These IDPs perform a wide range of biological functions such as signal transduction (Wright & Dyson, 2014), cell division, intracellular transport, posttranscriptional regulation, cell cycle control (Ou, Waddell, & Kriwacki, 2012; Yoon, Mitrea, Ou, & Kriwacki, 2012), regulation of transcription and translation, either in their disordered conformation or in their ordered state, induced by binding to different cellular partners (Dyson & Wright, 2005). Often, these IDPs tend to fold up into a variety of secondary structures upon binding to membranes or other cellular partners. Even in their bound states, the IDPs can remain partially disordered resulting in the formation of fuzzy complexes which allow these proteins to alternatively interact with multiple partners (Fuxreiter & Tompa, 2012; Gruet et al., 2016; Sharma, Raduly, Miskei, & Fuxreiter, 2015). An increasing body of work suggests many of these cellular functions are accomplished by the formation of membrane-less organelles comprising liquid droplets from IDPs and IDRs (Boeynaems et al., 2018; Courchaine, Lu, & Neugebauer, 2016). These liquid droplets are spherical supramolecular fluidic assemblies that are formed by weak intermolecular interactions leading to liquid–liquid phase separation of IDPs.

The inherent propensity of these IDPs to interact with multiple endogenous partners renders these proteins vulnerable to misfolding resulting in the formation of amyloidogenic conformers and oligomers that subsequently assemble into highly ordered amyloid fibrils. Amyloid formation from these IDPs is implicated in several neurodegenerative diseases (Knowles, Vendruscolo, & Dobson, 2014) such as Parkinson's disease (Baba et al., 1998; Wales, Lázaro, Pinho, & Outeiro, 2013), Alzheimer's disease (Attems, 2017; He et al., 2018; Spires-Jones & Hyman, 2014), Huntington's disease, and so forth. The abundance of IDPs in various proteomes along with their intricate involvement in several critical functions and diseases has sparked considerable interest in studying the structural attributes such as the intrinsic conformational flexibility, conformational heterogeneity, distribution, and their "promiscuous" binding to different cellular partners. Our current understanding of both disorder-to-function and disorder-to-disease is incomplete due to inherent challenges associated with the structural studies of IDPs. A growing body of studies have been devoted to the development of computational techniques (Burger, Gurry, & Stultz, 2014) such as molecular dynamics simulations (Henriques, Cragnell, & Skepö, 2015) as well as experimental techniques such as nuclear magnetic resonance (Konrat, 2014), electron paramagnetic resonance (Drescher, 2012), circular dichroism, triplet–triplet energy transfer, small-angle X-ray scattering, small-angle neutron scattering, fluorescence correlation spectroscopy (Szasz et al., 2011), single-molecule spectroscopy (Mukhopadhyay, Krishnan, Lemke, Lindquist, & Deniz, 2007), and time-resolved fluorescence (Jain et al., 2016; Pierce, Toptygin, & Wendland, 2013) in order to capture the structural ensembles sampled by the IDPs/IDRs. A combination of these ensemble and single-molecule methodologies has provided important insights into the structural and dynamical aspects of IDPs.

Fluorescence spectroscopy is a sensitive and multiparametric technique that can provide insights into the structural and dynamical aspects of biomolecules (Bhattacharya & Mukhopadhyay, 2016; Jain & Mukhopadhyay, 2015; Lakowicz, 2006). It covers a wide range of timescales from femtoseconds to seconds depending on the measurement format. The commonly used steady-state format allows one to record parameters such as fluorescence intensity and quantum yield, spectral shifts, quenching, and anisotropy of either intrinsic (Trp/Tyr) or extrinsic fluorophores in proteins (Jameson & Ross, 2010). These measurements are important in gaining preliminary insights into the local environment, solvent accessibility, and flexibility. Fluorescence (or Förster) resonance energy transfer (FRET) between two fluorophores covalently attached with the protein can report the average intermolecular distance that is an indicator of the extent of compaction/expansion of the polypeptide chains. A more advanced time-resolved format allows one to estimate fluorescence lifetimes, distance distributions, dynamics, and rotational correlation times. The technique of time-resolved fluorescence anisotropy decay has been widely used to probe the dynamics of different dyes, small molecules, peptides, and proteins in solution (Alexander Ross, Rousslang, & Brand, 1981; Christensen, Drake, & Phillips, 1986; Dutt, Doraiswamy, Periasamy, & Venkataraman, 1990; Lakowicz, Cherek, Maliwal, & Gratton, 1985; Prazeres, Fedorov, Barbosa, Martinho, & Berberan-Santos, 2008; Rich et al., 1993; Swaminathan, Periasamy, Udgaonkar, & Krishnamoorthy, 1994; Thevenin, Periasamy, Shohet, & Verkman, 1994). This technique has also been extensively used to study a broad spectrum of biomolecular interactions, including small-molecule protein (Heller, Aprile, & Vendruscolo, 2017; Mukherjee et al., 2013;

Scholtbach, Venegas, Bohne, & Fuentealba, 2015), protein–protein (De Beuckeleer, Volckaert, & Engelborghs, 1999; Paila, Kombrabail, Krishnamoorthy, & Chattopadhyay, 2011; Sharma et al., 2004), DNA–protein (Deprez et al., 2001; Millar, 2000), peptide–DNA (Liu, Liang, Liu, & Zu, 2014), small-molecule membrane (Ikeda, Lee, Yamaguchi, & Tazuke, 1990; Krishna, Das, Periasamy, & Nityananda, 2000; Krishna, Srivastava, & Periasamy, 2001; Maiti, Krishna, Britto, & Periasamy, 1997; Sarkar, Datta, Das, & Bhattacharyya, 1996), and protein–membrane interactions (Peng, Visser, van Hoek, Wolfs, & Hemminga, 1990). Analysis of time-resolved fluorescence depolarization kinetics has also provided detailed structural insights into the fundamental processes of protein folding (Canet, Doering, Dobson, & Dupont, 2001; Kim, Arjara, Richards, Gray, & Winkler, 2006; Saxena, Krishnamoorthy, Udgaonkar, & Periasamy, 2007; Weinreis, Ellis, & Cavagnero, 2010) and amyloid aggregation (Jha, Udgaonkar, & Krishnamoorthy, 2009; Mukhopadhyay, Nayak, Udgaonkar, & Krishnamoorthy, 2006; Sahay, Anoop, Krishnamoorthy, & Maji, 2014). Fluorescence depolarization kinetics measured using time-resolved fluorescence anisotropy decay measurements have the ability to discern different modes of protein dynamics within the excited state lifetime of the fluorophore involving the local-, segmental-, and global conformational dynamics. Therefore, time-resolved fluorescence anisotropy is an exceptionally robust tool to directly observe the molecular motions of the IDPs both in the free states (Jain et al., 2016) and in the assembled/bound states (Arya et al., 2016; Sahay et al., 2014). In this chapter, we will discuss the application of fluorescence depolarization kinetics methodology in probing the conformational plasticity, binding, and assembly of a variety of IDPs.

2. FLUORESCENCE DEPOLARIZATION KINETICS: TIME-RESOLVED FLUORESCENCE ANISOTROPY DECAY MEASUREMENTS

2.1 Fluorescence Polarization/Anisotropy

When a fluorophore is excited with linearly polarized light, the resultant fluorescence emission is also polarized and the extent of this polarization of emission is defined in terms of fluorescence anisotropy (*r*) (Eq. 1)

$$r_{ss} = \frac{I_{\parallel} - I_{\perp}}{I_{\parallel} + 2GI_{\perp}} \tag{1}$$

where r_{ss} represents the steady-state anisotropy, $I_{\|}$ and $I_{\perp}$ are the parallel and perpendicular polarized fluorescence emission intensities, respectively, and the *G*-factor is the ratio of the sensitivities of the detection system for horizontally and vertically polarized light.

The fluorescence anisotropy originates due to the existence of transition moments for both absorption and emission and involves the photoselection in the excitation of randomly oriented fluorophores. Fluorophores having their absorption transition dipoles oriented parallel to the vector of the incident polarized excitation light are preferentially excited in a $\cos^2\theta$ dependence in which θ is the angle between the absorption dipole moment and the electric field vector of the incident photon. This photoselection creates an instantaneous population of excited fluorophores along the transition dipole that equals to 2/5 (0.4); in other words, on an average 40% (2 out of 5) fluorophores are excited (Fig. 1).

This sets the maximum anisotropy at a value of 0.4 for one-photon excitation. This is also known as the fundamental or intrinsic (time-zero) anisotropy (r_0). Assuming that all fluorophore molecules are associated with only a single absorption transition moment, the theoretical value of the fundamental anisotropy, r_0, is related to the angle between the absorption and emission transition dipole moments according to the following equation:

$$r_0 = \frac{1}{5}\left(3\cos^2\xi - 1\right) \tag{2}$$

where ξ is the angle between the absorption and emission transition moments, in the absence of any rotational motion (i.e., under frozen/high viscosity/low-temperature conditions). The probability that a fluorophore will be excited from its ground state is proportional to $\cos^x\theta$, where $x = 0$ for excitation with unpolarized light, 2 for one-photon excitation, 4 for two-photon excitation, and 6 for three-photon excitation. The photoselection using multiphoton photon excitation creates a much narrower probability distribution, 4/7 and 2/3 for two-photon and three-photon excitation (Callis, 1997; Denk, Strickler, & Webb, 1990; Vogel, Thaler, Blank, & Koushik, 2009), respectively (Fig. 1A).

The lower probability leads to a much higher r_0 for multiphoton fluorescence (0.571 for two-photon and 0.667 for three-photon). The angular displacement of the transition dipole due to the rotational motion of the fluorophore during its excited state lifetime gives rise to the observed steady-state anisotropy that is determined using the following relationship given by Perrin's equation (Lakowicz, 2006) (Eq. 3)

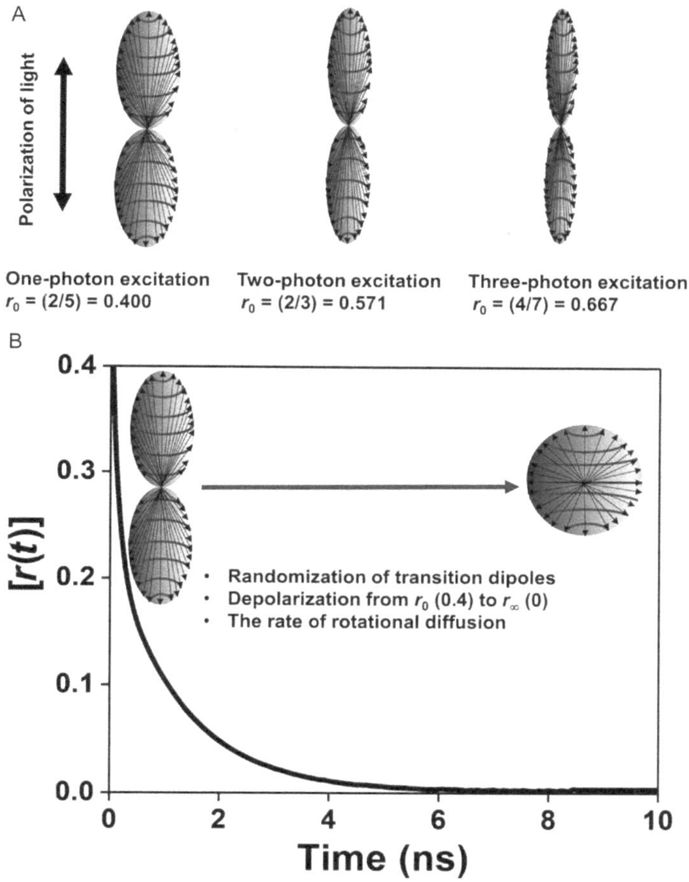

Fig. 1 (A) A schematic showing the photoselection indicating the orientational distribution of transition dipoles and the intrinsic fluorescence anisotropies (r_0) upon excitation with linearly polarized light for one-photon, two-photon, and three-photon excitations. (B) A schematic of a single-exponential anisotropy decay depicting a complete depolarization due to the randomization of transition dipoles upon rotational relaxation.

$$r_{ss} = \frac{r_0}{1 + \frac{\tau}{\theta}} \tag{3}$$

where r_{ss} is the steady-state emission anisotropy, r_0 is the intrinsic anisotropy, τ is the fluorescence lifetime, and θ is the rotational correlation time.

The steady-state anisotropy values provide information on the flexibility of a fluorophore in its immediate microenvironment. The anisotropy value of intrinsic Trp or fluorescently labeled IDPs will be low (near zero) due to fast rotational dynamics originating from the high conformational flexibility of IDPs. The chain compaction, binding-induced folding, and aggregation of IDPs will lead to an increase in the anisotropy. The steady-state anisotropy measurements provide preliminary information and represent only a time-averaged value of anisotropy and, therefore, lack the ability to discern the different modes of dynamics occurring on the picoseconds-to-nanoseconds (ps–ns) timescale (Lakowicz, 2006; Valeur, 2001). Fluorescence depolarization kinetics using the time-resolved anisotropy decay measurements allow us to dissect the different dynamic events occurring during the excited state lifetime of the fluorophore.

The time-dependent fluorescence anisotropy [$r(t)$] or the depolarization profile of a fluorophore at a particular emission wavelength is monitored by exciting the fluorophore with a vertically polarized laser pulse and recording the fluorescence emission through vertical (I_{VV} or $I_{||}$) and horizontal (I_{VH} or $I_{\perp}$) emission polarizers. This is followed by the global fitting and analysis of the decay profiles to recover the rotational correlation times and associated amplitudes (as described in Sections 2.3.1 and 2.3.2) (Cross & Fleming, 1984). For a freely rotating fluorophore, the depolarization of fluorescence can be expressed in terms of the following equation:

$$r(t) = r_0 e^{-\left(\frac{t}{\phi}\right)} \tag{4}$$

where r_0 is the intrinsic (time-zero) anisotropy, generally 0.4, ϕ is the rotational correlation time of the fluorophore which is also related to its hydrodynamic volume, V, by the Stokes–Einstein relationship:

$$\phi = \frac{\eta V}{k_B T} = \frac{\frac{4}{3}\pi R_h^3 \eta}{k_B T} \tag{5}$$

where η is the viscosity of the medium, k_B is the Boltzmann's constant, R_h is the hydrodynamic radius, and T is the absolute temperature.

Fig. 1B depicts a complete depolarization in the form of a single-exponential decay from the preferentially excited photoselected population to a random orientation upon tumbling of the fluorophore in solution. Free fluorophores in solution undergo a rapid tumbling that results in fast depolarization from r_0 to zero. On the other hand, when a fluorophore is attached to a (larger) macromolecule, such as a protein, the observed anisotropy exhibits a significantly slower depolarization rate. Additionally, a fluorophore attached to a macromolecule exhibits at least a biexponential decay function comprising the (fast) local "wobbling-in-cone" motion of the fluorophore and a (slow) global motion of the macromolecule (Fig. 2A) (Kinosita, Ikegami, & Kawato, 1982). Such a time-dependent anisotropy decay profile can be expressed in terms of a typical biexponential function (Eq. 6)

$$r(t) = r_0 \left[\beta_{\text{fast}} e^{\left(\frac{-t}{\phi_{\text{fast}}}\right)} + \beta_{\text{slow}} e^{\left(\frac{-t}{\phi_{\text{slow}}}\right)} \right] \tag{6}$$

where ϕ_{fast} represents the rotational correlation time corresponding to the local motion of the fluorophore attached to the protein, ϕ_{slow} is the corresponding correlation time associated with the global motion of the entire protein, and β_{fast} and β_{slow} represent the amplitudes of the local and global motions, respectively. A representative fluorescence anisotropy decay profile of a fluorophore, when attached to a protein, is shown in Fig. 2B.

More generally, the time-dependent fluorescence anisotropy can be written as follows:

$$r(t) = r_0 \sum_{0}^{i} \beta_i e^{\left(-t/\phi_i\right)} \tag{7}$$

where ϕ_i are the rotational correlation times, and β_i are the associated amplitudes contributing toward the anisotropy decay.

In Section 2.2, we will focus on the description of the time-correlated single-photon counting (TCSPC) setup and highlight the procedure for recording time-resolved fluorescence intensity decay curves.

2.2 Time-Correlated Single-Photon Counting

Time-resolved fluorescence measurements are generally carried out using a TCSPC setup. The sample is excited with a train of light pulses using high-repetition-rate lasers (femtoseconds/picoseconds) or laser diodes (picoseconds/nanoseconds). The emitted photons are detected at 90 degrees to the excitation beam using a microchannel plate photomultiplier tube and are sorted into different channels or time bins on the basis of the time difference between the sample irradiation and the photon emission. The detection rate for photons in the TCSPC setup is limited to around 1 photon per 100 excitation pulses (1%) such that the probability of detecting more than two photons per excitation pulse is negligible. In the forward mode of TCSPC, the

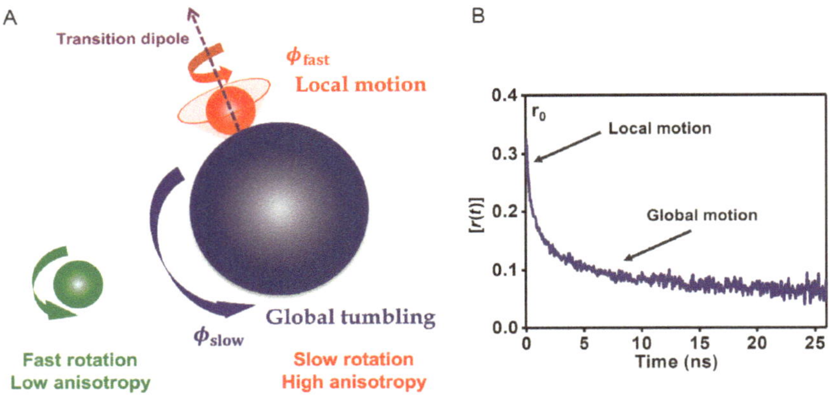

Fig. 2 (A) A schematic of the wobbling-in-cone model of the rotational motions of a fluorophore attached to a macromolecule. A free fluorophore undergoes a rapid tumbling and exhibits low anisotropy, whereas a covalently attached fluorophore undergoes two types of motions, namely, local/fast motion and global/slow tumbling. (B) A biexponential time-resolved anisotropy decay plot of a fluorescently labeled protein captures the fast (wobbling-in-cone) dynamics and the global rotational tumbling.

excitation pulse irradiates the sample and synchronously triggers the time-to-amplitude converter (TAC) which builds up a voltage ramp, linear with time, on the nanosecond timescale. Before reaching the TAC, the signal passes through a constant fraction discriminator (CFD) which accurately measures the arrival time of the start pulse. The second trigger pulse arises from the single detected photon and passes through another CFD before sending the electronic signal to TAC. Depending on the elapsed time between the start and stop electronic pulses, the TAC contains a voltage proportional to this time difference or time delay (Δt). This analog voltage is amplified using a programmable gain amplifier before being digitized to a numerical value by the analog-to-digital converter. These time differences are stored in a multichannel analyzer that allows one to construct the photon time distribution that is used to plot the time-resolved fluorescence decay profile.

Depending on the fluorophore under study, different pulsed-light sources including a tunable Ti:Sapphire laser, laser diodes, and light-emitting diodes are used. For a typical time-resolved fluorescence anisotropy decay measurement, the IDP under consideration is either site-specifically labeled with an extrinsic fluorophore or the anisotropy decay of an intrinsic fluorophore such as Trp is monitored. Before recording time-resolved anisotropy decays, it is necessary to record fluorescence intensity decay profiles of the sample as described in Section 2.2.1.

2.2.1 Time-Resolved Fluorescence Intensity Decay

Fluorescence lifetime is essentially the average length of time that a fluorophore remains in the electronic excited state following excitation before finally returning to the ground state. The excited state decays as a result of both radiative and nonradiative processes depending upon the environment around the excited state. The lifetime of this excited state is given by the following equation:

$$\tau = \frac{1}{k_{\mathrm{f}} + \sum k_{\mathrm{nr}}} \tag{8}$$

where k_{f} is the rate constant associated with fluorescence emission, and $\sum k_{\mathrm{nr}}$ is the summation of the rate constants associated with all other nonradiative processes. In the absence of nonradiative processes such as energy transfer, quenching, and intersystem crossing, the observable lifetime of the fluorophore approaches its natural lifetime. Since fluorescence emission is a random event, each fluorophore in the excited state has the same probability of emitting a photon in a given period of time. Therefore, in order to accurately measure the average time that a fluorophore spends in its excited state, it is necessary to record multiple fluorescence emission events. The recorded fluorescence intensity decay profile can essentially be visualized as a histogram made up of photon counts. The detected photons are distributed into different time bins according to the time that each of the respective fluorophores spends in its excited state, following excitation. For a single-exponential decay, the fluorescence lifetime is then determined from the slope of a plot of log $I(t)$ vs time. For a multiexponential model function (n decay times) the time-dependent intensity decay is expressed as the following:

$$I(t) = \sum_{1}^{n} \alpha_i \exp\left(-t/\tau_i\right) \quad n = 1, 2, \text{or } 3 \tag{9}$$

where α_i are the amplitudes of the different components associated with corresponding decay times, τ_i.

The first step toward recording a time-resolved fluorescence intensity decay data involves recording the instrument response function (IRF). The IRF (also known as prompt) may be defined as the response of the instrument to a zero-lifetime sample and is the shortest time profile that can be measured by the instrument. Since the IRF has a finite width, it causes a distortion of the measured decay curve as a result of which the measured function is essentially a convolution of the true fluorescence decay function with the IRF. To recover the true decay function of a fluorophore, the IRF has to be recorded.

The necessary steps required for recording the IRF are as follows:

2.2.1.1 Step 1: Recording Instrument Response Function

(i) An appropriate pulsed laser, laser-diode or light-emitting diode is used as the pulsed-light source, depending on the excitation maxima of the fluorophore under study.

(ii) A dilute solution (1%, v/v) of ludox (colloidal silica) in water is prepared to record the IRF of the instrument at that particular excitation wavelength.

(iii) The emission polarizer in the TCSPC instrument is set at a magic angle orientation of 54.7 degrees with respect to the direction of excitation polarization. The optimum photon counts to be recorded are generally set at 10,000 counts at the peak to obtain a good signal-to-noise ratio. A neutral density filter may be needed since there will be high-intensity Rayleigh scattering.

(iv) The emission monochromator is set at an identical wavelength as that of the excitation wavelength and the dilute solution of ludox is placed in the sample chamber. The alignment as well as the excitation and emission focus of the laser source is optimized to get optimal photon counts following which the IRF is recorded. If the scattering intensity is very high, an appropriate neutral density filter should be placed between the sample and the emission monochromator.

The necessary steps for recording a fluorescence intensity decay profile are described as follows:

2.2.1.2 Step 2: Recording Fluorescence Intensity Decay *I*(*t*) at a Magic Angle

(i) Keeping the alignment of the light source intact, the fluorescent sample is placed in the sample chamber and the emission monochromator is set at the desired emission wavelength. The intensity decays collected at a right angle to the incident excitation beam are then recorded at the magic angle polarizer condition ($\theta = 54.7$ degrees). The intensity decay for measuring fluorescence lifetime is always recorded with the emission polarizer at the magic angle orientation so as to remove any polarization artifact. The data analysis is described in Section 2.3.

2.2.2 *Time-Resolved Fluorescence Anisotropy or Fluorescence Depolarization Measurements*

Similar to the fluorescence intensity curves, the measured polarized intensity decays are also distorted due to convolutions with the IRF. The first step involves recording the IRF.

2.2.2.1 Step 1: Recording IRF for Fluorescence Depolarization Measurements

Using a 1% (v/v) ludox solution, the IRF is recorded similarly as described in Section 2.2.1, with the only difference being in the orientation of the emission polarizer, which is now at 0 degree (parallel) with respect to the excitation polarizer. It needs to be mentioned that the maximum photon counts fixed for recording the IRF must coincide with the maximum peak difference set for recording the polarized intensity decays. The parallel and perpendicular polarized fluorescence emission intensities upon excitation using a polarized light source are recorded as follows.

2.2.2.2 Step 2: Recording Fluorescence Intensity Decays $I_{II}(t)$ and $I_{\perp}(t)$

Both the excitation and emission polarizers in the TCSPC instrument are set at 0 degree orientations, and the maximum peak difference is generally set at 20,000 photon counts to get a better signal-to-noise ratio in the recorded decay curves. The slit width is kept at an appropriate bandpass. For time-resolved fluorescence anisotropy measurements, the polarized intensity decays $I_{II}(t)$ and $I_{\perp}(t)$ are collected with the excitation polarizer at 0 degree and emission polarizers at 0 and 90 degrees, respectively, with respect to the excitation polarizer. The excitation polarizer is fixed at an angle of 0 degree to ensure that the sample is irradiated with vertically polarized excitation light following which the $I_{II}(t)$ and $I_{\perp}(t)$ intensity decays are collected with alternate orientations of the emission polarizer at 0 and 90 degrees, with respect to the excitation polarizer, with identical accumulation times for the parallel and perpendicular polarized components. This is achieved by automated toggling of the emission polarizers between parallel and perpendicular orientations, every 60 s. We highly recommend toggling as opposed to a long acquisition of $I_{II}(t)$ and then switching to $I_{\perp}(t)$ in order to minimize the photobleaching artifact. At all points, care must be taken so that the ratio of stop to start pulses is always $<2\%$ for all possible combinations of the orientations of the excitation and emission polarizers. Neutral density filters should be used, if necessary.

The time-dependent fluorescence anisotropy $r(t)$ is expressed as:

$$r(t) = \frac{I_{||}(t) - GI_{\perp}(t)}{I_{||}(t) + 2GI_{\perp}(t)} \quad (10)$$

where $I_{II}(t)$ and $I_{\perp}(t)$ are the time-dependent parallel and perpendicularly polarized fluorescence emission intensities, respectively, and the *G*-factor is the ratio of the sensitivities of the detection system for horizontally and vertically polarized light (Lakowicz, 2006). The next step involves the determination of the geometry-factor or the *G*-factor which accounts for the differential sensitivity of the detector toward vertically and horizontally polarized light. The *G*-factor is estimated as follows.

2.2.2.3 Step 3: Estimation of the Geometry-Factor or the *G*-Factor

Two different methods for the *G*-factor estimation are described as follows (we recommend the use of both methods):

(i) The first method involves the measurement of two additional intensity decay components of the fluorescent sample, where the excitation polarizer remains fixed at a 90 degrees orientation, while the emission polarizer alternately shifts between 0 degree (I_{HV}) and 90 degrees (I_{HH}). The corresponding intensity decays are recorded in four cycles, each of 60 s duration. Then, the *G*-factor is estimated from the ratio of the obtained intensity decays ($\int I_{HV}(t)dt / \int I_{HH}(t)dt$) and used later for data analysis.

(ii) The second method involves tail matching of the vertically and horizontally polarized intensity decays [$I_{II}(t)$ and $I_{\perp}(t)$] (Cross & Fleming, 1984; Rachofsky & Laws, 2000; Rice & Kenney-Wallace, 1980). Tail matching is achieved by recording $I_{II}(t)$ and $I_{\perp}(t)$ intensity decays of a highly depolarized sample, such as the free dye in a low viscous solvent. The excitation/emission parameters (wavelength/bandpass, etc.) and the acquisition time for the estimation of the *G*-factor should be exactly the same as that of the samples under study. A free dye in the low viscous solvent will exhibit (fast) subnanosecond depolarization, therefore, the multiplication factor that is used to bring the tail of the anisotropy decay to zero by multiplying $I_{\perp}(t)$ is the inverse of the *G*-factor. A representative time-resolved anisotropy decay profile of NATA in water (Jain et al., 2016) is shown in Fig. 3A. After the *G*-factor correction is incorporated, the noise after the anisotropy has decayed to zero should be randomly distributed over the time axis.

Next, it is imperative to determine the intrinsic (time-zero) anisotropy (r_0).

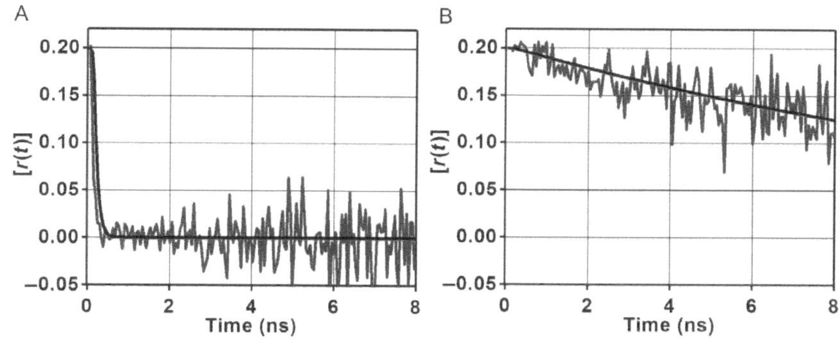

Fig. 3 (A) A representative time-resolved anisotropy decay curve of the Trp analog NATA in a nonviscous solvent (water) showing fast (single-exponential) depolarization kinetics that can be used to estimate the *G*-factor. (B) A representative time-resolved anisotropy decay curve of the Trp analog NATA in a viscous solution (80% glycerol–water) that allows an accurate estimating of r_0. *Adapted with permission from Jain, N., et al. (2016). Direct observation of the intrinsic backbone torsional mobility of disordered proteins.* Biophysical Journal, 111, *768–774.*

2.2.2.4 Step 4: Determining Intrinsic Anisotropy (r_0)

The r_0 is a fundamental property of a fluorophore at a given excitation wavelength. The maximum value of r_0 could be 0.4, but in a majority of cases, fluorophores exhibit a lower value. There are multiple reasons for the lower value of r_0. The reasons that contribute toward the lower values of r_0 could be due to the overlapping of two or more noncollinear, low-lying absorption bands (as seen in the case of Trp), the differences in geometries of the Frank–Condon and emissive states leading to noncollinear absorption and emission transition dipole moments, torsional vibrations of the emission dipoles around their equilibrium positions, and radiative or nonradiative energy transfer processes. In addition to these factors, instrumental artifacts also pose a major bottleneck in the accurate estimation of fluorescence emission anisotropy. Large cone angles associated with the incident/emitted beams or effects from polarizers, misaligned by even a few degrees, might contribute toward the inaccurate estimation of r_0 (Prazeres et al., 2008; Valeur, 2001). Therefore, we highly recommend estimating the r_0 of the fluorophore using the same excitation/emission parameters and the same TCSPC setup. In order to determine r_0, it is necessary to freeze the fast rotational motion that rapidly depolarizes the fluorescence. This is achieved by using a highly viscous medium such that the fluorophores become nearly immobilized.

A solution of the free fluorophore is made in 80% glycerol–water mixture. The $I_{II}(t)$ and $I_{\perp}(t)$ decays are collected as previously described in Section 2.2.2 (step 2), and a time-resolved anisotropy decay profile $[r(t)]$ is constructed and analyzed (for analysis, see Section 2.3). The anisotropy decay profile in a highly viscous medium shows much slower depolarization kinetics as compared to the decay obtained for fluorophores in a low viscous medium, allowing a better estimation of the r_0. A representative time-resolved anisotropy decay plot of the free tryptophan analog (NATA) in 80% glycerol–water mixture (Jain et al., 2016) is shown in Fig. 3B.

Section 2.3 deals with the analysis of time-resolved fluorescence anisotropy decay profiles to recover the rotational correlation times and associated amplitudes.

2.3 Time-Resolved Fluorescence Anisotropy Decay Data Analysis

There are different ways to analyze the anisotropy decay profiles (Cross & Fleming, 1984): (a) by fitting the $r(t)$ using single/multiexponential decay models; (b) by fitting the sum and the difference of $I_{II}(t)$ and $I_{\perp}(t)$; (c) by globally fitting $I_{II}(t)$ and $I_{\perp}(t)$. The global fitting of $I_{II}(t)$ and $I_{\perp}(t)$ involves the simultaneous fitting of the measured intensity decays $I_{II}(t)$ and $I_{\perp}(t)$ that yields the best results (Ameloot, vandeVen, Acuña, & Valeur, 2013; Cross & Fleming, 1984; Feinstein et al., 2003; Kapusta, Erdmann, Ortmann, & Wahl, 2003), and therefore we will discuss only this methodology in the following steps.

2.3.1 Step 1: Fitting the Magic Angle Intensity Decay I(t)

The fluorescence intensity decay $I(t)$ is recorded at a magic angle orientation with respect to the excitation. Initially, a model decay function is assumed for the fluorescence intensity decay using Eq. (9). This function is then convoluted with the IRF and used to fit the experimentally recorded intensity decay (Cross & Fleming, 1984; Lakowicz, 2006). For fitting the decay profiles, the nonlinear least squares fitting algorithm (Wahl, 1979), with iterative reconvolution, is used to recover the fluorescence lifetime parameters. The goodness of the fit is judged on the basis of reduced χ^2 values (χ^2_R) (should be close to 1) and the randomness in the plots of the weighted residuals against the time axis. The χ^2_R value is given by the following relation:

$$\chi_R^2 = \frac{\chi^2}{N} = \sum_{k=1}^{N} \left[\frac{N_k^m - N_k^c}{\sigma_k} \right]^2 \quad (11)$$

where k denotes the total number of channels or data points, N_k^m, N_k^c are the photon counts measured and calculated, respectively, and σ_k is the standard deviation in the counts in the kth channel, given by $\sigma_k = \sqrt{N_K}$ (number of counts in kth channel). The fluorescence lifetime parameters τ_i and α_i recovered from the best fits are further used in the next step.

2.3.2 Step 2: Global Fitting of $I_{//}(t)$ and $I_{\perp}(t)$ and Constructing r(t)

The global fitting of $I_{II}(t)$ and $I_{\perp}(t)$ involves simultaneous analysis of related experimental data by linking common parameters to different fitting functions. The robustness of this fitting lies in the fact that all the parameters associated with the intensity and anisotropy decays are simultaneously optimized in a single-step analysis.

The IRF, $I_{II}(t)$ and $I_{\perp}(t)$, are given as inputs for the analysis. Using the determined G-factor (step 3 of Section 2.2.2), the perpendicular decay component $I_{\perp}(t)$ is corrected. The next step in this analysis is to assume a decay function for $r(t)$ according to Eq. (7). Depending on whether the assumed decay function is single or multiexponential, initial guess values for the rotational correlation times, ϕ_i, the corresponding amplitudes, β_i, and the corresponding error windows associated with each of the parameters are used in the analysis. The error windows for these parameters are initially kept open and after successive iterations, the error windows are narrowed down depending on the relevance of the model. The values of the fluorescence lifetimes (τ_i) and the associated amplitudes (α_i), recovered from the magic angle intensity decay analysis, are provided to the analysis software as the initial guess values, with narrow error windows. The value of r_0 (determined in Section 2.2.2, step 4) is given as the initial guess value for r_0 with a small error window of about 5%. Now, the IRF recorded for fluorescence anisotropy decay measurements (Section 2.2.2, step 1) remains convoluted with the measured $I_{II}(t)$ and $I_{\perp}(t)$ polarized intensity decays (Section 2.2.2, step 2). Using Eqs. (12) and (13), the assumed anisotropy decay and the magic angle intensity decay functions generate nondistorted polarized intensity decay functions $I_{II}(t)$ and $I_{\perp}(t)$ (Cross & Fleming, 1984). The nondistorted decay functions are convoluted with the IRF and then, compared with the experimentally recorded $I_{||}(t)$ and $I_{\perp}(t)$ decays to get the best fit.

$$I_{\parallel}(t) = {}^{1}/_{3} I(t)[1 + 2r(t)] \tag{12}$$

$$I_{\perp}(t) = {}^{1}/_{3} I(t)[1 - r(t)] \tag{13}$$

The calculated fit of the raw anisotropy decay profile, determined from the fits to $I_{\parallel}(t)$ and $I_{\perp}(t)$, is overlaid with the raw anisotropy decay profile, calculated from the corresponding $I_{\parallel}(t)$ and $I_{\perp}(t)$ curves.

Generally, while fluorophores, assumed to be spherical, in solution display a single-exponential decay, a fluorophore attached to a protein displays two rotational correlation times corresponding to the local mobility of the fluorophore and the overall rotation of the protein. Therefore, a prior knowledge of the system under study is necessary before assuming a physically relevant model decay equation. The appropriateness of the chosen form of $r(t)$ is then assessed by the goodness of the fit on the basis of reduced χ^2 values (χ^2_{R}), the randomness in the plots of the weighted residuals as a function of time, and the autocorrelation analysis (Fig. 4).

To check for the robustness of the recovered rotational correlation times, one can fix one of the rotational correlation times (say ϕ_1) at different values and vary the other rotational correlation time (ϕ_2). A plot of the goodness of the fit χ^2_{R} vs ϕ_{slow} values exhibits a minimum $(\chi^2_{\mathrm{R}})_{\mathrm{min}}$. Such a representative plot is shown in Fig. 4H. A complete summary of the strategy for recording time-resolved fluorescence anisotropy decay is shown as a schematic in Fig. 5.

An accurate estimation of a rotational correlation time is limited by the lifetime of the fluorophore, used for the experiments. Generally, a rotational correlation time component may be resolved only if it falls within the range of ~0.1 to ~10 times the lifetime of a fluorophore (Wahl, 1979). A general analytical method of constrained analysis can be used to extend the range of rotational correlation times recoverable from a particular value of fluorescence lifetime. In this method (Feinstein et al., 2003; Rachofsky, Wolf, Bialik, Ross, & Laws, 1999), the experimentally recorded steady-state anisotropy (r_{ss}) is introduced as a constrained parameter in the global analysis algorithm. The calculated value of r_{ss}, recovered from the time-resolved fluorescence anisotropy measurements, is compared with the experimentally obtained value of steady-state emission anisotropy, to assess the quality of the fit of the time-resolved anisotropy decays.

In the next section, we will describe a few case studies from our laboratory. These studies include the applications of fluorescence depolarization kinetics measurements to investigate the structural and dynamical behavior of a number of amyloidogenic IDPs.

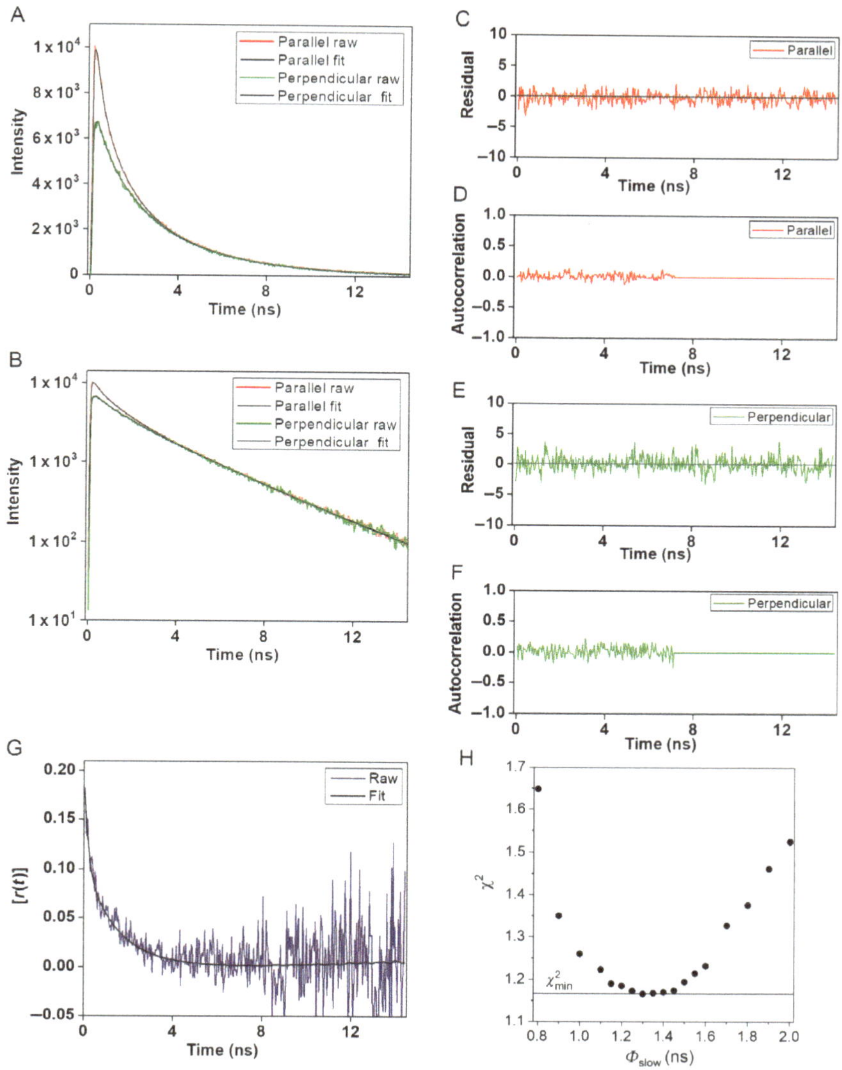

Fig. 4 The global analysis of a representative time-resolved fluorescence anisotropy decay data for the Trp-39 variant of α-synuclein. (A) The raw parallel [$I_{||}(t)$] and perpendicular [$I_{\perp}(t)$] polarized intensity decays are shown in *red* and *green*, respectively, and the *black lines* are the corresponding global fit using Eqs. (12) and (13). (B) The decay plots and fits are shown in logarithmic intensity scale. Panels (C) and (D) represent the residual plot and the autocorrelation function for $I_{||}(t)$, respectively. Panels (E) and (F) represent the residual plot and the autocorrelation function for $I_{\perp}(t)$, respectively. (G) The raw anisotropy decay plot and the fit obtained from the global analysis shown in *blue* and in *black*, respectively. (H) A representative plot of the rotational correlation time ϕ *vs* the goodness of fit (χ^2_R) obtained from forced-fits. This plot is used to assess the robustness of the recovered rotational correlation times. *Adapted with permission from Jain, N., et al. (2016) Direct observation of the intrinsic backbone torsional mobility of disordered proteins.* Biophysical Journal, 111, *768–774.*

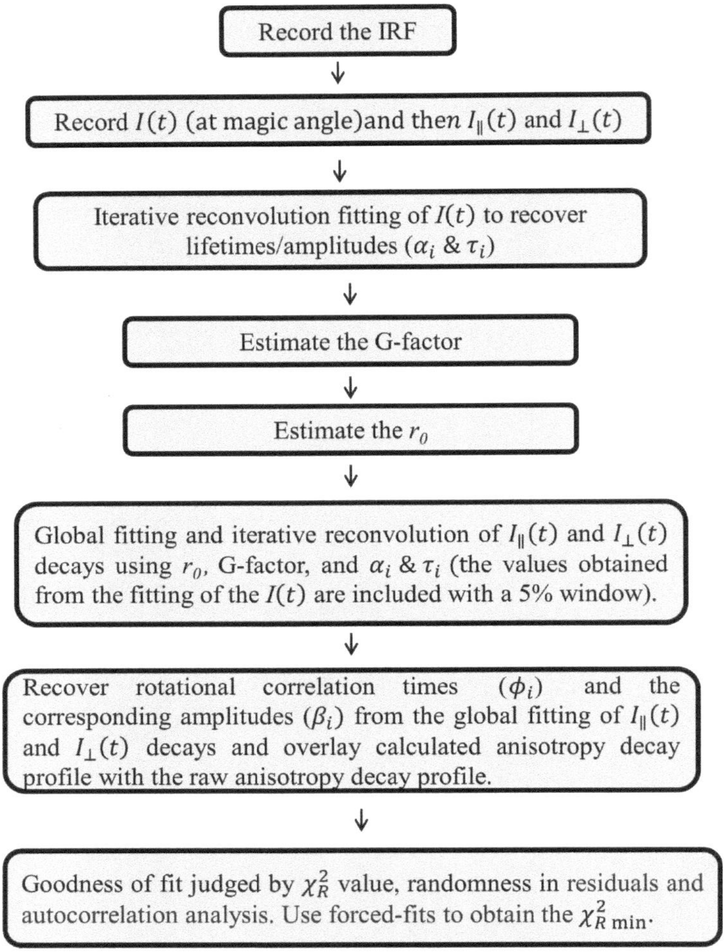

Fig. 5 The steps involved in acquisition and analysis of time-resolved fluorescence anisotropy decay data.

3. APPLICATION OF TIME-RESOLVED FLUORESCENCE ANISOTROPY MEASUREMENTS TO PROBE THE CONFORMATIONAL PREFERENCE, DYNAMICS, AND BINDING IN IDPs

3.1 Characterization of the Conformational Preference of IDPs: Collapsed or Extended

IDPs are known to adopt various conformations including a premolten globule, a structured molten globule, an extended random coil, and a transiently structured conformation (Dunker et al., 2001). In this section, we highlight the utility of this robust technique of time-resolved fluorescence anisotropy decay in characterizing the conformational preference of IDPs. In this section, we focus on two archetypal IDPs, κ-casein and α-synuclein (α-syn), and demonstrate how signature time-resolved anisotropy decay profiles have provided insights into the conformational preference of these IDPs.

κ-Casein is an 18.9 kDa, 169-residue protein predominantly expressed in bovine mammary glands. The whole length of the polypeptide chain of κ-casein is divided into two distinct domains: a positively charged N-terminal segment (1 − 100) that is enriched in glutamine/arginine residues and a negatively charged C-terminal domain (101–169) (Arya & Mukhopadhyay, 2014; Jain, Bhattacharya, & Mukhopadhyay, 2011). A representative diagram of the collapsed κ-casein bearing a Trp buried in the core is shown in Fig. 6A. α-Syn, on the other hand, is a 14.5-kDa, 140-residue protein abundantly expressed in the brain and presynaptic terminals of dopaminergic neurons. The polypeptide chain of α-syn consists of three domains: the positively charged N-terminal domain (1–60), the central hydrophobic NAC (non-Abeta component) domain, and the negatively charged C-terminal domain. Fig. 6B shows a representative diagram of the expanded α-syn polypeptide chain bearing a single Trp exposed to the solvent environment.

Initial picosecond time-resolved anisotropy decay measurements were made on native monomerized κ-casein containing a single Trp (Trp-76) (Arya & Mukhopadhyay, 2014). The analysis revealed a (fast) sub-ns rotational correlation time ($\phi_{fast} \sim 0.2$ ns) and a slow component ($\phi_{slow} \sim 7$ ns). Moreover, the hydrodynamic radius of refolded κ-casein estimated (~1.9 nm) from the value of ϕ_{slow} corroborated the hydrodynamic radius obtained from the dynamic light scattering measurements. This demonstrated the utility of time-resolved anisotropy decay measurements in clearly identifying the preference of native monomeric κ-casein to adopt a collapsed globule conformation. Now, in order to demonstrate the differences in the fluorescence depolarization kinetics of a collapsed IDP (κ-casein) and an expanded IDP (α-syn), we overlay the decays (Fig. 6C). The global analysis of the time-resolved anisotropy decay profile of a single-Trp variant of α-syn revealed a sub-ns component ($\phi_{fast} \sim 0.2$ ns), representing the local mobility of the Trp residue, which was similar to that observed in case of κ-casein. This pointed toward similar motional dynamics in the immediate microenvironments of the Trp residues in both the IDPs. On the other hand, the longer nanosecond rotational correlation time component, ϕ_{slow}, estimated to be ~1.4 ns for α-syn, was much faster than its global

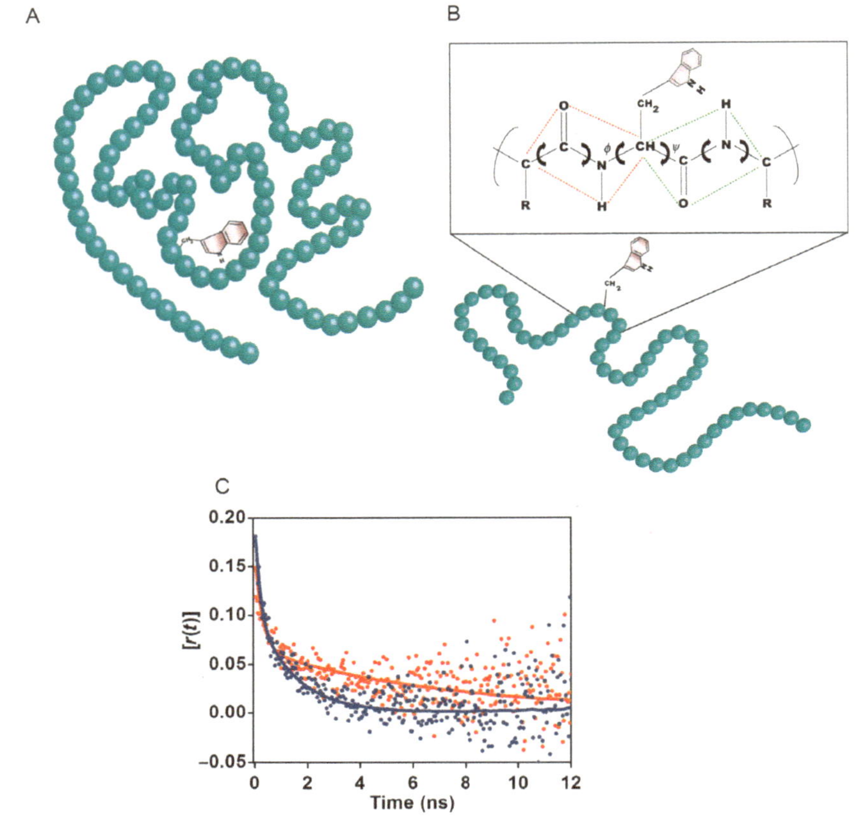

Fig. 6 (A) A model collapsed IDP, κ-casein, showing Trp residue in the core. The slower component of the anisotropy decay for such a collapsed globule would represent a global tumbling. (B) An expanded IDP, α-syn, having a single Trp. The slower component of the anisotropy decay for such an expanded coil would represent the intrinsic backbone torsional mobility in the ϕ–ψ Ramachandran dihedral space. (C) Representative time-resolved anisotropy decay curves and fits for a collapsed globule, κ-casein (shown in *red*) and for an expanded coil, α-syn (shown in *blue*). The fast decay components in both the curves represent the (fast) local motion of Trp, whereas the slow decay component represents the global motion in the case of κ-casein and the backbone dihedral dynamics in the case of α-syn. *Adapted with permission from Jain, N., et al. (2016) Direct observation of the intrinsic backbone torsional mobility of disordered proteins.* Biophysical Journal, 111, *768–774.*

rotational correlation time. This clearly suggested that this nanosecond rotational correlation time cannot represent the global rotational motion of the polypeptide backbone. This nanosecond component for α-syn was attributed to the intrinsic backbone segmental mobility that depolarizes the fluorescence anisotropy much more rapidly than the global rotational motion of the protein. This study reveals that the signature time-resolved anisotropy decay profiles can tell us about the conformational preferences of IDPs.

The slow rotational correlation time recovered in the case of denatured κ-casein ($\phi_{slow} \sim 2.7$ ns) could also correspond to backbone segmental mobility, even though the value is nearly twice as much slower than the ϕ_{slow} observed for α-syn. Therefore, it becomes imperative to understand that there are several additional factors that hinder the accurate estimation of rotational correlation time corresponding to the segmental dynamics in the denaturant-induced expanded state of any protein. Denaturant molecules are already known to bind to proteins (Möglich, Krieger, & Kiefhaber, 2005) and result in an alteration of the polypeptide chain dynamics, an enhancement in the apparent dimensions of the polypeptide chains, as well as in the modification of the bulk- and immediate-solvent viscosities around the polypeptide chain. These effects might lead to an overestimation of the measured timescale of the segmental mobility in the case of denaturant-induced extended protein conformations.

On the contrary, α-syn, being largely devoid of any stable local intrachain interactions, serves as an excellent candidate for capturing short-range backbone segmental mobility, under native conditions. In order to extract the site-specific information from the fluorescence depolarization kinetic measurements, we took advantage of the nonexistence of Trp in the polypeptide sequence of α-syn and created 10 single-Trp variants along the whole length of the polypeptide chain (residues positions 4, 27, 39, 56, 69, 78, 90, 124, and 140) (Jain et al., 2016). The time-resolved anisotropy decay curves for the three representative single-Trp variants of α-syn are shown in Fig. 7. The global fitting of the parallel and perpendicular polarized fluorescence intensity decays for the various single-Trp α-syn constructs revealed two well-resolved rotational correlation times. As mentioned previously, the sub-ns component ($\phi_{fast} \sim 0.2$ ns) represented the local mobility of the Trp and the slower nanosecond component ($\phi_{slow} \sim 1.4$ ns) was postulated to arise due to the intrinsic backbone segmental mobility of the disordered polypeptide chain of α-syn. In order to further validate this hypothesis, we performed time-resolved fluorescence anisotropy decay measurements using a preproline, Trp-107 variant of α-syn. Proline imparts restriction on the ψ dihedral angle rotation of the preceding carbon atom (Schimmel & Flory, 1968). Invariably, the Trp-107 variant exhibited a ϕ_{slow} of ~ 1.8 ns, which pointed toward a slower torsional mobility due to the limited accessibility in the Ramachandran dihedral space, as compared to all the other residue positions on α-syn (Fig. 7C).

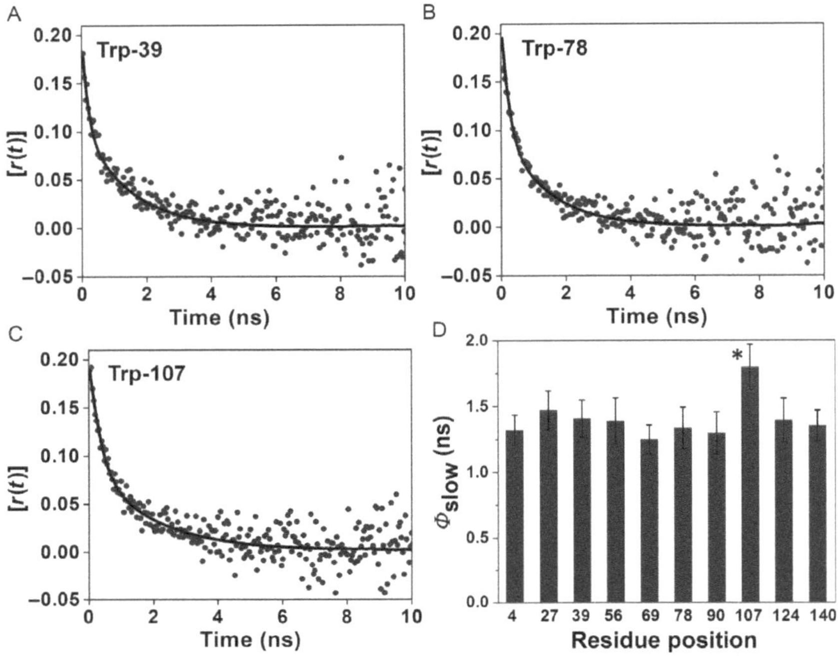

Fig. 7 Representative time-resolved anisotropy decay curves for (A) Trp-39, (B) Trp-78, and (C) Trp-107 variants of α-syn. (D) A plot of the slower rotational correlation time (ϕ_{slow}) vs residue position. The preproline residue Trp-107 (asterisked) exhibits a higher value of slow correlation time due to the rigidity imparted by a proline residue at position 108 of α-syn. *Adapted with permission from Jain, N., et al. (2016) Direct observation of the intrinsic backbone torsional mobility of disordered proteins.* Biophysical Journal, 111, *768–774.*

The constancy in the characteristic timescale of backbone torsional mobility (~1.4 ns), irrespective of the residue position, except the preproline residue, clearly indicated that ϕ_{slow} is independent of the overall tumbling of the protein (Jain et al., 2016). Interestingly, the nanosecond rotational correlation time component (~1.4 ns) also coincided well with the estimated reorientation timescale of a 14-residue statistical (Kuhn's) segment of α-syn (Parigi et al., 2014). This powerful fluorescence depolarization technique can be used to construct the backbone dynamical map of various IDPs.

Apart from its ability to discern the various conformations adopted by the IDPs, in their denatured and native forms, the fluorescence depolarization kinetics technique can also be useful in probing the transition of an IDP from its disordered monomeric form to its amyloid fibril form, as will be discussed subsequently in Section 3.2.

3.2 Investigation of the Disorder-to-Order Amyloid Transition

After elaborating on the utility of time-resolved fluorescence anisotropy measurements in discerning the conformational preference of two amyloidogenic IDPs, κ-casein and α-syn, we utilized the same technique to probe the disorder-to-order transitions of κ-casein from its disordered collapsed globule conformation to its β-sheet-rich amyloid fibril state (Arya et al., 2016). For our picosecond time-resolved fluorescence measurements, κ-casein was labeled with a small, solvent-sensitive dye, acrylodan. Our time-resolved fluorescence anisotropy decay measurements revealed the differences in the rotational dynamics of the fluorophore in both the monomeric and amyloid forms. The fluorescence depolarization profiles recorded for monomeric κ-casein could be fitted to a typical biexponential decay equation, resulting in a $\phi_{fast}(\sim 2$ ns) corresponding to the local mobility of the fluorophore and a $\phi_{slow}(\sim 8$ ns) corresponding to the global tumbling of the collapsed globular κ-casein, corroborating our Trp anisotropy decay results (Jain et al., 2011). For (ordered) β-rich amyloid fibrils of κ-casein, we observed significant dampening in both the local ($\phi_{fast} \sim 1$ ns) and global motions ($\phi_{slow} \geq 20$ ns). The global rotational motion of such a higher order assembly might be much slower on the fluorescence timescale. The representative fluorescence anisotropy decay profiles for monomeric κ-casein and its β-sheet-rich amyloid state are shown in Fig. 8. Time-resolved fluorescence depolarization measurements neatly captured the transition of a disordered, monomeric IDP to its ordered, amyloid form.

As an extension of this concept, we next discuss how time-resolved fluorescence anisotropy measurements have revealed the differences in the structural attributes of two amyloid folds that are associated with two distinct yeast prion strains of Sup35 (Narang, Swasthi, Mahapatra, & Mukhopadhyay, 2017). Yeast prions act as protein-only elements of inheritance in yeast (Serio & Lindquist, 2000; Shorter & Lindquist, 2005), and their prion domains give rise to diverse new beneficial phenotypes. The translation termination factor in *Saccharomyces cerevisiae*, known as Sup35, is a fungal prion protein which converts into a self-perpetuating prion state from its soluble functional nonprion form. The Sup35 protein has three distinct domains: the N-terminal domain $(1-123)$, a middle (M) domain (124–253), and a C-terminal domain (254–685). While the N-terminal forms the main amyloid core in the prion form [PSI^+] of Sup35, the highly charged M domain confers the solubility of the nonprion form [PSI^-] of Sup35. The C-terminal domain modulates the translational termination activity of Sup35. The NM domain of Sup35 (Sup35NM) exists as an IDP

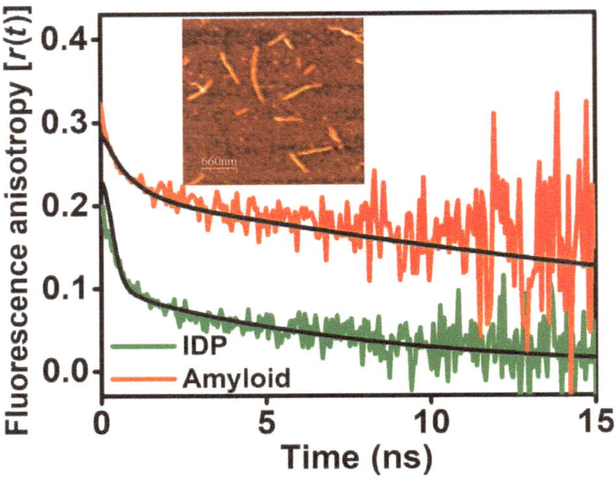

Fig. 8 The time-resolved fluorescence anisotropy decay plots with fits of acrylodan-labeled κ-casein, in the collapsed premolten globule state (*olive*) and in the ordered amyloid state (*red*). The slower component of the rotational correlation time is much longer due to the formation of highly ordered aggregates that exhibit much slower rotational diffusion on the fluorescence timescale. The *inset* shows an atomic force microscopy image of amyloid fibrils. *Adapted with permission from American Chemical Society Arya, S., et al. (2016). Water rearrangements upon disorder-to-order amyloid transition.* The Journal of Physical Chemistry Letters, *7(20), 4105–4110.*

in its monomeric state and aggregates into a self-propagating amyloid-like prion state. It was previously reported that two distinct forms of amyloid fibrils, known as Sc4 and Sc37, can be generated in vitro from Sup35NM of *S. cerevisiae* at two temperatures, namely, 4°C and 37°C, respectively (Marcelino-Cruz, Bhattacharya, Anselmo, & Tessier, 2011; Tanaka, Chien, Naber, Cooke, & Weissman, 2004). To discern the differences in the structural and dynamical attributes associated with the Sc4 and Sc37 amyloid fibrils, we took advantage of the nonexistence of Trp in the Sup35NM domain and created 18 single-Trp variants. The fluorescence depolarization kinetics for all the Trp residues within the amyloid fibrils of Sc37 and Sc4 could be fitted using a model biexponential decay function, giving rise to two rotational correlation times, ϕ_{fast} and ϕ_{slow}. Representative time-resolved anisotropy decay profiles of single-Trp variants of Sup35, within Sc37 amyloid fibrils, are shown in Fig. 9A. None of the anisotropy decay profiles decayed to zero within the Trp fluorescence lifetime, but rather displayed a residual anisotropy (r_∞) due to the large size of the amyloid aggregates. The variation in the values of r_∞ for different

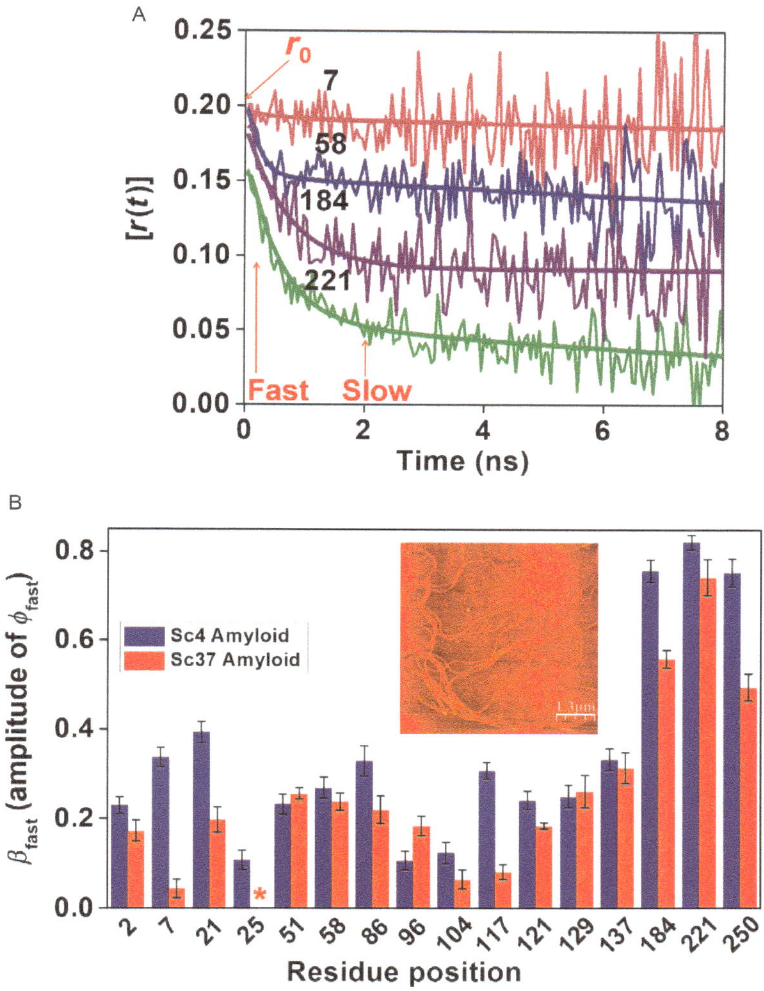

Fig. 9 (A) The time-resolved fluorescence anisotropy decay curves with fits of single Trp variants (7, 58, 184, and 321) of Sup35NM in the Sc37 amyloid state displaying a typical biexponential decay kinetics. These decays profiles show the different amplitudes of the local dynamics indicating varied conformational mobility in the amyloid state. (B) The conformational mobility map of two distinct types of amyloids, Sc4 (*blue*) and Sc37 (*red*), constructed from the recovered values of the amplitudes of the local dynamics (no amplitude of the local motion at residue 25 is indicated by an asterisk). These results suggest different local chain packing in the two types of amyloids responsible for the formation of two types of prion strains. *Adapted with permission from American Chemical Society Narang, D., Swasthi, H. M., Mahapatra, S., & Mukhopadhyay, S. (2017). Site-specific fluorescence depolarization kinetics distinguishes the amyloid folds responsible for distinct yeast prion strains.* The Journal of Physical Chemistry B, 121*(36), 8447–8453.*

residue positions was attributed to the differential extents of the fast depolarization. The amplitudes (β_{fast}) of the fast rotational correlation times reported on the residue-specific motional constraints within the different amyloid folds. Fig. 9B shows the residue-specific conformational mobility map containing the β_{fast} plotted as a function of the residue position for both the Sc37 and Sc4 fibrils. The fluorescence depolarization kinetics clearly revealed more order in the Sc37 fibrils as compared to the Sc4 fibrils (Narang et al., 2017). A distinct variation in the conformational dynamics was seen only at the N-terminal end and at the interface of the N- and M-domains. It was evident that residue positions 7, 21, and 25 at the N-terminal end and the residues 104, 117, and 121 at the N/M interface region exhibited lower values of β_{fast} in the Sc37 fibrils when compared to the Sc4 fibrils, which indicated higher order within the Sc37 amyloid fold. The time-resolved fluorescence anisotropy served as a useful tool to distinguish between the two distinct forms of amyloid fibrils of Sup35NM. The use of this technique could also be extended to distinguish the packing differences between the polymorphic amyloids of other IDPs.

3.3 Characterization of IDP-Membrane Interactions

The time-resolved anisotropy will be useful to study different types of dynamics of aggregates and lipid membrane-induced folded states of IDPs. To estimate the slow rotational correlation time associated with the global motion of large amyloid aggregates or protein–lipid complexes, the choice of the fluorophore should be such that it exhibits a long fluorescence lifetime. The accurate estimation of the slow rotational correlation time component is limited to values corresponding to ~10 times greater than the fluorescence lifetime value (Wahl, 1979). Using anisotropy decay measurements on the protein–lipid complexes, it might be possible to extract a variety of motional dynamics of proteins within the membrane. The long rotational correlation time in such systems could provide important insights into the translational diffusion of proteins on curved the membrane surface as described previously for dye–micelle complexes (Krishna et al., 2000).

3.4 Ultrafast Time-Resolved Fluorescence Anisotropy Decay Measurements

Ultrafast femtosecond time-resolved fluorescence anisotropy is often useful in the study of hydration dynamics. In order to probe the femtosecond hydration dynamics during the conformational conversion of κ-casein to

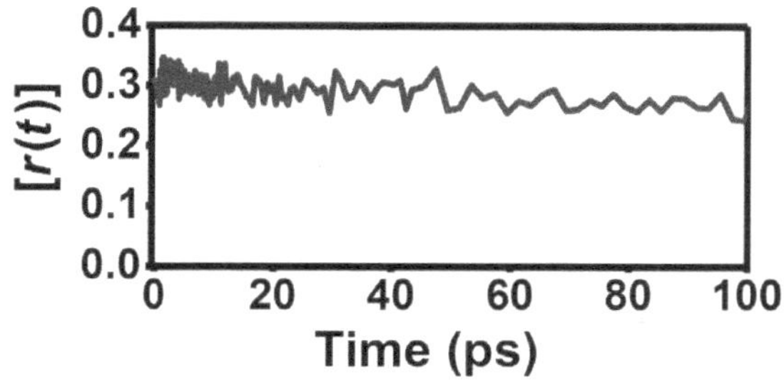

Fig. 10 A femtosecond time-resolved anisotropy decay curve of acrylodan labeled to κ-casein indicating negligible rotational dynamics on the picosecond timescale. *Adapted with permission from American Chemical Society Arya, S., et al. (2016). Water rearrangements upon disorder-to-order amyloid transition.* The Journal of Physical Chemistry Letters, *7(20), 4105–4110.*

the amyloid state, it was necessary to establish the absence of any chain motion occurring on the ultrafast timescale. In order to do so, we performed ultrafast femtosecond time-resolved fluorescence anisotropy decay measurements on κ-casein amyloid fibrils using the femtosecond upconversion technique. No significant rotational mobility of the fluorophore was observed on the sub-100 ps timescale, as shown in Fig. 10 (Arya et al., 2018, 2016). These studies confirm that fluorescence depolarization originating from the rotational motion is a much slower event compared to ultrafast solvation that relaxes the excited state (Lakowicz, 2006).

4. CONCLUSION AND FUTURE DIRECTIONS

In this chapter, we have discussed the principle and practice of fluorescence depolarization kinetics to study the conformational attributes of IDPs by monitoring a variety of dynamical events. We have also shown how the fluorescence depolarization measurements allowed us to directly capture the dynamics associated with the intrinsic torsional mobility of IDPs occurring on a timescale that is not easily accessible by other relaxation measurements. This methodology can be used in the future to discern different structural transitions including coil-to-globule and coil-to-amyloid transitions, in novel IDPs/IDRs using site-specific backbone segmental mobility dynamics as an important readout. This might allow us to quantify the contribution of internal friction from the dihedral dynamics. The fluorescence depolarization kinetics is also useful to probe the disorder-to-order amyloid transition. In addition, the site-specific fluorescence depolarization measurements have allowed us to dissect the differences in the local dynamical signatures in distinct amyloid strains that have altered chain packing. The use of this site-specific fluorescence depolarization technique can very well be extended to distinguish the differences between the polymorphic amyloids of various other IDPs. We also believe that this methodology will be useful to discover the dynamic events of lipid–protein complexes. Looking into the future, the time-resolved fluorescence anisotropy decay measurement can be coupled with a single-molecule setup to record FRET and anisotropy simultaneously. This single-molecule measurement will be useful for gaining insights into the conformational subpopulations, size distributions and structural dynamics of different IDPs in the (monomeric) disordered form, membrane-bound form, amyloid state, and in the liquid–liquid phase-separated droplets. Taken together, this technique will provide unique insights into various structural attributes of these conformationally heterogeneous IDPs and will enhance our understanding of the various dynamical signatures associated with different structural transitions of IDPs implicated in a variety of physiological functions and human diseases.

ACKNOWLEDGMENTS

We thank IISER Mohali, the Department of Science and Technology and the Ministry of Human Resource Development, Govt. of India for the financial support (National Postdoctoral Fellowship to A.M.; Centre of Excellence grant to S.M.). The work described here has been contributed by several former and present members of the Mukhopadhyay laboratory including Drs. M. Bhattacharya, N. Jain, D. Narang, S. Arya, K. Bhasne, Ms. A. Singh, and Ms. D. Das, and has been supported by several research grants from the Council of Scientific and Industrial Research, the Department of Science and Technology, and the Department of Biotechnology, Govt. of India (to S.M.). The analyses of all of the time-resolved fluorescence anisotropy decay profiles have been carried out using the data analysis software developed by Prof. N. Periasamy at the Tata Institute of Fundamental Research (TIFR) Mumbai, India. We thank Prof. N. Periasamy (Retd. TIFR Mumbai) for providing us with the data analysis software.

REFERENCES

Alberts, B., Johnson, A., Lewis, J., Morgan, D., Raff, M., Roberts, K., et al. (2014). *Molecular biology of the cell* (5th ed.). New York: Garland Science.

Alexander Ross, J. B., Rousslang, K. W., & Brand, L. (1981). Time-resolved fluorescence and anisotropy decay of the tryptophan in adrenocorticotropin-(1-24). *Biochemistry*, *20*, 4361–4369.

Ameloot, M., vandeVen, M., Acuña, A. U., & Valeur, B. (2013). Fluorescence anisotropy measurements in solution: Methods and reference materials (IUPAC technical report). *Pure and Applied Chemistry*, *85*(3), 589–608.

Arya, S., & Mukhopadhyay, S. (2014). Ordered water within the collapsed globules of an amyloidogenic intrinsically disordered protein. *Journal of Physical Chemistry B*, *118*, 9191–9198.

Arya, S., Singh, A. K., Bhasne, K., Dogra, P., Datta, A., Das, P., et al. (2018). Femtosecond hydration map of intrinsically disordered α-synuclein. *Biophysical Journal, 114*, 2540–2551.

Arya, S., Singh, A. K., Khan, T., Bhattacharya, M., Datta, A., & Mukhopadhyay, S. (2016). Water rearrangements upon disorder-to-order amyloid transition. *The Journal of Physical Chemistry Letters*, 7(20), 4105–4110.

Attems, J. (2017). Alzheimer's disease pathology in synucleinopathies. *The Lancet Neurology, 16*(1), 22–23.

Baba, M., Nakajo, S., Tu, P. H., Tomita, T., Nakaya, K., Lee, V. M., et al. (1998). Aggregation of alpha-synuclein in Lewy bodies of sporadic Parkinson's disease and dementia with Lewy bodies. *The American Journal of Pathology, 152*(4), 879–884.

Bhattacharya, M., & Mukhopadhyay, S. (2016). Studying protein misfolding and aggregation by fluorescence spectroscopy. In C. Geddes (Ed.), *Vol. 8. Reviews in Fluorescence 2015* (pp. 1–27). Switzerland: Springer International Publishing.

Boeynaems, S., Alberti, S., Fawzi, N. L., Mittag, T., Polymenidou, M., Rousseau, F., et al. (2018). Protein phase separation: A new phase in cell biology. *Trends in Cell Biology, 28*(6), 420–435.

Burger, V. M., Gurry, T., & Stultz, C. M. (2014). Intrinsically disordered proteins: Where computation meets experiment. *Polymer, 6*(10), 2684–2719.

Callis, P. R. (1997). Two-photon-induced fluorescence. *Annual Review of Physical Chemistry, 48*, 271–297.

Canet, D., Doering, K., Dobson, C. M., & Dupont, Y. (2001). High-sensitivity fluorescence anisotropy detection of protein-folding events: Application to α-lactalbumin. *Biophysical Journal, 80*(4), 1996–2003.

Christensen, R. L., Drake, R. C., & Phillips, D. (1986). Time-resolved fluorescence anisotropy of perylene. *The Journal of Physical Chemistry, 90*(22), 5960–5967.

Courchaine, E. M., Lu, A., & Neugebauer, K. M. (2016). Droplet organelles? *The EMBO Journal, 35*(15), 1603–1612.

Cross, A. J., & Fleming, G. R. (1984). Analysis of time-resolved fluorescence anisotropy decays. *Biophysical Journal, 46*(1), 45–56.

Das, R. K., Ruff, K. M., & Pappu, R. V. (2015). Relating sequence encoded information to form and function of intrinsically disordered proteins. *Current Opinion in Structural Biology, 32*, 102–112.

De Beuckeleer, K., Volckaert, G., & Engelborghs, Y. (1999). Time resolved fluorescence and phosphorescence properties of the individual tryptophan residues of barnase: Evidence for protein-protein interactions. *Proteins, 36*(1), 42–53.

Denk, W., Strickler, J. H., & Webb, W. W. (1990). Two-photon laser scanning fluorescence microscopy. *Science, 248*(4951), 73–76.

Deprez, E., Tauc, P., Leh, H., Mouscadet, J.-F., Auclair, C., Hawkins, M. E., et al. (2001). DNA binding induces dissociation of the multimeric form of HIV-1 integrase: A time-resolved fluorescence anisotropy study. *Proceedings of the National Academy of Sciences of the United States of America, 98*(18), 10090–10095.

Drescher, M. (2012). EPR in protein science: Intrinsically disordered proteins. *Topics in Current Chemistry, 321*, 91–119.

Dunker, A. K., Babu, M. M., Barbar, E., Blackledge, M., Bondos, S. E., Dosztányi, Z., et al. (2013). What's in a name? Why these proteins are intrinsically disordered. *Intrinsically Disordered Proteins, 1*(1), e24157.

Dunker, A. K., Lawson, J. D., Brown, C. J., Williams, R. M., Romero, P., Oh, J. S., et al. (2001). Intrinsically disordered protein. *Journal of Molecular Graphics and Modelling, 19*, 26–59.

Dutt, G. B., Doraiswamy, S., Periasamy, N., & Venkataraman, B. (1990). Rotational reorientation dynamics of polar dye molecular probes by picosecond laser spectroscopic technique. *The Journal of Chemical Physics, 93*(12), 8498–8513.

Dyson, H. J., & Wright, P. E. (2005). Intrinsically unstructured proteins and their functions. *Nature Reviews Molecular Cell Biology*, *6*(3), 197–208.

Feinstein, E., Deikus, G., Rusinova, E., Rachofsky, E. L., Ross, J. B. A., & Laws, W. R. (2003). Constrained analysis of fluorescence anisotropy decay: Application to experimental protein dynamics. *Biophysical Journal*, *84*(1), 599–611.

Fuxreiter, M., & Tompa, P. (2012). Fuzzy complexes: A more stochastic view of protein function. *Advances in Experimental Medicine and Biology*, *725*, 1–14.

Gruet, A., Dosnon, M., Blocquel, D., Brunel, J., Gerlier, D., Das, R. K., et al. (2016). Fuzzy regions in an intrinsically disordered protein impair protein-protein interactions. *FEBS Journal*, *283*(4), 576–594.

Habchi, J., Tompa, P., Longhi, S., & Uversky, V. N. (2014). Introducing protein intrinsic disorder. *Chemical Reviews*, *114*(13), 6561–6588.

He, Z., Guo, J. L., McBride, J. D., Narasimhan, S., Kim, H., Changolkar, L., et al. (2018). Amyloid-β plaques enhance Alzheimer's brain tau-seeded pathologies by facilitating neuritic plaque tau aggregation. *Nature Medicine*, *24*(1), 29–38.

Heller, G. T., Aprile, F. A., & Vendruscolo, M. (2017). Methods of probing the interactions between small molecules and disordered proteins. *Cellular and Molecular Life Sciences*, *74*(17), 3225–3243.

Henriques, J., Cragnell, C., & Skepö, M. (2015). Molecular dynamics simulations of intrinsically disordered proteins: Force field evaluation and comparison with experiment. *Journal of Chemical Theory and Computation*, *11*(7), 3420–3431.

Ikeda, T., Lee, B., Yamaguchi, H., & Tazuke, S. (1990). Time-resolved fluorescence anisotropy studies on the interaction of biologically active polycations with phospholipid membranes. *BBA—Biomembranes*, *1021*(1), 56–62.

Jain, N., Bhattacharya, M., & Mukhopadhyay, S. (2011). Chain collapse of an amyloidogenic intrinsically disordered protein. *Biophysical Journal*, *101*(4), 1720–1729.

Jain, N., & Mukhopadhyay, S. (2015). Applications of fluorescence anisotropy in understanding protein conformational disorder and aggregation. In P. Misra (Ed.), *Applied spectroscopy and the science of nanomaterials* (pp. 41–57). Singapore: Springer.

Jain, N., Narang, D., Bhasne, K., Dalal, V., Arya, S., Bhattacharya, M., et al. (2016). Direct observation of the intrinsic backbone torsional mobility of disordered proteins. *Biophysical Journal*, *111*(4), 768–774.

Jameson, D. M., & Ross, J. A. (2010). Fluorescence polarization/anisotropy in diagnostics and imaging. *Chemical Reviews*, *110*(5), 2685–2708.

Jha, A., Udgaonkar, J. B., & Krishnamoorthy, G. (2009). Characterization of the heterogeneity and specificity of interpolypeptide interactions in amyloid protofibrils by measurement of site-specific fluorescence anisotropy decay kinetics. *Journal of Molecular Biology*, *393*(3), 735–752.

Kapusta, P., Erdmann, R., Ortmann, U., & Wahl, M. (2003). Time-resolved fluorescence anisotropy measurements made simple. *Journal of Fluorescence*, *13*(2), 179–183.

Kim, J. E., Arjara, G., Richards, J. H., Gray, H. B., & Winkler, J. R. (2006). Probing folded and unfolded states of outer membrane protein a with steady-state and time-resolved tryptophan fluorescence. *Journal of Physical Chemistry B*, *110*(35), 17656–17662.

Kinosita, K., Ikegami, A., & Kawato, S. (1982). On the wobbling-in-cone analysis of fluorescence anisotropy decay. *Biophysical Journal*, *37*(2), 461–464.

Knowles, T. P. J., Vendruscolo, M., & Dobson, C. M. (2014). The amyloid state and its association with protein misfolding diseases. *Nature Reviews Molecular Cell Biology*, *15*(6), 384–396.

Konrat, R. (2014). NMR contributions to structural dynamics studies of intrinsically disordered proteins. *Journal of Magnetic Resonance*, *241*(2014), 74–85.

Krishna, M. M. G., Das, R., Periasamy, N., & Nityananda, R. (2000). Translational diffusion of fluorescent probes on a sphere: Monte Carlo simulations, theory, and fluorescence anisotropy experiment. *Journal of Chemical Physics*, *112*(19), 8502–8514.

Krishna, M. M. G., Srivastava, A., & Periasamy, N. U. (2001). Rotational dynamics of surface probes in lipid vesicles. *Biophysical Chemistry*, *90*(2), 123–133.

Lakowicz, J. R. (2006). *Principles of fluorescence spectroscopy* (3rd ed.). New York: Springer.

Lakowicz, J. R., Cherek, H., Maliwal, B., & Gratton, E. (1985). Time-resolved fluorescence anisotropies of diphenylhexatriene and perylene in solvents and lipid bilayers obtained from multifrequency phase-modulation fluorometry. *Biochemistry*, *24*(2), 376–383.

Lin, Y., Currie, S. L., & Rosen, M. K. (2017). Intrinsically disordered sequences enable modulation of protein phase separation through distributed tyrosine motifs. *Journal of Biological Chemistry*, *292*(46), 19110–19120.

Liu, C. X., Liang, G. T., Liu, Z., & Zu, L. L. (2014). Time-resolved fluorescence anisotropy study of the interaction between DNA and a peptide truncated from the p53 protein core domain. *Journal of Fluorescence*, *24*(2), 533–539.

Maiti, N. C., Krishna, M. M. G., Britto, P. J., & Periasamy, N. (1997). Fluorescence dynamics of dye probes in micelles. *The Journal of Physical Chemistry B*, *101*(51), 11051–11060.

Mao, A. H., Crick, S. L., Vitalis, A., Chicoine, C. L., & Pappu, R. V. (2010). Net charge per residue modulates conformational ensembles of intrinsically disordered proteins. *Proceedings of the National Academy of Sciences of the United States of America*, *107*(18), 8183–8188.

Marcelino-Cruz, A. M., Bhattacharya, M., Anselmo, A. C., & Tessier, P. M. (2011). Site-specific structural analysis of a yeast prion strain with species-specific seeding activity. *Prion*, *5*(30), 208–214.

Marsh, J. A., & Forman-Kay, J. D. (2010). Sequence determinants of compaction in intrinsically disordered proteins. *Biophysical Journal*, *98*(10), 2382–2390.

Millar, D. P. (2000). Time-resolved fluorescence methods for analysis of DNA-protein interactions. *Methods in Enzymology*, *323*(2000), 442–459.

Möglich, A., Krieger, F., & Kiefhaber, T. (2005). Molecular basis for the effect of urea and guanidinium chloride on the dynamics of unfolded polypeptide chains. *Journal of Molecular Biology*, *345*(1), 153–162.

Mukherjee, M., Saha Sardar, P., Ghorai, S. K., Samanta, S. K., Roy, A. S., Dasgupta, S., et al. (2013). A comparative study of interaction of tetracycline with several proteins using time resolved anisotropy, phosphorescence, docking and FRET. *PLoS One*, *8*(4), e60940.

Mukhopadhyay, S., Krishnan, R., Lemke, E. A., Lindquist, S., & Deniz, A. A. (2007). A natively unfolded yeast prion monomer adopts an ensemble of collapsed and rapidly fluctuating structures. *Proceedings of the National Academy of Sciences of the United States of America*, *104*(8), 26–49.

Mukhopadhyay, S., Nayak, P. K., Udgaonkar, J. B., & Krishnamoorthy, G. (2006). Characterization of the formation of amyloid protofibrils from barstar by mapping residue-specific fluorescence dynamics. *Journal of Molecular Biology*, *358*(4), 935–942.

Narang, D., Swasthi, H. M., Mahapatra, S., & Mukhopadhyay, S. (2017). Site-specific fluorescence depolarization kinetics distinguishes the amyloid folds responsible for distinct yeast prion strains. *The Journal of Physical Chemistry B*, *121*(36), 8447–8453.

Ou, L., Waddell, M. B., & Kriwacki, R. W. (2012). Mechanism of cell cycle entry mediated by the intrinsically disordered protein p27(Kip1). *ACS Chemical Biology*, 7(4), 678–682.

Paila, Y. D., Kombrabail, M., Krishnamoorthy, G., & Chattopadhyay, A. (2011). Oligomerization of the serotonin1A receptor in live cells: A time-resolved fluorescence anisotropy approach. *The Journal of Physical Chemistry B*, *115*(39), 11439–11447.

Parigi, G., Rezaei-Ghaleh, N., Giachetti, A., Becker, S., Fernandez, C., Blackledge, M., et al. (2014). Long-range correlated dynamics in intrinsically disordered proteins. *Journal of the American Chemical Society*, *136*(46), 16201–16209.

Peng, K., Visser, A. J. W. G., van Hoek, A., Wolfs, C. J. A. M., & Hemminga, M. A. (1990). Analysis of time-resolved fluorescence anisotropy in lipid-protein systems—II. Application to tryptophan fluorescence of bacteriophage M13 coat protein incorporated in phospholipid bilayers. *European Biophysics Journal*, *18*(5), 285–293.

Pierce, B. D., Toptygin, D., & Wendland, B. (2013). Pan1 is an intrinsically disordered protein with homotypic interactions. *Proteins: Structure, Function, and Bioinformatics*, *81*(11), 1944–1963.

Prazeres, T. J. V., Fedorov, A., Barbosa, S. P., Martinho, J. M. G., & Berberan-Santos, M. N. (2008). Accurate determination of the limiting anisotropy of rhodamine 101. Implications for its use as a fluorescence polarization standard. *The Journal of Physical Chemistry A*, *112*(23), 5034–5039.

Rachofsky, E. L., & Laws, W. R. (2000). Kinetic models and data analysis methods for fluorescence anisotropy decay. *Methods in Enzymology*, *321*(2000), 216–238.

Rachofsky, E. L., Wolf, B., Bialik, C. N., Ross, J. B. A., & Laws, W. R. (1999). A general method for constrained analysis of fluorescence anisotropy decay: Application of the steady-state anisotropy. *Journal of Fluorescence*, *9*(4), 379–390.

Redfern, O. C., Dessailly, B., & Orengo, C. A. (2008). Exploring the structure and function paradigm. *Current Opinion in Structural Biology*, *18*(3), 394–402.

Rice, S. A., & Kenney-Wallace, G. A. (1980). Time-resolved fluorescence depolarization studies of rotational relaxation in viscous media. *Chemical Physics*, *47*(2), 161–170.

Rich, R. L., Chen, Y., Neven, D., Négrerie, M., Gai, F., & Petrich, J. W. (1993). Steady-state and time-resolved fluorescence anisotropy of 7-azaindole and its derivatives. *Journal of Physical Chemistry*, *97*(9), 1781–1788.

Sahay, S., Anoop, A., Krishnamoorthy, G., & Maji, S. K. (2014). Site-specific fluorescence dynamics of α-synuclein fibrils using time-resolved fluorescence studies: Effect of familial Parkinson's disease-associated mutations. *Biochemistry*, *53*(5), 807–809.

Sarkar, N., Datta, A., Das, S., & Bhattacharyya, K. (1996). Solvation dynamics of coumarin 480 in micelles. *Journal of Physical Chemistry*, *100*(38), 15483–15486.

Saxena, A. M., Krishnamoorthy, G., Udgaonkar, J. B., & Periasamy, N. (2007). Identification of intermediate species in protein-folding by quantitative analysis of amplitudes in time-domain fluorescence spectroscopy. *Journal of Chemical Sciences*, *119*(2), 61–69.

Schimmel, P. R., & Flory, P. J. (1968). Conformational energies and configurational statistics of copolypeptides containing L-proline. *Journal of Molecular Biology*, *34*(1), 105–120.

Scholtbach, K., Venegas, I., Bohne, C., & Fuentealba, D. (2015). Time-resolved fluorescence anisotropy as a tool to study guest-cucurbit[n]uril-protein ternary supramolecular interactions. *Photochemical & Photobiological Sciences*, *14*(4), 842–852.

Serio, T. R., & Lindquist, S. L. (2000). Protein-only inheritance in yeast: Something to get [PSI+]-ched about. *Trends in Cell Biology*, *10*(3), 98–105.

Sharma, R., Raduly, Z., Miskei, M., & Fuxreiter, M. (2015). Fuzzy complexes: Specific binding without complete folding. *FEBS Letters*, *589*(19), 2533–2542.

Sharma, P., Varma, R., Sarasij, R. C., Ira, Gousset, K., Krishnamoorthy, G., et al. (2004). Nanoscale organization of multiple GPI-anchored proteins in living cell membranes. *Cell*, *116*(4), 577–589.

Shorter, J., & Lindquist, S. (2005). Prions as adaptive conduits of memory and inheritance. *Nature Reviews Genetics*, *6*, 435–450.

Spires-Jones, T. L., & Hyman, B. T. (2014). The intersection of amyloid beta and tau at synapses in Alzheimer's disease. *Neuron*, *82*(4), 756–771.

Swaminathan, R., Periasamy, N., Udgaonkar, J. B., & Krishnamoorthy, G. (1994). Molten globule-like conformation of barstar: A study by fluorescence dynamics. *The Journal of Physical Chemistry*, *98*(37), 9270–9278.

Szasz, C., Alexa, A., Toth, K., Rakacs, M., Langowski, J., & Tompa, P. (2011). Protein disorder prevails under crowded conditions. *Biochemistry*, *50*(26), 5834–5844.

Tanaka, M., Chien, P., Naber, N., Cooke, R., & Weissman, J. S. (2004). Conformational variations in an infectious protein determine prion strain differences. *Nature*, *428*, 323–328.
Thevenin, B. J., Periasamy, N., Shohet, S. B., & Verkman, A. S. (1994). Segmental dynamics of the cytoplasmic domain of erythrocyte band 3 determined by time-resolved fluorescence anisotropy: Sensitivity to pH and ligand binding. *Proceedings of the National Academy of Sciences of the United States of America*, *91*(5), 1741–1745.
Tompa, P. (2005). The interplay between structure and function in intrinsically unstructured proteins. *FEBS Letters*, *579*(15), 3346–3354.
Uversky, V. N. (2013a). A decade and a half of protein intrinsic disorder: Biology still waits for physics. *Protein Science*, *22*(6), 693–724.
Uversky, V. N. (2013b). Unusual biophysics of intrinsically disordered proteins. *Biochimica et Biophysica Acta—Proteins & Proteomics*, *1834*(5), 932–951.
Uversky, V. N. (2014). Introduction to intrinsically disordered proteins (IDPs). *Chemical Reviews*, *114*(13), 6557–6560.
Uversky, V. N. (2017a). Intrinsically disordered proteins in overcrowded milieu: Membrane-less organelles, phase separation, and intrinsic disorder. *Current Opinion in Structural Biology*, *44*, 18–30.
Uversky, V. N. (2017b). Protein intrinsic disorder-based liquid–liquid phase transitions in biological systems: Complex coacervates and membrane-less organelles. *Advances in Colloid and Interface Science*, *239*, 97–114.
Uversky, V. N., Santambrogio, C., Brocca, S., & Grandori, R. (2012). Length-dependent compaction of intrinsically disordered proteins. *FEBS Letters*, *586*(1), 70–73.
Valeur, B. (2001). *Molecular fluorescence principles and applications*. New York: Wiley-VCH.
van der Lee, R., Buljan, M., Lang, B., Weatheritt, R. J., Daughdrill, G. W., Dunker, A. K., et al. (2014). Classification of intrinsically disordered regions and proteins. *Chemical Reviews*, *114*(13), 6589–6631.
Vogel, S. S., Thaler, C., Blank, P. S., & Koushik, S. V. (2009). Time-resolved fluorescence anisotropy. In A. Periasamy & R. M. Clegg (Eds.), *FLIM microscopy in biology and medicine* (pp. 245–288). New York: Chapman and Hall/CRC.
Wahl, P. (1979). Analysis of fluorescence anisotropy decays by a least square method. *Biophysical Chemistry*, *10*(1), 91–104.
Wales, P., Lázaro, D. F., Pinho, R., & Outeiro, T. F. (2013). Limelight on alpha-synuclein: Pathological and mechanistic implications in neurodegeneration. *Journal of Parkinson's Disease*, *3*(4), 415–459.
Weinreis, S. A., Ellis, J. P., & Cavagnero, S. (2010). Dynamic fluorescence depolarization: A powerful tool to explore protein folding on the ribosome. *Methods*, *52*(1), 57–73.
Wright, P. E., & Dyson, H. J. (1999). Intrinsically unstructured proteins: Re-assessing the protein structure-function paradigm. *Journal of Molecular Biology*, *293*(2), 321–331.
Wright, P. E., & Dyson, H. J. (2014). Intrinsically disordered proteins in cellular signalling and regulation. *Nature Reviews Molecular Cell Biology*, *16*(1), 18–29.
Yoon, M.-K., Mitrea, D. M., Ou, L., & Kriwacki, R. W. (2012). Cell cycle regulation by the intrinsically disordered proteins p21 and p27. *Biochemical Society Transactions*, *40*(5), 981–988.

CHAPTER THIRTEEN

Building, Characterization, and Applications of Cuvette-FCS in Denaturant-Induced Expansion of Globular and Disordered Proteins

Timir Baran Sil, Bankanidhi Sahoo, Kanchan Garai[1]
Tata Institute of Fundamental Research, Hyderabad, India
[1]Corresponding author: e-mail address: kanchan@tifrh.res.in

Contents

Abstract

Fluorescence correlation spectroscopy (FCS) is a single-molecule sensitive technique with widespread applications in biophysics. However, conventional microscope-based FCS setups have limitations in performing certain experiments such as those requiring agitations such as stirring or heating, and those involving measurements in solvents with the mismatch of refractive indices. We have recently developed an FCS setup that

Methods in Enzymology, Volume 611
ISSN 0076-6879
https://doi.org/10.1016/bs.mie.2018.08.027

is suitable for performing measurements inside regular cuvettes. The cuvette-FCS is suitable for performing single-molecule measurements in experiments that are regularly performed in spectrofluorometers but are generally avoided in microscope-based FCS. Here we describe building and characterization of the performance of the cuvette-FCS setup in detail. Finally, we have used a natively folded protein and an intrinsically disordered protein to demonstrate and describe how cuvette-FCS can be applied conveniently to measure urea-dependent expansion of hydrodynamic size of proteins.

1. INTRODUCTION

Fluorescence correlation spectroscopy (FCS) uses autocorrelation of the spontaneous fluctuations of fluorescence to characterize the timescales and the amplitudes of underlying processes such as Brownian diffusion and chemical kinetics in equilibrium (Digman & Gratton, 2009; Elson, 2011). A major advantage of FCS is that it can extract information on dynamics from equilibrium measurements without requiring to perturb the system (Elson, 2011). Consequently, FCS is used widely to study conformational fluctuations and hydrodynamic size of proteins, protein–ligand interactions, and mobility of protein molecules in complex environments such as on the plasma membrane, inside the cellular milieu, and within tissues (Cardarelli & Gratton, 2010; Guldbrand et al., 2013; Matayoshi & Swift, 2001; Petrasek, Ries, & Schwille, 2010; Wohland, 2014).

Recently, FCS has been proven to be an important tool in characterization of structure and conformational dynamics of unfolded and disordered proteins (Borgia et al., 2016; Crick, Jayaraman, Frieden, Wetzel, & Pappu, 2006; Ferreon, Moran, Gambin, & Deniz, 2010; Sherman & Haran, 2006; Trexler & Rhoades, 2010). Single-molecule Forster resonance energy transfer (smFRET) is frequently used along with FCS to characterize, respectively, the radius of gyration (R_g) and the hydrodynamic radius (R_h) of the polypeptide chains (Borgia et al., 2016; Ferreon et al., 2010; Sherman & Haran, 2006; Trexler & Rhoades, 2010). Dimensions of the unfolded and the disordered proteins are known to be highly sensitive to the amino acid composition, temperature, and solvent quality (Nettels et al., 2009; Tran, Mao, & Pappu, 2008). Quantitative measurements of the temperature- and solvent-dependent dynamics of the unfolded and the disordered proteins have been used to evaluate the intrachain and chain–solvent interaction energies, and the internal friction (Hofmann et al., 2012; Nettels et al., 2009; Soranno et al., 2012). These measurements provide critical insights into the mechanism of early events of protein folding and the functions of intrinsically disordered proteins (IDPs) (Hofmann et al., 2012; Sherman & Haran, 2006). However, the temperature- and solvent-dependent FCS measurements are not straightforward; rather these experiments are fraught with several difficulties as will be discussed later (Chattopadhyay, Saffarian, Elson, & Frieden, 2005; Sherman et al., 2008).

Since FCS is a fluctuation spectroscopy technique, the sensitivity of FCS depends on the relative amplitude of the fluctuations, which is inversely proportional to the number of molecules in the observation volume. Macroscopic systems contain a large number of molecules; consequently, fluctuations of fluorescence in bulk measurements are negligible. Therefore, sensitive FCS measurements require a small (typically less than 1 fL) observation volume. The high spatial resolution required in FCS can be attained in confocal microscopy. Hence, FCS setups are generally part of confocal microscopes. Most manufacturers of microscopes such as Zeiss, Olympus, and Nikon provide FCS as an accessory to the confocal microscope. Confocal microscopes are highly popular instruments due to its applications in measurements of physical properties of the biomolecules in solutions, and in complex environments such as those present inside living cells and tissue samples.

However, traditional FCS setups have certain limitations peculiar to the structures of the microscopes. For example, experiments that require continuous mixing of samples by stirring cannot be monitored by FCS. Similarly, temperature-dependent experiments and experiments involving mismatch of refractive indices are generally avoided in conventional FCS. However, these experiments are performed routinely in spectrofluorometers. Spectrofluorometers are highly popular in biophysics and biochemistry laboratories for various types of experiments such as those involving protein aggregation, titration of macromolecules with ligands, and folding–unfolding of biomolecules in chemical denaturants. However, spectrofluorometer experiments provide only ensemble-level information. Therefore, an FCS setup capable of performing measurements in regular cuvettes can enable applications of FCS in a large array of experiments that are performed regularly in spectrofluorometers but are avoided in conventional FCS. We have recently built a setup capable of performing highly sensitive FCS measurements in regular cuvettes (Sahoo, Sil, Karmakar, & Garai, 2018). Here we provide detailed description of building and characterization of the cuvette-FCS setup. We also discuss the advantages and disadvantages of using the cuvette-FCS setup. Finally, we demonstrate how cuvette-FCS can be used to study unfolding and expansion of a natively folded protein and an IDP in a commonly used chemical denaturant such as urea.

2. WHEN IS CUVETTE-FCS ADVANTAGEOUS OVER MICROSCOPE-BASED FCS?

- Stirring of samples. Cuvette-FCS is useful for monitoring processes that require stirring of samples. For example, amyloid aggregation of proteins is an extremely slow process requiring several days or months for completion. These processes speed up in the presence of agitations such as continuous or intermittent stirring.
- Temperature-dependent experiments. Experiments that require temperatures below RT and above 37°C are generally avoided in microscopes. Since cuvette-FCS uses dry objective and the cuvette is thermally isolated from the rest of the FCS setup, it can be used to monitor processes over a large range of temperatures. We note here that Połatynska et al. have performed FCS measurements over a broad range of temperature using a specially designed cell and a dry and long working distance objective (Połatyńska et al., 2017).
- Corrosive solvents. Quartz cuvettes are resistant to most of the solvents; hence, cuvette-FCS can be used for measurements in most solvents including the corrosive solvents. For example, collapse and expansion of many polymers in organic solvents can be studied conveniently using cuvette-FCS.
- Solvents with the mismatch of refractive indices. Measurements in solvents such as guanidinium chloride (GdnCl) and urea are generally avoided in microscope-based FCS due to optical aberrations arising from the mismatch of refractive indices. We will demonstrate later that cuvette-FCS can be used conveniently to study protein folding and unfolding in chemical denaturants (Sahoo et al., 2018).
- Use of automatic titrators. Many experiments such as protein–ligand interactions and protein unfolding can be performed with better accuracies if the titrations are performed by autotitrators. Cuvette-FCS is compatible with the autotitrators (Sahoo et al., 2018). Furthermore, cuvette-FCS is compatible with most of the accessories commonly used with the spectrofluorometers.

3. THEORY

Here we briefly discuss the important parameters that characterize the resolution and the sensitivity of an FCS setup. FCS measures the autocorrelation of the fluctuations of fluorescence ($F(t)$) from a diffraction limited confocal observation volume. The autocorrelation function $G(\tau)$ is defined as (Elson & Magde, 1974)

$$G(\tau) = \frac{\langle F(t)F(t+\tau)\rangle}{\langle F(t)\rangle^2} - 1 \tag{1}$$

In case of confocal microscopy, the molecular detection function or the point spread function (PSF) is generally considered to be a 3D Gaussian given by Eq. (2)

$$\mathrm{PSF}(r, z) = I_0(0)\exp\left[-\left(\frac{2r^2}{\sigma_\mathrm{r}^2} + \frac{2z^2}{\sigma_\mathrm{z}^2}\right)\right] \tag{2}$$

where σ_r and σ_z are the radial and axial dimensions of the PSF, respectively. For a perfectly 3D Gaussian PSF, the analytical expression of $G(\tau)$ for a single diffusing particle is given by Eq. (3)

$$G(\tau) = \frac{1}{\langle N\rangle}\frac{1}{\left(1+\frac{\tau}{\tau_\mathrm{D}}\right)\left(1+\frac{\tau}{\omega^2\tau_\mathrm{D}}\right)^{0.5}} \tag{3}$$

The $\langle N\rangle$ is the average number of fluorophore molecules in the effective FCS observation volume (V_eff), ω is the axial ratio of the PSF (i.e., $\omega = \sigma_{\mathrm{z}/}\sigma_\mathrm{r}$), and τ_D is the diffusion time of the fluorophore in the solution. V_eff in this case is given by (Ivanchenko & Lamb, 2011)

$$V_\mathrm{eff} = \pi^{3/2}\sigma_\mathrm{r}^2\sigma_\mathrm{z} \tag{4}$$

The diffusion time τ_D depends on the radial dimension of the PSF as shown in Eq. (5)

$$\tau_\mathrm{D} = \frac{\sigma_\mathrm{r}^2}{4D} \tag{5}$$

where D is the diffusion coefficient of the molecule. Therefore, the analysis of the experimentally measured autocorrelation data $G(\tau)$ using Eq. (3) can yield the values of τ_D and ω. Then Eq. (5) and the known value of D can be used to estimate σ_r and σ_z. The V_eff can be determined experimentally from the estimated values of σ_r and σ_z using Eq. (4). It may be noted here that V_confocal and V_eff are not the same, rather $V_\mathrm{eff} = \gamma\ V_\mathrm{confocal}$. For a 3D Gaussian volume $\gamma = 2^{-3/2}$. The analysis of the $G(\tau)$ using Eq. (3) also yields the value of $\langle N \rangle$. Hence, V_{eff} can also be estimated using Eq. (6) from experimentally measured $\langle N \rangle$ if the concentration (C) of the fluorophore is known:

$$V_{\text{eff}} = \frac{\langle N \rangle}{C N_{\text{av}}} \tag{6}$$

Here N_{av} is the Avogadro number. Therefore, the analysis of the FCS autocorrelation data using Eq. (3) yields all the necessary parameters to characterize the effective observation volume in FCS.

Signal to noise in FCS: Signal to noise (S/N) in FCS was first calculated in a pioneering article by Koppel (1974). Subsequently, several articles have addressed the errors in FCS both theoretically and experimentally in detail (Kask, Gunther, & Axhausen, 1997; Qian, 1990; Saffarian & Elson, 2003; Wohland, Rigler, & Vogel, 2001). Qualitatively, the S/N in $G(\tau)$ depends on the number of photons detected per molecule per unit time (commonly known as counts per molecule or CPM) and the total acquisition time (T) (Ivanchenko & Lamb, 2011)

$$\frac{S}{N} \approx \text{CPM} \times T^{0.5} \tag{7}$$

Role of triplet state dynamics: In addition to the Brownian diffusion, another factor that may contribute significantly to $G(\tau)$ is the triplet state dynamics of the fluorophore. A fluorophore when excited from the ground state (S_0) to the singlet excited state (S_1) can cross to the triplet state (T_1) via intersystem crossing (ISC) instead of going back to S_0 state by emitting fluorescence. The effective rate of the ISC is typically low; however, the rate can increase significantly if the high excitation intensity is used. The lifetime of the T_1 state is typically a few μs. Hence the triplet state correlation time (τ_T) is significantly smaller than the diffusion time. In case of $\tau_T \ll \tau_D$, $G(\tau)$ can be expressed analytically as shown later (Schonle, Von Middendorff, Ringemann, Hell, & Eggeling, 2014; Widengren, Mets, & Rigler, 1995)

$$G(\tau) = \frac{1}{\langle N \rangle} \left(\frac{1 - F_T + F_T \exp\left(-\frac{\tau}{\tau_T}\right)}{1 - F_T} \right) \frac{1}{\left(1 + \frac{\tau}{\tau_D}\right)\left(1 + \frac{\tau}{\omega^2 \tau_D}\right)^{0.5}} \tag{8}$$

where F_T is the fraction of fluorophores in the triplet state in equilibrium. Detailed derivations of the F_T and the τ_T as a function of excitation intensity and the rate constants of transitions between $S_0 \rightarrow S_1$, $S_1 \rightarrow S_0$, $S_1 \rightarrow T_1$, and $T_1 \rightarrow S_0$ are presented elsewhere (Widengren et al., 1995). It may be mentioned here that the τ_T is inversely proportional to the sum of the effective rate of ISC and rate of deexcitation of the T_1 state. Hence τ_T decreases with increasing excitation power (Schonle et al., 2014). Furthermore, if the fluorophore fluctuates between a bright and a dark state due to other reasons such as due to conformational fluctuations in case of a fluorescently labeled protein, Eq. (8) can be used to estimate the timescales and the amplitudes of such processes (Ducas & Rhoades, 2014; Haupts, Maiti, Schwille, & Webb, 1998).

Notes

1. While most FCS autocorrelation data obtained from in vitro measurements are widely analyzed using Eq. (3) or (8), several recent studies have shown that most often the PSF in FCS deviates from a perfect 3D Gaussian. However, the deviations are significantly less if the back aperture of objective is underfilled (Hess & Webb, 2002). Detailed theoretical and experimental characterizations of the artifacts in FCS have been discussed in several pioneering articles (Hell, Reiner, Cremer, & Stelzer, 1993; Sheppard, 1989; Sheppard & Gu, 1992; Sheppard & Torok, 1997; Wilson & Carlini, 1989).

4. BUILDING THE CUVETTE-FCS SETUP

Cuvette-FCS setup uses essentially the same confocal detection scheme as the conventional microscope-based FCS setups with two main differences. First, the objective lens is held horizontally in cuvette-FCS. Such geometry is required for working with a cuvette as the sample cell. Second, here we use an extra long working distance (ELWD) objective lens with working distance greater than 1.8 mm, and with correction collar adjustments for 1.25 mm thick cover glasses. While the ELWD objectives are compatible with thick cuvette windows, these objectives have lower numerical aperture (NA) than the high NA water or oil immersion objectives used in the conventional FCS setups. The ELWD objective used here has an NA equal to 0.7. Furthermore, these objectives are classified as Plan Fluor not Plan Apochromat, which are carefully corrected for both the spherical and the chromatic aberrations. Plan Fluor objectives are also corrected for aberrations, but the corrections are less extensive. The long working distance objective lenses have been used in the past for performing FCS measurements through thick optical windows (Altamore, Lanzano, & Gratton, 2013;

Banachowicz, Patkowski, Meier, Klamecka, & Gapiński, 2014). However, the signal to noise (S/N) reported in these studies is much lower compared to conventional FCS setups. Below we describe the detailed procedure for building a highly sensitive cuvette-FCS setup.

4.1 Materials

- A low-noise stable single-mode (TEM_{00}) laser with an appropriate mount (HCM2/M, Thorlabs). We have used a green ($\lambda = 543$ nm) HeNe laser (model R-39581, Newport).
- Common optomechanical components such as optical posts, post holders, bases, table clamps, lens holder, and kinematic mirror mounts. All the optomechanical elements mentioned here are from Thorlabs.
- Optomechanical items for building the 30 mm cage system. Cage plates, cage rods, cage cube with a mounted 50/50 beam splitter (CCM1-BS013/M), filter mount (QRC1A), and iris diaphragms (CP20S). The iris diaphragms are used for the alignment of the laser. All these items are from Thorlabs.
- An aspheric lens with $f = 25$ mm (P/N-49660, Edmund Optics), a biconvex lens with $f = 200$ mm (LBF254-200-A, Thorlabs), and an achromatic lens with $f = 200$ mm (AC254-200-1-ML, Thorlabs).
- Silver-coated mirrors (PF10-03-P01, Thorlabs).
- Ultraflat dichroic mirror (zt488/543/635rpc, Croma Technology).
- Objective lens (Plan Fluor, ELWD, 60 ×, 0.7 NA, Nikon Instruments).
- Emission bandpass filter (ET605/70 m, Chroma Technology) and neutral density (ND) filters (Thorlabs).
- Multimode optical fibers (Ø50 μm, 0.22 NA, M42L01), fiber couplers (SM1/FC), and XY translational stages (ST1XY-D). All these are from Thorlabs.
- Translational stages (TS90-Mu05-01, Holmarc, India). Precision = 5 μm.
- Avalanche photodiodes or APD (detectors) (SPCM-AQRH-14-FC, Excelitas Technologies).
- Correlator card (Flex02-01 D/C, correlator.com) along with the software.
- Computer (Dell Precision Tower 3620).
- Quartz cuvette (10 mm light path, P/N—101-10-40, Hellma).
- Cuvette holder. The cuvette holder is custom designed in the machine shop of Tata Institute of Fundamental Research, Hyderabad. A schematic of the cuvette holder is shown in the inset of Fig. 1. The cuvette

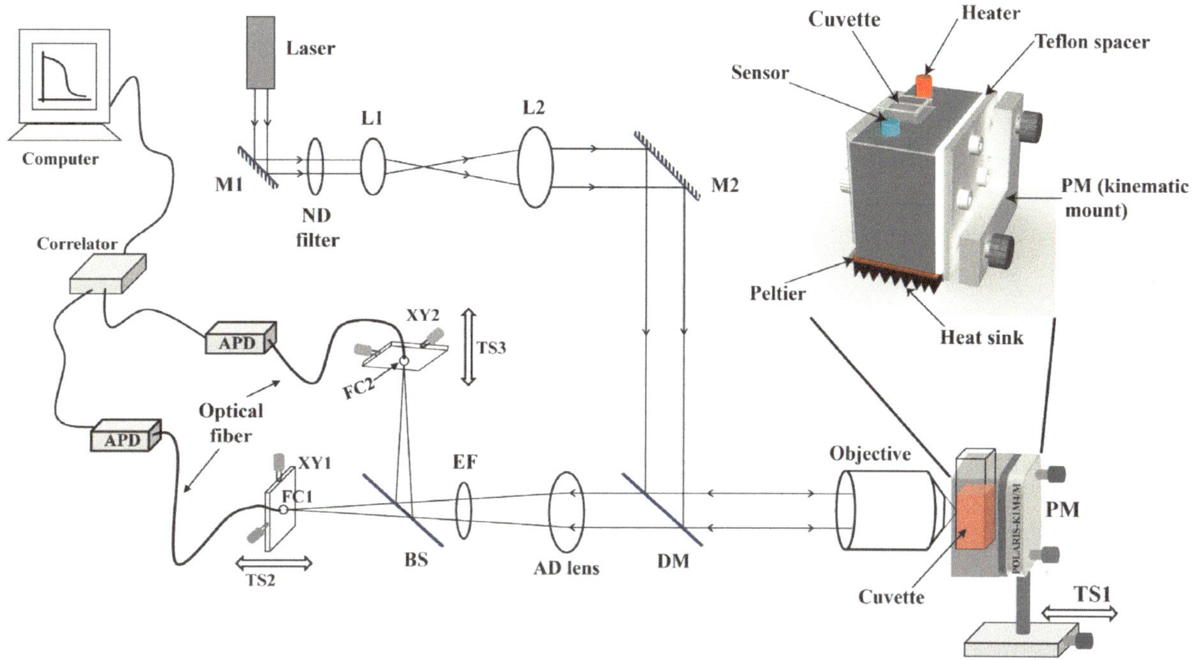

Fig. 1 Schematic of the cuvette-FCS setup. *AD*, achromatic doublet lens; *APD*, avalanche photodiode; *BS*, beam splitter; *DM*, dichroic on a kinematic mirror mount; *EF*, emission filter; *FC1 and FC2*, fiber couplers; *L1*, aspheric lens ($f = 25$ mm); *L2*, biconvex lens ($f = 200$ mm); *laser*, green He–Ne laser; *M1 and M2*, silver-coated mirrors on kinematic mirror mounts; *objective*, ELWD, 0.7 NA, dry; *TS1, TS2, and TS3*, micrometer translation stages; *XY1 and XY2*, XY stages. *Inset* shows the detailed design of the cuvette holder. PM: kinetic platform mount for angular alignment of the cuvette holder.

holder is designed such that it is compatible with the 30 mm cage system. The holder contains slots for inserting a heating element and a temperature sensor. A Peltier cooling element is attached at the bottom of the cuvette holder for cooling the holder to approximately 20°C. A heat sink is glued to the Peltier with thermal adhesive for radiating the heat from the heating side of the Peltier. The heating element and the sensor are connected to a temperature controller (model no. S.ELECT042, from local electronics market). The temperature controller maintains the temperature of the sample cell within ±0.5°C of the set temperature in our setup.

- Kinematic platform mount (PM) (POLARIS-K1M4/M, Thorlabs) for holding and adjusting the angular alignment of the cuvette holder. There is a Teflon spacer between the cuvette holder and the mirror mount for thermal isolation.
- A vacuum pump (Diaphragm pump, Chemvac 410, Tarsons, India) connected to a flexible PTFE tubing. This is for emptying of solutions from the cuvette and washing of the cuvette.

4.1.1 Notes

1. The specifications of the dichroic mirror need to be decided based on the laser(s) and the fluorophores of interest. However, the important point to consider is that the dichroic mirror must be "ultraflat" with <0.25 waves/in. Peak–Valley surface flatness for best performance.
2. We have tried cuvettes from several manufacturers and found that the quality of the cuvette is important. The cuvette from Hellma showed the best performance (Sahoo et al., 2018). If any user is interested to try a cuvette from a different manufacturer, then we recommend comparing it with the Hellma cuvette specified earlier.

4.2 Reagents and Buffers

- A concentrated solution of rhodamine B (10 μ*M*) in phosphate-buffered saline (PBS) at pH 7.4
- Rhodamine B solutions (35 and 100 n*M*) in PBS at pH 7.4

Note: The choice of the fluorophore must be based on the wavelength of the laser used in the setup. Since we are using a 543-nm laser here we have used solutions of rhodamine B. Several other fluorophores such as Cy3, Alexa 546, Atto 550, and HiLyte 555 can also be used with the 543-nm laser.

4.3 Building the Cuvette-FCS Setup

Fig. 1 shows the schematic of the cuvette-FCS setup. The overall design is the same as that of the cuvette-FCS setup reported by Sahoo et al. with slight modifications in the design of the cuvette holder (Sahoo et al., 2018). Below we provide a stepwise description of the procedure for building the setup using the components mentioned earlier.

1. First, the laser should be mounted rigidly. In our setup we maintain a height of 15 cm for the laser beam from the top surface of the optical table. This height is convenient to incorporate accessories such as the magnetic stirrer under the cuvette holder.
2. The laser beam is expanded using a pair of lenses L1 ($f=25$ mm) and L2 ($f=200$ mm). The L1 must be aspheric to avoid spherical aberration of the focal spot. If the setup is to be used with multiple lasers, then we recommend an achromatic aspherical L1 and an achromatic L2. We recommend mounting both L1 and L2 together in a cage system to improve the mechanical stability of the beam expander. The distance between L1 and L2 should be adjusted (~225 mm) such that the output beam is collimated or mildly diverging.
3. The expanded beam is reflected by a mirror M2 and the dichroic mirror DM. Both M2 and DM are mounted on kinematic mirror mounts. These two mirrors are used to align the laser beam to the central axis of the succeeding cage system holding the objective lens.

Note: This alignment is critical for obtaining diffraction-limited focusing of the laser beam.

4. The cuvette is inserted inside a home-built cuvette holder. Fig. 1 (inset) shows the design of the cuvette holder. The cuvette holder is designed to be compatible with the cage system for the ease of alignment of the cuvette as will be discussed later.
5. The cuvette holder is mounted on a kinematic PM. A Teflon spacer is used between the PM and the cuvette holder for thermal isolation. PM is mounted on a translation stage (TS1). TS1 is used to control the *Z*-position of the cuvette and, hence, the focal depth of the objective inside the sample. Align the cuvette holder, PM, and TS1 coaxial with the objective as shown in Fig. 1. PM is used for precise angular alignment of the cuvette. Step-by-step description of the alignment of the cuvette holder will be described in the following section.
6. The fluorescence from the observation volume collected by the objective lens is separated from the excitation laser beam by the DM and then

split into two mutually perpendicular channels using a 50:50 beam splitter. The fluorescence is focused using an AD lens ($f = 200$ mm) and collected by two multimode optical fibers mounted on *X–Y* stages in the two detection channels. The optical fibers with the mode field diameter equal to 50 μm act as the confocal pinholes here. The *X–Y* stages are further mounted on linear translation stages (TS2 and TS3) such that the pinholes can be positioned precisely at the focus of the AD lens in *X*, *Y*, and *Z* directions. The other ends of the optical fibers are connected to the APDs. The outputs of APDs are connected to the inputs of the correlator card, which is connected to the PC.

7. *X–Y–Z* alignment of the optical fiber is performed in the same manner as it is performed in the case of conventional FCS. Here we provide a brief description of the procedure of the alignment. First, the cuvette is filled with a concentrated (10 μ*M*) aqueous solution of rhodamine B. Initially, the cuvette is moved slightly away from the objective. Then the cuvette is moved slowly toward the objective. When the focal spot of the objective enters inside the solution, a bright fluorescence spot appears in front of the detection optical fibers. To see the fluorescence by eye the room needs to be somewhat dark. Then the TS1 can be moved to increase the focal depth to about 20 μm. The optical fibers in both the channels are then positioned approximately at the center of the fluorescence spot by eye estimation. This completes the coarse alignments of the pinholes. The final alignment must be performed by monitoring the photon count rates (CR) using the APDs. Hence, the concentrated rhodamine solution is decanted, and the cuvette is thoroughly washed with water and ethanol several times. Then the cuvette is filled with a dilute (~100 n*M*) solution of rhodamine B and the APDs are powered on. The CR and the autocorrelation curves ($G(\tau)$) are now displayed real time on the monitor of the computer. *X–Y* positions of the fiber couplers are then optimized by maximizing the CR. This procedure is performed at several *Z*-positions of the pinholes. Finally, *X–Y–Z*-position of the fiber coupler is decided based on the highest CR and the highest amplitude of the correlation curve $G(\tau)$.

4.3.1 Notes

1. The cuvette holder we are using is custom built. Alternatively, the cuvette holder from Thorlabs (CVH100/M) may also be used. However, this holder needs to be modified slightly to adapt to the cuvette-FCS setup. For example, the opening of the cuvette holder on the objective side

needs to be widened to allow the objective to come closer to the cuvette and a few groves must be made on the opposite side of the holder to attach it to the PM.

2. All the experiments reported here have been performed at 25°C. Hence the Peltier element is used to cool the holder to approximately 20°C. This requires a small amount of flow of current through the Peltier. Hence the passive cooling of the Peltier with a heat sink works fine. However, if a temperature below 20°C needs to be used, then the Peltier unit would require active cooling such as by flowing cold water through a jacket thermally attached to the Peltier. Interested readers may see the design of such a cuvette holder reported by Sahoo et al. (2018).
3. Stability of the position of the cuvette is highly important for the optimal performance of the cuvette. Therefore, the cuvette must sit tightly/rigidly inside the holder. If the cuvette sits loosely, then thin strips of papers or plastics may be inserted between the outer wall(s) of the cuvette and the inside wall(s) of the holder to hold the cuvette tightly in place.
4. The micrometer translation stage TS1 should be of high precision. The precision of TS1 used here is 5 μm.
5. To avoid optical aberration, the mode field diameter of the optical fiber should be smaller than the size of one Airy disc unit (AU), which is equal to $1.22 \times \lambda \times \text{magnification}/\text{NA}$. In our setup AU = 61 μm (Sahoo et al., 2018). Therefore, we have used optical fibers with mode field diameter = 50 μm.

4.4 Alignment of the Cuvette for Optimization of the Performance

Since we are using a dry objective and the sample is in water, spherical aberrations are expected to increase with the increase in the depth of the observation volume. We have observed that both the *Z*-alignment and the angular alignment of the cuvette are crucial for the sensitivity and the resolution of the cuvette-FCS (Sahoo et al., 2018). The *Z*-position of the cuvette can be controlled by moving the cuvette holder by using the micrometer stage TS1, and the angular alignment of the cuvette can be adjusted using the kinematic mount PM. Here we provide a detailed description of how the alignment of the cuvette is optimized to obtain the best performance.

4.4.1 Procedure

1. Fill the cuvette with 2 mL of 35 n*M* rhodamine B solution. Place the cuvette in the cuvette holder. Power the APDs. Open the appropriate program in the computer to monitor the CR and the $G(\tau)$ in real time.

2. Use TS1 to move the cuvette holder away from the objective until the focus of the objective leaves the solution and enters just inside the quartz window of the cuvette. At this point the CR decreases sharply to about 5 kHz. Record the position of the TS1. This position may be considered the approximate position of the inner surface of the optical window of the cuvette.
3. Move the TS1 in the direction of the objective slowly until the CR reaches a maximum value. The CR here is generally >50 kHz. Stop here and optimize the angular alignment of the cuvette by adjusting the knobs of the PM. Record the value of the CR. It may be seen from Fig. 2A that the CR is about 90 kHz after the first iteration.
4. Move the TS1 again to maximize CR. The TS1 should be moved in the direction of increasing CR. Move until the CR is maximum. Then adjust PM to maximize CR in the same way as mentioned earlier.

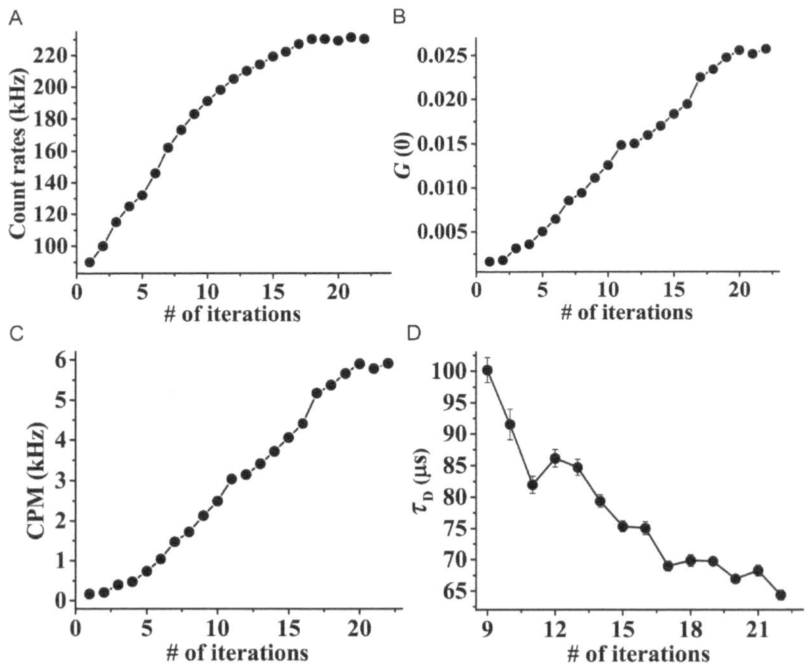

Fig. 2 Optimization of the alignment of the cuvette. Improvement of (A) count rate, (B) *G*(0), (C) counts per molecule (CPM) per second, and (D) diffusion time (τ_D) with number of iterations. Count rate reaches maximum by 17th iteration. However, *G*(0), CPM, and τ_D continue to improve till 22nd iterations. The *symbols* represent data and the *lines* are for guide to the eye.

5. Repeat Step 4 until CR does not increase any further. Fig. 2A shows how the CR increases with increasing iterations. The CR is shown to reach its maximum at 230 kHz after 17th iteration.
6. To optimize the alignment further we need to examine the $G(0)$ and the CPM. Move the TS1 away from the objective by 5 μm and record the $G(\tau)$. If the $G(0)$ increases from the previous measurement, then move the TS1 again by 5 μm in the same direction. If $G(0)$ is decreased, then move 10 μm in the opposite direction. We have observed that in this phase of the alignment, adjustment of the PM is not necessary.
7. The alignment is complete once CR, $G(0)$, and CPM have reached the corresponding maximum values. Fig. 2B and C shows how $G(0)$ and CPM reach maximum values with iterations. It may be seen from Fig. 2D that τ_D reaches minima at this position.

5. CHARACTERIZATION OF THE PERFORMANCE OF THE SETUP

5.1 Characterization of the $G(\tau)$ With Excitation Laser Power

Sensitivity of FCS measurements, i.e., S/N of the $G(\tau)$, is known to increase with the increase of CPM (see Eq. 7) (Ivanchenko & Lamb, 2011; Koppel, 1974; Saffarian & Elson, 2003; Sahoo et al., 2018), which can normally be increased with the increase of excitation power. However, the increase of excitation power can introduce artifacts related to the photophysical properties of the fluorophores. For example, at higher excitation intensities, contributions from triplet state dynamics of the fluorophore to the $G(\tau)$ may become significant. Furthermore, it can lead to saturation of fluorescence emission, altering the dimensions and the shape of the PSF. Therefore, $G(\tau)$ needs to be characterized as a function of excitation power before deciding what excitation power to be used in an experiment.

5.1.1 Equipment and Materials

- Power meter to measure laser power of a few μW (MODEL: PMKIT-05-01, Newport)
- ND filters with OD equal to 0.2, 0.4, 0.6, 0.8, 1.0, and 1.2
- A 35 nM rhodamine B solution in PBS buffer at pH 7.4

5.1.2 Procedure

1. Measure the power of the laser in front of the objective lens without using any ND filter. This is the highest excitation power at the sample. In our setup the highest power we can reach is 75 μW.
2. Take 2 mL of 35 n*M* rhodamine B solution in the cuvette. Make sure that the setup is aligned as discussed earlier.
3. Perform FCS measurements using the 1.2 OD filter first. Place the ND filter in front of the laser (see Fig. 1). Optimize the *X*–*Y* alignment of the multimode fibers (MF) each time the ND filter is changed. Generally, we record a total of 15 autocorrelation curves with acquisition time of 30 s each. Repeat the same with 1.0, 0.8, 0.6, 0.4, 0.2, and 0.0 OD filters. Use of these ND filters corresponds to 4.7, 7.5, 12, 19, 30, 47, and 75 μW of excitation power at the sample.

5.1.3 Analysis of the Autocorrelation Data G(τ)

(i) Plot all the $G(\tau)$ curves acquired with identical power in the same graph using any plotting software. We have used Origin 7.0 (Origin Lab). Examine if all the $G(\tau)$ data look similar.

(ii) Calculate the mean and the standard deviation (SD) of the $G(\tau)$ data at each τ. Plot the mean and SD of the $G(\tau)$ for each excitation power separately. The $G(\tau)$ data obtained using 7.5, 30, and 75 μW of excitation powers are shown in Fig. 3A and B.

(iii) Estimate S/N of the autocorrelation data and the CPM using the following:

$$\frac{S}{N} = \frac{\text{mean of } G(\tau)}{\text{SD of } G(\tau)} \tag{9}$$

$$\text{CPM} = \frac{CR}{\langle N \rangle} \tag{10}$$

(iv) Fit the $G(\tau)$ data using both Eqs. (3) and (8) for comparison. Ensure that the fitting program uses SD of the data to calculate the reduced χ^2 for optimization of the fitting. In origin if the $G(\tau)$ is plotted together with the SD, then the program takes the SD into consideration automatically.

(v) Generally, it is a good practice to initialize the fitting parameters reasonably close to the actual values to ensure that the fitting converges quickly.

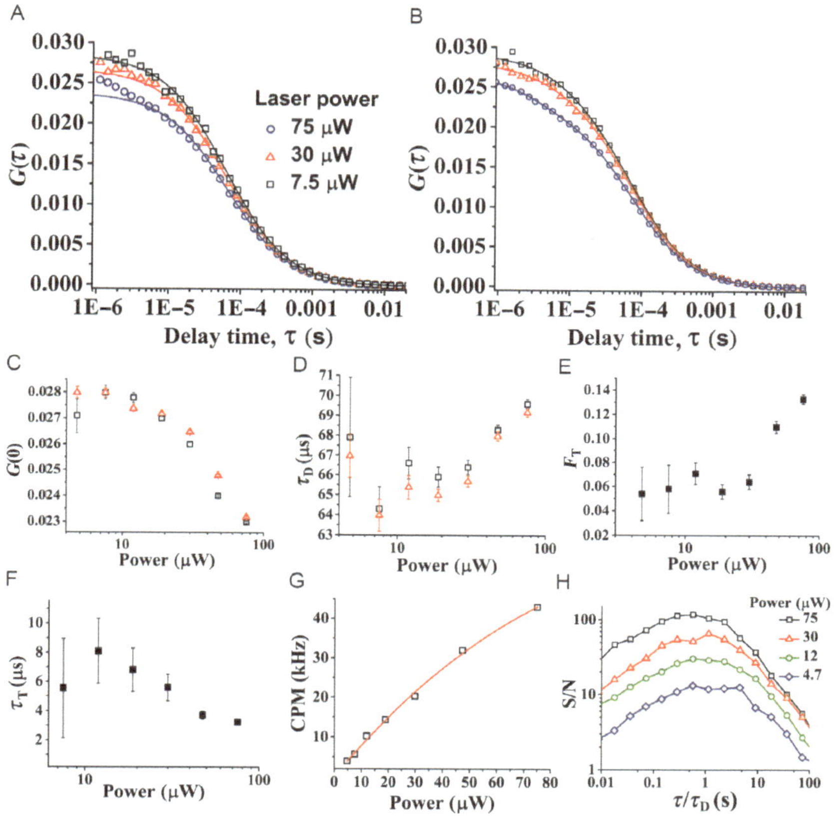

Fig. 3 Effect of excitation intensity on autocorrelation function. $G(\tau)$ data (*symbols*) with fits (*solid lines*) (A) using one-component diffusion and (B) using one-component diffusion plus triplet state dynamics. (C) $G(0)$ decreases and (D) τ_D increases with power indicating broadening of the PSF. (E) Triplet fraction (F_T) increases and (F) triplet correlation time (τ_T) decreases with power indicating increased rate of ISC. (G) Counts per molecule (CPM) and (H) S/N of $G(\tau)$ increases with power. CPM deviates from linearity above 40 μW indicating saturation effects at higher power. In (C) and (D), the *squares* represent results from fitting the $G(\tau)$ with one-component diffusion plus triplet state dynamics model, and *triangles* are from fitting the data from $\tau > 10\,\mu s$ using one-component diffusion model. The sample used is 35 n*M* rhodamine B in PBS at pH 7.4.

5.1.3.1 Notes

1. If the $G(\tau)$ data acquired under identical conditions do not overlap with each other, then there must be instabilities either in the setup or in the sample or both. Such things can happen, e.g., due to adsorption and photobleaching of the samples. Analysis of such data may give erroneous results.

5.1.4 Results

The $G(\tau)$ data obtained using a few representative powers of excitation have been compared in Fig. 3A and B. Clearly, noise in the $G(\tau)$ data is less at higher excitation powers, indicating improvement of S/N with power. Furthermore, $G(0)$ decreases at higher excitation power, indicating the increase of the V_{eff} with power. This happens because at high excitation intensity the fluorescence emission reaches saturation at the center of the PSF causing flattening of the peak and broadening of the PSF (Eggeling, Volkmer, & Seidel, 2005). The saturation of fluorescence emission at high intensity occurs mostly due to ISC from the S_1 to the T_1 state of the fluorophore. Since the lifetime of the triplet state is much longer than the S_1 state (a few μs vs a few ns), the fluorophore goes out of the excitation–emission cycle once it crosses to the T_1 state until it is deexcited back to the S_0 state. Therefore, fitting of the $G(\tau)$ data using a single diffusing species (see Eq. 3) works well for the data obtained at low power but not for those obtained at higher values of power (see Fig. 3A). Hence, fitting of the $G(\tau)$ data obtained using high power needs to include the contributions from the triplet state dynamics. Fig. 3B shows that all the $G(\tau)$ fit well when Eq. (8) is used. Fig. 3C shows that $G(0)$ remains almost constant at lower excitation power. However, $G(0)$ decreases monotonically for excitation power above 19 μW. As expected, the diffusion time (τ_D) also increases with power (Fig. 3D). Fig. 3E and F shows that the error bars in the values of the fraction of the triplet state population (F_T) and triplet correlation time (τ_T) are very high at lower excitation power. This is due to redundancy of the number of fitting parameters at low power. In fact the corresponding $G(\tau)$ data fit well with Eq. (3) (see Fig. 3A). Therefore, at low power contributions from the triplet state dynamics are insignificant. At excitation power >19 μW fraction of the triplet state population (F_T) increases and triplet correlation time (τ_T) decreases with excitation power. F_T increases due to the increased rate of ISC with the increase of excitation intensity. As discussed earlier, τ_T decreases with the increase of power due to the increase of effective rate of ISC at higher intensities (Schonle et al., 2014; Widengren et al., 1995). Fig. 3G and H shows that both CPM and S/N increase with the increase of excitation power. The CPM deviates from linearity for excitation power >40 μW, consistent with the saturation effects at high power.

Taken together, the data and the analysis presented earlier indicate that while the increase of excitation power helps to improve the S/N, high power also introduces additional components in the $G(\tau)$ due to contributions from the triplet state dynamics of the fluorophore. To avoid complications from the triplet state dynamics, we recommend using about 15 μW of excitation power in case of rhodamine B.

5.1.5 Notes

1. Here we have used excitation only up to 75 μW. Use of power even higher than this may lead to other effects such as photobleaching. Generally, it is advisable to use lesser excitation power to avoid such complications.
2. Dependence of the triplet state dynamics on the excitation power differs between fluorophores. For example, fluorescein exhibits more pronounced triplet state contributions to the $G(\tau)$ than Atto 488 (Blom, Chmyrov, Hassler, Davis, & Widengren, 2009). Hence, it is necessary to perform measurements of F_T and τ_T as a function of excitation power using the fluorophore of interest to decide the highest excitation power that may be used without significant contributions from the triplet state dynamics.
3. It may be noted that in Fig. 3E and F the error bars in F_T and τ_T are quite high at low excitation power primarily due to very small population in the triplet state. Therefore, it is better to use a one-component diffusion model to fit the $G(\tau)$ data obtained using low excitation power.
4. Many researchers prefer to use a one-component diffusion model (Eq. 3) to analyze the $G(\tau)$ data for simplicity of the analysis. This is generally done by truncating the $G(\tau)$ data below certain values of τ. For example, Fig. 3E and F shows that F_T is about 0.14 and the τ_T is about 3 μs at 75 μW excitation power. Therefore, contribution from the triplet state dynamics to the $G(\tau)$ is <1% for $\tau > 10$ μs. Hence the $G(\tau)$ data from $\tau > 10$ μs may be fit using Eq. (3). Fig. 3C and D shows that the values of $G(0)$ and τ_D obtained using this approach are almost the same as the values obtained from the analysis using Eq. (8).

5.2 Characterization of the FCS Observation Volume

Since the performance of an FCS setup is extremely sensitive to the spatial resolution of the PSF, it is important to characterize the FCS observation volume carefully. Below, we describe how V_{eff} can be determined from the values of $\langle N \rangle$ and the known concentration of the fluorophore (see Eq. 6). Alternatively, V_{eff} can also be determined from the τ_D and axial ratio (ω) (see Eqs. 4 and 5). Comparison of the values of V_{eff} obtained using these two different approaches can be used to test the validity of the 3D Gaussian approximation of the PSF.

5.2.1 Equipment and Materials

- A spectrophotometer (Jasco, V-670) for measurement of absorbance to determine concentration of the stock solution of rhodamine B
- The home-built cuvette-FCS
- A 50 μ*M* solution of bovine serum albumin (BSA) in PBS at pH 7.4
- A 4.3 μ*M* rhodamine B solution in PBS at pH 7.4 in the presence of 1 μ*M* BSA. Measure the concentration of the stock solution of rhodamine B using the absorbance at 555 nm. The extinction coefficient of rhodamine B in water is 106,000 (Dixon, Taniguchi, & Lindsey, 2005).

5.2.2 Procedure

1. Prepare a stock solution of rhodamine B in PBS buffer. Measure the absorbance in the spectrophotometer. Use the absorbance (A) of the solution at $\lambda = 555$ nm. Use the extinction coefficient (ε) of rhodamine B ($\varepsilon_{555} = 106{,}000$) to determine the concentration of the rhodamine B solution. The concentration of the stock solution measured in the experiments presented here is 4.3 μ*M*.
2. Prepare 2 mL solution of 1 μ*M* BSA in PBS buffer with no rhodamine B in it. Take this solution in the cuvette for FCS measurements.
3. Use a 0.7 OD ND filter in front of the laser. This gives about 15 μW of excitation laser power at the sample. Perform FCS measurements. Record the photon CR. This corresponds to the background counts ($\langle B \rangle$).
4. The background counts typically arise from the ambient light entering the detector, detector dark counts, and fluorescence impurities in the buffer. In our setup $\langle B \rangle$ is less than 0.4 kHz.
5. Add stock solution of rhodamine B into the 2 mL buffer to prepare a 15 n*M* rhodamine B with 1 μ*M* BSA in it. Perform FCS measurements. We record 10 autocorrelation traces of 30 s each. Note the average count rate ($\langle \mathrm{CR} \rangle$). Typically, we obtain $\langle \mathrm{CR} \rangle$ higher than 90 kHz per detection channel. Hence the contribution of the background ($\langle B \rangle$) in this is very low.
6. Add 5 μL of rhodamine B stock solution in the cuvette. Mix it thoroughly by using a 1-mL pipette. Do not touch the walls of the cuvette to avoid disturbing the alignment of the cuvette. Perform FCS measurements as before. Repeat the same with serial addition of 8, 10, and 20 μL of the 4.3 μ*M* stock solution of rhodamine B.

5.2.2.1 Notes

Since FCS experiments are performed using n*M* concentrations of fluorophores, the experimenter needs to be careful about the adsorption of the fluorophore molecules to the walls of the cuvette. To minimize adsorption, we have added 1 μ*M* BSA into the buffer. Some other researchers have used 0.001% Tween-20 (Hofmann et al., 2012). We have avoided detergents because trace amounts of it may affect the subsequent experiments involving proteins.

5.2.3 Analysis of the Autocorrelation Data G(τ)

(i) Plot all the $G(\tau)$ curves acquired under identical conditions together. Examine if all the $G(\tau)$ data look similar.

(ii) Calculate the mean and the SD of the $G(\tau)$ data at each τ. Plot the mean and the SD of the $G(\tau)$ for each concentration separately. We have used $G(\tau)$ data for $\tau > 10\,\mu s$ only for fitting the $G(\tau)$ data with the one-component diffusion model.

(iii) Fit the $G(\tau)$ data using Eq. (3). Ensure that the fitting program uses SD of the data to calculate the reduced χ^2 for optimization of the fitting.

(iv) Record the values of $\langle N \rangle$, τ_D, and ω obtained at each concentration of the fluorophore. Use Eq. (5) to calculate σ_r from τ_D, then σ_z using σ_r and ω. The diffusion coefficient (D) of rhodamine B is equal to 4.27×10^{-6} cm^2/s (Gendron, Avaltroni, & Wilkinson, 2008).

5.2.4 Results

It may be seen from Fig. 4A that the $\langle N \rangle$ increases linearly with the concentration (C) of rhodamine B. The slope obtained is 1.08 ± 0.01 molecule per n*M*. The V_{eff} calculated using Eq. (6) is found to be 1.8 fL. As discussed

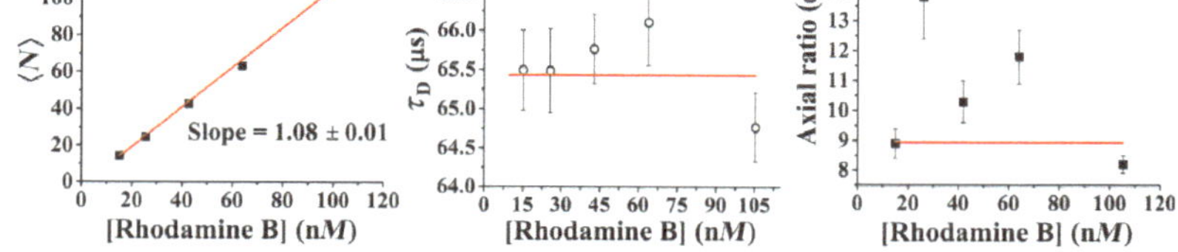

Fig. 4 Measurement of effective observation volume (V_{eff}) using known concentrations of rhodamine B in PBS. (A) $\langle N \rangle$ increases linearly with concentration. The slope of the line is 1.08 molecules/n*M*. From the slope the estimated $V_{eff} = 1.8$ fL (see Eq. 6). (B) τ_D and (C) ω at different concentrations of the fluorophore. The *solid lines* in (B) and (C) are weighted mean of τ_D and ω, respectively. The V_{eff} calculated using σ_r and σ_z obtained from the values of mean τ_D and ω (see Eqs. 3–5) is 1.9 ± 0.3 fL. The values of σ_r, ω, σ_z, and V_{eff} are summarized in Table 1.

Table 1 Summary of the Dimensions of the PSF in Cuvette-FCS

τ_D (μs)	ω	$\langle N \rangle/C$ (nM)$^{-1}$	σ_r (nm)	σ_z (μm)	V_{eff} (fL) (Using Eq. 4)	V_{eff} (fL) (Using Eq. 6)
65 ± 1	9.0 ± 0.9	1.08 ± 0.01	335 ± 6	3.01 ± 0.06	1.89 ± 0.3	1.79 ± 0.02

in the theory section that Eqs. (4) and (5) suggest an alternative approach to determine the V_{eff}. This approach does not require the knowledge of the concentration of the fluorophore; rather, σ_r and σ_z are calculated from the estimated values of τ_D and ω. It may be seen from Fig. 4B and C that the values of τ_D and ω are about 65.5 μs and 9.0, respectively. Hence the σ_r and σ_z are estimated to be about 335 nm and 3 μm, respectively. Using this approach, we find that the effective volume is about 1.9 fL. The values obtained for σ_r, ω, σ_z, and V_{eff} are summarized in Table 1. Clearly, V_{eff} estimated using two independent approaches is almost equal. Hence, 3D Gaussian approximation of PSF holds quite well in case of cuvette-FCS. Hess and Well have shown that small aberrations in the PSF can give rise to large, sometime diverging values of ω leading to unphysical values of the V_{eff} (Hess & Webb, 2002).

5.2.5 Notes

The linear relationship between $\langle N \rangle$ and V_{eff} (see Eq. 6) is valid only when the average photon counts from the fluorescence of the sample ($\langle F \rangle = \langle$total count rate$\rangle - \langle B \rangle$) is much higher than the average background counts ($\langle B \rangle$). For very low concentration (<1 nM) of the sample $\langle F \rangle$ can be comparable to $\langle B \rangle$. In such situations $\langle N \rangle$ needs to be corrected using the following formulae (Garai, Muralidhar, & Maiti, 2006; Koppel, 1974).

$$\langle N \rangle = \langle N \rangle_{\text{measured}} \left(\frac{\langle F \rangle}{\langle F + B \rangle} \right)^2 \tag{11}$$

The other problem that arises at very low concentrations is adsorption of the fluorophore to the walls of the sample cell. Hence, the minimum concentration of rhodamine B we have used here is 15 nM to avoid the above-mentioned complications. In our case the minimum $\langle F \rangle / \langle B \rangle$ is higher than 225. Therefore, the required correction factors under the experimental conditions reported here are less than 0.4%.

6. APPLICATION OF CUVETTE-FCS IN MEASUREMENT OF R_h OF PROTEINS IN UREA

A major advantage of cuvette-FCS is performing measurements in solvents such as the chemical denaturants, which have refractive indices different from water (Sahoo et al., 2018). Folding and unfolding of proteins using chemical denaturants such as urea and GdnCl are studied regularly in spectrofluorometers but rarely in FCS. However, experiments performed in spectrofluorometers provide only ensemble-level information, which are averages of the properties of the folded and the unfolded forms. On the other hand, single-molecule experiments can potentially investigate the property of the unfolded state under the native conditions and vice versa. For example, protein-unfolding experiments using single-molecule FRET (smFRET) and FCS have shown that the unfolded state of the protein may undergo expansion in chemical denaturant in GdnCl in a concentration-dependent manner (Hofmann et al., 2012; Sherman & Haran, 2006; Sherman et al., 2008). IDPs, which lack a well-defined structure, may also undergo collapse or expansion in chemical denaturants. Hofmann et al. used GdnCl-dependent expansion of prothymosin α (proTα) and activator for thyroid hormone and retinoid receptors (ACTR) to calculate the coil-to-globule (CG) transition and the solvation energy (Borgia et al., 2016; Hofmann et al., 2012). Soranno et al. determined the internal friction from solvent-dependent conformational dynamics and swelling or collapse of the polypeptide chains (Soranno et al., 2012). Trexler et al. have used smFRET and FCS to study the size and dynamics of α-synuclein to examine the correlations between the dimensions and the aggregation propensity of the protein (Trexler & Rhoades, 2010). Therefore, quantitative determination of the solvent-dependent hydrodynamic properties of the proteins can provide insights into the mechanisms of protein folding and the functions of the IDPs.

FCS measurements in chemical denaturants are fraught with difficulties due to the mismatch of refractive indices between the immersion water of the objective lens and the denaturant solutions. Several theoretical and experimental studies have characterized the aberrations arising due to the mismatch of refractive indices at the interface between the glass and the aqueous solution (Hell et al., 1993; Sheppard & Gu, 1992; Sheppard & Torok, 1997). The magnitudes of the aberrations increase dramatically with focal depth. For example, we have observed that the τ_D measured in FCS is extremely sensitive to the focal depth (Sahoo et al., 2018). Therefore, the distance between the focal spot of the objective and the inner surface of the cuvette needs to be kept constant at about 20 μm to avoid aberrations in denaturant solutions (Sahoo et al., 2018). A common procedure for FCS experiments with denaturation of proteins is the following. The solutions are prepared in different concentrations of urea typically in an eight-well sample chamber. Then each sample is measured independently in FCS. Hence, the distance between the glass and the objective needs to be adjusted for each sample to minimize aberrations making these experiments inconvenient and time consuming. On the other hand, the distance between the objective and the cuvette is fixed in cuvette-FCS. Therefore, adjustment of the distance for each sample is not required in cuvette-FCS. We have recently demonstrated that cuvette-FCS can be used successfully for measurements of diffusion times of proteins in urea (Sahoo et al., 2018). In the previous publication, we used an autotitrator for titration of urea into the native buffer. However, these experiments can also be performed easily by manual titrations. Below, we provide a step-by-step description of how such measurements could be performed. This section is broadly divided into two parts. In the first step, measurements on a rhodamine B solution are performed in different concentrations of urea to calibrate τ_D as a function of concentration of urea. Then we describe the procedure for the measurements on two fluorescently labeled proteins. Here we have used a natively folded protein, intestinal fatty acid-binding protein (IFABP), and an IDP α-synuclein to examine how these protein chains undergo expansions in urea. IFABP is a small 15 kDa protein, consisting of two β-sheets, each containing five β-strands and a small helical region (PDB id: 1AEL) (Hodsdon & Cistola, 1997; Sacchettini, Gordon, & Banaszak, 1989). The α-synuclein used in our experiments is 140-residues long consisting of (MW = 14.5 kDa) the N-terminal, the NAC (nonamyloid-β component), and the C-terminal regions. A cysteine mutant (A69C) of this protein has been used for TMR labeling. α-Synuclein does not form a stable secondary or tertiary structure although different regions are predicted to show different structural propensities (Marsh, Singh, Jia, & Forman-Kay, 2006). Below, we describe how cuvette-FCS can be used to measure hydrodynamic radii of the fluorescently labeled proteins as a function of concentrations of urea.

6.1 FCS Measurements on a Solution of Rhodamine B in Urea

6.1.1 Equipment and Materials

- A refractometer (DR-A1, Atago) to measure the refractive index of the urea solutions
- The home-built cuvette-FCS
- Rhodamine B solution (60 n*M*) in PBS, pH 7.4
- Urea (9 *M*) stock solution in PBS, pH 7.4

6.1.2 Procedure

(i) Measure the refractive index of the stock solution of urea. Use the measured refractive index to determine the concentration of urea in the solution. We use the online urea concentration calculator (http://sosnick.uchicago.edu/gdmcl.html). If the concentration is higher than 9 *M*, then add appropriate amount of PBS to adjust the concentration to 9 *M* urea.

(ii) Prepare a table for the titration of urea in the native buffer. Here we show the table that we have used (see Table 2).

(iii) Take a 2 mL solution of 60 n*M* rhodamine B in native buffer in the cuvette. Set the temperature of the cuvette holder to the desired temperature. Here we have used 25°C.

(iv) Use a 0.7 OD ND filter in front of the laser. This gives about 15 μW excitation power and 6 kHz of CPM per detection channel.

(v) If this experiment is being performed for the first time, then we recommend a *z*-scan, i.e., FCS measurements as a function of the focal depth (approximately between 0 and 70 μm). Fig. 5A shows the plot of CPM as a function of the focal depth. As expected the CPM is highest when the focal depth is about 25 μm. Optionally, one may also follow the τ_D as a function of focal depth. It may be seen from Fig. 5B that τ_D is the minimum at the position of the highest CPM. We will then compare this plot with a similar experiment at the end of the titration experiment to examine the stability of the setup and robustness of the measurements.

(vi) Record $G(\tau)$ data. We generally record 10 autocorrelation curves at each concentration of urea.

(vii) Follow Table 2 and withdraw the specified volume of the solution from the cuvette. Add the same volume of 9 *M* urea to keep the total

Table 2 Concentration of Urea in the Cuvette With Serial Addition of 9 *M* Urea

Sample Number	Total Volume (μL)	Volume Withdrawn From Cuvette (μL)	Volume of 9 *M* Urea Added to the Solution (μL)	[Urea] (*M*)	Relative Viscosity (cps)
1	2000	0	0	0.00	1.00
2	2000	100	100	0.45	1.02
3	2000	100	100	0.88	1.04
4	2000	100	100	1.28	1.05
5	2000	100	100	1.67	1.07
6	2000	200	200	2.22	1.10
7	2000	200	200	2.73	1.13
8	2000	200	200	3.20	1.16
9	2000	200	200	3.63	1.19
10	2000	300	300	4.17	1.23
11	2000	300	300	4.65	1.27
12	2000	300	300	5.09	1.31
13	2000	300	300	5.48	1.34
14	2000	400	400	6.01	1.40
15	2000	400	400	6.46	1.44
16	2000	400	400	6.84	1.49
17	2000	400	400	7.16	1.53

volume same. For example, row 2 in Table 2 shows that in the first step of the dilution series we withdraw 100 μL of the rhodamine B solution from the cuvette. Save this solution in a centrifuge tube clearly labeled with the sample number and the concentration of urea. This solution may be examined later using the refractometer if any confusion arises about the concentration of urea in the solution. Add equal volume (100 μL) of 9 *M* urea into the cuvette. Mix the solution with a 200 μL pipette gently. Take care that the pipette or the tip does not touch any part of the cuvette or the holder. Let the temperature of the sample equilibrate for ~2 min. Then perform the FCS measurements as before.

(viii) Repeat the withdraw/injection cycle as per Table 2 and perform the FCS measurements as described earlier.

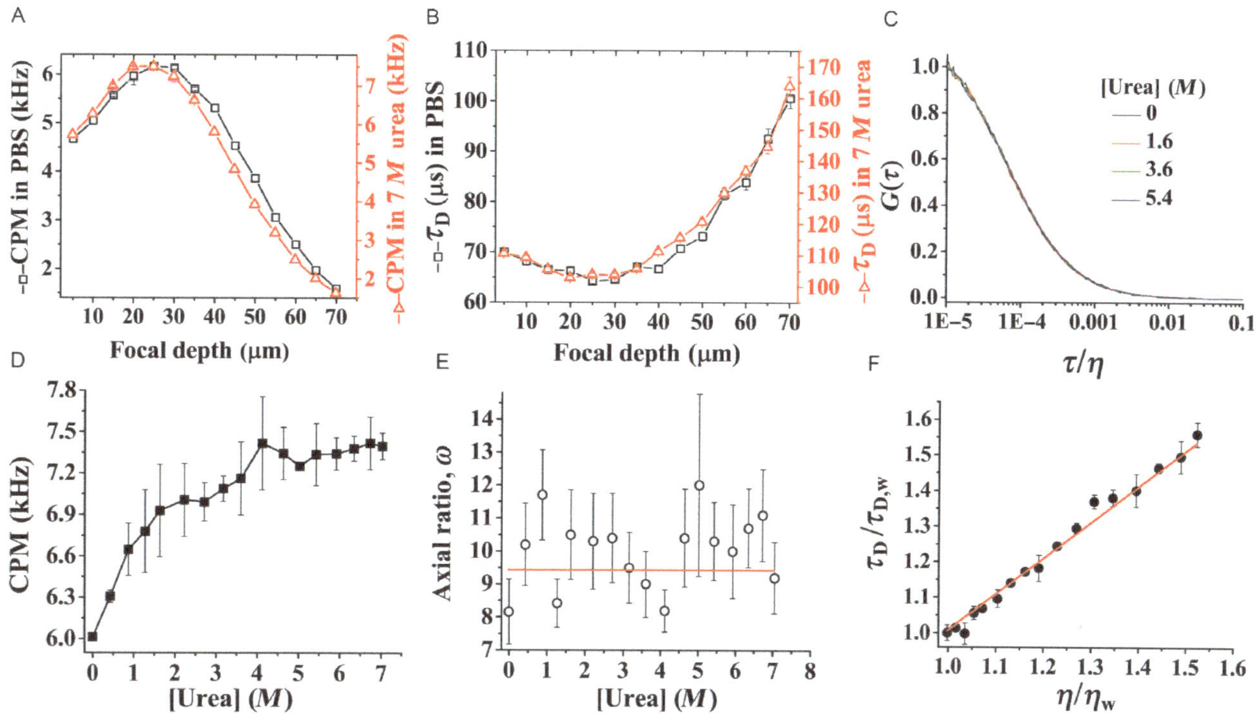

Fig. 5 Cuvette-FCS measurements on 60 n*M* rhodamine B in 0–7 *M* urea. (A) CPM and (B) τ_D vs focal depth in 0 *M* urea, i.e., at the beginning (*squares*) and in 7 *M* urea, i.e., at the end of the experiment (*triangles*). Clearly, the focal depth remains almost the same during the entire duration of the experiment indicating robustness of the setup. (C) Normalized $G(\tau)$ curves as a function of τ/η, where η is the viscosity of the solution. (D) CPM increases with the concentration of urea. An opposite trend would be expected if the observation volume suffers from aberrations in urea. (E) ω remains the same at all concentrations of urea. (D) Diffusion time increases linearly with the viscosity of urea consistent with the Stokes–Einstein relationship.

(ix) At the end of the measurement, i.e., at the highest concentration of urea perform FCS measurements as a function of sample *Z*-position as described earlier in Step (v). Record CPM and optionally, τ_D as a function of the focal depth. Compare this plot with the plot prepared at the beginning of the experiment. Data presented in Fig. 5A and B show that the plots of CPM and τ_D obtained in the beginning and at the end of the titration experiments agree quite well. This shows that the cuvette-FCS setup is quite stable.

(x) Repeat the entire experiment at least three times to test reproducibility of the measurements and to calculate the error bars.

6.1.2.1 Notes

1. Since the viscosity of water is extremely sensitive to the temperature, measurements of τ_D may be severely affected if the temperature of the cuvette holder is not maintained properly. Therefore, it is highly important that the temperature of the holder is maintained at a constant value. In our setup the temperature of the cuvette can be maintained within $\pm 0.5°C$ of the set temperature. Control of the temperature can be improved further by using PID-based temperature controllers.
2. To record CPM as a function of focal depth, it is not necessary to record, save, and analyze all the $G(\tau)$ data. At each position of the focal depth $G(0)$ may be noted from the $G(\tau)$ curve by eye estimation only. Mean CR can be noted from the time trace of CR by eye estimation. CPM can then be calculated by using the expression $CPM = CR \times G(0)$.
3. In the process of serial addition of urea the concentration of rhodamine B decreases serially. However, τ_D and CPM are intrinsic properties; hence these parameters are not affected.

6.1.3 Analysis of the FCS Data

(i) Calculate and plot both the mean and the SD of the $G(\tau)$ data from $\tau > 10\,\mu s$ at each concentration of urea as described in Section 5.2.3.

(ii) Fit the mean $G(\tau)$ data using Eq. (3) (one-component diffusion model). Make sure that the fitting program uses the SD to optimize the reduced χ^2 as discussed earlier.

(iii) Record CPM, τ_D, and ω at each concentration of urea.

(iv) A simple way to examine if the $G(\tau)$ data have undergone changes in urea is to plot all the normalized $G(\tau)$ curves together with the τ normalized by the viscosity (η) of the corresponding solution. For this plot the values of τ need to be divided by the η of the corresponding urea solution.

6.1.4 Results

Fig. 5C shows the viscosity normalized $G(\tau)$ curves obtained in different concentrations of urea. To plot these curves the τ values have been normalized by the viscosity of the corresponding solution and the $G(\tau)$ curves are normalized by the $G(0)$. It may be seen that all the $G(\tau)$ curves overlap with each other. In Fig. 5D and E we have plotted the CPM and the axial ratio (ω) as a function of concentration of urea. It may be seen that the CPM of rhodamine B increases in urea. This is due to the increase of quantum yield of the dye in urea (Sahoo et al., 2018). It may be noted here that CPM suffers significantly in case of optical aberrations in the presence of mismatch of refractive indices (Chattopadhyay et al., 2005; Hell et al., 1993). Therefore, this plot serves as a good indicator of the quality of the PSF and the robustness of the FCS measurements. Furthermore, it may be seen that ω remains almost constant within 10 ± 2 over the entire range of the concentration of urea. Once again it indicates minimal or no aberrations in the PSF. In Fig. 5F, we plot $\tau_D/\tau_{D,w}$ as a function of viscosity of urea (η/η_w). The $\tau_{D,w}$ and η_w are the diffusion time of rhodamine B in water and the viscosity of water, respectively. The values of η at different concentrations of urea are listed in Table 2 obtained from Kawahara and Tanford (1966). It may be seen here that the $\tau_D/\tau_{D,w}$ increases linearly with η/η_w, consistent with the Stokes–Einstein relationship (Sherman et al., 2008). Taken together results from these experiments indicate that the cuvette-FCS provides robust measurements of τ_D in urea.

6.1.5 Notes

1. Minimal aberrations due to refractive index mismatch in cuvette-FCS can be due to the following reason. We are using a low NA objective. Optical aberrations in such cases are shown to be low (Hell et al., 1993).
2. While fitting the $G(\tau)$ data with Eq. (3) here we have kept ω as a free parameter. However, sometimes more consistent values of τ_D may be obtained if the $G(\tau)$ curves at all concentrations of urea are fit using a fixed value of ω. For this purpose, ω may be fixed to its estimated average value (see Fig. 5E).
3. The Stokes–Einstein equation is the following:

$$D = \frac{k_B T}{6\pi\eta R_h} \tag{12}$$

Since τ_D is inversely proportional to D (see Eq. 5), the τ_D is expected to be proportional to η.

6.2 Measurements of Chain Expansion of a Natively Folded and a Disordered Protein in Urea

Here we describe the application of cuvette-FCS to study the unfolding of a natively folded protein IFABP and expansion of an IDP α-synuclein in urea.

6.2.1 Equipment and Materials

- The home-built cuvette-FCS
- Plasmid of IFABP (a gift from Carl Frieden, Washington University, St. Louis)
- Plasmid of α-synuclein (A69C) (a gift from Samrat Mukhopadhyay, IISER, Mohali, India)
- The stock solution of urea (9 *M*) in PBS at pH 7.4

6.2.2 Procedure

1. Expression and purification of IFABP and α-synuclein are performed using published protocols (Jiang & Frieden, 1993; Volles & Lansbury, 2007). IFABP is labeled with 5/6-carboxy-tetramethyl-rhodamine succinimidyl ester at the N-terminal amine and α-synuclein (A69C) is labeled with TMR-maleimide using standard protocols (Sahoo et al., 2018).
2. Take 60 n*M* TMR-IFABP or TMR-α-synuclein in PBS in the cuvette. Cuvette-FCS measurements of unfolding of TMR-IFABP and TMR-α-synuclein are performed using the same procedure described in Section 6.1.2.

6.2.2.1 Notes

1. Here we have performed serial dilution of the protein in urea. However, if the unfolding of the protein in urea is slow, taking longer than a few minutes then it is better to prepare the protein solutions in urea individually, incubate these at RT for overnight, and then measure them in FCS individually. In this case, the cuvette should be emptied using the diaphragm vacuum pump connected to the flexible PTFE tubing to avoid disturbance of the alignment of the cuvette. While adding new sample care should be taken to ensure that the pipette or the tip does not touch any part of the cuvette or the holder to avoid any disturbance of the alignment of the cuvette.

6.2.3 Analysis of Data

1. Calculate the normalized $G(\tau)$ curves with viscosity normalized τ for both the proteins at all the concentrations of urea.
2. Analyze the $G(\tau)$ data for both the proteins using the same procedure described in Section 6.1.3. Use data from $\tau > 10\,\mu s$ only. Calculate the viscosity normalized τ_D at each concentration of urea.
3. Calculate R_h of the proteins at each concentrations of urea using the following equation.

$$R_{h,\text{protein}} = \frac{\tau_{D,\text{protein}}}{\tau_{D,\text{rhodamine B}}} R_{h,\text{rhodamine B}} \tag{13}$$

The $R_{h,\text{rhodamine B}}$ is known to be 0.57 nm (Gendron et al., 2008). The τ_D of rhodamine B at a concentration of urea is taken from the fitted line presented in Fig. 5F.

6.2.4 Results

Fig. 6A and B shows the normalized $G(\tau)$ data with viscosity normalized τ for TMR-IFABP and TMR-α-synuclein, respectively. Unlike rhodamine B the $G(\tau)$ data for TMR-IFABP show considerable differences between 0 and 7 M urea indicating changes of R_h of the protein in urea. In contrast, the differences in $G(\tau)$ curves are very small in case of α-synuclein. Fig. 6C shows the plot of R_h of IFABP as a function of urea. It may be noted that the R_h of IFABP at 0 M urea is about 2.2 nm. Using the program Hydropro and the structure of apo-IFABP (PDB id: 1AEL) (Hodsdon & Cistola, 1997) the calculated R_h of IFABP is 2.15 nm (Ortega, Amorós, & García de la Torre, 2011). Therefore, R_h measured in cuvette-FCS is close to the estimated R_h. Furthermore, it may be seen from Fig. 6C that the R_h of IFABP undergoes an expansion from 2.2 to 3.4 nm between 0 and 7 M urea. The major extent of the expansion occurs between 5 and 7 M urea consistent with cooperativity of unfolding of the protein in urea. These results are consistent with the previous report that τ_D of IFABP increases by about 1.8-fold in GdnCl (Chattopadhyay et al., 2005). Fig. 6D shows the R_h of α-synuclein increases from 3.07 to 3.22 nm between 0 and 7 M urea, indicating a very small extent of expansion of the polypeptide chain. Unlike IFABP (see Fig. 6C) expansion of α-synuclein does not exhibit cooperativity consistent with its disordered nature under native conditions. It may be seen that while the molecular weights of IFABP and the α-synuclein are similar (~15 kDa), R_h of α-synuclein in the native buffer is significantly higher than that of

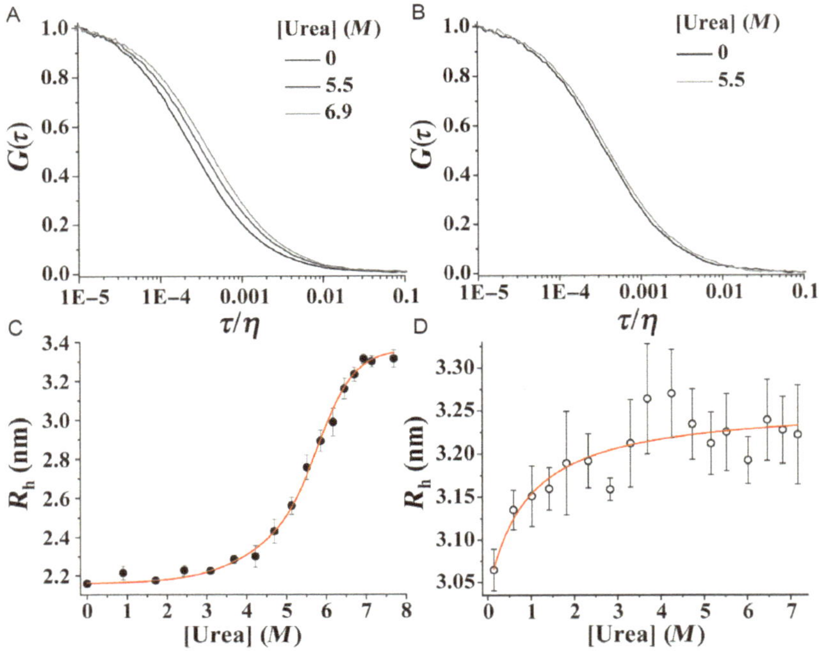

Fig. 6 Measurement of R_h of TMR-IFABP and TMR-α-synuclein in urea. Normalized $G(\tau)$ curves as function of τ/η obtained from TMR-IFABP (A) and TMR-α-synuclein (B) in different concentrations of urea. R_h as a function of concentration of urea for TMR-IFABP (C) and TMR-α-synuclein (D). While IFABP shows a large and cooperative expansion of size in urea, expansion observed in case of α-synuclein is small.

IFABP, indicating an already expanded conformation of α-synuclein in the native buffer. These results are consistent with the previously reported observations that α-synuclein is already expanded in neutral pH; hence it exhibits minimal or no expansion in GdnCl (Trexler & Rhoades, 2010). Taken together the results presented here indicate that cuvette-FCS can be used to measure effects of solvents such as urea on the R_h of both natively folded and disordered proteins.

6.2.5 Notes

1. Here we have discussed the effects of urea only on the R_h of the proteins. FCS measurements can also be used to extract information about the conformational dynamics both in the natively folded and in the disordered proteins (Chattopadhyay, Saffarian, Elson, & Frieden, 2002; Chen, Rhoades, Butler, Loh, & Webb, 2007; Ducas & Rhoades, 2014).

This is generally extracted from fitting the $G(\tau)$ with Eq. (8) and observing the dependence of F_T and τ_T on the concentration of urea. However, interpretation of such observations can be complex particularly in the presence of triplet state dynamics of the fluorophore. Therefore, we have not discussed the effects of urea on the conformational dynamics here. To quantify the amplitudes and the timescales of conformational dynamics use of two chromophores on the proteins is more appropriate (Chattopadhyay et al., 2002; Chen et al., 2007; Ducas & Rhoades, 2014).

7. SUMMARY

We have provided detailed instructions for building a highly sensitive cuvette-FCS setup. Furthermore, we have described the procedures for characterization of the resolution and the sensitivity of the setup from the analysis of $G(\tau)$ obtained using a standard fluorophore. Although the cuvette-FCS setup uses an objective lens with considerably lower NA than the regular microscope-based FCS, the resolution and the sensitivity achieved in this setup are comparable to some of the commercially available FCS (Ruttinger et al., 2008; Sahoo et al., 2018). For example, the radial resolution and the axial resolution obtained in cuvette-FCS are about 335 nm and 3 μm, respectively. Furthermore, CPM > 40 kHz and S/N > 100 can be obtained in cuvette-FCS using about 75 μW of excitation power. Sahoo et al. have compared the resolution of the PSF obtained in cuvette-FCS with the PSF obtained using a commercially available FCS from PicoQuant and found that the radial resolution of the FCS observation volume in the commercial FCS is about 1.4× better than that of the cuvette-FCS (Sahoo et al., 2018). However, the sensitivity (measured by the CPM and the S/N) of the cuvette-FCS is almost the same as the PicoQuant FCS (Sahoo et al., 2018). A few authors have previously used long working distance objectives to perform FCS measurements through thick glass windows. However, the CPM reported in these experiments has been much lower than the conventional FCS (Altamore et al., 2013; Banachowicz et al., 2014). The improved performance of our cuvette-FCS setup may be attributed to identifying and optimizing the factors that critically affect the PSF of the observation volume. For example, somewhat unexpectedly we find that both the quality and the alignment of the cuvette are extremely important determinants of the sensitivity of the cuvette-FCS.

We have then described the procedures to examine if 3D Gaussian approximation works suitably for the PSF in cuvette-FCS. The analytical forms of $G(\tau)$ presented in Eqs. (3) and (8) are based on 3D Gaussian approximation of the PSF. Several authors have characterized the confocal PSF in conventional FCS both theoretically and experimentally and concluded that the PSF deviates significantly from 3D Gaussian (Enderlein, Gregor, Patra, Dertinger, & Kaupp, 2005; Hess & Webb, 2002; Ivanchenko & Lamb, 2011). However, such deviations are less when low NA objectives are used (Hell et al., 1993). Here we have followed several approaches to examine the possible artifacts in cuvette-FCS. For example, we have examined whether the axial ratio (ω) of the PSF converges to a reasonable value from fitting of the $G(\tau)$ data (using Eq. 3 or 8). Figs. 4C and 5E show that ω in cuvette-FCS converges to around 9 ± 1. Typically, ω diverges to large values or even to infinity in the presence of even small deviations of the PSF from the 3D Gaussian (Hess & Webb, 2002). Additionally, we find that V_{eff} in cuvette-FCS measured using two independent approaches (see Eqs. 4 and 6) agrees well. Therefore, contributions from aberrations in cuvette-FCS seem to be small. We have also examined the effects of excitation laser power on the functional forms of $G(\tau)$ and described the effects of the triplet state dynamics of the fluorophore on $G(\tau)$ at higher excitation intensities.

Cuvette-FCS has several potential advantages over microscope-based FCS. The advantages are mostly about convenience of performing certain experiments mentioned in more detail in Section 2. It may be further emphasized that a major advantage of cuvette-FCS is that it can potentially be applied to most of the experiments performed in spectrofluorometers. Here we have described the effects of a regularly used chemical denaturant, viz., urea on the hydrodynamic size of a natively folded and a disordered protein. While conventional FCS has been used in such studies, we find that cuvette-FCS measurements are highly convenient in these experiments. Sahoo et al. have shown cuvette-FCS coupled with an automatic titrator can be used to monitor unfolding of an N-terminal fragment of apolipoprotein E4 in urea with high accuracy (Sahoo et al., 2018). Here we have shown that the titrations of urea can be performed manually as well. We think that the main reasons behind the robustness of the cuvette-FCS measurements in urea are the following: first, the focal depth remains constant because the distance between the cuvette and the objective remains fixed over the entire duration of the experiment; second, the use of low NA objective reduces optical aberrations in the PSF (Hell et al., 1993). Therefore, even though the use of low NA objectives reduces S/N of the data, it can improve the accuracy of the measurements by eliminating or reducing the effects of the optical aberrations in these experiments. Quantitative measurements of the hydrodynamic properties of the folded and the disordered proteins in chemical denaturants, in solvents with varying viscosities and at different temperatures, can reveal critical insights into the mechanisms of protein folding and the functions of the IDPs (Hofmann et al., 2012; Nettels et al., 2009; Soranno et al., 2012). We think that cuvette-FCS offers significant advantages over the microscope-based FCS in performing these experiments due to ease of use and the robustness of the measured values of the parameters. Furthermore, cuvette-FCS may also be used in smFRET measurements.

7.1 Current Limitations and Future Improvements of Cuvette-FCS

While cuvette-FCS has several advantages over conventional FCS, there are a few disadvantages too. The major drawback is that the CPM is limited by the low NA of the objective. Hence, cuvette-FCS may not be suitable for measuring fast (~few ns) conformational dynamics of biomolecules. The other disadvantage is due to the use of the dry objective. To avoid artifacts from optical aberrations the focal depth needs to be maintained carefully. However, we think that both the sensitivity and the ease of use of the cuvette-FCS can be improved further in the future. For example, currently commercially available choice of the ELWD objectives is limited to the dry objectives. In the future if the water immersion ELWD objectives are available commercially, then the alignment of the setup is expected to be easier and the sensitivity also can improve significantly.

ACKNOWLEDGMENT

K.G. received funding from Department of Science and Technology, India.

REFERENCES

Altamore, I., Lanzano, L., & Gratton, E. (2013). Dual channel detection of ultra low concentration of bacteria in real time by scanning FCS. *Measurement Science & Technology*, *24*(6), 065702. https://doi.org/10.1088/0957-0233/24/6/065702.

Banachowicz, E., Patkowski, A., Meier, G., Klamecka, K., & Gapiński, J. (2014). Successful FCS experiment in nonstandard conditions. *Langmuir*, *30*(29), 8945–8955. https://doi.org/10.1021/la5015708.

Blom, H., Chmyrov, A., Hassler, K., Davis, L. M., & Widengren, J. (2009). Triplet-state investigations of fluorescent dyes at dielectric interfaces using total internal reflection fluorescence correlation spectroscopy. *Journal of Physical Chemistry A*, *113*(19), 5554–5566. https://doi.org/10.1021/jp8110088.

Borgia, A., Zheng, W. W., Buholzer, K., Borgia, M. B., Schuler, A., Hofmann, H., et al. (2016). Consistent view of polypeptide chain expansion in chemical denaturants from multiple experimental methods. *Journal of the American Chemical Society, 138*(36), 11714–11726. https://doi.org/10.1021/jacs.6b05917.

Cardarelli, F., & Gratton, E. (2010). In vivo imaging of single-molecule translocation through nuclear pore complexes by pair correlation functions. *PLoS One, 5*(5), e10475. https://doi.org/10.1371/journal.pone.0010475.

Chattopadhyay, K., Saffarian, S., Elson, E. L., & Frieden, C. (2002). Measurement of microsecond dynamic motion in the intestinal fatty acid binding protein by using fluorescence correlation spectroscopy. *Proceedings of the National Academy of Sciences of the United States of America, 99*(22), 14171–14176. https://doi.org/10.1073/pnas.172524899.

Chattopadhyay, K., Saffarian, S., Elson, E. L., & Frieden, C. (2005). Measuring unfolding of proteins in the presence of denaturant using fluorescence correlation spectroscopy. *Biophysical Journal, 88*(2), 1413–1422. https://doi.org/10.1529/biophysj.104.053199.

Chen, H. M., Rhoades, E., Butler, J. S., Loh, S. N., & Webb, W. W. (2007). Dynamics of equilibrium structural fluctuations of apomyoglobin measured by fluorescence correlation spectroscopy. *Proceedings of the National Academy of Sciences of the United States of America, 104*(25), 10459–10464. https://doi.org/10.1073/pnas.0704073104.

Crick, S. L., Jayaraman, M., Frieden, C., Wetzel, R., & Pappu, R. V. (2006). Fluorescence correlation spectroscopy shows that monomeric polyglutamine molecules form collapsed structures in aqueous solutions. *Proceedings of the National Academy of Sciences of the United States of America, 103*(45), 16764–16769. https://doi.org/10.1073/pnas.0608175103.

Digman, M. A., & Gratton, E. (2009). Fluorescence correlation spectroscopy and fluorescence cross-correlation spectroscopy. *Wiley Interdisciplinary Reviews. Systems Biology and Medicine, 1*(2), 273–282. https://doi.org/10.1002/wsbm.5.

Dixon, J. M., Taniguchi, M., & Lindsey, J. S. (2005). PhotochemCAD 2: A refined program with accompanying spectral databases for photochemical calculations. *Photochemistry and Photobiology, 81*(1), 212–213. https://doi.org/10.1111/j.1751-1097.2005.tb01544.x.

Ducas, V. C., & Rhoades, E. (2014). Investigation of intramolecular dynamics and conformations of alpha-, beta- and gamma-synuclein. *PLoS One, 9*(1), e86983. https://doi.org/10.1371/journal.pone.0086983.

Eggeling, C., Volkmer, A., & Seidel, C. A. M. (2005). Molecular photobleaching kinetics of rhodamine 6G by one- and two-photon induced confocal fluorescence microscopy. *Chemphyschem, 6*(5), 791–804. https://doi.org/10.1002/cphc.200400509.

Elson, E. L. (2011). Fluorescence correlation spectroscopy: Past, present, future. *Biophysical Journal, 101*(12), 2855–2870. https://doi.org/10.1016/j.bpj.2011.11.012.

Elson, E. L., & Magde, D. (1974). Fluorescence correlation spectroscopy. 1. Conceptual basis and theory. *Biopolymers, 13*(1), 1–27. https://doi.org/10.1002/bip.1974.360130102.

Enderlein, J., Gregor, I., Patra, D., Dertinger, T., & Kaupp, U. B. (2005). Performance of fluorescence correlation spectroscopy for measuring diffusion and concentration. *Chemphyschem, 6*(11), 2324–2336. https://doi.org/10.1002/cphc.200500414.

Ferreon, A. C. M., Moran, C. R., Gambin, Y., & Deniz, A. A. (2010). Single-molecule fluorescence studies of intrinsically disordered proteins. In N. G. Walter (Ed.), *Methods in enzymology: Vol. 472* (pp. 179–204): Academic Press [Chapter 10].

Garai, K., Muralidhar, M., & Maiti, S. (2006). Fiber-optic fluorescence correlation spectrometer. *Applied Optics, 45*(28), 7538–7542. https://doi.org/10.1364/AO.45.007538.

Gendron, P.-O., Avaltroni, F., & Wilkinson, K. J. (2008). Diffusion coefficients of several rhodamine derivatives as determined by pulsed field gradient–nuclear magnetic resonance and fluorescence correlation spectroscopy. *Journal of Fluorescence, 18*(6), 1093. https://doi.org/10.1007/s10895-008-0357-7.

Guldbrand, S., Kirejev, V., Simonsson, C., Goksör, M., Smedh, M., & Ericson, M. B. (2013). Two-photon fluorescence correlation spectroscopy as a tool for measuring molecular diffusion within human skin. *European Journal of Pharmaceutics and Biopharmaceutics, 84*(2), 430–436. https://doi.org/10.1016/j.ejpb.2012.10.001.

Haupts, U., Maiti, S., Schwille, P., & Webb, W. W. (1998). Dynamics of fluorescence fluctuations in green fluorescent protein observed by fluorescence correlation spectroscopy. *Proceedings of the National Academy of Sciences of the United States of America, 95*(23), 13573–13578. https://doi.org/10.1073/pnas.95.23.13573.

Hell, S., Reiner, G., Cremer, C., & Stelzer, E. H. K. (1993). Aberrations in confocal fluorescence microscopy induced by mismatches in refractive index. *Journal of Microscopy, 169*(3), 391–405. https://doi.org/10.1111/j.1365-2818.1993.tb03315.x.

Hess, S. T., & Webb, W. W. (2002). Focal volume optics and experimental artifacts in confocal fluorescence correlation spectroscopy. *Biophysical Journal, 83*(4), 2300–2317. https://doi.org/10.1016/S0006-3495(02)73990-8.

Hodsdon, M. E., & Cistola, D. P. (1997). Ligand binding alters the backbone mobility of intestinal fatty acid-binding protein as monitored by N-15 NMR relaxation and H-1 exchange. *Biochemistry, 36*(8), 2278–2290. https://doi.org/10.1021/bi962018l.

Hofmann, H., Soranno, A., Borgia, A., Gast, K., Nettels, D., & Schuler, B. (2012). Polymer scaling laws of unfolded and intrinsically disordered proteins quantified with single-molecule spectroscopy. *Proceedings of the National Academy of Sciences of the United States of America, 109*(40), 16155–16160. https://doi.org/10.1073/pnas.1207719109.

Ivanchenko, S., & Lamb, D. C. (2011). *Fluorescence correlation spectroscopy: Principles and developments*. Dordrecht: Springer. https://doi.org/10.1007/978-94-007-0893-8_1.

Jiang, N., & Frieden, C. (1993). Intestinal fatty-acid-binding protein: Characterization of mutant proteins containing inserted cysteine residues. *Biochemistry, 32*(41), 11015–11021. https://doi.org/10.1021/bi00092a010.

Kask, P., Gunther, R., & Axhausen, P. (1997). Statistical accuracy in fluorescence fluctuation experiments. *European Biophysics Journal with Biophysics Letters, 25*(3), 163–169. https://doi.org/10.1007/s002490050028.

Kawahara, K., & Tanford, C. (1966). Viscosity and density of aqueous solutions of urea and guanidine hydrochloride. *Journal of Biological Chemistry, 241*(13), 3228–3232.

Koppel, D. E. (1974). Statistical accuracy in fluorescence correlation spectroscopy. *Physical Review A, 10*(6), 1938–1945.

Marsh, J. A., Singh, V. K., Jia, Z. C., & Forman-Kay, J. D. (2006). Sensitivity of secondary structure propensities to sequence differences between alpha- and gamma-synuclein: Implications for fibrillation. *Protein Science, 15*(12), 2795–2804. https://doi.org/10.1110/ps.062465306.

Matayoshi, E., & Swift, K. (2001). Applications of FCS to protein-ligand interactions: Comparison with fluorescence polarization. In R. Rigler & E. S. Elson (Eds.), *Fluorescence correlation spectroscopy: Theory and applications* (pp. 84–98). Berlin, Heidelberg: Springer Berlin Heidelberg.

Nettels, D., Muller-Spath, S., Kuster, F., Hofmann, H., Haenni, D., Ruegger, S., et al. (2009). Single-molecule spectroscopy of the temperature-induced collapse of unfolded proteins. *Proceedings of the National Academy of Sciences of the United States of America, 106*(49), 20740–20745. https://doi.org/10.1073/pnas.0900622106.

Ortega, A., Amorós, D., & García de la Torre, J. (2011). Prediction of hydrodynamic and other solution properties of rigid proteins from atomic- and residue-level models. *Biophysical Journal, 101*(4), 892–898. https://doi.org/10.1016/j.bpj.2011.06.046.

Petrasek, Z., Ries, J., & Schwille, P. (2010). Scanning FCS for the characterization of protein dynamics in live cells. *Methods in Enzymology, 472*, 317–343. https://doi.org/10.1016/S0076-6879(10)72005-X.

Połatyńska, A., Tomczyk, K., Pochylski, M., Meier, G., Gapinski, J., Banachowicz, E., et al. (2017). Temperature dependent FCS studies using a long working distance objective: Viscosities of supercooled liquids and particle size. *The Journal of Chemical Physics*, *146*(8), 084506. https://doi.org/10.1063/1.4977047.

Qian, H. (1990). On the statistics of fluorescence correlation spectroscopy. *Biophysical Chemistry*, *38*(1–2), 49–57. https://doi.org/10.1016/0301-4622(90)80039-A.

Ruttinger, S., Buschmann, V., Kramer, B., Erdmann, R., Macdonald, R., & Koberling, F. (2008). Comparison and accuracy of methods to determine the confocal volume for quantitative fluorescence correlation spectroscopy. *Journal of Microscopy*, *232*(2), 343–352.

Sacchettini, J. C., Gordon, J. I., & Banaszak, L. J. (1989). Refined apoprotein structure of rat intestinal fatty-acid binding-protein produced in *Escherichia coli*. *Proceedings of the National Academy of Sciences of the United States of America*, *86*(20), 7736–7740. https://doi.org/10.1073/pnas.86.20.7736.

Saffarian, S., & Elson, E. L. (2003). Statistical analysis of fluorescence correlation spectroscopy: The standard deviation and bias. *Biophysical Journal*, *84*(3), 2030–2042.

Sahoo, B., Sil, T. B., Karmakar, B., & Garai, K. (2018). A fluorescence correlation spectrometer for measurements in cuvettes. *Biophysical Journal*, *115*(3), 455–466. https://doi.org/10.1016/j.bpj.2018.05.038.

Schonle, A., Von Middendorff, C., Ringemann, C., Hell, S. W., & Eggeling, C. (2014). Monitoring triplet state dynamics with fluorescence correlation spectroscopy: Bias and correction. *Microscopy Research and Technique*, *77*(7), 528–536. https://doi.org/10.1002/jemt.22368.

Sheppard, C. J. R. (1989). Axial resolution of confocal fluorescence microscopy. *Journal of Microscopy*, *154*(3), 237–241. https://doi.org/10.1111/j.1365-2818.1989.tb00586.x.

Sheppard, C. J. R., & Gu, M. (1992). Axial imaging through an aberrating layer of water in confocal microscopy. *Optics Communications*, *88*(2), 180–190. https://doi.org/10.1016/0030-4018(92)90508-O.

Sheppard, C. J. R., & Torok, P. (1997). Effects of specimen refractive index on confocal imaging. *Journal of Microscopy-Oxford*, *185*, 366–374. https://doi.org/10.1046/j.1365-2818.1997.d01-627.x.

Sherman, E., & Haran, G. (2006). Coil–globule transition in the denatured state of a small protein. *Proceedings of the National Academy of Sciences of the United States of America*, *103*(31), 11539–11543. https://doi.org/10.1073/pnas.0601395103.

Sherman, E., Itkin, A., Kuttner, Y. Y., Rhoades, E., Amir, D., Haas, E., et al. (2008). Using fluorescence correlation spectroscopy to study conformational changes in denatured proteins. *Biophysical Journal*, *94*(12), 4819–4827. https://doi.org/10.1529/biophysj.107.120220.

Soranno, A., Buchli, B., Nettels, D., Cheng, R. R., Muller-Spath, S., Pfeil, S. H., et al. (2012). Quantifying internal friction in unfolded and intrinsically disordered proteins with single-molecule spectroscopy. *Proceedings of the National Academy of Sciences of the United States of America*, *109*(44), 17800–17806. https://doi.org/10.1073/pnas.1117368109.

Tran, H. T., Mao, A., & Pappu, R. V. (2008). Role of backbone—Solvent interactions in determining conformational equilibria of intrinsically disordered proteins. *Journal of the American Chemical Society*, *130*(23), 7380–7392. https://doi.org/10.1021/ja710446s.

Trexler, A. J., & Rhoades, E. (2010). Single molecule characterization of alpha-synuclein in aggregation-prone states. *Biophysical Journal*, *99*(9), 3048–3055. https://doi.org/10.1016/j.bpj.2010.08.056.

Volles, M. J., & Lansbury, P. T. (2007). Relationships between the sequence of α-synuclein and its membrane affinity, fibrillization propensity, and yeast toxicity. *Journal of Molecular Biology*, *366*(5), 1510–1522. https://doi.org/10.1016/j.jmb.2006.12.044.

Widengren, J., Mets, U., & Rigler, R. (1995). Fluorescence correlation spectroscopy of triplet-states in solution—A theoretical and experimental-study. *Journal of Physical Chemistry*, *99*(36), 13368–13379. https://doi.org/10.1021/j100036a009.
Wilson, T., & Carlini, A. R. (1989). The effect of aberrations on the axial response of confocal imaging-systems. *Journal of Microscopy*, *154*, 243–256. https://doi.org/10.1111/j.1365-2818.1989.tb00587.x.
Wohland, T. (2014). Imaging fluorescence correlation spectroscopy measures dynamics and structure in live samples. *Biophysical Journal*, *106*(2), 7a. https://doi.org/10.1016/j.bpj.2013.11.084.
Wohland, T., Rigler, R., & Vogel, H. (2001). The standard deviation in fluorescence correlation spectroscopy. *Biophysical Journal*, *80*(6), 2987–2999.

CHAPTER FOURTEEN

Stopped-Flow Kinetic Techniques for Studying Binding Reactions of Intrinsically Disordered Proteins

Michael D. Crabtree, Sarah L. Shammas[1]

Department of Biochemistry, University of Oxford, Oxford, United Kingdom
[1]Corresponding author: e-mail address: sarah.shammas@bioch.ox.ac.uk

Contents

Abstract

Intrinsically disordered proteins are abundant in signaling processes such as transcription. Suitable binding and unbinding rates of proteins with their partners are critical for allowing them to perform their biological roles. Understanding how these are achieved, and indeed designing strategies for intervening or modulating related biological processes, therefore requires kinetic studies. In this chapter, we describe

Methods in Enzymology, Volume 611
ISSN 0076-6879
https://doi.org/10.1016/bs.mie.2018.09.026

stopped-flow-based methods for determining association and dissociation rate constants for pairs of macromolecular binding partners. We describe how to select the simplest appropriate model to describe the interaction, and highlight cases where it is possible to distinguish between induced fit and conformational selection binding mechanisms. Finally, we go on to describe methods for examining the role of electrostatic forces in binding processes, and for describing the transition state for binding processes that have folding associated with them.

1. INTRODUCTION

In recent years it has become clear that a significant proportion of the eukaryotic proteome contains long intrinsically disordered regions. Protein disorder is by no means distributed equally throughout cellular functions. It is heavily overrepresented in classes of proteins involved in signaling, such as transcription factors (Ward, Sodhi, McGuffin, Buxton, & Jones, 2004). As a result, the interactions made by intrinsically disordered proteins with their binding partners, which are sometimes coupled with their folding, are of definite interest. Although the study of protein disorder is arguably most advanced in terms of bioinformatics, more recently, extensive experimental efforts are being made in describing individual IDPs. Unbound, bound, and in some cases intermediate states can be examined with structural methods such as NMR or SAXS. Advances in structural methods in particular have allowed the description of the structural ensembles formed by disordered proteins (Varadi et al., 2014). The stability of their complexes can be determined using equilibrium methods such as isothermal titration calorimetry, circular dichroism, and fluorescence-based studies. However, to examine the mechanism of these interactions kinetic methods are required. For example, this is the only way that induced fit and conformational selection mechanisms might be discriminated (Dogan & Jemth, 2014; Kiefhaber, Bachmann, & Jensen, 2012; Shammas, Crabtree, Dahal, Wicky, & Clarke, 2016).

Most kinetic methods rely on a simple principle of "mix and observe." For relatively slow reactions (15 min +), manual mixing of solutions within a quartz cuvette and observation within apparatus designed for equilibrium experiments, e.g., fluorimeter, CD spectrometer, arguably remains the best method. However, this approach comes with a significant dead time of at least 5–10 s, where no data are available, to allow for efficient mixing and instrumental startup. As this can be a critical time period for many reactions, alternative kinetic techniques, such as continuous-flow, stopped-flow, and relaxation methods, have been developed.

The stopped-flow technique involves rapid mixing of two solutions within a small reaction volume (cell), and with signal detection/recording being triggered by the mixing event by means of a stop syringe (Movie 1 in the online version at https://doi.org/10.1016/bs.mie.2018.09.026). The dead time of the experiment is determined by the mixing time of the apparatus and is typically between 1 and 10 ms. Above around 15 min there are likely to be problems with experimental drift, as a result of diffusion through the system, or fluctuations in lamp output. However, between these two timescales stopped-flow can be a powerful kinetic method. An appropriate kinetic technique is essentially dictated by the timescale of reaction. In practice, however, limitations may be imposed by the availability of apparatus. In some cases, the timescale of the reaction can be altered by choice of appropriate experimental conditions, e.g., temperature, pH, ionic strength. For association mixing reactions the reactant concentrations also have a direct impact on the timescale of the reaction, and depending upon the binding affinity, it may be possible to choose the desired range for the reaction timescale by altering the protein concentration. Where reactions remain too fast for stopped-flow, then continuous-flow methods can be useful since these have a somewhat reduced experimental dead time; however, these generally consume substantially larger sample volumes. Stopped-flow instruments are also more widely available than continuous-flow setups. This chapter therefore focuses on application of a stopped-flow methodology for binding studies; however, the data analysis is equally appropriate to describe continuous-flow or manual mixing approaches. Relaxation methods such as laser-induced temperature jump, which rely instead on rapid perturbations of experimental conditions to a system at equilibrium, are capable of monitoring at even shorter timescales, but the reaction kinetics can require different fitting equations. Finally, although we focus on describing coupled folding and binding protein–protein interactions, these methods are also appropriate for protein–nucleic acid interactions, and in the absence of any conformational changes.

2. PREPARATION FOR STOPPED-FLOW EXPERIMENTS

Kinetic experiments in general consume more material and research time than equilibrium experiments. In our experience, the key to performing efficient kinetic studies is suitable preparation in advance using knowledge obtained from equilibrium experiments. The methods described later help in this process.

2.1 Establishing a Suitable Probe

In principle a stopped-flow setup can be combined with many different types of detectors for electromagnetic radiation. However, protein–protein interactions rarely cause a change in absorbance, and circular dichroism stopped-flow has generally poor signal to noise without synchrotron radiation. As a result, most stopped-flow studies utilize fluorescence, i.e., fluorescence intensity, FRET, or fluorescence anisotropy.

2.1.1 Test for Intrinsic Change in Fluorescence

Tryptophan or tyrosine residues within one or more protein may provide an internal fluorescence probe with which to monitor the reaction without perturbing the system. However, fluorescence changes cannot be guaranteed and may not be large enough to be useful, even if the fluorescence probe becomes buried in the interface upon formation of the complex. Thus, a good initial step is to determine if a sufficient signal change already exists.

2.1.1.1 Equipment

Fluorimeter
Fluorescence quartz cuvette

2.1.1.2 Reagents

X μM solution of protein (L) (where $X > K_d$)
X μM solution of partner (P)

2.1.1.3 Protocol

1. Generate equimolar solutions of the protein (L) and partner (P).
2. Perform an emission scan for solutions of L and P, exciting at 280 nm (275 nm if only tyrosines are present) and measuring the emission from 300 to 400 nm.
3. Mix L and P in a 1:1 volume ratio.
4. Repeat the emission scan for the mixture.
5. Average the individual spectra obtained for L and P. If this differs from the mixture scan, then a signal change has occurred.

2.1.1.4 Notes

1. We recommend starting with a 10 nm excitation slit width and a 5 nm emission slit width.
2. If there are multiple stages to binding, each will have its own signal change and it is possible that these might cancel out for equilibrium experiments but be observable during kinetic experiments.

2.1.2 Labeling With an Extrinsic Fluorophore

We usually utilize a fluorescently labeled version of one protein for out-competition studies to determine the dissociation rate constant(s). This version can also be used to observe the binding process when there is no change in the intrinsic fluorescence upon binding, and provides a convenient method for determining the equilibrium dissociation constant (K_d). If one binding partner is a synthetic peptide, then it is generally most convenient and cost effective to purchase an N-terminally fluorophore-labeled version alongside the unlabeled version. If proteins are being recombinantly expressed, then the fluorophore can be incorporated with maleimide chemistry at an introduced cysteine residue as we now outline.

2.1.2.1 Equipment

- Ultrasonic bath
- Aluminum foil
- HiTrap® desalting, Hitrap® Q HP, or Hitrap® SP HP 5 mL column
- AKTA FPLC (GE Healthcare)
- Equipment for SDS-PAGE
- Fluorescence gel imager, e.g., Typhoon Trio, ORCA

2.1.2.2 Reagents and Buffers

- Plasmid construct for expression of wild-type protein
- Primers for introducing mutations
- Q5®—site-directed mutagenesis kit (or similar)
- Materials for expression and purification of target protein
- Alexa Fluor™ 594 C5-maleimide
- Dimethyl sulfoxide (DMSO)
- 20 m*M* pH 7.4 phosphate buffer
- Bond-Breaker™ TCEP solution (Neutral pH, 0.5 *M*)
- Liquid nitrogen
- Benchmark™ fluorescent protein standard

2.1.2.3 Procedure

1. Locate a suitable position for label incorporation with reference to any published final structures or simulations. Avoid residues known or assumed to be critical for complex formation such as those along binding interfaces.
2. Design primers for, and then perform, site-directed mutagenesis to introduce a cysteine residue at the identified position, and remove

other natural cysteine residues by mutating them to serine or alanine. Primer design can be assisted using NEBasechanger™ tool.

3. Express and purify the mutant protein.
4. For efficient maleimide reactions the protein should be buffer exchanged into a pH 7.0–7.5 buffer. Phosphate, Tris, or HEPES buffers without DTT or beta-mercaptoethanol are all suitable. Calculate the protein concentration and dilute or concentrate protein to around 50–100 μ*M*.
5. Dissolve 1 mg of AlexaFluor maleimide with 100 μL DMSO, sonicate in ultrasonication bath for ~3 min, spin down, and aliquot evenly over 10 aliquots. These can be lyophilized or flash frozen and stored at −80°C.
6. Calculate the amount of dye per aliquot, e.g., 110 nmol for AlexaFluor594 C5-maleimide, and then the volume of protein solution an aliquot can label to achieve a dye:protein ratio of 1.4:1, e.g., 80 nmol of protein needed for AlexaFluor594 C5-maleimide. Reduce this volume of protein solution for 1 h at 37°C with TCEP BondBreaker solution (0.5 *M*). Add 2 μL of TCEP solution for every 1 mL of protein solution.
7. Add the aliquot of dye to the protein 2 μL at a time while slowly rotating the tube. Rotate for 4 h at room temperature.
8. Protein and dye may be separated using a Hitrap® desalting column. However, for best results separation is performed using an appropriate ion exchange column. Disposable Hitrap® Q and Hitrap® SP columns are generally sufficient for this purpose and include literature for suggestions of suitable buffers and separation methods suitable for a range of proteins.
9. Analyze the collected fractions via SDS-PAGE, using a fluorescent protein ladder as a reference marker. Image the gel with a fluorescent imager immediately after running to determine which peaks contain labeled protein.
10. Calculate the labeling efficiency (=100 f/p). Fluorophore concentration (f) may be determined using the absorbance of the fluorophore. Protein concentration (p) is conveniently determined by quantitative amino acid analysis.

2.1.2.4 Notes

1. Fluorescein is a relatively cheap fluorophore that works well for short timescale reactions (<10 s). For longer timescales, we recommend fluorophores with improved photostability properties, such as the Alexa dye series.
2. Keep the samples protected from light as much as is reasonably possible, e.g., wrap containers and columns in foil.

2.2 Choosing a Suitable Concentration Range

As a general rule, the concentrations of unlabeled proteins used in association experiments should be at, or above, the K_d for the complex. By definition, below these concentrations the complexed state is not highly occupied, limiting observable signal changes. Higher concentrations can be used to achieve improved signal-to-noise or faster reactions. A first step in establishing a suitable concentration range for experiments is therefore to determine the K_d for the complex. Isothermal titration calorimetry (ITC) is commonly applied for this purpose if K_d is between 10 n*M* and 10 μ*M*. However, the availability of a fluorescently labeled ligand provides convenient alternatives, which may be more accurate, require less protein, and are suitable for determining a larger range of K_d values.

2.2.1 Determining K_d With a Fluorescently Labeled Ligand

2.2.1.1 Equipment

- Fluorescence spectrophotometer
- Excitation and emission polarizing filters (for anisotropy measurements)
- Fluorescence quartz cuvette

2.2.1.2 Buffers and Reagents

- Fluorescently labeled protein L
- Unlabeled protein partner P
- Biophysical buffer, e.g., 50 m*M* sodium phosphate pH 7

2.2.1.3 Procedure

1. Clean the cuvette by washing with water (×2), 2% Hellmanex III (×3), water (×3), and ethanol (×2), before drying with compressed air or nitrogen.
2. Generate an *X* μ*M* stock (a) of labeled protein L. *X* should be below the expected K_d.
3. Generate two to four stocks (b) containing both proteins. The concentration of protein L should be *X* μ*M* in all stocks. Concentrations of protein P should vary, e.g., 0.25, 5, and 100 μ*M* for an expected K_d of 1 μ*M*.
4. Mix the protein stocks (a and b) in various ratios to produce solutions with concentrations of unlabeled partner roughly equally separated on a log-scale (e.g., 24 concentrations between 10 n*M* and 100 μ*M* for a reaction with an expected K_d of 1 μ*M*).
5. Equilibrate the reaction mixtures at the desired temperature in the dark.
6. Use any leftover solution from stocks a and b to optimize the detector sensitivity (voltage), amount of time to record and average each point, and the number of repeat measurements.

7. Starting with the lowest, measure the fluorescence intensity and/or fluorescence anisotropy for each concentration of unlabeled protein.
8. Fluorescence anisotropy (R) may be calculated using Eq. (1):

$$R = \frac{I_{VV} - GI_{VH}}{I_{VV} + 2GI_{VH}} \quad (1)$$

where

$$G = \frac{I_{HV}}{I_{HH}} \quad (1a)$$

and the first subscript corresponds to the polarization of the excitation filter (vertical, V; or horizontal, H), the second subscript corresponds to the polarization of the emission filter, and I is the fluorescence intensity recorded.
9. Obtain the K_d by plotting the intensity/anisotropy signal as a function of the protein concentration and fitting the data to Eq. (2):

$$S = S_f + \Delta S \left(\frac{K_d + [L] + [P] - \sqrt{((K_d + [L] + [P])^2 - 4[L][P])}}{2[L]} \right) \quad (2)$$

where S_f is the signal of the free protein (A), ΔS is the difference in signal change between free and bound, and [L] and [P] are the total concentration of proteins L and P, respectively.

2.2.1.4 Notes

1. The midpoint of the binding isotherm only approximates the K_d when the concentration of labeled protein is $\leq K_d$ (Fig. 1).
2. To generate the reaction solutions, we have recommended diluting a stock of labeled protein with stocks containing unlabeled and labeled partners. This method allows the various concentrations to be produced by pipetting reasonable volumes and reduce random variation in labeled protein concentration between tubes. However, caution should be employed during stock preparation and analysis as inaccuracies in stock concentration may lead to systematic errors.
3. We recommend using small-volume cuvette (60 μL) for reactions with a $K_d > 100\,\mathrm{n}M$ and a large-volume cuvette (900 μL) for reactions with a $K_d < 100\,\mathrm{n}M$.

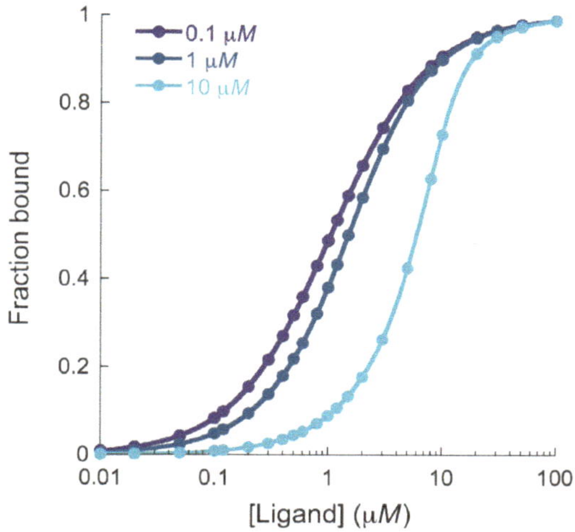

Fig. 1 Impact of ligand depletion on binding isotherms. Theoretical binding isotherms for a reaction with a K_d of 1 μ*M*. At dye-labeled protein concentrations below the K_d (0.1 μ*M*), the midpoint of the curve is equal to the K_d. For higher concentrations the midpoint shifts progressively to the *right* and the curve becomes less sigmoidal as complex formation significantly reduces the free ligand concentration, i.e., total protein concentrations plotted on the *x*-axis no longer correspond to the free protein concentration.

4. Equilibration should be for at least 10 half-lives of the dissociation rate constant. If this is not known, then initially equilibrate for 1 h. This should be sufficient for most reactions that have a $K_d > 1\,\text{n}M$.
5. As described previously (Dandliker, Hsu, Levin, & Rao, 1981), if there is a difference in fluorescence intensity between the free and bound states, then the anisotropy should be adjusted using Eq. (3):

$$R_{\text{adj}} = \frac{\dfrac{R - R_f}{R_b - R} \times \dfrac{Q_f}{Q_b} \times R_b + R_f}{1 + \dfrac{R - R_f}{R_b - R} \times \dfrac{Q_f}{Q_b}} \tag{3}$$

where Q is the measured fluorescence intensity, and *R* is the measured anisotropy. The subscripts f and b indicate values measured for the free or bound protein L, respectively.
6. The total fluorescence intensity of the free (Q_f) and bound (Q_b) peptide is calculated using Eq. (4):

$$Q = I_{VV} + 2I_{VH} \tag{4}$$

where the first subscript corresponds to the polarization of the excitation filter (vertical, V; or horizontal, H), the second subscript corresponds to the polarization of the emission filter, and I is the fluorescence intensity recorded.

2.2.2 Determining K_d With an Unlabeled Ligand

Labeling a protein with a fluorescent dye has the potential to change several of the reaction characteristics. Once the binding affinity of the labeled peptide has been determined, it is therefore prudent to check the K_d for the unlabeled version. This can be achieved through performing a competition binding assay, which provides the inhibition constant (K_i).

2.2.2.1 Equipment

- Fluorescence spectrophotometer
- Excitation and emission polarizing filters (for anisotropy measurements)
- Fluorescence quartz cuvette

2.2.2.2 Buffers and Reagents

- Fluorescently labeled protein (L)
- Unlabeled version of the protein (U)
- Unlabeled protein partner (P)
- Biophysical buffer

2.2.2.3 Procedure

1. Clean the cuvette by washing with water (×2), 2% Hellmanex III (×3), water (×3), and ethanol (×2), before drying with nitrogen.
2. Produce a stock (a) containing both X μM probe-labeled protein (L) and Y μM unlabeled partner (P).
3. Generate two to four stocks (b) of unlabeled protein U, containing X μM protein L and Y μM unlabeled partner P.
4. Mix the protein stocks (a and b) to produce the various concentrations of unlabeled protein U (with constant concentrations of probe-labeled protein and its partner).
5. Equilibrate the reaction mixtures at the desired temperature.
6. Use any leftover solution from stocks a and b to optimize the detector sensitivity (voltage), amount of time to record and average each point, and the number of repeat measurements.

7. Starting with the lowest, measure the fluorescence intensity and/or fluorescence anisotropy for each concentration of unlabeled competing protein.
8. Obtain the IC_{50} by fitting the data to Eq. (5):

$$Q = Q_f + \frac{\Delta Q}{\left(1 + \left(\frac{[U]}{IC_{50}}\right)^{-n}\right)} \quad (5)$$

where Q is the measured fluorescence intensity, Q_f is the fluorescence intensity of the free-labeled ligand, IC_{50} is the midpoint of the binding isotherm, [U] is the concentration of competing peptide (U), and n is the Hill coefficient.
9. The inhibition constant for the unlabeled protein (U) can then be calculated using Eq. (6) (Nikolovska-Coleska et al., 2004):

$$K_i = \frac{I_{50}}{\frac{L_{50}}{\frac{K_{d,lab} + P_0}{K_{d,lab} + 1}}} \quad (6)$$

where I_{50} is the concentration of free U at 50% inhibition, L_{50} is the concentration of the free L at 50% inhibition, P_0 is the concentration of free protein when no competitor is present, and $K_{d,lab}$ is the dissociation constant for the complex between the labeled peptide (L) and partner protein (P).

2.2.2.4 Notes

1. Notes from Section 2.2.1 are also relevant to these competition binding experiments.
2. If ITC proves problematic for determining K_d of an unlabeled ligand, then this comparative method can be useful provided that a labeled version can be generated.
3. We recommend starting with $X = K_{d,lab}$ and $Y \geq X$.
4. The K_i is equal to the $K_{d,unlab}$ but is named in reference of the experimental method used to determine it.

2.3 Preparing Protein Solutions for Stopped-Flow

Protein–protein interactions are by nature multimolecular, and this means that observed rates are also dependent upon reactant concentrations. For this reason, it is essential that concentrations in mixing reactions are accurate for determining rate constants. We recommend the following approach to solution preparation to maximize accuracy.

2.3.1 Equipment

- Quartz absorbance cuvette (1 cm pathlength)
- UV–visible spectrophotometer
- Fluorimeter
- Weighing balance

2.3.2 Buffer and Reagents

- Protein stock solutions
- Biophysical buffer (e.g., 50 mM sodium phosphate, pH 7)

2.3.3 Procedure

1. Perform an absorbance scan from 200 to 600 nm for all stock solutions.
2. Subtract from each the spectrum of a matched buffer control.
3. Check for aggregation by noting any absorbance at wavelengths away from protein chromophores. If present, then filter protein solutions and repeat.
4. Calculate the ratio of absorbance A_{260}:A_{280}. Pure protein samples have a ratio of approximately 0.6, whereas pure DNA and RNA have ratios of around 1.8 and 2, respectively. If the ratio is above 0.7, then consider using an additional purification method, such as ion exchange, to remove the nucleic acid contamination.
5. Calculate the stock concentration. For proteins containing tryptophan or tyrosine residues, then use the absorbance at 280 nm and an accurate extinction coefficient to determine the protein concentration. Extinction coefficients can be determined using the method described by Gill and von Hippel (1989). For synthetic fluorophore-labeled samples use the extinction coefficient of the fluorophore.
6. Eliminate surface absorption. At low concentrations protein, peptide, or nucleic acids may stick to the surfaces of reaction and mixing vessels. Perform a separate dilution series of each sample into buffer and check for linearity of buffer-subtracted fluorescence intensity. If lower concentrations give disproportionately lower responses, then this indicates sticking. Absorption can be reduced by switching type of microcentrifuge tubes, pipette tips, and drive syringes. Alternatively, addition of small

concentrations of detergent, e.g., 0.05% Tween-20, into buffer solutions often effectively eliminates sticking.

7. When performing a dilution to generate a stopped-flow solution, we recommend weighing the stock added and the total volume to determine an accurate dilution factor.

2.3.4 Notes

1. If possible perform absorbance scans of the stock solution directly. If the absorbance is too high for accurate assessment, i.e., above 1.0, then a dilution must be performed. Weigh stock solution and buffer upon addition to the mixing vessel to determine an accurate dilution factor.
2. In the absence of chromophores, the concentration should be determined using quantitative amino acid analysis (accurate within around 5%–10%).

3. PERFORMING STOPPED-FLOW EXPERIMENTS

In general, the setup of stopped-flow apparatus is dependent upon the model and described within the appropriate product manual. However, most of the following hints are generally applicable. The standard equipment required to set up the stopped-flow as we describe is shown in Fig. 2. Syringe volumes and emission filters are dependent on the experimental setup and examples are shown in Fig. 3.

3.1 Equipment

- SX20 Applied Photophysics stopped-flow
- Two FlexFluidics drive syringes
- One FlexFluidics 1-mL stop syringe
- Disposable plastic syringes
- Beaker of ddH_2O
- 2.2 cm diameter long-pass filter (optional)

3.2 Procedure

1. Switch on the nitrogen supply, electronics box, power supply, and computer.
2. Screw the two (dry) drive syringes into the adapter inside the sample handling unit.
3. Screw a (dry) stop syringe into the output (with the plunger fully up).

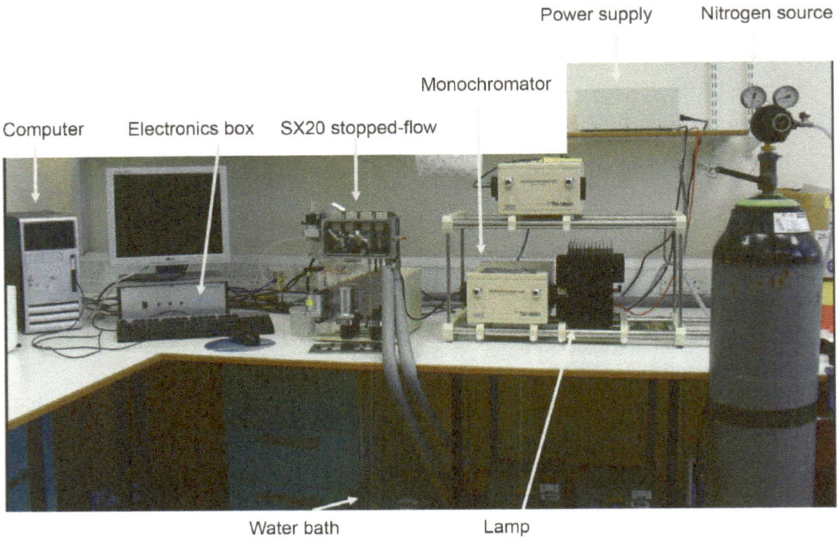

Fig. 2 Equipment for stopped-flow spectrometry. The stopped-flow spectrometer is shown in a disassembled state, with all of the basic hardware required.

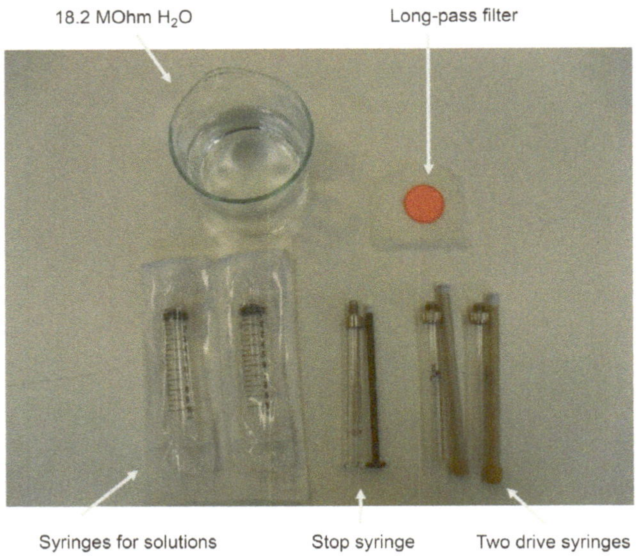

Fig. 3 Experiment-specific stopped-flow accessories. Two 2.5-mL drive syringes and a 1-mL stop syringe are shown, but other volumes are available to create various mixing ratios. The long-pass filter required is dependent on the fluorophore; a 570 nm filter is shown here.

4. Rotate the valves on the sample handling unit to isolate the drive syringes from the flow-path. Attach water-filled disposable plastic syringes to the top of the sample handling unit and use them to fill the two drive syringes with water. Empty the water back into the plastic syringes by manually pushing up the driving ram. Empty and fill the syringes until no bubbles remain in the drive syringes.
5. Rotate the valves to connect the drive syringes and the stop syringe to the rest of the system. Lift the driving ram to push water through and into the stop syringe. Rotate the valve above the stop syringe and push water out from the stop syringe to the waste. Reconnect the stop syringe into the flow-path and refill it with water from the drive syringes. Repeat until no bubbles remain. This generally takes at least four washes.
6. Perform a test injection into the system to test for leaks in the setup (for 5–10 s and with continual pressure applied to the syringes). If there is a leak, then this should be observed as a slow upward movement of the barrel throughout the pressure hold. Most commonly, the syringe plunger is not well fitting to the barrel, or the syringes are incorrectly connected to the apparatus. The location of the leak is most easily identified by careful drying of all visible parts and careful inspection for water leaks during repeat pressure tests. Action should then be taken to dismantle and then reattach at this point. If this does not fix the problem, then replacing parts can be considered.
7. If the apparatus uses a xenon flash lamp, then this should be switched on to equilibrate at least 30 min before experiments begin.
8. Unscrew the photomultiplier tube and install an appropriate long-pass emission filter before replacing it. Appropriate filter sets may generally be selected by examining the published spectra for the chosen fluorophore (Table 1).
9. Attach the Faceplate. The stopped-flow should now resemble Fig. 4.
10. Temperature control is generally supplied by means of a circulating water bath. If temperature is to be investigated as an independent variable, then a thermocouple should be used to accurately determine the temperature within the water bath or, if possible, within the sample handling unit.
11. Wash the system with buffer, and then with the appropriate protein solutions.
12. Use the "Auto PMT" function to select the detector voltage.

Table 1 Optical Characteristics of Fluorescent Probes

Fluorophore	Absorption Maximum (nm)	Emission Maximum (nm)	Extinction Coefficient (M^{-1} cm^{-1})	Recommended Long-Pass Filter (nm)
Tyrosine	275	305	1490	305
Tryptophan	280	320–360[a]	5500	320
Alexa Fluor 488	495	519	73,000	515
FITC	493	512	75,000	515
TAMRA	555	580	83,000	570
Alexa Fluor 546	556	573	112,000	570
Alexa Fluor 594	590	617	92,000	610

[a]The emission maximum of tryptophan is strongly dependent on its chemical environment. When exposed to water, the emission maximum of tryptophan tends toward 360 nm, whereas when shielded from water (e.g., the interior of a folded protein), the emission maximum is closer to 320 nm.

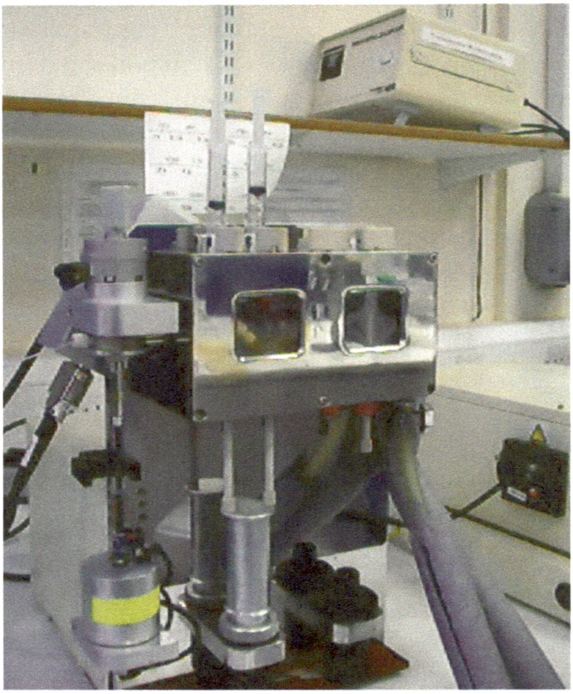

Fig. 4 Example of a stopped-flow instrument that has been fully assembled using components from Figs. 2 and 3.

13. Perform a test injection for 10 s (no pressure hold), recording data on a log scale. If the reaction involves a decrease in fluorescence, then the PMT voltage might need to be reduced manually so that the signal is not over range at the start of the trace. Note the PMT voltage used. For later experiments it may be advisable to use the same voltage so that absolute fluorescence values can be loosely compared.
14. A good starting point is to perform five replicate injections, recording for 50 s on a logarithmic scale (no pressure hold). Fit to an appropriate exponential decay function as outlined in the following sections. Use the fitted rate constant(s) to inform the timescale for data collection. As a rule of thumb each recording should be around ($10 \ln 2/k_{obs}$) seconds. If this required timescale is less than 50 ms, or more than 1000 s, then the solution conditions, reactant concentrations, or kinetic technique should be adjusted.
15. Continue making replicate shots, recording at the selected timescale on a linear scale, until the signal-to-noise ratio reaches a level where the residuals are less than 5% of the smallest signal change. If more than 100 replicates are required for this, then for time and material conservation it is wise to consider using an alternative reaction probe. Each replicate should display a similar starting point, amplitude, and rate constant, and at least five traces should be taken to confirm reproducibility. Occasionally, a bubble may pass into the cell chamber, resulting in a trace that does not overlay. These traces, which display clearly distinguishable signal profiles from the other replicates, should not be included in the analysis.

3.3 Notes

1. Bubbles in the flow-path are detrimental to data quality. They may also affect mixing volumes. It is essential to check regularly for bubbles and remove them. To remove bubbles that enter the cell chamber, we recommend running two to three shots on a short timescale (e.g., 0.2 s) with the pressure hold applied. This is usually sufficient to dislodge the bubble(s).
2. Extrinsic fluorophores can be prone to photobleaching. We recommend keeping solution containers and syringes wrapped with aluminum foil. Foil can also be taped above the viewing port so that this is covered during experiments but available for viewing when needed.
3. Exchanging solutions can introduce errors in sample concentrations. The following method provides accurate concentrations while minimizing

sample wastage (Movie 2 in the online version at https://doi.org/10.1016/bs.mie.2018.09.026):

- Wash syringes and system through with buffer several times and ensure that the stop syringe is filled with buffer.
- Empty as much buffer from the drive syringes as possible.
- Half-fill the drive syringe with sample. Manually push the plunger forward and back repeatedly to mix the sample with the remaining buffer.
- Remove as much sample as possible from the syringe and repeat.
- Fill the syringe with sample.

4. After the stop syringe hits the detector, there is typically a delay of approximately 30 ms before the pressure applied by the drive ram is released. This can lead to artifacts that are problematic for fast reactions. Enabling the pressure hold option can improve the data quality but should only be applied for 2 s or less.

5. We recommend fitting the data and plotting the residuals as the experiment is running. This confirms the use of the correct recording timescale and that the data align with the expected model.

6. The signal change becomes progressively smaller as reactions are converted to products. Increased sampling at the beginning of the reaction can therefore be of great benefit. Logarithmic sampling and split time bases are particularly useful when monitoring multiple phases that have rate constants that are more than an order of magnitude apart.

7. We recommend starting with a monochromator slit width of 2 mm (bandwidth of 9.3 nm). It may be necessary to reduce this if photobleaching is an issue.

8. For asymmetric mixing experiments, the drive pressure (set using the external nitrogen flow regulator) should not exceed 2 bar. For 1:1 mixing, a pressure of up to 4 bar may be used.

4. PERFORMING AND ANALYZING KINETIC BINDING STUDIES

Coupled folding and binding reactions may necessarily be split into at least two steps: folding and binding. The simplest schemes, known as induced fit and conformational selection, can be described as

$$\mathrm{P} + \mathrm{L} \underset{k_{-}}{\overset{k_{+}}{\rightleftharpoons}} \mathrm{P.L} \underset{k_{\mathrm{u}}}{\overset{k_{\mathrm{f}}}{\rightleftharpoons}} \mathrm{C}$$

and

$$\mathrm{P} + \mathrm{L} \underset{k_u}{\overset{k_f}{\rightleftharpoons}} \mathrm{P} + \mathrm{L}^* \underset{k_-}{\overset{k_+}{\rightleftharpoons}} \mathrm{C}$$

where P, L, P.L, L*, and C represent the folded protein, unfolded ligand, bound intermediate, prefolded ligand, and final complex, respectively.

Whether both steps are observed in kinetic experiments depends upon the relative rate constants and signal changes for the two phases. In many cases the rate constants are such that no intermediate is significantly populated so only a single phase is observed, and the reaction appears simply two-state. In other cases, it is possible that three phases may be observed (Dogan, Mu, Engström, & Jemth, 2013), though this appears less common. If the reaction is two-state under the chosen experimental conditions, as has been shown for many coupled folding and binding systems, then the rate equation describing the model can be solved analytically and mixing experiments can be easily interpreted for any range of reactant concentrations. For reactions of three or more states (two or more steps) the model can be solved analytically if certain approximations hold. In particular, if the folded protein partner is present at much higher concentrations than that of the unfolded ligand, then its concentration can be considered constant throughout the reaction. These conditions, known as pseudo-first-order conditions, greatly simplify the rate equations so that kinetic traces appear as multiexponential decay functions with observed rate constants and amplitudes that depend upon the fundamental rate constants. The numbers of states that are occupied therefore determine the fitting equations, and even the required reactant ratios in experiments, and this must be determined as early in the process as possible. A two-state model is not appropriate; if more than a single phase is observed during pseudo-first-order mixing experiments, observed rate constants depend upon the technique used, e.g., fluorescence intensity and anisotropy provide different estimates, or the K_d obtained from equilibrium and kinetic methods do not match, i.e., $K_d \neq k_{off}/k_{on}$. If a two-state model is appropriate, then it can be worth considering an equimolar (or near equimolar) mixing strategy for subsequent binding experiments. These strategies have several advantages including maximizing signal to noise for those experiments without extrinsic fluorophores, and reducing the required number of stopped-flow mixing experiments (Shammas, Rogers, Hill, & Clarke, 2012). Arguably the largest advantage is that for appropriate reagent concentrations, i.e., around the K_d, it is possible to extract both k_{off} and k_{on} from a single mixing experiment.

4.1 Association Mixing Experiments

4.1.1 Pseudo-First-Order Mixing Experiment

4.1.1.1 Additional Equipment

Two 2.5-mL FlexFluidics drive syringes
One FlexFluidics 1-mL stop syringe

4.1.1.2 Reagents

Fluorescently labeled disordered ligand (L)
Partner (P)
Biophysical buffer

4.1.1.3 Procedure

1. Fill the two drive syringes with ligand and protein solutions. The concentration of folded protein should be above the K_d to ensure sufficient complex formation. The concentration of disordered ligand should be at least 10-fold lower than folded protein after mixing. For example, for a reaction with 1 μM K_d, initial solutions could be 10 μM protein P and 0.5 μM disordered ligand (L).
2. Rapidly mix the solutions and observe the reaction to check for sufficient signal change and establish the rough timescale for the reaction.
3. If required, then reaction rates can be increased or decreased by increasing or decreasing the concentration(s), respectively. If the fluorescence intensity change is small, then consider swapping to fluorescence anisotropy if this is available.
4. Establish the number of observable phases. If a single exponential decay function (Eq. 7 with $i = 1$) fits well to the data without systematic deviations observable in the residuals plot, then continue collecting replicate measurements until the residuals represent $\leq 2\%$ of the total signal change:

$$[S] = S_{\text{final}} + \sum_i \Delta S_i e^{-k_{\text{obs},i} t} \quad (7)$$

where S is the signal, S_{final} is the final signal, ΔS_i are the signal amplitudes, and $k_{\text{obs},i}$ are the observed rate constants.
5. If a multiexponential decay function (Eq. 7 with $i = 1, 2, \ldots$) is required to fit the data, then this could indicate that an intermediate is being

populated. If one or more of the fluorescence amplitude changes are negative, then photobleaching needs to be ruled out as a cause. Collect kinetic traces for the mixing of protein with buffer, and for ligand with buffer using identical instrument settings to the mixing experiment. If both the collected traces are flat, then this suggests that photobleaching is not taking place.

6. Systematically exchange the partner solution (P) in the drive syringe for partner solutions of varying concentrations, performing 5–100 replicate injections for each sample. At least five relatively evenly spaced concentrations should be examined, e.g., 5, 10, 20, 30, and 40 μ*M*. The concentration of disordered ligand must always remain at least 10-fold lower than partner after mixing.
7. Fit each averaged kinetic trace with a multiexponential decay function (Eq. 7). Examine residuals plots to identify the simplest model consistent with the data for all concentrations.
8. Plot each observed association rate constant (k_{on}) against the protein concentration. If there is only a single phase observed for all concentrations, and k_{on} displays a linear dependence on protein concentration, then the reaction is best described by a two-state model (Fig. 5A). The k_+ can be identified as the gradient of the straight line fit (Fig. 5B).

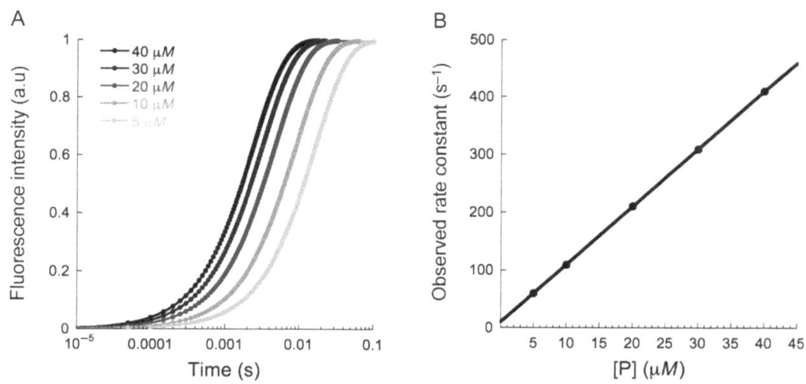

Fig. 5 Extracting the association rate constant of a two-state system under pseudo-first-order conditions. (A) Example kinetic data for reactions between protein L (0.5 μ*M*) and its partner P (5, 10, 20, 30, and 40 μ*M*) where $k_+ = 10\ \mu M^{-1}\ s^{-1}$ and $k_- = 10\ s^{-1}$. Fluorescence intensity changes are fit to a single exponential function (Eq. 7; $i = 1$) to obtain the observed rate constant. (B) For a two-state reaction, the observed rate constant is linearly dependent on the concentration of excess reactant (partner P). The k_+ can be extracted from the gradient of the straight line fit, while the k_- can be determined from the intercept. Note that since this method involves extrapolation it should be applied with caution to extract the k_-.

For two or more phases the dependence of both k_{on} on concentration is more complex (Shammas et al., 2016). We have provided two interactive programs to enable the reader to visualize how changes in the fundamental rate constants will alter observed rate constants and amplitudes for the simplest induced fit (Supplementary File 1 in the online version at https://doi.org/10.1016/bs.mie.2018.09.026) and conformational selection (Supplementary File 2 in the online version at https://doi.org/10.1016/bs.mie.2018.09.026) schemes. These programs can be opened by Wolfram CDF Player which may be downloaded for free at https://www.wolfram.com/cdf-player/.

9. Where two or more phases are present, then further mixing experiments with reversed pseudo-first-order conditions should be performed to identify whether the populated intermediate is a bound or an unbound species. In the example, 1 μM partner could be mixed with 10 μM ligand. At least two different concentrations should be examined. Plot both sets of k_{on} against the concentration of the protein that is in excess. If the two types of mixing experiment give consistent values for k_{on}, then this indicates that the intermediate is a bound species, i.e., induced fit scheme. If the values are clearly different, then this indicates that the intermediate is an unbound species, i.e., conformational selection scheme.

10. For two phases, the two k_{on} obtained with folded protein in excess can be used to obtain estimates for the various fundamental rate constants, by fitting both simultaneously to Eq. (8) or (9) for induced fit and conformational selection schemes, respectively (Fig. 6). For three-state reactions there is often difficulty converging to a unique solution. In this case consider global fitting of observed association and dissociation rates alongside equilibrium data, sharing the value of the four fundamental rate constants. Expressions for K_d and k_{off} in terms of the rate constants to be used for this purpose are shown in Table 2. If a unique solution is still not obtained, then it may be possible to extract information about some or all of the fundamental rate constants by fitting association kinetic data that was collected at high concentrations (where the high k_{on} appears linearly dependent on protein concentration, and the low k_{on} appears independent of concentration) (Dahal, Shammas, & Clarke, 2017):

$$k_{on,IF} = \frac{1}{2}\left(x \pm \sqrt{x^2 - 4([P]k_+k_f + [P]k_+k_u + k_-k_u)}\right) \quad (8)$$

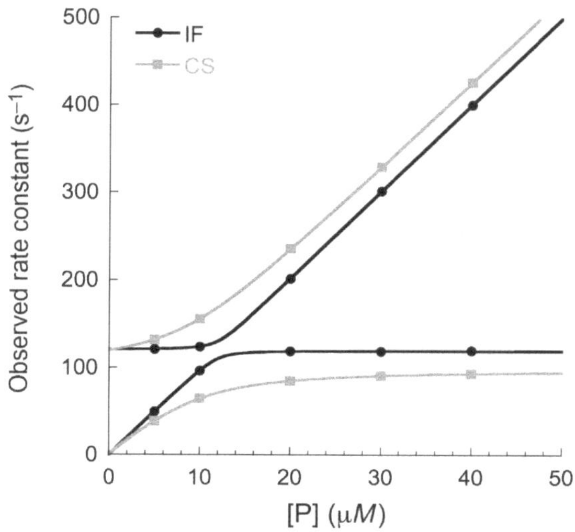

Fig. 6 Extracting the fundamental rate constants of a three-state system under pseudo-first-order conditions. Observed rate constants obtained for a reaction between protein L (0.5 μ*M*) and its partner P (5, 10, 20, 30, and 40 μ*M*) when the folding and binding mechanism follows either an induced fit (IF) or conformational selection (CS). The example data are fit to Eqs. (8) and (9) for an IF and CS mechanism (with folded protein in excess), respectively, and represent a reaction where $k_+ = 10\ \mu M^{-1}\ s^{-1}$, $k_- = 1\ s^{-1}$, $k_f = 100\ s^{-1}$ and $k_u = 20\ s^{-1}$. The goodness of fit to experimental data of the two models is often equivalent when values of all parameters are allowed to vary.

Table 2 Observed Equilibrium and Rate Constants for Induced Fit and Conformational Selection Mechanisms in Terms of the Fundamental Rate Constants

	Induced Fit	**Conformational Selection**
K_d	$\frac{k_- k_u}{k_+(k_u + k_f)}$	$\frac{k_u(k_- + k_+)}{k_f k_+}$
k_{off}	$\frac{1}{2}\left(k_- + k_u + k_f \pm \sqrt{-4k_- k_u + (k_- + k_u + k_f)^2}\right)$	k_- and $k_u + k_f$
$k_{on,high}$, $P \rightarrow \infty$	$k_+[P] + k_-$	$k_+[P] + k_- + k_u$
$k_{on,low}$, $P \rightarrow \infty$	$k_u + k_f$	k_f

$$k_{\mathrm{on,CS}} = \frac{1}{2}\left(x \pm \sqrt{x^2 - 4([\mathrm{P}]k_+k_\mathrm{f} + k_\mathrm{f}k_- + k_-k_\mathrm{u})}\right) \tag{9}$$

where

$$x = [\mathrm{P}]k_+ + k_- + k_\mathrm{f} + k_\mathrm{u} \tag{10}$$

4.1.1.4 Notes

1. Typical starting concentrations for folded partner protein are about 10 μM since average k_+ values are around $10^6\ M^{-1}\ \mathrm{s}^{-1}$; however, since this can vary by many orders of magnitude (Shammas et al., 2012), the concentration may need to be adjusted.
2. Low concentrations of fluorescently labeled ligands can be used, but we recommend using around 0.2 μM to achieve reasonable signal to noise.
3. Minor photobleaching over the timescale of the experiment can be accounted for by including a linear term (+ct) in Eq. (7). However, large changes in fluorescence require changes to the experimental protocol.
4. Conformational selection mechanisms may also be identified if one of the observed rate constants decreases with increasing protein concentration (Gianni, Dogan, & Jemth, 2014).
5. Where all observed kinetic rate constants and equilibrium data are used to estimate the fundamental rate constants, it can be difficult to confirm the applicability of the model. In this case, we recommend checking the concentration dependence of the amplitudes observed in association mixing experiments. Amplitudes for the two phases can change in relative size and also in sign and are determined by the values of the four fundamental rate constants and the changes in fluorescence corresponding to each step. Unlike observed rate constants, fluorescence intensity amplitudes are subject to day-to-day variation so wherever possible data should be collected in a single day or calibrated using some standard. Amplitudes can be checked for consistency using the interactive resources we created for the simplest induced fit (Supplementary File 1 in the online version at https://doi.org/10.1016/bs.mie.2018.09.026) and conformational selection (Supplementary File 2 in the online version at https://doi.org/10.1016/bs.mie.2018.09.026) schemes.

4.1.2 Equimolar and Near-Equimolar Mixing Experiments

4.1.2.1 Additional Equipment

Two 2.5-mL FlexFluidics drive syringes
One 1-mL FlexFluidics stop syringe

4.1.2.2 Buffers and Reagents

Y μM solution of protein L (where $Y\,\mu M \sim > K_d$)
Five solutions of partner P of $x_i Y\,\mu M$ (where $x_i = 1$, 2, 3, 4, and 5).
Biophysical buffer

4.1.2.3 Procedure

1. Fill the two drive syringes with $Y\,\mu M$ protein L and $5Y\,\mu M$ partner P and wash through the system.
2. Select the PMT voltage so that the fluorescence is in range. Note this voltage since it should be used for all subsequent data collection.
3. Perform 10–100 injections and average the traces.
4. Empty the drive solution of protein P.
5. Repeat steps 1–4 with $4Y$, $3Y$, $2Y$, and $Y\,\mu M$ protein P and then buffer.
6. Wash both drive syringes through with buffer, and the system with buffer. Mix buffer with buffer and record the signal.
7. If the K_d is unknown, then perform a preliminary fit for each kinetic trace using Eq. (7) ($i = 1$). Plot the amplitude against the concentration of partner protein and fit with Eq. (2) to obtain a preliminary estimate of K_d.
8. Attempt to fit each averaged kinetic trace to Eq. (11) using the estimated K_d, keeping [A] and x_i fixed to their calculated values (Fig. 7A). Examine the five estimated values for K_d and k_+ to check that they are consistent before averaging these. This acts as an internal check of model applicability. If this is successful, then future experiments may be restricted to performing only equimolar mixes to increase throughput:

$$F = F_0 + \Delta F \frac{(b - z)\left(1 - e^{(zk_+t)}\right)}{2\left(\left(\frac{b - z}{b + z}\right)e^{(zk_+t)} - 1\right)} \tag{11}$$

where

$$b = -(K_d + (1 + x)[\mathrm{L}]) \tag{11a}$$

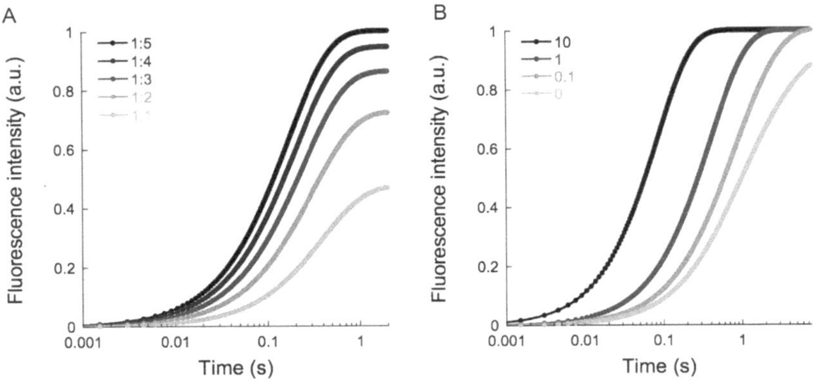

Fig. 7 Extracting the association rate constant and equilibrium binding constant of a two-state system under equimolar and near equimolar conditions. (A) When the concentration of protein L is at or near the K_d, an increase in signal change is observed upon mixing with increasing concentrations of partner P. The data can be either individually or globally fit to Eq. (11) to extract both the k_+ and K_d. Example data represent a reaction where [L] = 1 μ*M*, $k_+ = 1\ \mu M^{-1}\ s^{-1}$, $K_d = 1\ \mu M$, and the [L]:[P] concentration ratio is 1:1, 1:2, 1:3, 1:4, or 1:5. (B) Under conditions where the concentrations of L and P are >100-fold above K_d, the reaction becomes essentially irreversible. When [L] ≠ [P], Eq. (11) can be used to ascertain k_+ by setting the K_d to 0. Example data for reactions followed under near equimolar conditions, where [L] (1 μ*M*), [P] (0.99 μ*M*), and k_+ ($1\ \mu M^{-1}\ s^{-1}$) are the same, but the K_d varies (0, 0.1, 1, and 10 μ*M*).

and

$$z = \sqrt{K_d{}^2 + 2(1+x)K_d[\mathrm{L}] + (x^2 - 2x + 1)[\mathrm{L}]^2} \tag{11b}$$

and

$$x = \frac{[\mathrm{P}]}{[\mathrm{L}]} \tag{11c}$$

and [P] and [L] are equal to the concentration of proteins P and L, respectively.

9. If the solution does not converge, then this is likely because the signal to noise is not sufficient to extract all of the parameters independently. If the K_d is known accurately, then this can be used as a fixed parameter for fitting. Alternatively, the data for the five averaged kinetic traces should be simultaneously fit to Eq. (11), globally sharing the k_+, K_d, and ΔF.

4.1.2.4 Notes

1. At equimolar concentrations the kinetic traces deviate significantly from single exponentials, even under conditions where the reaction is essentially irreversible.
2. This method can also be used where $Y \gg K_d$, i.e., association is essentially irreversible if the K_d is fixed to zero for the purposes of fitting (unless [P] = [L]) (Fig. 7B).

4.2 Dissociation Experiments

Dissociation kinetics experiments can easily be performed incorrectly: if the complex does not fully dissociate, then the observed rate constant includes a contribution from the association rate constant. To ensure almost complete dissociation, we recommend an outcompetition approach, where a preformed complex of protein L and partner P is mixed with an additional protein U that competes for binding. The complex formed between P and U must have a different total fluorescence (intensity or anisotropy) to that between P and L. When using intrinsic tyrosine/tryptophan fluorescence, this typically means performing mutagenesis of protein L to create the competitor, or utilizing a different binding protein/partner. Using a fluorescent dye-labeled protein provides an easy solution: a dye-labeled protein L can be outcompeted using an excess of U. Below, we describe the method used to outcompete a preformed complex between a dye-labeled protein and its partner with an unlabeled competitor.

4.2.1 Additional Equipment

- 2.5-mL drive syringe
- 250-μL drive syringe
- 2.5-mL stop syringe

4.2.2 Buffers and Regents

- Dye-labeled protein (L)
- Partner (P)
- Unlabeled competitor (U)
- Biophysical buffer

4.2.3 Procedure

1. Mix the dye-labeled peptide (L) and partner (P) in a concentration ratio that ensures that the majority of L is in complex with P (e.g., 0.1 μ*M* L, 1 μ*M* P for a reaction with a K_d of 0.1 μ*M*).
2. Generate stocks of unlabeled competitor (e.g., 5, 10, and 50 μ*M*).

3. Set up the stopped flow as described in Section 3, ideally using one 250-μL drive syringe and one 2.5-mL drive syringe.
4. Wash the stock of dye-labeled protein (L) and partner (P) into the 250-μL syringe and the unlabeled competitor (C) into the 2.5-mL drive syringe.
5. Rapidly mix the complex and competitor, and follow the change in signal with time (see Sections 3 and 4).
6. Fit the data to an exponential function to obtain the observed rate constant, ensuring to collect at least 10 half-lives.
7. The observed rate constants from each mixing can usually be averaged to estimate the dissociation rate constant. However, if some complex remains after the reaction has reached equilibrium, the observed rate constant will decrease as the concentration of competitor is increased (Fig. 8). Concentration-dependent rates should not be included in the average.

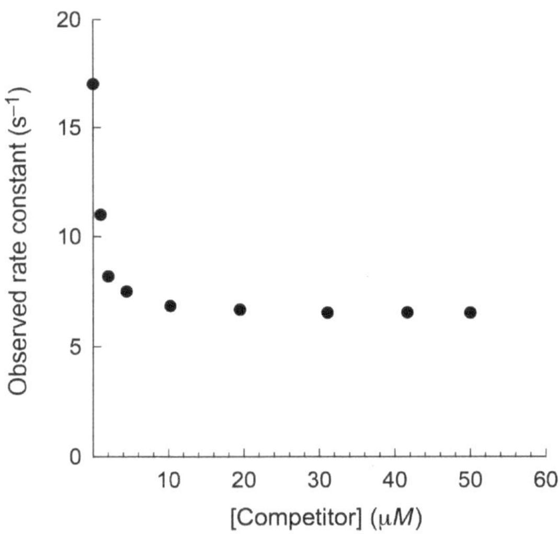

Fig. 8 Example kinetic data obtained from an outcompetition dissociation experiment. As the concentration of competitor is increased the observed rate constant approaches k_-. In this example, we would recommend excluding the values below 30 μ*M* of competitor. Alternatively, as described previously (Shammas, Travis, & Clarke, 2014), all the data can be fit to Eq. (12) to obtain the k_-:

$$k_{\mathrm{off}} = k_- + k_+[\mathrm{P}]\frac{1}{1+\frac{[\mathrm{U}]}{K_{\mathrm{d,U}}}} \tag{12}$$

where [U] is the concentration of competing peptide and $K_{\mathrm{d,U}}$ is the equilibrium dissociation constant for the interaction between the competitor and partner protein.

4.2.4 Notes

1. The required volumes of the complex and competitor vary depending on the number of repeat shots required. For a 1:10 mixing ratio, we recommend starting with 1 mL of complex and 4 mL of competitor. Including washing, this should provide enough material to complete three competitor concentrations, with six repeat shots for each.
2. As a dissociation event is effectively unimolecular in terms of the reactant (the complex), it will follow first order kinetics and can be described by an exponential function. For three-state processes there are strictly two characteristic rate constants (Table 2); however, the amplitude of one is usually so low that only one is observable.
3. The same method may be used to calculate $k_{off,unlab}$ if there is a large signal change by simply outcompeting U with a large excess of L. Alternatively if $K_{d,lab} \ll K_{d,unlab}$ outcompetition can utilize similar concentrations of L and U provided that $[L] > [P]$.

5. ROLE OF ELECTROSTATICS

Charge–charge interactions can be split into two categories—long range and short range—which have different kinetic signatures. Long-range interactions determine electrostatic driving forces that modulate encounter complex formation rates and help steer the reaction to its transition state. Short-range electrostatic interactions (e.g., salt-bridges) occur between residues at the reaction transition state and/or in the final bound state. They can therefore impact both association and dissociation rates depending upon contacts formed at the reaction transition state.

5.1 Ionic Strength Dependence of the Association Rate Constant

Below, we describe the method used to determine the impact of charge–charge interactions (electrostatic forces) on the k_{on} using pseudo-first-order conditions, but in principle any of the protocols described in this chapter can be similarly adapted by titrating the ionic strength of the buffer with salt.

5.1.1 Additional Equipment

- Two 2.5-mL drive syringes
- 1-mL stop syringe

5.1.2 Buffers and Reagents

- Protein (L)
- Partner (P)
- 10 m*M* MOPS Buffer with various concentrations of NaCl (e.g., 0, 32 m*M*, 48 mM, 125 mM, 500 m*M*, 2 *M*), all adjusted to pH 7

5.1.3 Protocol

1. Create a stock of protein (L) at a single concentration (e.g., 0.8 μ*M*) in MOPS 10 m*M* buffer pH 7.
2. Generate at least five stocks of the partner (P) at different concentrations (e.g., 8, 12, 16, 20, and 24 μ*M*) in MOPS 10 m*M* buffer pH 7.
3. Mix the protein (L) stock at a 1:1 volume ratio with the 10 m*M* MOPS 2 *M* NaCl buffer (final concentrations of 1 *M* NaCl and 0.4 μ*M* protein).
4. Mix the five partner (P) stocks at a 1:1 volume ratio with the 10 m*M* MOPS 2 *M* NaCl buffer (final concentrations of 1 *M* NaCl and 4, 6, 8, 10 and 12 μ*M* of partner).
5. Perform 1:1 pseudo-first-order mixing experiments and extract association rate constants using the method described in Section 4.
6. Repeat steps 3–5, mixing with 10 m*M* MOPS with different concentrations of NaCl (producing final NaCl concentrations of 0, 16, 24, 62.5, and 250 m*M*) to obtain the association rate constant at different ionic strengths.

5.1.4 Notes

1. We recommend using a monovalent ion to buffer as this limits contributions to solution ionic strength from the buffer itself and expands the available range for experiments, e.g., 10 m*M* MOPS at pH 7 (~4 m*M* ionic strength at 25°C). In comparison, a 10 m*M* phosphate buffer at pH 7 has an ionic strength of ~20 m*M*.
2. Altering the ionic strength of the buffer can alter the pH of the solution. It is important to check that the solutions are all of the same pH and alter them by adding (measured) quantities of concentrated HCl or NaOH if required.
3. The structure of the protein and partner should be checked at the different ionic strengths (e.g., using circular dichroism) as changes in structure can also modulate the reaction kinetics, complicating the analysis (Wicky, Shammas, & Clarke, 2017).

5.2 Extracting the Basal Association Rate Constant

The bimolecular (fundamental) association rate constant k_+ is expected to vary with ionic strength due to the involvement of long-range electrostatic forces, and an empirical model has been developed for fitting data. Once k_+

at each ionic strength has been determined, the protocol below can be used to extract the k_+ in the absence of any electrostatic steering (i.e., at infinite ionic strength).

5.2.1 Protocol

- Plot ln k_+ against $1/\sqrt{I}$.
- Fit the data to a Debye–Huckel-like model (Vijayakumar et al., 1998), as described in Eq. (13):

$$\ln k_+ = \ln k_+^{I=\infty} - \frac{AB}{Bd}\frac{I^{-1/2}}{\left(Bd + I^{-1/2}\right)} \tag{13}$$

where

$$AB = \left(\frac{Q_A Q_B}{k_B T_\varepsilon}\right) e\sqrt{\frac{8\pi N_A}{k_B T_\varepsilon}} \tag{13a}$$

and

$$Bd = e\sqrt{\frac{8\pi N_A}{k_B T_\varepsilon d}} \tag{13b}$$

where Q_A and Q_B represent the charges of the free proteins, k_B is the Boltzmann constant, N_A is Avogadro's number, d is the separation distance of the encounter complex, e is the elementary charge, ε is the permittivity of water, and T is the temperature.

5.2.2 Notes

1. High k_+ in the absence of electrostatic effects effectively rules out reactions involving a large degree of conformational selection (Kiefhaber et al., 2012; Shammas, Travis, & Clarke, 2013).

6. DESCRIBING THE TRANSITION STATE(S) FOR FOLDING AND BINDING

Kinetic experiments are the only available experimental method to offer information on the energy barriers between the transition state, and the free and bound states. Through studying how these free energy differences change upon mutation, Φ-value analysis provides a method to probe the interactions formed at the reaction transition state (Matouschek, Kellis, Serrano, & Fersht, 1989). It has been used extensively to characterize unimolecular folding pathways and mechanisms, and more recently has been applied to coupled folding and binding reactions of intrinsically disordered proteins (Crabtree, Mendonça, Bubb, & Clarke, 2018; Shammas et al., 2016). This can provide insight into the folding and binding mechanism.

6.1 Additional Equipment

- Two 2.5-mL drive syringes
- 1-mL stop syringe

6.2 Buffers and Reagents

- Wild-type and mutant protein (L)
- Partner (P)
- Buffer (e.g., 50 m*M* sodium phosphate pH 7)

6.3 Protocol

1. Design mutations along the binding interface. First, hydrophobic residues that contact the partner protein (e.g., that bind in a hydrophobic groove) are mutated to shorten the side chain (typically by mutation to alanine). The aim is to remove interactions without introducing new ones. Charged residues can also be mutated; however, this can alter long-range electrostatic effects, so a pseudo-wild type (with the charge already mutated out) must also be generated.
2. Design mutations along the solvent-exposed surface (that do not contact the partner) to probe intramolecular interactions, i.e., the amount of secondary structure formation. For example, for an intrinsically disordered protein that becomes helical upon binding, a pseudo-wild type is created by mutating to alanine, to remove the impact of the side chain, and then this is compared to a protein with a glycine in the same position.
3. Express and purify all of the mutant proteins.
4. For each mutant determine the association and dissociation rate constants using the method established with the wild-type protein.
5. Calculate the Phi-value (Φ) for each residue using Eq. (14):

$$\Phi = \frac{\ln\left(\frac{k_+^{\mathrm{WT}}}{k_+^{\mathrm{Mut}}}\right)}{\ln\left(\frac{K_\mathrm{d}^{\mathrm{Mut}}}{K_\mathrm{d}^{\mathrm{WT}}}\right)} \tag{14}$$

6. Examine the spatial distribution of Φ-values by referencing the structure for the final complex to visualize the formation of native contacts within the transition state.

6.4 Notes

1. A Φ of 1 suggests that all of the interactions that are formed in the bound state are also formed in the transition state. Φ of 0 indicates that none of the interactions that are formed in the bound state are present in the transition state. An intermediate value is harder to assess but it can mean that some of the interactions are formed at the transition state, or all of them are partially formed. Negative values occur if the mutation accelerates binding, or slows dissociation, and indicate the presence of nonnative interactions.
2. Caveats of Φ-value analysis have been previously discussed (Fersht & Sato, 2004). It is wise not to ascribe too much import to individual Φ-values but rather to consider them all collectively.
3. If the reaction is two-state, then $k_{-}/k_{+} = K_d$. We recommend performing equilibrium binding experiments for at least a few proteins to confirm that the K_d from kinetics matches the value from equilibrium.
4. Φ-Value analysis can also be performed on multistate reactions to probe each transition state by considering each step of the reaction in turn.
5. The level of residual structure within the unbound ensemble can be altered by mutation (Crabtree et al., 2017), additives such as urea, TFE or TMAO (Gianni, Morrone, Giri, & Brunori, 2012), or exchange of salt type (while keeping ionic strength constant) (Wicky et al., 2017). If increases in the residual structure observed by circular dichroism or NMR spectroscopy are not accompanied by an increased association rate constant, then this effectively rules out a conformational selection mechanism.

7. CONCLUSIONS

Association reactions of disordered proteins with their macromolecular partners often involve some folding. The in vitro kinetic methods we have described can reveal the rate constants describing (un)binding and (un)folding processes, the order in which they occur, and generate a picture of the native contacts present within the transition state(s).

ACKNOWLEDGMENT

S.L.S. is supported by an MRC career development award (MR/N024168/1).

REFERENCES

Crabtree, M. D., Borcherds, W., Poosapati, A., Shammas, S. L., Daughdrill, G. W., & Clarke, J. (2017). Conserved helix-flanking prolines modulate intrinsically disordered protein:target affinity by altering the lifetime of the bound complex. *Biochemistry*, *56*, 2379–2384.

Crabtree, M. D., Mendonça, C. A. T. F., Bubb, Q. R., & Clarke, J. (2018). Folding and binding pathways of BH3-only proteins are encoded within their intrinsically disordered sequence, not templated by partner proteins. *The Journal of Biological Chemistry*, *293*, 9718–9723.

Dahal, L., Shammas, S. L., & Clarke, J. (2017). Phosphorylation of the IDP KID modulates affinity for KIX by increasing the lifetime of the complex. *Biophysical Journal*, *113*, 2706–2712.

Dandliker, W. B., Hsu, M. L., Levin, J., & Rao, B. R. (1981). Equilibrium and kinetic inhibition assays based upon fluorescence polarization. *Methods in Enzymology*, *74*, 3–28.

Dogan, J., & Jemth, P. (2014). Only kinetics can prove conformational selection. *Biophysical Journal*, *107*, 1997–1998.

Dogan, J., Mu, X., Engström, Å., & Jemth, P. (2013). The transition state structure for coupled binding and folding of disordered protein domains. *Scientific Reports*, *3*, 2076.

Fersht, A. R., & Sato, S. (2004). Phi-value analysis and the nature of protein-folding transition states. *Proceedings of the National Academy of Sciences of the United States of America*, *101*, 7976–7981.

Gianni, S., Dogan, J., & Jemth, P. (2014). Distinguishing induced fit from conformational selection. *Biophysical Chemistry*, *189C*, 33–39.

Gianni, S., Morrone, A., Giri, R., & Brunori, M. (2012). A folding-after-binding mechanism describes the recognition between the transactivation domain of c-Myb and the KIX domain of the CREB-binding protein. *Biochemical and Biophysical Research Communications*, *428*, 205–209.

Gill, S. C., & von Hippel, P. H. (1989). Calculation of protein extinction coefficients from amino acid sequence data. *Analytical Biochemistry*, *182*, 319–326.

Kiefhaber, T., Bachmann, A., & Jensen, K. S. (2012). Dynamics and mechanisms of coupled protein folding and binding reactions. *Current Opinion in Structural Biology*, *22*, 21–29.

Matouschek, A., Kellis, J. T., Serrano, L., & Fersht, A. R. (1989). Mapping the transition state and pathway of protein folding by protein engineering. *Nature*, *340*, 122–126.

Nikolovska-Coleska, Z., Wang, R., Fang, X., Pan, H., Tomita, Y., Li, P., et al. (2004). Development and optimization of a binding assay for the XIAP BIR3 domain using fluorescence polarization. *Analytical Biochemistry*, *332*, 261–273.

Shammas, S. L., Crabtree, M. D., Dahal, L., Wicky, B. I. M., & Clarke, J. (2016). Insights into coupled folding and binding mechanisms from kinetic studies. *Journal of Biological Chemistry*, *291*, 6689–6695.

Shammas, S. L., Rogers, J. M., Hill, S. A., & Clarke, J. (2012). Slow, reversible, coupled folding and binding of the spectrin tetramerization domain. *Biophysical Journal*, *103*, 2203–2214.

Shammas, S. L., Travis, A. J., & Clarke, J. (2013). Remarkably fast coupled folding and binding of the intrinsically disordered transactivation domain of cMyb to CBP KIX. *Journal of Physical Chemistry B*, *117*, 13346–13356.

Shammas, S. L., Travis, A. J., & Clarke, J. (2014). Allostery within a transcription coactivator is predominantly mediated through dissociation rate constants. *Proceedings of the National Academy of Sciences of the United States of America*, *111*, 12055–12060.

Varadi, M., Kosol, S., Lebrun, P., Valentini, E., Blackledge, M., Dunker, A. K., et al. (2014). PE-DB: A database of structural ensembles of intrinsically disordered and of unfolded proteins. *Nucleic Acids Research*, *42*, 326–335.

Vijayakumar, M., Wong, K.-Y., Schreiber, G., Fersht, A. R., Szabo, A., & Zhou, H.-X. (1998). Electrostatic enhancement of diffusion-controlled protein-protein association: Comparison of theory and experiment on barnase and barstar. *Journal of Molecular Biology*, *278*, 1015–1024.

Ward, J. J., Sodhi, J. S., McGuffin, L. J., Buxton, B. F., & Jones, D. T. (2004). Prediction and functional analysis of native disorder in proteins from the three kingdoms of life. *Journal of Molecular Biology*, *337*, 635–645.

Wicky, B. I. M., Shammas, S. L., & Clarke, J. (2017). Affinity of IDPs to their targets is modulated by ion-specific changes in kinetics and residual structure. *Proceedings of the National Academy of Sciences*, *114*, 9882–9887.

CHAPTER FIFTEEN

The Use of Mass Spectrometry to Examine IDPs: Unique Insights and Caveats

Dale Stuchfield*, Aidan P. France*, Lukasz G. Migas*, Anja Thalhammer†, Anne Bremer†, Bruno Bellina*, Perdita E. Barran*,[1]

*The Michael Barber Centre for Collaborative Mass Spectrometry, The School of Chemistry, Manchester Institute of Biotechnology, The University of Manchester, Manchester, United Kingdom
†Max-Planck-Institut für Molekulare Pflanzenphysiologie, Potsdam, Germany
[1]Corresponding author: e-mail address: perdita.barran@manchester.ac.uk

Contents

Abstract

A sizeable proportion of active protein sequences lack structural motifs making them irresolvable by NMR and crystallography. Such intrinsically disordered proteins (IDPs) or regions (IDRs) play a major role in biological mechanisms. They are often involved in cell regulation processes, and by extension can be the perpetrator or signifier of disease. In light of their importance and the shortcomings of conventional methods of biophysical analysis to identify them and to describe their conformational variance, IDPs and IDRs have been termed "the dark proteome." In this chapter we describe the use of ion mobility-mass spectrometry (IM-MS) coupled with electrospray ionization to analyze the conformational diversity of IDPs. Using the LEA protein COR15A as an exemplar system and contrasting it with the behavior of myoglobin, we outline the methods for

Methods in Enzymology, Volume 611
ISSN 0076-6879
https://doi.org/10.1016/bs.mie.2018.09.038

analyzing an IDP using nanoelectrospray ionization coupled with IM-MS, covering sample preparation, purification; optimization of mass spectrometry conditions and tuning parameters; data collection and analysis. Following this, we detail the use of a "toy" model that provides a predictive framework for the study of all proteins with ESI-IM-MS.

ABBREVIATIONS

AIMS aspiration ion mobility spectrometry
ATDs arrival time distributions
CCS collision cross section
CCSD collision cross section distribution
CSD charge-state distribution
DTIMS drift time ion mobility spectrometry
ESI electrospray ionization
FAIMS field-asymmetric waveform ion mobility
GC gas chromatography
IDP intrinsically disordered protein
IDR intrinsically disordered region
IM-MS ion mobility-mass spectrometry
IMS ion mobility spectrometry
LC liquid chromatography
MS mass spectrometry
NMR nuclear magnetic resonance
TWIMS traveling wave ion mobility spectrometry
XRD X-ray diffraction

1. INTRODUCTION

1.1 Background

Over the past years, the study of intrinsically disordered proteins has gained more and more momentum (Uversky & Dunker, 2010). The growing interest in these proteins stems largely from their role in proteopathic diseases such as Parkinson's and Alzheimer's disease (Tompa, 2012) as well as the desire to develop a thorough understanding of the role of these conformationally diverse protein complexes (Uversky, Oldfield, & Dunker, 2008; Zubarev et al., 2000). A classic example of this is the tumor suppressor protein p53 which is involved in a high proportion of human cancers (Ananiev, Tchernev, Patterson, Gulubova, & Ganchev, 2011; Sormanni et al., 2017). IDP's differ significantly from ordered proteins, in that due to their conformational diversity it is not possible to characterize them by a single "native" structure; instead they are better considered to exist as conformational ensembles (Tompa, 2011). This conjecture helped to disprove the dogma that a proteins function is directly related to a resolvable single structure (Tompa, 2002; Wright & Dyson, 1999).

New definitions were developed following the discovery of IDPs that better described the protein paradigm. A "protein trinity" consisting of three structures: ordered, molten globule, and random coil was proposed by Dunker et al. (2001). This trinity was later expanded by Dyson and Wright (2005) to include a better description for the ordered state by partitioning it into two subsets of "mostly folded with localized disorder" and "linked folded domains." It became apparent that the study of IDPs by methods such as X-ray crystallography and NMR could not resolve valuable information on the inherent nature of the proteins (Daughdrill, Hanely, & Dahlquist, 1998; Tompa, 2011; Ward, Sodhi, McGuffin, Buxton, & Jones, 2004). As a result there arises an opportunity to develop alternative methods for the structural analysis of IDPs and the interactions they partake in with potential cofactors and substrates. One such method, which has grown in use (Beveridge, Chappuis, Macphee, & Barran, 2013; Borysik, Kovacs, Guharoy, & Tompa, 2015; D'Urzo et al., 2015; Faull et al., 2010; Jurneczko et al., 2012; Stuchfield & Barran, 2018), is ion mobility-mass spectrometry (IM-MS) coupled with electrospray ionization (ESI).

The origins of mass spectrometry (MS) came from the desire to identify atomic isotopes and it was initially used as a method for determining the molecular weight of small chemical compounds (Beveridge et al., 2013). However, with the development of soft ionization it became possible to transfer whole protein assemblies into the gas phase (Benesch, Ruotolo, Simmons, & Robinson, 2007; Fenn, Mann, Meng, Wong, & Whitehouse, 1989; Hernández & Robinson, 2007; Uetrecht et al., 2008). Once in the gas phase, the mass spectrometer can be used to determine the stoichiometry of complexes, as well as the strength of subunit interactions with exquisite detail (Ruotolo, Benesch, Sandercock, Hyung, & Robinson, 2008; Sahasrabuddhe et al., 2018; Young et al., 2018; Zhou, Dagan, & Wysocki, 2012).

MS is capable of trapping conformational intermediates, or manipulating structures and storing gas-phase ions over a microsecond to hour timescale, facilitating the investigation of protein folding and unfolding as well as the observation of conformational change(s) (Badman, Myung, & Clemmer, 2005; Breuker, Brüschweiler, & Tollinger, 2011; Freitas, Hendrickson, Emmett, & Marshall, 1999). This ability to probe conformational resilience of isolated complexes positions MS as an attractive biophysical technique for the study of conformationally diverse proteins (Uetrecht, Rose, van Duijn, Lorenzen, & Heck, 2010). In addition, MS is an extremely versatile technique which is often coupled to numerous separation methods in tandem, such as liquid chromatography (LC) and gas chromatography (GC). Following the development of ESI, the coupling of MS directly to the eluent from a chromatographic column was such a successful union that it became a standard procedure when beginning any analysis of proteins or other biomolecules (Chen & Pramanik, 2008) and had phenomenal power in top-down sequencing efforts (Vorontsov, Rensen, Prangishvili, Krupovic, & Chamot-Rooke, 2016; Zubarev et al., 2000). That being said, MS alone lacks the ability to provide further detail regarding structure and dynamics, hence when carrying out a study on a given protein, LC-MS is used primarily for the determination or confirmation of the molecular weight and for the assignment of posttranslational modifications (PTMS).

The development of ESI provided an extradimension to the analysis of proteins through the presentation of the protein in a charge-state distribution (CSD). It was observed very early on that when a protein was sprayed using ESI, it would present to the mass spectrometer in a range of charge states which primarily corresponded to the protein/protein complex with a neutral mass M, presenting as ions with several m/z values (m/z stands for the mass-to-charge ratio that is the unit of measurement for a mass spectrometer). Each of these m/z values corresponded to ions in positive ionization mode of the general form $[M+nH]^{nz+}$ (or in negative ionization mode as $[M-nH]^{nz-}$) where n increments by 1. Such CSDs can be deconvoluted to elucidate accurate mass and interpreted in terms of the conformational state of the protein under study (Beveridge et al., 2014; Dobo & Kaltashov, 2001; Frimpong, Abzalimov, Uversky, & Kaltashov, 2010; Konermann, Collings, & Douglas, 1997; Testa et al., 2013). The range of n is pertinent, and denatured proteins (for example, following LC separation) or even disordered proteins, present with wider CSDs, while globular proteins which have retained aspects of a stable "native" fold present with narrower CSDs (Konermann, 2007).

The term native MS has been coined to explain the careful use of ESI, primarily from salty aqueous solutions, followed by slow removal of solvent to best preserve the noncovalent interactions present in the native protein or protein complex (Leney & Heck, 2017). In IM-MS this tends to result in the production of a narrow CSD coupled with a narrow spread of conformations (Beveridge et al., 2014). For IDPs the use of native MS methods mostly provides a wide CSD and correspondingly wide range of conformational states, as shall be seen later in this chapter.

Following seminal work in the 1990s, by Clemmer, Jarrold, Bowers, Russell and others (Clemmer, Hudgins, & Jarrold, 1995; Gillig et al., 2000;

Mao, Woenckhaus, Kolafa, Ratner, & Jarrold, 1999; von Helden, Wyttenbach, & Bowers, 1995; Wyttenbach, Von Helden, & Bowers, 1996), as well as significant technical developments (Gerlich, 1992), IM-MS instruments started to become commercially available in the mid-2000s (Giles et al., 2004; Pringle et al., 2007). Adding ion mobility separation to MS adds an extradimension of analysis to MS experiments as proteins are separated based on their charge and size (Mason & Mason, 1988). As MS separates ions according to their movement due to an applied electric (or magnetic) field in a vacuum, ion mobility separates ions due to an applied electric field in the presence of a neutral buffer gas. In the case of drift tube ion mobility, DTIMS, the protein is introduced to a gas-filled chamber as a pulse of ions that traverse the cell due to a weak electric field (5–50 V cm^{-1}). The field causes the ions to move through a cell, and collisions with the inert drift gas impede the progress of the ions. Experiments are performed at the so-called low-field limit (Kemper & Bowers, 1990; Revercomb & Mason, 1975) such that the ions rapidly reach a constant drift velocity (ν_d) which is dependent on their charge and mass (as for MS), and also on the number and nature of the collisions with the buffer gas. The mobility (K) of a given ion is defined as the constant of proportionality between the drift velocity ν_d and the applied electric field (E).

Following data acquisition in an IM-MS experiment, an ion of a given m/z will have a corresponding arrival time distribution (ATD). This ATD can be used to calculate what is known as a rotationally averaged collisional cross section (CCS), which provides a direct measurement of the proteins' conformational diversity (Lanucara, Holman, Gray, & Eyers, 2014). It is also worth noting that it is possible to computationally calculate a CCS, allowing for the comparison between experiment and theory (Beveridge et al., 2014). IM-MS has been applied to the study of more than 1000 proteins or protein complexes,[a] and while most of these studies have focussed on folded proteins, there are an increasing number of examples where it has been applied to examine IDPs; such as the tumor suppressor protein p53 (Dickinson et al., 2015a; Dunker et al., 2001; Faull et al., 2010; Jurneczko et al., 2013; Pagel, Natan, Hall, Fersht, & Robinson, 2013; Sinz, Arlt, Chorev, & Sharon, 2015) and the highly disordered amyloid forming protein, α-synuclein (Bernstein et al., 2004; Phillips et al., 2015). Some systems, such as the family of cold regulated (COR) proteins, adopt a disordered or ordered conformation

[a] On June 1st 2018 a search of the terms "ion mobility" "mass spectrometry," and "protein" on PubMed returned 1113 hits.

depending on the environment they are in. These COR proteins are vital to the cold acclimation of plants (Thalhammer & Hincha, 2014). Specifically they are involved in the stabilization of chloroplast membranes in the leaf during freezing. COR15A (used as an example in this chapter) and COR5B are COR proteins, found in *Arabidopsis thaliana*, which are critical to the plants adaptation to cold temperatures and specifically to dehydration (Candat et al., 2013; Lin & Thomashow, 1992; Nakayama et al., 2007). The COR15 proteins are found to be relatively disordered in hydrophilic conditions and undergo conformational transition to helical forms upon dehydration.

A mechanism of action for the COR15 proteins was proposed by Thalhammer and Hincha (2014). Upon freezing, a cellular dehydration occurs which results in a crowding effect in the cell membrane; this is accompanied by a rise in the expression levels of COR15. As the water content decreases, the COR15 proteins are proposed to fold into a α-helix and then to associate with the inner envelope membrane of the chloroplast. The association further promotes helicity in COR15, and in this form it is found to stabilize the membrane against the formation of hexagonal II (H_{II}) phase lipid domains.

1.2 Overview of Practice

In this chapter we will explain our methodology for the analysis of IDPs using IM-MS, this will cover the initial sample preparation and purification procedures, the importance of running a standard when working with an unknown protein, how best to optimize spectrometer settings, the acquisition of consistent data, and postacquisition analysis. Following this we will detail the use of a predictive framework and "toy model" that provides additional theoretical data regarding a proteins CCS and the degree of disorder within its structure. Steps of the sample preparation and of the entire workflow are shown in Fig. 1A and B, respectively.

2. EXPERIMENTAL IM-MS

As well as the drift tube IMS (DTIMS) described earlier (Kanu, Dwivedi, Tam, Matz, & Hill, 2008), there are currently four other types of methods commonly available for ion mobility separation: aspiration ion mobility spectrometry (AIMS) (Juan & Simon, 2012), field-asymmetric waveform (or differential) ion mobility spectrometry (FAIMS/DIMS) (Buryakov, Krylov, Nazarov, & Rasulev, 1993), traveling wave ion mobility spectrometry (TWIMS) (Giles et al., 2004), and trapped ion mobility

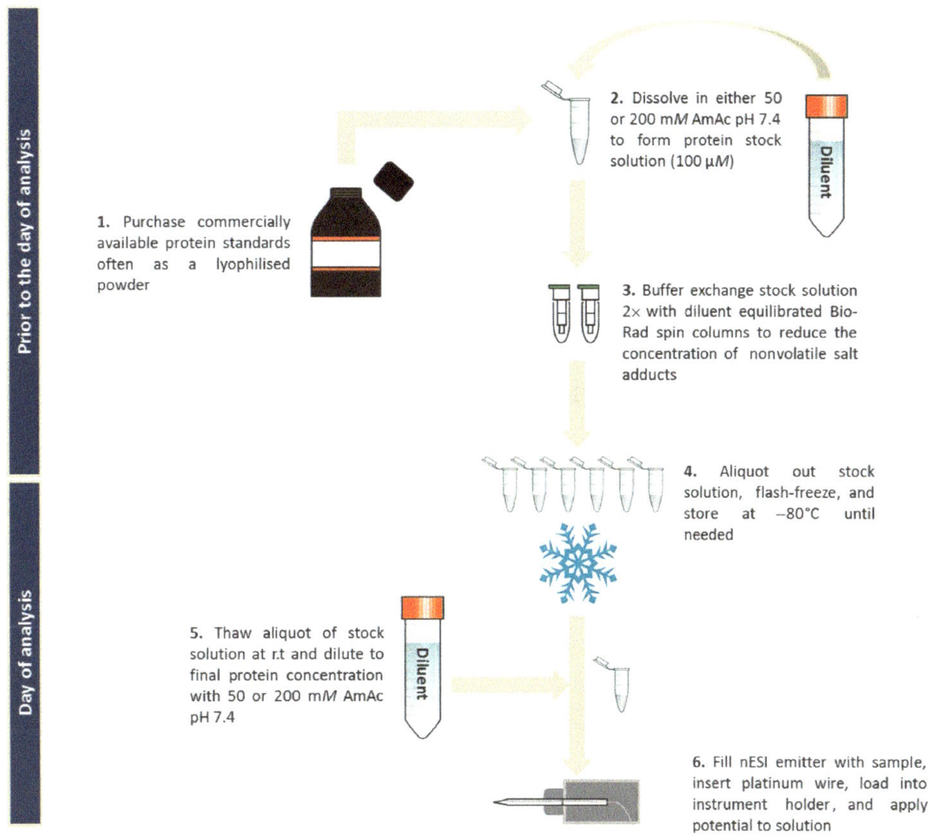
Prior to the day of analysis
Day of analysis
1. Purchase commercially available protein standards often as a lyophilised powder
2. Dissolve in either 50 or 200 mM AmAc pH 7.4 to form protein stock solution (100 µM)
Diluent
3. Buffer exchange stock solution 2× with diluent equilibrated Bio-Rad spin columns to reduce the concentration of nonvolatile salt adducts
4. Aliquot out stock solution, flash-freeze, and store at −80°C until needed
5. Thaw aliquot of stock solution at r.t and dilute to final protein concentration with 50 or 200 mM AmAc pH 7.4
Diluent
6. Fill nESI emitter with sample, insert platinum wire, load into instrument holder, and apply potential to solution

Fig. 1—Cont'd

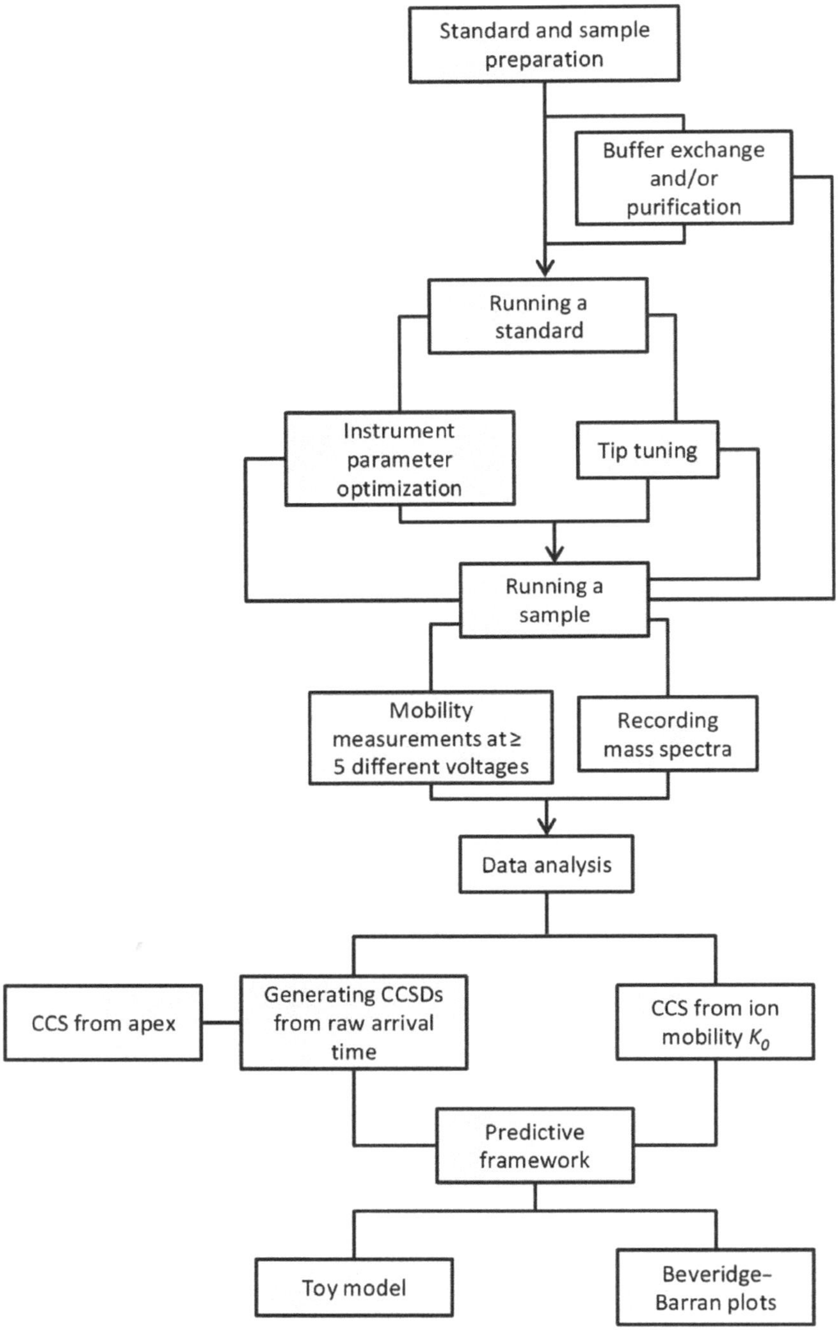

Fig. 1 **See legend on opposite page.**

spectrometry (TIMS) (Ridgeway, Lubeck, Jordens, Mann, & Park, 2018). This chapter will only cover DTIMS in detail, but similar approaches can be used for TWIMS and TIMS measurements albeit following calibration to obtain CCS values using well studied known proteins (Ridgeway et al., 2018; Stuchfield & Barran, 2018). The mechanism for how DTIMS separates ions will not be covered here, but the reader can find more detailed explanations in a number of papers (Kemper & Bowers, 1990; Lanucara et al., 2014; McCullough et al., 2008; Ujma, Giles, Morris, & Barran, 2016).

2.1 Sample Preparation and Purification

Sample preparation is the first step in the analysis of proteins by native IM-MS (Fig. 1A). This procedure typically involves the formulation of a solution at a pH appropriate for the protein under study, adjustment of the protein concentration (depending on ionization method) and at times, some purification steps (e.g., desalination). For the majority of native (IM)-MS experiments the low flow rate variant of ESI, namely, nano-ESI (nESI) is utilized. This uses borosilicate or quartz capillaries of ID $\sim$0.9 mm often home pulled to orifices 0.5–20 μM in diameter, with the ESI potential applied either to an inserted platinum wire or to a gold or palladium exterior. The principle advantage of this is that nESI uses cf. 100 × less sample than conventional ESI, and therefore is suited to low concentration samples. nESI from home-pulled tips also allows far better control of the size of the droplets, which in turn can influence the measured conformations of the protein as it alters the desolvation process.

Fig. 1 (A) Typical preparative procedure for commercially available protein "standards" prior to native (IM)-MS analysis. Majority of the "standard" protein/protein complexes we use for native (IM)-MS experiments are obtained as lyophilized powders (1). The formulation of the diluent, used to dissolve the lyophilized protein "standard" to form a stock solution (2), is described in Steps 1–4 in Section 2.1.1.2. The typical buffer exchange procedure (3) used to remove nonvolatile adducts from the protein sample is described in Steps 1–8 in Section 2.1.2.2. Following this process, the buffer-exchanged stock solution is aliquoted out, flash frozen, and stored at −80°C until the day of analysis (4). On the day of analysis, single aliquots are thawed (as required) at r.t. and diluted to their final concentration (5). Following this, the analyte solution is loaded into an "in-house" pulled nESI emitter (Tip pulling/tuning parameters on our tip puller are described in Section 2.2.2), platinum wire (if the tip is not coated) is inserted into the emitter so as to make contact with the solution and the instrument holder, and a potential is applied to the solution to permit sample ionization (6). (B) Workflow for all stages of IM-MS analysis of proteins including computational evaluation. This figure encompasses all steps described within this chapter.

For a typical nESI-(IM)-MS experiment, 20 μL of a 1–30-μ*M* solution of the protein of interest will last for c. 8 h of continual use (if aggregation does not occur).

2.1.1 Buffers and Solutions

A number of salt solutions are deemed suitable for the solvation of proteins which are subjected to native MS experiments. These solutions can also act as buffers if the pH needs to be adjusted, with the central aim of being volatile enough to depart the protein upon desolvation. Aqueous ammonium acetate (AmAc) with a pH range of ~6.8–7.4 is the "go to" salt for spraying intact proteins, as it is volatile and readily evaporates during electrospray. Furthermore, high concentrations of ammonium acetate also help to reduce the effects of nonvolatile buffer components (Hernández & Robinson, 2007). That being said, a number of groups have recently shown that it is possible to perform native (IM)-MS from highly salty solutions (Mortensen & Williams, 2016) and with great promise for the future from crude cell lysates (Gan et al., 2017; Raijmakers, Kraiczek, De Jong, Mohammed, & Heck, 2010; Visser, Scholten, Van Den Heuvel, & Heck, 2007).

For initial analysis (to check the protein mass) it may be appropriate to first examine the protein under conditions that will not preserve its native structure, for example, 50:50 H_2O:MeOH + 0.01% formic acid solution, which is similar to the make-up flow from an LC experiment (Hernández & Robinson, 2007).

For certain experiments other solvents can be used, for example, 2,2,2-trifluoroethanol (TFE) which is known to induce helical structures in proteins at ambient temperatures (Mohanta & Jana, 2018).

2.1.1.1 Equipment and Reagents

- pH meter
- Ammonium acetate
- MeOH:H_2O (50/50) + 0.01% formic acid
- Ammonium hydroxide solution (AmOH) (~20%)
- Acetic acid (AcOH)

2.1.1.2 Procedure

1. Dissolve solid ammonium acetate (in the required quantity) in 50 mL ultrapure water to yield a solution of ammonium acetate with the desired ionic strength.
2. Measure the pH of the solution.
3. Raise or lower the pH to the desired level.

- If the pH of the solution is below the desired level add ammonium hydroxide to raise it.
- If the pH of the solution is above the desired level use acetic acid to lower it.

4. Once the desired pH is reached, the solution is ready to be used.

2.1.1.3 Notes

- Generally not much AmOH or AcOH is needed to alter the pH. It is advisable to start with ~5 μL aliquots (however, these amounts will vary significantly depending upon the concentration of the AmOH/AcOH used, the ionic strength of the solution, and its target pH).
- Due to the small amount of solvent used in nESI, solution quantities are small. As such, it is advisable to calibrate pipettes every 6 months.

2.1.2 Buffer Exchange and Purification

Proteins often arrive in a solution that has been optimized for other biophysical experiments which are rarely suited to direct infusion MS. Even lyophilized proteins are often "enriched" in sodium or potassium salts, and often both powders and solutions can contain some impurities. As a result, it is necessary to carry out buffer exchange to transfer the protein to an MS compatible solvent (e.g., AmAc mentioned earlier).

2.1.2.1 Equipment and Reagents

- Micro Bio-Spin™ 6 Column (Bio-Rad, Hercules, CA, USA)
- Micro Centrifuge Collection Tubes (Bio-Rad, Hercules, CA, USA)
- MSE Micro Centaur Centrifuge
- Buffer solution
- Protein sample

2.1.2.2 Procedure

1. Shake the spin column (in order to resuspend the gel and remove any air bubbles) and break the bottom off, then place the spin column in the collection tube.
2. Remove the lid on top of the spin column and then centrifuge for 1 min at 5000 rpm in order to remove the packing buffer.
3. Remove any waste buffer collected in the bottom of the collection tube and rerun for another 1 min at 5000 rpm.
4. Remove any waste and if the powder (Bio-Gel P-6 gels) appears wet repeat step 3, else pipette 500 μL of the desired buffer to the spin column.

5. Centrifuge for 1 min at 5000 rpm. This will mark the first column equilibration step in the buffer exchange (BE) process. As such, after this first step, the resulting buffer will contain a 50:50 ratio of desired buffer: packing buffer.
6. Repeat step 5 again until a 99.99:0.01 ratio is reached:
   - Buffer composition after 2 × BE cycles—90:10
   - Buffer composition after 3 × BE cycles—99:1
   - Buffer composition after 4 × BE cycles—99.9:0.1
   - Buffer composition after 5 × BE cycles—99.99:0.01
7. Now place the spin column into a fresh collection tube, pipette between 20 and 75 μL of the sample into the spin column (larger volumes may affect the efficiency of the buffer exchange process), and centrifuge for 5 min at 5000 rpm.
8. Collect the filtrate from the bottom of the collection tube and repeat step 7 (using filtrate instead of sample) in a fresh, equilibrated spin column (i.e., 5 × BE). Repeat for the third time and collect your MS compatible sample solution.

2.1.2.3 Notes

- Another method for buffer exchange and/or desalination is dialysis. Typically, dialysis is performed in cassettes and this is feasible if larger amounts of protein (> 1 mg) are available.
- Buffer exchange utilizing spin columns are a good alternative to dialysis when sample quantity is an issue. However, sample loss and dilution during the purification step are not uncommon.
- Typically 2–3 buffer exchanges are carried out when purifying any given protein; this number should be minimized based on MS analysis.
- This process is also applicable to the purification of a protein obtained as a lyophilized powder, which can be dissolved in the desired buffer beforehand and buffer exchanged as described in step 7.

2.2 Ion Mobility and Mass Spectrometry

Each spectrometer is different and requires a certain level of finesse to operate in such a way as to obtain high-quality data. Accordingly, it is sensible to run a well characterized, cheap, readily available "standard" protein prior to analyzing an unknown sample. This preliminary analysis is used to check whether the nanospray emitters are optimal and whether the mass spectrometer is working correctly. Furthermore, this process allows for the preoptimization of instrument tuning parameters prior to the analysis of the subject protein. Standard proteins should ideally present with a similar m/z to the protein under investigation and be selected on the basis that they are of high-quality and low batch variability while presenting with available mass and CCS literature values. Commonly these standards are globular proteins rather than IDPs, but this does not influence their applicability within the preoptimization process. Following emitter and instrument preoptimization and buffer exchange of the subject protein, we can proceed to analyze the sample. From a samples mass spectrum we obtain a CSD from which we can infer the analytes experimental mass, which can be compared to its theoretical mass to ascertain if there are any PTMS or uncleaved tags present on the analyte. The CSD gives us valuable information about the proteins structure as mentioned in Section 1.1.

The next step is to acquire mobility data by carrying out an ion mobility experiment. In drift tube instruments which have a quadrupole prior to the IM cell, this can involve mass selection of specific charge states followed by IM analysis at a range of different drift voltages (typically $\geq$5). Following successful data acquisition, ATDs of specific charge states can be extracted and after further data processing collision cross section distributions (CCSDs) can be generated.

2.2.1 Equipment and Materials

- An ion mobility-mass spectrometer, some examples from our lab include: A Waters Synapt G2 fitted with an RF-confining drift cell across which is applied a static potential gradient (Bush et al., 2010), an Agilent 6560 DTIMS QToF (Kurulugama, Darland, Kuhlmann, Stafford, & Fjeldsted, 2015), and two home-built DT-IM-MS instruments capable of measurements at a range of drift gas temperatures (McCullough et al., 2008; Ujma et al., 2016)
- Thin-wall borosilicate glass capillaries (i.d. 0.9 mm, o.d. 1.2 mm, World Precision Instruments, Stevenage, UK) or equivalent
- Model P-1000 Flaming/Brown Micropipette Puller (Sutter Instrument Company, USA) or equivalent
- 10 μL gas tight syringe (Hamilton)
- Spool of platinum wire, 0.125 mm diameter (Goodfellow, Cambridge Ltd., UK)
- Helium/Nitrogen and Argon gas (>99.9%, vol/vol)

2.2.2 Tip Pulling/Tuning

In line with the majority of native MS studies, we ionize IDPs using nESI. In this study the solution is ionized via the application of a potential (which in

Table 1 Lists the Various Settings on the P-1000 Tip Puller We Use

Setting	Range	Description
Heat	0–999	Controls the level of electrical current supplied to the filament
Pull	0–255	Controls the force of the pull. The higher the PULL the longer the taper and smaller the diameter of the tip
Velocity	0–255	The adjustable velocity allows for a selection of a precise glass temperature as the trip point for the hard pull
Time	0–500	Controls the length of time the cooling air is active
Delay	0–500	Controls the delay time between when the heat turns off and when the hard pull is activated

More detail can be found in the operations manual which is available as a PDF online (https://www.sutter.com/manuals/P-1000_OpMan.pdf).

this case is positive) to a thin platinum wire inserted into the loaded nESI emitter. That being said, emitters can also be coated with gold or palladium and the source potential applied to this surface instead. All emitters utilized within this work were pulled in-house using a model P-1000 Flaming/Brown micropipette puller. This puller has various settings which can be tuned in order to produce emitters of varying taper and orifice morphology; these settings are listed in Table 1.

2.2.3 Mass Spectrometry and Ion Mobility

Our group has a variety of instruments to carry out (IM)-MS experiments, and we have analyzed IDPs on several of these (Stuchfield & Barran, 2018). For this chapter, data have been taken on the modified Synapt G2 (with an RF-confining drift cell) to highlight the steps undertaken to analyze IDPs by IM-MS.

Samples analyzed via nESI are typically sprayed from 10 to 500 m*M* ammonium acetate (the chosen salt concentration depends on the protein of interest, as nonnative self-interactions can occur if ionic strength is too low, and salt clustering can occur at high concentrations) solutions buffered to pH 6.8–7.4, in order to best mimic physiological conditions. We often apply capillary voltages of 0.7–1.2 kV, with (on Waters instruments) cone voltages ranging from 10 to 200 V, and source temperatures of 50–80°C. The pressure in the desolvation region of the source is also an important parameter to adjust since it alters how and when the solvent and salt departs the protein. This is affected either by throttling the flow on the roughing pump on the source and/or by adding more gas into the source (Sobott, Hernández, McCammon, Tito, & Robinson, 2002; Sobott & Robinson, 2002; Van Den Heuvel et al., 2006).

2.2.3.1 CCS From DTIMS

In order to obtain necessary data to derive CCSs of individual ions, we perform IM experiments at several (≥5) drift voltages while also recording the temperature and pressure of the buffer drift gas (N_2 or He) at each drift voltage. The ATDs of the entire mass spectrum are recorded by synchronization of the release of ions into the drift cell with the mass spectral acquisition. All MS and IM-MS data were analyzed using MassLynx v4.1 (Waters, Manchester, UK), Origin v8.5 (OriginLab Corporation, USA) software, as well as Microsoft Excel and ORIGAMI (Migas, France, Bellina, & Barran, 2017).

The CCS values are derived from raw arrival time data using the Mason–Schamp (Eq. 1) (Revercomb & Mason, 1975):

$$\Omega_{avg} = \frac{(18\pi)^{1/2}}{16}\left[\frac{1}{m_b}+\frac{1}{m}\right]^{1/2}\frac{ze}{(k_B T)^{1/2}\rho}\frac{1}{L^2}t_d V \quad (1)$$

where m_b and m represent the buffer gas and ion masses, respectively, z is the charge of the ion, e is the elementary charge, k_B is the Boltzmann constant, T is the gas temperature (in Kelvin), ρ is the buffer gas density, L is the drift tube length (in cm), V is the voltage across the drift tube, and t_d is the ion drift time. It should be noted that because the raw arrival time output, t_a, includes the time the ion spends outside of the drift cell, t_0, Eq. (2) has to be used to obtain the actual ion drift time, t_d. An ions dead time (t_0) is calculated from the y-intercept of the linear plot of the ion arrival time (t_a) extracted from the apex of the ATDs obtained at each drift voltage (DV) vs the reciprocal of the effective drift voltages ($1/V$). Accordingly we obtain a t_d which can be applied to Eq. (1) to calculate the CCS for the apex of an ion population.

$$t_d = t_a - t_0 \quad (2)$$

2.2.3.2 DTIMS Procedure

1. *Standard Preparation:* Begin by running a protein standard. We recommend myoglobin (*holo* = 17.6 kDa, *apo* = 16.9 kDa) or cytochrome *c* (12.4 kDa) for smaller IDPs. Prepare a solution of the protein standard in 10–50 m*M* ammonium acetate at ~1–5 μ*M* (use MeOH:H_2O:FA, 50:50:0.01 mixture if analyzing under denaturing conditions).
2. *Drift Gas:* Choose a drift gas for use in IM separation, the majority of experiments on proteins have used either N_2 or He, but other gases are viable. In Synapt instruments argon is the default gas employed in the Trap/Transfer regions.

3. *Instrument Settings:* Example tuning parameters employed for the native DTIM-MS analysis of *holo*-myoglobin on the modified Synapt G2 (with an RF-confining drift cell) are given in Table 2, and a schematic diagram of the modified Synapt G2 is shown in Fig. 2. The parameters outlined in Table 2 yielded a good compromise between MS quality and the minimization of gas-phase structural activation. This was concluded by comparison of our experimentally derived CCSs with previously reported literature values (Jurneczko, Kalapothakis, Campuzano, Morris, & Barran, 2012). As such, *holo*-myoglobin, sprayed from a nondenaturing solution (made up as described previously) utilizing the instrument tuning parameters described below should yield a similar MS to what is observed in Fig. 3.
4. *Sample Preparation:* Once the standard is spraying stably with good ion transmission (shot-to-shot of >1E3 from nondenaturing solutions), analysis of the subject protein can commence. Prepare a solution of the subject protein in ammonium acetate at ~5–20 μ*M* (higher concentrations of the subject will be used relative to the standard protein just to observe signal, after which the concentration can be reduced) following the procedures described in Section 2.1 and Fig. 1A.
5. *Instrument Settings:* With the same tuning parameters employed for the analysis of the standard, spray the subject protein (these parameters may be the same as in Table 2, or different depending on whether slight alterations were needed to improve signal intensity and/or the likelihood of structural activation).
6. *Judging Quality:* Mass spectral quality will differ depending upon the samples' solution composition (nonvolatile buffer components reduce mass spectral quality), homogeneity, and its propensity for salt adduction. As mentioned in Section 2.1 most IDP samples we obtain have been recombinantly expressed and as such are rarely homogenous in nature. This is why the buffer exchange procedure, outlined previously, is essential prior to MS analysis. An example of a spectrum containing insufficiently high amounts of nonvolatile adducts, where the analyte requires further purification, is depicted in Fig. 4. One way to solve this issue is to repeat the steps covered in Section 2.1.2.
7. *MS Analysis:* Improvement of the data quality is achieved, either by further purification or instrument parameter optimization, whereby source voltages within the instrument can be increased to effectively dissociate nonvolatile salt adducts from the analyte ions. After optimizing the preparative procedure and tuning parameters to obtain a

Table 2 Optimized Instrument Parameters Used to Run a Sample of 5 μ*M* Myoglobin in 50 m*M* Ammonium Acetate (pH 7.4) on the Modified Synapt G2 (With an RF-Confining Drift Cell)

Assignment in Fig. 2	Source Voltages (V) and Gas Flows (mL/min)	
1	Capillary voltage (kV)	0.7–1.2
Source	Source temperature (°C)	40.0
2	Sample cone	20
3	Extractor cone	1
4	Trap gas flow	2
5	Helium gas flow	100.0
DC voltages		
4	Trap cell energy (V)	0.0–4.0
4	Trap DC entrance	1.0
4*	Trap DC bias	1.0
4	Trap DC	0.0
4	Trap DC exit	1.0
4	Trap height (V)	10–15
5	IMS DC entrance	−6.5
5	Helium cell DC	Altered incrementally for DVs
5	Helium cell exit	−40
5	IMS bias	Altered incrementally for DVs
5	IMS DC exit	0.0
6	Transfer cell energy (V)	0.0–6.0
6	Transfer DC entrance	4.0
6	Transfer DC exit	15.0
Pressures		
Source	Backing (mbar)	2.7
4	Trap (mbar)	1.3×10^{-2}
5	IM (Torr)	1.94
6	Transfer (mbar)	1.4×10^{-2}
7	ToF (mbar)	7.4×10^{-7}

Voltages, gas flows, and typical pressures utilized and observed within the analysis of myoglobin (in this instance) are linked to regions within the modified Synapt G2 (outlined in Fig. 2) by the numbers in the left hand column.

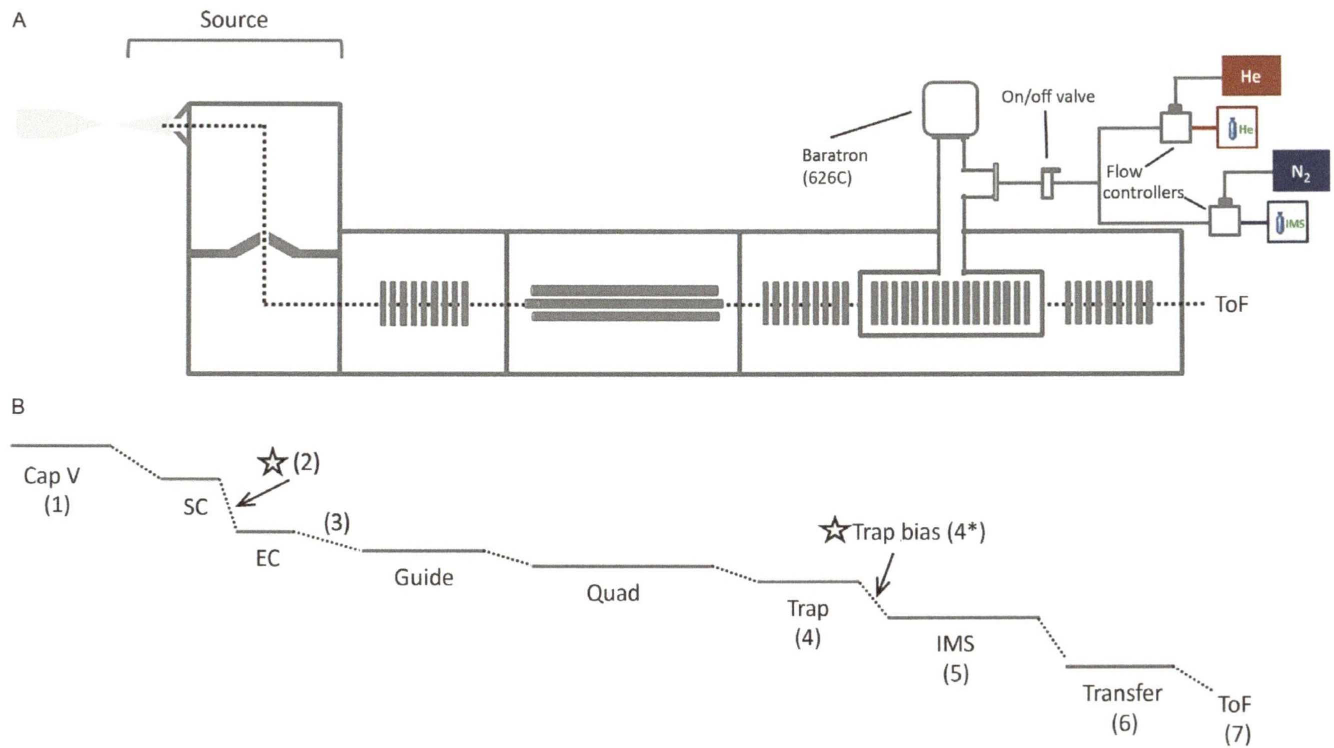

Fig. 2 (A) Schematic of the modified Synapt G2 (with an RF-confining drift cell), in which the *black dotted line* denotes the ion beam path through the instrument. Drift gases can be switched using the MassLynx software which is connected to a modified gas inlet system. Absolute measurements of the drift gas pressure were enabled by installation of a capacitance manometer (MKS Baratron type 626C) above the center of the drift cell. (B) Schematic of the potential gradient along the instrument, the numbers next to specific regions within the schematic correspond to their respective voltages(s)/gas flows and pressures outlined in Table 2. Voltages which impart the greatest energy upon ions, thus promoting their gas-phase activation, are highlighted by *gray stars*. The voltage gradient from the sample cone (SC) to the extractor cone (EC) can be utilized to promote in-source ion activation. Furthermore, the trap bias voltage, which acts to accelerate ions from a comparatively low-pressure environment (the trap cell) to a higher pressure one (the helium and IM cell), can also significantly perturb ion structures if not optimized for the system in question.

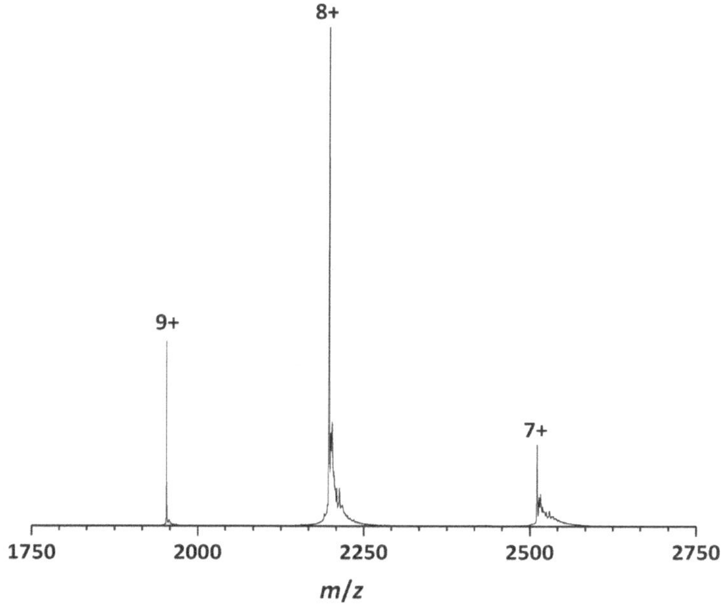

Fig. 3 Mass spectrum obtained following nESI of 5 μ*M* myoglobin in 50 m*M* ammonium acetate (pH 7.4). The numbers above the peaks denote the charge state *z* for the ion of the form $[M+nH]^{n+}$, where $n = z$. This spectrum was acquired on a modified Synapt G2 (with RF-confining drift cell).

high-quality mass spectrum, postacquisition analysis can commence. Initially the protein charge states can be found from the m/z peaks and their spacing within the mass spectrum (Eq. 3). After which the experimental molecular weight of the analyte can be determined (Eq. 4).

$$z = \frac{M_2 - mA}{M_1 - M_2} \tag{3}$$

$$\text{MW} = \frac{(M_1 - mA)(M_2 - mA)}{M_1 - M_2} \tag{4}$$

where z is the charge state of M_1, M_1 is the m/z of the lower charge state ion (higher m/z), M_2 is the m/z of the higher charge state ion (lower m/z), mA is the mass of a single proton (or any other observed adduct), and MW is the molecular weight of the analyte. Regarding the CSD, a large Δz observed for proteins analyzed from near-physiological solution pH, is indicative of at least partial disorder within the protein structure, albeit with some exceptions (Beveridge et al., 2015; Natalello, Santambrogio, & Grandori, 2017). It is worth noting

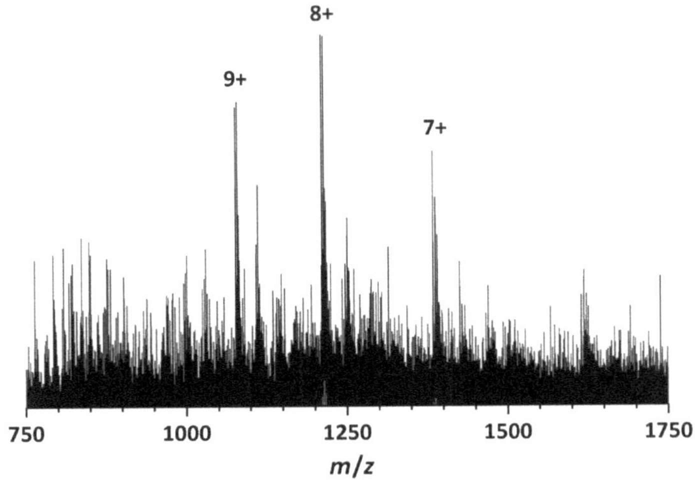

Fig. 4 Mass spectrum of COR15A 20 μ*M* in TFE. The numbers above the peaks denote the charge state *z* for the ion of the form $[M+nH]^{n+}$, where $n = z$. The low signal:noise ratio within the spectrum, as well as the comparatively wide mass spectral peaks within the charge state envelope, indicates that the sample solution still contains a high level of nonvolatile salts. This spectrum was acquired on a modified Synapt G2 (with RF-confining drift cell).

that another method for determining charge on a protein is using the Rayleigh Limit (Z_R) parameter which was developed by Fernandez De La Mora (2000) (Eq. 5). It allows the calculation of the maximum number of charges that a globular protein of a specific molecular weight can carry.

$$Z_R = 0.0778\sqrt{m} \quad (5)$$

where Z_R is the maximum charge and m is the mass of the protein. As mentioned in Section 1.1 the spread of z, *ergo* the CSD provides valuable information about the structure of the protein; wider CSDs generally suggest that the protein occupies a large range of conformers in solution, thus yielding a multitude of available protonation sites upon ionization (Konermann, 2007). The nESI mass spectrum from the LEA protein COR15A (Fig. 5) possesses a broad CSD with $\Delta z = 9$, as would be expected for an IDP. Interestingly and symptomatic for many proteins, the lower charge states ($z = 5$–7) still retain some salt adducts (see insert in Fig. 5) relative to $z \geq 8$, which would indicate that the lower charge states might sample more compact, globular structures with binding cavities which permit increased salt adduction.

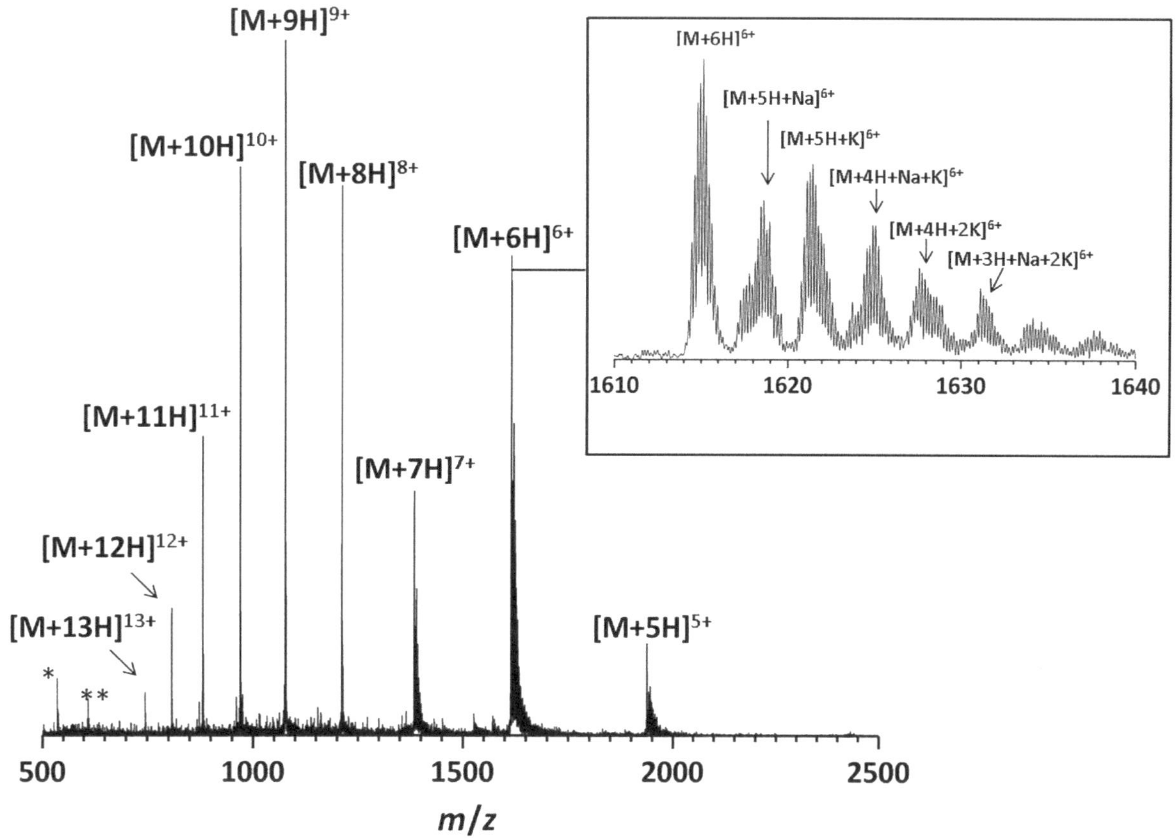

Fig. 5 Mass spectrum of COR15A sprayed from 200 m*M* ammonium acetate solution, pH 7.3. The experimentally determined MW is 9688.3 Da ± 3.4, and the theoretical MW is 9683 Da. Peaks denoted with "*" and "**" are contaminant species with *m/z* 527 and 610, respectively. The insert highlights the adducts seen for the charge states ranging from $z = 5$–7, specifically it shows the adducts for the $z = 6+$ charge state. This spectrum was acquired on a modified Synapt G2 (with an RF-confining drift cell).

8. *Acquiring Mobility Data:* The next step is to conduct IM experiments in order to obtain CCS data for the subject protein; for linear field instruments, this involves taking measurements at five or more drift voltages. The measurements taken at each voltage provide an ATD for each charge state. The apex values from these distributions give a time t_a (Eq. 2) at which the ion intensity is at a maximum. For each of the different voltages (V), these t_a values can be used to generate a plot of t_a vs $\frac{1}{V}$ with a gradient that is proportional to the inverse of the reduced mobility, $\frac{1}{K_0}$, and an intercept that corresponds to the dead time t_0, which is independent of the time spent in the drift cell. Linearity of these data points reflects an adherence to the Mason–Schamp equation (1) and indicates that all data are obtained in the low-field conditions. The slope of the plot can be used to determine the CCS value that corresponds to the apex of the ATD for that charge state using Eq. (6).

$$\Omega = \frac{3ze}{16N}\left(\frac{2\pi}{\mu k_B T}\right)^{1/2}\frac{1}{K_0} \tag{6}$$

where Ω is the CCS, z is the charge on the ion, e is the elementary electron charge, N is the gas number, μ is the reduced mass of the ion, k_B is the Boltzmann constant, T is the temperature in Kelvin, and K_0 is the reduced mobility. The plot stems from Eq. (7) which relates the mobility of an ion to its drift time.

$$KE = v_d = \frac{L}{t_d} \tag{7}$$

where K is the ion mobility, E is the electric field strength, v_d is the drift velocity, L is the drift tube length, and t_d is the drift time. Because low-field ion mobility measurements are dependent on pressure and temperature we use the reduced mobility, K_0, where pressure and temperature are normalized as shown in Eq. (8).

$$K_0 = K\left(\frac{T_0}{T}\right)\left(\frac{P}{P_0}\right) \tag{8}$$

Normalized temperature T_0 and pressure P_0 are 273.15 K and 760 Torr, respectively.

9. *Mass Select:* Tandem mass spectrometers allow any charge state of a given protein to be selected. In some IM-MS instruments (e.g., The Waters Synapt family) this takes place prior to the mobility separation, such that the ATD and the mass spectrum only arise from that parent ion.

This prevents contamination of the ATD for a given charge state, from higher charge state ions that have been "charge stripped" during or at the end of the drift cell, or from multimers that have dissociated in or after the drift cell to yield a product ion of the charge state of the ion of interest. If, mass selection prior to IM is not possible, then the analyst must be aware of these confounding factors when interpreting data. Irrespective of whether this is technically possible, it is always possible to extract a given ion from the acquired data, and this is a necessary step in data analysis. A zoomed-in region of the mass spectrum focusing on the $[M+10H]^{10+}$ ion of COR15A and corresponding ATD is shown in Fig. 6.

10. *Optimize:* Following mass selection we look at the mobiligram to judge the quality, usually a degree of tuning of the voltages and optimization of the gas flow around the drift cell is needed in order to improve the data quality (again consult Table 2 for good "start-point" settings; specifically the Trap DCs and IMS DCs). Fig. 7A and B shows examples of an acceptable ATD and an ATD where concentration-dependent non-native dimers are seen; it is worth noting that these dimers can be removed by serial dilution. Fig. 7C demonstrates the effect poor voltage optimization has on the mobiligram. In these steps, which adjust the passage of the ion into the drift cell, it is important to consider that we do not want to disrupt the structure of the ion. If a protein ion is injected into the drift tube with high acceleration voltages it is likely to deform either by first compacting and then (or directly) unfolding, thus distorting the subsequent interpretation of the ATD. With higher injection energies the transmission of ions through the drift cell becomes more favorable resulting in higher signal intensity, so it is important that the analyst tunes the signal while monitoring the ATD to prevent such distortion. At very high injection energies, the ion will penetrate far into the drift cell before it reaches the constant drift velocity that satisfies the Mason–Schamp equation (1) and this will effectively shorten the length of the drift region, causing errors in the CCS calculation. Conversely, if an ion is not given sufficient energy it will not enter the cell against the egressing drift gas, resulting in a loss of signal. When a pulse/packet of ions is introduced to the drift tube they begin to separate, therefore the measurement of the packet must be completed before the second packet is introduced otherwise overlap, known as "wrap around" occurs.
11. *Acquisition*: Once a good quality ATD is obtained, a run of data can be acquired. The analyst should take note of the drift cell DC, IMS bias, He cell exit, and transfer DC entrance (as these are used in combination to establish the drift voltage across the cell), as well as the temperature

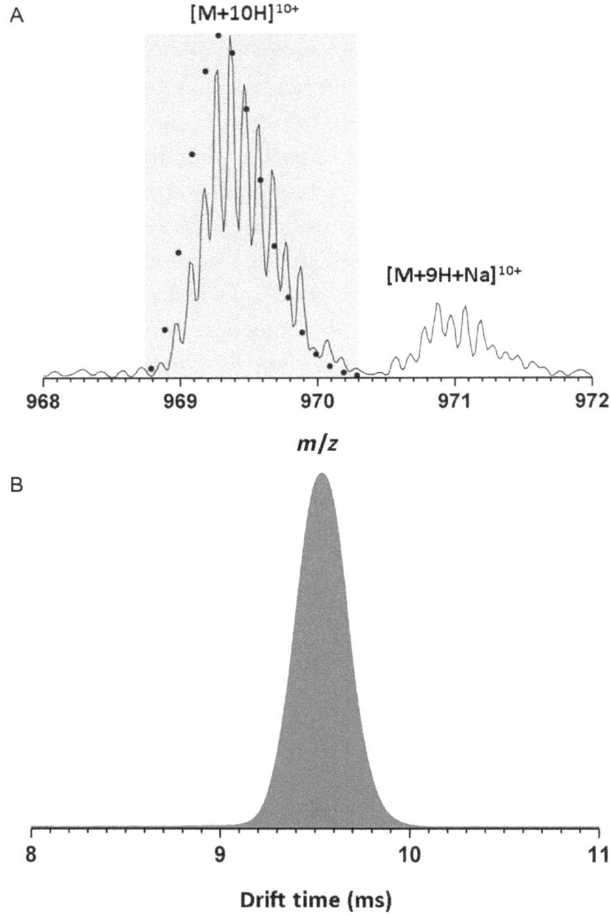

Fig. 6 (A) Isotopic distribution of the $[M+10H]^{10+}$ and the salt adduct $[M+9H+Na]^{10+}$ charge state of the COR15A protein. The *solid circles* represent theoretical isotopic distributions for the $[M+10H]^{10+}$. (B) The arrival time distribution for the COR15A, $z=10$ charge state at a drift voltage of 100 V. COR15A was sprayed from 200 m*M* AmAc, pH 7.3 at a concentration of 20 μ*M*. Data for (A) and (B) were acquired on a modified Synapt G2 (with an RF-confining drift cell).

and pressure within the drift cell. After acquisition of the measurement at a given voltage (for example, 100 V), the drift voltage is then incrementally increased by c. 25 V to begin the second measurement until we have six different drift voltage measurements. By the end of the experiment we have six ATDs at six different voltages as shown in Fig. 8. Using this drift data we can calculate the CCS of the ion.

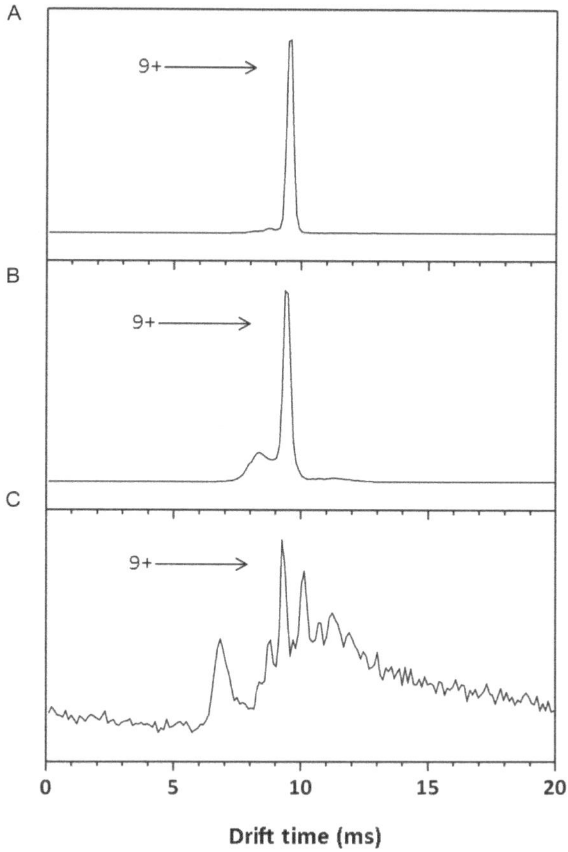

Fig. 7 (A) Shown is an example of a good ATD of the 9+ charge state of the COR15A protein. (B) Shown is an ATD of the 9+ charge state of COR15A, the shoulder seen to the left of the main peak likely corresponds to the 18+ dimer of COR15A which has the same nominal *m/z* and so will be selected along with the monomeric 9+ species. (C) Here is an example of really bad data where the clean peaks we saw previously are obscured by noise, this stems from poor voltage optimization. All ATDs were obtained on a modified Synapt G2 (with an RF-confining drift cell).

3. DATA ANALYSIS

3.1 CCS Direct From Ion Mobility

The measurements taken at different drift potentials are used to generate a plot of t_d vs $\frac{1}{V}$ with a gradient that is proportional to $\frac{1}{K_0}$, which can be used to determine an ions CCS. An example is shown in Fig. 9.

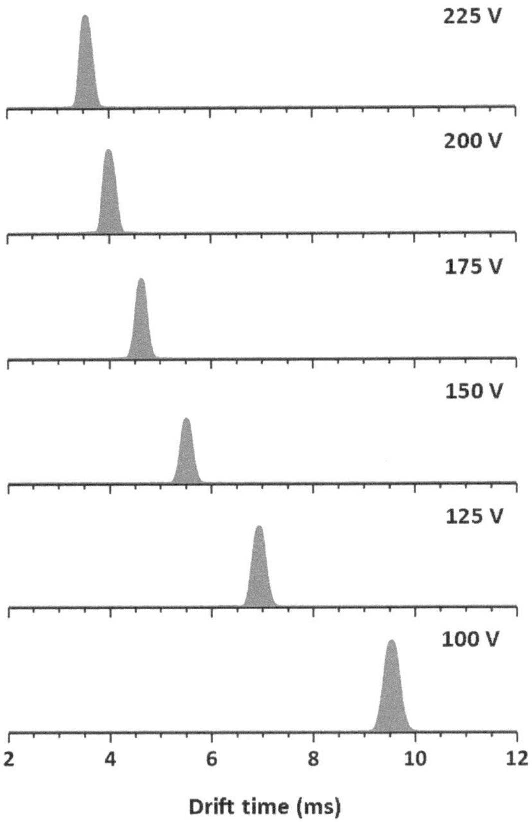

Fig. 8 Shown are six selected ion ATDs taken at six different drift voltages for the [M + 10H]$^{10+}$ ion of COR15A protein. As the drift voltage is increased, the ions drift velocity also increases, resulting in shorter residency times in the drift cell. For the modified Synapt G2 (with an RF-confining drift cell) used here, each ATD was acquired over c. 2 min.

The plot stems from Eq. (7) shown in Section 2.2. Because low-field ion mobility measurements are dependent on pressure and temperature we use the reduced mobility, K_0, where pressure and temperature are normalized as shown in Section 2.2 in Eq. (8).

3.2 Converting ATDs to CCSDs

In Section 1.1 we discussed that the most useful structural depiction of an IDP defines them in terms of conformational ensembles. In light of this, it is advantageous to think of the CCS as a distribution rather than a discreet value. We can derive a CCS distribution from the raw arrival time data using

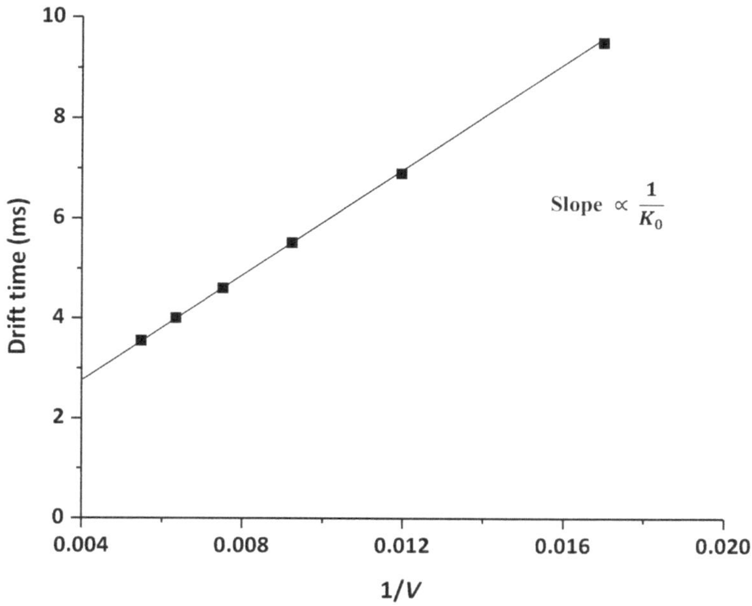

Fig. 9 Shown is a plot of t_d vs $^1/_V$ obtained using the six ATDs for COR15A as seen in Fig. 8. The slope of the plot is proportional to the inverse of the reduced mobility of the ion. Using this we can calculate the CCS of a specific ion. A similar approach is automated in the Agilent Software for use of their 6560 IM-QToF (Kurulugama et al., 2015).

Eq. (1), which directly relates drift time to the CCS. The process of doing this is as follows:

1. We obtain an ATD of a single charge state. In our case, we convert scan time (bins) to time (s) using Eq. (9):

$$\text{Arrival time (s)} = \text{scan number} \times \text{pusher time (s)} \tag{9}$$

2. For each charge state of our protein and at every drift voltage we must determine the apex of the major conformation in the ATD and use the plot shown in Fig. 9 to elucidate the intercept (dead time, t_0) and the slope ($1/K_0$) which is subsequently used to calculate the collision cross section using Eq. (6).
3. Application of the derived parameters for a single drift time value (i.e., apex of the distribution) gives us a single $^{DT}CCS_{He}$ value, however, as mentioned earlier on, IDPs are better explained using CCSDs which show the entire conformational ensemble of the ion. In order to obtain the CCSD, we use the Mason–Schamp equation (1) for each drift time value from the ATD, as shown in Fig. 10.

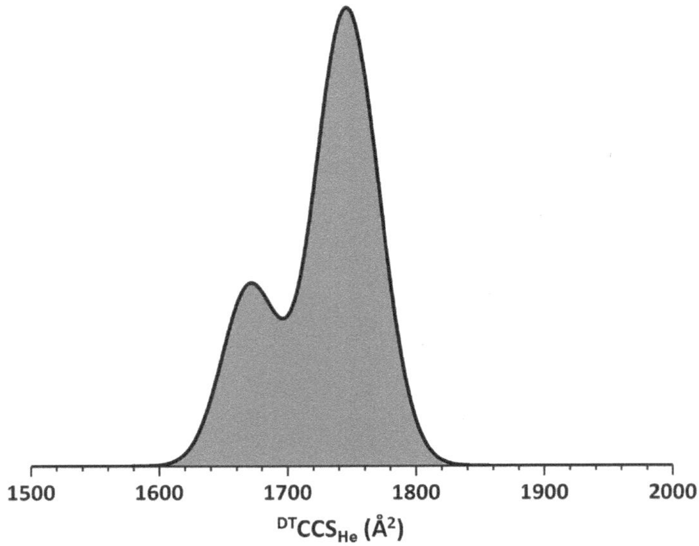

Fig. 10 Shown is the $^{DT}CCSD_{He}$ for the $[M+9H]^{9+}$ ion of COR15A. The CCSD is composed of two peaks, with the apex of the conformational families centered at 1670 and 1760 Å^2. The lack of baseline resolution may be instrumental, or due to interconversion between these two conformers on the timescale of the experiment. COR15A was sprayed from 200 m*M* AmAc, pH 7.3 at a concentration of 20 μ*M*. These data were acquired on a modified Synapt G2 (with an RF-confining drift cell).

When presenting IM-MS results, we tend to use the CCSD obtained from the lowest drift voltage as it tends to result in better separation of species with multiple conformations as the gradient in the drift tube is less steep and ions have more time to separate. It is worth noting, however, that as the ion velocity decreases (due to decreased electric field), the signal to noise of the ions of interest can decrease too meaning a higher drift voltage might be more appropriate.

3.3 Uncovering Conformational Diversity

The practical steps outlined earlier can be considered as guidelines for IM-MS analysis of proteins regardless of their inherent order. It is now pertinent to consider the difference seen in data obtained from ordered and disordered proteins. As well as CCS distributions for a specific charge state of a given protein (Fig. 10), it is possible to plot a total CCS distribution for a protein (Fig. 13). The process we use to obtain a total CCSD is given below:

Replicate Standardization: The first step is to fit Gaussian peaks so as to best reflect the experimental data. The cumulative fit of these peaks will have the same range and number of data points across each CCSD replicate for a given ion we do this in order to standardize the CCS bins across replicates. This because the replicate CCS axes will be slightly different, as the dead time and drift time will vary across experiments due to variables such as temperature and pressure.

Normalization: We need to normalize the intensities of the different charge states to one another. This is done as follows:

- For each charge state, we average the intensities of the replicates.
- For each charge state, we determine the area under the CCSD.
- For each charge state, we determine the peak height in the mass spectrum.
- We sum the areas for each charge state together.
- We sum the MS peak heights for each charge state together.
- We determine the fraction (area and MS) for each charge state using Eq. (10):

$$z^{\text{fraction}} = \left(\frac{A_z}{\sum_{z_{\min}}^{z_{\max}} A_z}\right)\left(\frac{h_z}{\sum_{z_{\min}}^{z_{\max}} h_z}\right) \tag{10}$$

  where A_z is the area under the CCSD for charge state z, and h_z is the peak height for charge state z.
- We then normalize to the highest fraction using Eq. (11):

$$z^{\text{normalized}} = \frac{z^{\text{fraction}}}{z^{\text{fraction}}_{\max}} \tag{11}$$

- Following this, we multiply the averaged intensities by $z^{\text{normalized}}$ to obtain normalized intensities for each charge state.
- Lastly we sum together the normalized intensities of each charge state to obtain the intensity for the total CCSD.

Following ESI from salty solutions, structured globular proteins present ATDs and corresponding CCSDs that are Gaussian-like and often represent single conformational families (Fig. 11B) (Beveridge et al., 2014, 2015; Jurneczko & Barran, 2011). However, it is not unusual for each charge state to have a slightly different CCSD, with higher values of charge corresponding to higher average CCS values (Pacholarz & Barran, 2015; Pacholarz et al., 2014, 2017). By contrast for IDPs, multiple resolved or partially resolved distributions are observed in the ATD even for single charge

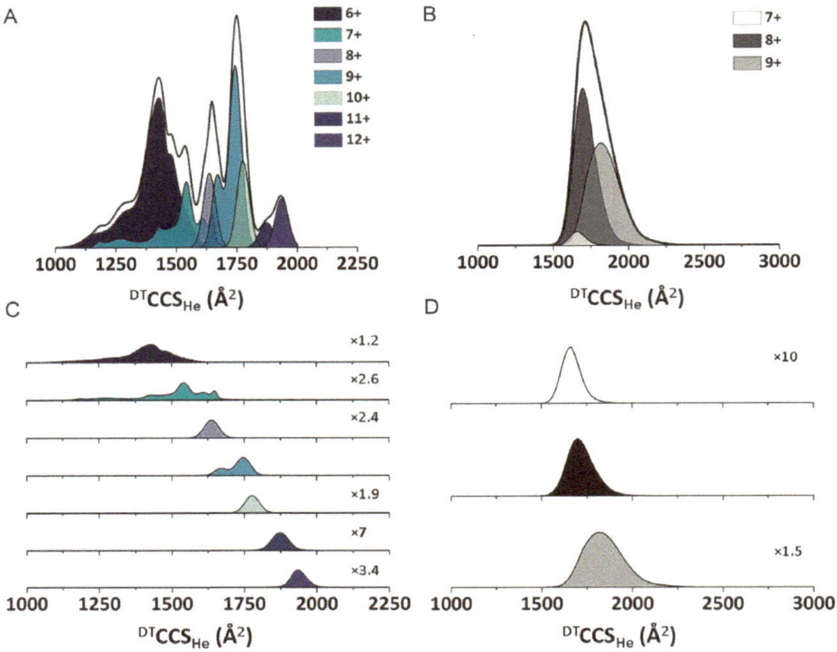

Fig. 11 (A) A total CCSD plot of the seven charge states of the COR15A protein, showing a typical spread in CCS values within and across charge states. The two lowest charge states, $z=6$ and $z=7$, have distinctively different distributions to the higher charge states; they are much broader with a number of partially resolved conformational ensembles. For $z=8$ only one conformational family can be observed. Interestingly the $z=9$ charge state appears to represent an intermediate species which is comprised of two partially resolved peaks each representing a distinct conformation. Regarding the higher charge states ($z=10$–12), the CCSDs are much narrower than $z=6$ and 7 and consist of a single observable conformational family. (B) A total CCSD plot of the three charge states of myoglobin, each charge state is seen to have a singular Gaussian-like distribution which is due to only one resolvable conformational family being present for each. (C) A stack plot of the CCSD for each of the seven charge states of COR15A. (D) A stack plot of the CCSD for the three charge states of myoglobin. Magnification values relative to the most abundant charge state are represented in the *right hand corner* of each row within the stacked plots.

states (Beveridge et al., 2014, 2015; Dickinson et al., 2015b). Under denaturing or partially denaturing conditions, structured proteins present with a greater Δz, often with higher charge states yielding larger CCSs (Clemmer et al., 1995; Dickinson et al., 2015b).

For myoglobin, $\Delta z=3$ and $\Delta\text{CCS}=\sim 400\ \text{Å}^2$ (ΔCCS = apex of the largest CCS—apex of the smallest CCS). The Gaussian-like distribution of the peaks indicates the presence of a single conformer for each charge state, although it can be noted that as the charge state increases so does the average CCS value. The low Δz and ΔCCS as well as the single Gaussian-like distribution of each charge state peak indicates that myoglobin occupies a narrow conformational landscape with few closely related conformational families, as might be expected for a well-ordered protein following ionization from native-like solution conditions (Beveridge et al., 2014). The total CCSD plot for COR15A, with $\Delta z = 7$ and ΔCCS of $\sim 1000\,\text{Å}^2$, supports the assertion made from the broad CSD (Fig. 5) that this protein presents to the mass spectrometer in numerous conformations. Additionally, for several charge states ($z = 6$, 7, and 9) there are multiple conformational families in the CCSD, indicative that COR15A occupies a conformational landscape which has many low energy conformations, which are close in energy to one another, but distinct in structure.

The charge states of $z = 6$, 7, and 9 of COR15A (Fig. 11A) give rise to complex ATDs with numerous partially resolved Gaussian-like signatures attributable to the presence of multiple conformers for the same net charge (Fig. 11C). In comparison, for myoglobin a Gaussian-like distribution is found for each charge state (Fig. 11D) indicative of single closely related conformational families. Comparison between the IM-MS data for myoglobin and COR15A (Fig. 11), which was taken on the same instrument with very similar conditions, reveals how dramatic the difference in structures between the two proteins are. This work highlights how we can use IM-MS to discern the conformational heterogeneity of any given protein/protein complex.

The total CCSD also allows us to define potential conformational families that exist within the conformational landscape of the IDP in question; this is shown in Fig. 12.

By simply fitting Gaussian peaks to the total CCSD of COR15A we are able to identify possible conformational families within the overall gas-phase landscape of the protein. This shows that in comparison to a globular protein, like myoglobin, COR15A most likely adopts numerous conformations in solution which can be sampled in vacuo.

3.4 A Predictive Framework for Disorder

We now understand that CSDs and CCSDs can be used to provide structural information on the conformational diversity and level of disorder within a protein. As we mentioned previously (Section 3.3), proteins have distinct features in their CSDs and CCSDs that can be used to categorize them as principally ordered or disordered when sprayed from certain solution conditions.

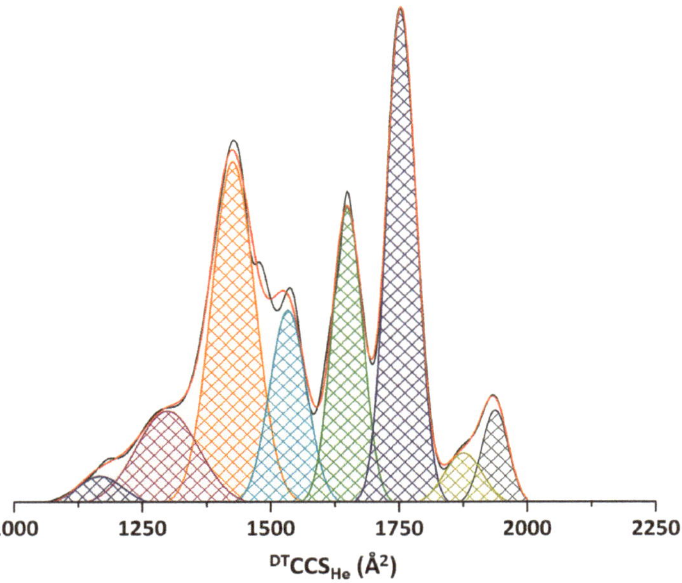

Fig. 12 Gaussian peaks were fitted to the total CCSD of COR15A and represent possible conformers present within the gas-phase conformational landscape of the protein. A total of eight peaks were fitted to the total CCSD and likely correspond to eight different conformational populations of the COR15A protein.

We have developed a predictive framework that uses a proteins CSD and CCSD to predict the level of disorder within it and a "toy model" that calculates a minimum and maximum boundary CCS for the protein. The framework and toy model have been written in python code and are available on GitHub for free use (https://github.com/ModronMan/Origami-Analyst).

3.4.1 Trends in Collected IM-MS Data: "Beveridge–Barran" Plots

Using data acquired from many proteins we have developed an empirical predictive framework that can be used to evaluate the level of disorder in an unknown protein (Beveridge et al., 2014). By plotting the range of CCS values (ΔCCS) and observed charge states (Δz) for proteins against their molecular weight, Beveridge et al. (2014) showed data plots and trends of the form below (Fig. 13). These plots indicate that some proteins have a narrow charge state and CCS range which hardly varies with molecular weight, while some have a far broader range of charge states and a correspondingly large range of CCSs. The former category is primarily made up of proteins that are known to be structured following careful native MS analysis and sample preparation as

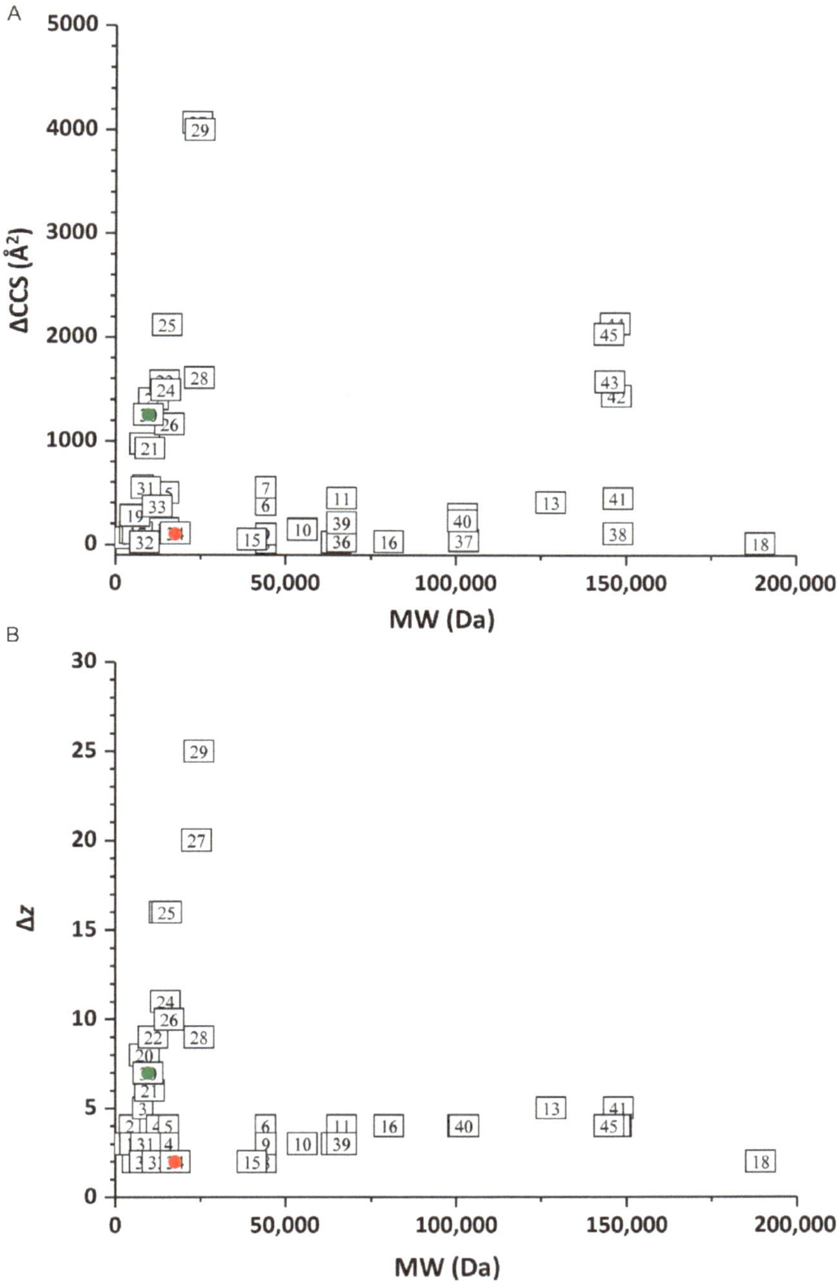

Fig. 13 Beveridge–Barran plots for a data set of proteins which can be found in Appendix. Each number corresponds to a protein, COR15A (*green dot*) and myoglobin (*red dot*) are included in the set. (A) A plot of ΔCCS vs molecular weight for a data set of proteins, including COR15A and myoglobin. (B) A plot of Δz vs molecular weight for a data set of proteins, including COR15A and myoglobin.

shown above, whereas the latter set, with large Δz and ΔCCS consist either of conformationally dynamic proteins, sprayed from pseudonative solutions, or from structured proteins sprayed from denaturing solutions.

Following IM-MS analysis, the analyst can obtain and plot ΔCCS and Δz vs molecular weight and estimate the structural propensity for a given analyte depending upon which region it resides in. For example, following the procedures described earlier, COR15A falls within the "disordered" region of the graphs, whereas myoglobin is found in the "ordered" region. This demonstrates how these plots can be used to give valuable insight into the structure of an unknown protein.

3.4.2 A Toy Model for Predicting CCS Boundaries

In addition to the Beveridge–Barran plots, we have developed a "toy model" which calculates a theoretical upper and lower boundary of a proteins CCS based off its amino acid sequence (Beveridge et al., 2014). The lower CCS boundary is calculated by assuming that a folded proteins' shape approximates to a sphere. This approach is representative of a globular highly compact form of a protein. We start by using a proteins molecular weight to calculate its volume via Eq. (12):

$$V = {}^{M_w}/_{\rho} \tag{12}$$

where M_w is molecular weight of the protein, and ρ is the density of the protein; the value of protein density used is $\rho = 0.904\,\mathrm{Da\,\AA^{-3}}$ (Fischer, Polikarpov, & Craievich, 2009). This volume can then be used to calculate the radius of a sphere as shown in Eq. (13):

$$r = \left(\frac{3V}{4\pi}\right)^{1/3} \tag{13}$$

The CCS of a sphere of this radius is then given by Eq. (14):

$$\mathrm{CCS}_{\mathrm{lower}}\left(\mathrm{\AA}^2\right) = \pi r^2 = \pi\left(\frac{3V}{4\pi}\right)^{2/3} \tag{14}$$

This equation gives the geometric size of the sphere of a compact protein. The value obtained by this equation, however, is exceeded by CCS numbers obtained in IM-MS using helium as a buffer gas. We therefore use a scaling parameter of 1.19 developed by Fernandez De La Mora (2000) to predict the smallest possible measured CCS. The scaling parameter is also applied to the upper boundary calculation.

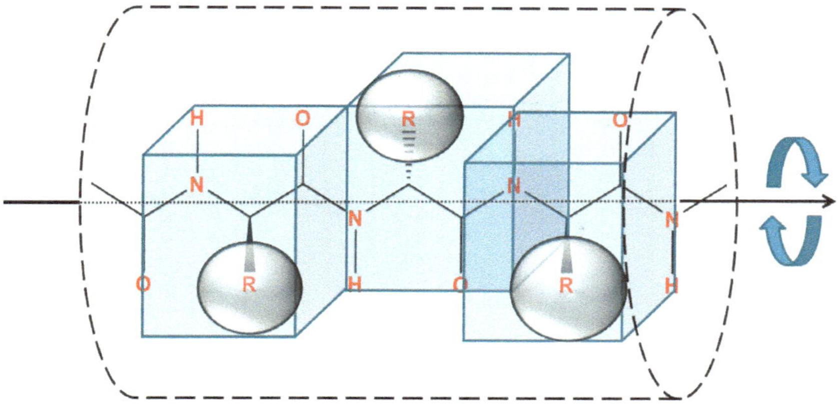

Fig. 14 The rotationally average cylindrical projection of a peptide chain, where each amino acid's volume is treated as a cube.

For the upper boundary we model a completely unfolded protein stretched out from end to end, as a cylinder as shown below, where each amino acids volume is treated as a cube that when rotationally averaged forms a cylindrical projection (Fig. 14).

The first step is to use Eq. (15) to determine the average volume of an amino acid in a protein's sequence:

$$\bar{V} = \frac{\sum_{i}^{n} V_i N_i}{n} \tag{15}$$

where V_i is the volume of a specific amino acid, N_i is the corresponding number of a specific amino acid, and n is the total number of amino acids in the sequence. Next we treat the radius of the cylinder to be equal to the length of the cube (Fig. 15).

Therefore, the radius of the cylinder is the cube root of the average amino acid volume. The CCS is then given by Eq. (16), since the protein "tumbles" in the drift tube:

$$\mathrm{CCS_{upper}}\left(\mathring{A}^2\right) = \left(\frac{4}{\pi}\right) rl + 2r^2 \tag{16}$$

where length, l, comes from knowing that the furthest distance between α-carbons in a protein chain is 3.63 Å (Heck, 2008). Therefore, the maximum length of a linear polypeptide chain with n residues is given by $l = n(3.63)$.

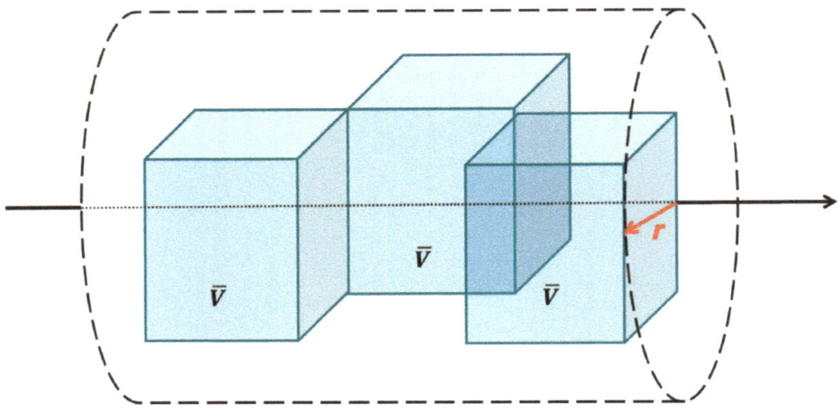

Fig. 15 The length of the cube is taken to be the radius of the cylinder.

4. SUMMARY AND CONCLUSION

In summary, our approach to the analysis of IDPs using native IM-MS can be simplified into three major steps: sample preparation and purification (Section 2.1), ion mobility and mass spectrometry experimental procedures (Section 2.2), and data analysis (Wu & Fuxreiter, 2016).

In the first step, sample preparation and purification, we have tried to highlight the importance of preparing a solution at a pH appropriate for the subject protein. A number of salts are commonly used within these solutions which can act as buffers upon pH adjustment. Furthermore, these salts need to be volatile enough to leave the protein upon desolvation. It is important to note that proteins are often dissolved in solutions that are not suitable for direct infusion MS experiments while even lyophilized samples are often saturated with salts and contain impurities. In light of this we buffer exchange sample solutions to transfer the protein to an MS compatible solvent, this process also helps to purify the protein. We use spin columns, as they are compatible with small sample volumes (20–75 μL).

Once sample preparation is complete we move onto the second step which is the running of a sample on an (IM)-MS instrument. We find it is important to initially run a "standard" protein sample on the instrument before tackling the IDP sample. This allows the analyst to check that the tip and instrument are both working correctly while permitting optimization of instrument settings (Table 2) to attain a good starting point for the analysis of the IDP. When the standard is spraying stably with good ion transmission, analysis of the subject IDP can commence. The optimized parameters used for the standard will become the starting parameters for the IDP; depending upon the quality of the mass spectrum and ATDs (which we can observe in real time), alterations may be needed to improve the signal intensity/instrument "softness." On the other hand poor spectral quality observed at this stage may be due to the sample; therefore, further purification may be required to resolve the issue. Once the IDP is running stably and at good quality, acquisition of the mass spectrum can be carried out. With the acquired mass spectrum we are able to use Eqs. (3) and (4) to determine the molecular weight and CSD of the subject protein, where for an IDP a broad CSD is likely observed and is representative of the numerous conformations occupied. Following this we move on to IM analysis of each charge state seen in the mass spectrum. After mass selecting the charge state of interest we can acquire IM data for the ion at ≥5 different drift voltages.

Upon acquisition of the ATDs we can begin the final step of data analysis; this involves using the ATDs alongside Eq. (1) to generate CCSDs for each charge state. In addition we can also produce a total CCSD that encompasses the whole conformational landscape of the IDP in vacuo. The total CCSD of IDPs are observed to be very different to globular proteins which exhibit Gaussian-like distributions (Fig. 11B) adhering to a small conformational landscape. Fitting Gaussian peaks to the IDPs total CCSD (Fig. 12) shows us that in comparison to globular proteins, more structures are observed within the conformational landscape of the IDP.

Lastly, we have designed a predictive framework (Section 3.4) to help define the level of disorder in the proteins we analyze. By plotting the ΔCCS and Δz of the subject protein against its molecular weight on a graph, with a large data set of other proteins (Fig. 13; Appendix), we are able to infer the level of disorder within the protein under study. Furthermore, we have developed a "toy model" (Section 3.4.2) that can calculate the maximum and minimum boundaries of a given proteins CCS which provides an additional dimension to this predictive framework.

ACKNOWLEDGMENTS

We acknowledge funding from The University of Manchester and the A*STAR program for the PhD studentship of D.S. and from EPSRC and the MRC for the studentships to L.G.M. and A.P.F., respectively. We also acknowledge the BBSRC for grants BB/L002655/1 and BB/M017702/1 which support our work, as well as thanking Waters Corp. and Agilent for support and help with our instrumentation.

APPENDIX

Table containing the proteins used in Beveridge–Barran plots in Fig. 13. The numbers correspond to those used in Fig. 13A and B.

Number	Name	Molecular Weight (Da)	dZ	dCCS (Å^2)
1	Melittin	2846	3	61
2	Human beta-defensin 2	4328	4	113
3	Lymphotactin 1–72	8111	5	560
4	Cytochrome *c* denatured	12,229	4	392
5	Hemoglobin-alpha *holo*-monomer	15,743	4	493
6	Ovalbumin reduced conformation 1	44,200	4	382
7	Ovalbumin reduced conformation 2	44,200	3	547
8	Ovalbumin intact conformation 1	44,287	2	19
9	Ovalbumin intact conformation 2	44,287	3	99
10	TTR tetramer	55,000	3	144
11	BSA conformation 2	66,430	4	446
12	Concanavalin A tetramer	102,000	4	292
13	SAP Pentamer	128,000	5	408
14	Lysozyme	14,314	3	148
15	Core UVR8 monomer	40,150	2	48
16	Core UVR8 dimer	80,300	4	29
17	Bovine pancreatic trypsin inhibitor (BPTI)	6531	2	107
18	MtATP-phosphoribosyltransferase (MtATP-PRT) he	189,374	2	14
19	Insulin	5730	3	274
20	Ubiquitin denatured	8556	8	970
21	Lymphotactin	10,175	6	928
22	N-terminal p53	11,162	9	1404
23	α-Synuclein	14,460	16	1577

24	N-terminal MDM2	14,790	11	1489
25	Hemoglobin-alpha apo monomer	15,126	16	2117
26	Hemoglobin-beta apo monomer	15,876	10	1161
27	β-Casein	23,944	20	4069
28	p53 DNA-binding domain	24,615	9	1609
29	p53 DNA-binding domain pH 1.5	24,616	25	4000
30	COR15A	9683	10	1254
31	Apolipoprotein C-II (ApoC-II)	8959	3	542
32	Ubiquitin	8566	2	21.66667
33	Cytochrome *c*	12,358	2	363.6667
34	Myoglobin	17,567	2	107.6667
35	Avidin tetramer	64,933	3	20
36	BSA	66,433	3	32
37	Concanavalin A	102,393	4	42
38	Alcohol dehydrogenase	147,659	5	111
mAbs				
39	IgG1 A	147,106.78	4	1440
40	IgG1 B	145,246.13	4	1580
41	IgG4 A	146,776.18	4	2140
42	IgG4 B	144,915.53	4	2040

REFERENCES

Ananiev, J., Tchernev, G., Patterson, J. W., Gulubova, M., & Ganchev, G. (2011). p53—'The guardian of genome'. *Acta Medica Bulgarica*, *38*, 72–82.

Badman, E. R., Myung, S., & Clemmer, D. E. (2005). Evidence for unfolding and refolding of gas-phase cytochrome c ions in a Paul trap. *Journal of the American Society for Mass Spectrometry*, *16*, 1493–1497.

Benesch, J. L. P., Ruotolo, B. T., Simmons, D. A., & Robinson, C. V. (2007). Protein complexes in the gas phase: Technology for structural genomics and proteomics. *Chemical Reviews*, *107*, 3544–3567.

Bernstein, S. L., et al. (2004). α-Synuclein: Stable compact and extended monomeric structures and pH dependence of dimer formation. *Journal of the American Society for Mass Spectrometry*, *15*, 1435–1443.

Beveridge, R., Chappuis, Q., Macphee, C., & Barran, P. (2013). Mass spectrometry methods for intrinsically disordered proteins. *Analyst*, *138*, 32–42.

Beveridge, R., et al. (2014). A mass-spectrometry-based framework to define the extent of disorder in proteins. *Analytical Chemistry, 86*, 10979–10991.
Beveridge, R., et al. (2015). Relating gas phase to solution conformations: Lessons from disordered proteins. *Proteomics, 15*, 2872–2883.
Borysik, A. J., Kovacs, D., Guharoy, M., & Tompa, P. (2015). Ensemble methods enable a new definition for the solution to gas-phase transfer of intrinsically disordered proteins. *Journal of the American Chemical Society, 137*, 13807–13817.
Breuker, K., Brüschweiler, S., & Tollinger, M. (2011). Electrostatic stabilization of a native protein structure in the gas phase. *Angewandte Chemie, International Edition, 50*, 873–877.
Buryakov, I. A., Krylov, E. V., Nazarov, E. G., & Rasulev, U. K. (1993). A new method of separation of multi-atomic ions by mobility at atmospheric pressure using a high-frequency amplitude-asymmetric strong electric field. *International Journal of Mass Spectrometry and Ion Processes, 128*, 143–148.
Bush, M. F., et al. (2010). Collision cross sections of proteins and their complexes: A calibration framework and database for gas-phase structural biology. *Analytical Chemistry, 82*, 9557–9565.
Candat, A., et al. (2013). Experimental determination of organelle targeting-peptide cleavage sites using transient expression of green fluorescent protein translational fusions. *Analytical Biochemistry, 434*, 44–51.
Chen, G., & Pramanik, B. N. (2008). LC-MS for protein characterization: Current capabilities and future trends. *Expert Review of Proteomics, 5*, 435–444.
Clemmer, D. E., Hudgins, R. R., & Jarrold, M. F. (1995). Naked protein conformations: Cytochrome c in the gas phase. *Journal of the American Chemical Society, 117*, 10141–10142.
D'Urzo, A., et al. (2015). Molecular basis for structural heterogeneity of an intrinsically disordered protein bound to a partner by combined ESI-IM-MS and modeling. *Journal of the American Society for Mass Spectrometry, 26*, 472–481.
Daughdrill, G. W., Hanely, L. J., & Dahlquist, F. W. (1998). The C-terminal half of the anti-sigma factor FlgM contains a dynamic equilibrium solution structure favoring helical conformations. *Biochemistry, 37*, 1076–1082.
Dickinson, E. R., et al. (2015a). The use of ion mobility mass spectrometry to probe modulation of the structure of p53 and of MDM2 by small molecule inhibitors. *Frontiers in Molecular Biosciences, 2*, 39.
Dickinson, E. R., et al. (2015b). Insights into the conformations of three structurally diverse proteins: Cytochrome c, p53, and MDM2, provided by variable-temperature ion mobility mass spectrometry. *Analytical Chemistry, 87*(6), 3231–3238.
Dobo, A., & Kaltashov, I. A. (2001). Detection of multiple protein conformational ensembles in solution via deconvolution of charge-state distributions in ESI MS. *Analytical Chemistry, 73*, 4763–4773.
Dunker, A. K., et al. (2001). Intrinsically disordered protein. *Journal of Molecular Graphics & Modelling, 19*, 26–59.
Dyson, H. J., & Wright, P. E. (2005). Intrinsically unstructured proteins and their functions. *Nature Reviews. Molecular Cell Biology, 6*, 197–208.
Faull, P. A., et al. (2010). Utilising ion mobility-mass spectrometry to interrogate macromolecules: Factor H complement control protein modules 10-15 and 19-20 and the DNA-binding core domain of tumour suppressor p53. *International Journal of Mass Spectrometry, 298*(1–3), 99–110.
Fenn, J. B., Mann, M., Meng, C. K. A. I., Wong, S. F., & Whitehouse, C. M. (1989). Electrospray ionisation for mass spectrometry of large biomolecules. *Science, 246*, 64–71 [80-.].
Fernandez De La Mora, J. (2000). Electrospray ionization of large multiply charged species proceeds via Dole's charged residue mechanism. *Analytica Chimica Acta, 406*, 93–104.

Fischer, H., Polikarpov, I., & Craievich, A. F. (2009). Average protein density is a molecular-weight-dependent function. *Protein Science, 13*, 2825–2828.

Freitas, M. A., Hendrickson, C. L., Emmett, M. R., & Marshall, A. G. (1999). Gas-phase bovine ubiquitin cation conformations resolved by gas-phase hydrogen/deuterium exchange rate and extent. *International Journal of Mass Spectrometry, 185*, 565–575.

Frimpong, A. K., Abzalimov, R. R., Uversky, V. N., & Kaltashov, I. A. (2010). Characterization of intrinsically disordered proteins with electrospray ionization mass spectrometry: Conformational heterogeneity of α-synuclein. *Proteins: Structure, Function, and Bioinformatics, 78*, 714–722.

Gan, J., et al. (2017). Native mass spectrometry of recombinant proteins from crude cell lysates. *Analytical Chemistry, 89*, 4398–4404.

Gerlich, D. (1992). Inhomogeneous RF fields: A versatile tool for the study of processes with slow ions. In C.-Y. Ng, M. Baer, I. Prigogine, & S. A. Rice (Eds.), *Advances in Chemical Physics: State-Selected and State-To-State Ion-Molecule Reaction Dynamics, Part 1. Experiment*, Vol. 82, Wiley. https://doi.org/10.1002/9780470141397.ch1. (Chapter 1).

Giles, K., et al. (2004). Applications of a travelling wave-based radio-frequency-only stacked ring ion guide. *Rapid Communications in Mass Spectrometry, 18*, 2401–2414.

K. J. Gillig et al., Coupling high-pressure MALDI with ion mobility/orthogonal time-of-flight mass spectrometry, *Analytical Chemistry* 72 (17), 2000, 3965–3971. https://doi.org/10.1021/AC0005619

Heck, A. J. R. (2008). Native mass spectrometry: A bridge between interactomics and structural biology. *Nature Methods, 5*, 927–933.

Hernández, H., & Robinson, C. V. (2007). Determining the stoichiometry and interactions of macromolecular assemblies from mass spectrometry. *Nature Protocols, 2*, 715–726.

Juan, D., & Simon, J. (2012). Revista mexicana de. *Revista Mexicana de Fisica, 77*, 1–26.

Jurneczko, E., & Barran, P. E. (2011). How useful is ion mobility mass spectrometry for structural biology? The relationship between protein crystal structures and their collision cross sections in the gas phase. *Analyst, 136*, 20–28.

Jurneczko, E., Kalapothakis, J., Campuzano, I. D. G., Morris, M., & Barran, P. E. (2012). Effects of drift gas on collision cross sections of a protein standard in linear drift tube and traveling wave ion mobility mass spectrometry. *Analytical Chemistry, 84*(20), 8524–8531.

Jurneczko, E., et al. (2012). Intrinsic disorder in proteins: A challenge for (un)structural biology met by ion mobility-mass spectrometry. *Biochemical Society Transactions, 40*, 1021–1026.

Jurneczko, E., et al. (2013). Probing the conformational diversity of cancer-associated mutations in p53 with ion-mobility mass spectrometry. *Angewandte Chemie, International Edition, 52*, 4370–4374.

Kanu, A. B., Dwivedi, P., Tam, M., Matz, L., & Hill, H. H. (2008). Ion mobility-mass spectrometry. *Journal of Mass Spectrometry, 43*, 1–22.

Kemper, P. R., & Bowers, M. T. (1990). A hybrid double-focusing mass spectrometer—High-pressure drift reaction cell to study thermal energy reactions of mass-selected ions. *Journal of the American Society for Mass Spectrometry, 1*, 197–207.

Konermann, L. (2007). A minimalist model for exploring conformational effects on the electrospray charge state distribution of proteins. *The Journal of Physical Chemistry. B, 111*, 6534–6543.

Konermann, L., Collings, B. A., & Douglas, D. J. (1997). Cytochrome c folding kinetics studied by time-resolved electrospray ionization mass spectrometry. *Biochemistry, 36*, 5554–5559.

Kurulugama, R. T., Darland, E., Kuhlmann, F., Stafford, G., & Fjeldsted, J. (2015). Evaluation of drift gas selection in complex sample analyses using a high performance drift tube ion mobility-QTOF mass spectrometer. *Analyst, 140*, 6834–6844.

Lanucara, F., Holman, S. W., Gray, C. J., & Eyers, C. E. (2014). The power of ion mobility-mass spectrometry for structural characterization and the study of conformational dynamics. *Nature Chemistry*, *6*, 281–294.

Leney, A. C., & Heck, A. J. R. (2017). Native mass spectrometry: What is in the Name? *Journal of the American Society for Mass Spectrometry*, *28*(1), 5–13. https://doi.org/10.1007/s13361-016-1545-3.

Lin, C., & Thomashow, M. F. (1992). DNA sequence analysis of a complementary DNA for cold-regulated Arabidopsis gene cor15 and characterization of the COR 15 polypeptide. *Plant Physiology*, *99*, 519–525.

Mao, Y., Woenckhaus, J., Kolafa, J., Ratner, M. A., & Jarrold, M. F. (1999). Thermal unfolding of unsolvated cytochrome c: Experiment and molecular dynamics simulations. *Journal of the American Chemical Society*, *121*, 2712–2721.

Mason, E. A., & Mason, E. W. (1988). *Transport properties of ions in gases*. John Wiley & Sons Inc.

McCullough, B. J., et al. (2008). Development of an ion mobility quadrupole time of flight mass spectrometer. *Analytical Chemistry*, *80*, 6336–6344.

Migas, L. G., France, A. P., Bellina, B., & Barran, P. E. (2017). ORIGAMI: A software suite for activated ion mobility mass spectrometry (aIM-MS) applied to multimeric protein assemblies. *International Journal of Mass Spectrometry*, *427*, 20–28.

Mohanta, D., & Jana, M. (2018). Can 2,2,2-trifluoroethanol be an efficient protein denaturant than methanol and ethanol under thermal stress? *Physical Chemistry Chemical Physics*, *20*, 9886–9896.

Mortensen, D. N., & Williams, E. R. (2016). Electrothermal supercharging of proteins in native MS: Effects of protein isoelectric point, buffer, and nanoESI-emitter tip size. *Analyst*, *141*, 5598–5606.

Nakayama, K., et al. (2007). Arabidopsis Cor15am Is a chloroplast stromal protein that has cryoprotective activity and forms oligomers. *Plant Physiology*, *144*, 513–523.

Natalello, A., Santambrogio, C., & Grandori, R. (2017). Are charge-state distributions a Reliable tool describing molecular ensembles of intrinsically disordered proteins by native MS? *Journal of the American Society for Mass Spectrometry*, *28*, 21–28.

Pacholarz, K. J., & Barran, P. E. (2015). Distinguishing loss of structure from subunit dissociation for protein complexes with variable temperature ion mobility Mass spectrometry. *Analytical Chemistry*, *87*, 6271–6279.

Pacholarz, K. J., et al. (2014). Dynamics of intact immunoglobulin G explored by drift-tube ion-mobility mass spectrometry and molecular modeling. *Angewandte Chemie, International Edition*, *53*, 7765–7769.

Pacholarz, K. J., et al. (2017). Hybrid mass spectrometry approaches to determine how L-Histidine feedback regulates the enzyzme MtATP-Phosphoribosyltransferase. *Structure*, *25*(5), 730–738.

Pagel, K., Natan, E., Hall, Z., Fersht, A. R., & Robinson, C. V. (2013). Intrinsically disordered p53 and its complexes populate compact conformations in the gas phase. *Angewandte Chemie, International Edition*, *52*, 361–365.

Phillips, A. S., et al. (2015). Conformational dynamics of α-synuclein: Insights from mass spectrometry. *Analyst*, *140*, 3070–3081.

Pringle, S. D., et al. (2007). An investigation of the mobility separation of some peptide and protein ions using a new hybrid quadrupole/travelling wave IMS/oa-ToF instrument. *International Journal of Mass Spectrometry*, *261*, 1–12.

Raijmakers, R., Kraiczek, K., De Jong, A. P., Mohammed, S., & Heck, A. J. R. (2010). Exploring the human leukocyte phosphoproteome using a microfluidic reversed-phase-TiO2 reversed-phase high-performance liquid chromatography phosphochip coupled to a quadrupole time-of-flight mass spectrometer. *Analytical Chemistry*, *82*, 824–832.

Revercomb, H. E., & Mason, E. A. (1975). Theory of plasma chromatography/gaseous electrophoresis. A review. *Analytical Chemistry*, *47*, 970–983.

Ridgeway, M. E., Lubeck, M., Jordens, J., Mann, M., & Park, M. A. (2018). Trapped ion mobility spectrometry: A short review. *International Journal of Mass Spectrometry*, *425*, 22–35.

Ruotolo, B. T., Benesch, J. L. P., Sandercock, A. M., Hyung, S.-J., & Robinson, C. V. (2008). Ion mobility–mass spectrometry analysis of large protein complexes. *Nature Protocols*, *3*, 1139–1152.

Sahasrabuddhe, A., et al. (2018). Confirmation of intersubunit connectivity and topology of designed protein complexes by native MS. *Proceedings of the National Academy of Sciences of the United States of America*, *115*, 1268–1273.

Sinz, A., Arlt, C., Chorev, D., & Sharon, M. (2015). Chemical cross-linking and native mass spectrometry: A fruitful combination for structural biology. *Protein Science*, *24*, 1193–1209.

Sobott, F., Hernández, H., McCammon, M. G., Tito, M. A., & Robinson, C. V. (2002). A tandem mass spectrometer for improved transmission and analysis of large macromolecular assemblies. *Analytical Chemistry*, *74*, 1402–1407.

Sobott, F., & Robinson, C. V. (2002). Protein complexes gain momentum. *Current Opinion in Structural Biology*, *12*, 729–734.

Sormanni, P., et al. (2017). Simultaneous quantification of protein order and disorder. *Nature Chemical Biology*, *13*, 339–342.

Stuchfield, D., & Barran, P. (2018). Unique insights to intrinsically disordered proteins provided by ion mobility mass spectrometry. *Current Opinion in Chemical Biology*, *42*, 177–185.

Testa, L., et al. (2013). Extracting structural information from charge-state distributions of intrinsically disordered proteins by non-denaturing electrospray-ionization mass spectrometry. *Intrinsically Disordered Proteins*, *1*, e25068.

Thalhammer, A., & Hincha, D. K. (2014). A mechanistic model of COR15 protein function in plant freezing tolerance: Integration of structural and functional characteristics. *Plant Signaling & Behavior*, *9*, e977722.

Tompa, P. (2002). Intrinsically unstructured proteins. *Trends in Biochemical Sciences*, *27*, 527–533.

Tompa, P. (2011). Unstructural biology coming of age. *Current Opinion in Structural Biology*, *21*, 419–425.

Tompa, P. (2012). On the supertertiary structure of proteins. *Nature Chemical Biology*, *8*, 597–600.

Uetrecht, C., Rose, R. J., van Duijn, E., Lorenzen, K., & Heck, A. J. R. (2010). Ion mobility mass spectrometry of proteins and protein assemblies. *Chemical Society Reviews*, *39*, 1633–1655.

Uetrecht, C., et al. (2008). Stability and shape of hepatitis B virus capsids in vacuo. *Angewandte Chemie (International Ed. in English)*, *47*, 6247–6251.

Ujma, J., Giles, K., Morris, M., & Barran, P. E. (2016). New high resolution ion mobility mass spectrometer capable of measurements of collision cross sections from 150 to 520 K. *Analytical Chemistry*, *88*, 9469–9478.

Uversky, V. N., & Dunker, A. K. (2010). Understanding protein non-folding. *Biochimica et Biophysica Acta, Proteins and Proteomics*, *1804*, 1231–1264.

Uversky, V. N., Oldfield, C. J., & Dunker, A. K. (2008). Intrinsically disordered proteins in human diseases: Introducing the D2 concept. *Annual Review of Biophysics*, *37*, 215–246.

Van Den Heuvel, R. H. H., et al. (2006). Improving the performance of a quadrupole time-of-flight instrument for macromolecular mass spectrometry. *Analytical Chemistry*, *78*, 7473–7483.

Visser, N. F. C., Scholten, A., Van Den Heuvel, R. H. H., & Heck, A. J. R. (2007). Surface-plasmon-resonance-based chemical proteomics: Efficient specific extraction and semi-quantitative identification of cyclic nucleotide-binding proteins from cellular lysates by using a combination of surface plasmon resonance, sequential elution and liquid chromatography-tandem mass spectrometry. *Chembiochem*, *8*, 298–305.

von Helden, G., Wyttenbach, T., & Bowers, M. T. (1995). Inclusion of a MALDI ion source in the ion chromatography technique: Conformational information on polymer and biomolecular ions. *International Journal of Mass Spectrometry and Ion Processes*, *146–147*, 349–364.

Vorontsov, E. A., Rensen, E., Prangishvili, D., Krupovic, M., & Chamot-Rooke, J. (2016). Abundant lysine methylation and N-terminal acetylation in sulfolobus islandicus revealed by bottom-up and top-down proteomics. *Molecular & Cellular Proteomics*, *15*, 3388–3404.

Ward, J. J., Sodhi, J. S., McGuffin, L. J., Buxton, B. F., & Jones, D. T. (2004). Prediction and functional analysis of native disorder in proteins from the three kingdoms of life. *Journal of Molecular Biology*, *337*, 635–645.

Wright, P. E., & Dyson, H. J. (1999). Intrinsically unstructured proteins: Re-assessing the protein structure-function paradigm. *Journal of Molecular Biology*, *293*, 321–331.

Wu, H., & Fuxreiter, M. (2016). The structure and dynamics of higher-order assemblies: Amyloids, signalosomes, and granules. *Cell*, *165*, 1055–1066.

Wyttenbach, T., Von Helden, G., & Bowers, M. T. (1996). Gas-phase conformation of biological molecules: Bradykinin. *Journal of the American Chemical Society*, *118*, 8355–8364.

Young, G., et al. (2018). Quantitative mass imaging of single biological macromolecules. *Science*, *360*, 423–427.

Zhou, M., Dagan, S., & Wysocki, V. H. (2012). Protein subunits released by surface collisions of noncovalent complexes: Nativelike compact structures revealed by ion mobility mass spectrometry. *Angewandte Chemie, International Edition*, *51*, 4336–4339.

Zubarev, R. A., et al. (2000). Electron capture dissociation for structural characterization of multiply charged protein cations. *Analytical Chemistry*, *72*, 563–573.

CHAPTER SIXTEEN

Characterizing Time-of-Day Conformational Changes in the Intrinsically Disordered Proteins of the Circadian Clock

Jacqueline F. Pelham*[,2], Alexander E. Mosier*[,2], Jennifer M. Hurley*[,†,1]

*Department of Biological Sciences, Rensselaer Polytechnic Institute, Troy, NY, United States
†Center for Biotechnology and Interdisciplinary Sciences, Rensselaer Polytechnic Institute, Troy, NY, United States
[1]Corresponding author: e-mail address: hurlej2@rpi.edu

Contents

Abstract

Circadian rhythms are 24-h oscillations conserved in nearly all living organisms that a low for the anticipation of daily environmental changes. These rhythms are maintained by a molecular clock comprised of a transcriptional/translational negative feedback loop. Many of the proteins that organize this feedback loop are intrinsically disordered proteins (IDPs), which lack a fixed or ordered three-dimensional structure. L ttle is known about the impact of intrinsic disorder in clock proteins and this lack of comprehension is compounded by the fact that sophisticated techniques to understand the inherent nature of IDPs are only now emerging. Here, we add to that conversation by describing our novel protocol to track the conformation of a core clock protein (FREQUENCY) in a vital clock model organism (*Neurospora crassa*). Our protocol,

[2] J.F.P. and A.E.M. contributed equally to this publication.

Methods in Enzymology, Volume 611
ISSN 0076-6879
https://doi.org/10.1016/bs.mie.2018.08.024

*CiR*cadian *nA*tive *FasT* parallel proteol*Y*sis (CRAFTY), utilizes a parallel proteolysis approach in native conditions to determine the conformational shifts in FREQUENCY over time, providing biologically relevant information and contributing to our understanding of the importance of disorder in the circadian clock.

1. INTRODUCTION

1.1 The Molecular Circadian Clock

Nearly all living organisms are subject to the 24-h day–night cycle of the planet. To better adapt to this day–night oscillation, most organisms have evolved 24-h physiological and behavioral rhythms to anticipate the environmental changes that are associated with the planet's rotation, dubbed circadian rhythms (circa = around, diem = day) (Albrecht, 2012; Dunlap, 1999; Montenegro-Montero & Larrondo, 2012). By definition, circadian rhythms must be (1) self-sustaining, meaning they are endogenous rhythms that allow an organism to anticipate its environment rather than be driven by responses to environmental cues; (2) entrainable, meaning circadian rhythms can be resynchronized by external environmental cues so that an organism can adjust to changing day/night cycles; and (3) compensated for a range of temperatures and other factors, meaning these rhythms can resist the thermodynamic drive to vary the circadian period based on environmental changes (Mori, Mchaourab, & Johnson, 2015; Pittendrigh, Bruce, Rosenweig, & Rubin, 1959). The combination of these circadian attributes enables an organism to recognize and integrate circadian cues at both the cellular and systemic levels.

Due to the ubiquitous importance of adapting to diurnal changes in the environment, circadian timekeeping has been conserved across all domains of life and regulates an extensive array of biological processes, from complex behaviors to cellular maintenance (Correa et al., 2003; Dong et al., 2008; Dunlap & Loros, 2004, 2016; Hardin, 2005; Kondo & Ishiura, 2003; Lowrey & Takahashi, 2004; Panda, Hogenesch, & Kay, 2002; Yakir, Hilman, Harir, & Green, 2007; Zhang, Lahens, Ballance, Hughes, & Hogenesch, 2014). This regulation imparts an ability to anticipate daily changes and the maintenance of circadian rhythms has been shown to directly impact organismal fitness and survival, granting the organisms that maintain circadian rhythms a competitive advantage over organisms that cannot predict these changes (DeCoursey, Krulas, Mele, & Holley, 1997; Klarsfeld & Rouyer, 1998; Outang, Andersson, Kondo, Golden, & Johnson, 1998). An excellent example of the circadian competitive advantage, circadian regulation of heliotropism in the sunflower has proven to be a beneficial reproductive mechanism. The immature sunflower has evolved to orient its face to track the sun, including the nighttime repositioning of the flowers face toward where the flower anticipates the sun will rise. By predicting sunrise, the sunflower gains an extra few hours of exposure to the sun, raising the temperature of the flower's face, attracting an increased number of pollinators, and enhancing reproduction rates (Atamian et al., 2016). As an illustration of this principle in humans, perhaps the best-known example of circadian regulation is the regulation of sleep. Sleep provides the body an opportunity for metabolic restoration by the removal of cellular waste and the proper timing of sleep in conjunction with the day/night cycle by circadian regulation is profoundly important to human health (Imeri & Opp, 2010; Karatsoreos, 2012; Kim, Jeong, & Hong, 2015; Sack et al., 2007a, 2007b; Stevens et al., 2007; Xie et al., 2013).

In fungal and animal cells, circadian rhythms are created and maintained by a molecular circuit commonly referred to as the circadian clock. The circadian clock comprises a transcriptional/translational negative feedback loop with two key components (Fig. 1A). The first component, the positive arm, consists of a heterodimeric transcription factor complex that transcriptionally activates the expression of a central constituent in the second component of the clock, the negative arm. After activation from the positive arm, the central constituent of the negative arm is transcribed, translated, and forms a complex with other key negative arm proteins (e.g., kinases). The negative arm then interacts with the positive arm and represses positive arm activity, most likely through the phosphorylation of the positive arm. The central constituent of the negative arm is then subject to a strictly timed half-life, regulated in part by phosphorylation, before it is marked for degradation and removed from the system (reviewed in Eide et al., 2005; Grima et al., 2002; Hurley, Loros, & Dunlap, 2015, 2016a; Ko, Jiang, & Edery, 2002; Lamia et al., 2009; Panda et al., 2002). Once the central constituent in the negative arm has been degraded, the positive arm is reactivated by dephosphorylation via phosphatases, allowing for the reinitiation of transcription of the core negative arm constituent and beginning the cycle anew (reviewed in Albrecht, 2012; Bell-Pedersen et al., 2005; Cha, Zhou, & Liu, 2015a; Dunlap, 1999; Montenegro-Montero & Larrondo, 2012).

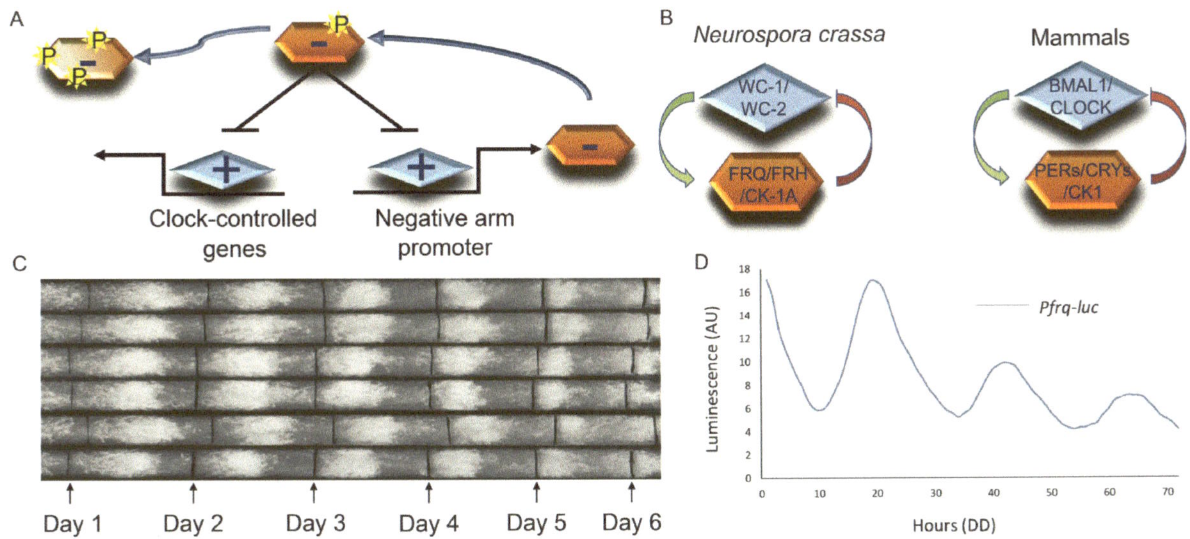

Fig. 1 Evolutionary conservation of the molecular architecture of the circadian clock enables the use of *Neurospora crassa* as a circadian model organism. (A) A representation of the basic timekeeping mechanism that underlies circadian rhythms in fungal and animal cells, a transcriptional/translational negative feedback loop. (B) A simplified schematic highlighting the conservation of the architecture of the circadian clock between fungal and animal species. The positive arm comprises WC-1/WC-2 and BMAL1/CLOCK in *Neurospora* and mammals, respectively. The negative arm comprises FRQ/FRH/CK-1A and PERs/CRYs/CK1 in *Neurospora* and mammals, respectively. (C) *Neurospora* banding-mutant strains grown on race tubes demonstrate periodic conidial formation, which is under the direct regulation of the circadian clock. *Black lines* demarcate each new day of growth. (D) Luciferase traces from a strain where luciferase is fused to the promoter of *frq* demonstrate the oscillation in the activation at the *frq* promoter via the WCC at the molecular level in vivo and in real time. *Blue line* = luciferase levels driven by the *frq* promoter. *Blue diamonds* = positive arm proteins. *Orange hexagons* = negative arm proteins. *Yellow stars* = phosphorylation events. *Images provided courtesy of Meaghan Jankowski and Joshua Thomas.*

Beyond the transcriptional promotion of the core negative arm constituent, the positive arm also regulates promoters of genes that do not participate in the core timekeeping loop, dubbed *clock control genes* (*ccgs*). The transcriptional activation of these *ccgs* via the positive arm is believed to be the primary mechanism by which the clock imparts its effects on cellular physiology (reviewed in Bell-Pedersen et al., 2005). Circadian positive arm promoter activation has broad effects in the cell; as many as 80% of transcripts have been determined to be oscillating with a 24-h rhythm (and therefore are believed to be regulated by the circadian clock) in at least one tissue type in simian models (Mure et al., 2018). This extensive regulation encompasses metabolic pathways, cell cycle control, immunomodulatory mechanisms, and a plethora of other cell regulatory systems (Dunlap & Loros, 2016; Hunt & Sassone-Corsi, 2007; Hurley et al., 2014; Hurley, Loros, & Dunlap, 2016b; Keller et al., 2009; Kim et al., 2015; Menet, Rodriguez, Abruzzi, & Rosbash, 2012; Rodriguez et al., 2013). Due to the circadian systems extensive transcriptional regulation, interconnection with a vast number of cellular pathways, and importance in organismal fitness, the disruption of the circadian clock by activities such as shift work and the exposure to light at night is accompanied by significant increases in a host of human diseases. Chronic disruptions to an organism's circadian clock have been associated with an increased risk of diabetes, obesity, depression, heart disease, cancer, and many more negative health consequences (reviewed in Bechtold, Gibbs, & Loudon, 2010; Evans & Davidson, 2013; Rajaratnam & Arendt, 2001).

1.2 *Neurospora crassa* as a Model for the Circadian Clock

One of the earliest and most widely used model organisms for studying circadian biology is *N. crassa*, a filamentous fungus (bread mold) whose well-defined clock shares a highly similar architecture with clocks in higher eukaryotic organisms (Fig. 1B) (Davis & Perkins, 2002; Dunlap et al., 2007; Fuller, Hurley, Loros, & Dunlap, 2014). *Neurospora* has long been used as a genetic model organism and has been a keystone of biochemical research for the better part of a century, gaining notoriety for work validating the one gene–one enzyme hypothesis (Beadle & Tatum, 1941; Davis & Perkins, 2002; Roche, Loros, McCluskey, & Glass, 2014). Many of the fundamental principles discovered in *Neurospora* that inform basic biology, including research in epigenetics, cell division and polarity, metabolism, and circadian clocks, are widely applicable in higher eukaryotes (reviewed in Roche et al., 2014). As a community, *Neurospora* researchers have sequenced its genome, created a near-complete genetic knockout library, established protocols for transformations, and generated a variety of regularly used selectable markers, all of which makes *Neurospora* an excellent model organism in which to investigate the molecular underpinnings of the circadian clock (Colot et al., 2006; Galagan et al., 2003; Ninomiya, Suzuki, Ishii, & Inoue, 2004; Roche et al., 2014).

Neurospora has been integral to circadian research for nearly six decades, following the description of a circadian rhythm in the production of its asexual spores, called conidia, in a mutant (band or *bd*) strain (Pittendrigh et al., 1959). This strain, which produces "bands" of conidia once every 24 h when run on media poured in a long glass tube, called a "race tube," allows for the overt analysis of this clock "band" phenotype (Fig. 1C). More recently, the fusion of a codon-optimized firefly luciferase gene to genes involved in, or regulated by, the circadian clock, allows for the tracking of luciferase protein levels in real time to examine circadian rhythms in *Neurospora* at the molecular level (Fig. 1D) (Gooch et al., 2008; Morgan, Greene, & Bell-Pedersen, 2003).

In *Neurospora*, the essential proteins which comprise the core clock include WHITE COLLAR-1 (WC-1), WHITE COLLAR-2 (WC-2), FREQUENCY (FRQ), FREQUENCY-Interacting RNA Helicase (FRH), and Casein Kinase-1 (CK-1) (Fig. 1B). In the subjective circadian morning, WC-1 and WC-2 form the White Collar Complex (the WCC or the positive arm of the clock) and facilitate the production of *frq* via transcriptional activation (Crosthwaite et al., 1997; Dunlap & Loros, 2004). *frq* is transcribed, translated, and then dimerizes with FRH to form the FRQ-FRH Complex (the FFC or the negative arm of the clock), with FRH acting to stabilize FRQ (Cheng, He, He, Wang, & Liu, 2005; Guo, Cheng, & Liu, 2010; Hurley, Larrondo, Loros, & Dunlap, 2013). As the day progresses, the FFC recruits a variety of kinases, including *Casein kinase-1* and *-2*, and interacts with the WCC, leading to the progressive hyperphosphorylation of both the WCC and FRQ (Baker, Kettenbach, Loros, Gerber, & Dunlap, 2009; Jinhu & Yi, 2010; Schafmeier et al., 2005; Tang et al., 2009). The extensive phosphorylation of the WCC is hypothesized to repress WCC transcriptional activity at the *frq* promoter, leading to a halt in the production of FRQ. In parallel, the hyperphosphorylation of FRQ leads to FRQ degradation (Baker et al., 2009; Jinhu & Yi, 2010; Tang et al., 2009). Once FRQ is degraded and the FFC is no longer present, the WCC is dephosphorylated by several different phosphatases, leading to WCC transcriptional reactivation, restarting the circadian cycle (Dunlap & Loros, 2004). Beyond its regulation at the *frq* promoter, the WCC also transcriptionally activates a host of gene promoters (*ccgs*). Approximately 40% of all transcripts in *Neurospora* fall under the regulation of the circadian clock, including many genes involved in metabolic, developmental, and stress response pathways (Correa et al., 2003; Hurley et al., 2014; Sancar, Sancar, Ha, Cesbron, & Brunner, 2015; Shinohara, Correa, Bell-Pedersen, Dunlap, & Loros, 2002).

The general principles which underlie timekeeping in *Neurospora* are paralleled in higher eukaryotes. For example, the mammalian circadian clock is also timed by a transcription/translation negative feedback loop. In mammals, the positive arm components BMAL1 and CLOCK bind together and transcriptionally activate the negative arm components PERIODs (PERs) and CRYPTOCHROMEs (CRYs), as well as a host of other *ccgs* which comprise the clock's output. As in *Neurospora*, the negative arm components in mammals (PERs/CRYs) form a complex with a variety of kinases to repress BMAL1-CLOCK activity, are subject to high levels of phosphorylation, and it is the half-life of these negative arm complex proteins that dictates the period of the clock (Fig. 1A and B) (reviewed in Bell-Pedersen et al., 2005; Buhr & Takahashi, 2013; Hurley et al., 2016a; Partch, Green, & Takahashi, 2014). Such effective usability coupled with the conservation of the core architecture of the circadian clock in higher eukaryotes makes *Neurospora* an exceptional model for circadian research (Fig. 1B) (Bell-Pedersen et al., 2005; Lowrey & Takahashi, 2004; Reddy & Rey, 2014).

1.3 Intrinsic Disorder and the Circadian Clock

A long-held tenant in biology has been that DNA is transcribed to RNA and translated into protein, which then takes on a specific conformation encoded by its amino acid composition that dictates the function of the protein (Dunker, Brown, Lawson, Iakoucheva, & Obradovic, 2002; Tompa, 2012; Uversky, 2011). However, a deeper understanding of a novel class of proteins, intrinsically disordered proteins (IDPs), is challenging this paradigm. IDPs are proteins that do not fold into stable, well-defined, three-dimensional structures upon their translation (Wright & Dyson, 2015). Instead, the conformation of an IDP, which is generally defined as a protein with regions that contain a string of at least 30 amino acids that lack a predetermined structure, remains variable and can be directly impacted by external factors, including, but not limited to, interacting partners, environmental changes within the cell, and posttranslational modifications (e.g., phosphorylation) (Tompa, 2012).

While a limited number of IDPs and IDP regions with biological significance have been recognized since the 1950s, a broader understanding of the prevalence and function of these proteins has only developed more recently (Oldfield & Dunker, 2014). What is known is that IDPs have many transient, yet highly specific binding partners, are commonly posttranslationally modified, and fall into a diverse array of functional categories. By current estimates, between 10%–35% of prokaryotic and 15%–45% of eukaryotic proteins contain intrinsically disordered regions (Tompa, 2012). The prevalence of IDPs and their involvement in many cellular regulatory pathways highlights the importance of this poorly understood class of proteins (Dunker et al., 2002; Uversky, 2011; Wright & Dyson, 2015).

One cellular process in which IDPs play a central role is the circadian clock. Computational analysis of clock proteins has shown that based on the primary sequence, many are predicted to be highly disordered (e.g., Fig. 2A and B) (Dosztányi, Csizmok, Tompa, & Simon, 2005; Hurley et al., 2013). Moreover, the IDP nature of FRQ, the core constituent of the *Neurospora* clock, has been validated experimentally. Our work demonstrated that FRQ (1) is more sensitive to protease activity under proteolytic treatment than proteins with defined structures, (2) stays in solution after heat treatment, and (3) behaves as an IDP under circular dichromatic analysis (Hurley et al., 2013). In line with its biochemically validated IDP nature, FRQ is highly phosphorylated, with over 75 phosphorylation sites validated, and has many specific binding partners which interact only at specific times over the circadian day (Fig. 2C) (Baker et al., 2009; Cha et al., 2015a; Guo et al., 2010; Larrondo, Olivares-Yañez, Baker, Loros, & Dunlap, 2015). In addition, like many IDPs, FRQ necessitates a partner protein for stability and immediately upon translation FRQ is bound by FRH, which stabilizes FRQ through its lifecycle (Cheng et al., 2005; Guo et al., 2010; Hurley et al., 2013; Uversky, 2015).

Beyond FRQ, many other core circadian clock proteins have been investigated for intrinsic disorder. FRH in *Neurospora* as well as PER1, PER2, CLOCK, and BMAL1 in mammals have been shown to be either intrinsically disordered or to contain disordered regions that are essential to the proper function of the protein (e.g., Fig. 2B) (Conrad et al., 2016; Gustafson & Partch, 2015; Hurley et al., 2013). While the prevalence of IDPs in the clock is well established, the underlying reason for the conservation of the IDP nature in clock proteins is unknown.

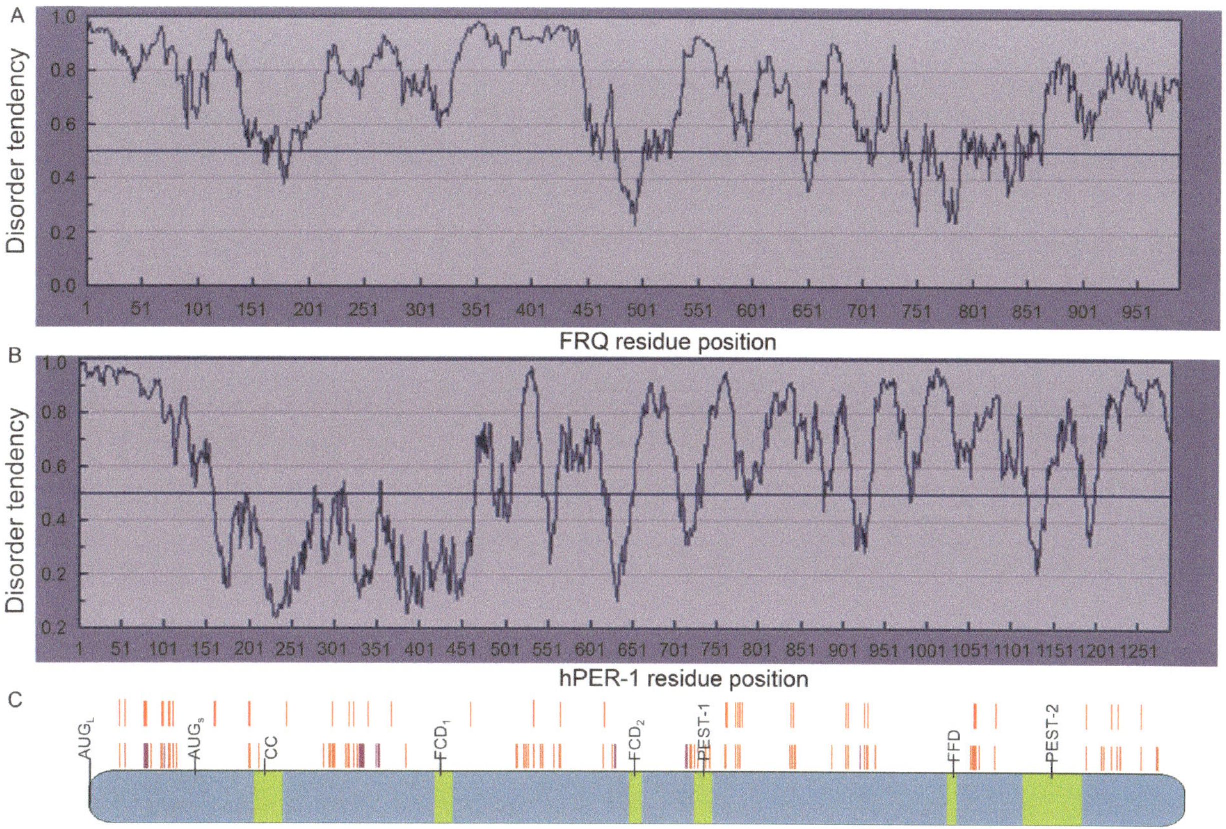

Fig 2 See figure legend on next page.

2. INVESTIGATING NEGATIVE ARM CONFORMATIONAL CHANGES OVER THE CIRCADIAN DAY

2.1 Conformational Determination of IDPs

A key element of the nature of IDPs is their ability to change conformations in different conditions, an important characteristic which has been theorized to underlie several aspects of circadian regulation (Conrad et al., 2016; Gustafson & Partch, 2015; Hurley et al., 2016a). However, changes in the conformation of the negative arm clock proteins, as it is with changes in the conformations of IDPs in general, are poorly understood. This is due in part to the limited number of methods to analyze IDP conformational changes on a global scale, as traditional mechanisms of protein structural analysis cannot define the global conformational changes of an IDP (e.g., NMR and crystallography) (Gsponer & Babu, 2009; Oldfield & Dunker, 2014; Tompa, 2011; Uversky, 2011). One of the few effective methods for studying the conformational changes of an IDP is limited proteolysis, protein degradation through enzymatic digestion (Oldfield & Dunker, 2014). This technique yields conformational details by relating the degradation rate to the degree of protein structure; in theory protein structure provides stereochemical barriers to protease access and the more flexible and "open" protein regions (or potentially whole proteins) are more quickly degraded. Put simply, this means that the less structured a protein is, the more readily available specific sites of protease cleavage are for digestion, leaving the more structured regions/proteins unaffected (Cha, Zhou, & Liu, 2015b).

Fig. 2 Proteins in the negative arm of the circadian clock are predicted to be intrinsically disordered. (A) The amino acid-based IUPred prediction of the tendency of FRQ toward disorder. A value above 0.5 in the IUPred plots suggests a high likelihood of disorder. (B) The amino acid-based IUPred prediction of the tendency of human PER1 (hPER1) toward disorder. (C) A representation of the phosphorylation sites on FRQ, with known protein–protein interaction domains highlighted in *green*. AUG_L = long isoform start position, AUG_S = short isoform start position, CC = coiled–coiled domain, FCD_1 = FRQ-CK-1A interacting domain, FCD_2 = FRQ-CK-1A interacting domain, FFD = FRQ-FRH interacting domain, PEST-1 and PEST-2 = Proline (P), Glutamic Acid (E), Serine (S), and Threonine (T) (PEST) domains (Baker et al., 2009; Querfurth et al., 2011). *Lines above* are experimentally verified phosphosites; the *red lines* are S/T residues unambiguously identified as phosphorylated; the *purple lines* were identified with low confidence and are labeled ambiguous in Baker et al. (2009). The *orange lines* were identified as phosphorylated in Tang et al. (2009).

Perhaps the most efficient proteolytic technique to analyze the conformation of an IDP is fast parallel proteolysis (FASTpp), which measures protein domain thermal melting in lysates (Minde, Maurice, & Rüdiger, 2012). The principle underlying this approach is that when the temperature to which a protein is exposed to is increased in the presence of the thermostable/thermoactivated protease Thermolysin, a protein's thermal melting point can be determined with high sensitivity. When coupled with SDS-PAGE and western blotting for target protein visualization, this technique can determine the folded fraction of a protein of interest in both purified samples and complex lysates, giving an estimate of the conformation and stability of an IDP of interest. FASTpp has been validated on proteins across a range of size and functional categories, used to probe the stability of the interaction between mutants of troponin and tropomyosin, and applied to understand ligand stabilization (Minde et al., 2012; Robaszkiewicz, Ostrowska, Cyranka-Czaja, & Moraczewska, 2015).

While the parallel aspect of FASTpp is an efficient and useful method to compare several samples at the same time, the thermal gradient implementation of FASTpp does not allow insights into an IDP's conformation in native conditions, which is important as environmental conditions are known to affect the conformation of IDPs. This may be particularly true of those IDPs involved in the temperature-responsive circadian clock, making the use of FASTpp to study circadian IDPs limited (Uversky, 2009).

In our previous study of FRQ, we were able to determine that FRQ was less stable under proteolytic cleavage than its partner protein FRH. Specifically, we performed limited digestion via proteinase K treatment on *Neurospora* protein lysates in near-native conditions, quantifying the remaining amounts of FRQ and FRH after a defined time of exposure (Hurley et al., 2013). Unlike FASTpp, this assay yielded a relative comparison of the conformation of these two core clock proteins in near-native conditions. However, this approach was neither temporal nor parallelized and therefore could not account for how the conformations of these proteins changed over circadian time, important as an IDP's conformation can be impacted by several temporal circadian factors (Baker et al., 2009; Querfurth et al., 2011; Tang et al., 2009; Uversky, 2009).

To adapt to the need for both native conditions and a time-sensitive analysis, we have developed a method for comparing protein conformation over circadian time in native conditions: *C*i*R*cadian n*A*tive *F*as*T* parallel proteol*Y*sis (CRAFTY). CRAFTY utilizes the parallel aspect and temperature-regulated protease activity of FASTpp and the native temperatures of a more typical proteolysis assay to probe circadian proteins for conformational dynamics in their native environment over time. In general, CRAFTY, specifically detailed in Section 2.2, utilizes Thermolysin (TL) added to lysates sampled at different circadian times to digest the protein of interest in near-native conditions from these samples in parallel. In theory, a protein in a more compact conformation in its native condition will be more resistant to degradation than one in a more open conformation, and therefore degrades more slowly in the proteolysis assay. Moreover, as many samples can be run in this parallel platform, differences in conformation from samples taken at different time points over the circadian day are more readily related. We believe this novel approach to proteolysis can be employed to compare conformational dynamics in near-native conditions in a variety of organisms, environments, and time-dependent processes.

2.2 Circadian Sample Preparation Protocol

To understand how protein conformations can change over time, one must first sample an organism in a circadian manner. The methodology of employing *Neurospora* mycelial mats (pads of actively growing fungal tissue) to create circadian time courses was developed in 1989 and has been widely adopted by the *Neurospora* community to track temporal cellular behavior to understand clock functions (Caster, Castillo, Sachs, & Bell-Pedersen, 2016; Hurley et al., 2014; Loros, Denome, & Dunlap, 1989; Tang et al., 2009). The basic structure and principle of this age-matched time course can be employed to study any other circadian model organism. For *Neurospora*, a variety of culturing media have been used, depending on experimental needs, and information on media used previously in *Neurospora* is available at the Fungal Genetics Stock Center website (McCluskey, Wiest, & Plamann, 2010). Various strains of *Neurospora* which have been mutated, tagged, and knocked out at the loci of interest have been cultured using this protocol; the choice of strain is based on experimental goals. For example, if there is no specific antibody for the protein of interest, a strain with an epitope tag fused to the protein of interest may be investigated using this technique. Below, we describe the basic culturing technique employed in our lab.

As an overview: age-matched *Neurospora* cultures are grown in the light for 24 h before being shifted to the dark to entrain (set) the clock. At the desired circadian time, the culture is flash frozen, after which samples can be extracted and downstream analyses can be performed. When this process

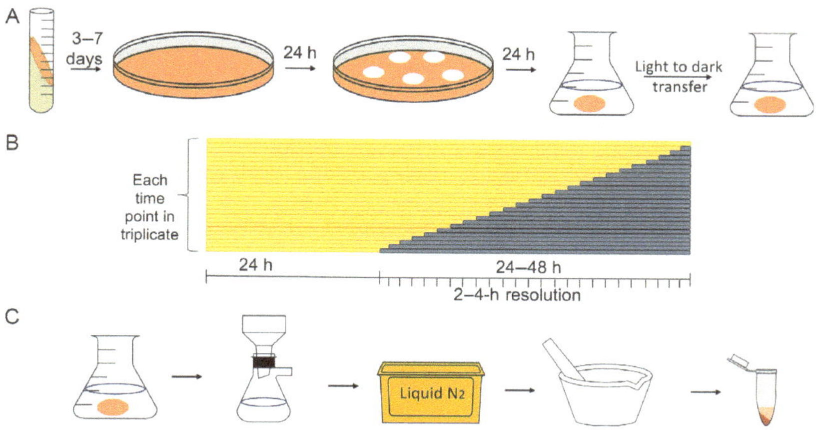

Fig. 3 A schematic of the *Neurospora* circadian time course culturing/sampling protocol. (A) *Tissue growth*: a conidial suspension inoculated from a 3–7 day-old slant is incubated for 24 h in constant light to yield a mycelial mat. Next "plugs" from this mat are cut (in triplicate for each time point) and individually placed in 125 mL flasks. These flasks are shaken orbitally in continuous light (LL) conditions for 24 h. Cultures are then transferred to the continuous darkness (DD) incubator in a staggered fashion. (B) *Sample transfer protocol*: A schematic of when to transfer samples from light to dark to achieve an age-matched circadian time course. *Yellow* represents duration in LL conditions and *gray* represents duration in DD conditions. (C) *Tissue harvest and protein extraction*: After the sample has grown for the allotted time, the culture is vacuum filtered and flash frozen. Next, each sample is ground into a fine powder in LN_2 and protein is extracted.

is done in batch and samples are harvested over the circadian day, we refer to the resultant samples as a "circadian time course." The resolution at which samples are harvested is dependent on the experiment, though the circadian field recommends that an optimal time course entails triplicate sampling, harvested at a 2-h resolution, for at least 48 h (Fig. 3) (Hughes et al., 2017).

2.2.1 Essential Equipment

- Light-proof harvesting (dark) room with low-red light for visualization without natural, white, or blue light. Light conditions depend on the organism of interest.
- Two Percival light- and temperature-controllable incubators (I36LLC8), each containing a level orbital shaker (VWR 89032-104), all maintained in the dark room.
  - Settings: Incubators both set to 25°C, one set to continuous light (max setting) (LL) and the other set to continuous darkness (DD).

- Sterile wooden inoculation sticks (14.8 cm).
- Ethanol and Bunsen burner for sterilization.
- Cork borer (8 mm).
- Small forceps (16.5 cm).
- Sterile culturing containers: 100 mm Petri dishes, 13 × 100 mm culturing slants, and 125 mL Erlenmeyer flasks.
- 2 L Vacuum flask, 10 cm Buchner funnel, and 5 μm grade 413 filter paper.
- Liquid nitrogen.
- Sterile microcentrifuge tubes (2 and 1.5 mL).
- Mortar and pestle (8 cm diameter).
- Small metal scoopula (16.5 cm).
- Refrigerated Eppendorf centrifuge (5424R) chilled to 4°C.
- Paper towels.
- Semi-micro two-sided cuvette (VWR 97000-586).

2.2.2 Buffers and Reagents

- 13 × 100 mm Culturing slant containing solidified/slanted 2 mL Vogel's minimal media: 1 × Vogel's salts (see Fungal Genetics Stock Center website for recipe), 2% Sucrose, and 1.5% agar.
- Liquid Culture Media (LCM): 2% Glucose, 0.5% Arginine, 1 × Vogel's salts.
- Chilled Protein Extraction buffer: pH 7.4, 50 m*M* HEPES, 137 m*M* NaCl, 3 m*M* CaCl, 0.4% NP-40, 10% Glycerol.
- HALT protease inhibitor, EDTA-free (100 ×), Thermo Scientific (87785).
- Bradford Reagent.

2.2.3 *Neurospora* Culture and Protein Extraction Procedure

2.2.3.1 Tissue Growth

1. Using a sterile wooden stick, transfer a small number of conidia from the *Neurospora* strain of interest onto a Vogel's minimal medial slant and incubate in the LL Percival incubator at 25°C (Fig. 3A).
2. After a sufficient amount of conidia have developed (3–7 days, depending on the strain), add 2 mL of LCM to the *Neurospora* slant culture and vortex vigorously to create a conidial suspension.
3. Transfer this conidial suspension into a 2 mL microcentrifuge tube. Centrifuge at 25°C and 2500 rpm for 5 min.

4. Discard the supernatant, which should contain the nonconidial debris, and retain the conidial pellet.
5. Gently resuspend the conidial pellet, avoiding any remaining debris, in 500 μL of LCM.
6. Using the conversion of OD_{600} $1 = 3 \times 10^6$ conidia, obtain a dilution of 1×10^7 conidia/mL in LCM.
7. Add 25 mL of fresh, sterile, LCM into a sterile 100 mm Petri dish (Fig. 3A).
8. Transfer 1 mL of 1×10^7 conidia to the Petri dish containing LCM, stirring gently to distribute the conidia.
9. Incubate the Petri dish with the conidial suspension in the LL Percival incubator at 25°C for 24 h.
10. After 24 h the conidia will grow into a mycelial mat in the Petri dish. Remove the dish to a sterile surface. Sterilize an 8 mm Cork borer with ethanol and a Bunsen burner. Use the sterile borer to cut "plugs" of mycelia from the hyphal mat and place a single plug (with sterile forceps) into a 125 mL flask containing 50 mL of LCM. Continue to cut and inoculate these plugs until the desired number of time course samples is obtained (Fig. 3A).
11. Incubate the flasks at 25°C in LL for 24 h under continuous shaking at 125 rpm (Fig. 3B).
12. After 24 h, begin to transfer flasks to the DD incubator depending on the desired circadian time of the sampling, maintaining the flasks in DD at 25°C under continuous shaking at 125 rpm (Fig. 3B).

2.2.3.2 Tissue Harvest

1. When the desired circadian time is reached, in the "dark room" (with only a red light for visualization), remove the sample from the incubator. Place a fresh piece of 5 μm grade 413 filter paper into a Buchner funnel and attach it to a vacuum flask. Pour the entire contents of the 125 mL flask sample into the Buchner funnel and allow the vacuum to remove the excess media from the *Neurospora* plug. Allow the vacuum to dry the tissue for 30–45 s after the media have been removed from the tissue (Fig. 3C).
2. Once the media have been removed from the plug, sandwich the plug between two dry, clean, paper towels on a solid surface. Compress the plug with a flat, sturdy, object (e.g., a microfuge tube rack) to remove excess liquid.

3. Place the plug in a 2mL liquid nitrogen-rated microcentrifuge tube and immediately flash freeze in liquid nitrogen. Samples at this stage can be stored indefinitely at −80°C.

2.2.3.3 Protein Extraction

1. Remove *Neurospora* plugs stored at −80°C to liquid nitrogen. Chill a clean mortar and pestle with liquid nitrogen. Remove a single plug from its microfuge tube (keeping the empty microcentrifuge tube in liquid nitrogen) and grind the plug into a fine powder (Fig. 3C).
2. Transfer the powdered tissue back into its cold microcentrifuge tube and add protein extraction buffer at a mg of powdered tissue: μL of buffer ratio of 1:2. Plugs at this stage will vary in size and weight depending on the strain. Immediately add the protease and phosphatase inhibitor (HALT) and vortex until thoroughly mixed.
3. Incubate the tissue/buffer mixture on ice for 5min, vortexing occasionally.
4. Centrifuge the tissue/buffer mixture at 4°C and 14,000rpm for 10min.
5. Transfer the lysate to a fresh microfuge tube (1.5mL), keeping on ice, and discard the pellet.
6. Quantify the total protein levels using a Bradford Assay and standardize the lysate total protein levels to 5mg/mL using protein extraction buffer for each time course sample.

2.3 CRAFTY Protocol

Due to both their disordered and environmentally sensitive nature, it is essential to maintain as near a native environment as possible when analyzing the global conformation of circadian proteins. Moreover, to compare structures across time points, it is crucial to perform conformational analyses in parallel. Therefore, to directly compare protein conformations in near-native conditions across many time points in our proteolytic conformational analysis of circadian proteins, we have devised the CRAFTY approach. CRAFTY employs limited digestion techniques at native temperatures and can be scaled to perform multiple parallel analyses. In theory, an increase in the rate of degradation of a protein at different circadian times suggests a more open conformation at that circadian time (Fig. 4A).

To summarize the CRAFTY approach: sample lysate is mixed with Thermolysin on ice (to inhibit enzyme activity) before each sample is aliquoted into a unique row in a 96-well plate, with multiple samples on the same plate. All samples are then subject to the same conditions via a preprogrammed, automatically controlled, PCR machine, minimizing variations

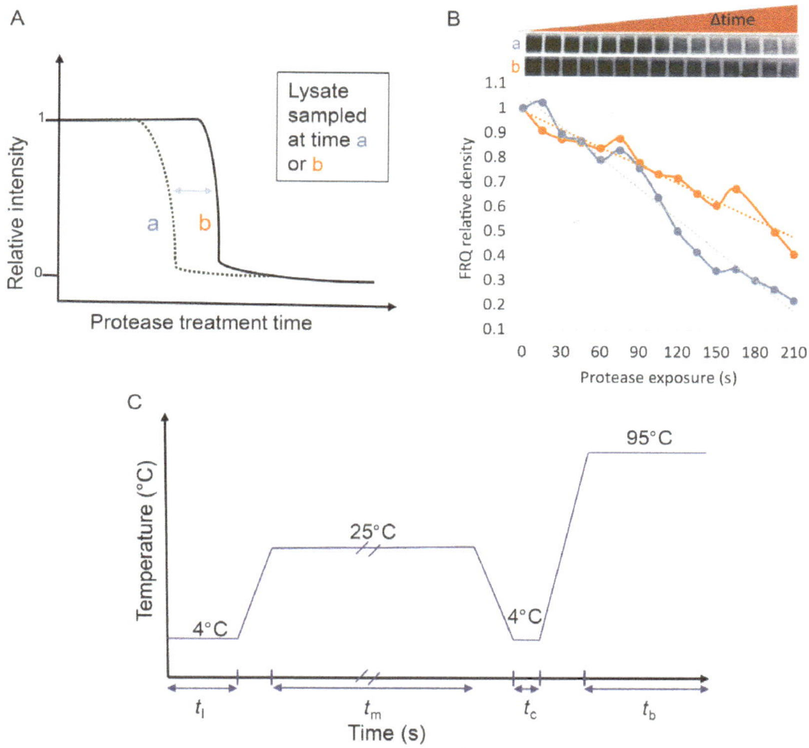

Fig. 4 CRAFTY analysis of proteins over circadian time: (A) A theoretical plot of the resultant data from CRAFTY, with the time of protease treatment on the *X*-axis and the relative intensity of the protein on the *Y*-axis. Proteins that are more resistant to protease cleavage are predicted to have a denser conformation and therefore would be more resistant to degradation over time (e.g., protein a is in a more open conformation than b). (B) An example of a western blot and quantified analysis of samples of FRQ from two distinct circadian times of day. Increasing time of exposure to the protease degrades FRQ at different rates depending on the time of day, suggesting circadian time-dependent conformational changes in FRQ. (C) A schematic of the CRAFTY protocol carried out in a gradient 96-well PCR setup; t_b = boiling time for SDS-PAGE preparation, t_c = cooling time, t_l = sample loading time, t_m = fractional melting time with interval quenching. *Panel (A and C): Adapted from Minde, D. P., Maurice, M. M., & Rüdiger, S. G. D. (2012). Determining biophysical protein stability in lysates by a fast proteolysis assay, FASTpp.* PLoS One, *7(10), e46147. https://doi.org/10.1371/journal.pone.0046147.*

in protease concentrations, temperature, and time to reaction quenching. CRAFTY samples are then western blotted and ImageJ software is used to quantify the relative density of the protein of interest (Fig. 4B). The below described protocol has been devised to specifically analyze the conformation of FRQ.

2.3.1 Equipment

- 96-Well PCR plate (Axygen PCR-96-SG-C) and thermal sealing film (VWR 82018-844).
- Thermocycler with gradient capability (such as Eppendorf 6331000025).
  - Preprogrammed with CRAFTY protocol (see Fig. 4C and Section 2.3.4).
- One heat-adjustable heating block (such as Eppendorf Thermomixer F1.5) set to 95°C.
- 8-Well multichannel pipette.
- Reagent Reservoir for SDS-PAGE EDTA protein loading buffer.
- SDS-PAGE: XCell4 SureLock Midi-Cell Electrophoresis System (WR0100).
- Western blot:
  - Bio-Rad Trans-Blot Turbo Transfer System (1704150).
  - Bio-Rad RTA PVDF Kit (1704273).
  - Chemiluminescent-capable imager or X-ray film.

2.3.2 Buffers and Reagents

- *Neurospora* lysate at a total protein concentration of 5 mg/mL from each sample of the circadian time course (from here on called "time points").
- Thermolysin (TL) solution: 0.19 m*M* Thermolysin, pH 7.4, 50 m*M* HEPES, 137 m*M* NaCl, 0.4% NP-40, 10% Glycerol.
- SDS-PAGE:
  - 5 × Protein loading buffer: 50 m*M* Tris–HCl pH 6.8, 2% SDS, 0.02% Bromophenol Blue, 10% Glycerol, 1% β-ME, 20 m*M* EDTA.
  - Invitrogen precast 4%–12% Bis-Tris gels (WG1402BOX).
  - MES SDS Running Buffer: 50 m*M* MES, 50 m*M* Tris base, 0.1% SDS, 1 m*M* EDTA.
- Western Blot:
  - 1 × Trans-Blot Turbo RTA Transfer Buffer (1704273).
  - Primary antibody: Anti-V5 Invitrogen (46-1157) at 1:5000 dilution.
  - Secondary antibody: Goat Anti-mouse IgG HRP conjugate Invitrogen (313430) at 1:25,000 dilution.
  - PBS 0.2% Tween-20, 5% milk (blocking).
  - PBS 0.2% Tween-20, 1% milk (primary incubation).
  - PBS 0.2% Tween-20 (washes and secondary incubation).
  - SuperSignal West Femto Maximum Sensitivity Substrate Thermo Scientific (34094).

2.3.3 Analysis of the Conformation of FRQ With CRAFTY

2.3.3.1 CRAFTY Protocol

1. Prepare PCR program as follows (Fig. 4C):
   a. Start with a 4°C hold for loading samples (t_l).
   b. Set the next step to a timed hold at 25°C and set the elapsed time for the hold as follows: number of PCR plate wells (from here on called "samples") per each circadian time point (generally 12) × the sampling resolution, e.g., 12 samples at a 30 s resolution means that the 25°C hold should be set for 6 min (t_m).
   c. Next, set the thermocycler to cool at 4°C for 1 min (t_c), following with a 95°C boiling step for 5 min (t_b).
2. Precool the thermocycler to 4°C in preparation for the lysate samples.
3. Prepare two aliquots for each circadian time point (285 μL of 5 mg/mL lysate each) and keep on ice (a Thermolysin (experimental) sample and a non-Thermolysin control).
4. While on ice, add 14 μL of Thermolysin to the experimental sample of each circadian time point and thoroughly mix by pipetting. Then add 14 μL of Thermolysin solution without Thermolysin to the non-Thermolysin control.
5. Aliquot 20 μL of the experimental and control samples for each circadian time point into their own unique row on the 96-well plate at 4°C, leaving at least 20 μL of each mixture on ice as a control. To get an accurate read on conformational change, we recommend utilizing all 12 wells in the PCR plate row, though the number of wells in the 96-well plate in which you place samples depends upon the number of samples you wish to take.
6. Place the 96-well plate into the precooled PCR machine and begin the preprogramed CRAFTY thermocycler protocol.
7. Quench samples at a given resolution across all time points by hand by simultaneously mixing the wells for a given sample thoroughly with 20 μL of the 5 × SDS-PAGE EDTA buffer in parallel at the desired time resolution using the multichannel pipettor (e.g., in the 6-min digestion window with 12 samples described above, a set of samples across each time point (± Thermolysin) are quenched in parallel every 30 s).
8. Once the total digestion time window has elapsed and quenching of all samples has occurred, the program will temporarily cool the plate. During this time place a thermal sealing film on the plate and close the PCR machine.

9. Allow the 96-well plate to boil the samples in preparation of SDS-PAGE analysis according to the preprogrammed CRAFTY PCR protocol.
10. In parallel, quench the ± Thermolysin control samples from each time point which were maintained at 4°C throughout the protocol and boil them at 95°C in a heat block for 5 min.
11. After all samples have been boiled, proceed to the western blot analysis protocol.

2.3.3.2 FRQ Western Blot Analysis Protocol

1. Assemble the XCell4 SureLock Midi-Cell Electrophoresis System containing the appropriate number of Invitrogen precast 4%–12% Bis-Tris gels and MES running buffer.
2. Load the CRAFTY samples so that the circadian time points to be compared are on the same gels if possible.
3. Separate the proteins with the Invitrogen precast 4%–12% Bis-Tris gels.
   a. Running conditions: 200 V for 1 h.
4. Perform a western blot transfer of the gels with RTA kit according to manufacturer's specifications.
5. Block the transferred membrane for 1 h with 50 mL 5% nonfat milk, 0.2% Tween PBS at 4°C while mixing on an orbital shaker.
6. Incubate the membrane with 25 mL at a dilution of 1:5000 anti-V5 in 1% milk, 0.2% Tween PBS overnight at 4°C while mixing on an orbital shaker.
7. Wash the membrane for 1 h, refreshing 30 mL of PBS 0.2% Tween-20 every 10 min at room temperature while mixing on an orbital shaker.
8. Incubate the membrane for 30 min with 25 mL of secondary Goat Anti-mouse IgG HRP conjugate antibody at a dilution of 1:25,000 in PBS 0.2% Tween-20 at room temperature while mixing on an orbital shaker.
9. Wash the membrane for 1 h, refreshing 30 mL of PBS 0.2% Tween-20 every 10 min at room temperature while mixing on an orbital shaker.
10. Expose the membrane to SuperSignal Femto Maximum Sensitivity Substrate according to manufacturer's specifications.
11. Image the blot using a chemiluminescent capturing device, e.g., gel imager or dark room X-film exposure.
12. Using ImageJ, or similar image quantification software, quantify the relative degradation rates of the protein of interest and compare the resultant data between samples.

2.3.4 Protocol Notes

Due to a given protein's difference in protease susceptibility, it is necessary to optimize each protocol with respect to the target protein. This may include the ratio of Thermolysin to lysate, the digestion resolution between samples (time between quenching), and the total time of Thermolysin exposure. In the above described experiment, we detailed the specific protocol used to analyze FRQ with CRAFTY. To do so, we used a frq::v5his103flag strain, in which a V510his3flag tag was fused to the C-terminus of FRQ for western blotting purposes.

3. CONCLUSION

We present here a novel method for the study of the time-dependent conformational changes of the IDPs in the circadian clock, CRAFTY. CRAFTY goes beyond previous proteolytic methods of IDP conformational definition using degradation by optimizing a proteolysis approach that includes the analysis of proteins in near-native conditions and the ability to compare conformational changes between time-resolved samples. Both of these elements are essential to the study of the circadian clock, a molecular circuit in which most of the proteins involved are, at least in part, intrinsically disordered (Gustafson & Partch, 2015; Hurley et al., 2013). Specifically, we summarized the protocol of this technique as we applied it to the key negative arm protein of the *N. crassa* circadian clock, FRQ, which has been shown to be almost completely lacking in predefined structure (Hurley et al., 2013). We indeed note time-of-day conformational changes (Fig. 4B) and it has not escaped our notice that this result has numerous implications in understanding the molecular mechanism of the circadian clock. As time of day is an essential factor for studying many cellular functions, we believe this technique is applicable to any processes where an IDP that regulates cellular physiology may change its conformation over time, e.g., the proteins involved in the cell cycle.

ACKNOWLEDGMENTS

We thank Meaghan Jankowski and Joshua Thomas for their contributions to Fig. 1 and Fenton Geoghegan for his contribution to Fig. 4. We thank the Cell and Molecular Biology Research Core at RPI for their assistance. We wish to thank the Fungal Genetics Stock Center at the University of Missouri for the resources they made available for working with *Neurospora*. This work was supported by the National Institutes of Health/ National Institute of Biomedical Imaging and Bioengineering (U01 EB022546), National Institute of General Medical Sciences (R35 GM128687), and RPI startup funds to J.M.H, and a SURP fellowship from RPI to J.F.P.

REFERENCES

Albrecht, U. (2012). Timing to perfection: The biology of central and peripheral circadian clocks. *Neuron*, *74*(2), 246–260. https://doi.org/10.1016/j.neuron.2012.04.006.

Atamian, H. S., Creux, N. M., Brown, E. A., Garner, A. G., Blackman, B. K., & Harmer, S. L. (2016). Circadian regulation of sunflower heliotropism, floral orientation, and pollinator visits. *Science*, *353*(6299), 587–590. https://doi.org/10.1126/science.aaf9793.

Baker, C. L., Kettenbach, A. N., Loros, J. J., Gerber, S. A., & Dunlap, J. C. (2009). Quantitative proteomics reveals a dynamic interactome and phase-specific phosphorylation in the *Neurospora* circadian clock. *Molecular Cell*, *34*, 354–363. https://doi.org/10.1016/j.molcel.2009.04.023.

Beadle, G. W., & Tatum, E. L. (1941). Genetic control of biochemical reactions in *Neurospora*. *Proceedings of the National Academy of Sciences of the United States of America*, *27*(11), 499–506. https://doi.org/10.1086/281267.

Bechtold, D. A., Gibbs, J. E., & Loudon, A. S. I. (2010). Circadian dysfunction in disease. *Trends in Pharmacological Sciences*, *31*(5), 191–198. https://doi.org/10.1016/j.tips.2010.01.002.

Bell-Pedersen, D., Cassone, V. M., Earnest, D. J., Golden, S. S., Hardin, P. E., Thomas, T. L., et al. (2005). Circadian rhythms from multiple oscillators: Lessons from diverse organisms. *Nature Reviews. Genetics*, *6*, 544–556. https://doi.org/10.1038/nrd1633.

Buhr, E. D., & Takahashi, J. S. (2013). Molecular components of the mammalian circadian clock. *Handbook of Experimental Pharmacology*, *217*(217), 3–27. https://doi.org/10.1007/978-3-642-25950-0-1.

Caster, S. Z., Castillo, K., Sachs, M. S., & Bell-Pedersen, D. (2016). Circadian clock regulation of mRNA translation through eukaryotic elongation factor eEF-2. *Proceedings of the National Academy of Sciences of the United States of America*, *113*(34), 9605–9610. https://doi.org/10.1073/pnas.1525268113.

Cha, J., Zhou, M., & Liu, Y. (2015a). Mechanism of the *Neurospora* circadian clock, a FREQUENCY-centric view. *Biochemistry*, *54*(2), 150–156. https://doi.org/10.1021/bi5005624.

Cha, J., Zhou, M., & Liu, Y. (2015b). Methods to study molecular mechanisms of the *Neurospora* circadian clock. *Methods in enzymology*. *Vol. 551* (1st ed.): Elsevier Inc. https://doi.org/10.1016/bs.mie.2014.10.002.

Cheng, P., He, Q., He, Q., Wang, L., & Liu, Y. (2005). Regulation of the *Neurospora* circadian clock by an RNA helicase. *Genes and Development*, *19*, 234–241. https://doi.org/10.1101/gad.1266805.234.

Colot, H. V., Park, G., Turner, G. E., Ringelberg, C., Crew, C. M., Litvinkova, L., et al. (2006). A high-throughput gene knockout procedure for *Neurospora* reveals functions for multiple transcription factors. *Proceedings of the National Academy of Sciences of the United States of America*, *103*(27), 10352–10357. https://doi.org/10.1073/pnas.0601456103.

Conrad, K. S., Hurley, J. M., Widom, J., Ringelberg, C. S., Loros, J. J., Dunlap, J. C., et al. (2016). Structure of the frequency-interacting RNA helicase: A protein interaction hub for the circadian clock. *The EMBO Journal*, *35*(15), e201694327. https://doi.org/10.15252/embj.201694327.

Correa, A., Lewis, Z. A., Greene, A. V., March, I. J., Gomer, R. H., & Bell-Pedersen, D. (2003). Multiple oscillators regulate circadian gene expression in *Neurospora*. *Proceedings of the National Academy of Sciences of the United States of America*, *100*(23), 13597–13602.

Crosthwaite, S. K., Dunlap, J. C., Loros, J. J., Science, S., Series, N., May, N., et al. (1997). *Neurospora* wc-1 and wc-2: Transcription, photoresponses, and the origins of circadian rhythmicity. *Science*, *276*(5313), 763–769.

Davis, R. H., & Perkins, D. D. (2002). *Neurospora*: A model of model microbes. *Nature Reviews. Genetics*, *3*(5), 397–403. https://doi.org/10.1038/nrg797.
DeCoursey, P. J., Krulas, J. R., Mele, G., & Holley, D. C. (1997). Circadian performance of suprachiasmatic nuclei (SCN)-lesioned antelope ground squirrels in a desert enclosure. *Physiology & Behavior*, *62*(5), 1099–1108. https://doi.org/10.1016/S0031-9384(97)00263-1.
Dong, W., Tang, X., Yu, Y., Nilsen, R., Kim, R., & Griffith, J. (2008). Systems biology of the clock in *Neurospora crassa*. *PLoS One*, *3*(8), e3105. https://doi.org/10.1371/journal.pone.0003105.
Dosztányi, Z., Csizmok, V., Tompa, P., & Simon, I. (2005). IUPred: Web server for the prediction of intrinsically unstructured regions of proteins based on estimated energy content. *Bioinformatics*, *21*(16), 3433–3434. https://doi.org/10.1093/bioinformatics/bti541.
Dunker, A. K., Brown, C. J., Lawson, J. D., Iakoucheva, L. M., & Obradovic, Z. (2002). Intrinsic disorder and protein function. *Biochemistry*, *41*(21), 6573–6582.
Dunlap, J. C. (1999). Molecular bases for circadian clocks. *Cell*, *96*, 271–290.
Dunlap, J. C., & Loros, J. J. (2004). The *Neurospora* circadian system. *Journal of Biological Rhythms*, *19*(5), 414–424. https://doi.org/10.1177/0748730404269116.
Dunlap, J. C., & Loros, J. J. (2016). Yes, circadian rhythms actually do affect almost everything. *Cell Research*, *26*(7), 759–760. https://doi.org/10.1038/cr.2016.65.
Dunlap, J. C., Loros, J. J., Colot, H. V., Mehra, A., Belden, W. J., Shi, M., et al. (2007). A circadian clock in *Neurospora*: How genes and proteins cooperate to produce a sustained, entrainable, and compensated biological oscillator with a period of about a day. *Cold Spring Harbor Symposia on Quantitative Biology*, *72*(1959), 57–68. https://doi.org/10.1101/sqb.2007.72.072.
Eide, E. J., Woolf, M. F., Kang, H., Woolf, P., Hurst, W., Camacho, F., et al. (2005). Control of mammalian circadian rhythm by CKIε-regulated proteasome-mediated PER2 degradation. *Molecular and Cellular Biology*, *25*(7), 2795–2807. https://doi.org/10.1128/MCB.25.7.2795.
Evans, J. A., & Davidson, A. J. (2013). Health consequences of circadian disruption in humans and animal models. *Progress in Molecular Biology and Translational Science*, *119*, 283–323. https://doi.org/10.1016/B978-0-12-396971-2.00010-5.
Fuller, K. K., Hurley, J. M., Loros, J., & Dunlap, J. C. (2014). Photobiology and circadian clocks in *Neurospora*. *Fungal Genomics*, *13*, 121–148. https://doi.org/10.1007/978-3-642-45218-5_6.
Galagan, J. E., Calvo, S. E., Borkovich, K. A., Selker, E. U., Read, N. D., Jaffe, D., et al. (2003). The genome sequence of the filamentous fungus *Neurospora crassa*. *Nature*, *422*(6934), 859–868. Retrieved from https://doi.org/10.1038/nature01554.
Gooch, V. D., Mehra, A., Larrondo, L. F., Fox, J., Touroutoutoudis, M., Loros, J. J., et al. (2008). Fully codon-optimized luciferase uncovers novel temperature characteristics of the *Neurospora* clock. *Eukaryotic Cell*, 7(1), 28–37. https://doi.org/10.1128/EC.00257-07.
Grima, B., Lamouroux, A., Chélot, E., Papin, C., Limbourg-Bouchon, B., & Rouyer, F. (2002). The F-box protein slimb controls the levels of clock proteins period and timeless. *Nature*, *420*(6912), 178–182. https://doi.org/10.1038/nature01122.
Gsponer, J., & Babu, M. M. (2009). The rules of disorder or why disorder rules. *Progress in Biophysics and Molecular Biology*, *99*(2–3), 94–103. https://doi.org/10.1016/j.pbiomolbio.2009.03.001.
Guo, J., Cheng, P., & Liu, Y. (2010). Functional significance of FRH in regulating the phosphorylation and stability of *Neurospora* circadian clock protein FRQ. *Journal of Biological Chemistry*, *285*(15), 11508–11515. https://doi.org/10.1074/jbc.M109.071688.

Gustafson, C. L., & Partch, C. L. (2015). Emerging models for the molecular basis of mammalian circadian timing. *Biochemistry, 54*(2), 134–149. https://doi.org/10.1021/bi500731f.
Hardin, P. E. (2005). The circadian timekeeping system of *Drosophila. Current Biology: CB, 15*(17), R714–R722. https://doi.org/10.1016/j.cub.2005.08.019.
Hughes, M. E., Abruzzi, K. C., Allada, R., Anafi, R., Arpat, A. B., Asher, G., et al. (2017). Guidelines for genome-scale analysis of biological rhythms. *Journal of Biological Rhythms, 32*(5), 380–393. https://doi.org/10.1177/0748730417728663.
Hunt, T., & Sassone-Corsi, P. (2007). Riding tandem: Circadian clocks and the cell cycle. *Cell, 129*(3), 461–464. https://doi.org/10.1016/j.cell.2007.04.015.
Hurley, J. M., Dasgupta, A., Emerson, J. M., Zhou, X., Ringelberg, C. S., Knabe, N., et al. (2014). Analysis of clock-regulated genes in *Neurospora* reveals widespread posttranscriptional control of metabolic potential. *Proceedings of the National Academy of Sciences of the United States of America, 111*(48), 16995–17002. https://doi.org/10.1073/pnas.1418963111.
Hurley, J. M., Larrondo, L. F., Loros, J. J., & Dunlap, J. C. (2013). Conserved RNA helicase FRH acts nonenzymatically to support the intrinsically disordered *Neurospora* clock protein FRQ. *Molecular Cell, 52*(6), 832–843. https://doi.org/10.1016/j.molcel.2013.11.005.
Hurley, J., Loros, J. J., & Dunlap, J. C. (2015). Dissecting the mechanisms of the clock in *Neurospora. Methods in Enzymology, 551*, 29–52. https://doi.org/10.1016/bs.mie.2014.10.009.
Hurley, J. M., Loros, J. J., & Dunlap, J. C. (2016a). Circadian oscillators: Around the transcription–translation feedback loop and on to output. *Trends in Biochemical Sciences, 41*(10), 834–846. https://doi.org/10.1016/j.tibs.2016.07.009.
Hurley, J. M., Loros, J. J., & Dunlap, J. C. (2016b). The circadian system as an organizer of metabolism. *Fungal Genetics and Biology, 90*, 39–43. https://doi.org/10.1016/j.fgb.2015.10.002.
Imeri, L., & Opp, M. R. (2010). How (and why) the immune system makes us sleep. *Nature Reviews. Neuroscience, 10*(3), 199–210. https://doi.org/10.1038/nrn2576.How.
Jinhu, G., & Yi, L. (2010). Molecular mechanism of the *Neurospora* circadian oscillator. *Protein & Cell, 1*(4), 331–341. https://doi.org/10.1007/s13238-010-0053-7.
Karatsoreos, I. N. (2012). Effects of circadian disruption on mental and physical health. *Current Neurology and Neuroscience Reports, 12*(2), 218–225. https://doi.org/10.1007/s11910-012-0252-0.
Keller, M., Mazuch, J., Abraham, U., Eom, G. D., Herzog, E. D., Volk, H.-D., et al. (2009). A circadian clock in macrophages controls inflammatory immune responses. *Proceedings of the National Academy of Sciences of the United States of America, 106*(50), 21407–21412. https://doi.org/10.1073/pnas.0906361106.
Kim, T. W., Jeong, J.-H., & Hong, S.-C. (2015). The impact of sleep deprivation on hormones and metabolism. *International Journal of Endocrinology, 2015*, 1–9. https://doi.org/10.1155/2015/591729.
Klarsfeld, A., & Rouyer, F. (1998). Effects of circadian mutations and LD periodicity on the life span of *Drosophila melanogaster. Journal of Biological Rhythms, 13*(6), 471–478. https://doi.org/10.1177/074873098129000309.
Ko, H. W., Jiang, J., & Edery, I. (2002). Role for Slimb in the degradation of *Drosophila* Period protein phosphorylated by Doubletime. *Nature, 420*(6916), 673–678. https://doi.org/10.1038/nature01272.
Kondo, T., & Ishiura, M. (2003). The circadian clock of cyanobacteria. *Chronobiology International, 20*, 775–794.
Lamia, K. A., Sachdeva, U. M., Di Tacchio, L., Williams, E. C., Alvarez, J. G., Egan, D. F., et al. (2009). AMPK regulates the circadian clock by cryptochrome phosphorylation and degradation. *Science, 326*(5951), 437–440. https://doi.org/10.1126/science.1172156.

Larrondo, L. F., Olivares-Yañez, C., Baker, C. L., Loros, J. J., & Dunlap, J. C. (2015). Decoupling circadian clock protein turnover from circadian period determination. *Science, 347*(6221), 1257277. https://doi.org/10.1126/science.1257277.

Loros, J. J., Denome, S. A., & Dunlap, J. C. (1989). Molecular cloning of genes under control of the circadian clock in *Neurospora. Science, 243*(4889), 385–388.

Lowrey, P. L., & Takahashi, J. S. (2004). Mammalian circadian biology: Elucidating genome-wide levels of temporal organization. *Annual Review of Genomics and Human Genetics, 5*(1), 407–441. https://doi.org/10.1146/annurev.genom.5.061903.175925.

McCluskey, K., Wiest, A., & Plamann, M. (2010). The fungal genetics stock center: A repository for 50 years of fungal genetics research. *Journal of Biosciences, 35*(1), 119–126. https://doi.org/10.1007/s12038-010-0014-6.

Menet, J. S., Rodriguez, J., Abruzzi, K. C., & Rosbash, M. (2012). Nascent-Seq reveals novel features of mouse circadian transcriptional regulation. *eLife, 2012*(1), 1–25. https://doi.org/10.7554/eLife.00011.

Minde, D. P., Maurice, M. M., & Rüdiger, S. G. D. (2012). Determining biophysical protein stability in lysates by a fast proteolysis assay, FASTpp. *PLoS One*, 7(10), e46147. https://doi.org/10.1371/journal.pone.0046147.

Montenegro-Montero, A., & Larrondo, L. F. (2012). *Circadian system: From genes to proteins and back in less than 24 hours.* Caister Academic Press.

Morgan, L. W., Greene, A. V., & Bell-Pedersen, D. (2003). Circadian and light-induced expression of luciferase in *Neurospora crassa. Fungal Genetics and Biology, 38*, 327–332. https://doi.org/10.1016/S1087-1845(02)00562-5.

Mori, T., Mchaourab, H., & Johnson, C. H. (2015). Circadian clocks: Unexpected biochemical cogs. *Current Biology, 25*(19), R842–R844. https://doi.org/10.1016/j.cub.2015.08.026.

Mure, L. S., Le, H. D., Benegiamo, G., Chang, M. W., Rios, L., Jillani, N., et al. (2018). Diurnal transcriptome atlas of a primate across major neural and peripheral tissues. *Science, 359*(6381). https://doi.org/10.1126/science.aao0318.

Ninomiya, Y., Suzuki, K., Ishii, C., & Inoue, H. (2004). Highly efficient gene replacements in *Neurospora* strains deficient for nonhomologous end-joining. *Proceedings of the National Academy of Sciences of the United States of America, 101*, 12248–12253.

Oldfield, C. J., & Dunker, A. K. (2014). Intrinsically disordered proteins and intrinsically disordered protein regions. *Annual Review of Biochemistry, 83*, 553–584. https://doi.org/10.1146/annurev-biochem-072711-164947.

Outang, Y., Andersson, C. R., Kondo, T., Golden, S. S., & Johnson, C. H. (1998). Resonating circadian clocks enhance fitness in cyanobacteria in silico. *Ecological Modelling, 221*(12), 1620–1629. https://doi.org/10.1016/j.ecolmodel.2010.03.015.

Panda, S., Hogenesch, J. B., & Kay, S. A. (2002). Circadian rhythms from flies to humans. *Nature, 417*(May), 329–335.

Partch, C. L., Green, C. B., & Takahashi, J. S. (2014). Molecular architecture of the mammalian circadian clock. *Trends in Cell Biology, 24*(2), 90–99. https://doi.org/10.1016/j.tcb.2013.07.002.

Pittendrigh, C. S., Bruce, V. G., Rosenweig, N. S., & Rubin, M. L. (1959). A biological clock in *Neurospora. Nature, 184*(4681), 169–170. https://doi.org/10.1038/184169a0.

Querfurth, C., Diernfellner, A. C. R., Gin, E., Malzahn, E., Höfer, T., & Brunner, M. (2011). Circadian conformational change of the *Neurospora* clock protein FREQUENCY triggered by clustered hyperphosphorylation of a basic domain. *Molecular Cell, 43*(5), 713–722. https://doi.org/10.1016/j.molcel.2011.06.033.

Rajaratnam, S. M. W., & Arendt, J. (2001). Health in a 24-h society. *Lancet, 358*, 999–1005.

Reddy, A. B., & Rey, G. (2014). Metabolic and nontranscriptional circadian clocks: Eukaryotes. *Annual Review of Biochemistry, 83*(1), 165–189. https://doi.org/10.1146/annurev-biochem-060713-035623.

Robaszkiewicz, K., Ostrowska, Z., Cyranka-Czaja, A., & Moraczewska, J. (2015). Impaired tropomyosin-troponin interactions reduce activation of the actin thin filament. *Biochimica et Biophysica Acta*, *1854*(5), 381–390. https://doi.org/10.1016/j.bbapap.2015.01.004.

Roche, C. M., Loros, J. J., McCluskey, K., & Glass, N. G. (2014). *Neurospora crassa*: Looking back and looking forward at a model microbe. *American Journal of Botany*, *101*(12), 2022–2035. https://doi.org/10.3732/ajb.1400377.

Rodriguez, J., Tang, C.-H. A., Khodor, Y. L., Vodala, S., Menet, J. S., & Rosbash, M. (2013). Nascent-Seq analysis of *Drosophila* cycling gene expression. *Proceedings of the National Academy of Sciences of the United States of America*, *110*(4), E275–E284. https://doi.org/10.1073/pnas.1219969110.

Sack, R. L., Auckley, D., Auger, R. R., Carskadon, M. A., Wright, K. P., Vitiello, M. V., et al. (2007a). Circadian rhythm sleep disorders: Part II, advanced sleep phase disorder, delayed sleep phase disorder, free-running disorder, and irregular sleep-wake rhythm. *Sleep*, *30*(11), 1484–1501. https://doi.org/10.1093/sleep/30.11.1484.

Sack, R. L., Auckley, D., Auger, R. R., Carskadon, M. a., Wright, K. P., Vitiello, M. V., et al. (2007b). Circadian rhythm sleep disorders: Part I, basic principles, shift work and jet lag disorders. *Sleep*, *30*(11), 1460–1483. https://doi.org/10.1136/bmj.e6664.

Sancar, C., Sancar, G., Ha, N., Cesbron, F., & Brunner, M. (2015). Dawn- and dusk-phased circadian transcription rhythms coordinate anabolic and catabolic functions in *Neurospora*. *BMC Biology*, *13*(1), 17. https://doi.org/10.1186/s12915-015-0126-4.

Schafmeier, T., Haase, A., Káldi, K., Scholz, J., Fuchs, M., & Brunner, M. (2005). Transcriptional feedback of *Neurospora* circadian clock gene by phosphorylation-dependent inactivation of its transcription factor. *Cell*, *122*(2), 235–246. https://doi.org/10.1016/j.cell.2005.05.032.

Shinohara, M. L., Correa, A., Bell-Pedersen, D., Dunlap, J. C., & Loros, J. J. (2002). *Neurospora* clock-controlled gene 9 (ccg-9) encodes trehalose synthase: Circadian regulation of stress responses and development. *Eukaryotic Cell*, *1*(1), 33–43. https://doi.org/10.1128/EC.1.1.33-43.2002.

Stevens, R. G., Blask, D. E., Brainard, G. C., Hansen, J., Lockley, S. W., Provencio, I., et al. (2007). Meeting report: The role of environmental lighting and circadian disruption in cancer and other diseases. *Environmental Health Perspectives*, *115*(9), 1357–1362. https://doi.org/10.1289/ehp.10200.

Tang, C.-T., Li, S., Long, C., Cha, J., Huang, G., Li, L., et al. (2009). Setting the pace of the *Neurospora* circadian clock by multiple independent FRQ phosphorylation events. *Proceedings of the National Academy of Sciences of the United States of America*, *106*(26), 10722–10727. https://doi.org/10.1073/pnas.0904898106.

Tompa, P. (2011). Unstructural biology coming of age. *Current Opinion in Structural Biology*, *21*(3), 419–425. https://doi.org/10.1016/j.sbi.2011.03.012.

Tompa, P. (2012). Intrinsically disordered proteins: A 10-year recap. *Trends in Biochemical Sciences*, *37*(12), 509–516. https://doi.org/10.1016/j.tibs.2012.08.004.

Uversky, V. N. (2009). Intrinsically disordered proteins and their environment: Effects of strong denaturants, temperature, pH, counter ions, membranes, binding partners, osmolytes, and macromolecular crowding. *Protein Journal*, *28*(7–8), 305–325. https://doi.org/10.1007/s10930-009-9201-4.

Uversky, V. N. (2011). Intrinsically disordered proteins from A to Z. *International Journal of Biochemistry and Cell Biology*, *43*(8), 1090–1103. https://doi.org/10.1016/j.biocel.2011.04.001.

Uversky, V. N. (2015). Proteins without unique 3D structures: Biotechnological applications of intrinsically unstable/disordered proteins. *Biotechnology Journal*, *10*(3), 356–366. https://doi.org/10.1002/biot.201400374.

Wright, P. E., & Dyson, H. J. (2015). Intrinsically disordered proteins in cellular signaling and regulation. *Nature Reviews. Molecular Cell Biology*, *16*(1), 18–29. https://doi.org/10.1038/nrm3920.Intrinsically.
Xie, L., Kang, H., Xu, Q., Chen, M. J., Liao, Y., Thiyagarajan, M., et al. (2013). Sleep drives metabolite clearance from the adult brain. *Science*, *342*(6156), 373–377. https://doi.org/10.1126/science.1241224.
Yakir, E., Hilman, D., Harir, Y., & Green, R. M. (2007). Regulation of output from the plant circadian clock. *FEBS Journal*, *274*(2), 335–345. https://doi.org/10.1111/j.1742-4658.2006.05616.x.
Zhang, R., Lahens, N. F., Ballance, H. I., Hughes, M. E., & Hogenesch, J. B. (2014). A circadian gene expression atlas in mammals: Implications for biology and medicine. *Proceedings of the National Academy of Sciences of the United States of America*, *111*(45), 16219–16224. https://doi.org/10.1073/pnas.1408886111.

CHAPTER SEVENTEEN

Assessing Allostery in Intrinsically Disordered Proteins With Ensemble Allosteric Model

Jing Li*, Vincent J. Hilser*,†,1
*T.C. Jenkins Department of Biophysics, Johns Hopkins University, Baltimore, MD, United States
†Department of Biology, Johns Hopkins University, Baltimore, MD, United States
[1]Corresponding author: e-mail address: hilser@jhu.edu

Contents

Methods in Enzymology, Volume 611
ISSN 0076-6879
https://doi.org/10.1016/bs.mie.2018.09.004

Abstract

Intrinsically disordered (ID) proteins have been shown to play a major role in signaling in a broad range of proteins, using a process known as allostery, wherein the protein can integrate one or a number of inputs to regulate its function. The disorder-mediated allostery can be understood energetically with ensemble allosteric model (EAM). In this model, the molecule without effectors is considered as an ensemble of preexisting conformations, and effector binding is treated as an energetic perturbation of the ensemble to redistribute the microstates that are favorable or unfavorable to the second binding partner. As it only considers the intrinsic energetics of the system and does not depend on a crystallographic structure, it can be applied to both structured proteins, ID proteins, and mixed proteins with both structured and ID domains. Simulation with EAM on the basis of experimental data can help quantitatively explain experimental observations, as well as to make predictions to direct future research. This has recently been illustrated with the case of human glucocorticoid receptor, a multidomain transcription factor that contains both structured and disordered regions. In this chapter, we describe the assays for measuring the transcriptional activity, binding affinity to cognate DNA, conformational stability, either on single domain or tandem coupled domains in the GR two-domain isoforms. We then explain how these data are utilized as input parameters or constraints in the EAM for quantitative estimates of stabilities and coupling energies for each domain through global minimization method.

1. INTRODUCTION

Intrinsically disordered regions (IDRs) of a protein lack a stable tertiary structure under physiological conditions and exist as dynamic ensembles of different conformational states. Coupled folding and binding are frequently observed in IDRs, the conditionally folded states are the functional states of ID proteins in most cases. As part of the binding energy compensates the folding, IDRs usually have high specificity and relatively low binding affinity to multiple binding partners (Uversky, Oldfield, & Dunker, 2005). These characteristics confer IDRs with malleability, making them sensitive regulators of protein functions. Indeed, more than 30% of the proteome is thought to consist of IDRs (Ward, Sodhi, McGuffin, Buxton, & Jones, 2004), with enrichment in posttranslational modification sites (Gao & Xu, 2012) and splicing sites (Romero et al., 2006). Of particular note is that

IDRs are hyperabundant in signaling proteins, RNA binding proteins, and transcription factors (Liu et al., 2006). Coincidentally, another common regulatory mechanism in transcription factors is allostery, which is the influence of a perturbation at one site of a molecule to a distant, often the functional, site of the same molecule. It plays important roles in signal transduction, gene transcription, and metabolic homeostasis by endowing cells with a sensitive responsiveness to environmental cues (Changeux & Edelstein, 2005). The convergence of allostery and IDRs in transcription factors suggests that allostery involving IDRs may represent a common regulatory paradigm.

Allostery is usually studied from structure point of view, represented by two classic models, the concerted MWC (Monod–Wyman–Changeux) model (Monod, Wyman, & Changeux, 1965) and the sequential KNF (Koshland–Nemethy–Filmer) model (Koshland, Nemethy, & Filmer, 1966). How then can we understand allostery in IDRs in a way that is divorced from a requirement of a purely structural representation? Recently, we showed that the ensemble allosteric model (EAM) (Hilser & Thompson, 2007; Motlagh & Hilser, 2012; Wrabl et al., 2011) can provide insights into these systems. In this model, the molecule without effectors is considered as an ensemble of preexisting conformations, and effector binding is treated as an energetic perturbation of the ensemble to redistribute the microstates that are favorable or unfavorable to the second binding partner. As it only considers the intrinsic energetics of the system and does not depend on a crystallographic structure, it can be applied to both structured proteins, ID proteins, and mixed proteins with both structured and ID domains. Practically, the model treats two or three tethered domains in an allosteric protein as being either folded (high-affinity conformation to ligand) or unfolded (low-affinity conformation to ligand), resulting in multiple states in the ensemble. Coupling between domains is expressed as interaction energy (Δg_{int}), which represents the statistical nature of the interaction. Depending on the energetic costs of interconverting between the folded state and unfolded state of each domain (ΔG_{i}) and of breaking the interaction between domains, each state in the ensemble has a different probability. Effector binding stabilizes the high-affinity state to effector and thus redistributes the ensemble. Allosteric coupling response is considered as the degree to which the probability of states in which functional sites are folded and is affected for a given perturbation to states that can bind effectors. More generally, allostery caused by different energetic perturbations (besides ligand binding) to the preexisting ensemble, such as posttranslational modifications, mutations, and truncations can all be understood in terms of regulating domain stability (ΔG_i) or modulating the interaction energy (Δg_{int}) between coupled domains in EAM (Hilser & Thompson, 2007). Due to its simplicity, simulation with EAM on the basis of experimental data can help quantitatively explain experimental observations, as well as to make predictions to direct future research. Indeed the EAM was used to investigate the origins of coupling in the IDR-mediated intradomain and interdomain allosteric coupling in a transcription factor, glucocorticoid receptor (GR) (Li et al., 2017).

GR, a transcription factor in the steroid hormone receptor (SHR) family, plays key roles in organ development, metabolite homeostasis, and the inflammatory response (Griekspoor, Zwart, Neefjes, & Michalides, 2007). Malfunction of GR leads to cancer, metabolic diseases, and inflammatory disorders (Oakley & Cidlowski, 2011), and GR-based pharmaceuticals are among the most commonly prescribed drugs. Full-length GR contains three major domains, the intrinsically disordered N-terminal domain (ID NTD), the DNA-binding domain (DBD), and the ligand-binding domain (LBD) (Hilser & Thompson, 2011). Allostery is known to govern the function of SHRs (Griekspoor et al., 2007; Gronemeyer & Bourguet, 2009). In GR, the intradomain and interdomain allosteric couplings are widely reported in the literature (Kumar et al., 1999; Meijsing et al., 2009; Pandit, Geissler, Harris, & Sitlani, 2002; Pfaff & Fletterick, 2010; Watson et al., 2013). However, the mechanism is elusive, especially how the long NTD is involved in the coupling. The major transactivation region of GR, activation function 1 (AF1) region, which contains various coregulator binding sites, is located in the NTD (Dahlman-Wright, Almlof, McEwan, Gustafsson, & Wright, 1994; Lavery & McEwan, 2005). In addition, the functional importance of GR's NTD is also suggested by the discovery of eight human GR translational isoforms with different tissue distributions, different transcriptional activities, and unique gene regulation sets (Lu & Cidlowski, 2005). Surprisingly, these translational isoforms differ only in the length of their NTD (Li, Motlagh, Chakuroff, Thompson, & Hilser, 2012; Lu & Cidlowski, 2005) (as shown in Fig. 1A). Truncation of the LBD removes GR's transcriptional activity dependence on hormone and generates constitutively active, two-domain constructs (Chen, Srinivasan, & Thompson, 1997) as shown in Fig. 1. Measurements of transcriptional activity and binding affinity to the cognate DNA of these two-domain transnational isoforms show no correlation (Fig. 1B).

Several features of the activities and DNA-binding properties of the different isoforms are noteworthy. First, the affinities of the isoforms for DNA vary, despite having identical DBD, indicating that the NTD of each

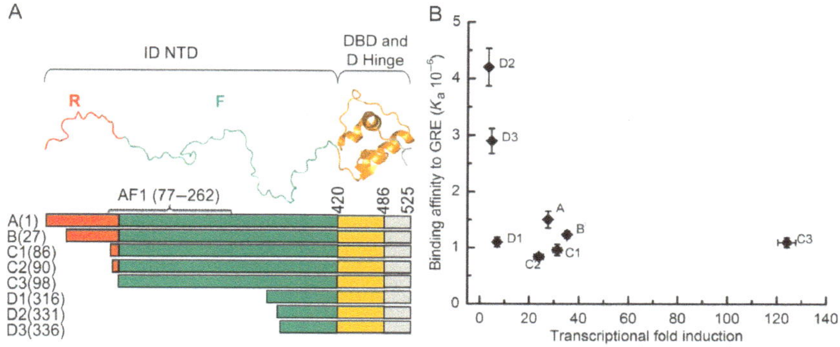

Fig. 1 Transcriptional activity and DNA-binding affinity are not correlated among GR translational isoforms. (A) Domain organization of the constitutively active GR constructs for translational isoforms, wherein the intrinsically disordered N-terminal domain (NTD) varies in length. Residues 1–97 (*red*) are labeled R (for Regulatory) and residues 98–420 (*Cyan*) are labeled F (for Functional). Also labeled are residues corresponding to the activation function 1 domain (AF1), which is required for transcriptional activity (Ford, McEwan, Wright, & Gustafsson, 1997). Error bars represent uncertainty of the individual fits. (B) Transcriptional activity, monitored by dual luciferase reporter assay (Fig. 2), and binding affinity to DNA, monitored by fluorescence anisotropy change (Fig. 3), show uncorrelated behavior.

isoform differentially communicates with its respective DBD (Fig. 1B). Second, the GR C3 isoform (i.e., 98–525) is almost five times more active than the full-length GR A isoform (i.e., 1–525), indicating that residues 1–97 somehow negatively regulate the activity of the remaining NTD residues, which contain the functionally important AF1 region (Fig. 1A). For this reason, we represent the full-length NTD as being composed of two distinct domains, a functional domain (F-domain) and a regulatory domain (R-domain) (Fig. 1A).

In this chapter, we describe the assays for measuring the transcriptional activity, binding affinity to cognate DNA, conformational stability, either on single domain or tandem coupled domains in the GR two-domain isoforms. We then explain how these data are utilized as input parameters or constraints in the EAM for quantitative estimates of stabilities and coupling energies for each domain through global minimization method.

2. TRANSCRIPTIONAL ACTIVITY ASSAY FOR DIFFERENT GR ISOFORMS

The transcriptional activity of the eight two-domain constructs was measured using a dual luciferase reporter assay. A luciferase reporter assay measures the activity of a transcription factor through the luminescence of luciferase. In our system, the expression vector of GR two-domain isoform was cotransfected with a *Gaussia* luciferase vector (Tannous, Kim, Fernandez, Weissleder, & Breakefield, 2005), in which GR response element (GRE) has been cloned in the regulatory region. To take into account the differences in cell density and transfection efficiency, we also cotransfected a control vector that encodes *Cypridina* luciferase (Nakajima, Kobayashi, Yamagishi, Enomoto, & Ohmiya, 2004) not under control of GR (as illustrated in Fig. 2A). To measure the transcriptional activity of transcription factors in a reporter gene assay, it is important to ensure that the reporter vector is fully bound by the transcription factor. Otherwise, the activity measured at non-saturating concentrations of transcription factor does not represent the relative activity of each transcription factor as they may have different binding affinities to the response element. To do this, a dosage curve was performed for each GR translational isoform.

2.1 Mammalian Expression Vectors for GR Translational Isoforms and Luciferase Reporters

Codons to express human GR A isoform NTD and DBD two-domain construct in U-2 OS cells were optimized for mammalian cell expression, synthesized by DNA 2.0 (Menlo Park, CA), and inserted into the PJ603 mammalian expression vector under CMV promoter control. Plasmids for the two-domain constructs of B, C1, C2, C3, D1, D2, and D3 were made by inserting the codons for each respective isoform amplified from A isoform into the *Nhe*I and *Xho*I sites of the PJ603 vector.

Plasmid GRE2-Gluc, to express secreted *Gaussia* luciferase under the control of two tandem full-length GREs in the promoter, was made by inserting an oligonucleotide containing two tandem full-length GREs, 5′-aattcAGAACAggaTGTTCTgagatccgtagc AGAACAggaTGTTC Tgagatccgtagcg-3′, into the *Eco*RI and *Bam*HI sites of the pGluc-miniTK vector (NEB). Plasmid pCluc-miniTK2 vector (NEB) to express *Cypridina* luciferase was utilized as an internal control in the cotransfection to account for cell density differences and transfection efficiency differences in each well.

2.2 Mammalian Cell Culture and Transfection

U-2 OS cells (American Type Culture Collection) were maintained in modified McCoy's 5a medium (Corning Cellgro) supplemented with

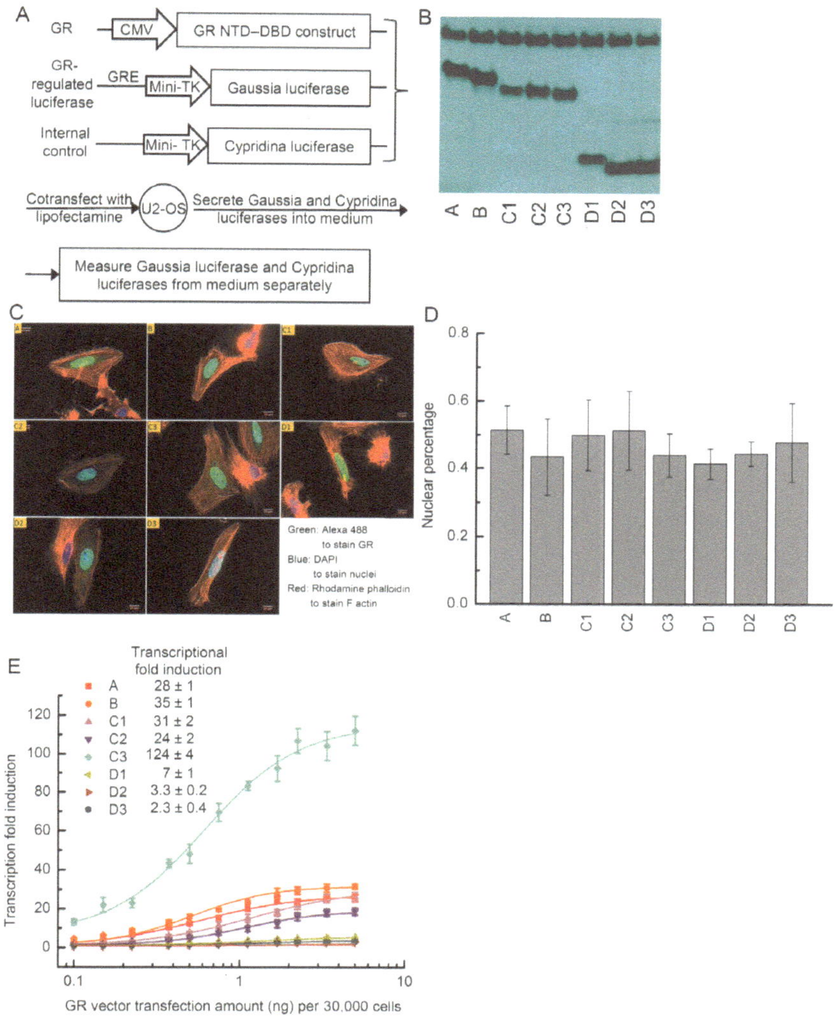

Fig. 2 Transcriptional activity assay on GR translational isoforms. (A) Cotransfection strategy in the dual luciferase reporter assay. (B) Western blot showing that U-2 OS cells transiently transfected with the expression plasmid for each isoform express each of them. (C) Multicolor immunostaining of U-2 OS cells transfected with A, B, C1, C2, C3, D1, D2, and D3 constructs. *Green*: Alexa 488 linked goat anti-mouse IgG staining GR. *Blue*: DAPI staining nuclei. *Red*: Rhodamine phalloidin staining F-actin. (D) Nuclear localization efficiency for the eight GR translational isoforms. Nuclear percentage is calculated by dividing the intensity of the *green* dye overlapped with *blue* dye with the total *green* dye intensity as shown in panel C. Three pictures were used for each isoform for the quantification. Average values and standard errors of the mean are reported in the graph. (E) Luciferase assay dosage curves for the eight GR translational isoforms. Per

(Continued)

10% fetal bovine serum and 100 U/mL penicillin and 100 μg/mL streptomycin. To transfect U-2 OS cells at about 80%–90% confluence, X-tremeGENE HP DNA transfection reagent (Roche) was used at 2 μL per 1 μg DNA according to the manufacturer's manual.

2.3 Western Blot to Check Expression

U-2 OS cells were plated on 6-well plates at a density of 5×10^5 cells/well. After 18 h, 50 ng of GR isoform expression vector with 450 ng Salmon Sperm DNA (Invitrogen) as transfection boosting reagent was transfected into each well with X-tremeGENE HP DNA transfection reagent (Roche) following the manufacturer's protocol. Twenty-four hours post-transfection the medium was changed once. After 48 h, the cells were collected for the assay.

Cells were removed from each well with PBS and pelleted by centrifuging at 1500 rpm for 10 min in a benchtop centrifuge. 50 μL of lysis buffer (8 *M* urea, 20 m*M* Tris–HCl, 500 m*M* NaCl, 1 m*M* Na_2EDTA, 1 m*M* EGTA, 1% Triton, 2.5 m*M* sodium pyrophosphate, 1 m*M* beta-glycerophosphate, 1 m*M* Na_3VO_4, and 1 μg/mL leupeptin, pH 7.5) was added into the cell pellet from each well. To reduce the viscosity, cells were passed through a 26G3/8 syringe needle 10 times and then the cell lysate was centrifuged at 14,000 rpm for 30 min. Supernatant was collected and the total protein concentration was measured by Bradford assay (BioRad). 5 μg of total protein was loaded onto each well of a 4%–15% Mini-PROTEAN TGX Precast Gel (BioRad) and separated in 1× Tris/glycine/SDS gel running buffer. Transfer of the protein from SDS page gel to PVDF film was done in transfer buffer (25 m*M* Tris–HCl, pH 8.3, 192 m*M* glycine, 20% methanol) under 120 V for 15 min. After blocking in 5% nonfat milk for 1 h, the PVDF film with protein transferred on it

Fig. 2—cont'd 30,000 cells, a constant 40 ng of GRE-driven luciferase vector was transfected, and the amount of GR vector cotransfected was increased from 0 to 5 ng. Errors are calculated from three samples. Curves are fitted to the data using the dose–response function, $F(C) = 1 + \frac{A_{max} - 1}{1 + (EC_{50}/C)^p}$, where $F(C)$ is the observed luciferase expression level when transfecting the plasmids coding GR isoforms at the amount of C; A_{max} represents the maximum transcriptional activity for each construct; EC_{50} represents the amount of GR construct transfected at the half-maximum transcriptional activity; *p* is an empirical value introduced in the fitting equation, to transform DNA vector amount to protein expression amount, to account for the possibility that different isoforms have different degradation rates, expression levels, nuclear localizations, and/or cooperativities.

was incubated with 5000-fold diluted primary antibody for GR (BD Transduction Laboratories, 611226) or 5000-fold diluted antibody for p150glued (BD Transduction Laboratories, 610473), which served as loading control, at 4°C overnight. The next morning, after washing with PBST buffer (PBS supplemented with 0.1% Tween 20) three times, the PVDF film was incubated in the 10,000-fold diluted HRP-linked anti-mouse IgG (GE healthcare, NA931). The detection was done with Amersham ECL Prime Western blotting reagent (GE healthcare, RPN2232). As shown in Fig. 2B, the eight GR translational isoforms have comparable expression level in U-2 OS cells.

2.4 Immunostaining to Check Nuclear Translocation

U-2 OS cells were plated on 6-well plates with 15 mm round German coverslips. All the culture and transfection procedures are the same as done for the cells treated for Western blot experiments. After 48 h, cells were rinsed with PBSM (PBS with 2 m*M* $MgCl_2$) three times and fixed with 4% paraformaldehyde in PBSM at room temperature for 10 min. Afterward, each coverslip was rinsed with PBSM three times again and quenched with 50 m*M* NH_4Cl in PBSM. The slide was then placed in PBSTB (PBS with 0.1% Triton X-100, 1%BSA) for 30 min at room temperature to permeabilize cells and block nonspecific binding. Thereafter, the slide was incubated with primary rabbit antibody for GR (cell signaling, #3660), which is 5000-fold diluted in PBSTB for 1 h at room temperature. The slide was then washed three times with PBSM and then incubated with the Alexa Fluor® 488 Goat Anti-Rabbit IgG, which is 600-fold diluted in PBSTB for 30 min at room temperature in the dark. Following that, the slide was incubated in PBSM with 0.2 μg/mL DAPI (Invitrogen) and 5 unit/mL rhodamine phalloidin (Invitrogen) for 10 min at room temperature in the dark to stain the nuclei and F-actin. Finally, each slide was washed with PBSM twice and mounted onto a microscope slide with Fluoromount (Sigma) and kept in the dark for drying.

Images were taken with an inverted light microscope (Axiovert 200, JHU Integrated Imaging Center). All images were taken the same day using the same gain, exposure times, and filter configurations (DAPI, FITC, and Texas Red filters of Sedat3 filter set).

As shown in Fig. 2C and D, the eight translational isoforms can all translocate to the nucleus, with indistinguishable nuclear translocation efficiency.

2.5 Luciferase Assay to Measure Transcriptional Activity of GR Translational Isoforms

U-2 OS cells were plated on 96-well plates at a density of 3×10^4 cells/well as measured by hemocytometer. After 18–24 h, when the cells reach about 80%–90% confluence, X-tremeGENE HP DNA transfection reagent (Roche) mediated chemical transfection was carried out. For the transcriptional activity dosage curve, 40 ng GRE2-Gluc, 40 ng pCluc-miniTK2, and up to 5 ng GR expression vector were cotransfected into U-2 OS cells on a 96-well plate. After 48 h, *Gaussia* luciferase activity and *Cypridina* luciferase activity were measured with the BioLux® *Gaussia* Luciferase Assay Kit (NEB) and the BioLux *Cypridina* Luciferase Assay Kit (NEB), respectively, on a TriStar^2 LB 942 Multidetection Microplate Reader (Berthold) according to the manufacturer's protocols. In each experiment, the *Gaussia* luciferase activity (normalized with the *Cypridina* luciferase activity) was measured in triplicate and averaged.

As shown in Fig. 2E, five GR translational isoforms A, B, C1, C2, and C3, which all contain the transactivation region, AF1, showed variable transcriptional activity; another three isoforms, D1, D2, and D3, which do not contain the AF1 region showed almost no significant transcriptional activation activity.

3. BINDING AFFINITY FOR COGNATE DNA OF DIFFERENT GR ISOFORMS

Allostery, by which remote sites of a protein are energetically coupled to elicit a functional response, manifests as action at distance. Thus, exploring one region's influence on a second regions energetics of binding to its ligands is critical to understanding the coupling mechanism. To gain insight into the allosteric coupling between the NTD and DBD of GR, binding affinities to the half-site of GRE were measured for the two-domain constructs of the eight GR translational isoforms. The binding was monitored in vitro by the change in the fluorescence anisotropy of the half-site GRE labeled with fluorescent dye 6FAM.

3.1 Design of Fluorescently Labeled GRE and Control Oligonucleotide

Oligonucleotides containing half-site GRE (5′-6FAM gcgcAGAACAggacgcg-3′ and 5′-cgcgtccTGTTCTgcgc-3′) were synthesized by Integrated DNA Technologies (IDT) with HPLC grade purification and annealed with each other to get the double-stranded 6FAM labeled half-site GRE.

Nucleotide G and C are included at the 5′ and 3′ ends to avoid the effect of fraying end on binding. Nonspecific control oligonucleotides (5′-GCGCCATATGATACGCG-3′ and 5′-CGCGTATCATATGGCGC-3′) were synthesized as complimentary strands by IDT and annealed to each other to get the double-stranded control oligonucleotide.

3.2 Protein Expression, Purification, and Storage

Expression and purification of single NTD and NTD–DBD conjoined two-domain constructs for either GR translational isoforms were carried out in *Escherichia coli*. As degradation and low yield are general problems in expression and purification of recombinant ID proteins from *E. coli*, we employed: (1) engineered plasmids with optimized codons for *E. coli* (DNA2.0), (2) affinity purification tags on both the N-terminus and C-terminus of each construct, (3) a protease-deficient competent expression strain (BL21(DE3) pLysS), (4) overnight protein expression at low temperature (15°C), (5) mild flash freezing and thawing to lyse cells, and (6) protease inhibitor cocktail (Roche). These strategies gave reasonable yields of about 10mg purified protein per liter *E. coli* cells and significantly countered degradation during expression and purification.

Optimized codons to express the NTD and DBD conjoined two-domain constructs for 8 GR translational isoforms with 10 tandem histidines tagged on both the N- and C-termini were synthesized by DNA 2.0 (Menlo Park, CA) and inserted into the pJ411 bacterial expression vector under T7 promoter control. BL21(DE3)pLysS competent cells (Novagen) transformed with expression plasmids were plated on LB plates with 30μg/mL kanamycin and incubated at 37°C for 24h. Single colonies were selected, inoculated into 50mL of LB medium with 30μg/mL kanamycin, and grown overnight at 30°C with shaking at 250rpm. 10mL of each overnight culture was transferred to 500mL of LB medium and grown at 37°C to an A_{600} of 0.8. The temperature was adjusted to 15°C, and after 1h, the culture was induced by 1m*M* isopropyl β-D-thiogalactopyranoside (American Bioanalytical). After 18h at 15°C with shaking at 250rpm, the *E. coli* cells were collected by centrifugation for 15min at 5000rpm at 4°C, and the pellets were washed with cold PBS (137m*M* NaCl, 7m*M* Na_2HPO_4, 3m*M* KCl, 1.4m*M* KH_2PO_4) and recentrifuged. To the cell pellets from a 2-L culture was added 50mL of lysis buffer (100m*M* monosodium phosphate, 10m*M* Tris, 20m*M* imidazole, pH 8.0) plus one complete EDTA-free protease inhibitor mixture tablet (Roche Applied Science). Cells were lysed by flash freezing twice in a dry ice and ethanol mixture for 20 min and then thawed for 10 min in a 42°C water bath. The cell lysate was centrifuged at 30,000 × *g* at 4°C for 1 h. The supernatant from 4 L of induced cells was loaded by gravity flow onto an FPLC column packed with 20 mL of nickel–nitrilotriacetic acid Superflow resin (Qiagen) preequilibrated with the lysis buffer. After thorough washing the column with the lysis buffer at a flow rate of 2 mL/min, protein was renatured on the column by flowing through lysis buffer (100 m*M* monosodium phosphate, 10 m*M* Tris, 20 m*M* imidazole, pH 8.0), target protein was eluted from the column by native lysis buffer containing 500 m*M* imidazole. Protein purity was checked by SDS-PAGE. The purified target protein was dialyzed against storage buffer (10 m*M* HEPES, 80 m*M* NaCl, 1 m*M* EDTA, 10% glycerol, pH 7.6). Protein concentration was then determined by A_{280} according to the Edelhoch method (Pace, Vajdos, Fee, Grimsley, & Gray, 1995).

3.3 Binding Affinity Measurement Monitored by Fluorescence Anisotropy Change

In the binding experiment, 6FAM labeled GRE was kept at constant concentration. And concentration of the GR construct was gradually increased. In the absence of GR, the GRE is in the free form and tumbling faster in the solution, so smaller fluorescence anisotropy is observed. In the presence of a saturating concentration of GR, the GRE is mostly in the bound complex and tumbles slower in solution with a larger anisotropy. Monitoring the fluorescence anisotropy change as increasing GR protein concentration allows us to monitor the binding process. The binding experiments were carried out in the following buffer: 10 m*M* HEPES (pH 7.6), 80 m*M* NaCl, 1 m*M* EDTA, 5 m*M* $MgCl_2$, 1 m*M* DTT, 200 μg/mL BSA, and 5 μ*M* control oligonucleotide. For each data point, 25 n*M* 6FAM labeled half-site GRE was incubated with 0 μ*M* up to 10 μ*M* of GR two-domain construct at 22°C for 30 min. Fluorescence anisotropy for mixtures at each GR concentration was measured using an Aviv ATF 105 fluorometer equipped with polarizers. The GR construct and 6FAM labeled half-site GRE mixture (150 μL) was allowed to rest in a "submicro" fluorometer cell (Starna Cells) at 22°C for 2 min to allow for temperature stabilization and then the mixture was excited at a wavelength of 495 nm. Anisotropy at 521 nm was recorded as a function of GR construct concentration.

As shown in Fig. 3, the eight GR translational isoforms have variable binding affinities to GRE, which varied approximately fivefold and ranged from 8.4×10^5 to 4.2×10^6 M^{-1} corresponding to the binding affinities of

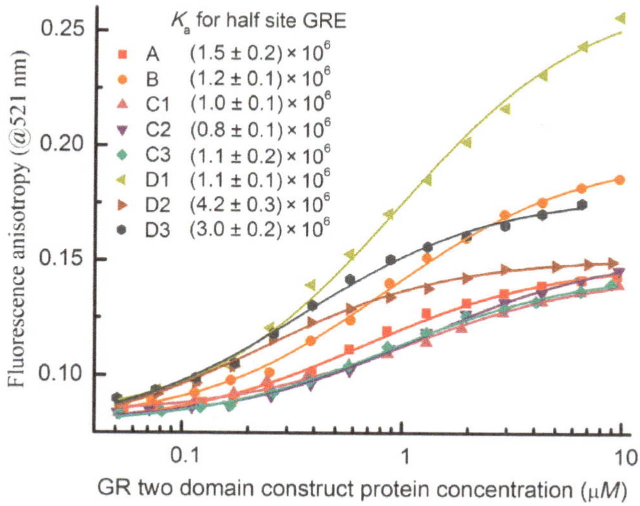

Fig. 3 Binding affinity of eight translational isoforms to half-site GRE by Fluorescence Polarization assay. Fluorescence anisotropy of the 6FAM labeled half-site GRE (5′-gcgcAGAACAggagcgc-3′) as a function of GR translational isoform concentration was measured for each construct. Curves represent fits to the data with a single site binding model. Binding affinity was fitted with nonlinear least squares method to the raw data with the following equation, $F(C_{GR}) = A_{free} + (A_{bound} - A_{free}) \cdot \frac{K_a \cdot C_{GR}}{1 + K_a \cdot C_{GR}}$, where $F(C_{GR})$ is the observed anisotropy value at certain GR two domain constructs concentration, C_{GR}; A_{free} is the anisotropy value of the free GRE oligonucleotide; A_{bound} is the anisotropy value of the GRE oligonucleotide in complex with GR; K_a is the binding affinity between the half-site GRE and GR construct. The fitting parameters are A_{free}, A_{bound}, and K_a.

C2 and D2 isoforms, respectively. The two-domain constructs of the eight translational isoforms all contain the same intact DBD to bind to the GRE. The variable binding affinity to the GRE indicates that the NTD of different translational isoforms allosterically modulates DBD-binding affinity to GRE. The coupling between NTD and DBD is consistent with previous reports that the GRE is an allosteric effector of GR (Geserick, Meyer, & Haendler, 2005; Gronemeyer & Bourguet, 2009; Meijsing et al., 2009; Watson et al., 2013).

4. STABILITY MEASUREMENT OF THE ID F-DOMAIN ALONE AND WITH EITHER THE R-DOMAIN OR THE DBD OR BOTH OF THEM

To determine the stability, i.e., the free energy difference between the unfolded and folded conformation, of the ID domain of the GR isoforms, we carried out thermodynamic characterizations. The naturally occurring osmolyte trimethylamine N-oxide (TMAO) was employed to induce the folding transitions. Organic osmolytes are small molecules in cells and function to stabilize and protect intracellular proteins against common denaturing environmental stresses (Bolen, 2001; Bolen & Baskakov, 2001). TMAO has been demonstrated to force thermodynamically unstable proteins to fold and regain high functional activities, provided there is a folded state in its functional conformational manifold (Baskakov et al., 1999; Kumar et al., 1999; Kumar, Lee, Bolen, & Thompson, 2001). We characterized the TMAO-induced folding transition of the intrinsically disordered F-domain, either in the presence or in the absence of its tandem coupled R-domain and the DBD.

4.1 Protein Expression, Purification, and Storage

The expression and purification of the single NTD of GR A isoform (i.e., R-F) and C3 isoform (i.e., F) are the same as described in (Section 3) for the NTD and DBD conjoined two-domain constructs, except that cell lysis was performed in the denaturing lysis buffer (6 *M* guanidine hydrochloride, 100 m*M* monosodium phosphate, 10 m*M* Tris, 20 m*M* imidazole, pH 8.0) and renatured on the Ni-column with the native lysis buffer.

4.2 Preparation of Trimethylamine N-oxide (TMAO) Buffers

Trimethylamine N-oxide dihydrate (TMAO, 98% pure, Acros Organics) was dissolved in 100 m*M* Tris, 200 m*M* NaCl, 50 m*M* arginine buffer to make the 0, 0.5, 1, 1.5, 2, and 2.5 *M* TMAO buffers. pH was adjusted to 7.4 for each buffer separately. Activated carbon, 12–20 mesh (Sigma-Aldrich) was added into each buffer solution and stirred for 30 min while protected from light to absorb small traces of fluorescent/fluorescence quenching impurities. The buffer was then filtered (0.22 μm filter, Millipore), aliquoted, and stored at −80°C. TMAO buffers with similar concentrations were mixed to obtain the target TMAO concentration used in the following fluorescence measurements, minimizing the pH changes due to mixing two TMAO buffers with equal pH but large concentration difference. Arginine, which is reported to aid protein solubility (Tsumoto et al., 2004), was included in the TMAO buffers at a concentration of 50 m*M* and efficiently prevented protein aggregation.

4.3 TMAO-Induced Protein Folding Transition Monitored by Tryptophan Emission Fluorescence Intensity

Fluorescence emission spectrum for each construct was measured using an Aviv ATF 105 fluorometer (Aviv Biomedical) in TMAO buffers of varying concentrations. Freshly prepared dithiothreitol 1 *M* stock solution was added to TMAO buffer to a final concentration of 1 m*M*. Protein at 0.5 μ*M* in a volume of 150 μL TMAO buffer at 22°C was allowed to rest in a "submicro" fluorometer cell for 5 min to allow for temperature stabilization and protein conformational equilibrium to be reached. Emission spectra were then recorded with excitation at 295 nm. All spectra were corrected for the contribution of its corresponding buffer. Fluorescence emission intensities at 338 nm were recorded as a function of TMAO concentration and normalized to the intensity at 0 *M* TMAO concentration. The resulting sigmoidal curve was fitted to a two-state cooperative folding transition with linear extrapolation method (LEM; Santoro & Bolen, 1988) to determine the stability ($\Delta G_{F \rightarrow U}$) and *m*-value for each construct. The LEM method has been shown to be valid for fitting the data of osmolyte induced folding of ID proteins (Wu & Bolen, 2006).

With excitation at 295 nm, tryptophan emission spectra were measured for F, R-F, F-DBD, and R-F-DBD (construct scheme shown in Fig. 1A) in buffers with different TMAO concentrations. All four constructs contain two tryptophans corresponding to the tryptophans in full-length NTD at residues 213 and 364. As shown in Fig. 4A inset, in the absence of TMAO, the maximal emission wavelength is at 348 nm with excitation at 295 nm for F construct (the same case was observed for the other constructs, data not shown), indicating the tryptophans are solvent exposed in these ID proteins. In the presence of 2.2 *M* TMAO, the emission wavelength maxima show a significant blue shift (338 vs 348 nm), indicating that TMAO induced a more ordered conformation in these disordered proteins. Fluorescence intensities at 338 nm as a function of TMAO concentration for each construct were fitted successfully to the cooperative two-state transition model using LEM (Santoro & Bolen, 1988), with F construct folding as an example shown in Fig. 4A. Free energies, $\Delta G_{F \rightarrow U}$, representing the stability of the unfolded state with folded state as reference, and *m*-value, representing the cooperativity of the transition, were tabulated in Fig. 4C. The population of the folded state (P_f) can be calculated from the distances of a measured data point to the unfolded state baseline and the folded state baseline (D_u, D_f), as illustrated in Fig. 4A. P_f values as a function of TMAO concentration for all constructs were shown in parallel in Fig. 4C.

A

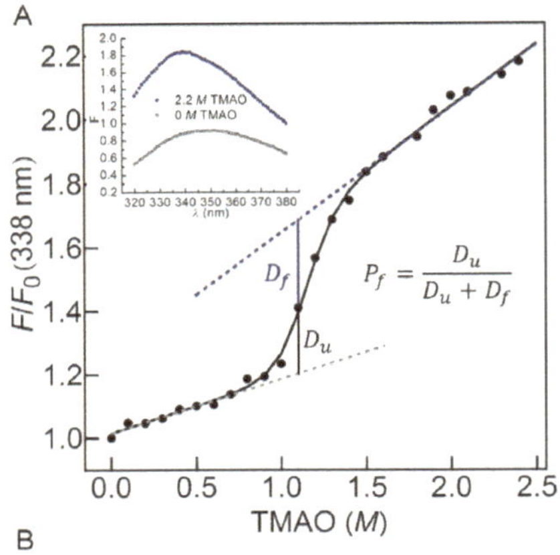

C

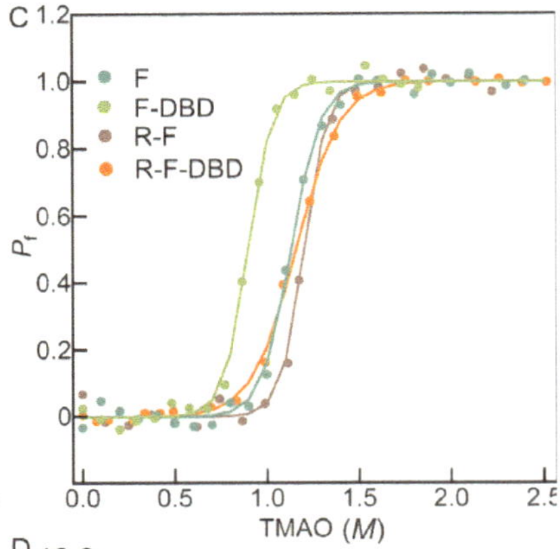

B

	$\Delta G_{F\rightarrow U}$ (kcal/mol)	*m*-value (kcal/(mol*M))
F	−7.6±0.3	6.7±0.3
F-DBD	−6.7±0.4	7.0±0.5
R-F	−10.1±0.9	8.8±0.4
R-F-DBD	−5.7±0.4	4.9±0.3

D

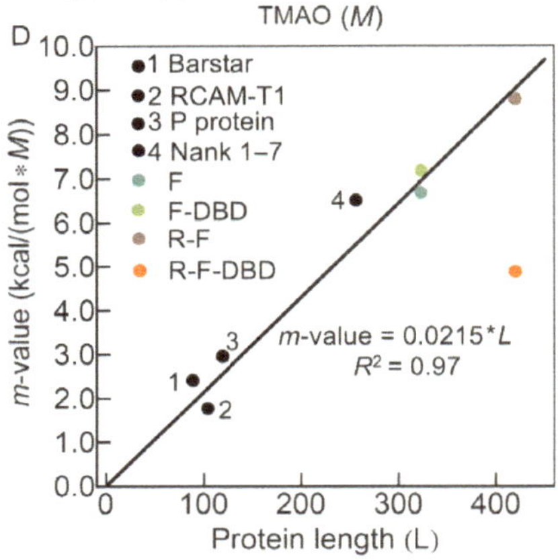

Fig. 4 TMAO-induced conformational transitions of GR F-domain, F-DBD, R-F, and R-F-DBD monitored by tryptophan emission fluorescence intensity. (A) TMAO-induced folding transition on F-domain. Tryptophan emission fluorescence intensity at 338 nm at each TMAO concentration was normalized to that at 0 *M* TMAO concentration. Nonlinear least square fit to two-state transitions using the linear extrapolation method was employed using the following fitting function:

$$F\left(C_{\mathrm{TMAO}}\right)=\frac{\left(b_F+s_F\cdot C_{\mathrm{TMAO}}\right)\cdot \mathrm{Exp}\left(\dfrac{\Delta G^0_{U\rightarrow F}+m\cdot C_{TMAO}}{R\cdot T}\right)+\left(b_U+s_U\cdot C_{\mathrm{TMAO}}\right)}{1+\mathrm{Exp}\left(\dfrac{\Delta G^0_{U\rightarrow F}+m\cdot C_{\mathrm{TMAO}}}{R\cdot T}\right)}$$

, where $F(C_{\mathrm{TMAO}})$ is the normalized fluorescence intensity at a certain TMAO concentration, C_{TMAO}; s_U and b_U are the slope and intercept of the baseline for the unfolded state, respectively; s_F and b_F are the slope and intercept of the baseline for the folded state, respectively; $\Delta G^0_{F\rightarrow U}$ is the stability of the unfolded state with folded state as reference in the absence of TMAO; *m* represents the cooperativity of the transition; *R* is the gas constant; and *T* is the temperature in Kelvin. Experimentally determined F (C_{TMAO}) values at different C_{TMAO} values were used to fit the following parameters: $\Delta G^0_{F\rightarrow U}$, m, s_U, b_U, s_F, b_F. Fitted baseline for the unfolded state (*dashed gray line*) and the folded state (*dashed blue line*)

(Continued)

One important thermodynamic parameter from the application of LEM (Santoro & Bolen, 1988) to osmolyte-induced protein folding is the *m*-value, which is defined as follows (Santoro & Bolen, 1988): $\Delta G_{F\rightarrow U}(C_{TMAO}) = \Delta G^0_{F\rightarrow U} + m \cdot C_{TMAO}$. $\Delta G_{F\rightarrow U}(C_{TMAO})$ is the free energy of the unfolded state with the folded state as reference at a certain TMAO concentration, and $\Delta G^0_{F\rightarrow U}$ is free energy of the unfolded state with the folded state as reference in the absence of TMAO. The *m*-value, which has the unit of kcal/(mol * *M*), represents the molar efficacy of osmolyte in folding a protein, and also measures the cooperativity of the transition between the unfolded and folded state (Auton & Bolen, 2007). For a cooperative two-state transition induced by a particular osmolyte, the *m*-value is proportional to the surface area buried upon protein folding, which, in turn, is generally proportional to the protein size (Auton & Bolen, 2007). The linear relationship between *m*-values and globular protein chain lengths has been reported for denaturant-induced protein unfolding (Ghosh & Dill, 2009). In Fig. 4D, the reported *m*-values in TMAO-induced folding for Barstar (Pradeep & Udgaonkar, 2004), RCAM-T1 (Baskakov & Bolen, 1998), P protein (Henkels, Kurz, Fierke, & Oas, 2001), and Nank1–7 (Mello & Barrick, 2003) are shown as a function of protein length. For comparison, *m*-values of TMAO-induced folding of F-domain, F-DBD, R-F, and R-F-DBD constructs determined in this study are also depicted. Seven data points, including Barstar, RCAM-T1, P protein, Nank1–7, F, F-DBD, and R-F could be fitted to a line with correlation coefficient $R^2 = 0.97$. Two important points emerge from this observation. First, it indicates that

Fig. 4—cont'd was shown. The distance of observable at certain TMAO concentration to the folded state baseline and the unfolded state baseline can be used to calculate the fraction of the folded state as illustrated. Inset, F-domain fluorescence emission spectrum with excitation at 295 nm in 0 *M* TMAO with λ_{max} at 348 nm (*gray dots*) and in 2.2 *M* TMAO with λ_{max} at 338 nm (*blue dots*). (B) Tabulation of the determined $\Delta G_{U\rightarrow F}$, representing the stability of the folded state with unfolded state as reference, and m-value, representing the cooperativity of the transition for each construct. Errors representing the errors from nonlinear least square fit. (C) Fraction of the folded state as a function of TMAO concentration was calculated from the *folding curve* as illustrated in (A) for each construct. (D) Dependence of *m*-values on protein chain length (*L*) for TMAO-induced folding transitions. Shown are the *m*-values fitted from TMAO-induced two-state folding transitions for Barstar (*black dot* 1), RCAM-T1 (*black dot* 2), P protein (*black dot* 3), Nank 1–7 (*black dot* 4), GR's F-domain (*cyan dot*), F-DBD (*green dot*), R-F (*gray dot*), and R-F-DBD (*orange dot*). All data points except for the values for R-F-DBD can be well fit to a line; i.e., *m*-value $= 0.0215 * L$ with a correlation coefficient of $R^2 = 0.97$.

GR's F-domain, F-DBD, and R-F undergo cooperative transitions with increasing TMAO concentration, only the fully unfolded state and the fully folded state are significantly populated during the transition. Second, the *m*-values for the F-domain, R-F, and F-DBD and *m*-values for the globular proteins Barstar, RCAM-T1, P protein, Nank 1–7, (which are in the unfolded state under physiological conditions by destabilizing mutations) fall on the same line, demonstrating that F-domain, R-F, and F-DBD are able to adopt native folds that are thermodynamically similar in terms of surface area burial (per unit length) to those adopted by globular protein. Significant deviation from the fitted linear line of the *m*-value for the R-F-DBD construct (Fig. 4D) suggests that it is not a cooperative two-state transition in TMAO-induced folding, some other states other than all the R-domain, F-domain, and DBD in the unfolded and all of them in the folded states are populated during folding.

5. DIRECT CHARACTERIZATION OF THE COUPLING BETWEEN R-DOMAIN AND DBD IN GR

The coupling between the R-domain and the DBD was assessed by comparing the DNA-binding affinity of various constructs that connect the DBD and the R-domain of each isoform. In order to only measure the coupling between the R-domain and the DBD, and to avoid the convoluting effects associated with the coupling between the F-domain and the R-domain and DBD, constructs were generated that utilized a flexible linker connecting the R-domain to the DBD (Fig. 5A, competitor). These constructs do not have the F-domain which is indispensable for transcription activation, thus they are not active transcription regulators; however, they have the DNA-binding domain, which could occupy the response element in the promoter of the luciferase codons and thus compete with the active transcriptional regulator for binding. As such, a competitive transcription assay was developed to compare the DNA-binding affinities of DBD when linked to the R-domains with different length.

5.1 Mammalian Expression Vectors for the R-Domain-Linked GR DBD and Gal4 DBD Construct and Luciferase Reporters

Amino acids 1–97 of GR were linked to GR DBD with an 11 amino acid, flexible linker (GTGGSGGSGGS). This plasmid was constructed by a PCR that deleted the codons for GR 98–420 from the plasmid coding A isoform NTD and DBD in PJ603 vector backbone (described in Section 2) with

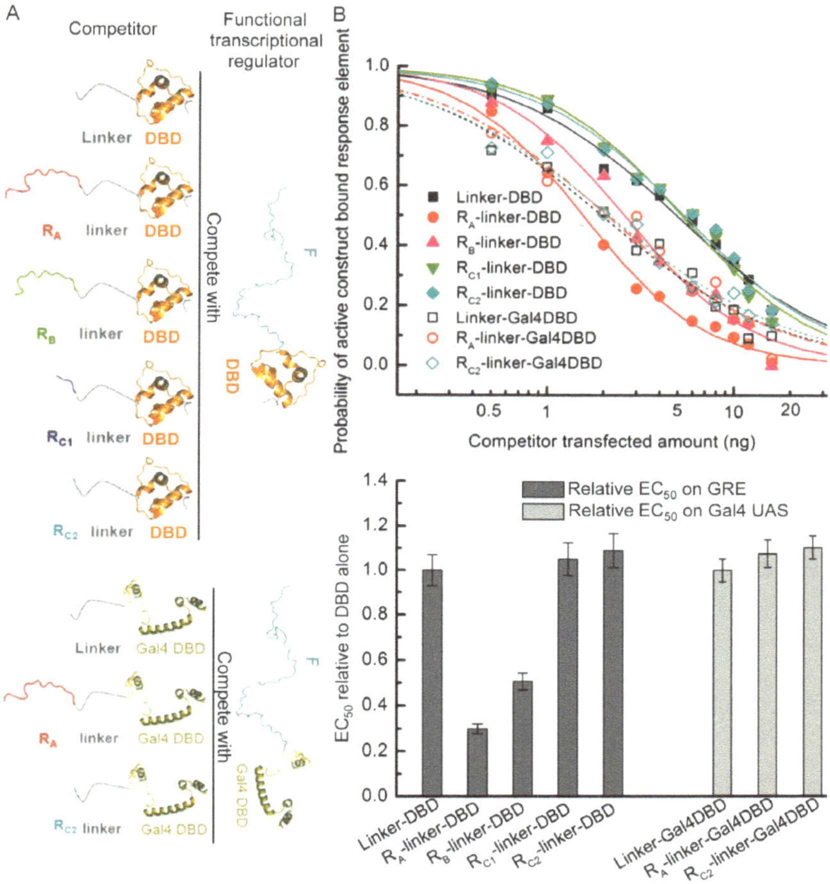

Fig. 5 Competitive transcriptional assay to check the coupling between R-domain and DBD in GR. (A) Schematic representation of competitive transfection assay design for constructs of the R-domains of isoforms A, B, C1, and C2 linked to GR DBD or Gal4 DBD. (B) Experimental competitive transfection assay for constructs shown in (A). The competitive binding data were fitted to dosage response curved represented by the following equation, $F\left(C_{\text{competitor}}\right) = A_{\text{Active}} - \frac{A_{\text{Active}} - A_{\text{Competitor}}}{1+\left(EC_{50}/C_{\text{competitor}}\right)^{p}}$, where $F(C_{\text{competitor}})$ is the observed luciferase expression level in transfection with competitor concentration $C_{\text{competitor}}$; A_{Active} is the expression level of Gaussia luciferase under the control of the functional transcriptional regulator as shown in (A); $A_{\text{Competitor}}$ is the expression level of Gaussia luciferase under the control of the competitor; EC_{50} is the amount of competitor vector transfected when the Gaussia luciferase expression level is half of the sum of A_{active} and $A_{\text{Competitor}}$; p is the Hill coefficient. In this equation, p is a practical value introduced in the fitting equation, and its purpose is the same as described in the caption of Fig. 1. In the data fitting, A_{active}, $A_{\text{Competitor}}$, EC_{50}, and p are fitting parameters. (C) The EC_{50} values of competitors shown in (A) were normalized to that of the linker-DBD construct or linker-Gal4 DBD construct and compared in the histogram.

*Bam*HI and *Kpn*I ends for ligation of an oligonucleotide coding for GTGGSGGSGGS. Plasmids coding R_B-linker-DBD, R_{C1}-linker-DBD, R_{C2}-linker-DBD, and linker-DBD were made by amplifying the appropriate codons from R_A-linker-DBD construct and inserting them into the *Nhe*I and *Xho*I site of PJ603 vector. Codons to express R_A-linker-Gal4DBD and GR F-Gal4 DBD constructs were optimized by DNA2.0 and inserted into the *Nhe*I and *Xho*I site of PJ603 vector. Plasmid coding R_{C2}-linker-Gal4DBD and linker-Gal4 DBD was made by amplifying the appropriate codons from R_A-linker-Gal4DBD plasmid.

Plasmid GRE4half-Gluc, to express secreted *Gaussia* luciferase under control of four tandem half-site GREs, was described in Section 2. Plasmid Gal4UAS-Gluc, to express secreted Gaussia luciferase under the control of two tandem Gal4 upstream activating sequences (Gal4 UAS), was made by inserting an oligonucleotide containing two tandem Gal4 UAS, 5′-aattCGGAGTACTGTCCTCCG aggaattcagCGGAGTACTGTCCTCCG-3′, into the *Eco*RI and *Bam*HI sites of the pGluc-miniTK vector (NEB). Plasmid pCluc-miniTK2 vector (NEB) to express *Cypridina* luciferase was utilized as an internal control in the cotransfection to account for cell density differences and transfection efficiency differences in each well.

5.2 Competitive Transfection Assay to Compare the Relative DNA-Binding Affinity

For the competitive transfection assay to compare the binding affinities of R-domain-linked GR DBD constructs, 40 ng of GRE4half-Gluc, 40 ng of pCluc-miniTK2, 3 ng of mammalian expression vector for NTD and DBD of C3 isoform (F-DBD), and up to 16 ng of plasmids coding for the competitors (including R_A-linker-DBD, R_B-linker-DBD, R_{C1}-linker-DBD, R_{C2}-linker-DBD, and linker-DBD shown in Fig. 5A) were cotransfected into U-2 OS cells on 96-well plates. After 48 h, the luciferase expression level was measured as described in Section 2.

As shown in Fig. 5B, linking the R domain of the A or B isoform to the DBD significantly increases the binding affinity of the DBD to GRE. The R_A-linker-DBD construct and R_B-linker-DBD constructs have significantly lower EC_{50} values compared to the linker-DBD construct in the competitive transfection assay. In contrast, linking the R domain of the C1 or C2 isoform to DBD does not significantly change the binding affinity of DBD to GRE, as the R_{C1}-linker-DBD and R_{C2}-linker-DBD constructs show similar EC_{50} values to the linker-DBD construct. As control, we also linked R-domains of A isoform and C2 isoform to the DBD of yeast transcription factor Gal4 and created R_A-linker-Gal4 DBD, R_{C2}-linker-Gal4 DBD, and linker-Gal4 DBD constructs. With these as competitors for the chimeric transcriptional regulator, GR F-domain-linked Gal4 DBD (GR F-Gal4 DBD), competitive transcription assay was carried out to monitor the luciferase expression level from Gal4UAS–Gluc vector (Fig. 5A, lower panel). There were indistinguishable capacities of the R_A-linker-Gal4 DBD, R_{C2}-linker-Gal4 DBD, and linker-Gal4 DBD constructs to compete off GR F-Gal4 DBD (Fig. 5B). These results support the notion that the R-domain affects DNA binding by the GR DBD specifically through stabilization of a high-affinity state of the GR DBD, and is not simply a consequence of a direct interaction between the R-domain and the DNA. In summary, the segment 1–85, which is unique to A and B isoforms, is involved in the favorable coupling network between R and DBD. Therefore, changing the length of the R domain in the transcriptionally active isoforms regulates the coupling energetics between the R domain and DBD.

6. UTILIZING THE EAM TO QUANTITATIVELY UNDERSTAND GR TRANSCRIPTION ACTIVATION

6.1 Building EAM for Each GR Isoform

An ensemble of states was constructed for each of the active isoform (A, B, C1, C2, and C3), enumerating all possible combinations of the DBD being in the high-affinity or low-affinity states, and the R- and F-domains being in their active (folded) or inactive (unfolded) states. For example, in A isoform (composed of the R-domain, the F-domain, and the DBD) and the most active C3 isoform (composed only of the F-domain and the DBD), the EAM produces 8 and 4 states in their respective ensembles, representing all combinations (Fig. 6A). Our choice of model is justified because the R- and F-domains can fold to globular protein-like structures (Fig. 4), consistent with the notion of coupled folding and binding for IDPs, where the folded conformation is the active form (Dyson & Wright, 2002).

In the absence of allosteric effector, which is the GRE in our case, the stability (ΔG_i) of the i state is determined by the intrinsic stability of each domain, ΔG_R, ΔG_F, and ΔG_D (the stability difference between unfolded state/low-affinity state and the folded state/high-affinity state), as well as and the interaction energy between the folded states (or high-affinity states) of the domains, Δg_{F-R}, Δg_{F-DBD}, Δg_{R-DBD} (shown in Fig. 6A) (Hilser & Thompson, 2007; Hilser, Wrabl, & Motlagh, 2012). Among these

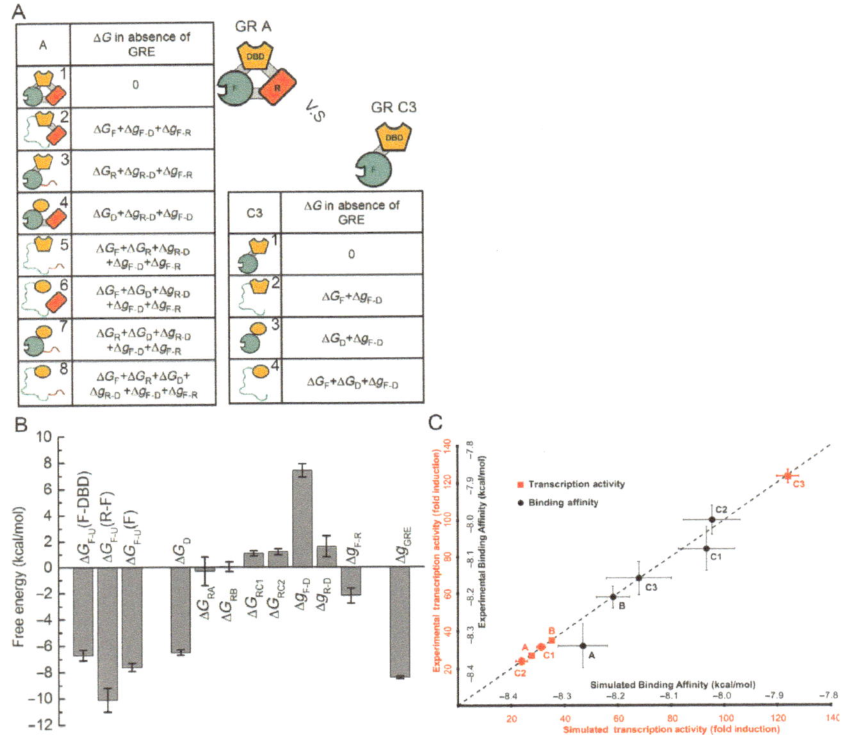

Fig. 6 EAM could quantitatively explain the uncorrelation between transcriptional activity and binding affinity. (A) EAM for GR A isoform and C3 isoform. (B) Histogram of the thermodynamic parameters from EAM that can recapitulate the DNA-binding affinity and transcriptional activity of GR A, B, C1, C2, and C3 isoforms. (C) The model recapitulates the DNA-binding affinity and relative transcriptional activity of A, B, C1, C2, and C3 isoforms. Error bars on experimental data represent uncertainty of the individual fits. Error bars on simulated data are average results from propagation of experimental error through the model.

parameters, ΔG_F, ΔG_D, and Δg_{F-DBD} are the same in all the five active isoforms as they have the same DBD and F-domain; ΔG_R, Δg_{F-R}, and Δg_{R-DBD} can be specific to A, B, C1, and C2 isoforms, as they have different length of the R-domain. Based on the free energy of the *i* state, ΔG_i, the statistical weight, SW, can be calculated using $SW_i = Exp(-\Delta G_i/RT)$. Sum of the statistical weights of all states in the ensemble results in the partition function, Q, for the system, which is defined by $Q = \sum_i SW_i$. Consequently, the probability of each state, P_i, can be calculated by dividing each state's statistical weight with the partition function Q for the ensemble, i.e., $P_i = SW_i/Q$.

In the presence of allosteric effector GRE, the state with high-affinity DBD state will be stabilized and the partition function of the ensemble will also be changed. The binding energy of DBD in high-affinity state to GRE is defined as $\Delta g_{\mathrm{GRE}} = -RT\ln(K_a^{\mathrm{DBD}} \cdot C_{\mathrm{GRE}})$, where K_a^{DBD} is the affinity of the high-affinity DBD conformation to GRE, and C_{GRE} is the concentration of the allosteric effector GRE. In such case, the statistical of state *i* with high affinity to DBD will be changed to, $\mathrm{SW}_i^{\mathrm{GRE}} = \mathrm{SW}_i \cdot (1 + \mathrm{Exp}(-\Delta g_{\mathrm{GRE}}/RT)$; and the statistical weight of the state with low-affinity DBD would not be altered. Consequently the partition function of the ensemble in the presence of GRE, Q^{GRE}, can be calculated as sum of the statistical weight of all the states; the probability of each state in the presence of GRE, P_i^{GRE}, can be calculated as its statistical weight divided by the partition function.

6.2 Constraints Utilized in the EAM

The following experimental constraints are utilized when building the EAM for the five active GR isoforms: (1) The measured TMAO refolding free energies ($\Delta G^0_{\mathrm{F}\rightarrow\mathrm{U}}$) for F, F-DBD, and R-F-DBD constructs (Fig. 4B) are related to the stability difference between the unfolded state (or low-affinity state) and folded state (or high-affinity state) of F-domain (ΔG_{F}), R-domain (ΔG_{R}), DBD (ΔG_{D}), and the interaction energy between the folded states (or high-affinity states) of them, $\Delta g_{\mathrm{F\text{-}R}}$, $\Delta g_{\mathrm{F\text{-}D}}$, $\Delta g_{\mathrm{R\text{-}D}}$ by the following equations:

$$\Delta G^0_{\mathrm{F}\rightarrow\mathrm{U}}(\mathrm{F}) = \Delta G_{\mathrm{F}} \quad (1)$$

$$\Delta G^0_{\mathrm{F}\rightarrow\mathrm{U}}(\mathrm{F\text{-}DBD}) = (\Delta G_{\mathrm{F}} + \Delta G_{\mathrm{D}} + \Delta g_{\mathrm{F\text{-}D}}) \quad (2)$$

$$\Delta G^0_{\mathrm{F}\rightarrow\mathrm{U}}(\mathrm{R\text{-}F}) = (\Delta G_{\mathrm{F}} + \Delta G_{\mathrm{R}} + \Delta g_{\mathrm{F\text{-}R}}) \quad (3)$$

(2) The interaction energies between the DBD and either the R-domain of C2 or C1 are equal to 0 kcal/mol ($\Delta g_{\mathrm{RC1\text{-}D}} = 0$, $\Delta g_{\mathrm{RC2\text{-}D}} = 0$), as the linking the respective R-domain to DBD did not significantly influence the binding affinity of DBD to the DNA (shown in Fig. 5B); (3) The coupling energy between the DBD and the R-domain of B isoform is 0.7 times the coupling energy between the DBD and the R-domain of A isoform,

$$\Delta g_{\mathrm{RB\text{-}D}} = 0.7\Delta g_{\mathrm{RA\text{-}D}} \quad (4)$$

based on the EC_{50} values of the RA-linker-DBD and the RB-linker-DBD constructs in the competitive transcription assay (Fig. 5B); (4) The coupling energy between the R-domain of A, B, C1, and C2 isoform and F are modeled as a single parameter, $\Delta g_{\mathrm{F\text{-}R}}$, as the eight amino acids between

90 and 98 (shared residues of A, B, C1, and C2 isoforms) seem to contribute most of the coupling between the R-domain and F-domain (Fig. S1 in the online version at https://doi.org/10.1016/bs.mie.2018.09.004).

6.3 Global Minimization Method to Simulate Experimental Data

EAM was employed to estimate reasonable parameters that reproduce experimental data for human GR transcriptional activity and DNA-binding affinity. The EAM is constrained by 10 data points measured from ex vivo transcription assays (Fig. 2E) and in vitro binding affinities (Fig. 3) measured on five active GR isoforms A, B, C1, C2, and C3. In the EAM, the experimentally observed transcriptional activity is represented by the summed probabilities of states whose folded F-domain cooccurs with the high-affinity DBD conformation (e.g., state 1 and 3 in A isoform ensemble, and state 1 in C3 isoform ensemble, Fig. 6A). Similarly, DNA-binding affinity is represented by the summed probabilities of states wherein the DBD is in the high-affinity conformation (e.g., state 1 and 3 in A isoform ensemble, and state 1 in C3 isoform ensemble, Fig. 6A).

Eight thermodynamic parameters are estimated from the global minimization: ΔG_{D}, $\Delta g_{\mathrm{RA\text{-}D}}$, Δg_{GRE}, ΔG_{RB}, ΔG_{RC1}, ΔG_{RC2}, $\Delta g_{\mathrm{F\text{-}R}}$, and a scaling parameter A (necessary to convert EAM population to the experimental observable of transcription activity). With the set of thermodynamic parameters satisfying the EAM (shown in Fig. 6B), simulated transcriptional activity and binding energy to GRE was calculated for each GR isoform and compared to the experimental data (Fig. 6C).

7. SUMMARY

This chapter suggests a framework in the methodology to study allostery, especially in IDPs. The combination of in vitro biophysical studies, cell-based functional assays, and EAM makes it possible to study a complicated allosteric system composed of long ID regions, as well as structured domains. The EAM quantitatively captures the variable transcriptional activity and binding affinity of the constitutively active constructs of the eight GR translational isoforms. The approach explained here offers a unifying strategy to investigate thermodynamics in ID region mediated allostery.

REFERENCES

Auton, M., & Bolen, D. W. (2007). Application of the transfer model to understand how naturally occurring osmolytes affect protein stability. *Methods in Enzymology, 428*, 397–418.

Baskakov, I., & Bolen, D. W. (1998). Forcing thermodynamically unfolded proteins to fold. *The Journal of Biological Chemistry, 273*, 4831–4834.

Baskakov, I. V., Kumar, R., Srinivasan, G., Ji, Y. S., Bolen, D. W., & Thompson, E. B. (1999). Trimethylamine N-oxide-induced cooperative folding of an intrinsically unfolded transcription-activating fragment of human glucocorticoid receptor. *The Journal of Biological Chemistry, 274*, 10693–10696.

Bolen, D. W. (2001). Protein stabilization by naturally occurring osmolytes. *Methods in Molecular Biology, 168*, 17–36.

Bolen, D. W., & Baskakov, I. V. (2001). The osmophobic effect: natural selection of a thermodynamic force in protein folding. *Journal of Molecular Biology, 310*, 955–963.

Changeux, J. P., & Edelstein, S. J. (2005). Allosteric mechanisms of signal transduction. *Science, 308*, 1424–1428.

Chen, H., Srinivasan, G., & Thompson, E. B. (1997). Protein-protein interactions are implied in glucocorticoid receptor mutant 465*-mediated cell death. *The Journal of Biological Chemistry, 272*, 25873–25880.

Dahlman-Wright, K., Almlof, T., McEwan, I. J., Gustafsson, J. A., & Wright, A. P. (1994). Delineation of a small region within the major transactivation domain of the human glucocorticoid receptor that mediates transactivation of gene expression. *Proceedings of the National Academy of Sciences of the United States of America, 91*, 1619–1623.

Dyson, H. J., & Wright, P. E. (2002). Insights into the structure and dynamics of unfolded proteins from nuclear magnetic resonance. *Advances in Protein Chemistry, 62*, 311–340.

Ford, J., McEwan, I. J., Wright, A. P., & Gustafsson, J. A. (1997). Involvement of the transcription factor IID protein complex in gene activation by the N-terminal transactivation domain of the glucocorticoid receptor in vitro. *Molecular Endocrinology, 11*, 1467–1475.

Gao, J., & Xu, D. (2012). Correlation between posttranslational modification and intrinsic disorder in protein. *Pacific Symposium on Biocomputing*, 94–103.

Geserick, C., Meyer, H. A., & Haendler, B. (2005). The role of DNA response elements as allosteric modulators of steroid receptor function. *Molecular and Cellular Endocrinology, 236*, 1–7.

Ghosh, K., & Dill, K. A. (2009). Computing protein stabilities from their chain lengths. *Proceedings of the National Academy of Sciences of the United States of America, 106*, 10649–10654.

Griekspoor, A., Zwart, W., Neefjes, J., & Michalides, R. (2007). Visualizing the action of steroid hormone receptors in living cells. *Nuclear Receptor Signaling, 5*, e003.

Gronemeyer, H., & Bourguet, W. (2009). Allosteric effects govern nuclear receptor action: DNA appears as a player. *Science Signaling, 2*, pe34.

Henkels, C. H., Kurz, J. C., Fierke, C. A., & Oas, T. G. (2001). Linked folding and anion binding of the *Bacillus subtilis* ribonuclease P protein. *Biochemistry, 40*, 2777–2789.

Hilser, V. J., & Thompson, E. B. (2007). Intrinsic disorder as a mechanism to optimize allosteric coupling in proteins. *Proceedings of the National Academy of Sciences of the United States of America, 104*, 8311–8315.

Hilser, V. J., & Thompson, E. B. (2011). Structural dynamics, intrinsic disorder, and allostery in nuclear receptors as transcription factors. *The Journal of Biological Chemistry, 286*, 39675–39682.

Hilser, V. J., Wrabl, J. O., & Motlagh, H. N. (2012). Structural and energetic basis of allostery. *Annual Review of Biophysics, 41*, 585–609.

Koshland, D. E., Jr., Nemethy, G., & Filmer, D. (1966). Comparison of experimental binding data and theoretical models in proteins containing subunits. *Biochemistry, 5*, 365–385.

Kumar, R., Baskakov, I. V., Srinivasan, G., Bolen, D. W., Lee, J. C., & Thompson, E. B. (1999). Interdomain signaling in a two-domain fragment of the human glucocorticoid receptor. *The Journal of Biological Chemistry*, *274*, 24737–24741.

Kumar, R., Lee, J. C., Bolen, D. W., & Thompson, E. B. (2001). The conformation of the glucocorticoid receptor af1/tau1 domain induced by osmolyte binds co-regulatory proteins. *The Journal of Biological Chemistry*, *276*, 18146–18152.

Lavery, D. N., & McEwan, I. J. (2005). Structure and function of steroid receptor AF1 transactivation domains: Induction of active conformations. *The Biochemical Journal*, *391*, 449–464.

Li, J., Motlagh, H. N., Chakuroff, C., Thompson, E. B., & Hilser, V. J. (2012). Thermodynamic dissection of the intrinsically disordered N-terminal domain of human glucocorticoid receptor. *The Journal of Biological Chemistry*, *287*, 26777–26787.

Li, J., White, J. T., Saavedra, H., Wrabl, J. O., Motlagh, H. N., Liu, K., et al. (2017). Genetically tunable frustration controls allostery in an intrinsically disordered transcription factor. *eLife*, *6*, e30688.

Liu, J., Perumal, N. B., Oldfield, C. J., Su, E. W., Uversky, V. N., & Dunker, A. K. (2006). Intrinsic disorder in transcription factors. *Biochemistry*, *45*, 6873–6888.

Lu, N. Z., & Cidlowski, J. A. (2005). Translational regulatory mechanisms generate N-terminal glucocorticoid receptor isoforms with unique transcriptional target genes. *Molecular Cell*, *18*, 331–342.

Meijsing, S. H., Pufall, M. A., So, A. Y., Bates, D. L., Chen, L., & Yamamoto, K. R. (2009). DNA binding site sequence directs glucocorticoid receptor structure and activity. *Science*, *324*, 407–410.

Mello, C. C., & Barrick, D. (2003). Measuring the stability of partly folded proteins using TMAO. *Protein Science*, *12*, 1522–1529.

Monod, J., Wyman, J., & Changeux, J. P. (1965). On the nature of allosteric transitions: A plausible model. *Journal of Molecular Biology*, *12*, 88–118.

Motlagh, H. N., & Hilser, V. J. (2012). Agonism/antagonism switching in allosteric ensembles. *Proceedings of the National Academy of Sciences of the United States of America*, *109*, 4134–4139.

Nakajima, Y., Kobayashi, K., Yamagishi, K., Enomoto, T., & Ohmiya, Y. (2004). cDNA cloning and characterization of a secreted luciferase from the luminous Japanese ostracod, Cypridina noctiluca. *Bioscience, Biotechnology, and Biochemistry*, *68*, 565–570.

Oakley, R. H., & Cidlowski, J. A. (2011). Cellular processing of the glucocorticoid receptor gene and protein: New mechanisms for generating tissue-specific actions of glucocorticoids. *The Journal of Biological Chemistry*, *286*, 3177–3184.

Pace, C. N., Vajdos, F., Fee, L., Grimsley, G., & Gray, T. (1995). How to measure and predict the molar absorption coefficient of a protein. *Protein Science*, *4*, 2411–2423.

Pandit, S., Geissler, W., Harris, G., & Sitlani, A. (2002). Allosteric effects of dexamethasone and RU486 on glucocorticoid receptor-DNA interactions. *The Journal of Biological Chemistry*, *277*, 1538–1543.

Pfaff, S. J., & Fletterick, R. J. (2010). Hormone binding and co-regulator binding to the glucocorticoid receptor are allosterically coupled. *The Journal of Biological Chemistry*, *285*, 15256–15267.

Pradeep, L., & Udgaonkar, J. B. (2004). Osmolytes induce structure in an early intermediate on the folding pathway of barstar. *The Journal of Biological Chemistry*, *279*, 40303–40313.

Romero, P. R., Zaidi, S., Fang, Y. Y., Uversky, V. N., Radivojac, P., Oldfield, C. J., et al. (2006). Alternative splicing in concert with protein intrinsic disorder enables increased functional diversity in multicellular organisms. *Proceedings of the National Academy of Sciences of the United States of America*, *103*, 8390–8395.

Santoro, M. M., & Bolen, D. W. (1988). Unfolding free energy changes determined by the linear extrapolation method. 1. Unfolding of phenylmethanesulfonyl alpha-chymotrypsin using different denaturants. *Biochemistry*, *27*, 8063–8068.

Tannous, B. A., Kim, D. E., Fernandez, J. L., Weissleder, R., & Breakefield, X. O. (2005). Codon-optimized Gaussia luciferase cDNA for mammalian gene expression in culture and in vivo. *Molecular Therapy: The Journal of the American Society of Gene Therapy*, *11*, 435–443.

Tsumoto, K., Umetsu, M., Kumagai, I., Ejima, D., Philo, J. S., & Arakawa, T. (2004). Role of arginine in protein refolding, solubilization, and purification. *Biotechnology Progress*, *20*, 1301–1308.

Uversky, V. N., Oldfield, C. J., & Dunker, A. K. (2005). Showing your ID: Intrinsic disorder as an ID for recognition, regulation and cell signaling. *Journal of Molecular Recognition*, *18*, 343–384.

Ward, J. J., Sodhi, J. S., McGuffin, L. J., Buxton, B. F., & Jones, D. T. (2004). Prediction and functional analysis of native disorder in proteins from the three kingdoms of life. *Journal of Molecular Biology*, *337*, 635–645.

Watson, L. C., Kuchenbecker, K. M., Schiller, B. J., Gross, J. D., Pufall, M. A., & Yamamoto, K. R. (2013). The glucocorticoid receptor dimer interface allosterically transmits sequence-specific DNA signals. *Nature Structural & Molecular Biology*, *20*, 876–883.

Wrabl, J. O., Gu, J., Liu, T., Schrank, T. P., Whitten, S. T., & Hilser, V. J. (2011). The role of protein conformational fluctuations in allostery, function, and evolution. *Biophysical Chemistry*, *159*, 129–141.

Wu, P., & Bolen, D. W. (2006). Osmolyte-induced protein folding free energy changes. *Proteins*, *63*, 290–296.

CHAPTER EIGHTEEN

A Tethered Vesicle Assay for High-Throughput Quantification of Membrane Fission

Wilton T. Snead*, Jeanne C. Stachowiak*,†,1

*Department of Biomedical Engineering, The University of Texas at Austin, Austin, TX, United States
†Institute for Cellular and Molecular Biology, The University of Texas at Austin, Austin, TX, United States
1Corresponding author: e-mail address: jcstach@austin.utexas.edu

Contents

Abstract

Membrane fission, which divides membrane surfaces into separate compartments, is essential to diverse cellular processes including membrane trafficking and cell division. Quantitative assays are needed to elucidate the physical mechanisms by which proteins drive membrane fission. Toward this goal, several experimental tools have been developed, including visualizing fission products using electron microscopy, measuring membrane shedding from a lipid reservoir, and observing fission of individual membrane tubes pulled from giant vesicles. However, no existing assay of membrane fission provides a quantitative, high-throughput measure of the distribution of vesicle curvatures generated by fission-driving proteins. Toward addressing this challenge, here we describe a novel approach that uses confocal fluorescence imaging to quantify the diameter distribution of membrane vesicles that have been tethered to a coverslip surface following exposure to fission-driving proteins. We employ this assay to measure the progressive appearance of high curvature fission products upon exposure of vesicles to increasing protein concentration. Results from this approach are in quantitative

Methods in Enzymology, Volume 611
ISSN 0076-6879
https://doi.org/10.1016/bs.mie.2018.08.014

agreement with measurements from electron microscopy, but can be collected with considerably greater throughput, enabling examination of a broad range of experimental conditions. Using the tethered vesicle approach, we have recently found that membrane-bound intrinsically disordered proteins are surprisingly potent drivers of membrane fission. The capacity to drive fission arises from steric pressure generated when disordered domains with large hydrodynamic radii bind to membranes at high local densities. More broadly, the experimental tools described here have the potential to improve our mechanistic understanding of membrane fission in diverse biophysical contexts.

1. INTRODUCTION

The process of membrane fission, which divides a continuous membrane surface into two or more separate compartments, is essential for cellular life (McMahon & Gallop, 2005). Cellular functions which rely on membrane fission include membrane trafficking (Conner & Schmid, 2003), cellular division (Mierzwa & Gerlich, 2014), and organelle biogenesis (Hoppins, Lackner, & Nunnari, 2007), among others. Harmful pathogens such as viruses and bacterial toxins also require membrane fission to enter and exit cells (Hurley, Boura, Carlson, & Rozycki, 2010). Therefore, detailed characterization of the physical mechanisms that drive membrane fission is important for understanding normal cell function as well as the physiology of disease.

Membrane fission is an energetically costly process (Helfrich, 1973), which relies upon significant energetic contributions from diverse protein machines (Zimmerberg & Kozlov, 2006). Quantitative experimental tools are needed to understand how individual proteins and molecular mechanisms overcome the energetic barriers to membrane fission. A significant body of work from multiple research groups has led to the development of a variety of assays for studying membrane fission (Fig. 1). Among the most common techniques is negative stain transmission electron microscopy (TEM) of vesicles, which enables direct visualization of the fission products generated upon mixing of membrane vesicles and proteins (Boucrot et al., 2012; Ford et al., 2002) (Fig. 1A). More recently, supported bilayers with extra membrane reservoir have been developed. These "SUPER" templates consist of glass beads coated with a low-tension membrane bilayer (Neumann, Pucadyil, & Schmid, 2013; Pucadyil & Schmid, 2008). Exposure of SUPER templates to fission-driving proteins leads to quantifiable

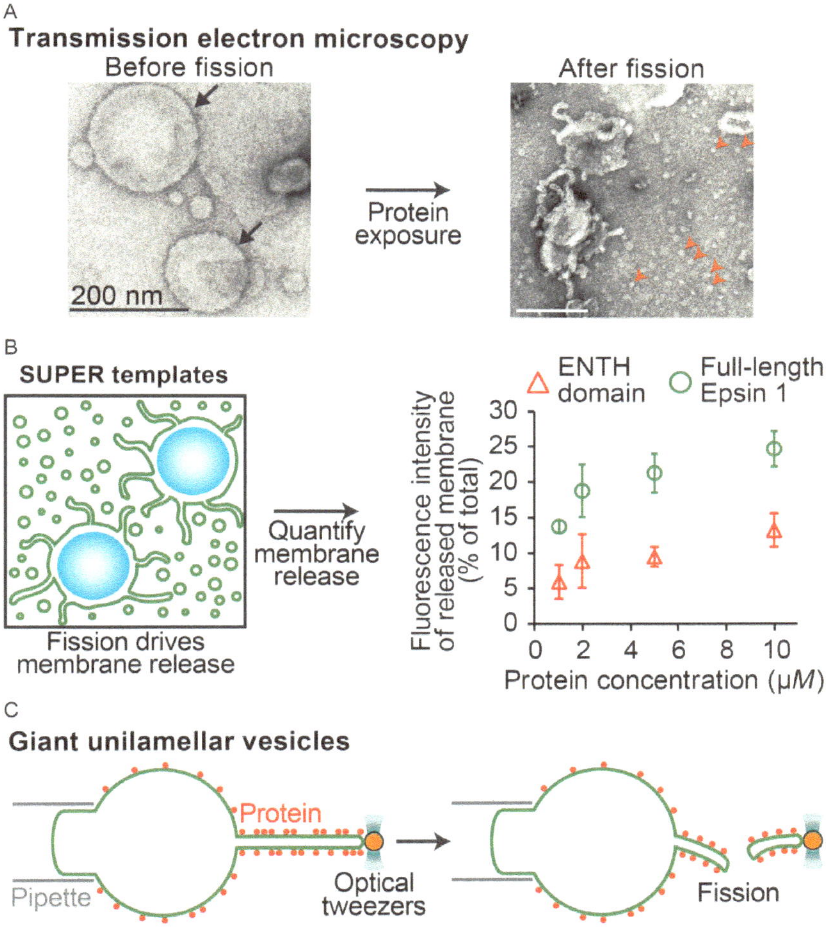

Fig. 1 Some existing assays of membrane fission. (A) Negative stain transmission electron microscopy allows direct visualization of vesicles before (*left*) and after (*right*) exposure to fission-driving proteins. Highly curved vesicles (*red arrowheads*) are generated from the initially low-curvature vesicles (*black arrows*). Vesicle diameters must be measured manually. (B) SUPER templates, which are glass beads coated in low-tension membrane, undergo membrane release upon exposure to fission-driving proteins. Fission is quantified as the bulk fluorescence of the released membrane in the supernatant after sedimenting the beads. An example plot of membrane release is shown on the *right*. Full-length Epsin 1 drives greater membrane release compared to the Epsin 1 N-terminal homology (ENTH) domain alone, owing to enhanced crowding by the bulky disordered domain of full-length Epsin 1. SUPER template membrane-release experiments do not allow precise quantification of vesicle curvatures. (C) Pulling membrane nanotubes from giant unilamellar vesicles using optical tweezers allows visualization of protein recruitment to high curvature membranes. Membrane tension is controlled via

(Continued)

release of membrane vesicles (Fig. 1B). Finally, lipid tubules pulled from giant unilamellar vesicles (GUVs) (Prevost, Tsai, Bassereau, & Simunovic, 2017) enable simultaneous measurement of lipid tubule diameter and protein binding, providing a highly controlled setting for observing membrane fission events (Fig. 1C).

While experimental systems like negative stain TEM, SUPER templates, and GUVs have enabled the investigation of membrane fission mechanisms, these tools generally required researchers to choose between the ability to make highly quantitative measurements and the ability to examine multiple conditions with high throughput. For example, SUPER templates provide a measure of the relative extent of membrane fission, quantified as the bulk fluorescence of membrane released by the addition of proteins (Neumann et al., 2013). While these experiments can achieve high throughput, they do not quantify the diameters of the fission products. In contrast, GUV tube-pulling assays provide highly quantitative estimates of membrane curvature, but can examine only a single lipid tubule at a time (Prevost et al., 2017), greatly limiting throughput. Finally, while TEM experiments quantify the diameters of fission products, they typically rely on manual analysis of vesicles in electron micrographs (Boucrot et al., 2012). This approach greatly limits throughput and can be biased by the selection of the field of view as well as insufficient understanding and control of vesicle adsorption and staining on TEM grids (Franken, Boekema, & Stuart, 2017). In light of these limitations, the development of novel assays that provide quantitative results in a high-throughput format is needed.

Here we describe a high-throughput assay for quantifying the distributions of vesicle diameters resulting from membrane fission. This procedure uses confocal fluorescence imaging to measure the brightness of individual, immobilized fluorescent vesicles. These measurements are then converted to distributions of vesicle diameter by calibrating the measured brightness distribution against an independent measure of the distribution of vesicle diameter from dynamic light scattering (DLS) (Fig. 2). This assay builds upon earlier work from the Stamou group (Hatzakis et al., 2009; Kunding,

Fig. 1—Cont'd pipette aspiration of the vesicle, which provides control over the diameter of the pulled nanotube. Membrane fission results in tube breaking (*right*). These experiments are low-throughput, as only one vesicle and pulled tube can be examined at a time. *Images (panel A) and data (panel B) from the same experiments reported in Snead, W., Hayden, C., Gadok, A., Zhao, C., Lafer, E., Rangamani, P., et al. (2017). Membrane fission by protein crowding.* Proceedings of the National Academy of Sciences of the United States of America, 114*(16), E3258–E3267.*

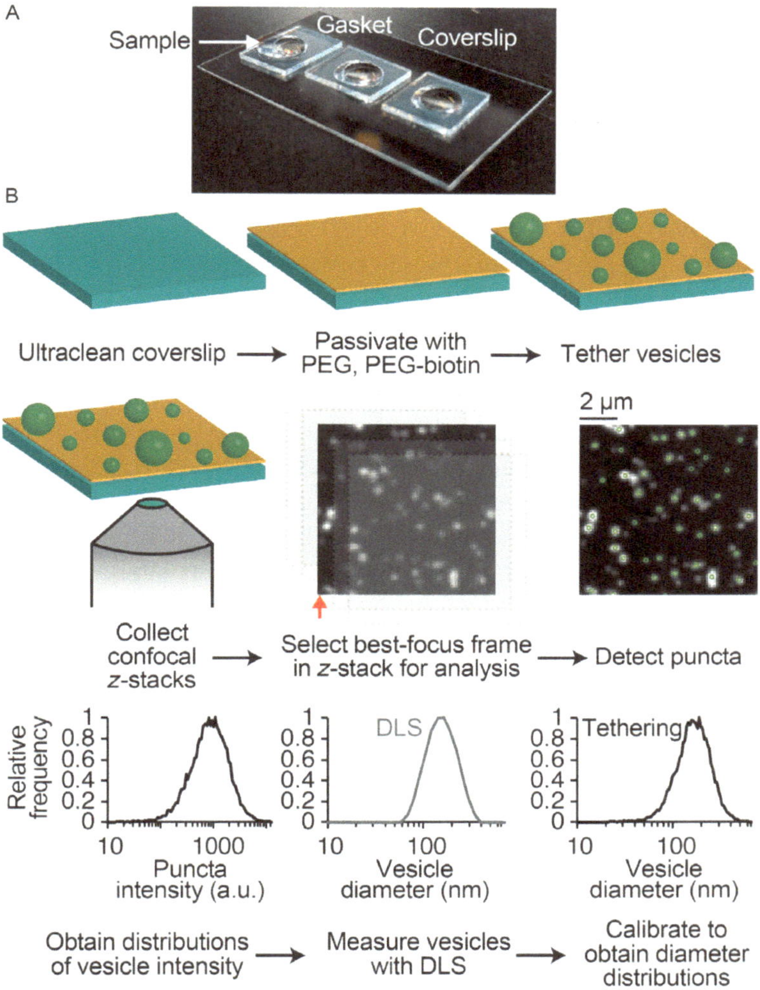

Fig. 2 Tethered vesicle assay of membrane fission. (A) Photo of three sample wells on an ultraclean coverslip. The gaskets were made from silicone sheets, and will hold 40 μL (see Section 2.2 for well preparation). (B) Workflow of tethered vesicle assay and image analysis to obtain distributions of vesicle diameter. This procedure is performed for a protein-free vesicle control, and the scaling factor calculated from calibration is used to generate vesicle diameter distributions for the fission samples.

Mortensen, Christensen, & Stamou, 2008; Lohr, Kunding, Bhatia, & Stamou, 2009), which utilized immobilized vesicles to measure the sensitivity of proteins to membrane curvature. However, our procedure, first reported in a recent study (Snead et al., 2017), is the first to use this experimental approach for quantifying the vesicle diameter distribution created during membrane fission. We validated our procedure by comparing it to traditional negative stain TEM of vesicles. Both approaches yield similar estimates of average vesicle diameter (see Section 4). However, the tethered vesicle assay provides much more detail about the distribution of vesicle diameters with considerably less experiment time in comparison to TEM. As such, the tethered vesicle assay facilitates study of membrane fission across a broad range of experimental parameters such as protein concentration and membrane composition. Therefore, the assay described here may be utilized to gain a more detailed understanding of how proteins drive membrane fission in a variety of biophysical contexts.

Based on studies that have employed combinations of the assays described above, the prevailing view is that membrane fission is mediated primarily by structured proteins or protein assemblies (Jarsch, Daste, & Gallop, 2016; McMahon & Gallop, 2005), including the ESCRT machinery (Schoneberg, Lee, Iwasa, & Hurley, 2017) or the dynamin helical scaffold (Antonny et al., 2016). In contrast, recent work from our group found that protein crowding can drive membrane fission independent of protein structural or assembly properties (Snead et al., 2017). Specifically, molecular crowding among proteins attached to membrane surfaces at high density generates steric pressure that can bend membranes and ultimately overcome the energetic barrier to membrane fission (Busch et al., 2015; Snead et al., 2017; Stachowiak et al., 2012). In support of this mechanism, we found that bulky, intrinsically disordered proteins, which occupy significantly larger membrane footprints in comparison to well-folded proteins of similar mass (Hofmann et al., 2012), can be potent drivers of membrane fission. Specifically, we examined membrane fission by Epsin 1, an adaptor protein of the clathrin-mediated endocytic pathway which contains a well-folded epsin N-terminal homology (ENTH) domain and a bulky, disordered domain. As shown below, both SUPER template experiments (Fig. 1B) and tethered vesicle experiments (Fig. 6) reveal that full-length Epsin 1 drives membrane fission more efficiently in comparison to the isolated ENTH domain (Snead et al., 2017).

2. TETHERED VESICLE FISSION ASSAY

In this section we describe procedures for preparing the vesicles and tethering substrates and performing the fission experiment. In the next section we describe the image analysis procedure used to quantify the distribution of vesicle diameters before and after fission. Fig. 2 shows an overall schematic of this workflow.

2.1 Preparing Vesicles

When designing a vesicle composition for fission studies, the lipid components should be chosen to adequately represent the membrane environment of interest. Factors to consider when designing a membrane composition include the binding lipid for the protein, the membrane bending rigidity, and the net charge of the membrane, among others. The lipid components can be adjusted to capture a variety of membrane physical and chemical properties.

In the experiments described here, we examine membrane fission by the ENTH domain of the clathrin adaptor protein Epsin 1. ENTH binds preferentially to the lipid phosphatidylinositol-4,5-bisphosphate (PIP_2), and inserts an N-terminal, amphipathic helix into the membrane surface (Ford et al., 2002). To study membrane fission by the ENTH domain, we use membranes composed primarily of 1,2-dioleoyl-*sn*-glycero-3-phosphocholine (DOPC). PIP_2 is incorporated to facilitate ENTH binding, as well as the negatively charged lipid 1,2-dioleoyl-*sn*-glycero-3-phospho-L-serine (DOPS) to further enhance ENTH recruitment via electrostatic interactions. The fluorescently labeled lipid Oregon Green 488 1,2-dihexadecanoyl-*sn*-glycero-3-phosphoethanolamine (Oregon Green 488-DHPE) is included to enable vesicle imaging, and the biotinylated lipid dipalmitoyl-decaethylene glycol-biotin (DP-EG10-biotin (Momin et al., 2015)) enables vesicle tethering. While the procedures described here focus on the ENTH domain, the membrane composition can readily be adapted to examine a wide variety of fission-driving proteins that bind to membrane surfaces.

2.1.1 Equipment

- Glass syringes for mixing lipids in solvent (Hamilton)
- Rotary evaporator (optional)
- Nitrogen gas
- Vacuum chamber and pump
- Mini extruder system (Avanti Polar Lipids, Alabaster, AL)

2.1.2 Reagents

- Lipids: PIP_2, DOPC, and DOPS (Avanti). PEG-biotin lipid (we use a custom-synthesized DP-EG10-biotin lipid (Momin et al., 2015), but others are available from Avanti). Fluorescent lipid (Oregon Green 488-DHPE; Thermo Fisher)
- Chloroform, methanol, ultrapure water
- MOPS, HEPES, NaCl, EDTA, EGTA (Sigma or Fisher)

2.1.3 Procedure

1. The lipid composition for experiments with ENTH is: 77.5 mol% DOPC, 15 mol% DOPS, 5 mol% PIP_2, 0.5 mol% Oregon Green 488-DHPE, and 2 mol% DP-EG10-biotin. In a clean glass conical tube, mix lipids in chloroform at the specified molar ratios.
   a. When working with PIP_2, the solvent mixture must be adjusted with methanol and water prior to adding PIP_2 such that the volume ratio is 20:9:1 chloroform:methanol:water, to ensure that PIP_2 remains soluble.
2. Dry the lipid into a film via rotary evaporation or under a stream of nitrogen. Dry further under vacuum for at least 2 h, preferably overnight.
3. Hydrate the lipid film in experiment buffer of 20 m*M* MOPS or 25 m*M* HEPES pH 7.35 and 150 m*M* NaCl. Vortex well.
4. Extrude using a 200 nm pore membrane in a mini extruder system.

2.1.4 Notes

- If needed, colorimetric assays for quantifying phospholipid concentration may be used after preparing the vesicles (Stewart, 1980; Takayama, Itoh, Nagasaki, & Tanimizu, 1977). The phosphatidylcholine assay described in Takayama et al. is available as a kit from several vendors.
- Store vesicles at 4°C and use within 48 h, as PIP_2 will degrade. Vesicles without PIP_2 can be stored for up to a week before degradation becomes a concern.
- Include 0.5 m*M* EDTA and EGTA in the buffer to prevent clustering of PIP_2 by divalent metal cations (Ellenbroek et al., 2011), which are present as trace contaminants in most buffers and salts.
- Our previous work has also utilized the synthetic lipid 1,2-dioleoyl-*sn*-glycero-3-[(*N*-(5-amino-1-carboxypentyl)iminodiacetic acid)succinyl], nickel salt (DOGS-NTA-Ni) in order to examine membrane fission by hexahistidine (6his)-tagged proteins. For example, our recent work showed that membrane-bound green fluorescent protein drove fission

when recruited to DOGS-NTA-Ni membranes by a 6his tag (Snead et al., 2017). In these experiments, PIP_2 and DOPS are replaced with 20 mol% DOGS-NTA-Ni. EDTA and EGTA should not be used in the buffer here, as they will bind and sequester nickel ions needed to promote 6his–NTA interactions.

2.2 Preparing Tethering Substrates

In order to tether the vesicles to a coverslip surface for imaging, the surface must first be passivated and functionalized with biotin. Here we use PEG and PEG-biotin for passivating the glass. Once passivated, the biotinylated vesicles are tethered to the PEG-biotin molecules by binding to neutravidin (see Section 2.2.6). Glass passivation can be achieved in two ways: (1) coating the glass with poly-L-lysine (PLL) conjugated to PEG and PEG-biotin, or (2) directly conjugating PEG and PEG-biotin to the glass via silanization. Here we describe procedures for cleaning and passivating glass using both of these approaches.

2.2.1 Equipment

- Glass coverslips, 24 × 40 mm (VWR cat. no. 48393-230)
- Glass carafes with slots for individual coverslips
- Glove bag
- Clean oven (for drying coverslips and performing the silane reaction)
- Princeton CentriSpin-20 size exclusion spin columns (Princeton Separations)
- Silicone gaskets
  - To make gaskets, we purchase 0.8-mm-thick silicone sheets (Grace Bio-Labs cat. no. 664172), cut out small pieces, and punch holes for imaging wells using a 0.5 mm biopsy punch. The wells hold approximately 40 μL.

2.2.2 Reagents

- RCA glass cleaning: 100% ethanol, ultrapure water, 30% H_2O_2, KOH, 37% HCl (Sigma or Fisher)
- PLL (Sigma)
- For PLL-PEG-biotin: mPEG-SVA (mPEG-succinimidyl valerate) and biotin-PEG-SVA (both having PEG chains of 5000 Da; Laysan Bio, Arab, AL)
- For silanization reaction: mPEG-silane and biotin-PEG-silane (both having PEG chains of 5000 Da; Laysan Bio)
- Sodium tetraborate (Fisher)

- Anhydrous isopropanol (Fisher)
- Glacial acetic acid (Fisher)
- Neutravidin (Thermo Fisher cat. no. 31000)
  - We purchase lyophilized neutravidin powder, create a 1 mg/mL solution in PBS or 25 m*M* HEPES pH 7.35, 150 m*M* NaCl, and store aliquots at −80°C.

2.2.3 RCA Glass Cleaning

Precautions: Appropriate personal protective equipment must be used when performing RCA glass cleaning, including chemical gloves, apron, and face shield. The procedure must be performed in a chemical fume hood, as the acid and base solutions with hydrogen peroxide generate caustic fumes. Prepare the acid and base solutions separately, and do not mix them together. Dilute the HCl by slowly pouring the concentrated HCl into water. Do not add water directly to concentrated HCl.

1. Place coverslips in carafes and add 100% ethanol to cover. Sonicate in a bath sonicator for 10 min. Rinse thoroughly with ultrapure water to remove all ethanol.
2. In a separate beaker, prepare a solution of 1.5 *M* KOH in ultrapure water and heat to 70–80°C. Slowly add H_2O_2 to a final concentration of 7.5% and stir well. Bubbles will appear spontaneously as the solution reacts. Pour the reactive mixture into the carafes to cover the slides. Let stand for 10 min and then rinse the carafe thoroughly with ultrapure water until all reactants are removed.
3. In a separate beaker, prepare a solution of 2.4 *M* HCl in ultrapure water and heat to 70–80°C. Slowly add H_2O_2 to a final concentration of 6.3% and stir well. Bubbles will appear spontaneously as the solution reacts. Pour the reactive mixture into the carafes to cover the slides. Let stand for 10 min and then rinse the carafe thoroughly with ultrapure water until all reactants are removed. Store clean slides in ultrapure water and use them within 2 weeks.

2.2.4 Preparing PLL-PEG-Biotin

1. Dissolve PLL at a concentration of 40 mg/mL in 50 m*M* sodium tetraborate at pH 8.5.
2. Calculate the amount of PEG-SVA needed to achieve a molar ratio of one PEG-SVA per five lysine subunits. Include biotin-PEG-SVA at a ratio of 1:50–1:20 with PEG-SVA (2%–5% biotinylated PEG).

3. In a glove bag filled with nitrogen gas, measure out PEG-SVA and biotin-PEG-SVA powders. Close and seal stock containers in the glove bag, as the PEGs are hygroscopic and should be stored in a low moisture environment.
   a. We like to pour out more of the PEG powder than needed in the glove bag, and then take the powder out of the glove bag and precisely measure the desired amount using a scale. Discard the excess powder that is not used. Once the powders are taken out of the glove bag, work quickly to weigh the powders and proceed to the reaction step.
4. Combine the PEG powders and mix well.
5. Add the PLL solution to the amine-reactive PEG-SVA powder, vortex well, and transfer to a test tube with a small stir bar. Stir continuously for 6 h at room temperature.
6. Separate PLL-PEG-biotin from the unconjugated molecules using CentriSpin-20 size exclusion spin columns hydrated with PBS or 25 m*M* HEPES pH 7.35 and 150 m*M* NaCl. Store at 4°C until use.

2.2.5 Passivating Glass With PEG-Silane

1. Dry ultraclean coverslips under a stream of nitrogen and further dry in a clean oven at 70–100°C for at least 30 min.
2. Create a 0.67% solution of PEG-silane in anhydrous isopropanol, with biotin-PEG-silane comprising 2%–5% of the total PEG amount, as described above. Sonicate in a bath sonicator for 10–15 min to further dissolve the powder.
3. Add acetic acid to a final concentration of 1%.
4. Immerse the dry, ultraclean coverslips in the reactive PEG-silane solution.
5. Incubate at 70°C for 30–60 min.
6. Remove the coverslips from the solution, rinse them well with ultrapure water, dry them under a stream of nitrogen, and store them dry until use.

2.2.6 Preparing Wells for Vesicle Tethering

This procedure is performed while vesicle and protein samples are incubating (see next section).

1. Place a silicone gasket, with a well volume of approximately 40 μL, onto an ultraclean coverslip or silane-PEG-biotin-coated coverslip.
2. If using a silane-PEG-biotin-coated coverslip, hydrate the substrate with 20 μL of buffer for approximately 5 min.

3. If using an uncoated ultraclean coverslip, add 20 μL of PLL-PEG-biotin solution diluted 10-fold in the experiment buffer of 20 m*M* MOPS pH 7.35, 150 m*M* NaCl. Coat the well for at least 20 min at room temperature. Longer times of up to 60 min are preferable to ensure complete coating.
4. Thoroughly wash to remove excess PLL-PEG-biotin by adding 80 μL of experiment buffer to the well, removing, and discarding. Repeat at least six times.
   a. While the final volume in the well will be 40 μL prior to imaging, the well can hold up to 100 μL, thereby allowing for 80 μL of wash buffer to be added to 20 μL in the well.
5. Add 1 mg/mL neutravidin solution to the well for a final concentration of approximately 0.2 mg/mL. Mix well and incubate for 10 min to provide sufficient time for neutravidin to bind to the biotin on the coverslip surface. Wash thoroughly with 20 m*M* MOPS pH 7.35, 150 m*M* NaCl following a similar procedure described in step 4.
   a. To ensure that the volume in the well remains 20 μL, add the neutravidin solution to the well, mix thoroughly, and remove the same volume that was added.
   b. After washing, the well volume should be 20 μL. Adding 20 μL of vesicle and protein sample to the well will bring the final volume to 40 μL (see Section 2.3).

The well is now ready for vesicle tethering (see Section 2.3).

2.2.7 Notes

Silicone gaskets can be cleaned and reused multiple times. Clean the gaskets by stirring them in a solution of 1–2 vol% Hellmanex detergent in ultrapure water at approximately 70–100°C for at least 15 min. Following exposure to detergent, wash the gaskets thoroughly in ultrapure water and dry them well under a stream of nitrogen gas before using.

2.3 Fission Reaction and Vesicle Tethering

2.3.1 Equipment and Reagents

- Low-adhesion microcentrifuge tubes (USA Scientific cat. no. 1415-2600)
- Water bath at 37°C
- Purified membrane-binding proteins
- Vesicles (prepared in Section 2.1)
- Coverslip ready for tethering (prepared in Section 2.2)
- Experiment buffer: 20 m*M* MOPS pH 7.35, 150 m*M* NaCl

2.3.2 Procedure

1. Create a 20 μL solution of vesicles and protein with a total lipid concentration of 10 μ*M* in a low-adhesion tube. The protein concentration can vary as determined by the experimental goals. Typically, a range of protein concentrations should be tested spanning from the nanomolar to micromolar range to determine the concentration that corresponds to the onset of membrane fission.
2. Incubate the solution of vesicles and protein at 37°C for 30–60 min to allow the fission reaction to take place. We chose 60 min for experiments with ENTH to ensure that the fission reaction had reached completion.
3. During incubation, prepare the imaging substrate as described in Section 2.2.
4. After the fission reaction, add the 20 μL solution of vesicles and protein to the imaging substrate and mix well by pipetting up and down gently several times. Wait 10 min for vesicles to reach the substrate surface and become tethered.
   a. Adding 20 μL of sample to the 20 μL already in the well (see Section 2.2) will bring the volume to 40 μL.
5. Wash to remove untethered vesicles by first removing 20 μL from the well, then pipetting in 80 μL of experiment buffer, removing, and discarding. Repeat at least six times. After washing, add 20 μL of experiment buffer to the well to bring the well volume back to 40 μL.

2.3.3 Notes

- Cover the wells with a Petri dish during incubations to avoid sample evaporation.
- Do not touch the pipet tip to the coverslip surface during washing steps.
- Image within 30 min to avoid vesicle unbinding or rupturing on the substrate.

2.4 Imaging Tethered Vesicles

Image the tethered vesicles using a confocal microscope. Confocal microscopy helps to minimize the background fluorescence from untethered vesicles in solution. Wide field and TIRF microscopy may also be used, as described in protocols from the Stamou group (Kunding et al., 2008), though we only utilize confocal microscopy in this procedure. We use a spinning disc confocal microscope (Zeiss Axio Observer Z1 with Yokogawa CSU-X1M), with a 488 nm excitation laser and emission filter centered at 525 nm with a 50 nm width. The objective is a Plan-Apochromat 100 ×, 1.4 numerical aperture (NA) oil immersion objective. Images are collected on a cooled (−70°C) EMCCD iXon3 897 camera (Andor Technology; Belfast, UK).

Acquire *z*-stacks (i.e., image stacks spaced at fixed perpendicular distances above and below the coverslip surface) that begin below the brightest, in-focus frame and extend a few frames above the brightest frame, with a spacing of 0.1 μm between frames. Acquire *z*-stacks of at least 10 fields of view within each sample to ensure an adequate number of vesicles are imaged. If possible, use the same imaging settings for all samples. For the protein-free sample, we detect approximately 200 vesicles per frame that are suitable for analysis (see next section for acceptable puncta criteria). Therefore, 10 *z*-stacks should yield approximately 2000 vesicles. The number of vesicles per frame in samples with protein may deviate from the protein-free condition, as proteins may coat the substrate and hinder vesicle tethering to some degree. Collect more than 10 *z*-stacks if the number of tethered vesicles is visibly lower than in the protein-free sample. If there are large differences in vesicle brightness between samples, for example, before and after membrane fission, parameters such as camera exposure time may need to be adjusted to improve signal quality. A linear correction for exposure time can be applied during analysis.

3. PARTICLE DETECTION AND QUANTIFICATION OF VESICLE DIAMETERS

Here, particle detection software is used to fit 2D Gaussian profiles to the diffraction-limited puncta generated in fluorescence images by individual tethered vesicles. The integrated intensities of the vesicles are proportional to the amplitudes of the Gaussian fits above the local background. These integrated intensities are also proportional to the surface areas of the vesicles. Therefore, the amplitudes of the Gaussian fits can be used to measure the relative differences in surface area among tethered vesicles. We perform particle detection using cmeAnalysis software, which was developed by the Danuser and Schmid groups to detect low signal-to-noise puncta of proteins associated with clathrin-coated endocytic structures (Aguet, Antonescu, Mettlen, Schmid, & Danuser, 2013). The software can be downloaded from the Danuser lab website. Carefully read the accompanying user guide to become familiar with how to organize data and use the software.

After detecting puncta, the distributions of vesicle brightness are converted to distributions of vesicle diameter by calibrating against an independent measurement of vesicle diameter. Specifically, we scale the median brightness value of the protein-free vesicles against the intensity-weighted average diameter of the same vesicles measured by DLS. Importantly, this scaling assumes that vesicle tethering did not alter the diameter distribution measured in DLS. This assumption was tested and shown to be valid in work from the Stamou group (Kunding et al., 2008; Lohr et al., 2009). We further verify this procedure by comparing the mean diameters determined using the tethered vesicle approach to the mean diameters assessed using electron microscopy, as described below.

3.1 Cropping Images, Selecting Best-Focus Frame, and Detecting Puncta

3.1.1 Procedure

1. To avoid variations in illumination across the image, crop the images to a central region where vesicles are uniformly illuminated. Specifically, crop the images to the central square of a 3×3 grid. Cropping can be performed in MATLAB®, using the Bio-Formats software package to import Zeiss .czi files or other file types not supported by MATLAB®.
2. Calculate the mean intensity of each frame in the stack.
3. Select the frame with the greatest mean brightest as the best-focus frame for particle detection analysis.
   a. The mean intensity of the brightest frame should be a local maximum in the *z*-stack. In other words, the frames above and below should have lower mean intensities to ensure that the chosen frame is actually the best-focus frame.
4. Save the best-focus frame in a directory labeled "ch1" within a parent directory named "Celli," where *i* is the *z*-stack number.
   a. See the cmeAnalysis user guide for more information on data organization.
5. Save the frames above and below the best-focus frame in separate directories labeled ch2 and ch3 within the same parent directory as the best-focus frame.
   a. The best-focus frame in the ch1 directory is the "master" channel in cmeAnalysis. The frames above and below in the ch2 and ch3 directories are the "slave" channels, in which the program uses the centroids of the detected puncta to search for corresponding puncta. The search radius is three times the standard deviation of the Gaussian fit to the microscope point spread function (PSF).

6. Load the dataset for an experimental condition using the command: data = loadConditionData. The loadConditionData script will open a user interface for selecting the directory containing the data to be analyzed.
   a. Before running analysis for the first time, enter values for the NA of the objective, total magnification (M), and physical pixel size of the camera (in m) into the "Parameters" vector in loadConditionData. The software uses these inputs (along with the fluorophore specified below in step 6d) to estimate the ideal standard deviation of the microscope PSF. If the true PSF is expected to differ significantly from the ideal case, then a direct estimate of the PSF standard deviation may be loaded, as described in the cmeAnalysis user guide.
   b. A prompt will appear in the MATLAB® command window asking for the number of channels. Type "3."
   c. Select ch1 as the master channel and ch2 and ch3 as the slave channels.
   d. A prompt will appear asking for a fluorescent marker for each channel. Since Oregon Green 488 is not a recognized fluorophore in cmeAnalysis, type "Alexa488" for each channel.
7. Pass the data to the runDetection script for particle detection with the command: runDetection(data).
8. After running detection, a folder entitled "Detection" is saved within each ch1 directory. This folder contains a file entitled "detection_v2.mat." Opening this file imports a structured array into MATLAB® called "frameInfo," which contains the puncta information for its respective image.
9. The "*A*" values in frameInfo are the amplitudes of the Gaussian fits to each puncta, which correspond to the intensities of individual puncta that will be used for further analysis.
10. Only accept *A* values that are significantly above the noise threshold in all three frames. Check this criterion with the "hval_Ar" logical array within frameInfo. Puncta which have a value of 1 in hval_Ar for all three channels can be used for further analysis.
11. To further exclude puncta close to the noise threshold, only accept *A* values that are greater than two standard deviations of the distribution of "c" values in each image. The c values are the local background intensities of each Gaussian fit.

12. Save the A values which meet the above criteria in a spreadsheet for further analysis.

Fig. 3 shows example images of vesicles before and after exposure to the ENTH domain. Membrane fission by ENTH resulted in a clear reduction in puncta brightness. The images are shown again to the right with the detected puncta marked.

3.2 Quantifying Vesicle Diameter

The final analysis step involves converting the distribution of Gaussian amplitudes (A values) obtained in the previous section to distributions of vesicle diameter. As noted above, the A values are proportional to the integrated intensity of each vesicle, and are therefore also proportional to the surface area of each vesicle. As such, the diameter of each vesicle is

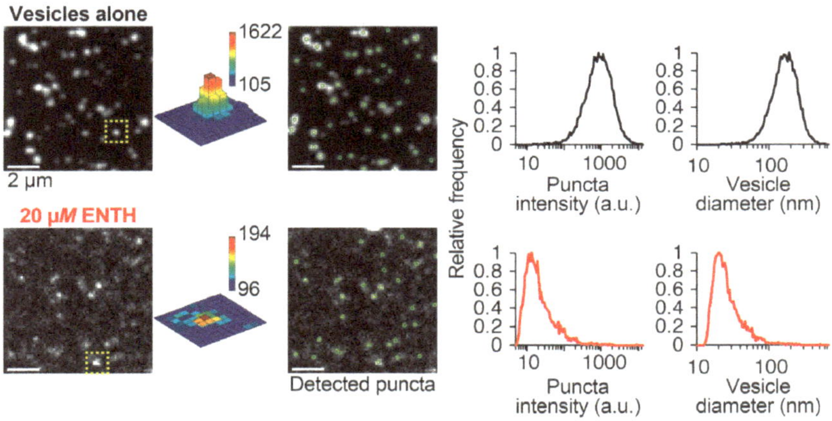

Fig. 3 Tethered vesicle assay reveals potent membrane fission by the ENTH domain. *Top row*: Before protein exposure. *Bottom row*: After exposure to 20 μ*M* ENTH. Contrast settings in the *top* and *bottom images* are different to show puncta clearly. The intensities of vesicle puncta are reduced after protein exposure, indicating fission. The *boxed region* in each image indicates vesicle intensity profile to the *right*, where the bar heights in both plots are scaled between 90 and 1622 brightness units and the *color map* corresponds to the specified range. The same images are shown again with the detected puncta from cmeAnalysis marked. The puncta intensity distributions indicate a strong reduction in vesicle intensity after protein exposure. The corresponding vesicle diameter distributions after calibration (*far right*) reveal a shift to a high curvature vesicle population after fission by ENTH. Scale bars: 2 μm. *Data from the same experiments reported in Snead, W., Hayden, C., Gadok, A., Zhao, C., Lafer, E., Rangamani, P., et al. (2017). Membrane fission by protein crowding.* Proceedings of the National Academy of Sciences of the United States of America, 114*(16), E3258–E3267.*

proportional to the square root of its A value. The use of the distribution of A values to estimate the distribution of vesicle diameters is achieved by computing a linear scaling factor that centers the square root of the median of the distribution of the A values on the intensity-weighted average diameter measured by DLS. This scaling is performed for the initial, protein-free vesicles, yielding a constant scaling factor that is then used to convert the other populations after protein exposure.

3.2.1 Dynamic Light Scattering

Here we perform DLS measurements of the protein-free vesicles to facilitate calibration. We use a Zetasizer Nano ZS (Malvern Instruments). Note that the vesicle concentration in DLS measurements is higher than in the tethered vesicle experiments to ensure sufficient signal. We load the vesicles at approximately 500 μ*M* into a cuvette suitable for DLS, and collect at least 14 traces for 10 s each. We use the average, intensity-weighted vesicle diameter distribution from these measurements for calibration, as described next.

Note: Owing to the large sample volume and vesicle concentration required for DLS-based measurements, the use of DLS to directly measure the diameter distribution after protein-driven fission is not generally feasible. In particular, a large volume of highly concentrated purified protein solution would be required, greatly limiting the number of experiments that could be performed.

3.2.2 Converting Vesicle Intensity Distributions to Vesicle Diameter Distributions

1. Compute a scaling factor, f, for converting A values to diameter values: $f = d/\sqrt{A_{\text{median}}}$, where d is the intensity-weighted average diameter of the initial, protein-free vesicles measured by DLS, and A_{median} is the median of the A values of the initial, protein-free vesicles.
2. Create distributions of vesicle diameter for each experimental condition by multiplying the square root of the respective A values by f.
3. Create histograms of vesicle diameter for each condition to assess shifts in diameter populations after protein exposure.

See Fig. 3 for examples of puncta intensity distributions before and after membrane fission. Converting these puncta intensity distributions to vesicle diameter distributions reveals that the ENTH domain drove the formation of a high curvature vesicle population centered near 20 nm diameter.

3.2.3 Examining Membrane Fission by the ENTH Domain

Using the procedure described above, we examined membrane fission by ENTH over a range of protein concentrations (Fig. 4). Experiments revealed that with increasing protein concentration, exposure to ENTH drove a progressive reduction in vesicle diameter, arriving at a high curvature population centered near 20 nm.

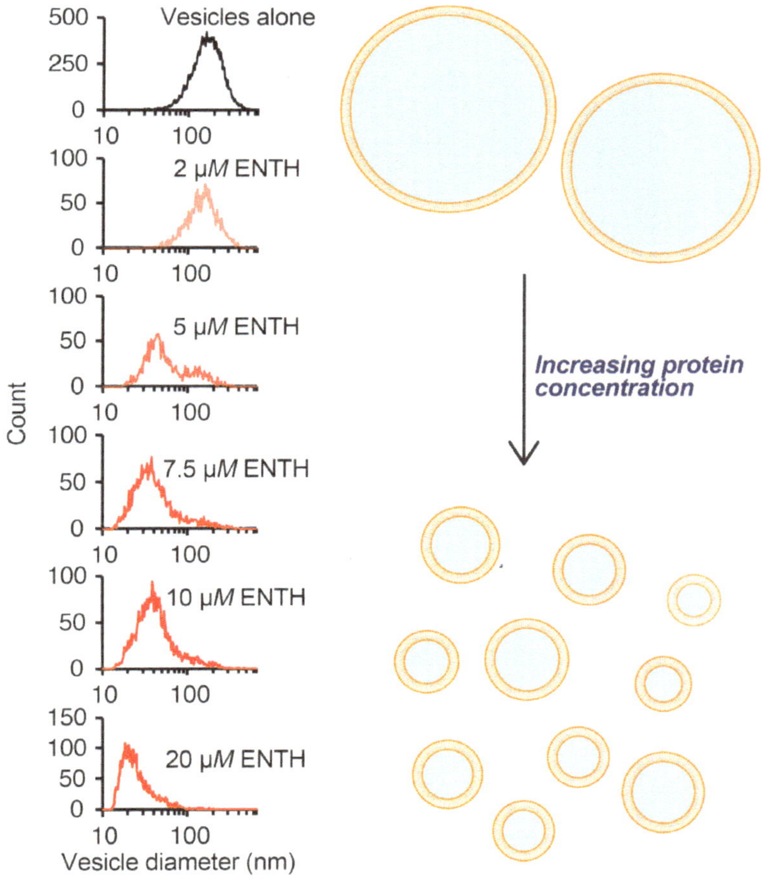

Fig. 4 Progressive reduction in vesicle diameter revealed by tethered vesicle assay. *Left*: Histograms of vesicle diameter after exposure to the indicated concentrations of ENTH. *Right*: Vesicles are refined to progressively higher curvature populations as the concentration of protein increases. *Data from the same experiments reported in Snead, W., Hayden, C., Gadok, A., Zhao, C., Lafer, E., Rangamani, P., et al. (2017). Membrane fission by protein crowding.* Proceedings of the National Academy of Sciences of the United States of America, 114*(16), E3258–E3267.*

4. COMPARISON TO AN ELECTRON MICROSCOPY-BASED ASSAY OF MEMBRANE FISSION

To verify that the tethered vesicle approach provides consistent results with established membrane fission assays, we compared the results of the 20 μ*M* ENTH condition in Fig. 4 with the results of a TEM assay performed under similar conditions. Fig. 5A shows electron micrographs of vesicles

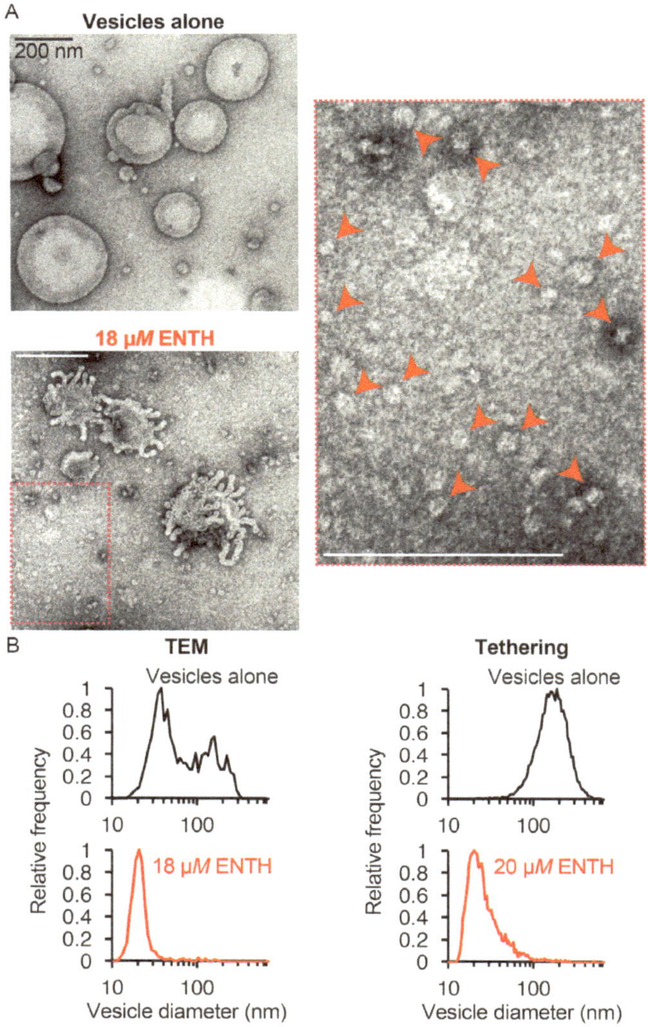

Fig. 5 Comparison of tethered vesicle assay with transmission electron microscopy (TEM). (A) Electron micrographs of vesicles before protein exposure (*top*) and after *(Continued)*

before and after exposure to ENTH, demonstrating the appearance of high curvature fission products. Measuring the vesicle diameters in these TEM images revealed that ENTH generated a vesicle population centered near 20 nm diameter not observed before protein exposure (Fig. 5B). This finding is in agreement with results from the tethered vesicle assay, in which ENTH formed populations of high curvature fission vesicles centered near 20 nm (Fig. 5B). This comparison indicates that the tethered vesicle approach provides consistent results with established procedures for assessing membrane fission.

Importantly, the family of distributions determined using the tethered vesicle assay (Fig. 4) can be measured in less than 1 day, but would likely be prohibitively laborious if performed using electron microscopy. Specifically, TEM image analysis required manual measurement of individual vesicles, a process that was both time consuming and somewhat subjective, owing to the low level of contrast associated with small vesicles in TEM. Therefore, the tethered vesicle assay facilitated the analysis of more experimental conditions compared to TEM.

5. CONCLUSIONS

Here we describe an assay for assessing membrane fission using fluorescence imaging of tethered vesicles. By quantifying the distribution of vesicle diameters after fission, this approach enables examination of progressive changes in fission efficiency as a function of experimental parameters such as protein concentration and lipid composition. We utilized this assay to demonstrate that bulky, intrinsically disordered protein domains can be potent drivers of membrane fission (Snead et al., 2017). Specifically, when we examined membrane fission by full-length Epsin 1, which contains both a well-folded N-terminal domain (ENTH), and a large, intrinsically disordered domain (Fig. 6A) (Busch et al., 2015; Kalthoff, Alves, Urbanke,

exposure to 18 μ*M* ENTH (*bottom*). Vesicles were adsorbed onto glow-discharged, carbon-coated TEM grids and stained with 2% uranyl acetate before imaging. *Dashed magenta box* indicates zoomed region to the *right*. *Red arrowheads* indicate high curvature fission products. Scale bars: 200 nm. (B) Distributions of vesicle diameter before protein exposure (*top*) and after exposure to ENTH (*bottom*) as quantified from TEM images (*left*) and the tethered vesicle assay (*right*). *Data from the same experiments reported in Snead, W., Hayden, C., Gadok, A., Zhao, C., Lafer, E., Rangamani, P., et al. (2017). Membrane fission by protein crowding.* Proceedings of the National Academy of Sciences of the United States of America, 114*(16), E3258–E3267.*

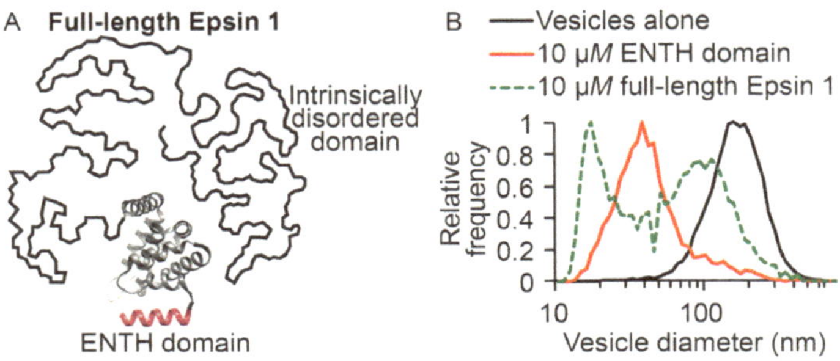

Fig. 6 The intrinsically disordered domain of Epsin 1 enhances membrane fission by the ENTH domain. (A) Schematic of full-length Epsin 1, which contains a large, intrinsically disordered domain and the folded ENTH domain (PDB 1H0A). Full-length Epsin 1 occupies a significantly larger projected area on the membrane surface compared to the ENTH domain alone, thereby enabling full-length Epsin 1 to reach a crowded coverage of the membrane surface with fewer membrane-bound protein copies compared to ENTH. (B) Tethered vesicle assay reveals that at the same concentration, full-length Epsin 1 generates a high curvature fission vesicle population not observed with the isolated ENTH domain. *Data from the same experiments reported in Snead, W., Hayden, C., Gadok, A., Zhao, C., Lafer, E., Rangamani, P., et al. (2017). Membrane fission by protein crowding.* Proceedings of the National Academy of Sciences of the United States of America, 114*(16), E3258–E3267.*

Knorr, & Ungewickell, 2002), we observed the appearance of a high curvature vesicle population at lower concentration of full-length Epsin 1 in comparison to the isolated ENTH domain (Fig. 6B). This finding was in agreement with results from SUPER templates (Fig. 1B), which showed that full-length Epsin 1 drove greater membrane release than the ENTH domain alone. Collectively, these results suggest that the disordered domain, which occupies a large membrane footprint (Busch et al., 2015), enabled the full-length protein to reach a crowded environment with fewer membrane-bound protein copies compared to the smaller ENTH domain. Our tethered vesicle assay facilitated this important observation, and helped to elucidate a new paradigm for membrane fission in which bulky, disordered proteins may be among the most potent drivers of fission. While the procedures described here focus primarily on study of the Epsin 1 protein, this assay can be readily adapted to study other fission-driving proteins. Ultimately, this assay may help elucidate new and interesting behaviors of essential fission proteins, improving our understanding of the molecular mechanisms responsible for fission.

ACKNOWLEDGMENTS

J.C.S. acknowledges funding from the National Institutes of Health (R01GM120549). W.T.S. acknowledges the support of a Ruth L. Kirschstein NRSA Predoctoral Fellowship from the National Institutes of Health (F31GM121013), as well as a fellowship from the Graduate School at UT Austin. The authors declare no conflict of interest.

REFERENCES

Aguet, F., Antonescu, C., Mettlen, M., Schmid, S., & Danuser, G. (2013). Advances in analysis of low signal-to-noise images link dynamin and AP2 to the functions of an endocytic checkpoint. *Developmental Cell*, *26*(3), 279–291.

Antonny, B., Burd, C., De Camilli, P., Chen, E., Daumke, O., Faelber, K., et al. (2016). Membrane fission by dynamin: What we know and what we need to know. *The EMBO Journal*, *35*(21), 2270–2284.

Boucrot, E., Pick, A., Camdere, G., Liska, N., Evergren, E., McMahon, H., et al. (2012). Membrane fission is promoted by insertion of amphipathic helices and is restricted by crescent BAR domains. *Cell*, *149*(1), 124–136.

Busch, D., Houser, J., Hayden, C., Sherman, M., Lafer, E., & Stachowiak, J. (2015). Intrinsically disordered proteins drive membrane curvature. *Nature Communications*, *6*, 7875.

Conner, S., & Schmid, S. (2003). Regulated portals of entry into the cell. *Nature*, *422*(6927), 37–44.

Ellenbroek, W., Wang, Y., Christian, D., Discher, D., Janmey, P., & Liu, A. (2011). Divalent cation-dependent formation of electrostatic PIP2 clusters in lipid monolayers. *Biophysical Journal*, *101*(9), 2178–2184.

Ford, M., Mills, I., Peter, B., Vallis, Y., Praefcke, G., Evans, P., et al. (2002). Curvature of clathrin-coated pits driven by epsin. *Nature*, *419*(6905), 361–366.

Franken, L., Boekema, E., & Stuart, M. (2017). Transmission electron microscopy as a tool for the characterization of soft materials: Application and interpretation. *Advanced Science*, *4*(5), 1600476.

Hatzakis, N., Bhatia, V., Larsen, J., Madsen, K., Bolinger, P., Kunding, A., et al. (2009). How curved membranes recruit amphipathic helices and protein anchoring motifs. *Nature Chemical Biology*, *5*(11), 835–841.

Helfrich, W. (1973). Elastic properties of lipid bilayers—Theory and possible experiments. *Zeitschrift Fur Naturforschung C*, *28*(11), 693–703.

Hofmann, H., Soranno, A., Borgia, A., Gast, K., Nettels, D., & Schuler, B. (2012). Polymer scaling laws of unfolded and intrinsically disordered proteins quantified with single-molecule spectroscopy. *Proceedings of the National Academy of Sciences of the United States of America*, *109*(40), 16155–16160.

Hoppins, S., Lackner, L., & Nunnari, J. (2007). The machines that divide and fuse mitochondria. *Annual Review of Biochemistry*, *76*, 751–780.

Hurley, J., Boura, E., Carlson, L., & Rozycki, B. (2010). Membrane budding. *Cell*, *143*(6), 875–887.

Jarsch, I., Daste, F., & Gallop, J. (2016). Membrane curvature in cell biology: An integration of molecular mechanisms. *Journal of Cell Biology*, *214*(4), 375–387.

Kalthoff, C., Alves, J., Urbanke, C., Knorr, R., & Ungewickell, E. (2002). Unusual structural organization of the endocytic proteins AP180 and epsin 1. *Journal of Biological Chemistry*, *277*(10), 8209–8216.

Kunding, A., Mortensen, M., Christensen, S., & Stamou, D. (2008). A fluorescence-based technique to construct size distributions from single-object measurements: Application to the extrusion of lipid vesicles. *Biophysical Journal*, *95*(3), 1176–1188.

Lohr, C., Kunding, A., Bhatia, V., & Stamou, D. (2009). Constructing size distributions of liposomes from single-object fluorescence measurements. *Methods in Enzymology Liposomes, 465*, 143–160. Pt. G.

McMahon, H., & Gallop, J. (2005). Membrane curvature and mechanisms of dynamic cell membrane remodelling. *Nature, 438*(7068), 590–596.

Mierzwa, B., & Gerlich, D. (2014). Cytokinetic abscission: Molecular mechanisms and temporal control. *Developmental Cell, 31*(5), 525–538.

Momin, N., Lee, S., Gadok, A., Busch, D., Bachand, G., Hayden, C., et al. (2015). Designing lipids for selective partitioning into liquid ordered membrane domains. *Soft Matter, 11*(16), 3241–3250.

Neumann, S., Pucadyil, T., & Schmid, S. (2013). Analyzing membrane remodeling and fission using supported bilayers with excess membrane reservoir. *Nature Protocols, 8*(1), 213–222.

Prevost, C., Tsai, F., Bassereau, P., & Simunovic, M. (2017). Pulling membrane nanotubes from giant unilamellar vesicles. *Journal of Visualized Experiments, 130*, e56086.

Pucadyil, T., & Schmid, S. (2008). Real-time visualization of dynamin-catalyzed membrane fission and vesicle release. *Cell, 135*(7), 1263–1275.

Schoneberg, J., Lee, I., Iwasa, J., & Hurley, J. (2017). Reverse-topology membrane scission by the ESCRT proteins. *Nature Reviews Molecular Cell Biology, 18*(1), 5–17.

Snead, W., Hayden, C., Gadok, A., Zhao, C., Lafer, E., Rangamani, P., et al. (2017). Membrane fission by protein crowding. *Proceedings of the National Academy of Sciences of the United States of America, 114*(16), E3258–E3267.

Stachowiak, J., Schmid, E., Ryan, C., Ann, H., Sasaki, D., Sherman, M., et al. (2012). Membrane bending by protein-protein crowding. *Nature Cell Biology, 14*(9), 944–949.

Stewart, J. (1980). Colorimetric determination of phospholipids with ammonium ferrothiocyanate. *Analytical Biochemistry, 104*(1), 10–14.

Takayama, M., Itoh, S., Nagasaki, T., & Tanimizu, I. (1977). A new enzymatic method for determination of serum choline-containing phospholipids. *Clinica Chimica Acta, 79*(1), 93–98.

Zimmerberg, J., & Kozlov, M. (2006). How proteins produce cellular membrane curvature. *Nature Reviews Molecular Cell Biology*, 7(1), 9–19.

CHAPTER NINETEEN

Generating Novel Materials Using the Intrinsically Disordered Protein Ubx

Gabriela Geraldo Mendes*,†, Rebecca M. Booth*,†, Donna L. Pattison‡, Adrian J. Alvarez*, Sarah E. Bondos*,†,1

*Texas A&M Health Science Center, College Station, TX, United States
†Texas A&M University, College Station, TX, United States
‡University of Houston, Houston, TX, United States
[1]Corresponding author: e-mail address: sebondos@medicine.tamhsc.edu

Contents

Abstract

The development of functionalized materials is needed to enable diverse applications. Protein-based materials are typically biocompatible and biodegradable and can exhibit a wide variety of useful mechanical properties. Most importantly, gene fusion enables facile incorporation of active proteins into the materials. However, many protocols rely on denaturing conditions to stimulate materials formation. These conditions would be expected to inactivate any appended functional proteins. This chapter describes methods to create protein fibers and films in a mild aqueous buffer near neutral pH. This facile, inexpensive single-pot approach to materials assembly does not require

Methods in Enzymology, Volume 611
ISSN 0076-6879
https://doi.org/10.1016/bs.mie.2018.08.007

any special equipment. Also included in this chapter are methods to fuse fibers to form fiber bundles, and to use fibers for cell culture. Although these methods were developed to generate materials from the *Drosophila* Hox transcription factor Ultrabithorax, they may also work for other self-assembling proteins, many of which have sequence features in common with Ubx.

1. INTRODUCTION

A major goal in materials engineering is the development of functionalized materials for use in diverse applications, including tissue engineering, biosensing, drug delivery, and photonics (Costa et al., 2018; Elsharkawy et al., 2018; Howell, Duran, Tsai, Bondos, & Bayless, 2016; Wei et al., 2017; Wolinsky, Colson, & Grinstaff, 2012). Realizing the potential of this exciting field requires the development of novel materials with a broad range of structural, mechanical, and functional properties optimized for each application. The inspiration for such materials is often drawn from biology, including the concept of hierarchical self-assembly of biopolymers to generate the materials' structure. Furthermore, embedding active biomolecules—often proteins—that are capable of functioning with unparalleled affinity and specificity has the potential to create bioactive materials (Huang et al., 2011; Nilebäck et al., 2017; Tsai et al., 2015; Zhou et al., 2015). Proteins offer many advantages in forming the scaffold, or basic structure, in these materials. First, recombinant protein technology allows facile production of monomers with uniform length and composition. Unlike organic polymers, the amino acid sequence of proteins can be easily engineered using standard molecular biology techniques to tune the properties of the resulting materials (Howell et al., 2015; Kim, Wong, Abou Neel, Chrzanowski, & Chan, 2014). Proteins are naturally biodegradable, and many protein-based materials are also biocompatible. Protein-based materials can exhibit a wide variety of useful mechanical properties: the tensile strength of spider dragline silk can rival that of structural steel, materials composed of elastin are extremely extensible, and collagen can form stiff three-dimensional gels (Koebley, Vollrath, & Schniepp, 2017; Li & Daggett, 2002; Wang, Lai, & Yang, 2016).

A key advantage of protein-based materials is the capacity to incorporate a specific protein, and thus its function, into materials by simple molecular biology approaches (Girotti, Orbanic, Ibáñez-Fonseca, Gonzalez-Obeso, & Rodríguez-Cabello, 2015; Huang et al., 2011; Jansson et al., 2014). In gene fusion, the DNA sequence encoding the self-assembling protein is attached to the gene encoding the functional protein without intervening stop codons. When this DNA sequence is transformed into a cell, the cell will produce a single polypeptide containing the amino acid sequences of both proteins (Fig. 1). Ideally, the materials-forming protein will retain the ability to self-assemble, and the activity of functional protein will be preserved in the resulting materials. This approach has numerous advantages for functionalizing materials. First, the functional protein does not have to be separately produced, reducing cost. Second, the material is uniformly and completely functionalized: every self-assembling protein monomer is attached to a functional protein. Third, only a single attachment site—a peptide bond—is formed between the functional protein and the self-assembling protein, thus uniformly orienting the functional protein. Fourth, the assembly and functionalization of the materials occur in a single step, eliminating the extra time and cost required for a multicomponent, multistep process. Finally, functional proteins, once embedded in protein materials, are often substantially stabilized, preserving their activity in otherwise hazardous conditions (Boothby et al., 2017; Tsai et al., 2015).

Several factors have prevented the widespread application of functionalization by gene fusion. The size, charge, or quaternary structure of the functional protein may misposition the self-assembling protein, slow assembly kinetics, or even prevent materials assembly (Jansson, Courtin, Sandgren, & Hedhammar, 2015; Tsai et al., 2015). Conversely, the function of an appended protein can be compromised either by the materials structure, or by harsh environmental conditions (e.g., heat, low pH, organic solvents) often encountered during materials assembly (Falconnet, Csucs, Grandin, & Textor, 2006; Hahn, Miller, & West, 2005; Huang et al., 2011; Tsai et al., 2015). This challenge has limited many self-assembling proteins to functionalization via peptides, which

Fig. 1 Primary sequence schematic of fusion proteins used to create functionalized materials, depicting a purification tag (*orange*), flexible linkers (*blue*), the functional protein sequence (*green*), and the self-assembling protein (*purple*). The functional protein can potentially be attached N-terminal or C-terminal of the self-assembling protein.

can still be a powerful tool. A subset of engineered proteins has successfully incorporated single-domain monomeric proteins via gene fusion, such as Enhanced Green Fluorescent Protein-Ultrabithorax (EGFP-Ubx) (Huang et al., 2011) and Z domain-4RepCT (Jansson et al., 2014). Multidomain, multimeric proteins have been incorporated into materials composed of the Hox transcription factor Ubx (Tsai et al., 2015).

Intrinsically disordered protein sequences play a major role in the design, formation, properties, and functionalization of protein-based materials. In many proteins that form materials, regions of intrinsic disorder alternate with structured protein regions, mimicking the arrangement of alternating hydrophobic–hydrophilic regions in block copolymers (Rabotyagova, Cebe, & Kaplan, 2011; Wesley et al., 2005). Intrinsically disordered proteins with low sequence complexities and multiple short-linear motifs may drive self-association and ultimately formation of phase-separated assemblies (Chavali, Gunnarsson, & Babu, 2017; Simon, Carroll, Rubinstein, Chilkoti, & López, 2017). Sequence enrichment in the structure-breaking amino acids proline and glycine is an integral feature of intrinsically disordered proteins that form materials, while the hydropathy and charge of the polypeptide determine the dependence of self-association on temperature (Quiroz & Chilkoti, 2015). Intrinsically disordered protein regions contribute to the mechanical properties of many protein-based materials, including silk, resilin, and Ubx (Balu et al., 2015; Giesa, Schuetz, Fratzl, Buehler, & Masic, 2017; Huang et al., 2010). Typically, materials composed of regularly arranged structured proteins are stiff (Young's modulus ~GPa) and less than 20% extensible (Guthold et al., 2007). In contrast, inclusion of intrinsically disordered regions in a protein sequence generates soft, extensible protein-based fibers (Young's modulus ~MPa, $\varepsilon_{max} > 100\%$) (Guthold et al., 2007). Indeed, the elastomeric properties of many proteins, including elastin, resilin, Ubx, and mussel adhesive protein, have been attributed to the presence of highly disordered sequences in these proteins (Huang et al., 2010; Qin, Hu, Cebe, & Kaplan, 2012; Tamburro, Bochicchio, & Pepe, 2005). In addition, the disordered regions of silk are required for supercontraction, the shrinkage of silk fibers along the main axis at high humidity (Giesa et al., 2017). Mutation of residues in the disordered regions can prevent or even reverse this behavior. Finally, the correlation of intrinsic disorder with extensibility suggests that these regions are relatively compact when the materials are not under tension. For the self-assembling protein Ubx, we previously hypothesized that these disordered regions can extend to accommodate the presence of a functional protein without sacrificing the interprotein contacts that stabilize the materials (Tsai et al., 2015). In support of this hypothesis, fibers composed of GFP-Ubx fusion proteins were less extensible than fibers composed only of the Ubx protein (Tsai et al., 2015).

A subset of proteins naturally self-assembles into useful materials. Common examples include silks, elastin, mussel adhesive proteins, bacterial S-layer proteins, and virus particles (Bellas et al., 2015; Bidwell, Fokt, Priebe, & Raucher, 2007; Dinjaski et al., 2017; Kim et al., 2011; Lee et al., 2009; Pum, Toca-Herrera, & Sleytr, 2013). This molecular toolbox is constantly being expanded by the enhancement of natural proteins sequences or the creation of novel proteins that form nanoscale to macroscale scaffolds (e.g., Yang, Holmberg, & Olsen, 2017). Herein, we describe facile methods to assemble materials using an unexpected protein source, the *Drosophila* Hox transcription factor Ubx, in gentle, aqueous conditions. During animal development, Ubx binds DNA to activate or repress the transcription of nearby genes. Although capable of cooperative DNA binding (Beachy et al., 1993), Ubx does not form materials as part of its natural function. Despite the lack of evolutionary selection for self-assembly, Ubx monomers hierarchically form ordered materials on a variety of length scales in vitro (Fig. 2). In buffers near neutral pH, Ubx coacervates at the air–water interface, generating disorganized clusters (aggregates) of molecules (Greer et al., 2009; Majithia, Patterson, Bondos, & Meissner, 2011). These clusters reshape themselves into lines, or protofibrils. Lateral interactions between protofibrils condense them into fibrils, which are similar in diameter to amyloid fibrils but lack amyloid structure. Fibrils interact laterally to form film, and film can be drawn into fibers that are meters long, with diameters ranging from 2 to 50 μm, depending on the thickness of the film. The strength of Ubx materials is derived from the rapid and spontaneous formation of dityrosine bonds, and mutation of tyrosine residues can either weaken or strengthen the resulting materials (Huang et al., 2010). The large intrinsically disordered regions within the Ubx sequence (Fig. 3) enable the successful incorporation into materials of a wide variety of proteins (Fig. 4), including fusions nearly 3 × the size of Ubx (Tsai et al., 2015). Thus, Ubx is a useful platform for generating active biomaterials. Furthermore, sequence similarities between Ubx and other self-assembling proteins suggest that this approach for generating fibers and film in aqueous conditions near neutral pH might be successfully applied to other protein systems.

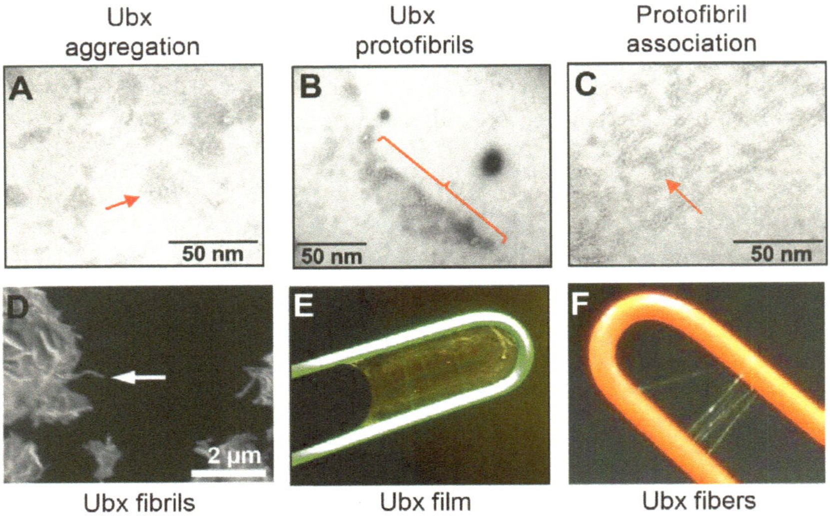

Fig. 2 Hierarchical assembly of Ubx materials. (A) About 15 min after dilution into a tray, Ubx begins to form irregular aggregates (*red arrow*). (B) Within 1 h, these aggregates begin to spontaneously reshape into lines of Ubx proteins, termed protofibrils. (C) After 1–2 h, the protofibrils align in parallel and begin to form side-to-side contacts. (D) After 2 h multiple protofibrils condense to form fibrils about 50 nm in diameter (*white arrow*). Fibrils also interact laterally to form films which can be lifted from the surface (E). Alternately, films on the air–water interface can be drawn into fibers composed of aligned fibrils (F). (A–C) 100-nm thick carbon films were placed on the surface of the Ubx solution, lifted using TEM grids, stained with 2% w/v phosphotungstic acid and imaged by TEM. (D) A film was transferred to a metal stub, sputter coated in gold, and imaged by scanning electron microscopy. *Panels (A)–(C): Reprinted from Majithia, R., Patterson, J., Bondos, S. E., & Meissner, K. E. (2011). On the design of composite protein-quantum dot biomaterials via self-assembly.* Biomacromolecules, 12*(10) 3629–3637. https://doi.org/10.1021/bm200889k with permission. Panel (D): Reprinted from Greer, A. M., Huang, Z., Oriakhi, A., Lu, Y., Lou, J., Matthews, K. S., et al. (2009). The Drosophila transcription factor Ultrabithorax self-assembles into protein-based biomaterials with multiple morphologies.* Biomacromolecules, 10*(4), 829–837. https://doi.org/10.1021/bm801315v with permission.*

2. EQUIPMENT

- A Teflon-coated 21.5 cm × 16.5 cm × 1.6 cm tray (Nordic Ware Toaster Oven Baking Sheet Cat No:43010)
- A disposable aluminum foil steam table pan, large enough to cover the Teflon-coated tray without touching it (for example, eHomeA2Z Heavy Duty Full-Size Deep Disposable Aluminum Foil Steam Table Pans for Cooking, Roasting, Broiling, Baking—21 × 13 × 3)

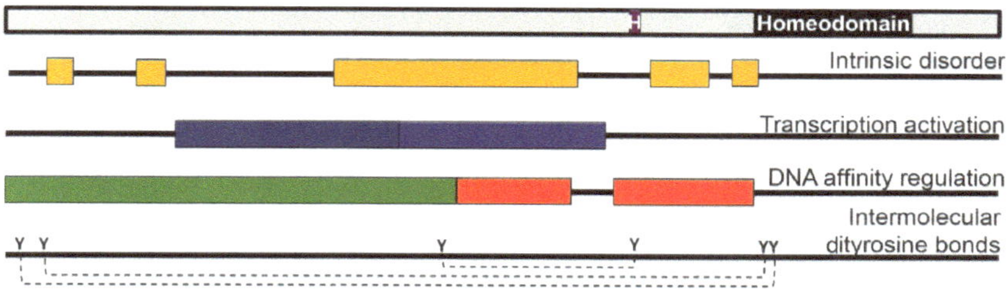

Fig. 3 Sequence schematic of the Ubx protein (*top, gray box*) depicting location of two characteristic features of Hox proteins, a hexapeptide motif (H, *purple box*), and the DNA binding homeodomain (*black box*). Bars below depict the location within the Ubx sequence of relevant features. *Yellow boxes* represent the location of regions that are intrinsically disordered, as determined by the average score from the VXLT-PONR, IUPRed, and Disembl—loops/coils algorithms and verified by native state proteolysis (Liu, Matthews, & Bondos, 2008). The transcription activation core domain (*dark blue*, 36% glycine) and the activation enhancing domain (*bright blue*, 21% glycine) together form the complete activation domain which is enriched in glycine (27%) (Tan, Bondos, Li, & Matthews, 2002). Regions of the Ubx sequence that either inhibit DNA binding (*red boxes*) or enhance DNA binding (*green boxes*) by the homeodomain are marked (Liu et al., 2008; Liu, Matthews, & Bondos, 2009). The location of tyrosines that form dityrosine bonds in materials is shown, with *dashed lines* connecting the tyrosines that form specific bonds (Howell et al., 2015). Although Ubx forms intermolecular dityrosine bonds in materials, they are depicted as intramolecular bonds for clarity.

- An 80-well microcentrifuge tube rack (e.g., VWR Cat No: 21150-294). A Teflon bar (1.3 cm × 1.3 cm × 24 cm) will also work
- Plastic-coated paper clips
- 10 μL yellow disposable polystyrene inoculating loops (VWR Cat No:12000-810)
- Siliconized glass microscope slides or glass microscope slides, Sigmacote® (Sigma-Aldrich SL2), fume hood, and Kimwipes
- A cap from a 50-mL centrifuge tube
- An empty styrofoam box
- A VWR® Multiwell Cell Culture Plate, 24 wells (VWR Cat No: 10062-896, for cell culture applications only)
- A biosafety cabinet (for cell culture applications only)

3. REAGENTS AND BUFFERS

- Buffer G (50 m*M* sodium phosphate monobasic, 500 m*M* NaCl, 5% glucose, pH 8.0), 600 mL
- Purified Ubx protein (2 mg, typically used as 2 mL of 1 mg/mL Ubx)
- Appropriate cell culture media

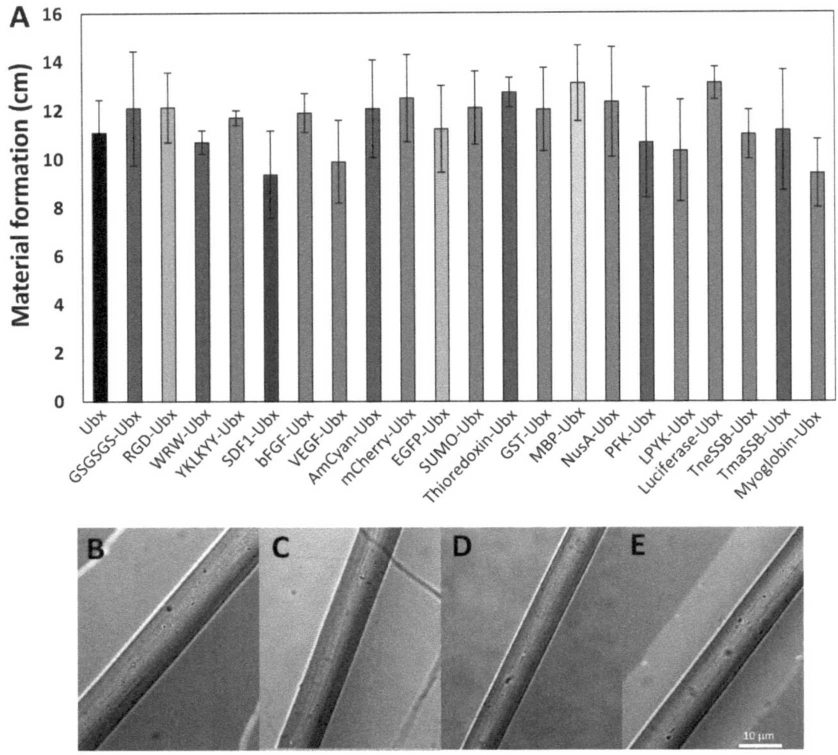

Fig. 4 The ability of materials to assemble is not significantly impacted by protein fusion. (A) Ubx variants (50 n*M*) were incubated for 16 h at room temperature. Materials formation was measured as length of fiber that can be drawn from the surface of the assembly solution. (B–E) Differential interference contrast microscopy of fibers formed by Ubx, VEGF-Ubx, MBP-Ubx, and Phosphofructokinase-Ubx, respectively. *Reprinted with permission from Tsai, S-. P., Howell, D. W., Huang, Z., Hsiao, H-. C., Lu, Y., Matthews, K. S., et al. (2015). The effect of protein fusions on the production and mechanical properties of protein-based materials.* Advanced Functional Materials, 25*(9), 1442–1450. http://doi.org/10.1002/adfm.201402997.*

4. PROCEDURES

A key advantage of the materials production approaches described below is that they are robust to a variety of environments and buffers. Although we typically perform these procedures at room temperature, they are also successful at elevated temperature and at 4°C. In addition, buffers other than Buffer G are compatible with the tray method for producing fiber and film. For instance, we have also successfully used phosphate-buffered saline. These procedures will tolerate pH values ranging from 6 to 8, NaCl concentrations ranging from 150 to 750 m*M*, and a variety of buffering agents including Tris, phosphate, and HEPES. A key limitation is that the Ubx variant being assembled into materials must be soluble in the selected buffer. A filter-based test for protein solubility (Bondos & Bicknell, 2003) can be used to rapidly screen buffer conditions if the buffer needs to be altered. These procedures are also sensitive to a few conditions. For both the tray method and the slide methods, large building vibrations can prevent film from forming and/or floating at the air–water interface. Because Ubx self-purifies as it assembles into materials (Greer et al., 2009), Ubx materials can be formed from crude cell lysate that has been centrifuged to remove solid cell debris.

Ubx materials derive their mechanical strength from the formation of dityrosine bonds between specific amino acids (Howell et al., 2015). Consequently, conditions that perturb bond formation also hamper materials assembly. Since these bonds form under oxidizing conditions, fibers drawn in the absence of oxygen are fragile, and lack the characteristic blue fluorescence that characterizes these bonds. Ubx assembly is also sensitive to the presence of Ubx proteolytic products. We hypothesize that these Ubx fragments act as chain terminators during assembly: these Ubx fragments include enough tyrosines to attach to the materials, but lack the tyrosines required for the next protein to bind. Consequently, a key aspect of the expression and purification of Ubx is the prevention of Ubx proteolysis and removal of any existing proteolytic products. Previous publications have described the purification of his-tagged and untagged Ubx protein (Greer et al., 2009; Huang et al., 2011; Liu et al., 2008).

We have developed two general methods for the production of Ubx films and fibers. In the first method, Ubx is diluted into buffer in a shallow tray, and the materials are drawn from the air–water interface. This most commonly used method provides an opportunity to change the buffer that hydrates the materials, and is the best approach when very long fibers are needed. For medium-scale experiments, trays holding smaller volumes (e.g., a whoopie pie tray) can be used. In the second (slide) method, a drop of concentrated protein rests on a siliconized surface where it assembles into materials. This is the fastest approach, and materials can be drawn after only 2 h of assembly time (Greer et al., 2009). Using higher protein concentrations can reduce the assembly time still further.

4.1 Tray Method for Production of Ubx Fibers

1. Place a metal tray on the lab bench, with the smaller side facing you, and fill with 600 mL of Buffer G[a]. The tray should be full, with the buffer domed slightly above the edges of the tray (Fig. 5A and B).
2. Slowly pipette 2 mL[b,c] of Ubx over the buffer, adding drop by drop. Move the pipette around the tray as you add protein to distribute the protein without stirring. Our average concentration of Ubx is 1 mg/mL.
3. Cover the tray containing Buffer G and Ubx with an upside-down aluminum foil steam table pan. Be careful not to touch the pan to the tray, or the liquid will spill.
4. Incubate the covered tray at room temperature from 6 h to 4 days.[d,e] We typically incubate for 16 h.

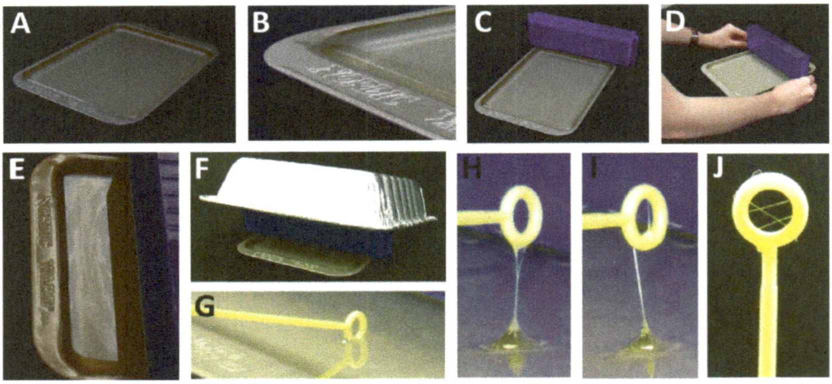

Fig. 5 Steps in Ubx fiber production. (A) Ubx is diluted into a shallow, Teflon-coated tray such that the liquid domes above the tray rim (B). (C) After incubation for at least 16 h under an aluminum pan, a microcentrifuge tube rack is placed on its side at the back of the tray (C) and slowly drawn forward (D) until it is roughly 5 cm from the front of the tray. (E) At this stage the film is visible as bright reflective areas on the film surface which resemble an oil slick. (F) The aluminum pan is placed on top of the tray and rack to prevent drafts from disturbing the film. (G) A *yellow inoculation loop* is touched to the surface and withdrawn (H) to reshape the film into a fiber. (I and J) Rotating the loop wraps the nascent fiber around the loop.

---

[a] Other buffers, e.g., PBS, also work.

[b] Add at least 1 mg of protein to the tray to ensure materials formation; more may be added to increase the yield of materials.

[c] Proteolytic products inhibit materials formation. Even adding a higher concentration of protein with proteolytic products can lead to fewer materials or low-quality materials with physical defects.

[d] Using longer incubation times creates more robust films which produce thicker and longer fibers.

[e] Very thick fibers (>35 μm diameter) are less uniform. The diameter can vary, and the film may occasionally fail to completely reshape into a cylinder, creating structural defects in the resulting fibers.

5. Remove the aluminum pan cover.
6. Gently place the long, narrow edge of the 80-well microcentrifuge tube rack on the far rim of the tray with the wells facing away from the tray (Fig. 5C). If the buffer has evaporated such that the rack is not flush against the surface of the buffer, gently insert a serological pipette into the buffer at the back edge of the pan and slowly add buffer such that the liquid is added below the surface so as not to disrupt any aggregates of biomaterial that have formed on the surface.
7. Slowly pull forward so the edge of the rack makes contact with the surface of the buffer, and then pull the rack forward until 2–5 cm remain between the front edge of the rack and the front of the tray (Fig. 5D). This procedure collects all of the film in a smaller area of the liquid surface.
8. When overhead lights reflect off the liquid, film, which looks like oil on water, can be seen on the surface of the buffer (Fig. 5E).
9. Recover the tray and rack with the aluminum foil container (Fig. 5F), being careful to not move the rack or touch the liquid in the tray. Incubate for at least an hour to allow the different "islands" of film time to interact.
10. Remove the aluminum pan cover.
11. Hold an inoculation loop above the surface with the loop facing you, and one curved side facing the liquid surface. Gently touch a yellow inoculation loop to the surface of the buffer where there is film, pushing the leading edge of the loop just below the surface. Slowly move the loop away from the surface to pull a fiber (Fig. 5G–J). You may initially draw film instead of fiber, but the film will collapse into a fiber as you continue to pull. Wrap the fiber around the loop to support the fiber.[f]
12. There are two ways to finish drawing a fiber and attaching it to a loop (or other support):
    i. To separate the fiber from the tray, quickly jerk the loop upward to break the fiber. If this procedure results in a dangling end, it can either be wrapped around the loop or removed with scissors.
    ii. To transfer the fiber to another loop, touch a new loop to the fiber below the first loop and pull the old loop away with a small, but quick jerk motion.

[f] Paper clips, reshaped into a candy cane structure (Fig. 7C) or other structures, can also be used for this procedure to support fewer, but longer fibers. Ubx materials will stick to plastic, Teflon, metal, and glass (Greer et al., 2009).

13. Although most fibers are produced in the first 24 h after setting up the tray, the tray will continue to yield fibers for several days, until the protein film is depleted.
14. To reset the tray such that fibers can be drawn on the following day, gently remove the rack, being very careful not to spill buffer. You may have to add additional buffer behind the rack to counteract evaporation. As shown in (Fig. 5A and B), the buffer should form a dome above the tray rim. The tray can then be covered as in step 3, and subsequent steps repeated the following day.
15. When you are finished making materials, the Teflon-coated tray, the aluminum pan, and the microcentrifuge tube rack can all be washed and reused.

4.2 Tray Method for Production of Unstabilized Ubx Film

1. Follow steps 1–8 of Section 4.1. If the film is visible on the liquid surface, then film can be lifted from the tray. Longer incubation times, both before and after squeezing the surface with the rack, will produce more robust films that are easier to handle.
2. Fold a plastic-coated paper clip (or floral wire) into a candy cane or shephard's crook shape such that there is a long and a short parallel arms.
3. Hold the reshaped paper clip by the long arm. Touch the entire short edge of the paper clip to the liquid surface just adjacent to where you can see film (Fig. 6A).
4. Slowly move the paper clip into the area with film, and then away from the surface to lift the film. Simultaneously rotate the paper clip along the axis of the long arm (Fig. 6B) such that the long arm is now facing down.
5. Allow the long arm of the paper clip to contact the film. The film should now be stretched between the long and short arms of the paper clip (Fig. 6C).
6. The indentations caused by packing of the nanoscale fibrils in the film can be seen in the reflection of the overhead lights. They should be oriented perpendicular to the two arms of the paperclip, in the direction in which the film was drawn.
7. The film will bow in and try to collapse along its short edges. Continuing to wrap the film around the paper clip will produce a thicker, more stable structure.

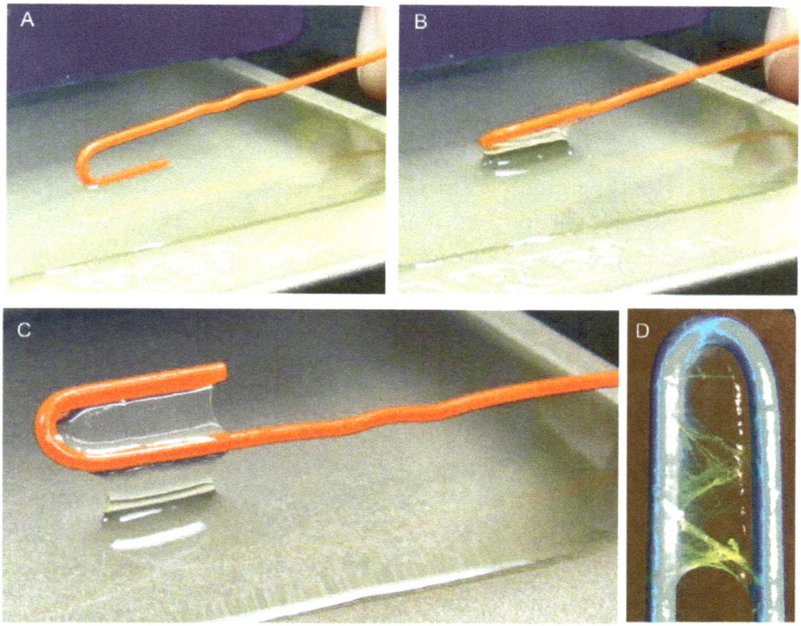

Fig. 6 The procedure to lift Ubx film from a tray. (A) The short arm of a reshaped paper clip is touched to the surface adjacent to an area with film. (B) The paper clip is rotated. As the short arm begins to lift film, the long arm contacts and supports the film. (C) Continued rotation of the paper clip allows multiple films to meld to one another, creating a thicker, more robust structure. (D) Alternately, a paper clip already wrapped in fibers can be used. Structural support provided by the fibers (*green*) can support a single layer of film (clear and transparent, but reflects light). This procedure is compatible with all Ubx variants able to form materials. EGFP-Ubx was used for these panels to increase fiber visibility.

4.3 Tray Method for Production of Stabilized Ubx Film

1. A second way to stabilize a film is to first wrap the paper clip with fibers to support the film structure.
2. Once the paper clip is wrapped in fibers, follow the directions in Section 4.2 to add film. A fiber-stabilized film is shown in Fig. 6D.

4.4 Slide Method for Production of Ubx Fibers

The slide method is a faster approach to materials assembly, but can be challenging in very dry air because dehydration can be faster than materials assembly. If the drop is drying before fibers can be drawn, place a small (e.g., 10 mL) beaker about half full of water next to the drop, and cover the drop and small beaker with a larger beaker instead of the 50-mL centrifuge tube. It is important to place the drop on a siliconized surface to maximize water beading. The deeper water bead is more resistant to dehydration.

Although the slide method can also be used to produce film, these films tend to be much thicker (termed sheets in Greer et al., 2009) and they lack the ridged surface, derived from fibrils packing in parallel, that is found in thin films.

1. Place the glass slide on the lab bench. If the slide has already been siliconized, then skip steps 2 through 4. Alternately, a siliconized cover slip can be affixed to a regular slide using clear nail polish.
2. If the slide has not been siliconized, open a bottle of Sigmacote in a fume hood. Wad three Kimwipe sheets together and press against the bottle opening such that the entire opening is covered. Invert the bottle until the Kimwipes are soaked about halfway through with Sigmacote.
3. Immediately (before the Kimwipes dry) rub the Sigmacote on the surface of the microscope slide. Let the slide air dry.
4. Rinse the slide with water and dry with clean Kimwipes.
5. Pipette 200 μL of Ubx into a single drop on the slide. Up to two drops may be placed on a single slide. Multiple slides may be used to increase the amount of materials collected (Fig. 7A).
6. Leave the protein drop exposed to air for 2–4 h.[g] Alternately, cover the drop with the cap from a 50-mL centrifuge tube (Fig. 7B) and leave at room temperature overnight.
7. Gently touch a plastic inoculation loop to the surface of the drop and slowly move the loop away to pull a fiber (Fig. 7C and D). Wrap the fiber around the loop to hold the fiber.
8. If the fiber breaks, you start a new one by touching this loop or a new loop to the drop surface.
9. To finish drawing a fiber and attaching it to a loop (or other support):
   i. To separate the fiber from the tray, quickly jerk the loop upward to break the fiber. If this procedure results in a dangling end, it can either be wrapped around the loop or removed with scissors.
   ii. To transfer the fiber to another loop, touch a new loop to the fiber below the first loop and pull the old loop away with a quick motion.

[g] Drops can be left for a longer period of time, but if too much liquid evaporates crystals will form and harvesting the materials becomes difficult.

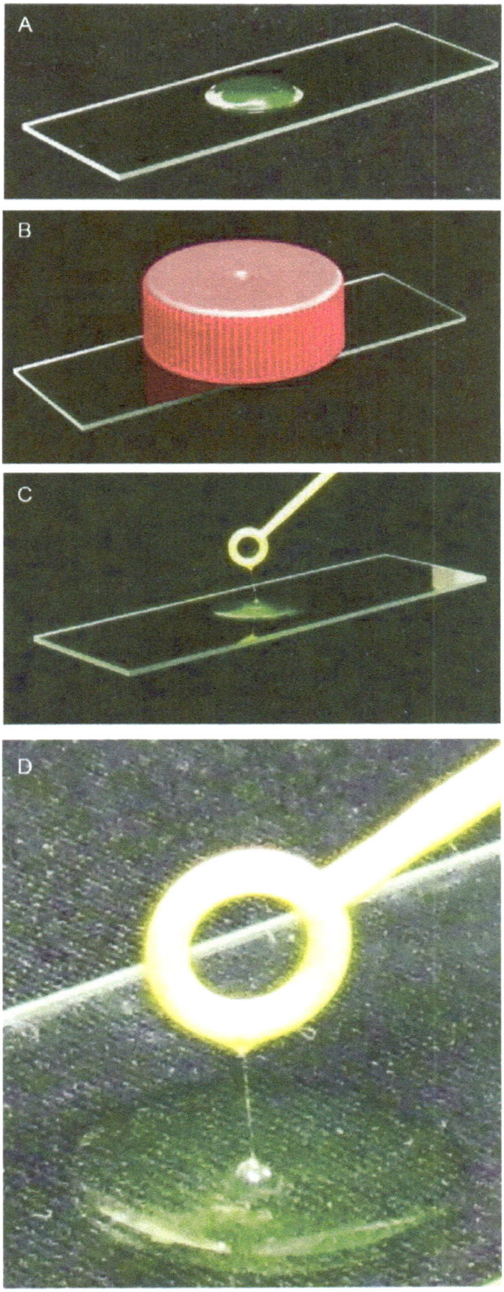

Fig. 7 Drawing fibers from a resting drop. (A) 200 μL of EGFP-Ubx was pipetted on a siliconized microscope slide and (B) covered with the cap of a 50-mL disposable centrifuge tube and incubated overnight. (C) The edge of an inoculation loop was touched to the film surface and (D) withdrawn to form a fiber. This procedure is compatible unmodified Ubx, Ubx mutants, and Ubx fusion proteins. EGFP-Ubx was used for these panels to increase fiber visibility.

4.5 Production of Ubx Fiber Bundles

Fibers continue to form dityrosine bonds just after they are drawn, allowing fibers in bundles to covalently bond to one another. Freshly drawn fibers, when bundled, will completely fuse to form a single cylinder, whereas fibers exposed to air for >5 min only partially fuse, leaving a small crevice visible at the fiber–fiber interface (Greer et al., 2009). Fibers that have completely bonded (after ~30 min) will not be able to form bundles, and will not adhere to themselves or other Ubx materials.

Bundles can be produced using fibers supported by loops or by a paper clip. However, it is easier to place the fibers in parallel and in close proximity when they are drawn with a paper clip.

1. Begin by drawing fibers with a paper clip bent to form a candy cane shape such that the fibers that will be fused in step 5 are in close proximity and roughly parallel to one another (Fig. 8A). The more distant the fibers are initially, the less long the fused region will be.
2. Place a drop (~200 μL) of Buffer G[h] on a siliconized microscope slide. The drop should form a deep dome, and not be spread into a thin layer. Instructions for siliconizing a slide can be found in Section 4.4, steps 2 through 4.
3. Hold the paper clip by the long arm over the drop. The plane that includes the long and short arms of the paper clip should be parallel to the lab bench.
4. Slowly lower the paper clip until the fibers are submerged in the drop, and the middle of the fiber group aligns with the center of the drop.
5. Raise the paper clip slowly (Fig. 8B). The fibers should stretch toward the midline of the drop, thus remaining submerged. The fibers will come into contact and form a bundle as they exit the liquid. The bundles may need to dry to remove trapped water (Fig. 8C).

4.6 Drying and Rehydrating Fibers

Dehydrated fibers are easy to store, and can be safely shipped in their Styrofoam box if needed. Although dried fibers are not flexible, they regain flexibility upon rehydration. Dried fibers appear to retain their activity. For instance, we rehydrated a fiber composed of EGFP-Ubx fusion protein that had been stored dry for 9 years at room temperature, and examination by microscopy revealed that it still fluoresced green.

[h] If fibers were drawn using the tray method using another buffer, then this other buffer should be used to create fiber bundles.

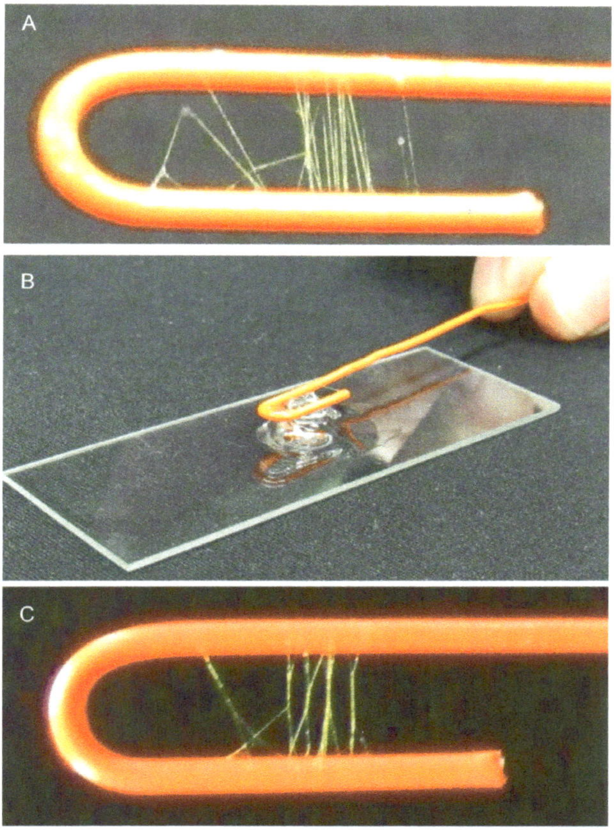

Fig. 8 Fusing fibers into fiber bundles. (A) A fiber was wrapped multiple times around a paper clip and (B) dipped into a 200-μL drop of buffer on a siliconized microscope slide. (C) The fibers melded to form five fiber bundles. Although this procedure is compatible with multiple Ubx variants, EGFP-Ubx was used for these panels to increase fiber visibility.

1. To dry fibers in air:
   i. For fibers wrapped around paper clips, insert the long arm of the paper clip into the inside face of the lid of a Styrofoam container.
   ii. For fibers wrapped around inoculation loops, use scissors to cut the loop such than a handle of convenient length (~7 cm) remains attached to the loop. Insert the handle into the lid of a Styrofoam container.
   iii. Leaving the lid on the bench, gently place the Styrofoam box over the top of the lid being careful not to bump the samples. Allow the fiber to dry while protected from drafts and accidental contact.

2. To rehydrate the fibers:
   i. Thoroughly wet a paper towel. Do not ring out the paper towel or it will be too dry. If the towel is dripping it is too wet.
   ii. Place the paper towel in the box such that the towel is not in contact with any loops, paper clips, or Ubx materials (Fig. 9A).
   iii. Incubate for 24h at room temperature. The fibers will be rehydrated.

4.7 Preparing Fibers for Cell Culture

The ability of most eukaryotic cells to adhere to Ubx fibers and films (Howell et al., 2016; Hsiao et al., 2016; Patterson, Abbey, Bayless, & Bondos, 2014; Patterson et al., 2015), combined with the ability to produce materials composed of active Ubx fusion proteins (Huang et al., 2011; Tsai et al., 2015), makes Ubx materials a powerful tool for cell culture studies and potentially a useful scaffold for tissue engineering. The fact that Ubx materials float makes maintaining contact between Ubx materials and cells more difficult. To accomplish this, media levels must be low enough to prevent the loop from floating and the bottom layer of materials will stay in contact with the cells.

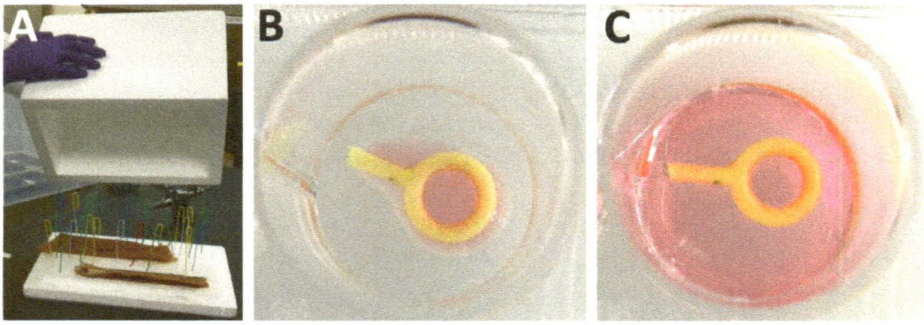

Fig. 9 (A) Rehydrating fibers supported by paper clips. (B) To prepare fibers for cell culture, an inoculation loop can be placed in a clean well of a 24-well plate, followed by the addition of cells and media inside the loop and some media along the outer edge of the loop to prevent dehydration. (C) Alternately, for migration assays cells are first be cultured in a well of a 24-well plate before placing the fibers in the well. After placing the loop on the floor of the well, 125 μL of cells and media are pipetted inside the loop and 125 μL of media only are pipetted around the outer rim of the loop such that the media cover the bottom of the well but the loop does not float.

1. For a cell migration assay, seed cells in media in 24-well plates at the desired concentration (usually 60k cells per 100 μL).[i]
2. Use media to bring the total volume to 1 mL.
3. Culture for 1–2 days until there is a confluent layer of cells.[j]
4. Remove old media and add 500–750 μL of new media.
5. Incubate for 6 h.
6. Remove media.
7. Very gently tilt the plate so that media collect at one edge, then aspirate.
8. If the fiber-wrapped loops need to be brought to the tissue culture area from another room, they should transported in a sterile closed container as follows:
   i. In a sterile biosafety hood, sterilize a Styrofoam box with ethanol.
   ii. Once the box is dry, insert the loop handle in the Styrofoam and close the box for transport.
9. Add to the well fibers pulled in a sterile biosafety hood to lower risk of contamination and wrapped around an inoculation loop. You will need to cut the loop handle shorted to fit in the well.
10. Place plastic inoculation loops wrapped with two to three fibers each in each well.
11. Gently pipet more media until loop is submerged but not floating. We usually pipette ~50 μL of media inside the loop and 200 μL outside (but touching) the loop bringing the total media volume to 250 μL. (Fig. 9A)
12. Incubate and perform desired analysis.

5. SUMMARY AND CONCLUSION

The amino acid sequence of Ubx has features in common with those of other proteins that phase separate and form materials. Elastin-like proteins are enriched in P-X_n-G motifs, where n ranges from 0 to 4 (Quiroz & Chilkoti, 2015; Rauscher, Baud, Miao, Keeley, & Pomès, 2006). In the Ubx sequence 50% of the prolines (7 of 14) fit this motif sequence. Proteins enriched in this motif also contain regions enriched in glycine (>30%) (Quiroz & Chilkoti, 2015). The Ubx transcription activation domain (amino acids 68–242) contains 27% glycine (Fig. 3) (Tan et al., 2002). Within the Ubx activation

[i] For other assays, the loop may be placed in a clean well, and cells and media may only be pipetted into the center of the loop as in Fig. 9B.

[j] The optimal cell density will depend on the cell line used and the assay used to analyze cell behavior on the fibers (Howell et al., 2016).

domain, the activation enhancing region corresponds to amino acids 159–242 and is over 36% glycine. In part due to this enrichment in glycine, Ubx contains 13 GGX motifs, which are found in both elastin and silk (Giesa et al., 2017). These sequence similarities suggest that, although the methods described in this chapter were developed for Ubx materials, these approaches might also be useful for other protein-based materials.

These methods described in this chapter have several advantages. First and foremost, these methods allow monomers to assemble into materials in mild aqueous buffers at neutral pH, enabling a wide variety of fused proteins to retain their activity (Howell et al., 2016). Second, these approaches only require a single buffer and inexpensive supplies and equipment, thus reducing the cost of materials production. Third, although fibers that are meters long can be drawn using the tray method, the approach can be easily scaled-up. Fourth, functionalized materials can be easily generated using this one-pot, single-step approach.

REFERENCES

Balu, R., Knott, R., Cowieson, N. P., Elvin, C. M., Hill, A. J., Choudhury, N. R., et al. (2015). Structural ensembles reveal intrinsic disorder for the multi-stimuli responsive bio-mimetic protein Rec1-resilin. *Scientific Reports*, *5*, 10896. https://doi.org/10.1038/srep10896.

Beachy, P. A., Varkey, J., Young, K. E., von Kessler, D. P., Sun, B. I., & Ekker, S. C. (1993). Cooperative binding of an Ultrabithorax homeodomain protein to nearby and distant DNA sites. *Molecular and Cellular Biology*, *13*(11), 6941–6956. https://doi.org/10.1128/MCB.13.11.6941.

Bellas, E., Lo, T. J., Fournier, E. P., Brown, J. E., Abbott, R. D., Gil, E. S., et al. (2015). Injectable silk foams for soft tissue regeneration. *Advanced Healthcare Materials*, *4*(3), 452–459. https://doi.org/10.1002/adhm.201400506.

Bidwell, G. L., 3rd, Fokt, I., Priebe, W., & Raucher, D. (2007). Development of elastin-like polypeptide for thermally targeted delivery of doxorubicin. *Biochemical Pharmacology*, *73*(5), 620–631. https://doi.org/10.1016/j.bcp.2006.10.028.

Bondos, S. E., & Bicknell, A. (2003). Detection and prevention of protein aggregation before, during, and after purification. *Analytical Biochemistry*, *316*(2), 223–231. https://doi.org/10.1016/S0003-2697(03)00059-9.

Boothby, T. C., Tapia, H., Brozena, A. H., Piszkiewicz, S., Smith, A. E., Giovannini, I., et al. (2017). Tardigrades use intrinsically disordered proteins to survive desiccation. *Molecular Cell*, *65*(6), 975–984. https://doi.org/10.1016/j.molcel.2017.02.018.

Chavali, S., Gunnarsson, A., & Babu, M. M. (2017). Intrinsically disordered proteins adaptively reorganize cellular matter during stress. *Trends in Biochemical Sciences*, *42*(6), 410–412. https://doi.org/10.1016/j.tibs.2017.04.007.

Costa, S. A., Simon, J. R., Amiram, M., Tang, L., Zauscher, S., Brustad, E. M., et al. (2018). Photo-crosslinkable unnatural amino acids enable facile synthesis of thermoresponsive nano- to microgels of intrinsically disordered polypeptides. *Advanced Materials*, *30*(5), 1704878. https://doi.org/10.1002/adma.201704878.

Dinjaski, N., Plowright, R., Zhou, S., Belton, D. J., Perry, C. C., & Kaplan, D. L. (2017). Osteoinductive recombinant silk fusion proteins for bone regeneration. *Acta Biomaterialia*, *49*, 127–139. https://doi.org/10.1016/j.actbio.2016.12.002.

Elsharkawy, S., Al-Jawad, M., Pantano, M. F., Tejeda-Montes, E., Mehta, K., Jamal, H., et al. (2018). Protein disorder-order interplay to guide the growth of hierarchical mineralized structures. *Nature Communications*, *9*, 2145. https://doi.org/10.1038/s41467-018-04319-0.

Falconnet, D., Csucs, G., Grandin, H. M., & Textor, M. (2006). Surface engineering approaches to micropattern surfaces for cell-based assays. *Biomaterials*, *27*(16), 3044–3063. https://doi.org/10.1016/j.biomaterials.2005.12.024.

Giesa, T., Schuetz, R., Fratzl, P., Buehler, M. J., & Masic, A. (2017). Unraveling the molecular requirements for macroscopic silk supercontraction. *ACS Nano*, *11*(10), 9750–9758. https://doi.org/10.1021/acsnano.7b01532.

Girotti, A., Orbanic, D., Ibáñez-Fonseca, A., Gonzalez-Obeso, C., & Rodríguez-Cabello, J. C. (2015). Recombinant technology in the development of materials and systems for soft-tissue repair. *Advanced Healthcare Materials*, *4*(16), 2423–2455. https://doi.org/10.1002/adhm.201500152.

Greer, A. M., Huang, Z., Oriakhi, A., Lu, Y., Lou, J., Matthews, K. S., et al. (2009). The Drosophila transcription factor Ultrabithorax self-assembles into protein-based biomaterials with multiple morphologies. *Biomacromolecules*, *10*(4), 829–837. https://doi.org/10.1021/bm801315v.

Guthold, M., Liu, W., Sparks, E. A., Jawerth, L. M., Peng, L., Falvo, M., et al. (2007). A comparison of the mechanical and structural properties of fibrin fibers with other protein fibers. *Cell Biochemistry and Biophysics*, *49*(3), 165–181. https://doi.org/10.1007/s12013-007-9001-4.

Hahn, M. S., Miller, J. S., & West, J. L. (2005). Laser scanning lithography for surface micropatterning on hydrogels. *Advanced Materials*, *17*(24), 2939–2942. https://doi.org/10.1002/adma.200500184.

Howell, D. W., Duran, C. L., Tsai, S.-P., Bondos, S. E., & Bayless, K. J. (2016). Functionalization of Ultrabithorax materials with vascular endothelial growth factor enhances angiogenic activity. *Biomacromolecules*, *17*(11), 3558–3569. https://doi.org/10.1021/acs.biomac.6b01068.

Howell, D. W., Tsai, S.-P., Churion, K., Patterson, J., Abbey, C., Atkinson, J. T., et al. (2015). Identification of multiple dityrosine bonds in materials composed of the Drosophila protein Ultrabithorax. *Advanced Functional Materials*, *25*(37), 5988–5998. https://doi.org/10.1002/adfm.201502852.

Hsiao, H.-C., Santos, A., Howell, D. W., Patterson, J. L., Fuchs-Young, R. S. L., & Bondos, S. E. (2016). Culture of tumorigenic cells on protein fibers reveals metastatic cell behaviors. *Biomacromolecules*, *17*(11), 3790–3799. https://doi.org/10.1021/acs.biomac.6b01311.

Huang, Z., Lu, Y., Majithia, R., Shah, J., Meissner, K., Matthews, K. S., et al. (2010). Size dictates mechanical properties for protein fibers self-assembled by the *Drosophila* Hox transcription factor Ultrabithorax. *Biomacromolecules*, *11*(12), 3644–3651. https://doi.org/10.1021/bm1010992.

Huang, Z., Salim, T., Brawley, A., Patterson, J., Matthews, K. S., & Bondos, S. E. (2011). Functionalization and patterning of protein-based materials using active Ultrabithorax chimeras. *Advanced Functional Materials*, *21*(14), 2633–2640. https://doi.org/10.1002/adfm.201100067.

Jansson, R., Courtin, C. M., Sandgren, M., & Hedhammar, M. (2015). Rational design of spider silk materials genetically fused with an enzyme. *Advanced Functional Materials*, *25*(33), 5343–5352. https://doi.org/10.1002/adfm.201501833.

Jansson, R., Thatikonda, N., Lindberg, D., Rising, A., Johansson, J., Nygren, P. A., et al. (2014). Recombinant spider silk genetically functionalized with affinity domains. *Biomacromolecules*, *15*(5), 1696–1706. https://doi.org/10.1021/bm500114e.

Kim, C. S., Choi, Y. S., Ko, W., Seo, J. H., Lee, J., & Cha, H. J. (2011). A mussel adhesive protein fused with the BC domain of protein a is a functional linker material that efficiently immobilizes antibodies onto diverse surfaces. *Advanced Functional Materials*, *21*(21), 4101–4108. https://doi.org/10.1002/adfm.201100710.

Kim, S. Y., Wong, A. H., Abou Neel, E. A., Chrzanowski, W., & Chan, H. K. (2014). The future perspectives of natural materials for pulmonary drug delivery and lung tissue engineering. *Expert Opinion on Drug Delivery*, *12*(6), 869–887. https://doi.org/10.1517/17425247.2015.993314.

Koebley, S. R., Vollrath, F., & Schniepp, H. C. (2017). Toughness-enhancing metastructure in the recluse spider's looped ribbon silk. *Materials Horizons*, *4*(3), 377–382. https://doi.org/10.1039/C6MH00473C.

Lee, Y. J., Yi, H., Kim, W. J., Kang, K., Yun, D. S., Strano, M. S., et al. (2009). Fabricating genetically engineered high-power lithium-ion batteries using multiple virus genes. *Science*, *324*(5930), 1051–1055. https://doi.org/10.1126/science.1171541.

Li, B., & Daggett, V. (2002). Molecular basis for the extensibility of elastin. *Journal of Muscle Research and Cell Motility*, *23*(5–6), 561–573. https://doi.org/10.1023/A:1023474909980.

Liu, Y., Matthews, K. S., & Bondos, S. E. (2008). Multiple intrinsically disordered sequences alter DNA binding by the homeodomain of the Drosophila Hox protein Ultrabithorax. *The Journal of Biological Chemistry*, *283*(30), 20874–20887. https://doi.org/10.1074/jbc.M800375200.

Liu, Y., Matthews, K. S., & Bondos, S. E. (2009). Internal regulatory interactions determine DNA binding specificity by a Hox transcription factor. *The Journal of Molecular Biology*, *390*(4), 760–774. https://doi.org/10.1016/j.jmb.2009.05.059.

Majithia, R., Patterson, J., Bondos, S. E., & Meissner, K. E. (2011). On the design of composite protein-quantum dot biomaterials via self-assembly. *Biomacromolecules*, *12*(10), 3629–3637. https://doi.org/10.1021/bm200889k.

Nilebäck, L., Hedin, J., Widhe, M., Floderus, L. S., Krona, A., Bysell, H., et al. (2017). Self-assembly of recombinant silk as a strategy for chemical-free formation of bioactive coatings: A real-time study. *Biomacromolecules*, *18*(3), 846–854. https://doi.org/10.1021/acs.biomac.6b01721.

Patterson, J. L., Abbey, C. A., Bayless, K. J., & Bondos, S. E. (2014). Materials composed of the Drosophila melanogaster protein Ultrabithorax are cytocompatible. *Journal of Biomedical Materials Research Part A*, *102*(1), 97–104. https://doi.org/10.1002/jbm.a.34675.

Patterson, J. L., Arenas-Gamboa, A. M., Wang, T.-Y., Hsiao, H.-C., Howell, D. W., Pellois, J.-P., et al. (2015). Materials composed of the Drosophila Hox protein Ultrabithorax are biocompatible and nonimmunogenic. *Journal of Biomedical Materials Research Part A*, *103*(4), 1546–1553. https://doi.org/10.1002/jbm.a.35295.

Pum, D., Toca-Herrera, J. L., & Sleytr, U. B. (2013). S-layer protein self-assembly. *International Journal of Molecular Sciences*, *14*(2), 2484–2501. https://doi.org/10.3390/ijms14022484.

Qin, G., Hu, X., Cebe, P., & Kaplan, D. L. (2012). Mechanism of resilin elasticity. *Nature Communications*, *3*, 1003. https://doi.org/10.1038/ncomms2004.

Quiroz, F. G., & Chilkoti, A. (2015). Sequence heuristics to encode phase behavior in intrinsically disordered protein polymers. *Nature Materials*, *14*, 1164–1171. https://doi.org/10.1038/nmat4418.

Rabotyagova, O. S., Cebe, P., & Kaplan, D. L. (2011). Protein-based block copolymers. *Biomacromolecules*, *12*(2), 269–289. https://doi.org/10.1021/bm100928x.

Rauscher, S., Baud, S., Miao, M., Keeley, F. W., & Pomès, R. (2006). Proline and glycine control protein self-organization into elastomeric amyloid fibrils. *Structure*, *14*(11), 1667–1676. https://doi.org/10.1016/j.str.2006.09.008.

Simon, J. R., Carroll, N. J., Rubinstein, M., Chilkoti, A., & López, G. P. (2017). Programming molecular self-assembly of intrinsically disordered proteins containing sequences of low complexity. *Nature Chemistry*, *9*, 509–515. https://doi.org/10.1038/nchem.2715.

Tamburro, A. M., Bochicchio, B., & Pepe, A. (2005). The dissection of human tropoelastin: From the molecular structure to the self-assembly to the elasticity mechanism. *Pathologie Biologie*, *53*(7), 383–389. https://doi.org/10.1016/j.patbio.2004.12.014.

Tan, X.-X., Bondos, S., Li, L., & Matthews, K. S. (2002). Transcription activation by Ultrabithorax Ib protein requires a predicted α-helical region. *Biochemistry*, *41*(8), 2774–2785. https://doi.org/10.1021/bi011967y.

Tsai, S.-P., Howell, D. W., Huang, Z., Hsiao, H.-C., Lu, Y., Matthews, K. S., et al. (2015). The effect of protein fusions on the production and mechanical properties of protein-based materials. *Advanced Functional Materials*, *25*(9), 1442–1450. https://doi.org/10.1002/adfm.201402997.

Wang, T., Lai, J. H., & Yang, F. (2016). Effects of hydrogel stiffness and extracellular compositions on modulating cartilage regeneration by mixed populations of stem cells and chondrocytes in vivo. *Tissue Engineering Part A*, *22*(23–24), 1348–1356. https://doi.org/10.1089/ten.tea.2016.0306.

Wei, G., Su, Z., Reynolds, N. P., Arosio, P., Hamley, I. W., Gazit, E., et al. (2017). Self-assembling peptide and protein amyloids: From structure to tailored function in nanotechnology. *Chemical Society Reviews*, *46*(15), 4661–4708. http://doi.org10.1039/C6CS00542J/.

Wesley, R. D., Dreiss, C. A., Cosgrove, T., Armes, S. P., Thompson, L., Baines, F. L., et al. (2005). Structure of a hydrophilic-hydrophobic block copolymer and its interactions with salt and an anionic surfactant. *Langmuir*, *21*(11), 4856–4861. https://doi.org/10.1021/la046830y.

Wolinsky, J. B., Colson, Y. L., & Grinstaff, M. W. (2012). Local drug delivery strategies for cancer treatment: Gels, nanoparticles, polymeric films, rods, and wafers. *Journal of Controlled Release*, *159*(1), 14–26. https://doi.org/10.1016/j.jconrel.2011.11.031.

Yang, Y. J., Holmberg, A. L., & Olsen, B. D. (2017). Artificially engineered protein polymers. *Annual Review of Chemical and Biomolecular Engineering*, *8*, 549–575. https://doi.org/10.1146/annurev-chembioeng-060816-101620.

Zhou, S., Huang, W., Belton, D. J., Simmons, L. O., Perry, C. C., Wang, X., et al. (2015). Control of silicification by genetically engineered fusion proteins: Silk-silica binding peptides. *Acta Biomaterialia*, *15*, 173–180. https://doi.org/10.1016/j.actbio.2014.10.040.

FURTHER READING

Breitwieser, A., Iturri, J., Toca-Herrera, J. L., Sleytr, U. B., & Pum, D. (2017). In vitro characterization of the two-stage non-classical reassembly pathway of S-layers. *International Journal of Molecular Sciences*, *18*(2), 400. https://doi.org/10.3390/ijms18020400.

CHAPTER TWENTY

Challenges in the Structural–Functional Characterization of Multidomain, Partially Disordered Proteins CBP and p300: Preparing Native Proteins and Developing Nanobody Tools

Angela Bekesi[*,†], Sara Abdellaoui[*,†], Natalie Holroyd[‡], Wouter Van Delm[§], Els Pardon[*,†], Jarne Pauwels[¶,||], Kris Gevaert[¶,||], Jan Steyaert[*,†], Stefaan Derveaux[§], Antoni Borysik[‡], Peter Tompa[*,†,#,1]

[*]VIB-VUB Center for Structural Biology (CSB), VIB, Brussels, Belgium
[†]Structural Biology Brussels (SBB), Vrije Universiteit Brussel (VUB), Brussels, Belgium
[‡]King's College London, Department of Chemistry, London, United Kingdom
[§]VIB Nucleomics Core, VIB, Leuven, Belgium
[¶]VIB Center for Medical Biotechnology, VIB, Ghent, Belgium
[||]Department of Biochemistry, Ghent University, Ghent, Belgium
[#]Institute of Enzymology, Research Centre for Natural Sciences of the Hungarian Academy of Sciences, Budapest, Hungary
[1]Corresponding author: e-mail address: peter.tompa@vub.vib.be

Contents

Methods in Enzymology, Volume 611
ISSN 0076-6879
https://doi.org/10.1016/bs.mie.2018.09.032

Abstract

The structural and functional characterization of large multidomain signaling proteins containing long disordered linker regions represents special methodological and conceptual challenges. These proteins show extreme structural heterogeneity and have complex posttranslational modification patterns, due to which traditional structural biology techniques provide results that are often difficult to interpret. As demonstrated through the example of two such multidomain proteins, CREB-binding protein (CBP) and its paralogue, p300, even the expression and purification of such proteins are compromised by their extreme proteolytic sensitivity and structural heterogeneity.

In this chapter, we describe the effective expression of CBP and p300 in a eukaryotic host, Sf9 insect cells, followed by their tandem affinity purification based on two terminal tags to ensure their structural integrity. The major focus of this chapter is on the development of novel accessory tools, single-domain camelid antibodies (nanobodies), for structural–functional characterization. Specific nanobodies against full-length CBP and p300 can specifically target their different regions and can be used for their marking, labeling, and structural stabilization in a broad range of in vitro and in vivo studies. Here, we describe four high-affinity nanobodies binding to the KIX and the HAT domains, either mimicking known interacting partners or revealing new functionally relevant conformations. As immunization of llamas results in nanobody libraries with a great sequence variation, deep sequencing and interaction analysis with different regions of the proteins provide a novel approach toward developing a panel of specific nanobodies.

1. INTRODUCTION

CREB-binding protein (CBP) and its paralogue p300 are multidomain signaling proteins involved in basic biological functions, such as cell proliferation, differentiation (Teo & Kahn, 2010), immunity (Revilla & Granja, 2009), DNA damage response (Dutto, Scalera, & Prosperi, 2018), and apoptosis (Lazarova et al., 2013; Lee et al., 2013; Yuan et al., 1999). Due to interacting with more than 400 of interaction partners (Bedford, Kasper, Fukuyama, & Brindle, 2010) and carrying intrinsic histone acetyltransferase (HAT) activity (Ogryzko, Schiltz, Russanova, Howard, & Nakatani, 1996), they primarily act as transcription coactivators that modify chromatin structure and target a broad range of partner proteins (Dancy & Cole, 2015). Their function and regulation are extremely complex, as they are targeted by multiple cellular signaling pathways and undergo a bewildering variety of posttranslational modifications (PTMs). PhosphoSite Plus database cites 92 and 140 PTM sites for human CBP and p300, respectively, including lysine acetylation, sumoylation, ubiquitylation, arginine methylation, and phosphorylation. Due to their critical integrator function(s) in cellular signaling, their mutations are implicated in diseases, such as cancer (Wang, Marshall, & Ikura, 2013) and Rubinstein–Taybi syndrome (Lopez-Atalaya, Valor, & Barco, 2014), and they are targets of drug development (Janknecht, 2002; Shaknovich & Melnick, 2011), which traditionally requires their detailed high-resolution structural characterization.

This, however, is a task much more complex than solving the structure of a folded enzyme, as both CBP and p300 are about 2400 residues in length and contain multiple folded domains connected by long intrinsically disordered linker regions (IDRs) (Fig. 1). The two proteins show high overall sequence identity and share the same domain architecture (Fig. 1); hence we refer to them as CBP/p300 when the data are applicable for both. For all of the folded domains, detailed structural–functional studies have been published and high-resolution structures are also available, providing essential insight into the function of protein. It should not be missed though that flexibility has a significant role even in the folded domains.

CBP/p300 has several protein interaction domains (TAZ1, KIX, ZZ, TAZ2, and NCBD) and epigenetic reader modules (bromodomain and PHD) essential in their transcriptional regulator function (Fig. 1). The catalytic core encodes the acetyltransferase activity (HAT domain, Fig. 1) modulated by a disordered autoregulatory loop (AL) prone to be heavily autoacetylated (Fig. 1) and also by immediate flanking regions (Bromo, PHD, RING, ZZ, and maybe TAZ2, Fig. 1) (Delvecchio, Gaucher, Aguilar-Gurrieri, Ortega, & Panne, 2013; Park et al., 2017). The two structures of the core region (4BHW: p300 Bromo-HAT, and 5U7G: CBP Bromo-ZZ) show basically the same domain architecture where the position of RING and ZZ is less defined, indicating possible conformational changes. Regulatory interdomain interactions are also suggested, like between the Bromo and the acetylated AL (Park et al., 2017), the RING and the L1 loop of the HAT domain or the AL, and the protein substrate-binding site of the HAT domain (Delvecchio et al., 2013).

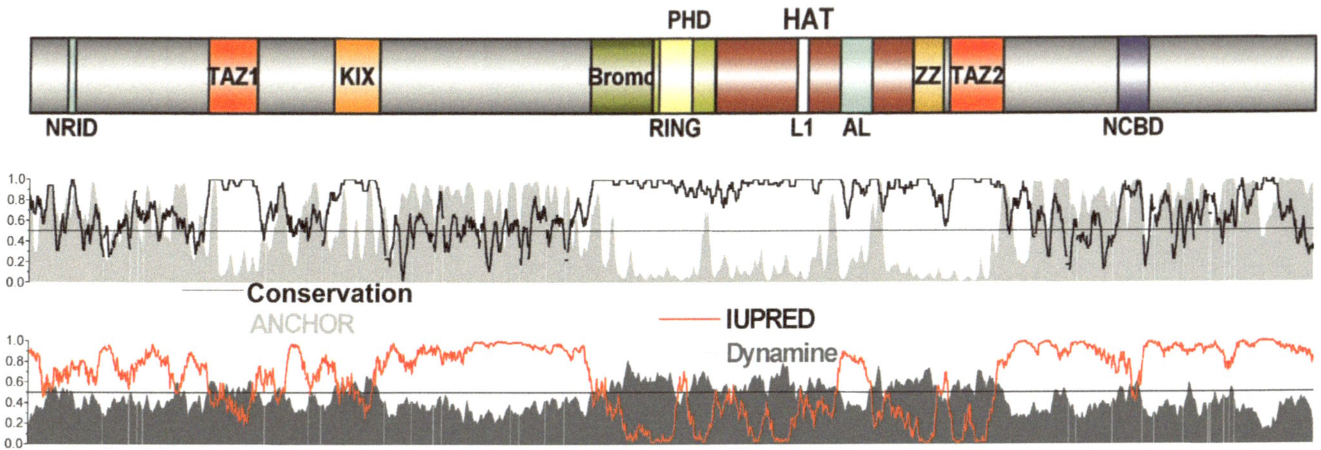

Fig. 1 Domain architecture and sequence analysis of p300 and CBP. Both p300 and CBP contains the same nine structured domains: the histone acetyltransferase (HAT) domain, five Zn-binding domains (TAZ1, TAZ2, ZZ, PHD, and a nonconventional RING domain inserted into the PHD domain), the epigenetic reader bromodomain, the CREB/KID interaction domain (KIX), and the nuclear receptor coactivator binding domain (NCBD). These domains show high sequence identity between CBP and p300 (*black trace* on the *upper graph*, conservation is scored from 0 to 1, plotted against residue numbers of the p300 sequence). A conserved motif for nuclear receptor interaction (LxxLL, NRID) is also common in CBP and p300, as well as the intrinsically disordered autoregulatory loop (AL) within the HAT domain. The rest of the proteins (~60% of the sequences, *gray*) are predicted to be intrinsically disordered regions (IDRs). IUPRED (Dosztányi, Csizmok, Tompa, & Simon, 2005) results for p300, above 0.5 it is more disordered (*bottom graph*, *red trace*). Prediction by DynaMine (Cilia, Pancsa, Tompa, Lenaerts, & Vranken, 2014) also for p300, above 0.75 it is more rigid (*bottom graph, gray chart*). These long IDRs are less conserved between the two proteins (*upper graph, black trace*), still, they might have functionality like comprising binding sites for protein partners (*upper graph*, *gray chart* shows ANCHOR (Mészáros, Simon, & Dosztányi, 2009) prediction for p300). The *graphs* and the domain organization figure are strictly proportional and aligned.

During acetyl transfer, the acetyl group is transferred from the cosubstrate acetyl-coenzyme A (AcCoA) to a Lys side chain of the protein substrate. The mechanism of the acetyl transfer by CBP/p300 was described as "hit-and-run," where a preformed enzyme–AcCoA complex interacts with the protein substrate only transiently (Liu et al., 2008). All available crystal structures comprising p300 HAT domain (3BIY (Liu et al., 2008), 4PZS, 4PZR, 4PZT (Maksimoska, Segura-Peña, Cole, & Marmorstein, 2014), 4BHW (Delvecchio et al., 2013)) contain ligands (AcCoA, Lys-CoA, acetonyl-CoA, and CoA) and lack the AL eliminated by either limited digestion or replaced during the cloning. In these structures, all CoA analogues have equivalent positions and are quite buried by the L1 loop. Interestingly, the CBP Bromo-ZZ structure, 5U7G, also lacking the AL, does not contain any substrate analogues, still the L1 loop has the same "closed" position (Park et al., 2017). Though the CoA as one of the products needs to leave and be replaced by acetyl-CoA before the next catalytic cycle, significant conformational changes within the HAT domain were not suggested by either of these structural–functional studies. Flexibility and chemical heterogeneity (c.f. autoacetylated AL) present within the otherwise structured core region obviously challenge most of the structural biology methods.

Flexibility and intrinsic disorder present much more in the full-length CBP/p300 due to its five long IDRs. Traditionally, folded domains of CBP/p300 are thought as functional regions mediating protein–protein interactions, but recently it was shown that IDRs might also be involved in binding protein partners (Contreras-Martos et al., 2017). The many possible PTMs combined with the structural flexibility of IDRs result in a large number of possible conformational and functional states of the protein. For the global structural and functional analysis of such proteins, one needs to consider this fact, as it may confuse the output of many traditional structural biology experiments. Whereas heterogeneity in traditional systems may be a sign of inadequate sample quality, in the case of proteins like CBP/p300 it may be linked with the real physiological state of the protein. This behavior, actually, may be a general structural biology problem, as structural disorder is prevalent in signaling and regulatory proteins (Van Der Lee et al., 2014), which often function by the structural and functional interplay of folded domains and disordered linker regions (Buday & Tompa, 2010). A unique functional capacity of such proteins is that they can assemble large signaling complexes of complicated architecture (Hegyi, Schad, & Tompa, 2007; Panne, Maniatis, & Harrison, 2007).

Even the conceptual foundation of how to describe the structure of such proteins is not yet settled. In solution, they may assume an ensemble of distinct conformations that have been described for a few fully disordered proteins (Varadi et al., 2014), but not for proteins with a mixture of folded domains and disordered linkers (Yang, Blachowicz, Makowski, & Roux, 2010). These proteins are too large for studying by NMR, whereas their successful atomic resolution structural characterization by X-ray crystallography or electron microscopy (EM) limits the conformational space and selects and stabilizes one or a few dominant—and functionally not necessarily the most relevant—conformations. Interestingly, due to the large size and extreme flexibility of these proteins, low-resolution single-molecule methods, such as smFRET (McCann et al., 2012) or AFM (Ando, Uchihashi, & Scheuring, 2014), can provide valuable insight into their structural organization.

Our premise in this work is that due to the structural complexity and heterogeneity of these proteins, their successful characterization requires specific labeling of their different parts. In addition, crystallization of their bigger elements (let alone the whole protein) requires molecular stabilization of functionally important conformations. In addition, dissecting allosteric functional regulation within the context of the full-length protein also requires molecules that can specifically interact with different parts of the protein and potentially compete with its regulatory intra- and intermolecular interactions.

The aim of this chapter is to demonstrate the problems of protein heterogeneity and provide improved purification strategies with multiple quality checks as well as molecular aids, nanobodies (NBs) for structural–functional studies. Nanobodies are small (~15 kDa) recombinant proteins derived from the variable region of "heavy chain only" natural antibodies (HH antibodies) of camelids (Ghassabeh, Devoogdt, De Pauw, Vincke, & Muyldermans, 2013; Hamers-Casterman et al., 1993; Muyldermans, 2013). Such nanobodies selected from a cDNA library derived from immunized llamas usually encompass the full antigen-binding ability of the original HH antibody. They have very diverse applications, such as crystallization aids of complex and flexible proteins (Rasmussen et al., 2011), markers in live-cell imaging and tracking (Traenkle & Rothbauer, 2017) and in assessing conformational dynamics (Dmitriev, Lutsenko, & Muyldermans, 2016), agents for in vivo diagnostics (Huang, Muyldermans, & Saerens, 2010), and biologics of potential therapeutic application (Bannas, Hambach, & Koch-Nolte, 2017). Their broad potential prompted us to develop nanobodies against both CBP and p300, along a novel path combining deep-sequencing, affinity-based pull-down, and mass spectrometry-based identification.

2. CLONING, EXPRESSION, AND PURIFICATION OF LARGE, PARTIALLY DISORDERED PROTEINS CBP/p300

Production of full-length CBP and p300 in large quantities and of quality appropriate for structural–functional studies is not trivial, although early papers reported successful expression in baculovirus-infected Sf9 cells and subsequent purification via single Ni^{2+}-based immobilized metal ion-affinity chromatography (IMAC) (Kraus & Kadonaga, 1998). We wanted to improve this procedure, so both N-terminal His- and C-terminal FLAG-tag were inserted into the constructs; hence Ni^{2+}-affinity and anti-FLAG immunoaffinity chromatography could be combined in the purification protocol. As expression and purification of intrinsically disordered proteins (IDPs)/IDRs are often compromised by their proteolytic sensitivity, such a tandem affinity purification approach is required to ensure that the final protein product is of full length, having both its N- and C-terminus present. In the case of such long proteins, premature termination of translation is also relatively frequent event causing the N-terminal fragments to be present (Chiabudini et al., 2014; Svidritskiy, Demo, & Korostelev, 2018). Purification via the C-terminal tag could help to get rid of these side products.

2.1 Cloning of the Full-Length CBP and p300

Cloning of cDNAs with length 7326 and 7243 nt, respectively, challenges common protocols involving restriction digestion, annealing, and ligation. Even sequencing of the clones is not trivial, as six to seven overlapping fragments should be sequenced in two directions to have a full coverage. For these reasons, highly efficient Gateway cloning was applied: (1) the BP reaction results in entry clones that contain the C-terminal FLAG-tag; then (2) LR clonase reaction is used to transfer the construct to several different destination vectors including donor vectors for baculovirus systems, like the N-terminal His-tag containing pDEST10 (Thermo Fisher Scientific). By transforming DH10Bac cells (Thermo Fisher Scientific) with pDEST10 constructs, recombinant bacmids can easily be produced by induced transposition.

2.1.1 Gateway Cloning

Full-length human CBP (UniProt Q92793) with C-terminal Myc-DDK-tag (DDK = FLAG-tag) was subcloned from CREBBP (NM_004380) Human cDNA ORF Clone (Origene) to Gateway vector pDONR221 (Thermo Fisher Scientific) using BP clonase reaction (according to the protocol provided by Thermo Fisher Scientific) after PCR amplification with attB1 and attB2 sites containing specific primers (forward primer: attB1 + CBP 5′: 5′-GGGGACAAGTTTGTACAAAAAAGCAGGCTCAATGGCTGA-GAACTTGCTGGACG-3′, reverse primer: reverse complementer of DDK 3′ + Stop + attB2: 5′-GGGGACCACTTTGTACAAGAAAGCTGGGTG TTAAACCTTATCGTCGTCATCCTTG-3′).

Full-length human p300 (UniProt Q09472, original vector was a kind gift from James Kadonaga) without a tag was subcloned into pDONR221, and then the same Myc-DDK-tag was inserted to have equivalent constructs of p300 and CBP (forward primer: attB1 + p300 5′: 5′-GGGGACAAGTTT GTACAAAAAAGCAG GCTCAATGGCCGAGAATGTGGTGGA-3′, reverse primer: reverse complementer of p300 3′ + *Not*I site + *Pme*I site including Stop + attB2: 5′-GGGGACCACTTTGTACAAGAAAGCTGG GTTCGTTTAAACCTGCGGCCGCGTGTGTATGTCTAGT GTAC TCTGTGAGAG-3′. This reverse primer allowed cloning the Myc-DDK-tag using the restriction sites present in the CBP construct).

Using the LR clonase reaction (according to the protocol of Thermo Fisher Scientific), entry clones were then transferred to the pDEST 10 (Thermo Fisher Scientific) destination vector encoding an N-terminal His-tag.

2.1.2 Generation of Mutants With In-Fusion Technique

All mutations were prepared at the level of entry clones. For inactivating the HAT domain (hereafter referred to as HATn), double mutations were introduced: WY(1466–67)AS into p300 and the corresponding WY(1502–03)AS into CBP using the In-Fusion technique (Clontech). This strategy is particularly relevant for constructs too long for the most often used QuickChange mutagenesis approach (Agilent). This technique allows to join two or more fragments (e.g., a vector and an insert), as long as they share homology of 15 bases at each end. For mutagenesis, a segment containing the mutation site and located between two unique restriction sites should be amplified in PCR. Mutations are introduced into this segment using mutagenesis primers B and C (for CBP, primer B: 5′-CCCAAGCCAAAACGACTG CAGGAGGCGTCCAAAAAGATGCTGGACAAGGCGTTTGCAG-3′, primer C: 5′-CTGCAAACGCCTTGTCCAGCATCTTTTTGGACGCC TCCTGCAGTCGTTTTGGCTTGGG-3′; for p300, primer B: 5′-GAAG ATACCCAAGCCCAAGCGACTGCAGGAAGCGTCCAAAAAAATG CTTGACAAGGCTG-3′, primer C: 5′-CAGCCTTGTCAAGCATTTT TTTGGACGCTTCCTGCAGTCGCTTGGGCTTGGGTATCTTC-3′).

In-Fusion primers A and D (for CBP primer A: 5′-CTATAAGTTTTGCAG TAAGCTTGCAGAGG-3′, primer D: 5′-GAGCAGGGGGTCGGGGTC GACGAT-3′; for p300 primer A: 5′-CGGAAGAAAGATGCATCAGAT CTGTG-3′, primer D: 5′-AGGAGGGGTTGGAGGGGTCACCTGG-3′) are designed to match the ends of this segment with a 15-nt extension over the restriction sites matching the rest of the wildtype (wt) construct. For details of primer design, visit Clontech's page http://bioinfo.clontech.com/infusion/convertPcrPrimersInit.do and https://www.takarabio.com/learning-centers/cloning/in-fusion-cloning-tools. The mutant segment is produced in two steps: first, primers A and C and primers B and D are used in two separate PCR reactions, to amplify the two halves of the segment carrying the desired mutation. Then, the products are purified from an agarose gel, and both are used as templates in a template-switching PCR amplification with primers A and D. The amplified mutant PCR product is then purified from an agarose gel. The wt plasmid construct is digested with the two restriction enzymes and is purified from the PCR mixture. Then, the mutant DNA segment and the plasmid fragment are mixed equimolarly and treated with the In-Fusion enzyme mix that digests the 3′ ends producing 5′-overhangs. The two DNAs anneal, resulting in a nicked plasmid that can be transformed into *Escherichia coli*, where it is repaired by the cellular DNA repair system. For further details, see the protocols of Clontech.

Note

(1) Due to the length of the constructs, the fidelity of PCR is critical. Several high-fidelity polymerases were tried with variable success. Still, checking the entry clones (pDONR constructs) with full sequencing is indispensable. For pDEST10 constructs, it is sufficient to check by restriction digestion, while recombinant bacmids are better be sequenced again, covering the full protein-coding region including the tags. For sequencing, overlapping PCR products with about 1300–1600 nt length are prepared. Sequencing data are easy to be aligned with the reference sequence in CLC Main Workbench (Qiagen).

2.1.3 Bacmid Generation

pDEST10 constructs are transformed into DH10Bac cells (Thermo Fisher Scientific) that already contain the bacmid with a mini-*att*Tn7 target site and a helper plasmid-encoding transposition proteins. With this help, the mini-Tn7 element on the pDEST10 donor plasmid can transpose to the mini-*att*Tn7 target site on the bacmid disrupting the *LacZα* gene. Positive clones can then be selected by their white color. The recombinant bacmid is isolated with a protocol optimized for large DNA. These protocols are slightly modified from the original ones of the manufacturer.

Materials

LB Broth with agar
$NH_4Fe(III)$ citrate
Antibiotics: kanamycin, tetracycline, gentamicine
IPTG
S-gal (Sigma-Aldrich, S-9811)
DH10Bac competent cells (Invitrogen, 10359-016)
S.O.C. medium
Solution I: 15 m*M* Tris pH 8.0, 10 m*M* EDTA, 100 μg/mL RNase
Solution II: 0.2 *M* NaOH, 1% SDS
KAc buffer: 3 *M* K-acetate pH 5.5
TE buffer: 10 m*M* Tris pH 8.0, 1 m*M* EDTA
Abs. isopropanol
70% ethanol

Equipment and Disposables

Autoclave
Centrifuge
Block thermostat
Plastic Petri dishes, sterile
Centrifuge tubes 50 and 1.5 mL
Culture tubes, plastic

Procedure

(1) Prepare LB agar medium containing 300 mg/L S-gal. Autoclave and cool down (to hand warm) the medium and add the following sterile-filtered components: (Work sterile!) 500 mg/L $NH_4Fe(III)$ citrate, 50 μg/mL kanamycin, 10 μg/mL tetracycline, 7 μg/mL gentamicine, 40 μg/mL IPTG. Prepared and solidified plates wrapped with parafilm and protected from light can be stored at 4°C for several weeks.
(2) Thaw the DH10Bac™ competent cells on ice.
(3) Dispense 100 μL of the cells into 1.5-mL round-bottom polypropylene tubes.
(4) Add approximately 1 ng recombinant donor plasmid (in 5 μL) and gently mix the DNA into the cells by tapping the side of the tube.
(5) Incubate the cells on ice for 30 min.
(6) Heat shock the cells by transferring the tubes to 42°C for 90 s.
(7) Chill the tubes on ice for 2 min.
(8) Add 900 μL S.O.C. medium to the transformed cells.

(9) Place the tubes in a shaking incubator horizontally at 37°C with medium agitation (225 rpm) for 4 h.
(10) Serially dilute the cells, using S.O.C. medium, to 10^{-1}, 10^{-2}, 10^{-3}.
(11) Place 100 μL of each dilution on a plate and spread evenly over the surface.
(12) Incubate for 24–48 h at 37°C (colonies are very small and black colonies may not be discernible prior to 24 h).
(13) Select about 10 white colonies from a plate with approximately 100–200 colonies and streak them to fresh plates to verify the phenotype. Incubate 18–24 h at 37°C.
(14) Inoculate single colonies with white phenotype to 3–6 mL LB medium supplemented with 50 μg/mL kanamycin, 7 μg/mL gentamicin, and 10 μg/mL tetracycline in a sterile 50-mL falcon tube. We typically select three to four colonies per constructs for separate bacmid preparations, as transfection and/or expression efficiency may vary from clone to clone even if the coding sequence is without mutations.
(15) Grow at 37°C to stationary phase (up to 24 h) shaking at 250–300 rpm.
(16) Pellet the cells with centrifuging them at 14,000 × *g* for 1 min.
(17) Remove the supernatant accurately by pipetting.
(18) Resuspend the bacterial pellet in 300 μL Solution I, and place to a 2-mL tube.
(19) Add 400 μL of the Solution II, gently mix, and incubate for 5 min at RT.
(20) Add 400 μL of the Kac buffer, mixing gently during addition. A thick white precipitate of protein and *E. coli* genomic DNA will form. Place the sample on ice for 5–10 min.
(21) Centrifuge 10 min at 14,000 × *g*, 4°C.
(22) During centrifugation, add 0.8 mL absolute isopropanol to another microcentrifuge tube.
(23) Gently transfer the supernatant to this tube. Avoid any white precipitate material. Mix by gently inverting tube a few times and place on ice for 5–10 min.
(24) Centrifuge at 14,000 × *g*, 4°C, for 15 min.
(25) Remove the supernatant, and add 0.5 mL 70% ethanol to the pellet (that might be invisible). Invert the tube several times to wash the pellet.
(26) Centrifuge at 14,000 × *g* RT, for 5 min.
(27) Remove as much of the supernatant as possible by pipetting, and dry the pellet at room temperature (RT) or in a thermo-block at 50°C for 5–10 min. Proper evaporation of the ethanol is as crucial as avoiding overdrying.

(28) Resolve the pellet in 100–200 μL TE buffer. Allow the solution to sit in the tube with occasional gentle tapping of the bottom of the tube. The DNA is generally ready for use within 10 min, as long as the pellets are not overdried.

(29) Determine the DNA concentration with the Nanodrop. It could be fairly high, above 1 mg/mL.

(30) Store the bacmid DNA at −20°C in aliquots to avoid a drastic reduction in transfection efficiency due to repeated freeze–thaw cycles.

Notes

(1) Screening for *LacZα* activity is easier by S-gal and $NH_4Fe(III)$ citrate that will result in an intense black color, instead of the less-intense blue color produced by X-gal or Bluo-gal.

(2) When Solution II is added and the sample is mixed, the suspension should change from very turbid to almost translucent.

(3) At this level, PCR analysis is recommended to verify successful transposition to the bacmid. However, in case of such long constructs as CBP/p300, we recommend full sequencing also in the same way as described in Section 2.1.2.

2.2 Expression of Full-Length Proteins in Sf9 Cells

Culturing eukaryotic cells requires a dedicated laboratory equipped with at least a sterile hood, appropriate incubators, centrifuges, a freezer, and a fridge. Growing Sf9 cells is possible in T-flasks, 6-well plates, as well as in shaking flasks without requiring CO_2 supplement. Shaking the cultures allows the production of higher amounts of recombinant protein. We used the following materials and equipment for culturing cells that will be referred to as "cell-culture lab materials and equipment" later on.

Materials

Sf9 cells (Gibco, Thermo Fisher Scientific)
ESF921 medium (Expression Systems, 96-001)
Penicillin–Streptomycin–Neomycin 100 × (Invitrogen) (PSN)
Fetal Bovine Serum, Certified, Heat-Inactivated (Invitrogen) (FBS)
CellFectin II reagent (Thermo Fisher Scientific)

Equipment and Disposables

Cell-culture hood, Class II
Shaking incubator, at 120 rpm and 27°C, humidified
Cell-culture flasks, Corning® Erlenmeyer baffled, vent cap, sterile
6-Well plates, sterile

Serological pipettes, sterile
Pipette controller
Sterile tubes: 50, 15, and 2 mL
Centrifuge for low speed for 2, 15, and 50 mL tubes and 350 mL flasks
Pipettes
Sterile pipette tips with filters
Filter units, sterile, disposable
Syringe filter, sterile
Pasteur pipettes, glass, sterilized
VACUSAFE™ Vacuum Aspiration System (INTEGRA Biosciences)
Sterile PBS
RBS or bleach for disinfection

2.2.1 Transfection and Virus Amplification

Usually, two to three purified bacmids per construct with the right sequence are transfected into Sf9 cells cultured in ESF921 medium using CellFectin II reagent, according to the supplier's protocol (https://tools.thermofisher.com/content/sfs/manuals/bactobac_topo_exp_system_man.pdf), with slight modifications, i.e., we use ESF921 medium instead of Grace, decrease the initial amount of Sf9 cells to 4.5×10^5 per well, and reduce PSN concentration to 0.25%.

After harvesting the P1 virus generation (500 × *g*, 20°C, 5 min), we usually do not make a virus plaque assay and instead continue with virus amplification in a second round of infection resulting in P2 virus generation. Hence, all the ~1.8 mL P1 is used to prepare 2 × 30 mL P2 virus generation. For this, 0.9 mL P1 is added to 30 mL 2×10^6 cells/mL culture and incubated for an additional 48 h.

Next, P2 is harvested, mixed with 2% FBS, and filtered through a 0.45-μm filter for long-term storage. From the amplified virus in the P2 generation, we usually make a trial expression, as well as prepare P3 virus generation that means rather scaling up, than further amplification, of the viruses.

For the P3 generation, 210 μL ofP2 is added to the 70 mL 2×10^6 cells/mL culture and is incubated for further 48 h. Harvested P3 is supplemented with 2% FBS, filtered on a 0.45-μm filter, aliquoted to 50 mL in sterile Falcon tubes, and stored at −80°C.

Notes

(1) We found that the evaporation of the medium from the 2 mL cultures during the 72 h incubation of the transfection might cause problems.

Therefore, we suggest to seal the 6-well plates with Parafilm, and to make sure that the incubator is appropriately humidified.

(2) Although baculovirus stocks can be stored at −80°C in the presence of 2% FBS, when they are protected from light, it is recommended that they be used fresh.

(3) In our experience, viruses stored for a long time can lose their ability to express the recombinant protein.

2.2.2 Trial Expression and Western Blot Analysis

We found that the oversized, partially disordered CBP and p300 frequently suffer from severe degradation already in the cell in which they are expressed. The baculovirus encodes for several carboxyl- and cysteine proteases (Gotoh, Miyazaki, Sato, Kikuchi, & Bentley, 2001) that might be active on overexpressed recombinant proteins too. Therefore, we applied a cocktail of cell-penetrable nontoxic proteinase inhibitors, adding them to the cultures 24 h before harvesting.

Materials

Growing Sf9 cells in ESF921 0.5% PSN medium
Recombinant P2 or P3 virus stocks
Histone Acetyltransferase (HAT) inhibitor II (Millipore, VWR)
Leupeptin (CarlRoth)
Pepstatin A (CarlRoth)
AEBSF (CarlRoth)
E64 (CarlRoth)
DMSO (Sigma-Aldrich)
5 × SDS loading dye
PageRuler Plus Prestained Protein Ladder (Thermo Fisher Scientific)
PBS (phosphate-buffered saline)
Low-fat milk powder
Bovine Serum Albumin (BSA)
Tween 20 (Sigma-Aldrich)
Monoclonal ANTI-FLAG® M2 antibody produced in mouse (Sigma-Aldrich, F1804)
Anti-Mouse IgG (whole molecule)-Peroxidase antibody produced in rabbit (Sigma-Aldrich, A9044)
Monoclonal Anti-polyHistidine-Peroxidase antibody, mouse, clone HIS-1 (Sigma-Aldrich, A7058)
Clarity Western ECL Substrate (Bio-Rad)

Equipment and Disposables

Cell-culture lab materials and equipment
MiniProtean Tetra Cell 2 Gel system (Bio-Rad) for gel electrophoresis
8% SDS-PAGE, 1 mm thick
Trans-Blot Turbo semidry blotting system (Bio-Rad)
Trans-Blot Turbo Transfer Pack, Mini PVDF (Bio-Rad)
Molecular Imager ChemiDoc XRS+ (Bio-Rad)

Procedure

(1) For expression of HATn mutants, prepare 10 m*M* stock of HAT inhibitor II in DMSO (10 mg HAT inhibitor II, dissolved in 2.15 mL DMSO), aliquot and store at −20°C protected from light .

(2) *Prepare cell-penetrable nontoxic protease inhibitor mix*: This protease inhibitor mix is 200× concentrated and has to be diluted 200 times when used. Prepare the following stock solutions first:
- 1 mg/mL Leupeptin in ESF921 medium (containing 0.5% PSN) and sterilize by filtering
- 1 mg/mL Pepstatin A in DMSO
- 0.25 *M* AEBSF in ESF921 medium (containing 0.5% PSN) and sterilize by filtering
- 5 mg/mL E64 in DMSO

Mix 3 mL Leupeptin, 6 mL Pepstatin A, 6 mL AEBSF, and 3 mL E64 stocks, and dilute it with ESF921 medium (containing 0.5% PSN) up to 30 mL. This mixture can be stored at −20°C in aliquots.

(3) Take 3- to 4-day growing cultures of Sf9 cells in ESF921 0.5% PSN medium. Prepare 30 mL cultures at 2×10^6 cells/mL in sterile, plastic 125-mL cell-culture flasks.

(4) Take 1 mL sample from this culture, use as negative control, spin it down, and store the pellet at −20°C till analysis.

(5) Start expression by infecting 30 mL cultures typically with 5% or 10% of P2 or P3 (e.g., adding 1.5 or 3 mL of viruses to the 30 mL cultures). *Further optimization* of the applied virus concentration might also be necessary.

(6) Grow the cells at 27°C shaking incubator with agitation at 120 rpm.

(7) After 24 h, take 1 mL sample, centrifuge down, and store the pellet at −20°C till analysis.

(8) Add the protease inhibitor mix at 200× dilution.

(9) If nonacetylated HAT-inactive CBP/p300 constructs are desired to be produced, or if a putative acetylation-related cytotoxicity/loss of expression is to be tested, HAT inhibitor should be added also at this step in 1000× dilution of the 10 m*M* stock solution.

(10) Continue culturing and take 1 mL samples at 48, 60, and 72 h of expression. Centrifuge them down and store the pellets at −20°C till analysis. The rest of the 30 mL expressions can also be harvested for further purification tests (details see later in Sections 2.2.3 and 2.3).

(11) Prepare samples for SDS-PAGE according to the following steps:
- Add 40 μL PBS to each sample
- Mix it in a vortex
- Incubate for 5 min at 99°C
- Mix it in a vortex
- Add 50 μL of 5 × SDS loading dye
- Mix it in a vortex
- Incubate for 5 min at 99°C
- Centrifuge down briefly

(12) Load 15 μL of samples from the clear supernatant to the wells of an 8% SDS-PAGE (prepared 1 mm thick, and with 10-well comb).

(13) Gel electrophoresis is done at constant 200 V until the blue front reaches the bottom of the gel.

(14) Transfer the SDS-PAGE in Trans-Blot Turbo system to Trans-Blot Turbo Mini PVDF membranes using 2 times the 7 min protocol. Transferring full-length CBP/p300 is not trivial even from an 8% gel. Doubling the time of the standard protocol is usually enough to detect expression, but it has to be noted that majority of the proteins at this size remain in the gel.

(15) Develop Western blot with the following incubation steps on a shaking platform at RT:
- Block for 1 h in 5% milk/0.1% Tween20/PBS (blocking solution)
- Rinse 3 times in PBS
- Incubate for 2 h with 2500 × diluted anti-FLAG antibody in blocking solution
- Rinse 3 times in PBS
- Wash 2 times 20 min in 0.1% Tween 20/PBS
- Rinse 3 times in PBS
- Incubate for 1 h with 80,000 × diluted antimouse Ig HRP conjugated in 5% milk/PBS
- Rinse 3 times in PBS
- Wash 2 times 20 min in 0.1% Tween 20/PBS
- Rinse 3 times in PBS

(16) Pipette 4 mL 1:1 mixture of ECL substrates to the membrane surface and for 5 min rinse it by hand to make sure the reagent reaches the surface evenly.

(17) Take out the membrane from the solution. Place the wet membrane on a glass plate to the white screen of the ChemiDoc for imaging. Use the application Blots/colorimetric with White Epi light source, and without filter, to adjust the zoom and make a photo of the prestained marker on the membrane. Do not change the position of the membrane, and use the application Blots/chemiluminescent with no light source and no filter.

Notes

(1) We found that performing trial expression at the level of P2 or P3 virus generations is more efficient, than determining the virus titer, and then optimizing the multiplicity of infection (MOI) for expression of each protein. In a trial expression, we optimize the amount of infecting P2/P3 virus, make a time course, and also get some insight about the intactness of the protein. The results of the trial expression are checked by Western blot using anti-FLAG and/or anti-His antibodies. Based on the results of the trial expression, the best-expressing clones can be chosen, and optimal virus concentration and expression time can be defined.

(2) Exposition time and the number of acquisitions in detection of Western blot have to be defined. In our experience, it is better to define long exposition (like 600 s) and ask for 15 pictures during this time. Usually one of these will produce an optimal signal-to-noise ratio.

2.2.3 Large-Scale Expression

To produce recombinant proteins enough for structural–functional studies, the expression needs to be scaled up. To this end, Sf9 cells are cultured in 480 mL medium in 2-L shaking flasks. 2 L cell culture typically yields 0.5–1 mg purified fl CBP/p300. P3 virus should be used for infection according to the results of the trial expression. We typically use 5%–10% virus and continue the expression for 48 h. In the case of an N-terminally truncated mutant, we experienced clearance/degradation by 48 h, but harvesting at 36 h helped. In contrast, another deletion mutant, the AL-deleted CBP/p300, showed delayed expression; thus the optimal time for its harvesting was about at 60 h. Expression pellets can be stored at −80°C until extraction and purification.

Materials and Equipment

Growing Sf9 cells in ESF921 0.5% PSN medium
P3 virus expressing the recombinant protein

Cell-penetrable protease inhibitor cocktail 200× (see Section 2.2.2, Procedure: (2))

Cell-culture lab materials and equipment (described in Section 2.2)

Procedure

(1) Dilute growing Sf9 cells to 2×10^6 cells/mL concentration into fresh 480 mL ESF921 medium supplemented with PSN in 2-L shaking flasks.
(2) Add P3 virus in a concentration determined in the trial expression.
(3) Continue culturing at 27°C shaking with 120 rpm
(4) Add the cell-penetrable protease inhibitor cocktail in 200× dilution 24 h before harvesting.
(5) After the time period determined in the trial expression, take 1 mL samples from the culture to test the actual expression efficiency on Western blot (according to the protocol in Section 2.2.2)
(6) Harvest the cells by centrifuging the culture for 15 min, 500 × *g*, at 20°C.
(7) Remove the supernatant with VacuSafe.
(8) Resuspend pellet in sterile PBS and pour into 50-mL falcon tubes (one to two tubes per 480 mL culture).
(9) Centrifuge: 8 min, 500 × *g*, 20°C.
(10) Remove the supernatant with the VacuSafe.
(11) Repeat the washing (last three steps) once more.
(12) Store the pellet at −80°C until extraction and purification.

Notes

(1) Instead of disposable, plastic cell-culture flasks, clean, sterilized glass flasks closed with strong autoclavable aluminum foil can also be used, and reused after disinfection with bleaching, proper cleaning, and sterilization.
(2) In our experience, using sterile hood is not necessary for harvesting, but the prompt work and applied washing steps detailed here clearly minimize uncontrolled degradation and improve the quality of protein preparation.

2.3 Purification of the Full-Length Proteins

CBP/p300 protein constructs were purified by consecutive Ni^{2+}-affinity and anti-FLAG immunoaffinity chromatography with extensive protease inhibition and also avoiding EDTA and other chelating agents to preserve native state of the Zn-finger domains. All steps are carried out in a cold room (4–8°C). 2 L Sf9 culture typically yielded 0.5–1 mg protein. Here, a detailed protocol is given.

Materials

Protease inhibitors: Leupeptin, Pepstatin A, Bestatin, E64, AEBSF
Roche protease inhibitor tablet EDTA-free Sigma protease inhibitor cocktail
Buffer components: Tris, NaCl, glycerol, imidazole
Additives: $MgCl_2$, Triton X-100, $ZnSO_4$, TCEP, RNase A, DNase I
FLAG®-peptide (Sigma-Aldrich)
ANTI-FLAG® M2 Affinity Gel (Sigma-Aldrich, A2220)
Ni-NTA or other Ni^{2+}-resin, like His-select, or Ni^{2+}-agarose

Equipment and Disposables

Dounce homogenizer
Sonication bath
Cold room
Centrifuge
Sterile filter units
Syringes
Pipettes
Falcon tubes (50 and 15 mL)
PD10 columns and tubing, closures
Dialysis tubing (~12 kDa cutoff)
Vivaspin concentrator with 100 kDa cutoff
SDS-PAGE apparatus
Superdex 200 and/or Superose 6B size-exclusion columns (GE Healthcare)
AKTA (GE Healthcare) or other available chromatographic system with cooling
UV/Vis spectrophotometer (i.e., NanoDrop or NanoVue)

Buffers to Be Prepared Before Start

- *500× protease inhibitor mixture in DMSO (PI mix for lysis)*: 1 mg/mL leupeptin, 0.5 mg/mL pepstatin A, 0.5 mg/mL bestatin, 2.5 mg/mL E64, 1 *M* AEBSF
- *1000× protease inhibitor mixture in DMSO (PI mix for purification)*: 1 mg/mL leupeptin, 0.5 mg/mL pepstatin A, 0.5 mg/mL bestatin, 1 mg/mL E64, 0.1 *M* AEBSF
- *5× basic buffer*: 250 m*M* Tris pH 7.5, 1.5 *M* NaCl, 25% glycerol
- *Lysis buffer*: 1× basic buffer, +5% glycerol, 2 m*M* $MgCl_2$, 0.2% Triton X-100, 1 m*M* $MgCl_2$, 0.1 μ*M* $ZnSO_4$, 50 μ*M* TCEP, 5 μg/mL RNase A,5 μg/mL DNase I, 1 tablet of the Roche protease inhib. for 50 mL, 1000× diluted Sigma protease inhibitor cocktail, 1× "*PI mix for lysis*"

- *Wash buffer for Ni^{2+}-affinity chromatography (wash)*: 1 × basic buffer, 1 × "*PI mix for purification*," 50 μ*M* TCEP
- *1 M salt buffer for Ni^{2+}-based IMAC (salt wash)*: 1 × basic buffer, 0.7 *M* NaCl, 1 × "*PI mix for purification*," 50 μ*M* TCEP
- *1 M imidazole buffer for Ni^{2+}-based IMAC (1 M imid. elution)*: 1 × basic buffer, 1 *M* imidazole, 1 × "*PI mix for purification*," 50 μ*M* TCEP. The pH should be checked and reset after dissolving the imidazole!
- *10 mM imidazole buffer for Ni^{2+}-based IMAC (10 mM imid. Wash)*: Dilute "1 *M* imid. elution" buffer 100 times in "wash" buffer.
- *Dialysis buffer for anti-FLAG immunoaffinity purification*: 1 × basic buffer, 1 × "*PI mix for purification*."
- *5 × dialysis buffer*: 250 m*M* Tris pH 7.5, 750 m*M* NaCl, 25% glycerol
- *1 M salt-wash buffer for anti-FLAG immunoaffinity (salt-wash FLAG)*: 1 × dialysis buffer, 850 m*M* NaCl, 1 × "*PI mix for purification*," No TCEP!
- *Wash buffer for FLAG affinity (wash FLAG)*: 1 × dialysis buffer, 1 × "*PI mix for purification*," No TCEP!
- *Glycine buffer for regeneration of immunoaffinity beads*: 100 m*M* glycine buffer pH 3.0
- *Final dialysis or gel-filtration buffer used also for storage at −80°C*: 1 × dialysis buffer, 1 × "*PI mix for purification*," 50 μ*M* TCEP

Notes

(1) Addition of 0.1 μ*M* $ZnSO_4$ to the lysis buffer is thought to support the integrity of the Zn-finger domains (Nyborg & Peersen, 2004), but on the other hand more degradation occurs due to activity of metalloproteases.

(2) The pH of the lysis buffer should be checked and reset, if necessary, after adding all additives!

(3) All buffers need to be sterile filtered for long-term usage not to contaminate sensitive protein preparation!

(4) Do not add TCEP for the buffers used in immunoaffinity, as it might interfere with the anti-FLAG antibody integrity.

2.3.1 Extraction of Cellular Proteins From Frozen Sf9 Cell Pellet

Efficient extraction can be helped by applying rigorous homogenization, as well as by many additives, such as detergents, DNase, RNase, reducing agents, and protease inhibitors, including EDTA against metalloproteases. On the other hand, it is also important not to interfere with structural–functional integrity of the recombinant protein to be produced. In our case, many opportunities had to be disclosed. EDTA could efficiently compete with

CBP/p300 Zn-fingers for Zn^{2+} (K_d for the EDTA: Zn^{2+} complex is around 10^{-15} *M*) that might cause the loss of function and/or denaturation of Zn-finger domains within the full-length protein. Detergents could modify the conformation of the protein that is hard to control due to significant intrinsic disorder, large size, and high inherent flexibility of the protein. The most common reducing agent, DTT, can denature Zn-fingers and has some disadvantages: (1) its oxidized form absorbs light at around 275 nm strongly influencing the protein concentration measurements later (Getz, Xiao, Chakrabarty, Cooke, & Selvin, 1999), and (2) it can form precipitates with Zn^{2+} or Ni^{2+} ions that might interfere the Ni^{2+}-based IMAC (Kim et al., 2017). Therefore, TCEP was used finally everywhere as a reducing agent instead of DTT.

Procedure

(1) Resuspend the cells from ~2 L of cultures in 100 mL ice-cold lysis buffer on ice. For HAT-inactive mutants, add 10 μ*M* HAT inhibitor (Histone acetyltransferase inhibitor II, Millipore) (1000 × dil. of the 10 m*M* stock)!

(2) Homogenize it in a Dounce homogenizer manually!

(3) 10 min sonication in a sonication bath (!) in the cold room!

(4) Homogenize in a Dounce homogenizer again

(5) Centrifuge at 21,000 × *g* at 4 °C for 45 min (if it is necessary, do another 30 min centrifugation step)

(6) Filter the supernatant through a 0.45 μ*M* pore-size filter (might be difficult even after the second centrifugation step)

Notes

(1) Check the pH during and after homogenization: if it is shifted, set it back.

(2) Rigorous homogenization methods like sonication directly with a sonication head at high power (direct sonication) might result in efficient disruption of inner membranes/protein networks, but may result in the release of more contaminating proteins into the extract. In our experience direct sonication had severe effects on purity; two extra protein bands appeared on the SDS-PAGE as copurified contaminants. These could be CBP/p300 fragments, but tubulin and ATP synthase were also detected from these gel bands by MS.

2.3.2 Ni^{2+}-Based Immobilized Metal Ion-Affinity Chromatography

In our experience, stepwise manual chromatography was better for this step as compared to a continuous method in an automatized chromatographic system. We found that Ni^{2+}-based IMAC gives extremely high background in both methods, probably because Zn-fingers frequently bind to the Ni^{2+}-resins on their own (Voráčková, Suchanová, Ulbrich, Diehl, & Ruml, 2011), and degradation products (fragments) of the overexpressed CBP/p300 might dominate the extract and also the Ni^{2+}-eluate (Fig. 2A). The stepwise manual procedure has at least two advantages over the automatized one: (1) the resin can be incubated with the protein extract that helps binding, and (2) the whole process is faster, decreasing the chance of uncontrolled degradation. We used the following protocol.

Procedure

(1) Load ~5 mL Ni-NTA resin in a fresh empty PD-10 column assembled with tubing and proper closure in the cold room.

(2) Wash the resin with wash buffer (3 column volume (CV)), then with 1 *M* imidazole (2 CV), and then equilibrate with wash buffer (5 CV).

(3) Take out the prewashed and equilibrated resin from the PD10 column, and add to the sterile-filtered extract in a Falcon tube!

(4) Incubate it on the rotator for 1–1.5 h at 4°C.

(5) Fill it back into the same column for fractionation! Take the following fractions into 15-mL falcon tubes labeled appropriately: flow-through (FT), "wash" (3 CV), "salt wash" (1 CV), "wash" (2 CV), "10 m*M* imid. wash" (1 CV), 1 *M* imid. elution (2 CV), washing back to the "wash" buffer (4 CV). For buffer composition, see buffer list in Section 2.3.

(6) Take 50 μL samples from each fraction, including the extract, for further SDS-PAGE analysis. It is better to mix these samples with SDS loading dye immediately and store them on ice until the analysis.

(7) Dialyze the "1 *M* imid. elution" fraction 2 times for 1.5 h in 1 L of "dialysis buffer for anti-FLAG immunoaffinity chromatography" in the cold room.

Note

(1) Imidazole and reducing agents interfere with the efficiency of the anti-FLAG immunoaffinity chromatography! Therefore last dialysis step is crucial. However, too long incubation of the highly contaminated Ni^{2+}-eluate might allow more degradation.

2.3.3 Anti-FLAG Immunoaffinity Chromatography

As noted, a critical element of the purification protocol is the combination of two affinity chromatography steps that rely on the two intact termini of the protein. Therefore, purification continues via the FLAG-tag (Brizzard, Chubet, & Vizard, 1994) located on the C-terminus of the protein (Fig. 2B).

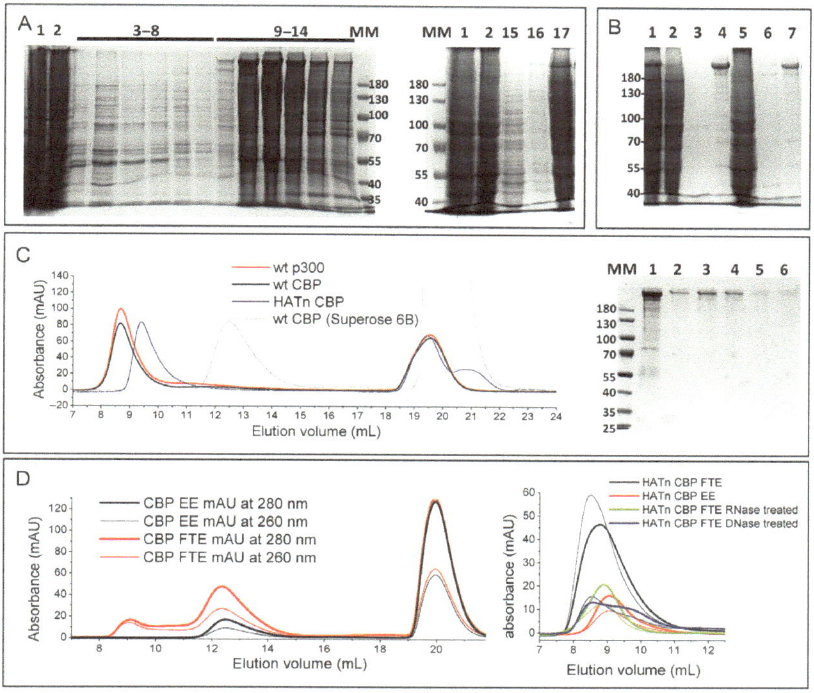

Fig. 2 Purification of full-length CBP and p300. (A) Comparison of gradient and step elution in Ni^{2+}-based IMAC by SDS-PAGE of CBP fractions: Extract (1), flow-through (2), buffer and salt gradient wash (3–8), elution by imidazole gradient (9–14), PageRuler prestained molecular marker (MM), washing steps with 1 *M* NaCl (15), with 10 m*M* imidazole (16), and with 1 *M* imidazole (17) containing buffer. (B) Fractions from anti-FLAG immunoaffinity chromatography of fl CBP on SDS-PAGE. Dialyzed Ni-affinity elution (1), anti-FLAG flow-through (2), 1 *M* salt wash (3), eluted fraction by 0.5 mg/mL FLAG peptide, "EE" (4), and the same fractions from dialyzed Ni^{2+}-affinity flow-through: anti-FLAG flow-through, "FTFT" (5), 1 *M* salt wash (6), eluted fraction by 0.5 mg/mL FLAG peptide, "FTE" (7). (C) Size-exclusion chromatography. Chromatograms of different constructs using Superdex 200 Increase or Superose 6B SEC column. The first peak corresponds to CBP/p300, the second peak at around 20 mL to FLAG peptide used for elution. At *right*, SDS-PAGE shows injected CBP (1) and consecutive gel-filtration fractions (2–6) corresponding to the first peak. (D) Comparison of CBP preparations purified from Ni^{2+}-affinity elution (EE) and flow-through (FTE). Gel-filtration on Superose 6B Increase (*left*) shows higher amount of aggregates, and/or presence of nucleic acid contaminants in the case of CBP FTE. On Superdex 200 Increase, high nucleic acid contamination of HATn CBP FTE was detected (*right panel, black curve*), while HATn CBP EE sample (*red*) shows pure protein peak. By RNase treatment, this nucleic acid contamination was efficiently removed (*green*), while DNase treatment caused less pronounced effect (*blue*). *Thick* and *thin lines* indicate the absorbance at 280 and 260 nm, respectively.

Procedure

(1) Prepare 2 mL of anti-FLAG affinity gel (mixed properly before taking the necessary amount) according to the supplier's protocol:
   i. Fill 3 mL 50% suspension into a fresh column
   ii. Wash it "wash FLAG" buffer to remove glycerol
   iii. Wash with 100 m*M* glycine buffer pH 3.0
   iv. Equilibrate with "wash FLAG" buffer

(2) Add the resin to the dialyzed "1 *M* imid elution" fraction and incubate in a rotator at 4°C for at least 4 h or overnight

(3) Fill the suspension back to the PD10 column and take the flow-through

(4) Wash with "wash FLAG" buffer (~10 mL). Do the whole fractionation in the cold room!

(5) Wash with "salt wash FLAG" buffer (~5 mL)

(6) Wash extensively (>40 mL) with "wash FLAG" buffer

(7) Elute the bound FLAG-tagged protein with 5 × 700 μL 0.5 mg/mL FLAG peptide dissolved in "wash FLAG" buffer, with 15 min incubation in each elution step in the cold room. The elute from these steps can be collected to the same 15-mL Falcon tube.

(8) Complete elution by washing with 1.5–2 mL "wash FLAG" buffer.

(9) Take 50 μL samples from the flow-through, the salt wash (optional), and the elution for further SDS-PAGE analysis.

(10) Analyze the fractions of the whole purification on 8% SDS-PAGE.

(11) The eluted protein preparation now contains a high concentration of FLAG peptide that might interfere with further applications, and for some reason also interfere with concentration measurements based on UV spectroscopy. In theory, FLAG peptide can be removed by dialysis using 12 kDa cutoff membrane, as well as by concentration on 100 kDa cutoff Vivaspin concentrator. In our experience, these are not really efficient: the Vivaspin concentrator could concentrate FLAG peptide too for some reason. To completely remove FLAG peptide, size-exclusion chromatography is the best (see Section 2.3.4)

(12) Concentrating up to 0.4 mg/mL is normally safe. Higher concentration should be carefully monitored for precipitation!

(13) Concentration can be determined by several methods. We routinely use UV absorbance at 280 nm in a NanodDrop or NanoVue. In NanoVue, use the "wavelength scan" function in the range of 220–400 nm. As another method, Qubit assay is quite reliable for IDPs (Contreras-Martos et al., 2018), and we found it was good for CBP/p300 too. Concentration measurements by Bio-Rad Protein Assay using BSA as standard usually underestimated the real amount 2–3 times for CBP and p300.

(14) Check the final prep on 8% SDS-PAGE! Add loading control also (dilution series of BSA in the range 0.5–4 μg) to estimate the protein concentration by densitometry.

(15) For short-term (maximum 3 days) storage, store the purified CBP/p300 on ice in the cold room. For a longer period (maximum 2 weeks), add Na-azide and "PI mix for purification" also and store it on ice. For storing it even longer (more than 2 weeks), the purified protein is better be frozen in liquid N_2 and stored at −80°C in its storage buffer.

Notes

(1) In our experience the sensitivity of NanoVue is lower, and the results are much less reproducible and more easily influenced by viscous materials (e.g., glycerol or detergent content, or higher protein concentration) than in the case of the NanoDrop.

(2) Please note that absorbance at 280 nm is highly influenced by oxidized DTT!

(3) In Ni^{2+}-affinity flow-through, there is always a high amount of apparently full-length CBP/p300 remaining (Fig. 2B), though purification from this fraction results in more contaminated/aggregated samples (Fig. 2D).

2.3.4 Size-Exclusion Chromatography

Gel-filtration (size-exclusion chromatography, SEC) was used to improve homogeneity, remove aggregates, and estimate the apparent molecular mass indicative of protein disorder also. SEC also serves as quality check for aggregation and/or degradation. Superdex 200 Increase (S200, GE Healthcare Life Sciences) or Superose 6 Increase (GE Healthcare Life Sciences) columns were used in Akta systems to analyze full-length wt CBP, wt P300, and HATn CBP (Fig. 2C). On S200, the first peak at 8.8 mL (wt) or 9.4 mL (HATn) elution volume corresponds to the protein, and the second peak around 20 mL probably appears due to the FLAG peptide applied in the previous step for elution (c.f. Section 2.3.3). On Superose 6B, the elution profile is quite similar, but the separation range is more optimal for proteins of this size: aggregates, if there are some, could be separated better. Using a gel-filtration standard (Bio-Rad), the column can be calibrated: log M vs elution volume plots then can be fitted linearly, and the apparent molecular mass of the protein in question can be calculated based on its elution volume.

This value represents an apparent size that is highly influenced by the shape and flexibility of the molecule (La Verde, Dominici, & Astegno, 2017; Tompa, 2009). The apparent size of full-length CBP and p300 in SEC (860 ± 70 and 950 ± 70 kDa, respectively) is much larger than the one expected based on their amino acid sequence (273 and 272 kDa, respectively) confirming an extended, partially disordered character (Fig. 2C). HATnCBP shows a somewhat more compact character, its apparent Mw being 710 ± 45 kDa. The chromatographic peak usually shows asymmetry toward smaller sizes that might be due to either disorder character or protein degradation (Fig. 2C).

As quantity is limiting, we were interested if the CBP/p300 fraction isolated from Ni^{2+}-affinity flow-through has the appropriate quality: gel-filtration on Superose 6B clearly shows a higher amount of aggregates and/or the presence of nucleic acid contaminants (Fig. 2D). Interestingly, with RNase treatment, this nucleic acid contamination was efficiently removed, while DNase treatment caused a less pronounced effect (Fig. 2D, right graph). The copurifying RNA might also have functional significance in light of recent findings that noncoding RNAs bind to and activate CBP (Bose et al., 2017). Based on above results, Ni^{2+}-based IMAC is necessary, but not sufficient in purifying CBP/p300.

2.4 Quality Control

Ensuring good quality of a recombinant protein is always a sensitive question. In the case of a smaller protein of well-defined structure and function, there are many possibilities to verify purity, intactness, and functionality. In the case of a multidomain, hence multifunctional, protein, functional assays should be multiplied too. The effect of the loss of function of some domains, even at a low-level, could be combinatorial. For example, in the case of nine functional domains, if all of them are 90% functional, it still can happen that only 10% of the protein prep is fully functional. If the protein has a significant intrinsic disorder, it is very hard, if not impossible, to judge whether its conformational ensemble is the same as in its native state. If the protein is very large, and has many possible PTMs (including autoacetylation) and interacting partners, like in our case, it is inherently inhomogeneous and the expected behavior in any quality check is unclear. Here, we show some examples of how to approach this problem. In Section 3, we describe what kind of molecular tools we could develop thus far to improve the "homogeneity" and/or to facilitate structural–functional studies of our recombinant CBP/p300.

2.4.1 Sodium Dodecyl Sulfate-Polyacrylamide Gel Electrophoresis

SDS-PAGE is the most frequently used technique to visualize protein degradation and the presence of contaminating proteins; however, in the case of CBP/p300, it has considerable limitations. We used MiniProtean Tetra Cell 2 Gel system (Bio-Rad), with 1 mm thick 8% PAGEs to analyze the full-length proteins (described in Section 2.2.2).

Given the huge number of known interacting partners (>400 (Bedford et al., 2010)), it can easily happen that several of them are present in a minor amount not visible on SDS-PAGE. Still, such invisible contaminations can influence structural–functional assays, especially at the single-molecule level.

In spite of extensive protease inhibition, degradation is frequently seen. Some extra bands are usually visualized at high-Mw positions, most probably representing "almost" full-length CBP/p300. However, smaller fragments must also be there in the background, as they might escape detection due to their smaller amount and high diversity in size. If unwanted degradation takes place, it frequently appears as a clearance of the full-length form, as encountered in limited proteolysis trials also. Uncontrolled degradation can be followed also on Western blots developed with antibodies against the terminal tags, as it is more sensitive to minor contaminations, although it is blind to the presence of inner fragments.

2.4.2 Acetylation State by Western Blot

Western blot can also be used for checking the acetylation state of the CBP/p300 constructs, as well as of other target proteins used in CBP/p300 HAT activity assays. The anti-Acetylated-Lysine (Ac-K2-100) antibody (Cell Signaling Technology) could be efficiently used according to the manufacturer's protocol. We could confirm that both full-length wild-type CBP and p300 have specific acetyltransferase activity on known substrates like the transcription factor p53 (Fig. 3A and B) and histone H4 (Fig. 3B). C-Terminally truncated p53 lacking the known p300 acetylation sites (Kim, Rivera, Coffman, & Haber, 2012; Wang, Tsay, Tan, Lo, & Lee, 2003) was not acetylated (Fig. 3A), indicating that the activity of recombinant p300 is specific. Inactive CBP/p300 mutants (HATn) have no measurable activity in these assays (Fig. 3A). Moreover, autoacetylation could not be detected on either of these HATn mutants, while wt proteins were always autoacetylated (Fig. 3C). We also tried to acetylate HATn CBP/p300 with the wt protein in vitro, without success (Fig. 3D), suggesting either intramolecular autoacetylation mechanism or a conformation of HATn CBP/p300 that is not appropriate for being a substrate for the active enzyme.

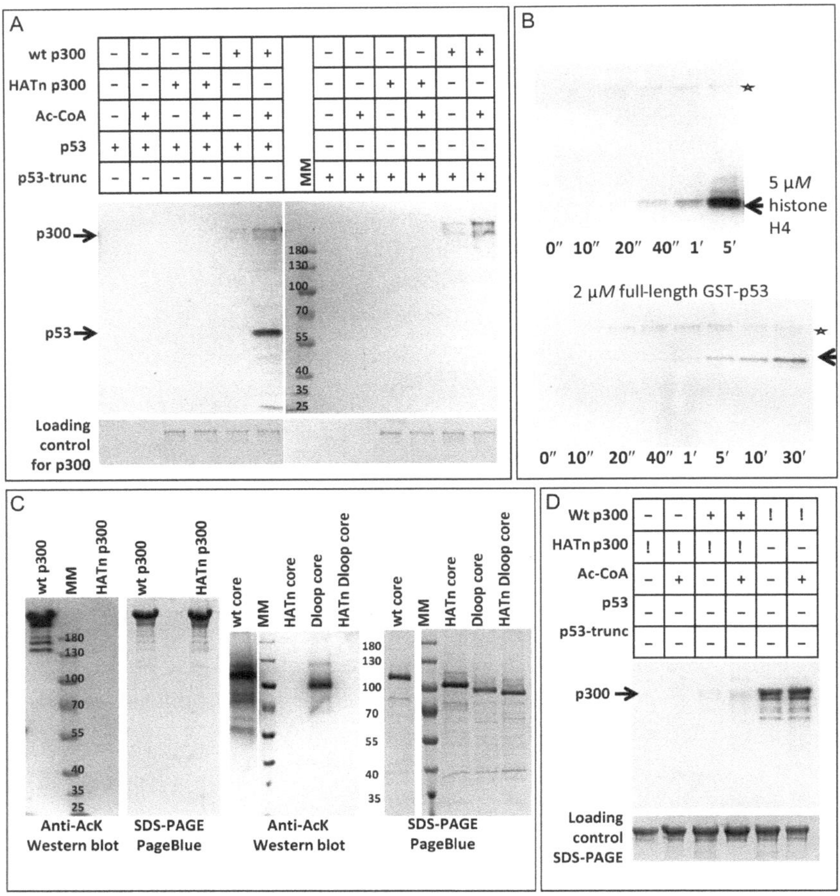

Fig. 3 Acetylation by full-length CBP and p300. (A) Acetylation of p53 by recombinant wt and HATn p300. Full-length and C-terminally truncated (lacking C-terminal acetylation sites) p53 (gift from Peter Wright) were tested in acetyltransferase reaction: 10 μg substrate protein in 50 μL reaction mixtures (73 n*M* (1 μg) wt or HATn p300, and 300 μ*M* Ac-CoA in 50 m*M* Tris pH 8.0, 100 m*M* NaCl, 10% glycerol) was incubated at 30°C for 30 min. 10 μL of the samples were analyzed on Western blot developed with anti-acetylated lysine (Ac-K2-100) antibody. Positions of p300 and p53 are indicated. Autoacetylation of the wt, but not of the HATn p300 is detected also. fl p300 remaining in the gel after transblotting served as loading control. (B) Acetylation of histone H4 (Sigma-Aldrich) and full-length p53 (GST-tagged, Enzo Life Sciences) by fl CBP. Reaction mixtures contained of 40 n*M* fl CBP, 800 μ*M* Ac-CoA, 0.02 mg/mL BSA and 5 or 2 μ*M* substrate protein, Histone H4 or fl p53, respectively, in buffer (50 m*M* Tris pH 7.5, 150 m*M* NaCl, 1000 × diluted PI mix, 50 μ*M* TCEP). Reactions were incubated at 20°C and 7.5 μL of samples were taken in the given time points and analyzed on the same way described for (A). Substrate and fl CBP positions are indicated by *arrows* and *stars*, respectively. Autoacetylation of fl CBP is detected also. (C) Autoacetylation of wt and HATn CBP/p300 constructs. Wt and HATn fl p300 (*left*) as well as different CBP core constructs (*right*, "Dloop core": CBP core lacking the AL, "HATn": inactive versions) were analyzed on anti-AcK Western blots and SDS-PAGEs. (D) wt p300 cannot acetylate the HATn p300 in in vitro autoacetylation. Reaction conditions were the same as in Fig. 3A. HATn or wt p300 were added either as substrate (!) in 10 μg or as enzyme (+) in 1 μg quantity.

2.4.3 Dynamic Light Scattering

Dynamic light scattering (DLS) is a state-of-the-art method to check mono/polydispersity of macromolecular samples. It is to be noted, however, that with a highly flexible system one should be aware that polydispersity can come from structural disorder also, and not only form the tendency of the protein to oligomerize/aggregate. We used DynaPro™ NanoStar™ DLS instrument from Wyatt with its software package Dynamics. Disposable plastic cuvettes were used allowing measurements in as little as 4 μL volume.

Our full-length CBP/p300 (after gel-filtration!) usually shows DLS autocorrelation curves and size distribution as shown in Fig. 4A, indicating 13.6 ± 2.1 nm radius, $25.6\% \pm 0.4\%$ polydispersity, and 1430 ± 500 kDa apparent Mw. A monomodal system with approximately 20%–30% polydispersity might be explained by the intrinsically disordered character too (Stetefeld, McKenna, & Patel, 2016). Interestingly, there is no significant difference between gel-filtered CBP peak fractions corresponding to Ni^{2+}-eluate (CBP EE GF peak) or flow-through (CBP FTE GF peak) (Fig. 4A). Upon in vitro autoacetylation, the radius, the corresponding apparent Mw, and polydispersity remain essentially the same (Fig. 4A). For the CBP Bromo-TAZ2 construct, static light scattering (SLS) measurement was also performed, and the results (6 nm hydrodynamic radius, 222 kDa apparent Mw, 22% polydispersity, and 109.0 kDa calculated Mw (vs 93.8 kDa based on the amino acid composition)) confirm a more extended conformation and an intrinsic disorder even within the core domains.

A thermal transition curve can also be recorded, which reports on the proper folding of a protein under different buffer conditions, and can be used also as quality check of different preparations. It can help to find optimal buffer conditions, checking effects of some additives on the protein fold, as well as it can be used for the determination of affinity toward ligands in a titration experiment.

We checked the stability in some buffer conditions, and the effect of some additives, like EDTA, DTT, TCEP, also the effect of freezing and quick- and slow thawing. We found that different CBP fractions of the purification process have altered heat stability (Fig. 4B). Autoacetylation increases the melting temperature of full-length CBP, resulting in essentially the same stabilized state, regardless of the initial state (Fig. 4B). Based on this, autoacetylation can be the main factor that influences the stability of our recombinant CBP protein preps.

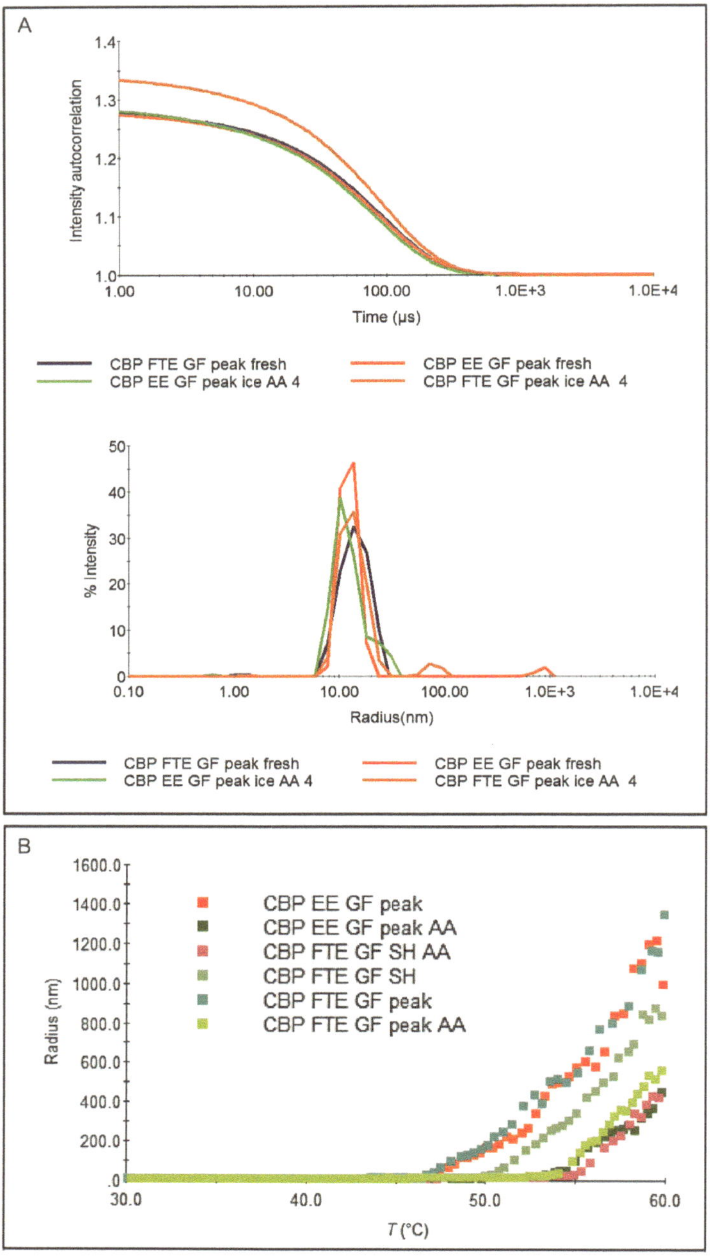

Fig. 4 Dynamic light scattering (DLS) on full-length CBP/p300. (A) Autocorrelation curves (*top graph*) for fl CBP peak fractions of gel-filtration, prepared from Ni^{2+}-affinity elution (EE GF peak) and flow-through (FTE GF peak). Addition of AcCoA allowing further autoacetylation (sample name labeled by AA 4) caused a little shift in the amplitude of

Notes

(1) For measuring SLS with NanoStar, one should be aware that in such experiments a true monodisperse sample, the exact protein concentration, and offset measurement with the buffer, are required.

(2) Large-size aggregates interfere with the DLS signal; therefore samples are recommended to be filtered right before the DLS measurement in a centrifuge-tube filter with 0.2 μm membrane pore size.

2.4.4 Posttranslational Modifications by Mass Spectrometry

As we were interested in possible PTMs occurring on the recombinant proteins purified from Sf9 cells, LC-MS/MS experiments on trypsin digests were performed using an LTQ Orbitrap Velos (Thermo Fisher Scientific) operating in HCD mode (see also Morais et al., 2014 for experimental settings). Several different PTMs were identified with acetylation of lysines being the predominantly observed modification (most of them as a result of autoacetylation (Contreras-Martos et al., 2017)). Also, different phosphorylation and some methylation events were detected (Table 1). As the protein sequences of the recombinant CBP and p300 protein were not completely covered by our MS/MS data, this resulted in an incomplete coverage of possible PTMs. Also, as PTMs quite often have a regulatory function, they are only present in about 10% of the proteins. This increases the heterogeneity of the samples and thereby reduces the possibility of detecting PTMs. Reducing the overall heterogeneity is complicated due to the high number of possible modifications and the different combinations (i.e., PTM crosstalk (e.g., see Venne, Kollipara, & Zahedi, 2014)), making it very troublesome to handle them in vitro. One possibility involves introducing specific enzymes, which completely remove or introduce certain specific modifications, but then again traces of these enzymes could potentially influence further structural or functional studies of our recombinant proteins. In all, this is a difficult issue to tackle.

---

the CBP FTE sample only. Intensity distributions of the same samples are shown in the *bottom graph*. (B) Thermal shift curves in DLS on different CBP fractions prepared from Ni^{2+}-affinity elution (EE GF peak) and flow-through (FTE GF peak, FTE GF SH). Gel-filtration main peak (peak) and its shoulder (SH) toward smaller sizes are also compared. Samples were incubated with AcCoA allowing further autoacetylation (AA). There is an increase in radius with the elevation of temperature, probably indicating aggregation upon thermal denaturation.

Table 1 List of the Identified PTMs on Recombinant CBP/p300

Phosph.Sites	Wtp300	HATnp300	Phosph.Sites	WtCBP	HATnCBP	Acetyl.Sites	Wtp300	HATnp300	Acetyl.Sites	WtCBP	HATnCBP
S285p	+	+	S124p	+	+	K17a	+		K13a	+	+
T841p	+		S745p	+		K79a	+		K663a	+	
T845p	+		S765p	+	+	K977a	+		K1014a	+	
S848p	+		T1006p	+		K1336a	+	+	K1372a[c]	+	
T887p	+	+	S1076p	+		K1340a		+	K1376a	+	
T890p	+		S1755p		+	K1542a	+		K1583a	+	
T891p	+	+	S1763p	+	+	K1546a	+		K1586a	+	
S1038p	+	+	S2063p	+	+	K1549a	+		K1587a	+	
S1044p	+		S2065p		+	K1550a	+		K1588a	+	
S1329p		+	S2079p	+	+	K1551a	+		K1591a	+	
T1332p		+	S2351p	+		K1554a	+		K1592a	+	
S1726p	+	+	S2356p	+		K1555a	+		K1595a	+	
T1851p	+					K1558a	+		K1597a	+	
T1854p	+					K1560a	+		K1605a	+	
T1856p	+					K1568a	+		K1607a	+	
T1906p	+					K1569a	+		K1741a	+	
T1909p	+					K1570a	+		K1762a	+	
T1913p	+					K1900a	+		K1937a	+	+
S2279p	+					**Other PTMs**			**Other PTMs**		
S2325p	+					R237met	+		R2353met	+	
S2326p	+					R754met[a]	+				
						Y611AEBSF[b]	+				
						R2317met	+				

[a]It was not yet found for p300, just for the equivalent site of CBP.
[b]This site was found to be phosphorylated in the literature.
[c]It was found only for mouseCBP.
PTMs at *gray background* were not yet found in the literature.

3. NANOBODY DEVELOPMENT AGAINST PURIFIED FULL-LENGTH CBP AND p300

According to the foregoing notes, we should say that true homogeneity is hardly achievable in the case of such large, partially disordered proteins such as CBP and p300, with numerous PTMs and interacting partners. Traditional biophysical and structural techniques provide mixed results on these proteins, which prompted us to find novel tools to improve homogeneity and/or help identification in structural and functional studies. Nanobodies (NBs) are found to be efficient tools as specific molecular labels, purification aids, molecular switches, and/or crystallization chaperons (c.f. Introduction). Such versatile molecular tools must be especially useful in handling very large and structurally heterogeneous proteins, like CBP/p300. Thus, we aimed to develop novel nanobodies ideally with one of the following properties: (1) specific for different parts including IDRs of CBP/p300, (2) selective for CBP or p300, (3) selective for autoacetylated or non-acetylated forms, (4) specific for other PTMs, (5) specific for certain conformations corresponding to regulations or putative bound states, etc. Although we have not met all these goals, our characterization of a panel of selected NBs suggests that these can be achieved.

3.1 Immunization of Two llamas With p300 and CBP, Respectively

Two llamas were immunized over a period of 6 weeks, one with 600 μg CBP and other with 620 μg p300 in total (Fig. 5A). Lymphocytes isolated from the anticoagulated blood of the immunized llama were used to prepare cDNA, which served as a template to amplify the open reading frames coding for the variable domains of the heavy chain only antibodies. The PCR fragments were ligated into the pMESy4 phage-display vector and transformed in *E. coli* TG1 cells. Nanobody libraries of >1E8 transformants were obtained. After super infection with helper phage, the nanobody repertoire of each library was expressed on the tip of the phage (Pardon et al., 2014).

3.2 Selection and Screening Using Phage Display and Solid-Surface Coating With Purified Full-Length Proteins

Selection of phage particles expressing nanobodies that specifically bind to one of the target proteins was performed. Subsequent rounds of selection and screening were done using the same full-length proteins which were

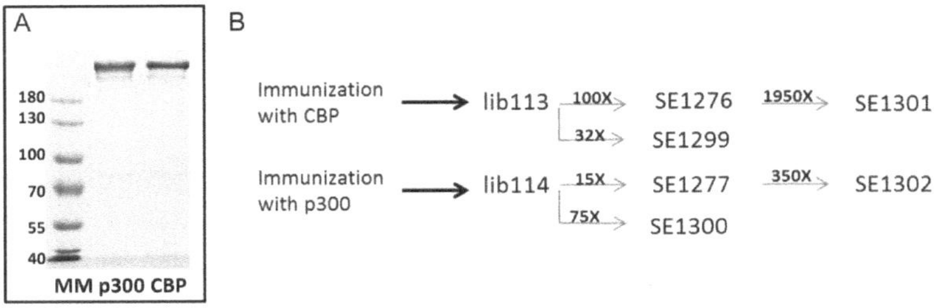

C

	1	2	3	4	5	6	7	8	9	10	11	12
A	~~C6	Failed	A3	Failed	C6	C6		A8			A11	
B				~G4, C6	Failed	**B6**			C12	C12	Failed	Failed
C	Failed	C6	G4,~C6		C6	C6	Failed	A8			F12	C12
D	~~C6	Failed		A3	~G4,~C6	Failed					C12	**E12**
E	~~C6	Failed	Failed		C6	**C6**	E7	E7	Failed	Failed	Failed	**E12**
F	C6	C6		C6	**F5~B6**				Failed	**E12**	Failed	F12
G	C6	Failed	C6	G4~C6	**G5**	Neg. ctrl						Neg. ctrl
H	~~C6	Failed	C6	C6	Failed	Blank				A11		Blank

SE1299: Selection round I on lib113 — SE1301: Selection round II on SE1276 (previous round I on lib113) — SE1300: Selection round I on lib114 — SE1302: Selection round II on SE1277 (previous round I on lib114)

Positive on ELISA

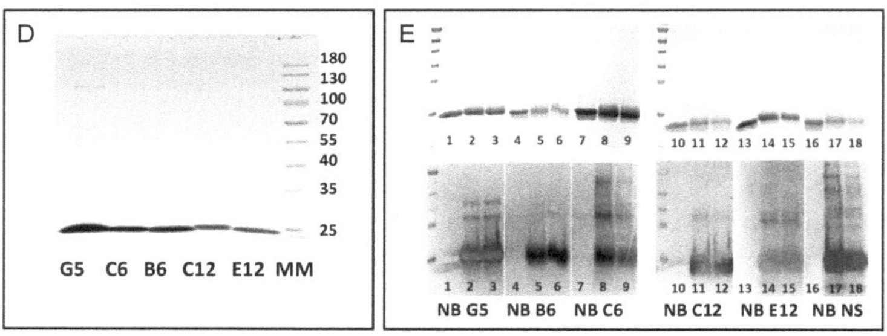

Fig. 5 Immunization, selection, and screening for NBs with full-length CBP and p300. (A) Purified recombinant CBP and p300 used for immunization and screening. (B) Scheme of immunization and selection indicating the codes of libraries and selected sublibraries and the enrichment above the *arrow* representing the selection procedure. (C) Summary of results: sequencing and ELISA of screen MP524. All samples were sequenced. Identical sequences are indicated by the well number of one representative. Close relative sequences to the overrepresented C6 are also indicated with ~C6 or ~~C6. Failed sequencing reactions putatively due to double-infected colonies are also indicated in *gray*. (D) SDS-PAGE of purified NBs. (E) Biotinylation of NBs. The *upper panels* show PageBlue stained SDS-PAGE, the *bottom panels* show the same on Western blot developed by Streptavidin HRP. The sample order follows the scheme: before biotinylation, biotinylated, and biotinylated, desalted. *NB NS*: nonspecific NB as negative controls.

nonspecifically adsorbed (solid-surface coating) on ELISA plates according to the protocol described in Pardon et al. (2014). Although the selection resulted in good enrichment in both cases (Fig. 5B), only five nanobodies (B6, C6, F5, G5, E12) gave positive results in the ELISA screen (Fig. 5C) probably due to the limited space to coat with these high-Mw samples.

Still, the whole screen was sent for sequencing revealing that C6 and related sequences are overrepresented in CBP-selected samples (14 out of 46 are identical, and further 8 are close relative, see also in Fig. 5C). The fact that ELISA screens produced only one positive signal for these otherwise identical nanobodies suggested that the applied conditions were suboptimal. In the sequencing data, further seven nanobodies were found in duplicates or multiplicates on the plate (C12, A11, F12, A8, E7, A3, and G4, see also in Fig. 5C). Although these were not positive under the applied conditions in ELISA, their multiple occurrences make them also candidates, especially the C12 that was found 4 times. Sequencing also revealed that B6 and F5 are close relatives (95% sequence identity, differing only in six amino acids in the whole sequence).

3.3 Expression, Purification, and Biotinylation of Five Selected Nanobodies

Based on both ELISA and sequencing results, nanobodies B6, C6, G5, C12, and E12 were expressed and purified via their His-tags to apparent homogeneity for further characterization (Fig. 5D) using a protocol described elsewhere in detail (Pardon et al., 2014). These nanobodies were biotinylated using Ez-Link NHS-PEO_4-Biotin reagent according to the manufacturer's protocol (Fig. 5E).

3.4 Pull-Down of CBP Domain Constructs With Biotinylated Nanobodies to Map the Epitopes

To address the possible epitope sites of these NBs, pull-down experiments were performed. As a first attempt, CBP domain constructs were used also for the p300-selected nanobodies. CBP domain constructs covering the whole sequence (Fig. 6A) were cloned into pET200D Topo plasmids (also encoded an N-terminal His-tag) and expressed together in BL21 *E. coli* cells. Cloning and bacterial expression was performed according to Suppliers' and standard protocols, detailed description of which is beyond the scope of this chapter.

Using biotin-labeled NBs and Streptavidin MagSepharose beads (GE Healthcare), a pull-down experiment was made from the extract of

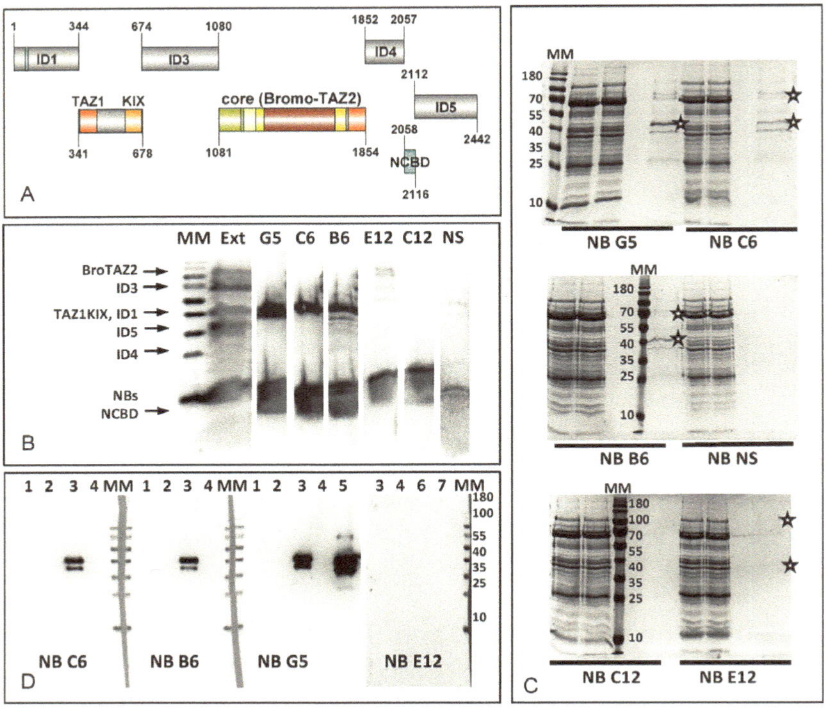

Fig. 6 Pull-down of CBP domain constructs with biotinylated nanobodies to estimate the epitopes. (A) CBP domain constructs designed for epitope mapping. The *color code* follows that in Fig. 1. All the constructs were fused to N-terminal His-tag. (B) Pull-down by biotinylated NBs and Streptavidin Magnetic Sepharose beads from extracts of coexpressed His-tagged CBP domain constructs. Western blots are developed by HRP-conjugated anti-His antibody. (C) Coexpression extract of CBP domain constructs was enriched in His-tagged proteins by Ni-affinity, and then pulled down with biotinylated NBs conjugated to magnetic Sepharose beads. Samples were analyzed by SDS-PAGE and MS protein identification. Sample order follows the scheme: enriched extract, flow-through, last wash, elution. MM: PageRuler prestained molecular marker. The bands labeled by a star were sent for MS analysis (see the peptide hits in Table 3). (D) Western blot on purified FLAG-tagged CBP domain constructs developed with the indicated nanobodies, followed by HRP-conjugated anti-His antibody. The sample order is the following: 1, ID1-FLAG; 2, ID3-FLAG; 3, TAZ1-KIX-FLAG; 4, BromoTAZ2-FLAG; 5, TAZ1-KIX-His as positive control; 6, HAT-inactive BromoTAZ2-FLAG; 7, loop-deleted BromoTAZ2-FLAG. Prestained PageRuler molecular marker (MM).

coexpressed CBP domains and the results were analyzed by anti-His-tag Western blotting and by LC/MS-MS (Fig. 6B and C, the list of peptides identified is given in Table 2). Based on these results, we could ascertain that three nanobodies, G5, B6, and C6, bind to the TAZ1-KIX region, while

Table 2 Pull-Down of CBP Domain Constructs With Biotinylated Nanobodies, Identified Peptides by Mass Spectrometry From SDS-PAGE Bands Indicated in Fig. 6C

Sample	Mw App (kDa)	CBP Domain Construct	Start	End	Sequence	Mean (Score)	Mean (Threshold)
NB G5 band 1	45	TAZ1-KIX	608	622	LVQAIFPTPDPAALK	27	25
			608	624	LVQAIFPTPDPAALKDR	40	28
			625	634	RMENLVAYAK	34	27
			626	634	MENLVAYAK	55	28
			635	647	KVEGDMYESANSR	74	24
			661	668	IQKELEEK	48	28
NB C6 band 1	65	ID3	743	768	AASPMNHSVQMNSMGSV PGMAISPSR	27	16
NB C6 band 2	42	TAZ1-KIX	625	634	RMENLVAYAK	33	27
						50	26
			626	634	MENLVAYAK	59	27
			626	635	MENLVAYAKK	30	26
			635	647	KVEGDMYESANSR	33	24
			661	668	IQKELEEK	46	28
NB E12 band 1	119	BromoTAZ2	1224	1233	DAAYYSYQNR	31	16
			1334	1341	LGNHLEDR	41	27
			1349	1360	QNHPEAGEVFVR	48	27
			1506	1514	MLDKAFAER	62	26
			1525	1535	QATEDRLTSAK	65	29

Continued

Table 2 Pull-Down of CBP Domain Constructs With Biotinylated Nanobodies, Identified Peptides by Mass Spectrometry From SDS-PAGE Bands Indicated in Fig. 6C—cont'd

Sample	Mw App (kDa)	CBP Domain Construct	Start	End	Sequence	Mean (Score)	Mean (Threshold)
NB E12 band 2	54	BromoTAZ2	1525	1535	QATEDRLTSAK	35	29
NB B6 band 1	65	TAZ1-KIX	625	634	RMENLVAYAK	39	28
NB B6 band 2	40	TAZ1-KIX	352	366	LIQQQLVLLLHAHK	54	13
			608	622	LVQAIFPTPDPAALK	33	25
			608	624	LVQAIFPTPDPAALKDR	36	28
			625	634	RMENLVAYAK	32	28
						41	26
			626	634	MENLVAYAK	36	26
						61	28
			635	647	KVEGDMYESANSR	65	24
			636	657	VEGDMYESANSRDEYYHLLAEK	46	20
			661	668	IQKELEEK	45	28

E12 binds somewhere to the core domains from Bromo to TAZ2. Nanobody C12 could not pull out any of these constructs efficiently that might be due to low affinity, or p300-preference, or the low expression level of the interesting CBP construct. The pull-down experiment was made according to the following steps.

Materials

Biotin-labeled NBs (C6, G5, B6, E12, C12, and a nonspecific one as negative control)
MagSepharose beads (GE Healthcare)
Buffer components: Tris, NaCl, "PI mix for purification" (c.f. Section 2.3)
Lysozyme
DNase I
Kanamycin
BL21(DE3) competent cells (*Thermo Fisher Scientific*)
Plasmid mixture of the domain constructs
LB medium
Materials for SDS-PAGE and Western blot (c.f. Section 2.2.2)

Equipment and Disposables

Shaking incubator
Centrifuge
Sonicator
Rotator
Equipment for SDS-PAGE and Western blot (c.f. Section 2.2.2)

Procedure

(1) Prepare a mixture of the seven CBP domain constructs (c.f. Fig. 6A) in pET200D Topo plasmid each at 2 fmol/μL concentration. Transform 50 μL of BL21(DE3) competent cells with 1 μL of this mixture according to the standard transformation protocol.

(2) Instead of plating, dilute the cells into 3 mL liquid LB broth supplemented with kanamycin and let them grow overnight at 37°C, 200 rpm to have a preculture.

(3) Inoculate 1 mL of overnight culture to 1 L LB and let it grow at 37°C, 200 rpm until logarithmic phase (OD ~ 0.7). Then cool the culture down to 16°C.

(4) Induce coexpression of the domain constructs with 1 m*M* IPTG and continue incubation at 16°C, 200 rpm, for overnight. Then pellet the cells by centrifuging at 3500 rpm, 4°C for 10 min.

(5) Prepare lysis buffer: 50 m*M* Tris pH 7.5, 150 m*M* NaCl, 5 × "PI mix for purification," Roche protease inhibitor tablet (EDTA-free, 1–50 mL), 5 μg/mL DNase, 15 μg/mL lysozyme.

(6) Resuspend bacterial pellet in 50 mL lysis buffer. Homogenize by direct sonication for 5 min by 50% power, 5 s on, 5 s off and then centrifuge down at 4°C, 14,000 rpm for 30 min. Filter extract through 45 μm cutoff filter.
(7) Add 200 μL of 1 mg/mL biotinylated nanobodies to 10 mL of protein extract in 15-mL Falcon tubes and rotate them for 5 h at 4°C.
(8) Add 200 μL Streptavidin Magnetic Sepharose (prepared according the manufacturer's instruction) to each sample and incubate for 30 min at RT.
(9) Collect magnetic beads with the help of magnetic rack, remove supernatant, and place the magnetic beads to a new 2-mL tube.
(10) Wash with 1 mL wash buffer (50 m*M* Tris pH 7.5, 150 m*M* NaCl, 1 × "PI mix for purification").
(11) Wash with 0.5 mL "salt wash" (50 m*M* Tris pH 7.5, 1 *M* NaCl, 1 × "PI mix for purification").
(12) Elute with 150 μL 0.1 *M* glycine pH 3.5 buffer. Add 10% 1 *M* Tris pH 7.5 to the fractions immediately to set back the pH.
(13) Make a second elution using 150 μL of 2 × SDS loading dye and heat the samples at 99°C for 5 min.
(14) Analyze samples on high purity SDS-PAGE appropriate for MS and Western blot.

3.5 Epitope Characterization by Western Blot

Using NBs in Western blotting can help to decide if they bind to a linear or conformational epitope, since denaturing SDS-PAGE disrupts the native structure of proteins. However, the method is only indicative, as in some cases the protein can partially refold on the membrane (Zhou, Chen, Purcell, & Emerson, 2007). Here, FLAG-tagged CBP constructs were loaded on denaturing and reducing SDS-PAGE and then transferred to a PVDF membrane. His-tagged NBs were used as first antibodies (in concentrations of 1 μg/mL for C6, B6, and C12, 0.3 μg/mL for G5, 0.5 μg/mL for E12 in blocking solution; c.f. Section 2.2.2), and HRP-conjugated anti-His-tag antibody was used for detection according to the Western-blot protocol described in Section 2.2.2 (Fig. 6D). The putative TAZ1-KIX binders, NB G5, C6, and B6 gave a nice signal on CBP TAZ1-KIX-FLAG sample, indicating SDS resistant and/or linear epitope for these nanobodies. However NB E12, the putative BromoTAZ2 binder, did not give a positive signal, indicating either lower affinity or a conformational epitope.

3.6 Affinity Determination and Characterization of the Epitopes by BioLayer Interferometry

BioLayer interferometry (BLI), similar to SPR, is a state-of-the-art label-free method for real-time studies of biomolecular interactions. In addition, BLI enables to study crude samples, even in a semihigh-throughput (HTP) manner. These make the technique ideal for epitope binning (Abdiche et al., 2014), and also for HTP kinetic studies (Yang, Singh, Wu, & Kroe-Barrett, 2017). In our lab, an Octet® RED96 System (Pall, ForteBio) is available, in which eight samples can be analyzed in parallel.

In screening and epitope-binning studies, the antigens are immobilized on the biosensor surface, and NBs (in periplasmic extracts or purified NBs) are tested in solution. We also tried this arrangement coupling biotinylated p300 to streptavidin biosensors for screening periplasmic extracts (PEs) derived from MP524 plate (c.f. Fig. 5C) instead of ELISA. The observed binding signals were usually too low, less than 0.2 nm (E12: 0.08 nm, C12: 0.04 nm, A8: 0.16 nm, F12: 0.08 nm, E7: 0.04 nm, and A11: 0.03 nm). We think that such arrangements might be suboptimal, as interferometry is hardly predictable upon the addition of a small and compact 15 kDa nanobody to a large, 273 kDa protein that looks even bigger due to intrinsic disorder allowing significant changes in conformation upon binding. Probing of such p300-coupled sensors with purified NBs resulted in better signals, still binding curves suffered from signs of heterogeneous binding making quantitative evaluation difficult. As an example, NB C12 probed with biotin-p300 sensors (Fig. 7A) shows that fitting with a 1:1 model gives a large deviation, and also the K_d values calculated from individual curve fitting and from steady-state analysis are orders of magnitude different (K_d values were calculated as 14 ± 4, 5.9 ± 0.2, 133 ± 8, and 250 ± 14 nM from individual curve fitting at 3.8, 38, 950 nM ligand concentration, and from steady-state analysis, respectively).

To overcome these limitations, the opposite arrangement is a logical choice: probing immobilized NBs with fl CBP/p300 in solution. This is not compatible with screening, but for epitope binning of a limited number of NBs and for affinity determination it should work. Concentration-dependent binding of all five nanobodies to CBP and/or p300 was confirmed using biotinylated NBs conjugated to biosensors, though heterogeneous binding still occurred. Moreover, K_d values were frequently below 10^{-12} M, due to nondetectable dissociation. Indeed, BLI off rates are known to be frequently overestimated (Yang et al., 2017) that can be real problem in the case of large, disordered proteins like CBP/p300. However, the signal-to-noise

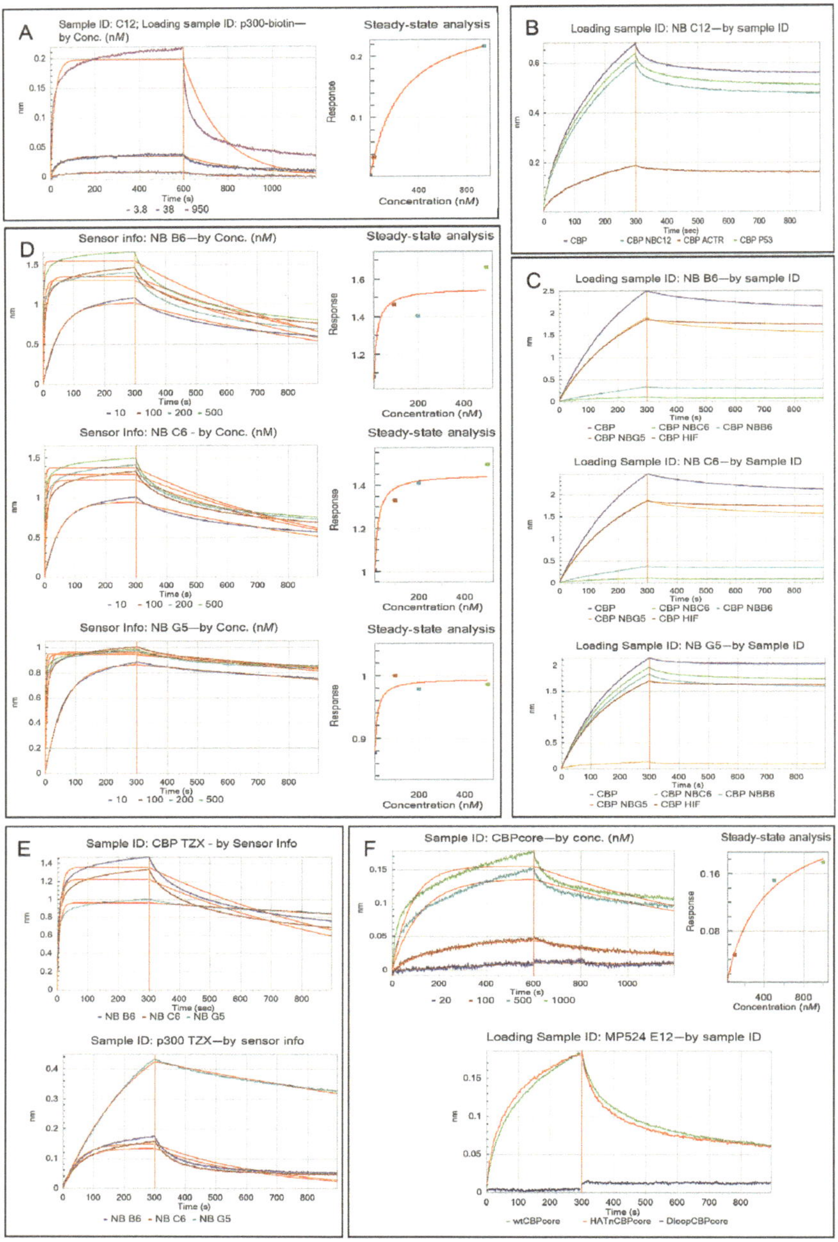

Fig. 7 Affinity determination and further characterization of the epitopes by BLI. (A) Biotinylated fl p300 immobilized on Streptavidin (SA) biosensors binds to NB C12 in a concentration-dependent manner. Fitting (*red curves*) of individual binding curves (*colored ones* on *left panel*) and steady-state analysis (*red curve* on *right panel*). (B) Epitope mapping of NB C12 using transactivation domains of known CBP-interacting

ratio was impressive; therefore a kind of epitope binning experiment was performed with this setup using fl CBP in the presence of other NBs and/or known interacting partners. We could show that NB C12 competes with MBP-ACTR(TAD) (gift from Peter Wright) (Fig. 7B) a known interacting partner of the NCBD domain (K_d = 34 n*M*) (Demarest, Deechongkit, Dyson, Evans, & Wright, 2004). Signal intensity was drastically dropped in the presence of ACTR, while a similar construct of p53(NTAD) (gift from Peter Wright) known to bind primarily to TAZ2 (26 n*M*), and with lower affinity to TAZ1, NCBD, and KIX domains (0.9, 3.1, 22 μ*M*, respectively (Ferreon et al., 2009)) did not cause a significant change (Fig. 7B). Based on this result, we can ascertain that the epitope of NB C12 is somewhere in the vicinity of the NCBD domain and ACTR binding cause a steric hindrance for the NB C12 binding. However, it can also happen that affinity is not changed toward the NB C12, and just some altered conformation of fl CBP bound to ACTR(TAD)-MBP caused a drastically decreased signal intensity. Moreover, interaction could not be detected between NB C12 and isolated CBP NCBD construct using either BLI or HDX/MS. Further description of NB C12 is not in the scope of this chapter; rather, we focus on the other four NBs. In competition BLI assays, we could further narrow down the epitope sites of the TAZ1-KIX binders: C6 and B6 bind to their epitope in a competitive way, while G5 binding does not have significant influence on C6 or B6 binding and vice versa (Fig. 7C). Interestingly, the

---

partners, ACTR and p53. Biotinylated NB C12 is coupled to SA biosensors and probed in 100 n*M* of fl CBP (*blue*) in the absence or presence of 10 × molar excess of NB C12 (*light blue*), or ACTR (*orange*), or p53 (*green*) constructs. (C) Epitope mapping of the three TAZ1-KIX binders by competition experiments. Biotin NB B6, C6, and G5 were conjugated to SA biosensors and probed in 100 n*M* of fl CBP in the presence or absence of 10 × molar excess of one of the three TAZ1-KIX binder nanobodies or the known TAZ1 binder, MBP-Hif1α(TAD) (*orange*). (D) Affinity determination of TAZ1-KIX binding NBs. Concentration-dependent binding of CBP TAZ1-KIX-His was measured with biotinylated NB conjugated SA biosensors. Fitting the individual curves with 1:1 model (*left*) and steady-state analysis (*right*) resulted in somewhat different K_d values (c.f. Table 3). (E) Comparison of NB binding to p300 and CBP TAZ1-KIX constructs. All three TAZ1-KIX binders immobilized on SA biosensors were probed in 100 n*M* CBP TAZ1-KIX-His construct (*top graph*) and in 100 n*M* p300 TAZ1-KIX-FLAG construct (*bottom graph*). For p300, binding signals are less intense and result in higher K_d values (c.f. Table 3). (F) NB E12 binds to the CBP core. Biotinylated NB E12 was immobilized on SA biosensors, and measured against different concentrations of wtCBP core (*top left graph*). Data were analyzed by single curve fitting as well as steady-state analysis (*top right graph*) (c.f. also Table 3). Binding of NB E12 to wt, HATn, and AL-deleted CBP core was compared (*bottom graph*).

known TAZ1 binder Hif1α(TAD)-MBP construct (gift from Peter Wright) does not significantly influence the binding of either of the NBs (Fig. 7C).

For detailed kinetic analysis, smaller CBP domain constructs, known to comprise the epitopes (see Section 3.4), were probed against biotinylated NBs immobilized on Streptavidin biosensors. With this, signs of heterogeneous binding decreased and affinity could have been analyzed better; however, K_d values derived from individual fitted curves and steady-state analysis still show some deviation. Concentration-dependent binding of NBs G5, B6, C6 was measured to CBP TAZ1-KIX (Fig. 7D) and p300 TAZ1-KIX constructs (Fig. 7E). The apparent K_d of binding to CBP TAZ1-KIX is about 1–4 n*M* for NB B6 and C6 and only 0.1–1 n*M* for NB G5 due to its extremely low k_{off} (Fig. 7D and Table 3). All the three NBs show significantly lower affinity toward p300 TAZ1-KIX (Table 3).

Concentration-dependent binding of NB E12 to wt CBP core was also detected and analyzed (Fig. 7F upper panels and Table 3) though heterogeneous binding was also present. Both binding signal intensity and affinity are somewhat smaller as compared to the TAZ1-KIX binders. While signal intensity might also be influenced by apparent size/orientation/conformation of the bound macromolecule, the K_d value reports on affinity only.

Table 3 Dissociation Constants Derived From BLI Experiments: Individual Curve Fitting With 1:1 Model and Steady-State Analysis

	NB B6		NB C6		NB G5			NB E12	
	K_d (n*M*)	Error (n*M*)	K_d (n*M*)	Error (n*M*)	K_d (n*M*)	Error (n*M*)		K_d (n*M*)	Error (n*M*)
10 n*M*	17	0.008	0.55	0.008	0.157	1.00E-03	100 n*M*	31.1	0.7
100 n*M*	1.27	0.04	1.25	0.04	0.211	0.005	250 n*M*	38.3	1.2
200 n*M*	1.5	0.07	1.46	0.07	0.206	0.006	500 n*M*	48.9	1.1
500 n*M*	2.01	0.14	1.94	0.15	0.233	0.009	1000 n*M*	70.8	1.8
100 n*M*[a]	16.4	0.7	17.3	0.8	13.35	0.14	250 n*M*[b]	30.5	1.1
Steady-state analysis									
Req	4.2	1.3	4.1	0.7	0.97	0.15		310[c]	64[c]
Response	4.5	1.3	4.5	0.7	1.3	0.17		380	80

[a]Data for p300 TAZ1-KIX binding.
[b]Data for HATn CBP core binding.
[c]Data corresponding to for 20 n*M* curve was omitted as Req could not be determined.

Note that E12 was originally selected by p300 from lib114 and it might prefer p300 core over CBP core. BLI experiments also revealed that NB E12 could bind to either the wt or the inactive CBP core domains with similar affinity, but binding was not detected with an AL-deleted mutant (Fig. 7F bottom). Therefore, NB E12 might bind to somewhere around the AL, but not to those regions that are either heavily autoacetylated or not, in the wt or the inactive CBP, respectively (Thompson et al., 2004; Wang, Thompson, Cole, & Cotter, 2005).

Notes

To design a BLI experiment it is hard to provide a general detailed protocol, as proteins in question largely determine the optimal experimental setup. Here we note some critical aspects that need to be considered before start.

(1) Selection of an appropriate sensor. It is said that sensor selection should be driven by the aim of the experiment. If detailed kinetic analysis is required, stable immobilization is absolutely necessary, so either amine reactive (ARG), or streptavidin (SA), or super-streptavidin (SSA) sensors can be used. However, immobilization via either reactive-amine coupling or biotinylated Lys residues on the protein surface might lead to steric hindrance that is hard to control. One solution might be enzymatic biotin ligation (AviTag technology, Avidity) at the N- or C-terminus (de Boer et al., 2003; Gautier & Hinner, 2015). In our experience, optimizing immobilization is critical for any of the applications, even for efficient screening. The most irreversible and the simplest layer structure is recommended (avoiding large sandwiches, such as protein A-coupled specific antibody that captures protein 1 for test with protein 2).

(2) It also has to be considered that the BLI signal is proportional to the mass and density of the biolayer. To increase signal intensity, the smaller protein should be immobilized on the sensor (bait) and the larger one be tested in solution (analyte).

(3) Loading scout is also recommended to find the optimal concentration of the bait protein for immobilization, where the binding of the analyte gives an optimal signal-to-noise ratio.

(4) The success of regenerating biosensors strongly depends on the nature of the protein immobilized. It is advised to try different regeneration methods, and if the signal intensity drops in the subsequent binding step, it is better to avoid regeneration as it will severely influence kinetic analysis later on. In our experience, regeneration hardly worked either because of not removing the bound analyte, or damaging the structure of the bait protein.

(5) Also related to regeneration is that using the "regeneration" step built in the Data Acquisition Software (ForteBio) leads to a kind of autozero effect, starting new baselines from zero again even if proteins were not removed from the sensors. We recommend to use the "wash" option instead of "regeneration," merging sensors into regeneration and then into neutralization buffers repeatedly (5 s in Reg., 5 s in Neut., repeated 3 times).

(6) Choosing an appropriate buffer is also critical. Basically, it has to be optimal for both the bait and the analyte; moreover it requires some detergent and/or blocking agent that minimizes nonspecific binding. We routinely used the "final dialysis or gel-filtration buffer" (c.f. Section 2.3) completed with 0.02% Tween 20 and 0.02 mg/mL BSA.

(7) The possible occurrence of nonspecific interactions also needs to be controlled by using reference sensors in a separate sensor row in the same experiment, and with exactly the same steps (same order, same length) as the measured ones, otherwise analysis in the ForteBio Analysis software will not work. It is also possible, though much more labor-intense, to export processed data to txt files and analyze curves in other software like MatLab or Origin. A good reference protein needs to be similar in size to the bait and appropriate for the same immobilization as well as compatible with the applied buffer conditions. We recommend to test several different reference proteins, in order to not overestimate the nonspecific interactions because of accidental choice of a sticky one. Empty blocked sensors can also be used, but in our experience the binding signal intensity is strongly dependent on the initial thickness of the biolayer (BLI signal upon loading the Bait; if it is already too large, the binding signal will be smaller, so an empty sensor can result in larger nonspecific binding signals).

(8) Subtraction of a buffer reference (called "reference wells," where one of the eight sensors in the row is not exposed to the analyte just to the buffer(s)) can help if signal is not stable due to inefficient capture of the bait protein. Double reference subtraction is also possible (reference sensors and reference wells).

(9) Measurement time in BLI is limited by the evaporation of samples. In order to get reproducible results, it is recommended to limit the time of an experiment to 5 h.

(10) For detailed kinetic analysis, it is recommended to measure several (three to five) different concentrations of the analyte ranging from 0.1 × to 10 × of the approximate K_d. The curves can then be fitted individually using different methods built in the analysis software, resulting in kinetic parameters for the binding. A steady-state analysis is also possible, plotting either "response" or "Req" parameters vs different analyte concentrations (c.f. Fig. 7A, D, and F upper panels, and also Table 3). In our experience, with flexible proteins like CBP and p300, heterogeneity of the binding cannot be fully avoided, resulting in large deviation of data derived by the two analysis methods. Binding heterogeneity always increases with the concentration of the Analyte.

3.7 Detailed Epitope Mapping by Hydrogen Deuterium Exchange With Mass Spec

Hydrogen-deuterium exchange mass spectrometry (HDX-MS) is an advanced analytical technique that reports on protein conformations and dynamics (Oganesyan, Lento, & Wilson, 2018). It monitors the chemical exchange of covalently bound hydrogen atoms in proteins for deuterium in D_2O solvents providing a structural probe at almost every residue across the protein backbone. Differences in the HDX-MS signatures of bound and unbound proteins can be used in footprinting to rapidly identify associating regions between proteins including antibody–antigen interactions (Deng, Lento, & Wilson, 2016; Oganesyan et al., 2018). Although the chemical information of HDX-MS is only spatially resolved to the level of individual peptides, it has many advantages over other methods including NMR, particularly with regard to throughput and sensitivity. Moreover, the ability of MS to characterize proteins significantly larger than the upper mass limit of NMR has made the technique particularly appealing for research on conventional antibodies (Majumdar, Middaugh, Weis, & Volkin, 2015).

3.7.1 TAZ1-KIX Binders: NB B6 and C6 as Molecular Labels, NB G5 as Molecular Switch

Given the high affinities of nanobodies G5, C6, and B6, we wanted to map their epitopes more precisely. HDX-MS experiments revealed that all these NBs bind to the KIX domain (Fig. 8A). However, protection from HDX is

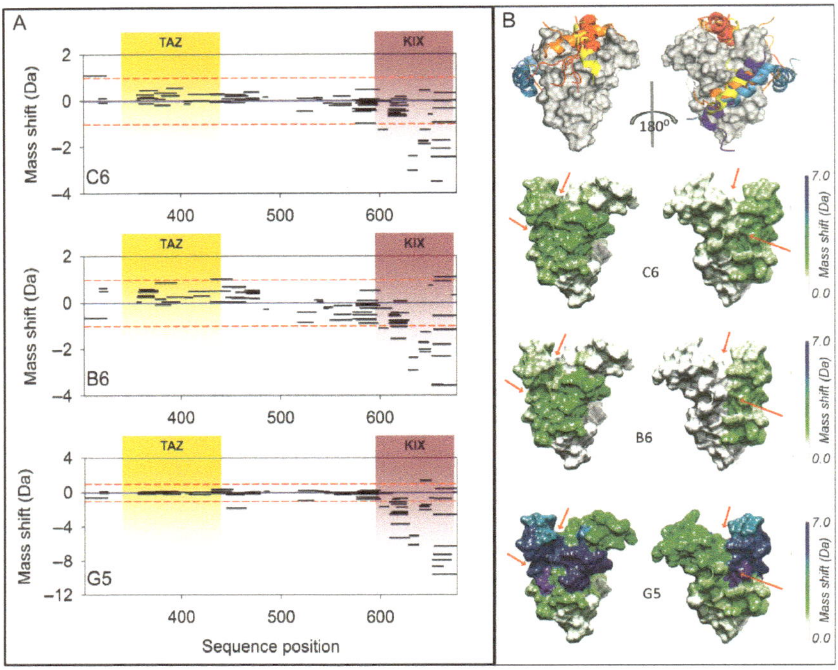

Fig. 8 Epitope mapping for TAZ1-KIX binders using HDX-MS. (A) HDX-MS Wood's plots showing the mass shifts of all peptides in response to the binding of NB C6, B6, and G5 to CBP TZX. The mass shifts of each peptide are plotted vs their position in the protein sequence. Negative shifts below the 95% confidence interval threshold (*red lines*) indicate slower isotope uptake and therefore greater protection in the NB bound form. The position of the TAZ1 and the KIX domains are indicated by *colored backgrounds*. (B) Comparison of the HDX-MS results with the known binding sites of KIX domain. *Upper*: 1KDX.pdb with surface representation, superimposed with other available high-resolution KIX structures (2AGH, 5SVH, 2LQH, 2LQI, 1KDX, 2LXS, 2LXT, 2KWF, not shown). All protein partners from these structures are shown with cartoon representation indicating the two binding surfaces on the KIX, MLL (*red*), ITF2 TF4 (pink), CREB pKID (*light blue*), Myb (*blue*), and FoxO3A TAD segments (*orange* and *yellow*) can bind to both sites in a promiscuous way. The structural representation was made in PyMOL (The PyMOL Molecular Graphics System, Schrödinger, LLC.). Panels below show HDX-MS difference data for the effect of NB binding projected to 1KDX structures in the same orientation than in the two upper panels. Protection by NB C6, NB B6, and NB G5 is shown according to the same color scale. These structural representations were done in UCSF Chimera (Pettersen et al., 2004). Known protein binding sites are indicated by *red arrows*. Note the mass shifts for surface representation were prepared by averaging the change in mass across all peptides intersecting each residue hence the total mass shift is lower in the surface representation than in the Wood's plots.

significantly larger in the case of nanobody G5, which suggests that the binding of this nanobody also induces large conformational changes. It is worth mentioning that there is no high-resolution structure available for the apo form of the KIX. However, several NMR structures report on its bound state, in interaction with partners (CREB: 1KDX.pdb, MLL: 2LXS.pdb, MLL and CREB: 2LXT.pdb, cMyb: 1SBO.pdb, MLL and cMyb: 2AGH.pdb, ITF2 TF4: 2KWF.pdb, FoxO3A TAD: 2LQH and 2LQI.pdb) revealing two interacting surfaces (Fig. 8B top panels) that recognize basically the same conserved motif of the partners (ΦxxΦΦ (Dyson & Wright, 2016; Wang et al., 2012)). The first crystal structure was published recently in a complex with two known interacting partners fused together artificially (5SVH.pdb, Langelaan, D. N., Smith, S. P. to be published on design of nanomolar affinity ligand to the KIX domain of CBP).

Projecting the HDX-MS results onto a structure of CBP KIX derived from 1KDX.pdb (Radhakrishnan et al., 1997) reveals that NB B6 and C6 protect approximately the same surfaces, most probably not overlapping with known binding sites of the KIX domain (Fig. 8B second and third rows of vs top panels). Therefore, we propose that NB C6 and B6 can be used as molecular labels and/or crystallization aids in studies addressing the structure of larger part of CBP/p300 and/or putative new conformations of the KIX domain, respectively.

NB G5 protects a larger surface, at least partially overlapping with both known binding sites of KIX domain (Fig. 8B bottom panels vs top panels). Indeed, the known KIX-binding motif can be identified in the CDR3 region of NB G5 as segment of "IPTML" residues. Hence, NB G5 might efficiently mimic the binding of known protein partners and can be potentially used as a kind of molecular switch in in vitro (Dmitriev et al., 2016) and in-cell (Traenkle & Rothbauer, 2017) functional studies. Blocking binding functions of the KIX domain might interfere with the assembly of transcription signaling complexes at certain promoters that could be addressed with the help of the NB G5 molecular switch.

3.7.2 Core Binder: NB E12 as a Potential Crystallization Aid

With HDX-MS, two regions of protection were detected in both wt and HATn CBP core (Bromo-TAZ2-FLAG) constructs upon NB E12 binding (P1373-V1380, and F1523-E1562: Fig. 9A and B), though with the inactive core, the protection was apparently weaker (Fig. 9B). The second site partially overlaps with the N-terminus of the 62-residue long AL (L1556–L1617). It is important to mention that from residue A1571, the AL region

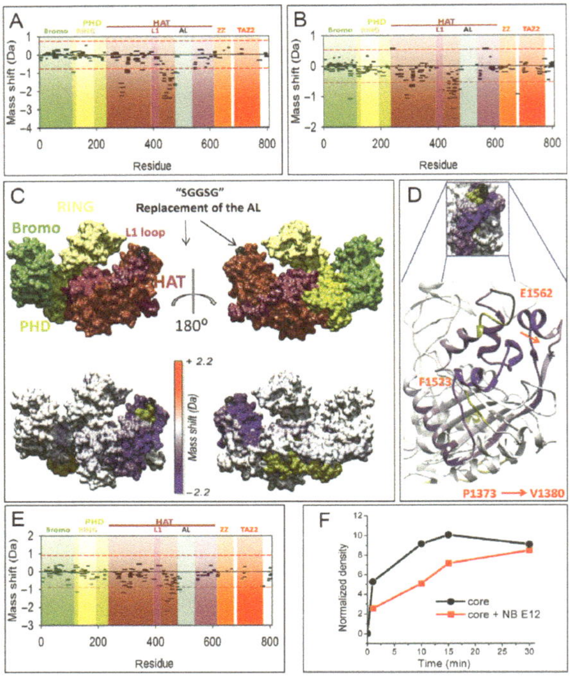

Fig. 9 Epitope mapping for NB E12 on CBP core constructs using HDX-MS. (A and B) Wood's plot showing the effect of NB E12 binding to wt (A) and HATn (B) CBP core. Negative shifts below the 95% confidence interval threshold (*red lines*) indicate slower isotope uptake and therefore greater protection in the NB-bound form. Subdomains within the Bromo-TAZ2 core region are indicated with background colors following the color code in Fig. 1 and (C). (C) Projection of HDX-MS results to a CBP Bromo-HAT structure. Homology model for CBP Bromo-HAT (*upper panels*) was made by Swiss Model Homology Modeling (Arnold, Bordoli, Kopp, & Schwede, 2006) using the ligand bound p300 Bromo-HAT structure, 4BHW, in which the 62-residue AL is exchanged for a 5-residue linker (SGGSG, *black*). The domain representation (*top*) and the projection of HDX-MS data (*bottom*) were done in UCSF Chimera (Pettersen et al., 2004). (D) Structure of NB E12 epitope in CBP HAT domain. Cartoon representation of the previous projection focuses on the regions protected by NB E12 binding. The 5-residue linker replacing AL is shown in *black*. The two protected segments are indicated by naming the first and the last residues of the track (*red*). The F1523-E1562 protected region partially overlaps with the N-terminus of the AL; therefore the overlapping 6-residue track cannot be shown in this representation. The most-protected segment (*dark blue*) is embedded within the crystal structure and locates close to the substrate-binding site. (E) Wood's plot on NB E12 and wt CBP Bromo-TAZ2 binding in the presence of CoA. Nb E12 binding shows similar protection pattern in the presence of CoA, but is somewhat less pronounced suggesting possible competition between NB E12 and CoA binding. (F) NB E12 inhibitory effect on HAT activity of CBP core. Acetylation of histone H4 substrate by CBP core in the presence and absence of NB E12 was followed on Western blot and evaluated by densitometry. NB E12 shows slight inhibition of the HAT activity, confirming the putative competition with CoA binding. Note regions for which no peptides were observed for p300 are colored *tan* on the structural representation.

was not covered in HDX-MS (Fig. 9A, B, and E). These results were projected to a model structure of CBP Bromo-HAT that was created based on the available p300 Bromo-HAT structure (4BHW), where the disordered AL is replaced by a linker (SGGSG) (Fig. 9C). Recently, the structure of apo CBP Bromo-ZZ (5U7G) was also published (Park et al., 2017) that shows quite good agreement with 4BHW and our model structure also, regarding PHD, and HAT domains. Bromodomain is slightly shifted toward the HAT, and RING is not visible due to its flexibility. As we found that ZZ is not affected by either of the tested conditions in HDX-MS, the model structure was chosen to project the HDX-MS results.

The two protected regions are close to each other, providing one single conformational epitope for NB E12 (Fig. 9C and D). It seems interesting that the protection is quite extended and the major part of the protected site is buried in the structure (Fig. 9C and D). It suggests that conformation(s) existing in solution might be considerably different from the published crystal structures where either substrate analogues or crystal contacts could stabilize a more closed conformation (c.f. Section 1).

Moreover, this putative embedded binding site of NB E12 is close to the known CoA-binding pocket in these structures, suggesting that NB E12 and CoA might compete with each other. To test this by HDX-MS experiments, we compared the CoA-bound CBP core in the presence and the absence of NB E12. Indeed, NB E12-induced protection was apparently weaker than in the case of the apo CBP core; still the protection pattern remained similar (Fig. 9E vs A). Moreover, in the presence of NB E12, acetyltransferase activity of CBP core on histone H4 was decreased (Fig. 9F) also confirming the putative competition between CoA and NB E12 binding.

Based on this, we propose that NB E12 could serve as a crystallization aid stabilizing a conformation more relevant for solution structure, having a potential impact also on dissecting the catalytic mechanism of acetyl transfer by CBP/p300.

3.8 Estimating the Potential of the NB Libraries

It was surprising that despite the good enrichment in the first NB selection, not too many positives could be detected in an ELISA screen (c.f. Fig. 5). Given the size and structural complexity of CBP/p300, we are convinced that much more specific and useful NBs can be isolated from the two libraries (lib113 and lib114) using alternative approaches. We assumed that the solid-surface coating using such a big and partially disordered protein as p300 and CBP might hinder large part of the protein surface from interaction with the NBs. The other concern was that CBP/p300, due to their flexibility, and multiple possible PTMs, might represent a conformationally/chemically very heterogeneous protein population that can decrease the detection potential for NBs specific for one or another state. To overcome these limitations, other immobilization methods need to be tried as well as smaller CBP/p300 domain constructs should be used as baits addressing more specific questions. Given that HTP sequencing recently became more accessible and has been efficiently used for characterizing NB libraries (Miyazaki et al., 2015; Turner et al., 2016), we wanted to use this technique for assessing the potential of the two NB libraries, and also for establishing a sequence database making future NB isolation more plausible using protein pull-down and mass spec identification (Fridy et al., 2014).

3.8.1 Systematic Selections With the Full-Length Proteins Immobilized via the C-Terminal FLAG-Tag

In a new selection, full-length proteins were immobilized on anti-FLAG magnetic beads through their C-terminal FLAG-tag, making more protein surface available. A series of selection was performed with wt and HATn full-length CBP and p300 against both libraries according to the scheme in Fig. 10A, and sublibraries were analyzed by HTP sequencing. With such selection, we attempted to cover the pool of potential binders as much as possible; therefore maximizing the enrichment was not an express goal. With blank samples, sticky and nonspecific hits could also be filtered out.

Here we provide a protocol only for antigen presentation and phage selection using anti-FLAG magnetic beads, as other steps follow the protocol given in Pardon et al. (2014).

Materials

NB library in infective phages (cfu $> 10^{12}$ mL^{-1}, see details in Pardon et al., 2014)
Preculture of TG1 cells in exponential growing phase (OD_{600nm} = 0.5–0.6)
Anti-FLAG® M2 magnetic beads (Sigma-Aldrich)
FLAG®-peptide (Sigma-Aldrich)
Culturing media: LB and 2xTY
Ampicillin

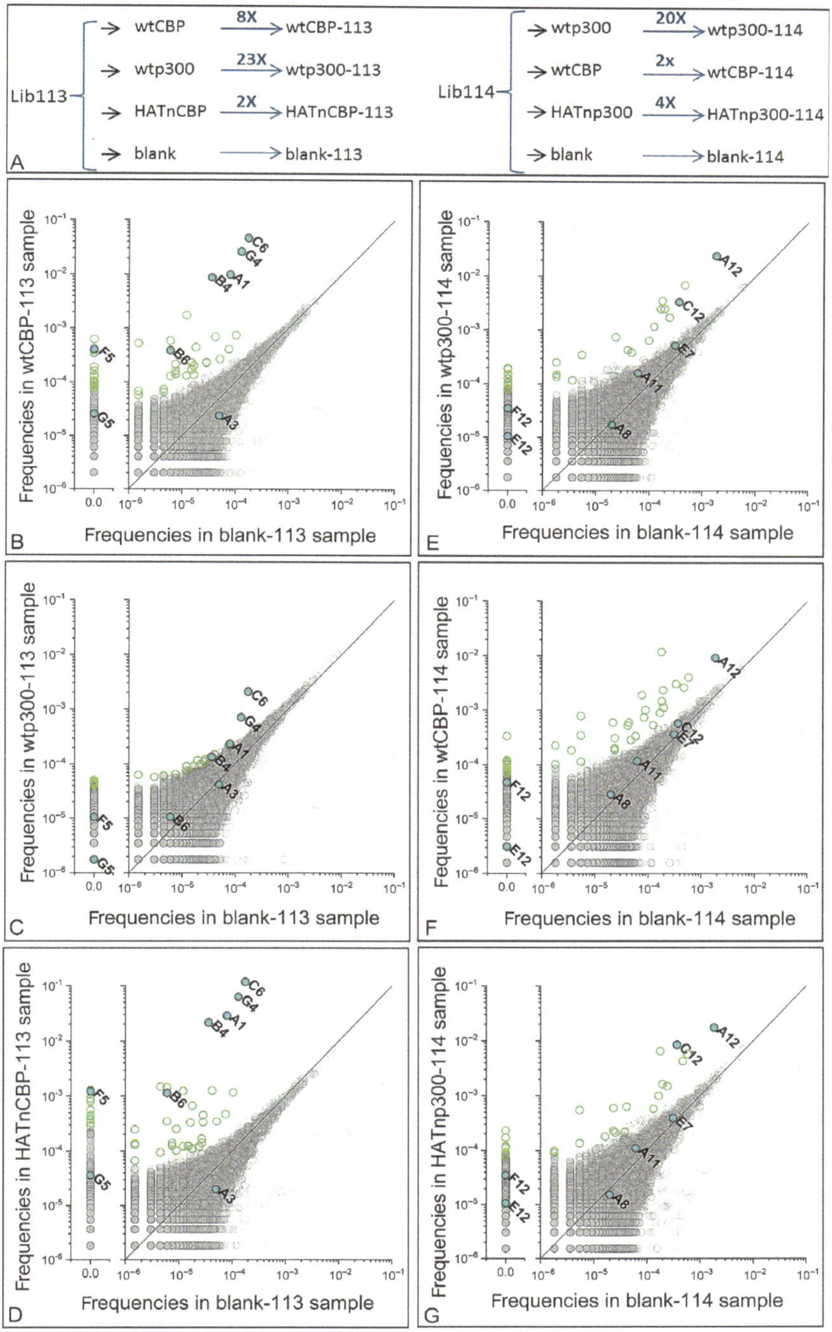

Fig. 10 See legend on next page.

Glucose

0.1 *M* glycine buffer pH 3.0

Wash buffer: 50 m*M* Tris pH 7.5, 150 m*M* NaCl 5% glycerol, 10 × "PI mix for purification" (c.f. Section 2.3)

Low-fat milk powder for blocking

Extracts of Sf9 cells expressing CBP/p300 constructs (fresh or stored at − 80°C in aliquots, c.f. Section 2.3.1).

Trypsin

PBS

4 mg/mL AEBSF stock solution

Standard kits for plasmid miniprep, like PureYield Plasmid Miniprep from Promega

Equipment and Disposables

Shaking incubator at 37°C, 170 rpm

Magnetic rack for microtubes

Rotator for microtubes

Pipettes and sterile filter tips

Polypropylene microcentrifuge tubes

50-mL Falcon tubes

Procedure

(1) Dilute TG1 preculture 100-fold into 50 mL LB and let them grow at 37°C and 170 rpm until exponential phase (OD_{600nm} = 0.5–0.6) and keep on ice until used.

Fig. 10 Systematic selection with full-length constructs and analysis of sublibraries by HTP sequencing. (A) Scheme of systematic selections with full-length wtCBP, wtp300, and HATn CBP in CBP immunized library (Lib113), and with wt p300, wtCBP, and HATn p300 in p300 immunized library (Lib114). Full-length proteins were immobilized on anti-FLAG magnetic beads directly from Sf9 cell extracts. Blank selections with FLAG peptide conjugated anti-FLAG magnetic beads were also performed in both libraries. The enrichment is indicated above the *arrow* representing the selection procedure. The names of the resulting sublibraries (the same for the corresponding samples for HTP sequencing) are also indicated. (B–G) HTP sequencing results represented on frequency-correlation plots. Selected NB sublibraries were analyzed by HTP sequencing. After processing, clonotypes were defined using VDJtools (Shugay et al., 2015), and their frequencies were calculated. Frequency–correlation plots (Origin 8.0, OriginLab, Northampton, MA) show frequency values in the given sample vs the blank. For hits with blank frequency = 0, an additional panel was inserted. Given the large number of overlapping data points, partially transparent symbols were used. Promising candidates (*green circles*) and NBs from the first screening (*cyan dots*) are also shown.

(2) Preparation of antigen conjugated magnetic resin. Pipette 20 μL of anti-FLAG magnetic beads (50% suspension) in separate polypropylene microcentrifuge tubes for each selection sample and label them accordingly.
- Activate beads by washing them with 2 × 1 mL 0.1 *M* glycine buffer pH 3.5. Equilibrate the beads with 3 × 1 mL "wash buffer."
- Add 4 mL of Sf9 extract and incubate for 4 h at 4°C in a rotator.
- As a control for blank selection, one aliquot of the beads is incubated with 600 μL of 0.25 mg/mL FLAG peptide in wash buffer.
- Remove the extract and wash with 3 × 1 mL "wash buffer."
- Resuspend beads in 40 μL wash buffer and take 10 μL for SDS-PAGE to check the efficiency of antigen capture.

(3) Phage selection
- Block the beads in 500 μL of blocking solution: 1% milk in "wash buffer" for 2 h at RT.
- Wash the beads with 3 × 1 mL "wash buffer."
- Add 300 μL of blocking solution and 20 μL of phage to the beads, and incubate for 2 h at RT.
- Wash the beads with 8 × 1 mL "wash buffer."
- Wash the beads with 2 × 1 mL "wash buffer" but without "PI mix."
- Resuspend the beads in 100 μL of "wash buffer" without "PI mix" and transfer the beads to a new polypropylene microcentrifuge tube.
- Elute phages with 34 μL of 0.25 mg/mL trypsin in PBS by incubation for 30 min at RT.
- Stop trypsin by addition of 5 μL of 4 mg/mL AEBSF.

(4) Phage repertoire rescue and titration are done according to the protocol given in (Pardon et al., 2014)
- Infect 3 mL of exponentially growing TG1 cells from step (1) with 50 μL of the eluted phage in a 50-mL Falcon tube and incubate for 30 min at 37°C without shaking.
- Add 15 mL of 2xTY supplemented with 100 μg/mL ampicillin and 2% glucose, and incubate overnight at 37°C and 170 rpm.
- 1 mL aliquots are to be prepared as glycerol stocks (20%, v/v) and stored at −80°C for years.
- For further analysis by deep sequencing, plasmid preps are to be done also from this infected TG1 culture, using standard kits like PureYield Plasmid Miniprep.

Notes

(1) TG1 cells are easily infected by phages; therefore it is advisable to test them by overnight growing on a solid selective medium.

(2) Right practice of handling beads in a magnetic rack is critical; otherwise beads can easily be lost or washing/incubation can be inefficient due to residual solution from the previous step. Right practice follows the steps: (1) Add solution to the beads in the tubes. (2) Mix them by inverting the tubes. (3) Place the tubes into a magnetic rack, and invert them again several times to make sure that all beads are collected even from the cap of the tubes. (4) Remove the supernatant by pipetting even from the cap.

(3) Incubation steps are made in a rotator to ensure appropriate mixing.

(4) Considering the degradation-prone character of CBP/p300 antigen, we decided to elevate the amount of the proteinase inhibitors and use 10 times more "PI mix for purification" as usually (c.f. Section 2.3). To avoid inhibition of the trypsin used for the elution of the phages, in the last washing steps PI mix was omitted.

(5) For work with phages, a separate room is recommended to avoid unintended phage infection of bacterial cultures. Use filtertips for pipetting, and discard all disposables to 1% sodium hypochlorite to inactivate contaminating phage particles.

3.8.2 Characterization of Sublibraries by HTP Sequencing

cDNA of nanobodies were amplified by PCR on the minipreps prepared from selected sublibraries (c.f. Section 3.8.1) using SuperFi polymerase (Thermo Fisher Scientific) and primers specific for pMESy4 phage display vector (forward: 5′-ATACCTATTGCCTACGGCAGC-3′, reverse: 5′-CGGGTACTAGGCTTCAGGTTC-3′). From purified PCR products, sequencing libraries were established with NEBNext Ultra DNA Library Prep Kit for Illumina (NEB) adding sample-specific tags, and sequenced together on a MiSeq platform with MiSeq Reagent kit v3 2× 300bp (Illumina). Sequencing was done from two directions to cover the full NB-coding region. Base-calling accuracy was measured by the Phred quality score (Q-score) (Ewing & Green, 1998). About 83% and 75% of the bases were above Q20 (~99% accuracy) in the case of forward and reverse runs, respectively.

In raw data we obtained 16,541,037 fragments. After trimming and filtering the corresponding reads, we kept 13,100,167 fragments. After stitching the paired reads, translating to protein sequences, and applying further quality filters, 559,988 individual amino acid sequences remained. This data set is used to establish a database for future nanobody identification by MS.

To get a better view on this NB repertoire, clonotypes were identified and quantified per sample with MiXCR v2.1 software package (Bolotin et al., 2015). We defined clonotypes only by their CDR3 amino acid sequence and did not use the read correction functionality of MiXCR to avoid reporting of nonexisting clonotypes. We also removed out-of-frame clonotypes or clonotypes with stop codons. The clonotype frequencies in different samples were subsequently joined with the VDJtools v.1.1.5 software package (Shugay et al., 2015), again only based on the CDR3 amino acid sequence. In the 8 samples, overall 355,376 individual CDR3 sequences were found in this way. In 353 cases out of the 355,376, GQGT segment was also found in the CDR3, mainly in single hits, that might be due to either library cloning, or sequencing or clonotyping errors.

Potential positive hits can be visualized well on frequency–correlation plots (Fig. 10B–G). In theory, every data point above the diagonal might be positive, but several hits look obviously outstanding from the majority (green circles in Fig. 10B–G). Also, numerous clonotypes were not found in the blank at all, and these are supposed to be potential binders, though it is worth to define top hits in each sample (green circles in Fig. 10B–G). Based on this ranking, we defined a group of 162 individual CDR3 sequences that are expected to be strong and specific binders to CBP and/or p300 (Fig. 10B–G). Within this group of promising candidates, 91 clonotypes represent unique sequences. The remaining 71 CDR3 could be clustered by homology: they defined 17 clusters, which might represent families or superfamilies including the one we found already for NB C6 (c.f. also Fig. 5C).

3.8.2.1 Nanobodies From the First Selection Appearing in HTP Sequencing Data

All five NBs from the first selection using solid-surface coating are identified also in HTP sequencing data, where selection was performed with anti-FLAG antibody conjugated CBP or p300 (cyan dots in Fig. 10B–G). On frequency plots, the KIX binders NB B6, C6 are abundant and highly enriched, showing some preference toward CBP over p300 (Fig. 10B vs C) as it was expected by BLI results (c.f. Fig. 7E and Table 3). Comparing frequency plots, we can assume that autoacetylation of CBP might have a negative effect on their affinity, as they were enriched more in the selection with the nonautoacetylated

HATn than with the acetylated wt CBP (Fig. 10B vs D). The other KIX binder, NB G5, is less abundant and is not present in the blank selection at all (Fig. 10B–D). It might prefer CBP over p300 (Fig. 10B vs C) and might not be affected by autoacetylation of CBP (Fig. 10B vs D). NB G5 is a good example that phage display has the potential to select high affinity, but low-frequency nanobodies. The p300-selected NB C12 is highly enriched (Fig. 10E and G). Comparing frequency plots, NB C12 might show stronger p300 preference (Fig. 10E vs F), and acetylation might have a negative effect on the binding (Fig. 10E vs G). Interestingly, the well-characterized core binder, NB E12, was not found in the blank, and in the samples it is not abundant at all (Fig. 10E–G). NB E12 shows some p300 preference (Fig. 10E vs F) and is not significantly affected by the acetylation state (Fig. 10E vs G), in good agreement with the binding data in Section 3.6 (c.f. Fig. 7F and Table 3).

We also searched for those NBs that were found in multiplicates in the first screening (c.f. Fig. 5C). Interestingly, these are not necessarily highly enriched according to the HTP sequencing data. NB A3, A11, A8, and E7 are rather nonspecific hits, while the NB B6 relative NB F5 and the NB C6 relative NB G4 are among the most outstanding hits. NB F12 also seems to be positive (Fig. 10B–G) that might bind to p300 and CBP equally, and disregards acetylation state (Fig. 10E–G). The NB C6 relative NB G4 behaves similar to NB C6, as well as the NB B6 relative NB F5 behaves similar to NB B6, as expected (Fig. 10B–D). Another interesting observation is that one of the most abundant and most enriched nanobody in lib114, NB A12, was found only once in the first screening plate, as well as the abundant and highly enriched NB C6 relatives, NB A1 and NB B4 (Fig. 10B–G vs Fig. 7). These data suggest that screening of 46 clones might not be fully representative for a highly variable NB library.

3.8.2.2 Comparison of Low-Scale Screening With HTP Sequencing Data

Parallelly, overall 96 clones from the 8 selections were individually sequenced to have a comparison with the deep-sequencing data. These sequences were identified in the frequency–correlation plots (Fig. 11), indicating that such low-scale screening cannot be fully representative for selected libraries that show extreme variability. It is in agreement of recently published results (Turner et al., 2016).

3.8.2.3 NB Families in HTP Sequencing Data

Though it is impressive that so many new nanobodies might emerge in future selections, it might be intriguing how these clonotypes can be

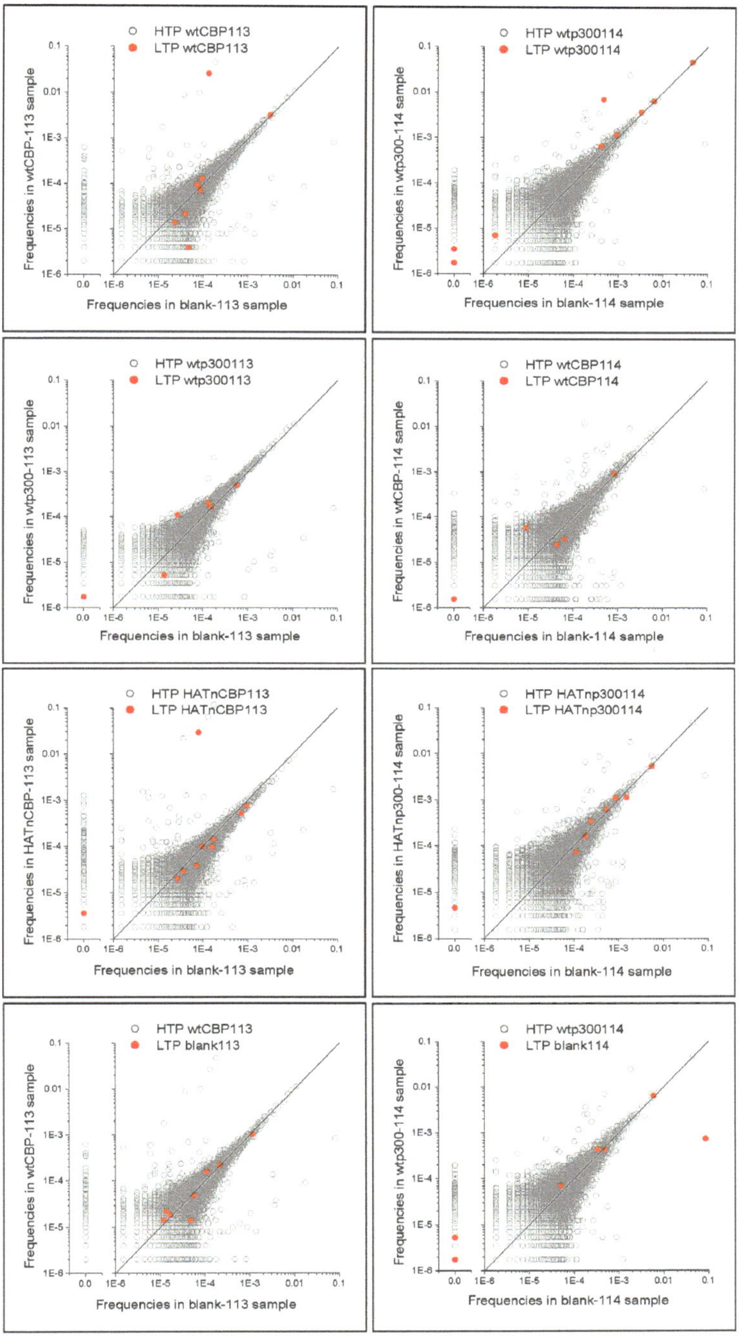

Fig. 11 96 Individually sequenced clones on frequency–correlation plots. From the selection (c.f. Fig. 10A) 96 individual clones were isolated, and sequenced. The corresponding clonotypes were searched for in HTP sequencing results (*red dots*). Frequency-correlation plots were done in Origin 6.0 (OriginLab, Northampton, MA) applying *gray circle symbols*.

clustered into NB families simplifying the picture. As a first attempt, we established a BLAST database from the sequences of clonotypes and performed a homology search with the CDR3 sequences of the five characterized NBs, and NB F12 and the NB A12 showing high enrichment in the frequency plots (c.f. Fig. 10E–G). Homologues were defined as having identical length and at least 80% identity in CDR3 (Pardon et al., 2014). With this threshold, NB C6 and NB A12 were found as members of the largest families with 182 and 106 clonotypes, respectively. NB C12 and NB B6 also belong to families (41 and 20 members, respectively), while NB F12 has two relatives and NB G5 has only one. Interestingly, NB E12 represents a unique sequence in the HTP sequencing data. The members of the families C6, A12, C12, and B6 are indicated in the frequency plots (Fig. 12). Eleven of C6 family members and four of A12 relatives were among the "promising" candidates (Fig. 12 and c.f. Fig. 10B–G). In the group of promising candidates, further nine clonotypes show lower but still significant homology to NB C6. Therefore, the C6 family might belong to a superfamily overrepresented in lib113.

3.8.3 Attempt for Further Systematic Selections and Screening With Disordered Regions and Different Core Constructs

Besides the four to five characterized NBs, we attempted to search for others, asking more specific questions around CBP/p300 using smaller CBP domain constructs as baits and immobilization via their FLAG-tag. In a more systematic selection and screen, we asked (1) specific questions around the core domain (Bromo-TAZ2), and also (2) if IDRs can have some transient structured motifs or regions immunogenic enough to be recognized by specific NBs. The first experiments were already performed with traditional phage display from Library 113 (CBP immunization c.f. also Fig. 5A) searching for specific NBs for CBP domains ID1, ID4, and different versions of the core in different conditions. The results were analyzed by both screening on ELISA and low-scale sequencing. For disordered regions (ID1 and ID4) thus far there was no specific NB isolated. For CBP core, some weaker binders were found primarily to the BromoPHD regions. Interestingly, from these screening trials, only one sequence (MP972 G2) was found among "promising" candidates on frequency plots (c.f. Fig. 10B–D). Most of the other sequences were found, but hidden, in the mass in the HTP sequencing data.

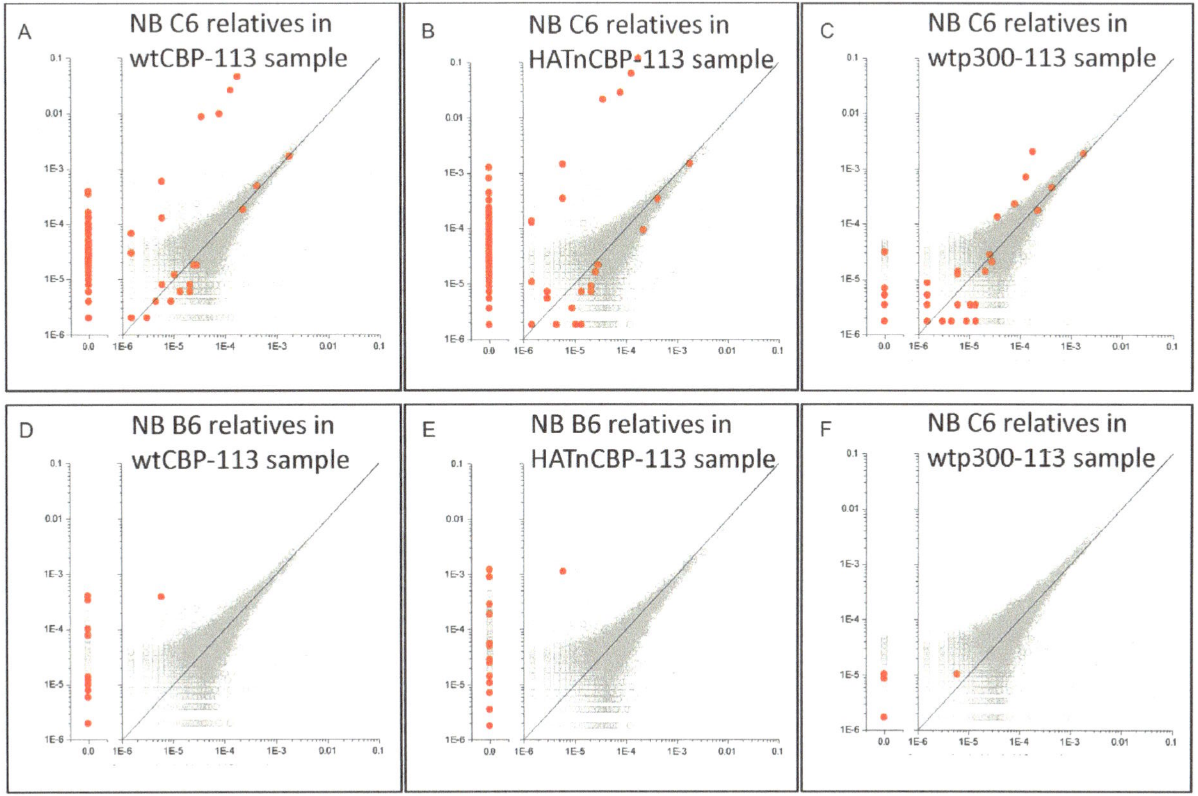

Fig. 12 Nanobody families in frequency-correlation plots. Blast database was created from the clonotype sequences using standalone BLAST (Camacho et al., 2009). Sequences homologous to several NBs of interest were then searched. Frequency data corresponding to each homologue were collected from the HTP sequencing data using PowerShell (Microsoft), and plotted on frequency-correlation plots. Here, members of NB C6 (A–C) and B6 (D–F) from lib113 and NB A12 (G–I) and C12 (J–L) from lib114 are shown (*red dots*) in HTP sequencing results. Frequency–correlation plots were done in Origin 6.0 (OriginLab, Northampton, MA) applying *gray circle symbols*.

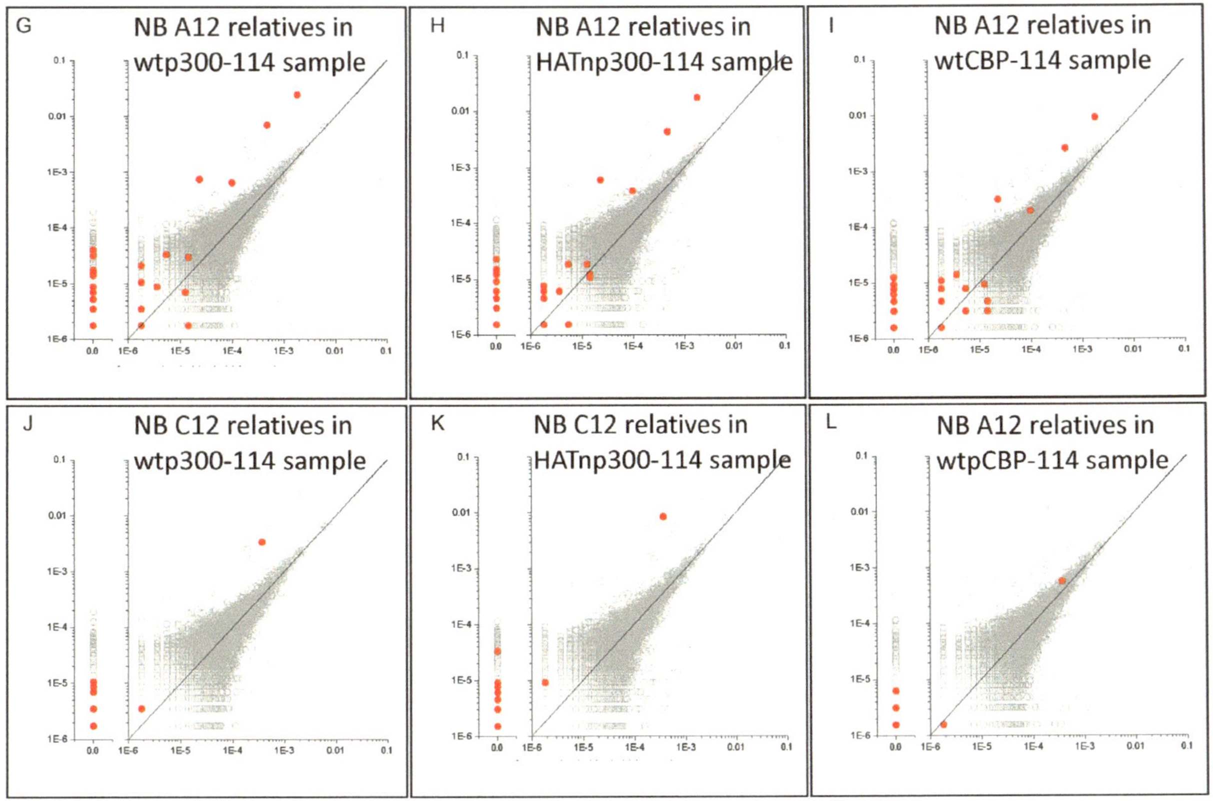
G
NB A12 relatives in wtp300-114 sample
H
NB A12 relatives in HATnp300-114 sample
I
NB A12 relatives in wtCBP-114 sample
J
NB C12 relatives in wtp300-114 sample
K
NB C12 relatives in HATnp300-114 sample
L
NB A12 relatives in wtpCBP-114 sample
0.1
0.01
1E-3
1E-4
1E-5
1E-6
0.0

Fig. 12—Cont'd

4. CONCLUSIONS AND FUTURE PERSPECTIVES

Conformational heterogeneity and complex PTM patterns of multi-domain, partially disordered proteins, represent a special challenge to their structural characterization, which we suggest to be mitigated by nanobodies directed against various segments of the protein. The development of such molecular tools may have a novel potential in studying highly dynamic proteins, opening unexplored novel avenues in structural-molecular biology. Due to the sheer size of such proteins, however, developing specific nanobodies by the conventional approaches of nanobody selection and screening in itself is a methodological challenge.

Here we suggest that methodological shortcomings emanate from several factors. (1) Given the huge number of potential epitopes within the full-length proteins, the variability of the emerging NB library must also be larger than in the case of a smaller, better defined, structured protein antigen. That is, screening of individual clones in a 96-well plate is not really representative for such variability. (2) Solid-surface coating might lead to a much larger steric hindrance than in the case of a small structured antigen. (3) Also, solid-surface coating might easily result in critical decrease of surface-bound antigen/epitope concentration in the case of such big, flexible, and extended protein as CBP/p300. These considerations might explain the contradicting results of ELISA screens.

We highlight in this chapter that in such a case, it is better to ask more specific questions regarding one domain, or segment, or specific PTM using smaller domain constructs manipulated in a specific way. Toward these points, we suggest an alternative procedure using deep sequencing to generate a sequence database of initial nanobody libraries. Then random expression of the whole library followed by protein pull-down with the given domain construct(s) of the antigen and mass spectrometry-based identification can be much less labor-intensive and give more representative results. The most interesting nanobody genes can then be either synthetized or cloned using inverted PCR primed at individual CDR3 sites. The effective generation of such dedicated nanobodies opens up novel possibilities for studying and functionally manipulating large multidomain proteins, such as CBP and p300.

ACKNOWLEDGMENTS

This work was supported by the Odysseus grant G.0029.12 from Research Foundation Flanders (FWO). We acknowledge the support and the use of resources of Instruct-ERIC, part of the European Strategy Forum on Research Infrastructures (ESFRI), and the Research Foundation Flanders (FWO) for their support to the Nanobody discovery. We are indebted to Peter Wright for constructs of fl and truncated p53, and the MBP-fusions of Hif1a(TAD), ACTR(TAD), and p53(NTAD), and also for inspiring discussions; Henri De Greve for useful advices in cloning; Alison Lundqvist for the technical assistance during Nanobody discovery; Mohamed Sallam for help in initial NB purification; Ivo Van Overstraeten and Karolien Van Belle for technical assistance; Mihaly Varadi for scores of homology between CBP and p300; VIB sequencing facility for low-scale sequencing. Molecular graphics were performed with the UCSF Chimera package, developed by the Resource for Biocomputing, Visualization, and Informatics at the University of California, San Francisco (supported by NIGMS P41-GM103311).

REFERENCES

Abdiche, Y. N., Miles, A., Eckman, J., Foletti, D., Van Blarcom, T. J., Yeung, Y. A., et al. (2014). High-throughput epitope binning assays on label-free array-based biosensors can yield exquisite epitope discrimination that facilitates the selection of monoclonal antibodies with functional activity. *PLoS One*, *9*(3), e92451. https://doi.org/10.1371/journal.pone.0092451.

Ando, T., Uchihashi, T., & Scheuring, S. (2014). Filming biomolecular processes by high-speed atomic force microscopy. *Chemical Reviews*, *114*(6), 3120–3188. https://doi.org/10.1021/cr4003837.

Arnold, K., Bordoli, L., Kopp, J., & Schwede, T. (2006). The SWISS-MODEL workspace: A web-based environment for protein structure homology modelling. *Bioinformatics*, *22*(2), 195–201. https://doi.org/10.1093/bioinformatics/bti770.

Bannas, P., Hambach, J., & Koch-Nolte, F. (2017). Nanobodies and nanobody-based human heavy chain antibodies as antitumor therapeutics. *Frontiers in Immunology*, *8*, 1–13. https://doi.org/10.3389/fimmu.2017.01603.

Bedford, D. C., Kasper, L. H., Fukuyama, T., & Brindle, P. K. (2010). Target gene context influences the transcriptional requirement for the KAT3 family of CBP and p300 histone acetyltransferases. *Epigenetics*, *5*(1), 9–15. Retrieved from http://www.pubmedcentral.nih.gov/articlerender.fcgi?artid=2829352&tool=pmcentrez&rendertype=abstract.

Bolotin, D. A., Poslavsky, S., Mitrophanov, I., Shugay, M., Mamedov, I. Z., Putintseva, E. V., et al. (2015). MiXCR: Software for comprehensive adaptive immunity profiling. *Nature Methods*, *12*, 380–381. https://doi.org/10.1038/nmeth.3364.

Bose, D. A., Donahue, G., Reinberg, D., Shiekhattar, R., Bonasio, R., & Berger, S. L. (2017). RNA binding to CBP stimulates histone acetylation and transcription. *Cell*, *168*(1–2). 135–149.e22. https://doi.org/10.1016/j.cell.2016.12.020.

Brizzard, B. L., Chubet, R. G., & Vizard, D. L. (1994). Immunoaffinity purification of FLAG epitope-tagged bacterial alkaline phosphatase using a novel monoclonal antibody and peptide elution. *BioTechniques*, *16*, 730–735.

Buday, L., & Tompa, P. (2010). Functional classification of scaffold proteins and related molecules. *FEBS Journal*, *277*(21), 4348–4355. https://doi.org/10.1111/j.1742-4658.2010.07864.x.

Camacho, C., Coulouris, G., Avagyan, V., Ma, N., Papadopoulos, J., Bealer, K., et al. (2009). BLAST+: Architecture and applications. *BMC Bioinformatics*, *10*, 421. https://doi.org/10.1186/1471-2105-10-421.

Chiabudini, M., Tais, A., Zhang, Y., Hayashi, S., Wolfle, T., Fitzke, E., et al. (2014). Release factor eRF3 mediates premature translation termination on polylysine-stalled ribosomes in Saccharomyces cerevisiae. *Molecular and Cellular Biology*, *34*(21), 4062–4076. https://doi.org/10.1128/MCB.00799-14.

Cilia, E., Pancsa, R., Tompa, P., Lenaerts, T., & Vranken, W. F. (2014). The DynaMine webserver: Predicting protein dynamics from sequence. *Nucleic Acids Research*, *42*, W264–W270. https://doi.org/10.1093/nar/gku270.

Contreras-Martos, S., Nguyen, H. H., Nguyen, P. N., Hristozova, N., Macossay-Castillo, M., Kovacs, D., et al. (2018). Quantification of intrinsically disordered proteins: A problem not fully appreciated. *Frontiers in Molecular Biosciences*, *5*, 83. https://doi.org/10.3389/fmolb.2018.00083.

Contreras-Martos, S., Piai, A., Kosol, S., Varadi, M., Bekesi, A., Lebrun, P., et al. (2017). Linking functions: An additional role for an intrinsically disordered linker domain in the transcriptional coactivator CBP. *Scientific Reports*, *7*, 4676. https://doi.org/10.1038/s41598-017-04611-x.

Dancy, B. M., & Cole, P. A. (2015). Protein lysine acetylation by p300/CBP. *Chemical Reviews*, *115*(6), 2419–2452. https://doi.org/10.1021/cr500452k.

de Boer, E., Rodriguez, P., Bonte, E., Krijgsveld, J., Katsantoni, E., Heck, A., et al. (2003). Efficient biotinylation and single-step purification of tagged transcription factors in mammalian cells and transgenic mice. *Proceedings of the National Academy of Sciences of the United States of America*, *100*(13), 7480–7485. https://doi.org/10.1073/pnas.1332608100.

Delvecchio, M., Gaucher, J., Aguilar-Gurrieri, C., Ortega, E., & Panne, D. (2013). Structure of the p300 catalytic core and implications for chromatin targeting and HAT regulation. *Nature Structural & Molecular Biology*, *20*(9), 1040–1046. https://doi.org/10.1038/nsmb.2642.

Demarest, S. J., Deechongkit, S., Dyson, H. J., Evans, R. M., & Wright, P. E. (2004). Packing, specificity, and mutability at the binding interface between the p160 coactivator and CREB-binding protein. *Protein Science: A Publication of the Protein Society*, *13*(1), 203–210. https://doi.org/10.1110/ps.03366504.

Deng, B., Lento, C., & Wilson, D. J. (2016). Hydrogen deuterium exchange mass spectrometry in biopharmaceutical discovery and development—A review. *Analytica Chimica Acta*, *940*, 8–20. https://doi.org/10.1016/j.aca.2016.08.006.

Dmitriev, O. Y., Lutsenko, S., & Muyldermans, S. (2016). Nanobodies as probes for protein dynamics in vitro and in cells. *Journal of Biological Chemistry*, *291*(8), 3767–3775. https://doi.org/10.1074/jbc.R115.679811.

Dosztányi, Z., Csizmok, V., Tompa, P., & Simon, I. (2005). IUPred: Web server for the prediction of intrinsically unstructured regions of proteins based on estimated energy content. *Bioinformatics*, *21*(16), 3433–3434. https://doi.org/10.1093/bioinformatics/bti541.

Dutto, I., Scalera, C., & Prosperi, E. (2018). CREBBP and p300 lysine acetyl transferases in the DNA damage response. *Cellular and Molecular Life Sciences: CMLS*, *75*(8), 1325–1338. https://doi.org/10.1007/s00018-017-2717-4.

Dyson, H. J., & Wright, P. E. (2016). Role of intrinsic protein disorder in the function and interactions of the transcriptional coactivators CREB-binding protein (CBP) and p300. *The Journal of Biological Chemistry*, *291*, 6714–6722. https://doi.org/10.1074/jbc.R115.692020.

Ewing, B., & Green, P. (1998). Base-calling of automated sequencer traces using phred. II. Error probabilities. *Genome Research*, *8*, 186–194. https://doi.org/10.1101/gr.8.3.186.

Ferreon, J. C., Lee, C. W., Arai, M., Martinez-Yamout, M. A., Dyson, H. J., & Wright, P. E. (2009). Cooperative regulation of p53 by modulation of ternary complex formation with CBP/p300 and HDM2. *Proceedings of the National Academy of Sciences of the United States of America*, *106*(16), 6591–6596. https://doi.org/10.1073/pnas.0811023106.

Fridy, P. C., Li, Y., Keegan, S., Thompson, M. K., Nudelman, I., Scheid, J. F., et al. (2014). A robust pipeline for rapid production of versatile nanobody repertoires. *Nature Methods*, *11*, 1253–1260. https://doi.org/10.1038/nmeth.3170.

Gautier, A., & Hinner, M. J. (2015). *Site-specific protein labeling: Methods and protocols: Vol. 1266*. In *Methods in Molecular Biology* (pp. 1–267). Springer Nature. https://doi.org/10.1007/978-1-4939-2272-7.

Getz, E. B., Xiao, M., Chakrabarty, T., Cooke, R., & Selvin, P. R. (1999). A comparison between the sulfhydryl reductants tris(2-carboxyethyl) phosphine and dithiothreitol for use in protein biochemistry. *Analytical Biochemistry*, *273*, 73–80.

Ghassabeh, G. H., Devoogdt, N., De Pauw, P., Vincke, C., & Muyldermans, S. (2013). Nanobodies and their potential applications. *Nanomedicine*, *8*(6), 1013–1026.

Gotoh, T., Miyazaki, Y., Sato, W., Kikuchi, K., & Bentley, W. E. (2001). Proteolytic activity and recombinant protein production in virus-infected Sf-9 insect cell cultures supplemented with carboxyl and cysteine protease inhibitors. *Journal of Bioscience and Bioengineering*, *92*(3), 248–255. https://doi.org/10.1016/S1389-1723(01)80258-9.

Hamers-Casterman, C., Atarhouch, T., Muyldermans, S., Robinson, G., Hammers, C., Songa, E. B., et al. (1993). Naturally occurring antibodies devoid of light chains. *Nature*, *363*, 446–448. https://doi.org/10.1038/363446a0.

Hegyi, H., Schad, E., & Tompa, P. (2007). Structural disorder promotes assembly of protein complexes. *BMC Structural Biology*, 7, 65. https://doi.org/10.1186/1472-6807-7-65.

Huang, L., Muyldermans, S., & Saerens, D. (2010). Nanobodies®: Proficient tools in diagnostics. *Expert Review of Molecular Diagnostics*, *10*(6), 777–785. https://doi.org/10.1586/erm.10.62.

Janknecht, R. (2002). The versatile functions of the transcriptional coactivators p300 and CBP and their roles in disease. *Histology and Histopathology*, *17*(2), 657–668. Retrieved from http://www.ncbi.nlm.nih.gov/pubmed/11962765/.

Kim, Y., Kaluz, S., Mehta, A., Weinert, E., Rivera, S., & Van Meir, E. (2017). Purifying properly folded cysteine-rich, zinc finger containing recombinant proteins for structural drug targeting studies: The CH1 domain of p300 as a case example. *Bio-Protocol*, 7(17), 1–14. http://doi.org/10.21769/BioProtoc.2537.

Kim, W. J., Rivera, M. N., Coffman, E. J., & Haber, D. A. (2012). The WTX tumor suppressor enhances p53 acetylation by CBP/p300. *Molecular Cell*, *45*(5), 587–597. https://doi.org/10.1016/j.molcel.2011.12.025.

Kraus, W. L., & Kadonaga, J. T. (1998). P300 and estrogen receptor cooperatively activate transcription via differential enhancement of initiation and reinitiation. *Genes & Development*, *12*(3), 331–342. Retrieved from http://www.pubmedcentral.nih.gov/articlerender.fcgi?artid=316488&tool=pmcentrez&rendertype=abstract.

La Verde, V., Dominici, P., & Astegno, A. (2017). Determination of hydrodynamic radius of proteins by size exclusion chromatography. *Bio-Protocol*, 7(8), 1–14. http://doi.org/10.21769/BioProtoc.2230.

Lazarova, D. L., Chiaro, C., Wong, T., Drago, E., Rainey, A., O'Malley, S., et al. (2013). CBP activity mediates effects of the histone deacetylase inhibitor butyrate on WNT activity and apoptosis in colon cancer cells. *Journal of Cancer*, *4*, 481–490. https://doi.org/10.7150/jca.6583.

Lee, Y. Y., Moujalled, D., Doerflinger, M., Gangoda, L., Weston, R., Rahimi, A., et al. (2013). CREB-binding protein (CBP) regulates β-adrenoceptor (β-AR)-mediated apoptosis. *Cell Death and Differentiation*, *20*, 941–952. https://doi.org/10.1038/cdd.2013.29.

Liu, X., Wang, L., Zhao, K., Thompson, P. R., Hwang, Y., Marmorstein, R., et al. (2008). The structural basis of protein acetylation by the p300/CBP transcriptional coactivator. *Nature*, *451*(7180), 846–850. https://doi.org/10.1038/nature06546.

Lopez-Atalaya, J. P., Valor, L. M., & Barco, A. (2014). Epigenetic factors in intellectual disability: The Rubinstein-Taybi syndrome as a paradigm of neurodevelopmental disorder with epigenetic origin. *Progress in Molecular Biology and Translational Science*, *128*, 139–176. https://doi.org/10.1016/B978-0-12-800977-2.00006-1.

Majumdar, R., Middaugh, C. R., Weis, D. D., & Volkin, D. B. (2015). Hydrogen-deuterium exchange mass spectrometry as an emerging analytical tool for stabilization and formulation development of therapeutic monoclonal antibodies. *Journal of Pharmaceutical Sciences*, *104*(2), 327–345. https://doi.org/10.1002/jps.24224.

Maksimoska, J., Segura-Peña, D., Cole, P. A., & Marmorstein, R. (2014). Structure of the p300 histone acetyltransferase bound to acetyl-coenzyme A and its analogues. *Biochemistry, 53*(21), 3415–3422. https://doi.org/10.1021/bi500380f.
McCann, J. J., Zheng, L., Rohrbeck, D., Felekyan, S., Kuhnemuth, R., Sutton, R. B., et al. (2012). Supertertiary structure of the synaptic MAGuK scaffold proteins is conserved. *Proceedings of the National Academy of Sciences of the United States of America, 109*(39), 15775–15780. https://doi.org/10.1073/pnas.1200254109.
Mészáros, B., Simon, I., & Dosztányi, Z. (2009). Prediction of protein binding regions in disordered proteins. *PLoS Computational Biology, 5*(5), e1000376. https://doi.org/10.1371/journal.pcbi.1000376.
Miyazaki, N., Kiyose, N., Akazawa, Y., Takashima, M., Hagihara, Y., Inoue, N., et al. (2015). Isolation and characterization of antigen-specific alpaca (Lama pacos) VHH antibodies by biopanning followed by high-throughput sequencing. *Journal of Biochemistry, 158*(3), 205–215. https://doi.org/10.1093/jb/mvv038.
Morais, V. A., Haddad, D., Craessaerts, K., De Bock, P. J., Swerts, J., Vilain, S., et al. (2014). PINK1 loss-of-function mutations affect mitochondrial complex I activity via NdufA10 ubiquinone uncoupling. *Science, 344*(6180), 203–207. https://doi.org/10.1126/science.1249161.
Muyldermans, S. (2013). Nanobodies: Natural single-domain antibodies. *Annual Review of Biochemistry, 82*, 775–797. https://doi.org/10.1146/annurev-biochem-063011-092449.
Nyborg, J. K., & Peersen, O. B. (2004). That zincing feeling: The effects of EDTA on the behaviour of zinc-binding transcriptional regulators. *The Biochemical Journal, 381*(Pt. 3), e3–e4. https://doi.org/10.1042/BJ20041096.
Oganesyan, I., Lento, C., & Wilson, D. (2018). Contemporary hydrogen deuterium exchange mass spectrometry. *Methods, 144*, 27–42.
Ogryzko, V. V., Schiltz, R. L., Russanova, V., Howard, B. H., & Nakatani, Y. (1996). The transcriptional coactivators p300 and CBP are histone acetyltransferases. *Cell, 87*(5), 953–959. Retrieved from http://www.ncbi.nlm.nih.gov/pubmed/8945521.
Panne, D., Maniatis, T., & Harrison, S. C. (2007). An atomic model of the interferon-beta enhanceosome. *Cell, 129*(6), 1111–1123. https://doi.org/10.1016/j.cell.2007.05.019.
Pardon, E., Laeremans, T., Triest, S., Rasmussen, S. G. F., Wohlkönig, A., Ruf, A., et al. (2014). A general protocol for the generation of nanobodies for structural biology. *Nature Protocols, 9*(3), 674–693. https://doi.org/10.1038/nprot.2014.039.
Park, S., Stanfield, R. L., Martinez-Yamout, M. A., Dyson, H. J., Wilson, I. A., & Wright, P. E. (2017). Role of the CBP catalytic core in intramolecular SUMOylation and control of histone H3 acetylation. *Proceedings of the National Academy of Sciences of the United States of America, 114*(27), E5335–E5342. https://doi.org/10.1073/pnas.1703105114.
Pettersen, E. F., Goddard, T. D., Huang, C. C., Couch, G. S., Greenblatt, D. M., Meng, E. C., et al. (2004). UCSF chimera—A visualization system for exploratory research and analysis. *Journal of Computational Chemistry, 25*(13), 1605–1612. https://doi.org/10.1002/jcc.20084.
Radhakrishnan, I., Pérez-Alvarado, G. C., Parker, D., Dyson, H. J., Montminy, M. R., & Wright, P. E. (1997). Solution structure of the KIX domain of CBP bound to the transactivation domain of CREB: A model for activator:coactivator interactions. *Cell, 91*(6), 741–752. Retrieved from http://www.ncbi.nlm.nih.gov/pubmed/9413984.
Rasmussen, S. G. F., Choi, H. J., Fung, J. J., Pardon, E., Chae, P. S., DeVree, B. T., et al. (2011). Structure of a nanobody-stabilized active state of the β(2) adrenoceptor. *Nature, 469*, 175–180. https://doi.org/10.1038/nature09648.
Revilla, Y., & Granja, A. (2009). Viral mechanisms involved in the transcriptional CBP/p300 regulation of inflammatory and immune responses. *Critical Reviews in Immunology, 29*(2), 131–154.

Shaknovich, R., & Melnick, A. (2011). Epigenetics and B-cell lymphoma. *Current Opinion in Hematology*, *18*(4), 293–299. https://doi.org/10.1097/MOH.0b013e32834788cf.

Shugay, M., Bagaev, D. V., Turchaninova, M. A., Bolotin, D. A., Britanova, O. V., Putintseva, E. V., et al. (2015). VDJtools: Unifying post-analysis of T cell receptor repertoires. *PLoS Computational Biology*, *11*(11), e1004503. https://doi.org/10.1371/journal.pcbi.1004503.

Stetefeld, J., McKenna, S. A., & Patel, T. R. (2016). Dynamic light scattering: A practical guide and applications in biomedical sciences. *Biophysical Reviews*, *8*(4), 409–427. https://doi.org/10.1007/s12551-016-0218-6.

Svidritskiy, E., Demo, G., & Korostelev, A. A. (2018). Mechanism of premature translation termination on a sense codon. *Journal of Biological Chemistry*, *293*(32), 12472–12479. https://doi.org/10.1074/jbc.AW118.003232.

Teo, J.-L., & Kahn, M. (2010). The Wnt signaling pathway in cellular proliferation and differentiation: A tale of two coactivators. *Advanced Drug Delivery Reviews*, *62*(12), 1149–1155. Retrieved from http://linkinghub.elsevier.com/retrieve/pii/S0169409X10001900.

Thompson, P. R., Wang, D., Wang, L., Fulco, M., Pediconi, N., Zhang, D., et al. (2004). Regulation of the p300 HAT domain via a novel activation loop. *Nature Structural & Molecular Biology*, *11*(4), 308–315. Retrieved from http://www.ncbi.nlm.nih.gov/pubmed/15004546.

Tompa, P. (2009). *Structure and function of intrinsically disordered proteins*. Boca Raton, FL: CRC Press. vol. 11.

Traenkle, B., & Rothbauer, U. (2017). Under the microscope: Single-domain antibodies for live-cell imaging and super-resolution microscopy. *Frontiers in Immunology*, *8*, 1–8. https://doi.org/10.3389/fimmu.2017.01030.

Turner, K. B., Naciri, J., Liu, J. L., Anderson, G. P., Goldman, E. R., & Zabetakis, D. (2016). Next-generation sequencing of a single domain antibody repertoire reveals quality of phage display selected candidates. *PLoS One*, *11*(2), 1–15. https://doi.org/10.1371/journal.pone.0149393.

Van Der Lee, R., Buljan, M., Lang, B., Weatheritt, R. J., Daughdrill, G. W., Dunker, A. K., et al. (2014). Classification of intrinsically disordered regions and proteins. *Chemical Reviews*, *114*(13), 6589–6631. https://doi.org/10.1021/cr400525m.

Varadi, M., Kosol, S., Lebrun, P., Valentini, E., Blackledge, M., Dunker, A. K., et al. (2014). PE-DB: A database of structural ensembles of intrinsically disordered and of unfolded proteins. *Nucleic Acids Research*, *42*(D1), 326–335. https://doi.org/10.1093/nar/gkt960.

Venne, A. S., Kollipara, L., & Zahedi, R. P. (2014). The next level of complexity: Crosstalk of posttranslational modifications. *Proteomics*, *14*(4–5), 513–524. https://doi.org/10.1002/pmic.201300344.

Voráčková, I., Suchanová, Š., Ulbrich, P., Diehl, W. E., & Ruml, T. (2011). Purification of proteins containing zinc finger domains using immobilized metal ion affinity chromatography. *Protein Expression and Purification*, *79*(1), 88–95. https://doi.org/10.1016/j.pep.2011.04.022.

Wang, F., Marshall, C. B., & Ikura, M. (2013). Transcriptional/epigenetic regulator CBP/p300 in tumorigenesis: Structural and functional versatility in target recognition. *Cellular and Molecular Life Sciences: CMLS*, *70*(21), 3989–4008. https://doi.org/10.1007/s00018-012-1254-4.

Wang, F., Marshall, C. B., Yamamoto, K., Li, G.-Y., Gasmi-Seabrook, G. M. C., Okada, H., et al. (2012). Structures of KIX domain of CBP in complex with two FOXO3a transactivation domains reveal promiscuity and plasticity in coactivator recruitment. *Proceedings of the National Academy of Sciences of the United States of America*, *109*(16), 6078–6083. https://doi.org/10.1073/pnas.1119073109.

Wang, D., Thompson, P., Cole, P. A., & Cotter, R. J. (2005). Structural analysis of a highly acetylated protein using a curved-field reflectron mass spectrometer. *Proteomics*, *5*(9), 2288–2296. https://doi.org/10.1002/pmic.200401167.

Wang, Y. H., Tsay, Y. G., Tan, B. C. M., Lo, W. Y., & Lee, S. C. (2003). Identification and characterization of a novel p300-mediated p53 acetylation site, lysine 305. *The Journal of Biological Chemistry*, *278*(28), 25568–25576. https://doi.org/10.1074/jbc.M212574200.

Yang, S., Blachowicz, L., Makowski, L., & Roux, B. (2010). Multidomain assembled states of Hck tyrosine kinase in solution. *Proceedings of the National Academy of Sciences of the United States of America*, *107*(36), 15757–15762. https://doi.org/10.1073/pnas.1004569107.

Yang, D., Singh, A., Wu, H., & Kroe-Barrett, R. (2017). Determination of high-affinity antibody-antigen binding kinetics using four biosensor platforms. *Journal of Visualized Experiments*, *122*, e55659. https://doi.org/10.3791/55659.

Yuan, Z. M., Huang, Y., Ishiko, T., Nakada, S., Utsugisawa, T., Shioya, H., et al. (1999). Function for p300 and not CBP in the apoptotic response to DNA damage. *Oncogene*, *18*(41), 5714–5717. https://doi.org/10.1038/sj.onc.1202930.

Zhou, Y. H., Chen, Z., Purcell, R. H., & Emerson, S. U. (2007). Positive reactions on Western blots do not necessarily indicate the epitopes on antigens are continuous. *Immunology and Cell Biology*, *85*(1), 73–78. https://doi.org/10.1038/sj.icb.7100004.

CHAPTER TWENTY-ONE

Characterization of the Binding of Small Molecules to Intrinsically Disordered Proteins

Veselin S. Dobrev*, Lisette M. Fred*, Kaitlyn P. Gerhart*, Steven J. Metallo*,†,1

*Department of Chemistry, Georgetown University, Washington, DC, United States
†Institute for Soft Matter Synthesis and Metrology, Georgetown University, Washington, DC, United States
[1]Corresponding author: e-mail address: sjm24@georgetown.edu

Contents

Abstract

Intrinsically disordered proteins (IDPs) comprise a large fraction of eukaryotic proteomes. IDPs are prevalent in cellular regulation, signaling networks, and disease pathways. The abundance and activity of IDPs is tightly controlled at multiple levels, and their dysregulation is associated with disease. Because of the importance of IDPs in both normal and disease states of the cell, IDPs are attractive targets for modulation by small molecules both to understand their biology and to provide potential drug leads. Multiple screens have successfully identified small molecules that bind to IDPs. Here, we describe how surface plasmon resonance, NMR, and fluorescence methods can be used to characterize the direct binding affinity between small molecules and IDPs. We describe how these techniques can contribute to identifying previously unknown small-molecule binding sites on IDPs.

Methods in Enzymology, Volume 611
ISSN 0076-6879
https://doi.org/10.1016/bs.mie.2018.09.033

1. INTRODUCTION

Intrinsically disordered proteins (IDPs) or intrinsically disordered protein regions (IDRs) lack stable secondary or tertiary structure, existing instead in an ensemble of conformations that rapidly interconvert. Disordered proteins are highly prevalent in eukaryotes and fulfill a range of functions (van der Lee et al., 2014). Many IDPs have crucial biological roles serving as hubs in signaling and regulatory networks, an activity that is facilitated by their ability to interact with multiple partners. Early in the development of our understanding of IDPs, the importance of coupled folding and binding in the recognition of IDP segments was noted (Dyson & Wright, 2002). However, some IDPs were found to form functional complexes while remaining in their dynamic, ensemble state via formation of "fuzzy" complexes (Mittag et al., 2008; Tompa & Fuxreiter, 2008). Recently one of these dynamic complexes was described in which the interacting proteins achieved picomolar affinity, while both partners remained disordered, indicating that structure is not a prerequisite for high affinity (Borgia et al., 2018). In addition, there has been rapid progress in understanding the nature and function of liquid–liquid phase separated systems in vivo and the importance of unstructured, transient interactions of IDPs to the generation and the activity of these systems (Banani, Lee, Hyman, & Rosen, 2017).

During the same period when our understanding of IDPs has expanded dramatically, the recognition and interest in protein–protein interactions as viable drug targets has also expanded (Arkin, Tang, & Wells, 2014; Oprea et al., 2018; Scott, Bayly, Abell, & Skidmore, 2016; Wells & McClendon, 2007). The number of human proteins targeted by FDA-approved drugs has roughly doubled over the last decade. While the expansion remains strongly biased toward previously successful classes of targets (GPCRs, ion channels, nuclear receptors, and kinases), there has been a substantial increase of first-in-class drugs targeting underrepresented families (Overington, Al-Lazikani, & Hopkins, 2006; Santos et al., 2017). An increased focus on target validation and biological understanding at one major pharmaceutical company has led to an increase in the inclusion of targets such as protein–protein interactions and transcription factors that typically have been seen as difficult to target (Morgan et al., 2018; Rees, Gribbon, Birmingham, Janzen, & Pairaudeau, 2016).

The combination of our growing knowledge about IDPs, their biological significance, and the expanding view of the range of proteins that can be targeted has led IDPs to be considered as potential drug targets (Metallo, 2010). Their many functions and ability to interact with multiple binding partners contribute to their importance as hub proteins, their overrepresentation in disease, and hence their desirability as drug targets (Uversky et al., 2014; Uversky, Oldfield, & Dunker, 2008; van der Lee et al., 2014). The tight regulation of IDP abundance and the propensity of changes in IDP concentration to have cellular effects indicate that modulation of IDP activity with small molecules would be likely to have biological effects (Gsponer, Futschik, Teichmann, & Babu, 2008; Vavouri, Semple, Garcia-Verdugo, & Lehner, 2009). Unsurprisingly, then, the search for small-molecule modulators of IDP function has been expanding.

2. SCREENING FOR BINDERS OF IDPs

The approach to finding small-molecule binders of disordered proteins, though clearly different from targeting structured proteins, follows an analogous process using similar techniques (Heller, Aprile, & Vendruscolo, 2017). The number of clear examples of small-molecule IDP-binding interactions in the literature is increasing but remains small. As with structured targets, a common approach to initial discovery of IDP binders involves activity-based screening for disruption of a specific binding event or detectable output. The approach commonly employs a diverse set of compounds, typically >10,000 molecules, with the reported screens so far coming from academic centers. While these screens may be comparatively small by pharma standards, they have yielded a variety of hits against IDPs (although the fraction of these hits that have the potential to progress to true lead compounds is uncertain). In the nascent field of small-molecule inhibitors of IDPs, discovering tool molecules that modulate IDP function via binding to the disordered sequence remains very valuable. First, the molecules help in understanding the biology of the given IDP and potentially validate the protein as a viable target, and second, they increase our understanding of the types of molecules and IDP sequences that can interact by increasing the limited set of examples available for comparison.

An IDP that has been targeted by us and several other groups is the oncogenic transcription factor c-Myc (Myc). The screening methods used in finding inhibitors against Myc demonstrate the similar approach to screening against both IDPs and structured protein–protein interactions. One early screen used purified proteins in a FRET-based approach in which Myc and its binding partner Max were labeled with CFP and YFP, respectively (Berg et al., 2002). A combinatorial library of 7000 compounds was screened to yield several hits that were characterized in additional functional assays. In another approach to targeting Myc, a cell-based assay was used in which Myc and Max were introduced as target and bait in a yeast two-hybrid system. The system was used to screen a diversity library of 10,000 compounds and again followed up with functional assays (Yin, Giap, Lazo, & Prochownik, 2003). Several screens used a fluorescence polarization assay in which the binding of fluorescently tagged *E*-box DNA by Myc–Max heterodimers was the primary readout (Hart et al., 2014; Kiessling, Sperl, Hollis, Eick, & Berg, 2006). A recent screen for Myc inhibitors was done on a true high-throughput scale with 400,000 drug-like molecules screened via a split-luciferase assay conducted in mammalian HEK293 cells (Choi et al., 2017). On-target effects of an active molecule from the screen, sAJM589, were shown through Myc–Max co-IP inhibition, disruption of Myc transcriptional targets, and reduced Myc half-life.

An alternative screening approach utilizes direct binding, label-free methods, most commonly nuclear magnetic resonance (NMR), or surface plasmon resonance (SPR), to detect directly the interaction between an IDP and small molecule. Because these assays are often lower throughput, they are typically conducted with small libraries such as a set consisting of approved drugs or approved drugs along with compounds that have entered some phase of clinical trial (these compounds have the advantage of known bioactivity or pharmacology). Direct screening assays were able to determine binders of the IDPs EWS-FLI1 and NUPR1 (Erkizan et al., 2009; Neira et al., 2017). Direct binding is also used to screen fragment libraries, sets whose components are smaller than those in a typical diversity library, with the potential of elaborating hits into higher affinity molecules. An NMR-based fragment screen found multiple binders to the disordered D2 subdomain of $p27^{Kip1}$ (Iconaru et al., 2015).

Computational modeling of ensembles of disordered sequences has been continuously improving (Das, Matysiak, & Mittal, 2018). Using the ability to generate ensembles computationally, an analogy to structure-based design is emerging for IDPs. This method entails generating ensembles of the target sequence, either purely computationally or by including experimental constraints (often from NMR or SAXS). The generated conformations are subjected to a clustering algorithm that groups similar conformations, with the most populated conformations then serving as the basis for in silico docking of small molecules (Convertino, Vitalis, & Caflisch, 2011; Matthes, Gapsys, Griesinger, & de Groot, 2017; Toth et al., 2014; Yu et al., 2016). This technique has succeeded in discovering IDP binders by experimental characterization of a limited set of molecules selected based on their docking scores. Targeting populated conformations should require less of a conformation shift for the targeted sequence, thereby avoiding a penalty in conformational energy. An alternative model is that small molecules actually expand the conformational ensemble rather than constrain it, resulting in stabilized binding interactions based on protein conformational entropy gain rather than solvent release (Heller, Sormanni, & Vendruscolo, 2015). Although it has not been determined that the process works simply as modeled (that is, the binders may not be interacting with the ensemble in the predicted manner), the process of computationally selecting potential binding molecules has proved a viable starting point in finding IDP binders and will improve as modeling and docking protocols improve.

3. CHARACTERIZATION OF SMALL-MOLECULE IDP BINDING

Initial hits from a screening protocol for IDP-binding molecules (as with hits from screens against structured targets) need subsequent validation and characterization. There are two primary questions regarding IDP-binding molecules: what is the affinity and what is the binding site? The first question is common to ligand binding and is typically assessed by titrating the ligand concentration while monitoring protein binding under equilibrium conditions to determine a dissociation constant (K_D). The question of binding site determination is more specific to IDP binding. As opposed to enzyme active sites or GPCR ligand-binding sites, IDPs do not have well-defined sites associated with small-molecule binding. Screens for IDP-binding molecules may provide only a general sense of where the ligand is binding—a domain or region important for interaction in the screen. If the assay was an SPR direct binding assay, the binding site is limited only by the size of the immobilized target, again often a whole disordered domain or region. The assay used to establish titratable binding can be used in experiments designed to determine the ligand interaction site. In the case of inhibitors binding to the basic helix–loop–helix leucine zipper (bHLHZip) of Myc, we used a fluorescence assay with a series of randomly mutated bHLHZip regions along with truncations to identify specific binding sites within the bHLHZip domain (Follis, Hammoudeh, Wang, Prochownik, & Metallo, 2008). When NMR analysis is available, chemical shift perturbations (CSPs) can be used to determine residues important for binding (De Mol et al., 2016; Iconaru et al., 2015; Krishnan et al., 2014). For IDPs, binding sites may not have sharp delimitations; there will be residues that make contact with the compound, but there may also be adjacent or surrounding residues that bias the conformation and thereby affect binding (Das, Crick, & Pappu, 2012). Both types of residues may be considered components of the binding site and definitively distinguishing between them may be difficult. Here, we describe three techniques (SPR, NMR, and fluorescence quenching) that have been useful in determining small-molecule IDP-binding affinity and binding sites.

3.1 Surface Plasmon Resonance

Current SPR (and surface interferometry technologies) has achieved very high sensitivity allowing them to detect binding of even small molecules (150 Da) to surface immobilized targets. This ability to quantitatively detect direct binding to a protein or peptide is a major advantage for the characterization of small-molecule IDP interactions. These binders often have weak affinity, and direct binding analysis eliminates the need to conduct competition assays and facilitates the quantification of the interaction.

3.1.1 Equipment

BiOptix 404pi SPR instrument (similar to Biacore T200)
NiHC 200M (XanTec bioanalytics GmbH) NTA sensorchips
96-well plates
0.2 μm polyethersulfone (PES) sterile filter
Volumetric flasks
Analytical balance

3.1.2 Buffers and Reagents

10% Polyoxyethylene (20) sorbitan monolaurate (P20)
Dimethylsulfoxide (DMSO)
10 × PBS EDTA (1.37 *M* NaCl, 27 m*M* KCl, 14 m*M* KH_2PO_4, 43 m*M* Na_2HPO_4, 1 m*M* EDTA)
Running buffer (1 × PBS pH 7.4, 0.005% polyoxyethylene (20) sorbitan monolaurate (P20), 5% DMSO, 0.1 m*M* EDTA)
Immobilization buffer (50 m*M* MES buffer pH 6.3)
0.5 *M* NaCl
0.5 *M* EDTA, pH 8.5
5 m*M* $NiCl_2$ in running buffer
Compound of interest dissolved in DMSO
Hexahistidine tagged IDP

3.1.3 Preparation of Running Buffer and Analyte

1. Prepare 10× PBS EDTA pH 7.4 and sterile filter the buffer using 0.2 μm filter.
2. Prepare 10% polyoxyethylene (20) sorbitan monolaurate (P20) by weighing 5.5 g of detergent (density 1.1 g/mL) and adding it to a 50-mL volumetric flask that already contains 15–20 mL of deionized (DI) water and gently mix. Carefully add DI water to the fill line and stir the mixture avoiding bubbles.
3. Prepare the running buffer in a 2 L volumetric flask by combining: 200 mL 10× PBS EDTA; 1000 μL of 10% P20; 110.00 g of DMSO (equivalent to 100 mL of DMSO for a final of 5% (v/v)); add deionized water to the 2 L mark and allow the flask to stir and mix for 1 h. Sterile filter the running buffer using a 0.2 μm filter. Degas the buffer under vacuum (following vacuum filtration) or by 1 h of sonication at room temperature.

   Note: In SPR proper refractive index matching in samples and running buffer is crucial for the data acquisition and reduction of data noise and error. Even a small mismatch between the concentration of buffer components, especially DMSO, in analyte samples and the running buffer can skew the data or render the experiment useless, especially when the expected maximal response is small (Frostell-Karlsson et al., 2000). Delivering exact, reproducible volumes of viscous liquids such as DMSO (or P20) is difficult. For this reason, the high accuracy and reproducibility of using an analytical balance to weigh the amount of DMSO needed to prepare a 5% (v/v) DMSO running buffer is preferred. To further improve reproducibility and avoid buffer mismatch, one can prepare running buffer containing all components except the DMSO and then remove an aliquot of that running buffer to be used in sample preparation. The amount of running buffer removed (typically 100 mL is sufficient) can be quantified on an analytical balance allowing the amount of DMSO required to complete the preparation of the running buffer to be calculated.
4. Prepare the analyte sample in a 1.5 mL polypropylene tube. To prepare 1 mL of 10 μ*M* analyte compound, here 10074-A4 (Hammoudeh, Follis, Prochownik, & Metallo, 2009), combine: 100 μL 10× PBS EDTA, 50 μL of 1% P20, and 800 μL deionized water in a 1.5 mL tube (or use running buffer minus DMSO removed in step 3). Lastly weigh 55.0 mg of 200 μ*M* 10074-A4 in DMSO and add slowly to the sample (final DMSO of 5% (v/v), matching the running buffer). The sample

can be sterile filtered; however, some compounds can adhere to filter membranes and should not be filtered.

5. Make serial dilutions required for the titration of the analyte using running buffer.

3.1.4 Protein Immobilization

1. Mount a NiHC 200M sensorchip into the SPR and wash and calibrate the chip per manufacturer instructions. The NiHC chip (XanTec) displays nitrilotriacetic acid (NTA) groups on a linear, synthetic polycarboxylate hydrogel. Carboxymethyl dextran hydrogel-based chips displaying NTA are widely available and can also be used.
2. Run 0.5 *M* EDTA, pH 8.5 for 5 min at 200 μL/min.
3. Equilibrate the chip with immobilization buffer for 2 min.
4. Apply 5 m*M* $NiCl_2$ in running buffer for 2 min then equilibrate the chip with immobilization buffer for another 2 min.
5. Inject 500 n*M* His-tagged protein in immobilization buffer for 2–15 min on two of the four detector channels. The channels without protein will act as reference channels during data analysis. The change in response units is proportional to the amount of ligand immobilized. To control the amount of protein ligand deposited on the chip the duration and ligand concentration can be changed.

   Note: Ligand concentration should be controlled to minimize aggregation on the chip surface and ligand–ligand interactions. Too little ligand immobilization, however, will make detection of weak signal from binding of low molecular weight analytes difficult (Giannetti, 2011).
6. Equilibrate the chip with running buffer until the baseline is stable to ensure protein is not leaching off the chip surface. We observed a stable baseline after 12–24 h. If protein continues to leach off the chip using a hexahistidine tag, increasing the tag to 10 or 12 histidines can provide stable binding (Fischer, Leech, & Hubbard, 2011).

3.1.5 Analyte Binding and Analysis

1. Equilibrate the chip with running buffer at 200 μL/min until the baseline is stable.
2. Calibrate the instrument response per the instrument manual.
3. Carefully prepare solutions of running buffer containing 3%, 3.5%, 4%, 4.5%, 5%, 5.5%, 6%, and 6.5% DMSO by adding the appropriate amount of DMSO by mass. This titration is used to correct for DMSO

mismatch between the sample and the running buffer during data analysis (Frostell-Karlsson et al., 2000).

4. Set up analyte, DMSO calibration, and running buffer injections in the SPR software and plate the appropriate samples on a 96-well plate per instrument specification. Seal plates to prevent evaporation.
5. Inject analyte at 150 μL/min for 60 s followed by 90 s dissociation. Between each analyte injection, running buffer (blank) samples should be injected not only to provide referencing but also to minimize carryover between analyte samples.
6. Analyte solutions are injected in ascending order from lowest to highest concentration. At the end of each titration set the DMSO standards are injected. This cycle is replicated two additional times. The chip surface can be regenerated between injections using 0.5 *M* NaCl to help remove any residual analyte.
7. Analyze data using the instrument's analysis package or import blank, binding, and DMSO titration into BioLogic Scrubber 2.0 software. Standard SPR analysis packages use the DMSO titration data to automatically correct for variations in DMSO concentration which can come from imprecisely matched buffers or from DMSO compound stocks that have absorbed water (Giannetti, 2011).
8. Zero the data by using the buffer signal from 5 to 15 s prior to injection. Crop and align data if needed.
9. Select the appropriate reference channel for your experiment and choose how blank injections are picked for data correction. Both closest and average blank corrections work very well when samples and running buffer are well matched.
10. Determine the corrected binding for analytes by selecting the last 5–10 s of the injection. This portion of the sensogram should be saturated and flat.
11. Use the Scrubber fitting routines, or another data analysis package, to calculate the analyte affinity (K_D) by fitting the corrected binding data. For single site, one-to-one binding Eq. (1) can be used.

$$R_{eq} = \frac{R_{max}[A]}{K_D + [A]} \tag{1}$$

The R_{max} is the maximum response achievable for the analyte when flowed over the immobilized ligand, R_{eq} is the equilibrium SPR response at a given analyte concentration, and $[A]$ represents the concentration of analyte. Occasionally (e.g., in some fragment screens) small molecules may bind independently to more than one site, and the binding will be fit more accurately by a multisite binding model (Giannetti, 2011; Rich, Day, Morton, & Myszka, 2001).

The data in Fig. 1 demonstrate the utility of SPR for binding characterization. We took advantage of the hexahistidine tag used for purification of Myc353-437 to immobilize the protein. A titration of the *R*-10074-A4

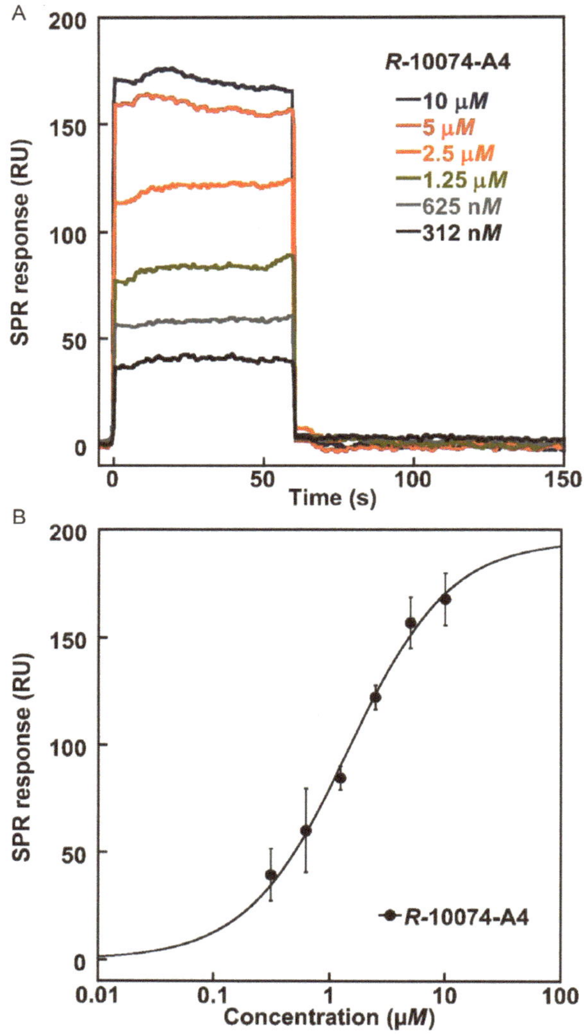

Fig. 1 (A) Full traces of the SPR analysis of *R*-10074-A4 binding to immobilized HexaHis-Myc353-437. Data were collected in 1 × PBS at pH 7.4 with 5% DMSO, 0.005% w/v P20, and 0.1 m*M* EDTA. Sensograms are representative of four independent trials. (B) Equilibrium binding data from sensograms in (A) fit to Eq. (1). *Error bars* represent the standard deviation of four independent trials.

enantiomer (Hammoudeh et al., 2009) yielded reproducible equilibrium binding data that were fit to Eq. (1) to provide a K_D for the interaction of $1.9 \pm 0.21\,\mu M$. As is expected for small-molecule IDP interactions, both the association and dissociation of the compound are too fast to determine the kinetics of the interaction by SPR. The ability to readily determine a K_D for a small-molecule interaction with an IDP can be enabling for characterization of the binding site or determination of buffer effects on the binding. We found His-tag immobilization useful in that it allowed us to easily strip the chip and immobilize fresh protein if we noted changes in binding such as reduced R_{max} values. All SPR makers also provide protocols for immobilization via thiol coupling, a strategy useful for stable surface binding and, in particular, for immobilization of peptides.

3.2 Nuclear Magnetic Resonance

NMR experiments are the most likely to provide molecular detail for small-molecule interactions with IDPs. Chemical shift changes can potentially be measured for all observable protein and ligand signals including Hα, N_H, Hβ, and others. For coupled binding and folding interactions such changes are usually on the order of 100–500 ppb (Williamson, 2013), while for small molecules binding to IDPs, as shown here, binding-induced changes can be an order of magnitude smaller. It should also be noted that not all small-molecule protein interactions result in detectable chemical shift changes.

Here, we demonstrate a fast, label-free method for measuring 1D ^{1}H chemical shift changes within the R-group protons of an IDP peptide upon interaction with a small molecule. We demonstrate a 10–26-ppb chemical shift change within the phenylalanine aromatic protons of the intrinsically disordered peptide Myc365-379 (Fig. 2) when interacting with the small-molecule NBZ-PABA (nitrobenzofurazan *para*-aminobenzoic acid, JY-3-094) (Yap et al., 2013).

3.2.1 Equipment

11.75 T NMR spectrometer
Shigemi NMR tubes
High-quality NMR tubes
Straight capillary tube
pH meter with small-volume pH electrode

3.2.2 Buffers and Reagents

500 m*M* Sodium phosphate buffer pH 6.5
DMSO-d_6
50 m*M* 2,2,3,3-*d*(4)-3-(trimethylsilyl)propionic acid (TMSP) in D_2O

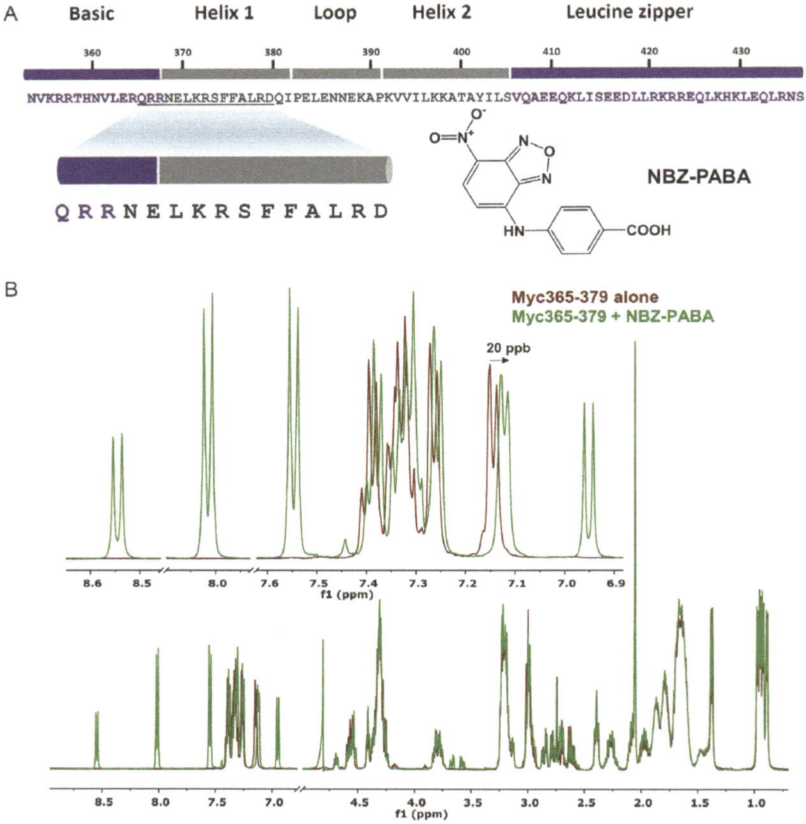

Fig. 2 (A) Sequence of c-Myc353-437 (highlighting Myc365-379) and the structure of NBZ-PABA. (B) 1D ^{1}H spectrum of peptide Myc365-379 alone (*red*) and peptide in the presence of equimolar NBZ-PABA (*green*). *Inset* enlarges the aromatic region, showing the changes in chemical shift for the phenylalanine protons indicating an interaction with the small molecule.

1 *M* DCl and NaOD
2.5 m*M* Myc365-379 in D_2O
20 m*M* NBZ-PABA in DMSO-d_6

3.2.3 Sample Preparation

1. Resuspend lyophilized aliquots of Myc365-379 in 99.9% D_2O and allow them to equilibrate for 4 h at room temperature. Measure the concentration of each stock using UV–vis (for Myc365-379 $\varepsilon_{274} = 1470\ M^{-1}\ cm^{-1}$). Depending on the sensitivity of the NMR used, the minimum final concentration of protein needed will be between 5 and 100 μ*M*.

2. Dissolve NBZ-PABA in DMSO-d_6 to obtain a 20 m*M* stock of the small molecule. The concentration and integrity of the compound should be confirmed by NMR.
3. Prepare two 500 µL samples of Myc365-379 in a 1.5 mL polypropylene tubes. The final samples contain 1.5 m*M* c-Myc365-379; 5 m*M* sodium phosphate, pH 6.5; 5.5 µ*M* TMSP from a D_2O stock; and 5% DMSO-d_6 either with or without compound. Add the peptide first followed by D_2O and TMSP, then buffer. Finally, into one tube deliver compound in DMSO-d_6 and into the other tube deliver an equivalent amount of DMSO-d_6 alone.

   Note: A high concentration of peptide and compound was used because both have good solubility at pH 6.5 and the high concentration facilitates NMR measurements. Often the usable concentration range of either the target protein (or peptide) or of the compound is substantially lower. Quantitative NMR (qNMR) (Malz & Jancke, 2005) is a useful technique to ensure that all the expected protein and compound are observed in your sample. In some instances, we have found that soluble peptide and soluble compound, both in final buffers, can be mixed in equal ratios and the resulting solution can lose material to nano-aggregates. These aggregates may leave observable chemical shifts unperturbed but will change the stoichiometry of the system. A 5-mm high-quality thin wall NMR tube and a sealed capillary tube filled with 0.11 m*M* TMSP in D_2O can be used as an external reference for qNMR. This setup ensures proper referencing, allows for qNMR, and guarantees no interaction with the standard.
4. Mix samples well, avoiding bubble formation. The pH of each sample is measured using a small-volume pH probe and adjusted with 1 *M* DCl or NaOD. Incubate samples at 25°C for a minimum of 1 h prior to measurement.

   Note: Synthesis of compounds and their isolation and purification can play a role in the pH of the stocks and samples made. Compounds that contain titratable groups such as amines and carboxylic acids can cause changes in sample pH and affect protein chemical shifts. For this reason, it is important to be mindful of the compound pK_a and protonation state which may vary by source or isolation technique. In our earlier work we observed larger chemical shift changes upon addition of 10058-F4 ($pK_a = 5.7$) and NBZ-PABA ($pK_a = 8.0$ for deprotonation of the anilinic NH, this unusually low value is due to the *para*-nitro group) that were likely influenced by pH perturbations from the compounds.

3.2.4 NMR Measurements

1. Transfer the equilibrated samples to 5 mm high-quality Shigemi NMR tubes.
2. Acquire ^{1}H NMR using the Agilent vNMRJ v4.0 PRESAT routine with a 2-s presaturation delay, transmitter offset and presaturation frequency centered on HOD, spectral width of 5000, and an acquisition time of 3.3 s.
3. Acquire data for 16–256 transients depending on sample concentration. The number of transients should be kept consistent for each sample set.
4. Process the acquired spectra using MestReNova 9.0. Baseline correct and apodize the data as needed.
5. Use TMSP to reference the spectra and overlay the data (Fig. 2).
6. Measure the alignment shift in the peptide phenylalanine peaks (or other perturbed peaks) using the MestReNova advanced analysis tool (Figs. 2 and 3).

   Note: Referencing biological and D_2O ^{1}H NMR is best done using internal or external chemical shift standards such as DSS (4,4-dimethyl-4-silapentane-1-sulfonic acid), TMSP, or formate. We reference

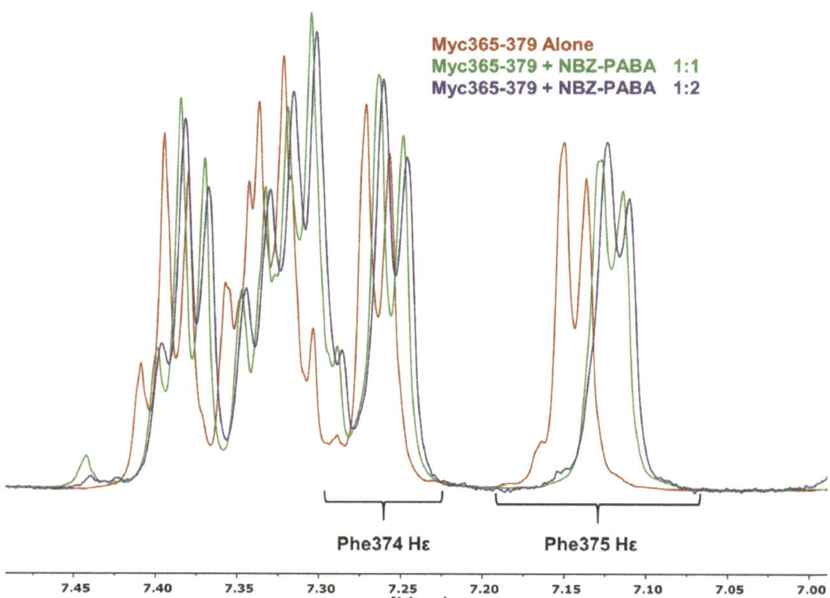

Fig. 3 NMR spectrum of Myc365-379 phenylalanine residues with and without NBZ-PABA. There is a 20-ppb shift upon addition of equimolar NBZ-PABA; similarly a twofold excess of NBZ-PABA shifts the phenylalanine 22 ppb.

our TMSP signal to 0.00 ppm (Wang, Filippov, Richter, Luo, & Kriwacki, 2005; Wishart et al., 1995). If the data are acquired over a pH range, Sykes and coworkers offer an equation to correct the TMSP chemical shift (Wishart et al., 1995). Precise chemical shift referencing is crucial for accurate measurement of the CSPs we see in IDP–small-molecule interactions. Use of chemical shift standards is preferable to using cosolvent chemical shifts. Note also that samples with varying excess of compound should be collected to confirm that binding saturates (Fig. 3).

Two-dimensional NMR can also be used to elucidate specific binding site residues. For EPI-001, ^{1}H–^{15}N HSQC was used to demonstrate interaction within the AF-1 domain of the androgen receptor and to observe CSPs for residues within 354–448 suggesting an interaction site (De Mol et al., 2016). Analogously, ^{1}H–^{15}N TROSY was used to examine the interaction between protein tyrosine phosphatase 1B (PTP1B) and the small-molecule MSI-1436 (Krishnan et al., 2014). Larger proteins ($\geq$35 kDa) have fast transverse relaxation rates which can lead to signal broadening and loss; TROSY experiments minimize this effect and enable researchers to acquire meaningful 2D NMR for large protein targets such as PTP1B (Pervushin, Riek, Wider, & Wuthrich, 1997; Wang et al., 2005).

In addition to chemical shift methods, saturation transfer difference (STD) (Mayer & Meyer, 1999; Yu et al., 2016) and water–ligand observed via gradient spectroscopy (WaterLOGSY) (Dalvit, Fogliatto, Stewart, Veronesi, & Stockman, 2001; Dalvit et al., 2000) are 1D NMR experiments that rely on the nuclear Overhauser effect and spin diffusion to facilitate magnetization transfer from protein to ligand or from bulk water to ligand, respectively. In STD, two spectra are acquired for a mixture of ligand(s) and protein: without protein saturation (off resonance) and while selectively saturating the protein peaks (on resonance). The subtractive difference of the two reveals bound ligands. Yu et al. used STD, in conjunction with ^{1}H–^{1}H TOCSY, to confirm the binding between PKUMDL-YC-1205 and Myc370-409 (Yu et al., 2016). WaterLOGSY is also able to distinguish between protein-bound and unbound ligands. Unbound ligands will have peaks 180 degrees out of phase from the complex. Both techniques are commonly used for library screens since they lend themselves to multiplexing and specific assignments of the ligands are not needed to obtain meaningful results. Kriwacki and coworkers carried out STD and WaterLOGSY screens to find nine preliminary hits that bound to p27-KID. Follow-up ^{1}H–^{15}N HSQC titration experiments with compounds structurally similar to the initial hits identified the compound SJ572403. Then, by titrating p27-KID with SJ572403, the authors were able to localize a binding site by measuring CSPs for several amino acid residues, especially W60 and W76 (Ban, Iconaru, Ramanathan, Zuo, & Kriwacki, 2017; Iconaru et al., 2015).

3.3 Intrinsic Fluorescence Quenching

Another technique that has been useful specifically for monitoring small-molecule binding to Myc is quenching of intrinsic protein fluorescence. While not universal in the way that SPR and NMR are, monitoring binding of small molecules to IDPs through changes in intrinsic protein fluorescence can provide an additional, orthogonal technique to observe the interaction. Advantages of the technique include the simplicity of the measurement and instrumentation (standard steady-state fluorimeter), its label-free nature, and sensitivity in the low micromolar to high nanomolar range. The clear drawback of the technique is its lack of generality: a Tyr or Trp is needed in or near the IDP-binding site and the small-molecule binder must perturb the fluorescence. While the requirements are restricting, within the examples where small-molecule IDP-binding sites have been localized aromatic residues have been found and are likely a common feature in small-molecule IDP-binding sites (Ban et al., 2017; De Mol et al., 2016; Follis et al., 2008; Neira et al., 2017).

Here, we describe the use of fluorescence quenching to quantify the binding interaction of the small-molecule 10058-F4 and the bHLHZip of Myc. The 10058-F4 binding site is localized within the short contiguous region Y402 to K412, a region that would be at the interface of helix 2 and the leucine zipper in the folded Myc–Max heterodimer (Fig. 4A). We and others have shown that 10058-F4 quenches the native tyrosine fluorescence within the Myc bHLHZip (Fig. 4B) (Follis et al., 2008; Heller, Aprile, Bonomi, et al., 2017). Although less fluorescent than tryptophan, tyrosine still provides a strong fluorescence signal at micromolar concentrations, within the affinity range common for small molecules binding to IDPs. At equimolar concentrations, the 10058-F4 small molecule quenches the Myc tyrosine fluorescence up to 86%; the quenching is titratable and can be used to determine the 10058-F4 binding affinity for Myc (Fig. 4B).

3.3.1 Equipment

Fluorimeter
UV–vis spectrometer
Quartz fluorimeter and UV–vis cuvettes

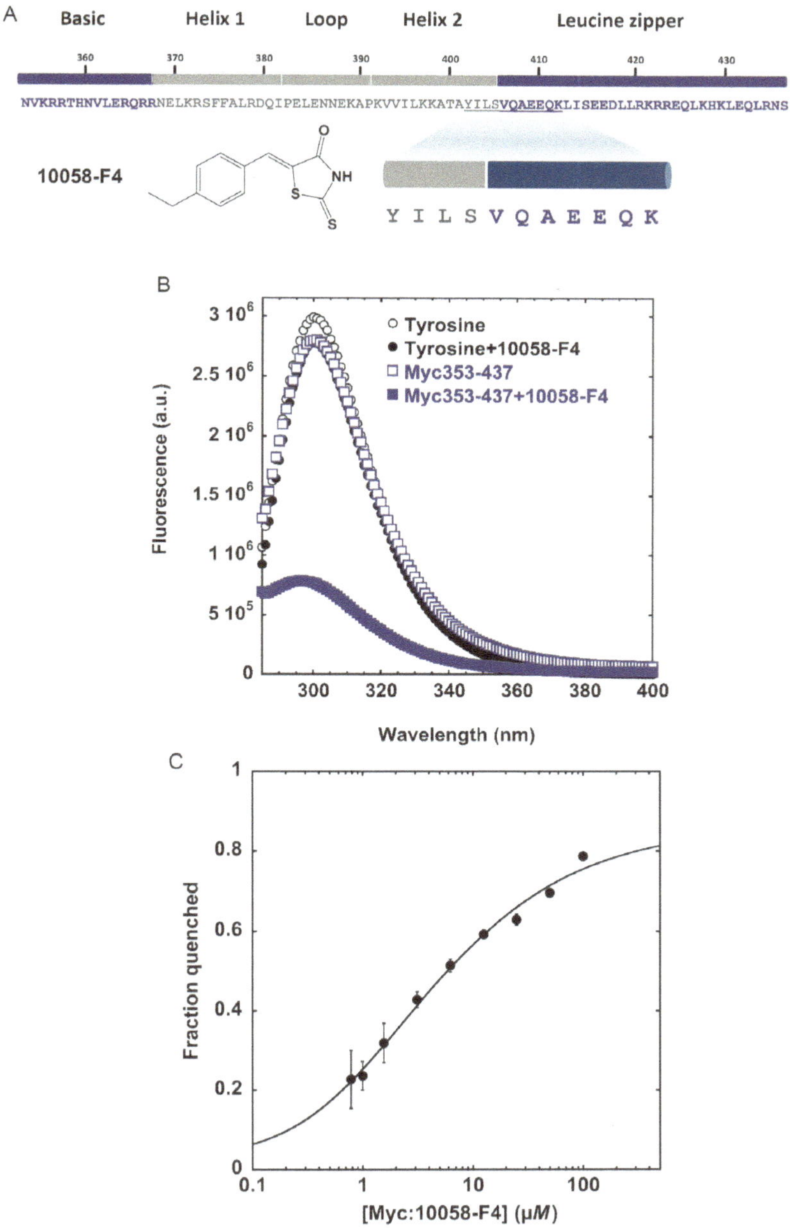

Fig. 4 (A) Myc bHLHZip domain with 10058-F4 binding site (Y402 to K412) highlighted and the 10058-F4 structure (neutral form). (B) Fluorescence emission scan of 50 μ*M* free tyrosine in the presence (*filled black circle*) and absence (*open black circle*) of equimolar 10058-F4 and 50 μ*M* wild-type Myc bHLHZip domain in the presence (*filled blue square*) and absence (*open blue square*) of equimolar 10058-F4 in 1 × assay diluent at 25°C. (C) Equilibrium quenching of an equimolar titration of Myc bHLHZip and 10058-F4 fit to a 1:1 binding model. *Error bars* represent the standard deviation of five independent trials.

Branson Ultrasonic bath sonicator
Centrifuge
0.2 μm PES sterile filter

3.3.2 Reagents and Buffers

Purified protein or peptide of interest
Small molecule in DMSO (or other compatible solvent)
DMSO or other high-purity organic solvent
100 m*M* dithiothreitol (DTT)
10 × PBS (1.37 *M* NaCl, 27 m*M* KCl, 43 m*M* Na_2HPO_4, 14 m*M* KH_2PO_4)
1 × Assay diluent for dilution series preparation (1 × PBS pH 7.4, 1 m*M* DTT, 5% DMSO)

3.3.3 Procedure

1. Determine the concentration of the IDP or IDR of interest by absorbance measurement and the protein molar extinction coefficient or by micro BCA (Thermo Scientific) assay. For the Myc bHLHZip domain, the protein was HPLC purified to >95% purity and then stored lyophilized at −20°C in aliquots of 2–4 mg/tube. The protein was reconstituted in 1 mL of ultrapure water and incubated at room temperature for at least 1 h. Subsequently, the protein was sterile filtered using a 0.2 μm PES filter and the concentration was determined using Beer's law from the UV absorbance at 274 nm and $\varepsilon_{274} = 1470\, M^{-1}\, cm^{-1}$. Final protein concentration generally ranged from 200 to 400 μ*M*. The reconstituted Myc bHLHZip solution was stored at −20°C for up to 1 month. Multiple freeze–thaw cycles were avoided in order to disfavor potential protein aggregation.
2. Prepare 20 × working stocks of small molecule in appropriate solvent. It is important to understand the solubility of the small molecule in the final assay diluent to avoid formation of insoluble nanoaggregates which will interfere with measurements. The 10058-F4 solubility in 1 × assay diluent was approximately 200 μ*M*. To maintain a final 5% DMSO in the sample, 20 × small-molecule working stocks were prepared in 100% DMSO.
3. Prepare the highest sample concentration with equimolar IDP and small molecule. Combine water (if necessary), disordered protein, 10 × PBS, 100 m*M* DTT, and 20 × small molecule in a microcentrifuge tube. At the same time prepare the blank sample containing IDP but no

compound (adding just solvent as the blank) as well as background samples of compound in diluent and diluent alone.

Note: Ionic strength and pH can affect protein solubility. The order of addition during sample preparation can be important to avoid conditions which could induce protein or small-molecule aggregation. In the case of the Myc bHLHZip domain, which is acidic (pH 3–4) after HPLC purification and reconstitution, the protein should be diluted to a lower concentration first and then buffered using 10 × PBS to a final 1 × PBS concentration. The protein solution should be buffered prior to addition of the small-molecule 10058-F4 because the small molecule precipitates at low pH.

4. Sonicate samples for 60 min at room temperature. For small molecules dissolved in organic solvent (and with low water solubility), the rate of dilution into aqueous solvent can affect the efficiency of the dispersion of the molecule. In this system, sonication helped ensure dispersion of the compound and increased the reproducibility of measurements. It is important to monitor and control the temperature of the sonicator bath to prevent heating the samples.
5. Remove samples from the sonicator and incubate at the desired temperature for 15 min. Centrifuge samples for at least 1 min at 13,000 rpm. Inspect samples for precipitation prior to fluorescence or UV–vis measurements. Even when both components (protein or peptide and small molecule) are fully soluble individually, aggregates may still form in the mixture and these samples cannot be used.
6. Prepare a serial dilution of the IDP–small-molecule solution, the IDP alone blank, and the compound alone background using 1 × assay diluent. A minimum eight-point dilution series over at least a 100-fold concentration range spanning the dissociation constant of the complex is recommended.
7. Measure the fluorescence of each sample. For tyrosine quenching of Myc bHLHZip with 10058-F4, samples were excited at 274 nm and emission was collected from 285 to 385 nm using 4 nm excitation slits and 8 nm emission slits.
8. Measure the UV–vis absorbance for all samples at the fluorescence excitation and emission wavelengths. This measurement is most convenient if done in the same cuvette used for fluorescence.
9. Correct each sample's emission intensity (IDP and compound or IDP alone) with the background intensity and then apply the inner filter effect correction specific to the sample (Eq. 2). Fluorescence intensity

can be suppressed by absorbance at both excitation and emission wavelengths. This phenomenon is called the inner filter effect. The absorbance values of each sample are used to calculate the correction (Eq. 2) for the specific sample (Lakowicz, 2006).

$$F_{\text{corr}} = F_{\text{obs}} \times 10^{\frac{(A_{\text{ex}} + A_{\text{em}})}{2}} \tag{2}$$

The corrected fluorescence is F_{corr}, the background-subtracted observed fluorescence is F_{obs}, and A_{ex} and A_{em} are the absorbance at the excitation wavelength and the emission wavelength, respectively (for the same path length as in the sample).

10. To obtain a dissociation constant from the corrected data, plot the fraction quenched, Q (Eq. 3) vs. concentration of IDP and small molecule (which are equimolar), where F_{corr} is the corrected fluorescence for a compound-IDP solution at a given concentration and F_{free} is the corrected fluorescence for the IDP alone at the same concentration (Fig. 4C).

$$Q = 1 - \frac{F_{\text{corr}}}{F_{\text{free}}} \tag{3}$$

The resulting titration curve of fraction quenched vs. concentration can be fit to a 1:1 binding model with equimolar components using Eq. (4), where Q_{max} is the fraction quenched for complete complex formation and the fraction of complex formed is expressed in terms of the K_{D} for binding and $[C]_{\text{tot}}$, the total concentration of IDP (or small molecule). The K_{D} and Q_{max} are fitted values (Follis et al., 2008).

$$Q = Q_{\text{max}} \times \frac{2 + K_{\text{D}}/[C]_{\text{tot}} - \sqrt{\left(-2 - K_{\text{D}}/[C]_{\text{tot}}\right)^2 - 4}}{2} \tag{4}$$

Control experiments in which the fluorescence intensity of free Tyr or Trp amino acid is measured with and without equimolar small molecule are useful. Quenching of free amino acids in the presence of the small molecule should be minimal and nontitratable. Comparison of the fluorescence intensity of the free Tyr or Trp with the fluorescence intensity exhibited by the IDP can also indicate if there are interactions within the IDP that may already be reducing the fluorescence and affecting its ability to act as a probe.

In Fig. 4B, we saw that the fluorescence of Y402 in Myc was nearly equivalent to the fluorescence from free Tyr.

Tryptophan's fluorescence intensity is substantially greater than tyrosine's, and the position of its emission maximum and intensity is more sensitive to the surrounding fluorophore environment (Lakowicz, 2006). Introduction of a Trp into a suspected binding site may provide a sensitive probe for small-molecule binding. To determine if Trp could act as a probe for 10058-F4 binding, the Myc bHLHZip mutant MycY402W was produced. The MycY402W tryptophan fluorescence was comparable to the fluorescence of free Trp; however, the emission maximum was blue-shifted 10 nm relative to the free Trp emission (Fig. 5). The addition of equimolar 10058-F4 strongly quenched the fluorescence of MycY402W but did not have the same effect on the free Trp (Fig. 5), indicating that, in this case, Trp could be introduced and act as a specific indicator of small-molecule interaction. Substitution with tryptophan within an IDP could change potential interaction sites as well as shift the protein's conformational ensemble and therefore not report on the wild-type protein's binding ability.

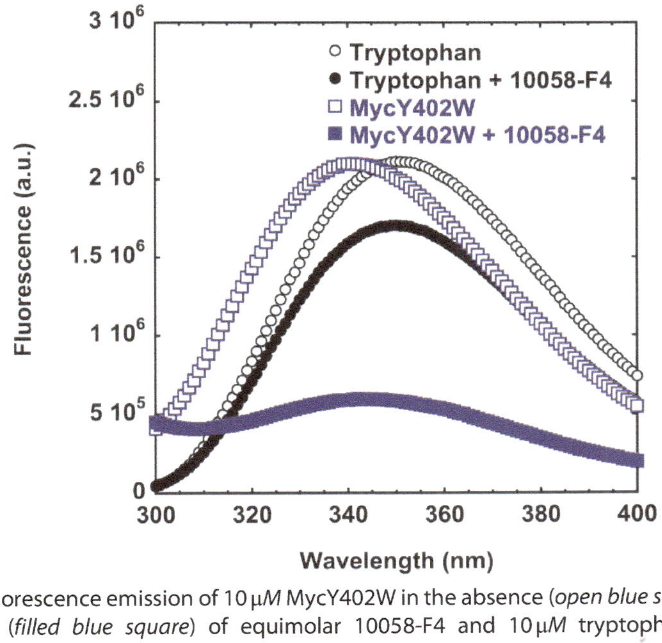

Fig. 5 Fluorescence emission of 10 μ*M* MycY402W in the absence (*open blue square*) and presence (*filled blue square*) of equimolar 10058-F4 and 10 μ*M* tryptophan in the absence (*open black circle*) and presence (*filled black circle*) of equimolar 10058-F4 in 1 × PBS, 5% DMSO, 1 m*M* DTT, pH 7.4 at 25°C.

However, if small-molecule binding is detectable with the Trp mutant, the mutant can then be used in a competition assay with wild-type protein (or other nonfluorescent mutants) and compound to assess the binding competence of the nonfluorescent sequence.

We and others have demonstrated that 10058-F4 will bind to the isolated minimal binding site peptide Myc402-412 with nearly the same affinity as the full Myc bHLHZip region (Follis et al., 2008; Heller, Aprile, Bonomi, et al., 2017). While the observation does indicate that binding recognition and affinity is dominated by the short recognition motif, the binding is not wholly independent of the context of the peptide. We found that when we extended the peptide sequence to Myc397-412, which placed two consecutive lysine residues at the N-terminus (Fig. 4A), the peptide did not display measurable binding to 10058-F4 in the fluorescence quenching assay. Peptides can be very useful in isolating the binding sites for small molecules on IDPs; however, the immediate local context of the binding site can strongly modulate binding behavior. In this case, selection of the native sequence to specifically include multiple charged residues within a short stretch rendered the peptide very soluble, but also changed aspects of the peptide's interaction such that it no longer bound to the small molecule, while both shorter (402-412) and longer sequences (353-437) did.

4. CONCLUSIONS

Given the range of examples for small-molecule binding to disordered proteins, it seems likely that many disordered regions that are involved in protein–protein interactions will support some level of small-molecule binding. There is no single method to go from an initial hit in an interaction screen to measuring a dissociation constant and defining a binding site on an IDP. The process uses the combination of techniques (e.g., NMR, SPR, fluorescence, competition) that are available and to which the particular IDP is amenable. Only a subset of initial hits is likely to exhibit specific binding vs. the intended target but what defines a useful level of specificity and affinity depends on the goal: a potential tool molecule or a hit for lead development. At this point we do not yet know the range of affinity and specificity achievable in these interactions. The characterization of additional small-molecule IDP-binding interactions will allow us to understand their commonalities, their differences, and the aspects that drive specificity, and to develop predictors of the most likely IDP-binding sites.

REFERENCES

Arkin, M. R., Tang, Y., & Wells, J. A. (2014). Small-molecule inhibitors of protein-protein interactions: Progressing toward the reality. *Chemistry & Biology*, *21*(9), 1102–1114.

Ban, D., Iconaru, L. I., Ramanathan, A., Zuo, J., & Kriwacki, R. W. (2017). A small molecule causes a population shift in the conformational landscape of an intrinsically disordered protein. *Journal of the American Chemical Society*, *139*(39), 13692–13700.

Banani, S. F., Lee, H. O., Hyman, A. A., & Rosen, M. K. (2017). Biomolecular condensates: Organizers of cellular biochemistry. *Nature Reviews. Drug Discovery*, *18*(5), 285–298.

Berg, T., Cohen, S. B., Desharnais, J., Sonderegger, C., Maslyar, D. J., Goldberg, J., et al. (2002). Small-molecule antagonists of Myc/Max dimerization inhibit Myc-induced transformation of chicken embryo fibroblasts. *Proceedings of the National Academy of Sciences of the United States of America*, *99*(6), 3830–3835.

Borgia, A., Borgia, M. B., Bugge, K., Kissling, V. M., Heidarsson, P. O., Fernandes, C. B., et al. (2018). Extreme disorder in an ultrahigh-affinity protein complex. *Nature*, *555*(7694), 61–66.

Choi, S. H., Mahankali, M., Lee, S. J., Hull, M., Petrassi, H. M., Chatterjee, A. K., et al. (2017). Targeted disruption of Myc-Max oncoprotein complex by a small molecule. *ACS Chemical Biology*, *12*(11), 2715–2719.

Convertino, M., Vitalis, A., & Caflisch, A. (2011). Disordered binding of small molecules to Abeta(12-28). *Journal of Biological Chemistry*, *286*(48), 41578–41588.

Dalvit, C., Fogliatto, G., Stewart, A., Veronesi, M., & Stockman, B. (2001). WaterLOGSY as a method for primary NMR screening: Practical aspects and range of applicability. *Journal of Biomolecular NMR*, *21*(4), 349–359.

Dalvit, C., Pevarello, P., Tato, M., Veronesi, M., Vulpetti, A., & Sundstrom, M. (2000). Identification of compounds with binding affinity to proteins via magnetization transfer from bulk water. *Journal of Biomolecular NMR*, *18*(1), 65–68.

Das, R. K., Crick, S. L., & Pappu, R. V. (2012). N-terminal segments modulate the alpha-helical propensities of the intrinsically disordered basic regions of bZIP proteins. *Journal of Molecular Biology*, *416*(2), 287–299.

Das, P., Matysiak, S., & Mittal, J. (2018). Looking at the disordered proteins through the computational microscope. *ACS Central Science*, *4*(5), 534–542.

De Mol, E., Fenwick, R. B., Phang, C. T., Buzon, V., Szulc, E., de la Fuente, A., et al. (2016). EPI-001, a compound active against castration-resistant prostate cancer, targets transactivation unit 5 of the androgen receptor. *ACS Chemical Biology*, *11*(9), 2499–2505.

Dyson, H. J., & Wright, P. E. (2002). Coupling of folding and binding for unstructured proteins. *Current Opinion in Structural Biology*, *12*(1), 54–60.

Erkizan, H. V., Kong, Y., Merchant, M., Schlottmann, S., Barber-Rotenberg, J. S., Yuan, L., et al. (2009). A small molecule blocking oncogenic protein EWS-FLI1 interaction with RNA helicase A inhibits growth of Ewing's sarcoma. *Nature Medicine*, *15*(7), 750–756.

Fischer, M., Leech, A. P., & Hubbard, R. E. (2011). Comparative assessment of different histidine-tags for immobilization of protein onto surface plasmon resonance sensorchips. *Analytical Chemistry*, *83*(5), 1800–1807.

Follis, A. V., Hammoudeh, D. I., Wang, H., Prochownik, E. V., & Metallo, S. J. (2008). Structural rationale for the coupled binding and unfolding of the c-Myc oncoprotein by small molecules. *Chemistry & Biology*, *15*(11), 1149–1155.

Frostell-Karlsson, A., Remaeus, A., Roos, H., Andersson, K., Borg, P., Hamalainen, M., et al. (2000). Biosensor analysis of the interaction between immobilized human serum albumin and drug compounds for prediction of human serum albumin binding levels. *Journal of Medicinal Chemistry*, *43*(10), 1986–1992.

Giannetti, A. M. (2011). Chapter eight—From experimental design to validated hits: A comprehensive walk-through of fragment lead identification using surface plasmon resonance. In L. C. Kuo (Ed.), *Vol. 493. Methods in enzymology* (pp. 169–218): Academic Press.
Gsponer, J., Futschik, M. E., Teichmann, S. A., & Babu, M. M. (2008). Tight regulation of unstructured proteins: From transcript synthesis to protein degradation. *Science*, *322*(5906), 1365–1368.
Hammoudeh, D. I., Follis, A. V., Prochownik, E. V., & Metallo, S. J. (2009). Multiple independent binding sites for small-molecule inhibitors on the oncoprotein c-Myc. *Journal of the American Chemical Society*, *131*(21), 7390–7401.
Hart, J. R., Garner, A. L., Yu, J., Ito, Y., Sun, M., Ueno, L., et al. (2014). Inhibitor of MYC identified in a Krohnke pyridine library. *Proceedings of the National Academy of Sciences of the United States of America*, *111*(34), 12556–12561.
Heller, G. T., Aprile, F. A., Bonomi, M., Camilloni, C., De Simone, A., & Vendruscolo, M. (2017). Sequence specificity in the entropy-driven binding of a small molecule and a disordered peptide. *Journal of Molecular Biology*, *429*(18), 2772–2779.
Heller, G. T., Aprile, F. A., & Vendruscolo, M. (2017). Methods of probing the interactions between small molecules and disordered proteins. *Cellular and Molecular Life Sciences*, *74*(17), 3225–3243.
Heller, G. T., Sormanni, P., & Vendruscolo, M. (2015). Targeting disordered proteins with small molecules using entropy. *Trends in Biochemical Sciences*, *40*(9), 491–496.
Iconaru, L. I., Ban, D., Bharatham, K., Ramanathan, A., Zhang, W., Shelat, A. A., et al. (2015). Discovery of small molecules that inhibit the disordered protein, p27Kip1. *Scientific Reports*, *5*, 15686.
Kiessling, A., Sperl, B., Hollis, A., Eick, D., & Berg, T. (2006). Selective inhibition of c-Myc/Max dimerization and DNA binding by small molecules. *Chemistry & Biology*, *13*(7), 745–751.
Krishnan, N., Koveal, D., Miller, D. H., Xue, B., Akshinthala, S. D., Kragelj, J., et al. (2014). Targeting the disordered C terminus of PTP1B with an allosteric inhibitor. *Nature Chemical Biology*, *10*(7), 558–566.
Lakowicz, J. R. (2006). *Principles of fluorescence spectroscopy* (3rd ed.). . United States: Springer.
Malz, F., & Jancke, H. (2005). Validation of quantitative NMR. *Journal of Pharmaceutical and Biomedical Analysis*, *38*(5), 813–823.
Matthes, D., Gapsys, V., Griesinger, C., & de Groot, B. L. (2017). Resolving the atomistic modes of Anle138b inhibitory action on peptide oligomer formation. *ACS Chemical Neuroscience*, *8*(12), 2791–2808.
Mayer, M., & Meyer, B. (1999). Characterization of ligand binding by saturation transfer difference NMR spectroscopy. *Angewandte Chemie, International Edition in English*, *38*(12), 1784–1788.
Metallo, S. J. (2010). Intrinsically disordered proteins are potential drug targets. *Current Opinion in Chemical Biology*, *14*(4), 481–488.
Mittag, T., Orlicky, S., Choy, W. Y., Tang, X., Lin, H., Sicheri, F., et al. (2008). Dynamic equilibrium engagement of a polyvalent ligand with a single-site receptor. *Proceedings of the National Academy of Sciences of the United States of America*, *105*(46), 17772–17777.
Morgan, P., Brown, D. G., Lennard, S., Anderton, M. J., Barrett, J. C., Eriksson, U., et al. (2018). Impact of a five-dimensional framework on R&D productivity at AstraZeneca. *Nature Reviews. Drug Discovery*, *17*(3), 167–181.
Neira, J. L., Bintz, J., Arruebo, M., Rizzuti, B., Bonacci, T., Vega, S., et al. (2017). Identification of a drug targeting an intrinsically disordered protein involved in pancreatic adenocarcinoma. *Scientific Reports*, 7, 39732.

Oprea, T. I., Bologa, C. G., Brunak, S., Campbell, A., Gan, G. N., Gaulton, A., et al. (2018). Unexplored therapeutic opportunities in the human genome. *Nature Reviews. Drug Discovery, 17*, 317.

Overington, J. P., Al-Lazikani, B., & Hopkins, A. L. (2006). How many drug targets are there? *Nature Reviews. Drug Discovery, 5*(12), 993–996.

Pervushin, K., Riek, R., Wider, G., & Wuthrich, K. (1997). Attenuated T2 relaxation by mutual cancellation of dipole-dipole coupling and chemical shift anisotropy indicates an avenue to NMR structures of very large biological macromolecules in solution. *Proceedings of the National Academy of Sciences of the United States of America, 94*(23), 12366–12371.

Rees, S., Gribbon, P., Birmingham, K., Janzen, W. P., & Pairaudeau, G. (2016). Towards a hit for every target. *Nature Reviews. Drug Discovery, 15*(1), 1–2.

Rich, R. L., Day, Y. S. N., Morton, T. A., & Myszka, D. G. (2001). High-resolution and high-throughput protocols for measuring drug/human serum albumin interactions using BIACORE. *Analytical Biochemistry, 296*(2), 197–207.

Santos, R., Ursu, O., Gaulton, A., Bento, A. P., Donadi, R. S., Bologa, C. G., et al. (2017). A comprehensive map of molecular drug targets. *Nature Reviews. Drug Discovery, 16*(1), 19–34.

Scott, D. E., Bayly, A. R., Abell, C., & Skidmore, J. (2016). Small molecules, big targets: Drug discovery faces the protein-protein interaction challenge. *Nature Reviews. Drug Discoverys, 15*(8), 533–550.

Tompa, P., & Fuxreiter, M. (2008). Fuzzy complexes: Polymorphism and structural disorder in protein-protein interactions. *Trends in Biochemical Sciences, 33*(1), 2–8.

Toth, G., Gardai, S. J., Zago, W., Bertoncini, C. W., Cremades, N., Roy, S. L., et al. (2014). Targeting the intrinsically disordered structural ensemble of alpha-synuclein by small molecules as a potential therapeutic strategy for Parkinson's disease. *PLoS One, 9*(2), e87133.

Uversky, V. N., Dave, V., Iakoucheva, L. M., Malaney, P., Metallo, S. J., Pathak, R. R., et al. (2014). Pathological unfoldomics of uncontrolled chaos: Intrinsically disordered proteins and human diseases. *Chemical Reviews, 114*(13), 6844–6879.

Uversky, V. N., Oldfield, C. J., & Dunker, A. K. (2008). Intrinsically disordered proteins in human diseases: Introducing the D2 concept. *Annual Review of Biophysics, 37*(1), 215–246.

van der Lee, R., Buljan, M., Lang, B., Weatheritt, R. J., Daughdrill, G. W., Dunker, A. K., et al. (2014). Classification of intrinsically disordered regions and proteins. *Chemical Reviews, 114*(13), 6589–6631.

Vavouri, T., Semple, J. I., Garcia-Verdugo, R., & Lehner, B. (2009). Intrinsic protein disorder and interaction promiscuity are widely associated with dosage sensitivity. *Cell, 138*(1), 198–208.

Wang, Y., Filippov, I., Richter, C., Luo, R., & Kriwacki, R. W. (2005). Solution NMR studies of an intrinsically unstructured protein within a dilute, 75 kDa eukaryotic protein assembly; probing the practical limits for efficiently assigning polypeptide backbone resonances. *Chembiochem, 6*(12), 2242–2246.

Wells, J. A., & McClendon, C. L. (2007). Reaching for high-hanging fruit in drug discovery at protein-protein interfaces. *Nature, 450*(7172), 1001–1009.

Williamson, M. P. (2013). Using chemical shift perturbation to characterise ligand binding. *Progress in Nuclear Magnetic Resonance Spectroscopy, 73*, 1–16.

Wishart, D. S., Bigam, C. G., Yao, J., Abildgaard, F., Dyson, H. J., Oldfield, E., et al. (1995). 1H, 13C and 15N chemical shift referencing in biomolecular NMR. *Journal of Biomolecular NMR, 6*(2), 135–140.

Yap, J. L., Wang, H., Hu, A., Chauhan, J., Jung, K. Y., Gharavi, R. B., et al. (2013). Pharmacophore identification of c-Myc inhibitor 10074-G5. *Bioorganic and Medicinal Chemistry Letters*, *23*(1), 370–374.

Yin, X., Giap, C., Lazo, J. S., & Prochownik, E. V. (2003). Low molecular weight inhibitors of Myc-Max interaction and function. *Oncogene*, *22*(40), 6151–6159.

Yu, C., Niu, X., Jin, F., Liu, Z., Jin, C., & Lai, L. (2016). Structure-based inhibitor design for the intrinsically disordered protein c-Myc. *Scientific Reports*, *6*, 22298.

CHAPTER TWENTY-TWO

Targeting the Intrinsically Disordered Proteome Using Small-Molecule Ligands

Sławomir Wójcik*,†, Melissa Birol‡, Elizabeth Rhoades‡, Andrew D. Miranker*,§, Zachary A. Levine*,¶,1

*Department of Molecular Biophysics and Biochemistry, Yale University, New Haven, CT, United States
†Department of Chemistry, Yale University, New Haven, CT, United States
‡Department of Chemistry, University of Pennsylvania, Philadelphia, PA, United States
§Department of Chemical and Environmental Engineering, Yale University, New Haven, CT, United States
¶Department of Pathology, Yale School of Medicine, New Haven, CT, United States
[1]Corresponding author: e-mail address: zachary.levine@yale.edu

Contents

Abstract

Intrinsically disordered proteins (IDPs) and regions (IDRs) make up a significant part of the proteome and facilitate a wide range of physiological and pathological functions that are only beginning to be understood. As such, they are highly attractive targets for drug development and bioengineering. However, their inability to adopt well-defined structures provides significant obstacles for developing ligands that regulate their behaviors. In this chapter, we review how the conformational flexibility of IDPs and their propensity to phase separate make them tractable targets for small-molecule manipulation. We also describe both theoretical and experimental approaches to characterize disordered proteins, including novel thermodynamic and single-molecule techniques that help identify complimentary partners of IDPs and their ability to shift protein ensembles toward preferred conformations.

Methods in Enzymology, Volume 611
ISSN 0076-6879
https://doi.org/10.1016/bs.mie.2018.09.036

ABBREVIATIONS

Aβ amyloid-β protein
CD circular dichroism
DEST dark-state exchange saturation transfer
FCCS fluorescence cross-correlation spectroscopy
FCS fluorescence correlation spectroscopy
FRET fluorescence resonance energy transfer
GPMV giant plasma membrane vesicle
HSQC heteronuclear single quantum coherence spectroscopy
IAPP islet amyloid polypeptide
IDP intrinsically disordered protein
IDR intrinsically disordered region
LLPS liquid–liquid phase separation
MD molecular dynamics
NMR nuclear magnetic resonance
PTM posttranslational modification
RDC residual dipolar coupling
REMD replica-exchange molecular dynamics
SAXS small-angle X-ray scattering
SELEX systematic evolution of ligands by exponential enrichment
smFRET single-molecule fluorescence resonance energy transfer
STD saturation transfer difference
ThT thioflavin T
T-REMD temperature replica-exchange molecular dynamics
trNOE transferred nuclear Overhauser effect
αSyn α-synuclein

1. INTRODUCTION

Proteins fulfill a wide variety of biological roles based on unique and well-defined structural components, spanning roles such as building blocks to molecular machines. By extension, one might expect that long chains of amino acids have evolved to adopt highly ordered structures that are optimized to carry out specific functions. However, nearly a third of the proteome completely or partially lacks this quality (Xue, Dunker, & Uversky, 2012) with some viruses exhibiting more than 75% disorder. At the sequence level, intrinsically disordered proteins (IDPs) and proteins containing disordered regions (IDRs) often exhibit some common traits, which include low complexity (abundance of a small subset of amino acids), a small number of hydrophobic residues, and a large net charge compared to globular proteins. Many of these traits are also naturally tunable through alternative splicing (sequence control) or posttranslational modifications (PTMs) (charge control) (Uversky, 2013).

As a result, these traits reduce the presence of stable three-dimensional structures and promote heterogeneous protein ensembles, where highly dynamic conformations are readily interconverted (Fisher & Stultz, 2011; Marsh & Forman-Kay, 2012). Such behaviors have historically made IDPs and IDRs elusive to characterize with traditional methods, since many of these techniques (e.g., X-ray crystallography) expect static structures and breakdown when dealing with poorly defined structural regions. In order to overcome these obstacles, techniques sensitive to dynamic protein interactions such as optical spectroscopy, nuclear magnetic resonance (NMR), small-angle X-ray scattering (SAXS), or solution-state techniques coupled with computational tools are preferred to adequately characterize IDPs. The persistent presence and evolution of IDPs in the proteome indicates that they impart important physiological functions associated with their conformational flexibilities. Some of these functions have been identified and include increased binding versatility, elevated susceptibility to PTMs, high specificity combined with low affinity, and elevated resistance to nonnative conditions among other factors (Liu & Huang, 2014).

Unlike globular proteins, IDPs achieve their specificity by adopting a broad spectrum of physiological structures that are capable of carrying out disparate signaling and regulatory functions (Wright & Dyson, 2014) through interactions with multiple partners (Li et al., 2012). Furthermore, their roles in reversible signal transduction appear to be attributed to their high dissociation rates that drive low affinity interactions. IDP interactions are multifaceted, and can be one-to-one, one-to-many, or many-to-one, which enables them to act as hubs in protein–protein interaction networks (Uversky, 2013). These behaviors are reflected in their anatomy, which are comprised of relatively short functional segments separated by flexible linkers. Another interesting property of IDPs is their unexpected reactions to environmental changes compared to globular proteins. For example, IDPs tend to be relatively insensitive to changes in temperature and pH, which can be exploited for biotechnological applications ranging from expression and purification tags to drug carriers (Uversky, 2015).

There is also considerable interest in the pathological roles of IDPs and disordered peptides, due to their marked association with multiple human diseases and disorders (Ambadipudi & Zweckstetter, 2016; Uversky, Oldfield, & Dunker, 2008). Given their ubiquitous presence in the proteome, it is not surprising then that proteins exhibiting substantial gains-of-function have the potential to severely impact cellular homeostasis. Prominent examples of IDPs involved in pathological processes include the islet amyloid polypeptide (IAPP) associated with type II diabetes (Asthana, Mallick, Alexandrescu, & Jha, 2018; Birol, Kumar, Rhoades, & Miranker, 2018), the amyloid-β (Aβ) protein associated with Alzheimer's disease (Ittner & Gotz, 2011; Lee, Nam, Lee, Savelieff, & Lim, 2017), the α-synuclein (αSyn) protein associated with Parkinson's disease (Breydo, Wu, & Uversky, 2012; Rosborough, Patel, & Kalia, 2017), and the tau protein (Bulic, Pickhardt, & Mandelkow, 2013; Ittner & Gotz, 2011; Kolarova, Garcia-Sierra, Bartos, Ricny, & Ripova, 2012), which contributes to a variety of neurodegenerative disorders. Other pathological proteins containing IDRs include prion protein (Ronga et al., 2006), which causes transmissible spongiform encephalopathy, and the tumor suppressor proteins p53 (Uversky, 2016; Wells et al., 2008) and BRCA1 (Mark et al., 2005).

Overall, there are a multitude of environmental factors that influence the conformational behaviors of IDPs, consequentially shifting their pathological behaviors. These factors include biological interfaces containing phospholipid membranes (Fusco, Sanz-Hernandez, & De Simone, 2018; Rawat, Langen, & Varkey, 2018; Sasahara, 2018). Binding of an IDP to lipid bilayers can often dramatically alter the structural properties of both protein and membrane (Ouberai et al., 2013), which can sometimes lead to membrane disruption (Dong, Qiao, Qian, & Wei, 2018; Zhang et al., 2017) and cytotoxicity (Bodner, Maltsev, Dobson, & Bax, 2010; Caillon, Hoffmann, Botz, & Khemtemourian, 2016; Fusco et al., 2017; Galvagnion et al., 2016). Such factors must be taken into account when constructing strategies to alter IDPs, including techniques to minimize interfacial interactions through, e.g., membrane displacement (Perni et al., 2017) or through targeting membrane-bound species (Birol et al., 2018; Kumar et al., 2015).

Such examples underscore the importance of molecular context, which add additional dimensions to the conformational landscape of IDPs. Inside the cell that context includes the effects of crowded, anisotropic environments comprised of macromolecules, metabolites, and partitioning compartments (Qin & Zhou, 2017). This complex network of interactions has a substantial impact on IDPs and IDRs, which is comprehensively summarized by Theillet et al. (2014). In addition to lipids and other biomolecules, IDPs are also affected by one another, and can even adopt liquid-ordered states in high concentrations similar to coacervating polymers (Overbeek & Voorn, 1957; Pak et al., 2016). In particular, there is growing attention among biophysicists in how IDPs microcompartmentalize in intracellular environments, facilitating the formation of membraneless organelles and Cajal bodies within the cytoplasm or nucleoplasm (Dundr, 2012).

These highly dynamic assemblies, enriched with macromolecules such as RNA and proteins, remain in direct contact with the surrounding fluid, and are considered a separate liquid phase (Brangwynne, Tompa, & Pappu, 2015). IDPs are thought to be important drivers of liquid–liquid phase separation (LLPS), based on the significant content of disordered structures that make up membraneless organelles. Similarly, the unique properties of disordered proteins make them ideal building blocks for macromolecular assemblies, since they can flexibly attach to ordered domains and slowly stabilize the interactions between them (Uversky, Kuznetsova, Turoverov, & Zaslavsky, 2015). Because IDPs exhibit low sequence complexity along with a large net charge, they can easily mimic polyelectrolytes known to participate in polymer LLPS. Another trait that drives sequestration of disordered regions is multivalency, which is enforced by multiple short repeats of weakly binding domains.

The aim of this chapter is to present the necessary context to better characterize proteins with disordered regions, and to highlight how small-molecule ligands can be utilized to address these behaviors. In the first part of this chapter, focus will be devoted to characterizing the types of states that are populated by IDPs. This will be followed by a review of computational and experimental methods for targeting IDPs, offering insights into the assemblies of intrinsically disordered structures. These techniques are intended to inform efforts to design, screen, and evaluate small-molecule ligands that favorably interact with IDPs, thereby harnessing their degenerate structures and functions.

2. DESIGNING LIGANDS TO SHIFT DISORDERED PROTEIN ENSEMBLES

IDPs can exhibit different levels of assembly and organization ranging from freely fluctuating chains to liquid droplets, in addition to pathological amyloid fibers. To some extent, these states resemble gas, liquid, and solid states of matter as an increasing function of order or interaction (Fig. 1A). Polypeptide chains neatly folded into cross β-sheets form amyloid structures that are no longer disordered, and which exhibit stabilizing interactions that make them very inert. Such structures are extraordinarily difficult to disassemble with small molecules once an insoluble amyloid is formed. Molecules that disassemble amyloid fibers have been reported (Haney et al., 2017; Li, Zhu, Manning-Bog, Di Monte, & Fink, 2004), however they are often not well-tolerated biologically or require large concentrations to function

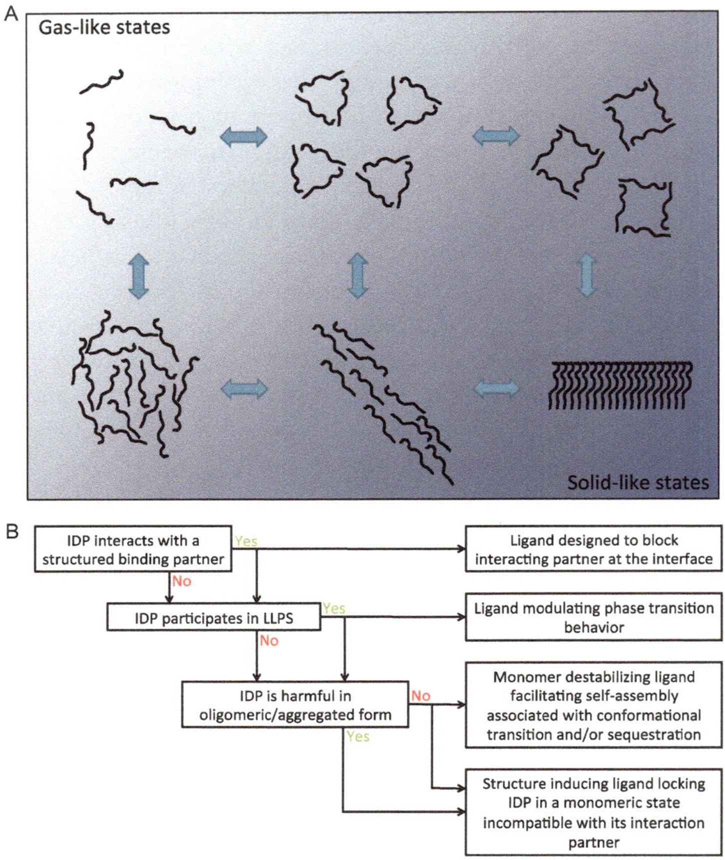

Fig. 1 (A) IDPs can exist in multiple forms and levels of assembly, ranging from loosely diffusing free monomers to highly organized and rigid aggregates. Several condensed phases may even coexist in the same system. (B) A flowchart illustrating different strategies to develop small molecules that manipulate IDPs based on their biophysical properties.

properly. In contrast, soluble IDPs that seed amyloid assembly are much more susceptible to therapeutic interventions.

The primary challenge posed by disordered sequences stems from their unique conformational flexibility and lack of a stable binding pocket to perform docking studies against. Furthermore, there is an unfavorable entropic cost of forming rigid complexes with disordered polypeptides that must be offset with relatively small enthalpic gains, which are characteristic of the low affinity interactions between IDPs (Metallo, 2010). This suggests that additional approaches beyond classical ligand screening should be considered to increase the likelihood of targeting a dynamic and demanding binding partner. Several techniques have been proposed to modulate the behavior of IDPs (Joshi & Vendruscolo, 2015; Marasco & Scognamiglio, 2015; Wang, Cao, Zhao, & Li, 2011). Properties of the disordered target must be considered first, while specific end goals will dictate which route is optimal for ligand development (Fig. 1B).

To avoid dealing with elusive binding sites, complimentary IDP partners can be screened for so that efforts can focus on exploiting known interaction interfaces. An example of a complementary IDP pair, where one of the targets was crystallized with a small-molecule inhibitor, is p53/MDM2 (Estrada-Ortiz, Neochoritis, & Dömling, 2016; Zhang, Golding, & Hardcastle, 2015). Upon interaction with various ligands, MDM2 forms a well-defined binding pocket that interacts with the tumor suppressor protein p53. This pocket became a natural target in designing small molecules that disrupted physiological p53/MDM2 interactions. However, when confronted with interactions primarily involving disordered regions, different strategies must be employed. One of the most common approaches is to use small molecules to stabilize IDP structures and thereby prevent them from further interactions. Locking proteins in particular conformations by pharmacological chaperones often makes them inaccessible for native binding partners or precludes their self-assembly into toxic oligomers and/or aggregates. Barecca et al. provides a thorough review of one of these agents that target the cellular form of the prion protein (Barreca, Iraci, Biggi, Cecchetti, & Biasini, 2018). αSyn has also been stabilized into a native and nonamyloidogenic state by two different small-molecule compounds (Lendel et al., 2009). Interestingly, in the latter case, self-association of the stabilizing compounds was key to their nonspecific inhibitory properties. It is of great importance to ensure that blocking one pathway does not open another equally harmful pathway. For example, past efforts to inhibit amyloid formation sometimes increased the population of cytotoxic oligomers, causing more damage than the amyloid fibers themselves (Sandberg et al., 2010). Similarly, stabilizing the central IDR in amyloid-beta monomers would likely inhibit its ability to form pathological amyloids, but may devastate its physiological behaviors. Therefore, great care must be taken to ensure that neighboring domains are not significantly perturbed in the presence of IDR-stabilizing ligands.

Strategies that promote disordered-to-ordered transitions of polypeptide chains often take advantage of the fact that unstructured regions can fluctuate around or between more ordered structures. Existence of these fluctuations

(Lee et al., 2012) is helpful for recognizing fragments that are susceptible to structure-inducing interactions with small molecules. Similarly, there have been attempts to identify common segments called molecular recognition features that fold upon binding (Mohan et al., 2006; Sharma, Bayarjargal, Tsunoda, Patil, & Sharma, 2018), which can greatly inform the development of efficient ligands.

The propensity of IDPs to participate in LLPS brings another set of challenges, but also enables new possibilities for small-molecule design. The selective concentration of proteins followed by their aggregation (Wegmann et al., 2018) or crystallization (Wang, Lomakin, McManus, Ogun, & Benedek, 2010) can contribute to pathological processes. Similarly, pathological processes can be triggered by the formation of "microreactors"—or droplets enriched in cofactors that initiate cytotoxic processes. Other pathogenic pathways involve the sequestration of functional proteins, thus rendering them unusable. Studies have recently found that the milieu of these droplets is quite different from the surrounding environment, enabling unique events such as the melting of DNA duplexes (Nott, Craggs, & Baldwin, 2016) or the restriction of proteolytic activities (Ge, Conley, Brandle, Truant, & Filipe, 2009), which likely interferes with the maintenance of intracellular proteins and nucleic acids. In more complex systems, disordered domains can also assemble into "selective phases." These phases function as a molecular filter, which has been documented in FG repeat-containing nucleoporins (Hülsmann, Labokha, & Görlich, 2012; Schmidt & Görlich, 2015). Another example of IDP gating behavior is the oligomeric assembly of IAPP on mitochondrial membranes, which are responsible for their subsequent cytotoxicity (Birol et al., 2018).

Distinct molecular environments within cellular microcompartments can be mimicked to localize metabolite (Frankel, Bevilacqua, & Keating, 2016) or macromolecule populations. This phenomenon can potentially direct externally introduced small molecules to their destination and increase their effective concentration. On the other hand, an elevated presence of enzymes inside of protein droplets could lead to the accelerated breakdown of ligands. Another inherent property of multiphase systems is the presence of boundaries. Small molecules with modular designs can be optimized to localize at an interface, and may become capable of screening proteins from interactions with the rest of the cell.

Building on these examples, it is possible to imagine the selective perturbation of IDPs without forming permanent complexes or structures, similar to inducing freezing-point depression in liquids by substoichiometric amounts of simple molecules. An example of "doping" protein structures to affect their phase behaviors is the LLPS of sickle cell hemoglobin, which is modulated by the addition of polyethylene glycol (Galkin, Chen, Nagel, Hirsch, & Vekilov, 2002). Successfully perturbing the internal dynamics of oligomeric IDP assemblies by small molecules has also been demonstrated for IAPP both in vitro in cell-derived vesicles and in vivo (Birol et al., 2018). A novel technique, known as "Diluted-FRET," which is described in detail in Section 8, indicates marked shifts between oligomeric populations as a result of small-molecule addition.

In all of these examples, the manipulation of IDPs by small molecules can potentially disrupt undesirable properties or the formation of pathological assemblies, highlighting the considerable need for approaches to effectively screen for IDP-specific ligands that modulate their physiological and pathological behaviors. The following sections provide a comprehensive overview of both the theoretical and experimental methods that can be utilized for this purpose.

3. THEORETICAL APPROACHES TO CHARACTERIZE AND TARGET PROTEIN DISORDER

In order to evaluate the efficacy of ligands designed to interact with unstructured targets, baseline fluctuations without ligands need to be established in order to identify quantitative controls. Theoretical models, including molecular dynamics (MD) simulations, are well suited for providing these controls since they can be used to collect structural observations and convert their statistics into thermodynamic signatures. MD simulations have also suggested that protein disorder results from a superposition of destabilizing forces (Levine, Larini, LaPointe, Feinstein, & Shea, 2015) and can be modulated in the presence of stabilizing environments or chemical agents. This is consistent with the hypothesis that IDPs inhabit multiple disordered states, capable of binding ligands of varying shape and size (Ma, Shatsky, Wolfson, & Nussinov, 2002). Wolynes and colleagues (Shoemaker, Portman, & Wolynes, 2000) have also suggested that IDPs can utilize a fly-casting mechanism in order to bind and capture targets while in their disordered states, followed by a slow folding as the target is reeled in. Taken together, these hypotheses suggest that IDPs functionalize their disorder, with the capability to inhabit conformations unique to their chemical environments. Thus, efforts to characterize IDPs using MD need to ensure that both unfolded and folded states (if any) are adequately sampled.

As noted earlier, IDPs typically exhibit a large net charge and low hydrophobicity (Uversky, 2017a; Uversky, Gillespie, & Fink, 2000; van der Lee et al., 2014), thus making it difficult to adopt stable three-dimensional structures. However, subgroups of disordered proteins with intermediate net charges and hydrophobicities can transiently sample well-defined secondary structures (van der Lee et al., 2014). These IDP ensembles are often the target of MD simulations, which can be used to inform the construction of IDP-targeting ligands. Therefore, theoretical and computational efforts are well suited for characterizing the thermodynamic shifts of IDPs in the presence of cosolvents, ligands, and other IDPs, which have historically been difficult to extract from experiments.

While the timescales of atomistic simulations are on the order of nano-to-microseconds, enhanced-sampling algorithms take advance of ergodicity, where the complete properties of a system can be evaluated either through longer timescales or through independent replicates that sample multiple folding pathways (Shields, 1983). As a result, theoretical techniques used to sample fluctuations in well-folded proteins can be repurposed for characterizing the preferred states of IDPs. An example is temperature replica-exchange molecular dynamics (T-REMD) simulations (Sugita & Okamoto, 1999), where replicates of a protein are heated and cooled in order to access states that are thermodynamically difficult to access. When this technique is successful, IDP ensembles can be mapped out at physiological temperatures and compared to those in the presence of ligands (Levine et al., 2015). Similar algorithms, such as Hamiltonian-REMD, can directly access folded and unfolded IDP states by modifying fractional α- and β-content (Comer, Roux, & Chipot, 2014; Laghaei, Mousseau, & Wei, 2011), however it can often be difficult to specify intermediate states without prior knowledge of the system. Efforts to sample more collective variables of IDPs have also been successful, including the use of metadynamics in well-tempered ensembles (Bonomi & Parrinello, 2010; Bussi, Gervasio, Laio, & Parrinello, 2006; Zerze, Miller, Granata, & Mittal, 2015). In many cases, however, force fields optimized for use on globular proteins tend to exhibit artifacts when modeling IDPs, often yielding a smaller radius of gyration than in experiments.

Detailed protocols have been published regarding the characterization of IDPs with T-REMD (Shea & Levine, 2016). This technique, broadly, goes through the following steps. First, a linear peptide chain is constructed from a disordered sequence using software such as Avogadro (Hanwell et al., 2012) or VMD (Humphrey, Dalke, & Schulten, 1996). This is because published structures are unlikely to exist in the absence of stabilizing agents, and because the structure is expected to be transient. After an explicit solvent is added, the system is energy-minimized and equilibrated for 10–20 ns to ensure that there are multiple states to seed REMD simulations. Next, a set of temperatures can be calculated from the REMD temperature generator at http://folding.bmc.uu.se/remd/, which provides a set of starting temperatures for optimizing REMD. Aside from the system size, one only needs to specify a preferred exchange rate between replicas, which is often near 25%, and the range of temperatures to sample, often between 290 and 500 K. Replicas of the equilibrated system are then heated to these temperatures for another 10–20 ns to stabilize their folding behaviors, with the goal of sampling folded and unfolded states at low and high temperatures, respectively. Multiple states from the equilibration trajectory can be used to seed these replicas to further diversify the starting structures used to populate REMD. Then, the true exchange rate must be measured between replicas after exchanges are turned on, with an attempt to swap states occurring at least every 3 ps. If the exchange rate is much higher than expected, then the difference in temperatures between specific pairs must be increased in order to lower their interconversion. Conversely, if the exchange rate is much lower than expected, the difference in temperatures between pairs much be decreased. The goal is to heuristically establish an ensemble that exchanges states as closely as possible to the ideal exchange rate. After a satisfactory ensemble is established, a production run can be carried out for at least 200 ns or greater, which is the time required to fold small disordered peptides (Levine et al., 2015). Convergence of T-REMD can be measured using multiple metrics, including the time required for the highest and lowest temperature replicates to make a round-trip journey (Prakash, Barducci, & Parrinello, 2011) or when the potential between IDP and solvent converges to a minimal value.

Analysis of T-REMD ensembles only requires a detailed inspection of one replicate, usually at room temperature (or at any temperature one wants to investigate). This is because each replicate samples states that is informed by the ensemble as a whole, despite being limited to a single temperature. This amounts to a Markov-Chain Monte Carlo model, in that each replicate represents a Markov state with a finite transition probability between states. The incorporation of a Monte Carlo step ensures that some transitions between unfavorable states remain, thereby reducing the likelihood of getting caught in a local free energy basin. Clustering of IDP structures in the analysis replicate is usually based on the Daura algorithm

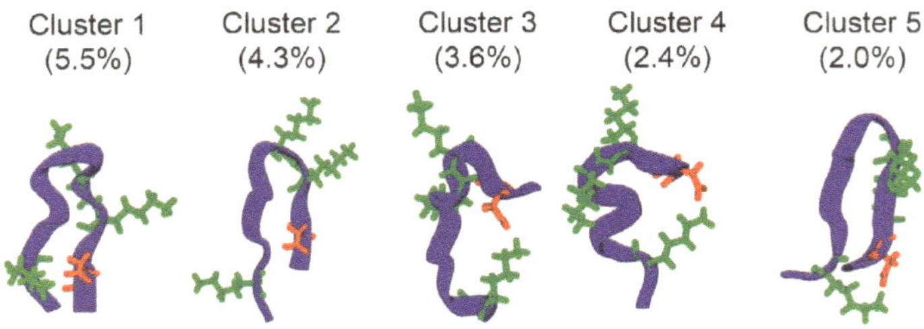

Fig. 2 T-REMD simulations provide a ranked list of preferred IDP states that are transiently sampled, forming the basis of their thermodynamic signatures. A normalized distribution of discrete IDP clusters can also be used to inform the free energy differences between states by taking the natural log of their probabilities, which can be done with and without ligands present. Here, *blue IDP backbones* are populated by numerous charged amino acids, where *green sidechains* represent lysine and *red sidechains* represent aspartic acid. *Figure adapted with permission from Shea, J. E., & Levine, Z. A. (2016). Studying the early stages of protein aggregation using replica exchange molecular dynamics simulations.* Methods in Molecular Biology, 1345, *225–250.*

(Daura, Suter, & van Gunsteren, 1999), which aggregates structures using an RMSD less than 1.4 Å. These clusters then provide a histogram of preferred IDP states (Fig. 2) that can be normalized and turned into free energy differences, which is the basis of their thermodynamic signatures (Levine et al., 2015). The introduction of chemical denaturants such as urea reinforces extended IDP populations, while protein stabilizers such as trimethylamine-*N*-oxide enhance compact states (Levine et al., 2015). These simulations highlight the use of multiple thermodynamic metrics for characterizing IDP ensembles that can be compared directly back to experiments. These include average end-to-end distances, radii of gyration, the frequency of transient secondary structures, and the number of unique protein conformations, all of which can be directly compared to fluorescence resonance energy transfer (FRET), SAXS, circular dichroism (CD), and NMR experiments, respectively. As a whole, these results reinforce the notion that ligands can perturb IDPs in subtle ways that affect their heterogeneities, especially in the context of complex physiological surfaces (Birol et al., 2018).

Theoretical models have also highlighted the increasingly important role of water in mediating protein disorder, which can greatly inhibit backbone hydrogen bonding (Best, Zheng, & Mittal, 2014; Schiro et al., 2015). Using MD simulations, water was recently shown to mediate single-component LLPS in simple polyampholytes (Rauscher & Pomès, 2017) and likely plays a role in biological LLPS where polyelectrolytes spontaneously form coexisting dilute and dense phases, similar to the coacervation of polymers (Overbeek & Voorn, 1957; Pak et al., 2016). Replica-exchange simulations have helped identify a number of water soluble conformations adopted by disordered oligomers (Rauscher & Pomès, 2017), and multiple novel force fields have been developed to increase the interaction between IDPs and water (Huang et al., 2017). This was due, in part, to inconsistencies in earlier force fields that artificially inflated the number of folded states IDPs could sample (Rauscher et al., 2015). Efforts to mitigate these problems involved multiple strategies, including manually patching the van der Waals interaction between IDPs and water (Best et al., 2014; Zerze, Best, & Mittal, 2015; Zerze, Mullen, Levine, Shea, & Mittal, 2015) though a complete reparameterization of existing force fields was also contemplated. These efforts are complicated, however, due to the simplicity of water in MD, which often adopts rigid tetrahedral structures containing three simple point charges (Ferguson, 1995). Four- and five-site water models such as the TIP4P-2005 force field exhibits considerably improved dynamics (Abascal & Vega, 2005), however the incorporation of these waters into modern force fields would require a complete reparameterization. Thus, bootstrapping of van der Waals interactions between water and IDPs is highly preferred and is heavily utilized in the literature (Henriques & Skepö, 2016), though there is evidence that older force fields (e.g., CHARMM22) may be better suited for IDP simulations (Rauscher et al., 2015; Weber & Uversky, 2017).

Simulations have also highlighted the effects of posttranslationally modified IDPs and their ability to outperform their wild-type analogs (Levine et al., 2016; Lyons, Gandhi, & Mancera, 2014; Ma, Wei, Zhen, & Nussinov, 2016). Small amino acid changes, including the conversion of tyrosine to dihydroxyphenylalanine (DOPA), had a significant impact on the available states that IDPs sampled (Levine et al., 2016). These simulations also found that while PTMs modulate the conformations of IDPs, they can also enhance binding to both organic and inorganic aqueous interfaces that are typically difficult to target due to competition with hydrating water molecules. Thus, as mentioned before, the roles of ligands are complex since they can inadvertently impart a gain-of-function onto disordered targets or destabilize the behaviors of IDRs.

As a whole, theoretical construction, characterization, and optimization of small-molecule ligands can be highly synergistic to experimental efforts that seek to efficiently target IDPs and IDRs. Decreasing hardware costs and increasingly accessible computing platforms have made the construction of ligand libraries easy to generate and cost efficient, especially when compared to the high costs of purifying and characterizing ligands in experiments. However, simulations require ongoing validation from experiments and should be used to inform and interpret experimental observations, rather than as a standalone method for ligand development. Characterization of IDP ensembles remains difficult, and simulations have helped fill this gap in order to identify the types of secondary structures that might be induced by potential ligands. However, theoretical models lack some of the necessary details required to describe IDP folding, though models continue to improve and are best coupled to experiments. For further information on theoretical efforts to characterize and target IDPs, the reader is encouraged to read a thorough review by Baker and Best (2014), which provides a comprehensive overview of the literature to date.

Using REMD, one can pair disordered targets with a diverse set of small peptides in order to quantify how specific side chains effect IDP folding and behavior. For example, if one finds that a disordered target populates fundamentally new states in the presence of polar amino acids, then more sophisticated ligands can be constructed (perhaps with a nonpeptidic backbone) that contain similar residues. REMD can be repeated and refined until the optimized ligands induce the desired effect across the most dominant protein clusters. The use of peptide intermediates as ligand substitutes offers a simple, and biophysically intuitive approach to link IDP behaviors to fundamental interactions between amino acids. Alchemical free energy perturbations and umbrella sampling can also be employed to inform the energetics of ligand binding. However, many ligands are capable of modulating IDP folding and aggregation in the absence of stable binding complexes.

4. EXPERIMENTAL IDENTIFICATION AND EVALUATION OF BINDING PARTNERS

Given historical omissions of systems involving intrinsically disordered structures, and the need to validate molecular models in multilevel studies, theoretical approaches alone should not be used to fully describe these poorly understood processes. Robust experimental methods to probe the transient dynamics of proteins and ligands at different levels are necessary, including at monomeric, oligomeric, and mesoscopic scales that capture the dynamics of assemblies and colloids. Integrating these observations enables ligands to be effective, while minimizing unwanted effects to other components. Experimental characterization of IDPs includes deducing their molecular weight, geometry, compactness, flexibility, mutual orientation, stoichiometry, association/dissociation rates, and diffusion properties. Many methods also measure the propensity of IDPs to aggregate as an indicator of their self-affinity. The end product of aggregation—higher-order assemblies—can be detected directly via electron microscopy or by exogenous probes such as thioflavin T (ThT) dyes. Inhibition of IDP aggregation in the presence of ligands is an indicator of their interaction strength. However, these approaches also have shortcomings. One is the requirement that the IDP is capable of aggregation. Second, external probes require a molecular recognition pattern to be present that does not cross-interact with ligands, suggesting that additional methods are necessary for evaluating generalized IDPs.

Other techniques include SAXS, which does not require chemical labeling. This method provides information about the radius of gyration, shape, maximum dimension, and molecular weight of the IDP in solution. Data collection can be automated and the analysis is not complicated as long as the sample is monodisperse. Unfortunately, this is not the case for many systems containing IDPs, but the ensemble approach and potential comparisons to modeling often lead to valuable insights into disordered states with and without ligands present (Kachala, Valentini, & Svergun, 2015; Kikhney & Svergun, 2015). Examples include a study of αSyn oligomers stabilized by the small-molecule flavonoid baicalein (Hong, Fink, & Uversky, 2008). Less sophisticated tools for the quick determination of macromolecular compactness in multicomponent solutions include chromatographic or electrophoretic separation. Both size-exclusion chromatography (Uversky, 2012) and gel electrophoresis (Tantos & Tompa, 2012) have also been used to study IDPs. However, many of the techniques described earlier offer only high-level and low-resolution snapshots.

To overcome these obstacles, we discuss two highly versatile spectroscopic methods and their abilities to extract molecular details from heterogeneous IDP assemblies. These techniques include NMR and single-molecule fluorescence measurements.

5. NMR TECHNIQUES

NMR spectroscopy has several distinct advantages over other techniques for evaluating IDPs and requires no external or attached molecular probes. Depending on the particular size of the interacting species, isotopic labeling may be advantageous, especially when residue- or atomic resolution is desired. NMR methods cover a wide range of timescales to investigate dynamic molecular events from picoseconds to hours (Kleckner & Foster, 2011) and can report transient, underpopulated states (Sekhar & Kay, 2013) often considered invisible by other methods. NMR can also probe interactions beyond IDPs and ligands, including those between IDP monomers and oligomers. The resolution of these studies can be distilled down to the residue level, as shown for Aβ (Fawzi, Ying, Torchia, & Clore, 2010). Saturation transfer difference (STD) techniques have also been employed to study the interactions between Aβ and peptidomimetic inhibitors. This technique allows epitope mapping indicating if inhibitors bind to oligomers but not monomers (Kaffy et al., 2016). The same method has helped localize binding epitopes in the ligand-stabilizing insulin oligomer, thus preventing IDP fiber formation (Banerjee et al., 2013).

Insight into bound ligand conformations can be obtained via transferred nuclear Overhauser effect (trNOE) (Post, 2003), while transferred residual dipolar coupling (RDC) provides orientation information (Chen & Tjandra, 2012). Both techniques have been used to study the inhibition of Aβ oligomerization by small peptides (Chen, Krause, & Reif, 2005) and molecules (Benaki et al., 2009). Transferred NOE also reveals interaction modes between amyloidogenic αSyn and its inhibitor dye Congo Red, especially in the micellar form. These results emphasize that IDP folding and aggregation is based on multiple transient interactions at big and small length scales, rather than solely on monomeric or dimeric behaviors (Maltsev, Grishaev, & Bax, 2012).

Other NMR methods focus solely on the disordered target, such as heteronuclear single quantum coherence spectroscopy (HSQC), which tends to highlight interactions between residues in monomers and small oligomers. In one study by Kumar et al., HSQC shifts were used to identify the interaction site of Aβ in the presence of a novel self-assembly inhibitor (Kumar, Henning-Knechtel, Chehade, Magzoub, & Hamilton, 2017). Dark-state exchange saturation transfer (DEST), described in detail by Fawzi, Ying, Torchia, and Clore (2012), has also provided a better understanding of IDP self-assembly and their elusive dynamic states, further identifying disordered regions for small molecules to target. Pithadia and coworkers evaluated various ligands of natural origin (polyphenols and their derivatives) based on their activity against IAPP toxicity (Pithadia, Brender, Fierke, & Ramamoorthy, 2016) demonstrating that data obtained from NMR is well suited for designing ligands to target IDPs. Ahmed and Melacini comprehensively reviewed solution-state NMR techniques using Aβ aggregates and the small-molecule inhibitor EGCG (Ahmed & Melacini, 2018). In addition to previously discussed approaches, the authors assessed the solvent accessibility of Aβ oligomers modified by inhibitors. The same system has been also characterized by solid-state NMR (Lopez del Amo et al., 2012), which is particularly useful for analyzing heterogeneous systems containing soluble and insoluble oligomers.

6. FLUORESCENT TECHNIQUES

The application of computational methods and conventional ensemble techniques such as NMR, CD, and fluorescence spectroscopy has brought about significant advancement in the understanding of IDP structural and functional properties (Wright & Dyson, 2009). Most common strategies using fluorescence for probing protein structure include single-molecule and super-resolution techniques, which directly probe the distribution of molecular events and can reveal important mechanisms that otherwise remain obscured. Single-molecule fluorescent approaches have allowed the ability to directly detect molecular processes without the loss of information due to ensemble averaging and the capacity to resolve complex structural distributions and dynamics (Iljina et al., 2016; Schuler & Eaton, 2008; Trexler & Rhoades, 2009). Combined with the sensitivity of fluorescence detection, the use of single-molecule fluorescence spectroscopy in the study of IDPs has provided important insights into the conformational properties of the disordered ensemble, and how these features are altered upon binding with cellular partners (Lee, Moran-Gutierrez, & Deniz, 2015; Melo et al., 2016). Detailed insights into the structural landscapes of IDPs have proven important for understanding the link between conformation and IDP functional activities, in addition to its pathological roles. According to the conformational nature of IDPs, these approaches have been informative in revealing additional techniques that move the design of small-molecule inhibitors forward (Fig. 1B).

Fluorescent approaches offer unique capabilities that are particularly well suited to studies of IDP structure and function. Single-molecule fluorescence resonance energy transfer (smFRET) measurements can provide detailed information about long-range structural properties of IDPs, and the coupling between folding and ligand binding (Ferreon, Gambin, Lemke, & Deniz, 2009; Melo et al., 2016). FRET signals arise upon the interaction of two fluorophores, whereby an excited donor probe transfers its excitation energy to a nearby acceptor probe (Schuler, 2013). This transfer of energy results in both a quenching of the fluorescence of the donor and an appearance of a fluorescence emission spectra of the acceptor.

Importantly, the efficiency of the energy transfer is strongly dependent on the distance between the donor and acceptor (in the range between 1 and 10 nm), thereby enabling FRET to effectively quantify molecular associations and conformation (Schuler & Eaton, 2008). Fluorescence correlation spectroscopy (FCS) is a near single-molecule resolution method to measure and analyze fluorescence fluctuations from molecules diffusing freely in solution (Haustein & Schwille, 2007). Fluorescence intensity data are subjected to correlation analysis to identify molecular events ranging from molecular diffusion to conformational dynamics at very rapid timescales. FCS measurements provide important information about conformational fluctuations and associated timescales for proteins and other biomolecules. This approach has been used for determining binding affinities of IDPs with binding partners as well as with small-molecule inhibitors (Kumar et al., 2016; Rhoades, Ramlall, Webb, & Eliezer, 2006).

To monitor in vitro amyloid fibril formation, a fluorescent probe, ThT, is commonly used. Upon binding to amyloid fibrils, ThT produces a strong fluorescence signal at approximately 482 nm when excited at 450 nm (Naiki, Higuchi, Hosokawa, & Takeda, 1989). The mechanism of fluorescence enhancement upon binding to amyloid has been attributed to the rotational immobilization of the central C—C bond connecting the benzothiazole and aniline rings (Srivastava et al., 2010; Voropai et al., 2003). It is commonly accepted that ThT binds to the side chain channels along the long axis of amyloid fibrils (Krebs, Bromley, & Donald, 2005).

7. Diluted-FRET

Single-molecule conditions are typically very diffuse and may reside below the critical concentration for monitoring IDP oligomerization. Studies using smFRET to report on aggregation, such as those probing the tau protein, require the use of heparin, a polyanion cofactor that accelerates aggregation at single-molecule concentrations (Shammas et al., 2015). However, the physiological relevance of this interaction in cells is unclear (Fichou, Vigers, Goring, Eschmann, & Han, 2018). NMR conditions are typically performed at high concentrations that are highly prone to aggregation, restricting its use in studying lower-order species formed in the early stages of assembly. Additionally, neither single-molecular approaches nor NMR allows for comparisons between intracellular oligomers and those formed in solution biophysical measures.

To overcome this issue, a more recent approach utilizing single-molecule concentrations of fluorescent probes (as low as 100 nM donor protein and 100 nM acceptor protein) was described that monitors the conformational fluctuations of IDPs using a high concentration of unlabeled proteins (donor:acceptor:unlabeled, 1:1:500) (Birol et al., 2018). This approach is well suited for characterizing large oligomeric assemblies and was demonstrated for IAPP, an IDP associated with type II diabetes (Hoppener, Ahren, & Lips, 2000). This structure-based fluorescence approach, termed diluted-FRET, was developed in order to focus on large, homooligomeric assemblies of IAPP (Birol et al., 2018). It was shown that oligomers contained up to hundreds of IAPP, populating several discrete states that were not canonically amyloid. A key advantage of this study was to apply diluted-FRET on biologically relevant membrane systems, namely on giant plasma membrane vesicles (GPMVs) (Schlamadinger & Miranker, 2014; Sezgin et al., 2012), recapitulating the oligomeric species found on the surface of biologically relevant membranes (Fig. 3A). Moreover, by performing diluted-FRET inside cells the molecular-scale detail

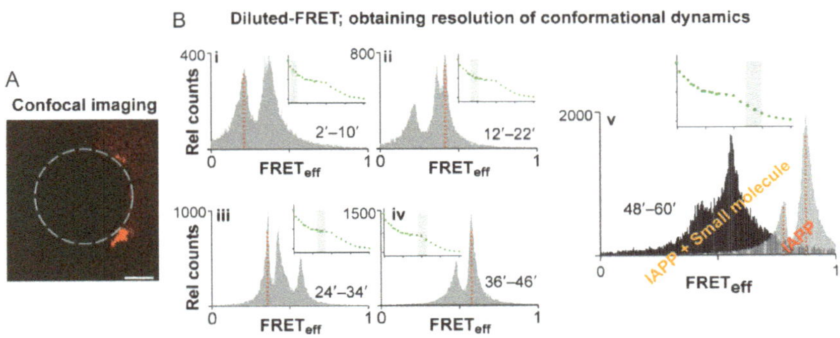

Fig. 3 Conformational dynamics depicted by diluted-FRET. (A) Confocal imaging of oligomeric membrane-bound protein. Image showing IAPP labeled with Alexa 594 bound to a GPMV. (B) Conformation changes within IAPP oligomers obtained by intermolecular diluted-FRET. Diluted-FRET histograms summed over a minimum of three GPMVs, with five optical sections for each GPMV over the indicated periods of time. For time points 48′–60′, IAPP in the presence of a small-molecule inhibitor is shown, capturing a distinct conformation from those formed in the absence of small-molecule inhibitors. Pixel statistics correspond to summation across the indicated time period. The *red dotted lines* indicate the different conformations sampled. (*Inset*) Schematic of the membrane leakage profile collected under matched conditions. *Gray rectangles* show the time period over which $FRET_{eff}$ is summed in the associated histogram. *Figure adapted with permission from Birol, M., Kumar, S., Rhoades, E., & Miranker, A. D. (2018). Conformational switching within dynamic oligomers underpins toxic gain-of-function by diabetes-associated amyloid.* Nature Communications, *9(1), 1312.*

obtained from GPMVs was comparable to intracellular assemblies of IAPP in association with depolarized mitochondrial membranes. Below, key findings using diluted-FRET are discussed that are otherwise neglected by other approaches.

Diluted-FRET can also report on the dynamics of IDPs within oligomeric assemblies. Using both GPMVs and cells, the fact that a FRET signal is obtained even in the presence of high concentrations of unlabeled protein, relative to labeled protein, signifies that the oligomeric species of IAPP is internally dynamic. Additionally, diluted-FRET reports on mechanistic models for oligomer growth. Possible mechanisms of oligomer growth may include monomer–monomer, monomer–dimer, monomer–small-N oligomer ($N < 10$), small-N oligomer–small-N oligomer, monomer–large-N oligomer, etc., which results in heterogeneous assemblies. By studying the progression in FRET efficiency over time, mechanisms driving oligomer growth can be identified.

Deducing the conformational heterogeneity of a system by diluted-FRET can greatly inform the design of small-molecule inhibitors. In addition, it is now clear that small-molecule strategies targeting oligomeric ensembles of IDPs are becoming increasingly necessary. By using diluted-FRET in the presence of small-molecule inhibitors, the conformation of IDPs responsible for other behaviors, such as membrane permeation or cellular toxicity can be readily revealed. As shown for IAPP, a small-molecule inhibitor was able to constrain the protein at a specific conformation. At this conformation IAPP was unable to induce large pore formation on the membrane and cause cellular toxicity (Fig. 3B). As such, diluted-FRET can be used as a screen to establish effective small-molecule inhibitors for oligomeric forms of IDPs that otherwise are not achievable by conventional methods limited to the monomeric form of IDPs.

Diluted-FRET requires the labeling of a batch of protein with a donor and another batch with an acceptor probe. The most notable limitations of diluted-FRET, like most fluorescent approaches, is the dependence on fluorescent probes that may result in signal contamination due to photophysical effects or that may affect protein function. As such, vigorous characterization of labeling, labeling positions, and fluorophores is required for the IDPs in question. It is important to consider a variety of different FRET donor and acceptor pairs, and to perform the experiments at multiple donor:acceptor: unlabelled protein ratios. A rigid protein conformation, e.g., protein fibrils in solution and in the presence of donor and acceptor probes, should be designed to establish a control structure lacking dynamic properties.

Additionally, when using this tool in cells, the possibility for specific protein and/or cell types to degrade the fluorescent probe must be accounted for. In order to obtain statistically significant image-based diluted-FRET measurements, a sufficiently large dataset must be utilized to draw conclusions. Potential improvements might include adapting diluted approaches with lifetime measurements that can neglect data contamination due to fluorescence bleed through and have the ability to infer from smaller datasets as this approach is based entirely on the lifetime of each pixel in the extracted images can be accounted for.

Despite these limitations, the information obtained from diluted-FRET can report on the conformational and mechanistic details of oligomer formation in complex systems and environments, which may not otherwise be available through conventional approaches. Most importantly, this tool is accessible to many as it can be performed on a conventional confocal microscope. Diluted-FRET can also be monitored across timescales in which the functional properties of proteins can be simultaneously studied. For example, it has been used to monitor IAPP-induced membrane leakage and cellular toxicity (Fig. 3B) (Birol et al., 2018) and has provided insight into the oligomeric conformation of proteins that induce these functions. Given that not all higher-order IDP assemblies are pathological, the ability to differentiate heterogeneous IDP oligomers is attractive and can inform a wide range of applications. These applications include designing and evaluating the effects of small-molecule inhibitors on dynamic protein assemblies, and span from systems that form soluble oligomers to gels and liquid-like organelles (Elbaum-Garfinkle et al., 2015; Fung, Birol, & Rhoades, 2018; Uversky, 2017b).

Extensions of this approach with other fluorescent methods, including FCS and fluorescence cross-correlation spectroscopy (FCCS) (Bacia, Kim, & Schwille, 2006), can also be used to inform the number of molecules within oligomers, and can provide insights into the dynamic nature of IDPs. Finally, a three-color diluted-FRET approach can be implemented by introducing an orthogonal labeling site via a genetically encoded unnatural amino acid into a third protein, or into an additional compartment of interest (Milles, Koehler, Gambin, Deniz, & Lemke, 2012).

8. ADDITIONAL CONSIDERATIONS

An important consideration when searching for ligands that target IDPs is the ability to test multiple compounds over short periods of time.

Whether one wishes to identify preliminary candidates in MD simulations, or test variations of an attractive candidate, it remains desirable to run many samples in quick succession or in parallel. Some of the NMR techniques, described in Section 6, allow simultaneous assessment of interactions with several compounds at once (Meyer & Peters, 2003). Other screening methods have been developed and tested on a variety of IDPs. Ligands inhibiting IAPP aggregation have been analyzed using in vivo assays based a β-lactamase tripartite fusion constructs (Saunders et al., 2016). Both IAPP and Aβ were used as targets in studies combining ion mobility spectrometry with mass spectrometry (IMS-MS), in order to identify interacting protein species, detect small molecules binding, and characterize inhibition modes (Young et al., 2015). Another study describes efforts to identify Aβ-recognizing aptamers through systematic evolution of ligands by exponential enrichment (SELEX)—which utilizes a combinatorial in vitro molecular-evolution technique that selects from libraries of oligonucleotide sequences (Rahimi, 2018). Conformations of αSyn and its associated aggregation propensities were successfully modulated by small molecules identified through second harmonic generation-based screening (Moree et al., 2015). This same IDP was the focus of a combined computational–experimental approach, yielding a small-molecule compound with substantial biological activity (Toth et al., 2014). For additional information about experimental methods to probe interactions between IDPs and small molecules, including low- and high-throughput methods not described in this chapter, a review by Heller et al. is strongly recommended (Heller, Aprile, & Vendruscolo, 2017).

Apart from developing techniques to evaluate small molecules, it is also important to consider strategies devoted to generating novel ligands. Significant efforts have been directed to repurpose existing bioactive compounds and natural products. Joshi and coworkers proposed how to extract fragments from structures identified in the literature, and then cross-reference small molecules from large databases in order to generate more specific IDP libraries for Aβ, αSyn, or Tau (Joshi et al., 2016). Promising results have been achieved using modular scaffolds with interchangeable functionalities when targeting IDPs with small molecules that extend beyond known bioactive structures. For example, Kumar and coworkers developed a series of rigid (Kumar, Henning, Magzoub, & Hamilton, 2018) and flexible (Kumar et al., 2016) backbones functionalized to target IDPs such as IAPP or Aβ. Control over the conformational flexibility of the scaffold imparted additional benefits, such as the ability to target IDPs in different environments like membranes or liquid protein phases.

Finally, in the presence of complex or multicomponent environments, there have been attempts to develop multiple ligands that target the same IDP using synergistic effects (where each ligand has its own function and potentially distinct interaction site). Insulin, for example, is a good inhibitor of IAPP aggregation in solution, but this behavior is lost in the presence of lipid bilayers. However, simultaneous addition of the peptidomimetic IS5, which reduces IAPP binding to membranes, resulted in reduced cytotoxicity (Hebda, Saraogi, Magzoub, Hamilton, & Miranker, 2009). Multiple functionalities in these scenarios could be delivered independently by disparate small molecules, or perhaps combined using a skillfully designed scaffold.

9. CONCLUSIONS

In this chapter, we introduced IDPs, described their features, and discussed how they diverge from structured proteins in the context of ligands that potentially modulate their behaviors. We also presented a variety of theoretical and experimental methods that are well suited for characterizing IDPs accompanied by examples of their practical implementation. Taken together, these examples highlight the ability of IDPs to function at multiple levels, from monomeric solutions to membrane-bound oligomers. These approaches enable new opportunities for small molecules to target unique IDP behaviors, including interconverting ensembles between ordered protein states. In particular, we encourage readers to take advantage of single-molecule fluorescence spectroscopy, diluted-FRET, exchange-based NMR methods, and T-REMD simulations, since they are particularly useful for characterizing dynamic and heterogeneous protein assemblies. However, the provided examples demonstrate that no single approach is sufficient to comprehensively characterize disordered targets. Instead, combining several approaches is recommended when attempting to characterize IDPs, especially in the context of small-molecule ligands that modulate disordered protein behaviors.

REFERENCES

Abascal, J. L., & Vega, C. (2005). A general purpose model for the condensed phases of water: TIP4P/2005. *The Journal of Chemical Physics*, *123*(23), 234505.

Ahmed, R., & Melacini, G. (2018). A solution NMR toolset to probe the molecular mechanisms of amyloid inhibitors. *Chemical Communications (Cambridge, England)*, *54*(37), 4644–4652.

Ambadipudi, S., & Zweckstetter, M. (2016). Targeting intrinsically disordered proteins in rational drug discovery. *Expert Opinion on Drug Discovery*, *11*(1), 65–77.

Asthana, S., Mallick, B., Alexandrescu, A. T., & Jha, S. (2018). IAPP in type II diabetes: Basic research on structure, molecular interactions, and disease mechanisms suggests potential intervention strategies. *Biochimica et Biophysica Acta (BBA)—Biomembranes*, *1860*(9), 1765–1782.
Bacia, K., Kim, S. A., & Schwille, P. (2006). Fluorescence cross-correlation spectroscopy in living cells. *Nature Methods*, *3*(2), 83–89.
Baker, C. M., & Best, R. B. (2014). Insights into the binding of intrinsically disordered proteins from molecular dynamics simulation. *Wiley Interdisciplinary Reviews: Computational Molecular Science*, *4*(3), 182–198.
Banerjee, V., Kar, R. K., Datta, A., Parthasarathi, K., Chatterjee, S., Das, K. P., et al. (2013). Use of a small peptide fragment as an inhibitor of insulin fibrillation process: A study by high and low resolution spectroscopy. *PLoS One*, *8*(8), e72318.
Barreca, M. L., Iraci, N., Biggi, S., Cecchetti, V., & Biasini, E. (2018). Pharmacological agents targeting the cellular prion protein. *Pathogens*, 7(1), 27.
Benaki, D., Stathopoulou, K., Leondiadis, L., Ferderigos, N., Pelecanou, M., & Mikros, E. (2009). Detection of interactions of the beta-amyloid peptide with small molecules employing transferred NOEs. *Journal of Peptide Science*, *15*(6), 435–441.
Best, R. B., Zheng, W. W., & Mittal, J. (2014). Balanced protein–water interactions improve properties of disordered proteins and non-specific protein association. *Journal of Chemical Theory and Computation*, *10*(11), 5113–5124.
Birol, M., Kumar, S., Rhoades, E., & Miranker, A. D. (2018). Conformational switching within dynamic oligomers underpins toxic gain-of-function by diabetes-associated amyloid. *Nature Communications*, *9*(1), 1312.
Bodner, C. R., Maltsev, A. S., Dobson, C. M., & Bax, A. (2010). Differential phospholipid binding of alpha-synuclein variants implicated in Parkinson's disease revealed by solution NMR spectroscopy. *Biochemistry*, *49*(5), 862–871.
Bonomi, M., & Parrinello, M. (2010). Enhanced sampling in the well-tempered ensemble. *Physical Review Letters*, *104*(19), 190601.
Brangwynne, C. P., Tompa, P., & Pappu, R. V. (2015). Polymer physics of intracellular phase transitions. *Nature Physics*, *11*, 899.
Breydo, L., Wu, J. W., & Uversky, V. N. (2012). Alpha-synuclein misfolding and Parkinson's disease. *Biochimica et Biophysica Acta*, *1822*(2), 261–285.
Bulic, B., Pickhardt, M., & Mandelkow, E. (2013). Progress and developments in tau aggregation inhibitors for Alzheimer disease. *Journal of Medicinal Chemistry*, *56*(11), 4135–4155.
Bussi, G., Gervasio, F. L., Laio, A., & Parrinello, M. (2006). Free-energy landscape for beta hairpin folding from combined parallel tempering and metadynamics. *Journal of the American Chemical Society*, *128*(41), 13435–13441.
Caillon, L., Hoffmann, A. R., Botz, A., & Khemtemourian, L. (2016). Molecular structure, membrane interactions, and toxicity of the islet amyloid polypeptide in type 2 diabetes mellitus. *Journal of Diabetes Research*, *2016*, 5639875.
Chen, Z., Krause, G., & Reif, B. (2005). Structure and orientation of peptide inhibitors bound to beta-amyloid fibrils. *Journal of Molecular Biology*, *354*(4), 760–776.
Chen, K., & Tjandra, N. (2012). The use of residual dipolar coupling in studying proteins by NMR. *Topics in Current Chemistry*, *326*, 47–67.
Comer, J., Roux, B., & Chipot, C. (2014). Achieving ergodic sampling using replica-exchange free-energy calculations. *Molecular Simulation*, *40*(1–3), 218–228.
Daura, X., Suter, R., & van Gunsteren, W. F. (1999). Validation of molecular simulation by comparison with experiment: Rotational reorientation of tryptophan in water. *The Journal of Chemical Physics*, *110*(6), 3049–3055.
Dong, X., Qiao, Q., Qian, Z., & Wei, G. (2018). Recent computational studies of membrane interaction and disruption of human islet amyloid polypeptide: Monomers, oligomers and protofibrils. *Biochimica et Biophysica Acta*, *1860*(9), 1826–1839.

Dundr, M. (2012). Nuclear bodies: Multifunctional companions of the genome. *Current Opinion in Cell Biology*, *24*(3), 415–422.

Elbaum-Garfinkle, S., Kim, Y., Szczepaniak, K., Chen, C. C., Eckmann, C. R., Myong, S., et al. (2015). The disordered P granule protein LAF-1 drives phase separation into droplets with tunable viscosity and dynamics. *Proceedings of the National Academy of Sciences of the United States of America*, *112*(23), 7189–7194.

Estrada-Ortiz, N., Neochoritis, C. G., & Dömling, A. (2016). How to design a successful p53-MDM2/X interaction inhibitor: A thorough overview based on crystal structures. *ChemMedChem*, *11*(8), 757–772.

Fawzi, N. L., Ying, J., Torchia, D. A., & Clore, G. M. (2010). Kinetics of amyloid beta monomer-to-oligomer exchange by NMR relaxation. *Journal of the American Chemical Society*, *132*(29), 9948–9951.

Fawzi, N. L., Ying, J., Torchia, D. A., & Clore, G. M. (2012). Probing exchange kinetics and atomic resolution dynamics in high-molecular-weight complexes using dark-state exchange saturation transfer NMR spectroscopy. *Nature Protocols*, 7(8), 1523–1533.

Ferguson, D. M. (1995). Parameterization and evaluation of a flexible water model. *Journal of Computational Chemistry*, *16*(4), 501–511.

Ferreon, A. C., Gambin, Y., Lemke, E. A., & Deniz, A. A. (2009). Interplay of alpha-synuclein binding and conformational switching probed by single-molecule fluorescence. *Proceedings of the National Academy of Sciences of the United States of America*, *106*(14), 5645–5650.

Fichou, Y., Vigers, M., Goring, A. K., Eschmann, N. A., & Han, S. (2018). Heparin-induced tau filaments are structurally heterogeneous and differ from Alzheimer's disease filaments. *Chemical Communications (Cambridge)*, *54*(36), 4573–4576.

Fisher, C. K., & Stultz, C. M. (2011). Constructing ensembles for intrinsically disordered proteins. *Current Opinion in Structural Biology*, *21*(3), 426–431.

Frankel, E. A., Bevilacqua, P. C., & Keating, C. D. (2016). Polyamine/nucleotide coacervates provide strong compartmentalization of Mg^{2+}, nucleotides, and RNA. *Langmuir*, *32*(8), 2041–2049.

Fung, H. Y. J., Birol, M., & Rhoades, E. (2018). IDPs in macromolecular complexes: The roles of multivalent interactions in diverse assemblies. *Current Opinion in Structural Biology*, *49*, 36–43.

Fusco, G., Chen, S. W., Williamson, P. T. F., Cascella, R., Perni, M., Jarvis, J. A., et al. (2017). Structural basis of membrane disruption and cellular toxicity by alpha-synuclein oligomers. *Science*, *358*(6369), 1440–1443.

Fusco, G., Sanz-Hernandez, M., & De Simone, A. (2018). Order and disorder in the physiological membrane binding of alpha-synuclein. *Current Opinion in Structural Biology*, *48*, 49–57.

Galkin, O., Chen, K., Nagel, R. L., Hirsch, R. E., & Vekilov, P. G. (2002). Liquid–liquid separation in solutions of normal and sickle cell hemoglobin. *Proceedings of the National Academy of Sciences of the United States of America*, *99*(13), 8479–8483.

Galvagnion, C., Brown, J. W., Ouberai, M. M., Flagmeier, P., Vendruscolo, M., Buell, A. K., et al. (2016). Chemical properties of lipids strongly affect the kinetics of the membrane-induced aggregation of alpha-synuclein. *Proceedings of the National Academy of Sciences of the United States of America*, *113*(26), 7065–7070.

Ge, X., Conley, A. J., Brandle, J. E., Truant, R., & Filipe, C. D. (2009). In vivo formation of protein based aqueous microcompartments. *Journal of the American Chemical Society*, *131*(25), 9094–9099.

Haney, C. M., Cleveland, C. L., Wissner, R. F., Owei, L., Robustelli, J., Daniels, M. J., et al. (2017). Site-specific fluorescence polarization for studying the disaggregation of alpha-synuclein fibrils by small molecules. *Biochemistry*, *56*(5), 683–691.

Hanwell, M. D., Curtis, D. E., Lonie, D. C., Vandermeersch, T., Zurek, E., & Hutchison, G. R. (2012). Avogadro: An advanced semantic chemical editor, visualization, and analysis platform. *Journal of Cheminformatics*, *4*(1), 17.

Haustein, E., & Schwille, P. (2007). Fluorescence correlation spectroscopy: Novel variations of an established technique. *Annual Review of Biophysics and Biomolecular Structure*, *36*, 151–169.

Hebda, J. A., Saraogi, I., Magzoub, M., Hamilton, A. D., & Miranker, A. D. (2009). A peptidomimetic approach to targeting pre-amyloidogenic states in type II diabetes. *Chemistry & Biology*, *16*(9), 943–950.

Heller, G. T., Aprile, F. A., & Vendruscolo, M. (2017). Methods of probing the interactions between small molecules and disordered proteins. *Cellular and Molecular Life Sciences*, *74*(17), 3225–3243.

Henriques, J. o., & Skepö, M. (2016). Molecular dynamics simulations of intrinsically disordered proteins: On the accuracy of the TIP4P-D water model and the representativeness of protein disorder models. *Journal of Chemical Theory and Computation*, *12*(7), 3407–3415.

Hong, D.-P., Fink, A. L., & Uversky, V. N. (2008). Structural characteristics of α-synuclein oligomers stabilized by the flavonoid baicalein. *Journal of Molecular Biology*, *383*(1), 214–223.

Hoppener, J. W., Ahren, B., & Lips, C. J. (2000). Islet amyloid and type 2 diabetes mellitus. *The New England Journal of Medicine*, *343*(6), 411–419.

Huang, J., Rauscher, S., Nawrocki, G., Ran, T., Feig, M., de Groot, B. L., et al. (2017). CHARMM36m: An improved force field for folded and intrinsically disordered proteins. *Nature Methods*, *14*(1), 71.

Hülsmann, B. B., Labokha, A. A., & Görlich, D. (2012). The permeability of reconstituted nuclear pores provides direct evidence for the selective phase model. *Cell*, *150*(4), 738–751.

Humphrey, W., Dalke, A., & Schulten, K. (1996). VMD: Visual molecular dynamics. *Journal of Molecular Graphics*, *14*(1), 33–38 [27-38].

Iljina, M., Garcia, G. A., Horrocks, M. H., Tosatto, L., Choi, M. L., Ganzinger, K. A., et al. (2016). Kinetic model of the aggregation of alpha-synuclein provides insights into prion-like spreading. *Proceedings of the National Academy of Sciences of the United States of America*, *113*(9), E1206–E1215.

Ittner, L. M., & Gotz, J. (2011). Amyloid-beta and tau—A toxic pas de deux in Alzheimer's disease. *Nature Reviews. Neuroscience*, *12*(2), 65–72.

Joshi, P., Chia, S., Habchi, J., Knowles, T. P., Dobson, C. M., & Vendruscolo, M. (2016). A fragment-based method of creating small-molecule libraries to target the aggregation of intrinsically disordered proteins. *ACS Combinatorial Science*, *18*(3), 144–153.

Joshi, P., & Vendruscolo, M. (2015). Druggability of intrinsically disordered proteins. *Advances in Experimental Medicine and Biology*, *870*, 383–400.

Kachala, M., Valentini, E., & Svergun, D. I. (2015). Application of SAXS for the structural characterization of IDPs. *Advances in Experimental Medicine and Biology*, *870*, 261–289.

Kaffy, J., Brinet, D., Soulier, J.-L., Correia, I., Tonali, N., Fera, K. F., et al. (2016). Designed glycopeptidomimetics disrupt protein–protein interactions mediating amyloid β-peptide aggregation and restore neuroblastoma cell viability. *Journal of Medicinal Chemistry*, *59*(5), 2025–2040.

Kikhney, A. G., & Svergun, D. I. (2015). A practical guide to small angle X-ray scattering (SAXS) of flexible and intrinsically disordered proteins. *FEBS Letters*, *589*(19 Pt. A), 2570–2577.

Kleckner, I. R., & Foster, M. P. (2011). An introduction to NMR-based approaches for measuring protein dynamics. *Biochimica et Biophysica Acta*, *1814*(8), 942–968.

Kolarova, M., Garcia-Sierra, F., Bartos, A., Ricny, J., & Ripova, D. (2012). Structure and pathology of tau protein in Alzheimer disease. *International Journal of Alzheimer's Disease*, *2012*, 731526.

Krebs, M. R., Bromley, E. H., & Donald, A. M. (2005). The binding of thioflavin-T to amyloid fibrils: Localisation and implications. *Journal of Structural Biology*, *149*(1), 30–37.
Kumar, S., Birol, M., Schlamadinger, D. E., Wojcik, S. P., Rhoades, E., & Miranker, A. D. (2016). Foldamer-mediated manipulation of a pre-amyloid toxin. *Nature Communications*, 7, 11412.
Kumar, S., Henning, A., Magzoub, M., & Hamilton, A. D. (2018). Peptidomimetic-based multi-domain targeting offers critical evaluation of Abeta structure and toxic function. *Journal of the American Chemical Society*, *140*(21), 6562–6574.
Kumar, S., Henning-Knechtel, A., Chehade, I., Magzoub, M., & Hamilton, A. D. (2017). Foldamer-mediated structural rearrangement attenuates Abeta oligomerization and cytotoxicity. *Journal of the American Chemical Society*, *139*(47), 17098–17108.
Kumar, S., Schlamadinger, D. E., Brown, M. A., Dunn, J. M., Mercado, B., Hebda, J. A., et al. (2015). Islet amyloid-induced cell death and bilayer integrity loss share a molecular origin targetable with oligopyridylamide-based alpha-helical mimetics. *Chemistry & Biology*, *22*(3), 369–378.
Laghaei, R., Mousseau, N., & Wei, G. H. (2011). Structure and thermodynamics of amylin dimer studied by Hamiltonian-temperature replica exchange molecular dynamics simulations. *Journal of Physical Chemistry B*, *115*(12), 3146–3154.
Lee, S. H., Kim, D. H., Han, J. J., Cha, E. J., Lim, J. E., Cho, Y. J., et al. (2012). Understanding pre-structured motifs (PreSMos) in intrinsically unfolded proteins. *Current Protein & Peptide Science*, *13*(1), 34–54.
Lee, T., Moran-Gutierrez, C. R., & Deniz, A. A. (2015). Probing protein disorder and complexity at single-molecule resolution. *Seminars in Cell & Developmental Biology*, *37*, 26–34.
Lee, S. J., Nam, E., Lee, H. J., Savelieff, M. G., & Lim, M. H. (2017). Towards an understanding of amyloid-beta oligomers: Characterization, toxicity mechanisms, and inhibitors. *Chemical Society Reviews*, *46*(2), 310–323.
Lendel, C., Bertoncini, C. W., Cremades, N., Waudby, C. A., Vendruscolo, M., Dobson, C. M., et al. (2009). On the mechanism of nonspecific inhibitors of protein aggregation: Dissecting the interactions of alpha-synuclein with Congo red and lacmoid. *Biochemistry*, *48*(35), 8322–8334.
Levine, Z. A., Larini, L., LaPointe, N. E., Feinstein, S. C., & Shea, J. E. (2015). Regulation and aggregation of intrinsically disordered peptides. *Proceedings of the National Academy of Sciences of the United States of America*, *112*(9), 2758–2763.
Levine, Z. A., Rapp, M. V., Wei, W., Mullen, R. G., Wu, C., Zerze, G. H., et al. (2016). Surface force measurements and simulations of mussel-derived peptide adhesives on wet organic surfaces. *Proceedings of the National Academy of Sciences*, *113*(16), 4332–4337.
Li, P., Banjade, S., Cheng, H.-C., Kim, S., Chen, B., Guo, L., et al. (2012). Phase transitions in the assembly of multivalent signalling proteins. *Nature*, *483*, 336.
Li, J., Zhu, M., Manning-Bog, A. B., Di Monte, D. A., & Fink, A. L. (2004). Dopamine and L-dopa disaggregate amyloid fibrils: Implications for Parkinson's and Alzheimer's disease. *The FASEB Journal*, *18*(9), 962–964.
Liu, Z., & Huang, Y. (2014). Advantages of proteins being disordered. *Protein Science*, *23*(5), 539–550.
Lopez del Amo, J. M., Fink, U., Dasari, M., Grelle, G., Wanker, E. E., Bieschke, J., et al. (2012). Structural properties of EGCG-induced, nontoxic Alzheimer's disease Abeta oligomers. *Journal of Molecular Biology*, *421*(4–5), 517–524.
Lyons, A. J., Gandhi, N. S., & Mancera, R. L. (2014). Molecular dynamics simulation of the phosphorylation-induced conformational changes of a tau peptide fragment. *Proteins: Structure, Function, and Bioinformatics*, *82*(9), 1907–1923.
Ma, B., Shatsky, M., Wolfson, H. J., & Nussinov, R. (2002). Multiple diverse ligands binding at a single protein site: A matter of pre-existing populations. *Protein Science: A Publication of the Protein Society*, *11*(2), 184–197.

Ma, B., Wei, G., Zhen, J., & Nussinov, R. (2016). Dancing with strings: The conformational dynamics of VQIXXK motifs within tau protein in monomer, fibril and hyper-phosphorylated filament states. *Biophysical Journal*, *110*(3), 553a–554a.

Maltsev, A. S., Grishaev, A., & Bax, A. (2012). Monomeric alpha-synuclein binds Congo red micelles in a disordered manner. *Biochemistry*, *51*(2), 631–642.

Marasco, D., & Scognamiglio, P. L. (2015). Identification of inhibitors of biological interactions involving intrinsically disordered proteins. *International Journal of Molecular Sciences*, *16*(4), 7394–7412.

Mark, W. Y., Liao, J. C., Lu, Y., Ayed, A., Laister, R., Szymczyna, B., et al. (2005). Characterization of segments from the central region of BRCA1: An intrinsically disordered scaffold for multiple protein–protein and protein–DNA interactions? *Journal of Molecular Biology*, *345*(2), 275–287.

Marsh, J. A., & Forman-Kay, J. D. (2012). Ensemble modeling of protein disordered states: Experimental restraint contributions and validation. *Proteins*, *80*(2), 556–572.

Melo, A. M., Coraor, J., Alpha-Cobb, G., Elbaum-Garfinkle, S., Nath, A., & Rhoades, E. (2016). A functional role for intrinsic disorder in the tau-tubulin complex. *Proceedings of the National Academy of Sciences of the United States of America*, *113*(50), 14336–14341.

Metallo, S. J. (2010). Intrinsically disordered proteins are potential drug targets. *Current Opinion in Chemical Biology*, *14*(4), 481–488.

Meyer, B., & Peters, T. (2003). NMR spectroscopy techniques for screening and identifying ligand binding to protein receptors. *Angewandte Chemie (International Ed. in English)*, *42*(8), 864–890.

Milles, S., Koehler, C., Gambin, Y., Deniz, A. A., & Lemke, E. A. (2012). Intramolecular three-colour single pair FRET of intrinsically disordered proteins with increased dynamic range. *Molecular BioSystems*, *8*(10), 2531–2534.

Mohan, A., Oldfield, C. J., Radivojac, P., Vacic, V., Cortese, M. S., Dunker, A. K., et al. (2006). Analysis of molecular recognition features (MoRFs). *Journal of Molecular Biology*, *362*(5), 1043–1059.

Moree, B., Yin, G., Lazaro, D. F., Munari, F., Strohaker, T., Giller, K., et al. (2015). Small molecules detected by second-harmonic generation modulate the conformation of monomeric alpha-synuclein and reduce its aggregation in cells. *The Journal of Biological Chemistry*, *290*(46), 27582–27593.

Naiki, H., Higuchi, K., Hosokawa, M., & Takeda, T. (1989). Fluorometric determination of amyloid fibrils in vitro using the fluorescent dye, thioflavine T. *Analytical Biochemistry*, *177*(2), 244–249.

Nott, T. J., Craggs, T. D., & Baldwin, A. J. (2016). Membraneless organelles can melt nucleic acid duplexes and act as biomolecular filters. *Nature Chemistry*, *8*(6), 569–575.

Ouberai, M. M., Wang, J., Swann, M. J., Galvagnion, C., Guilliams, T., Dobson, C. M., et al. (2013). Alpha-synuclein senses lipid packing defects and induces lateral expansion of lipids leading to membrane remodeling. *The Journal of Biological Chemistry*, *288*(29), 20883–20895.

Overbeek, J. T. G., & Voorn, M. (1957). Phase separation in polyelectrolyte solutions. Theory of complex coacervation. *Journal of Cellular and Comparative Physiology*, *49*(S1), 7–26.

Pak, C. W., Kosno, M., Holehouse, A. S., Padrick, S. B., Mittal, A., Ali, R., et al. (2016). Sequence determinants of intracellular phase separation by complex coacervation of a disordered protein. *Molecular Cell*, *63*(1), 72–85.

Perni, M., Galvagnion, C., Maltsev, A., Meisl, G., Muller, M. B., Challa, P. K., et al. (2017). A natural product inhibits the initiation of alpha-synuclein aggregation and suppresses its toxicity. *Proceedings of the National Academy of Sciences of the United States of America*, *114*(6), E1009–e1017.

Pithadia, A., Brender, J. R., Fierke, C. A., & Ramamoorthy, A. (2016). Inhibition of IAPP aggregation and toxicity by natural products and derivatives. *Journal of Diabetes Research*, *2016*, 2046327.

Post, C. B. (2003). Exchange-transferred NOE spectroscopy and bound ligand structure determination. *Current Opinion in Structural Biology*, *13*(5), 581–588.
Prakash, M. K., Barducci, A., & Parrinello, M. (2011). Replica temperatures for uniform exchange and efficient roundtrip times in explicit solvent parallel tempering simulations. *Journal of Chemical Theory and Computation*, 7(7), 2025–2027.
Qin, S., & Zhou, H. X. (2017). Protein folding, binding, and droplet formation in cell-like conditions. *Current Opinion in Structural Biology*, *43*, 28–37.
Rahimi, F. (2018). Aptamers selected for recognizing amyloid beta-protein—A case for cautious optimism. *International Journal of Molecular Sciences*, *19*(3), 668.
Rauscher, S., Gapsys, V., Gajda, M. J., Zweckstetter, M., de Groot, B. L., & Grubmüller, H. (2015). Structural ensembles of intrinsically disordered proteins depend strongly on force field: A comparison to experiment. *Journal of Chemical Theory and Computation*, *11*(11), 5513–5524.
Rauscher, S., & Pomès, R. (2017). The liquid structure of elastin. *eLife*, *6*, e26526.
Rawat, A., Langen, R., & Varkey, J. (2018). Membranes as modulators of amyloid protein misfolding and target of toxicity. *Biochimica et Biophysica Acta (BBA)—Biomembranes*, *1860*(9), 1863–1875.
Rhoades, E., Ramlall, T. F., Webb, W. W., & Eliezer, D. (2006). Quantification of alpha-synuclein binding to lipid vesicles using fluorescence correlation spectroscopy. *Biophysical Journal*, *90*(12), 4692–4700.
Ronga, L., Tizzano, B., Palladino, P., Ragone, R., Urso, E., Maffia, M., et al. (2006). The prion protein: Structural features and related toxic peptides. *Chemical Biology & Drug Design*, *68*(3), 139–147.
Rosborough, K., Patel, N., & Kalia, L. V. (2017). Alpha-synuclein and parkinsonism: Updates and future perspectives. *Current Neurology and Neuroscience Reports*, *17*(4), 31.
Sandberg, A., Luheshi, L. M., Sollvander, S., Pereira de Barros, T., Macao, B., Knowles, T. P., et al. (2010). Stabilization of neurotoxic Alzheimer amyloid-beta oligomers by protein engineering. *Proceedings of the National Academy of Sciences of the United States of America*, *107*(35), 15595–15600.
Sasahara, K. (2018). Membrane-mediated amyloid deposition of human islet amyloid polypeptide. *Biophysical Reviews*, *10*(2), 453–462.
Saunders, J. C., Young, L. M., Mahood, R. A., Jackson, M. P., Revill, C. H., Foster, R. J., et al. (2016). An in vivo platform for identifying inhibitors of protein aggregation. *Nature Chemical Biology*, *12*(2), 94–101.
Schiro, G., Fichou, Y., Gallat, F. X., Wood, K., Gabel, F., Moulin, M., et al. (2015). Translational diffusion of hydration water correlates with functional motions in folded and intrinsically disordered proteins. *Nature Communications*, *6*, 6490.
Schlamadinger, D. E., & Miranker, A. D. (2014). Fiber-dependent and -independent toxicity of islet amyloid polypeptide. *Biophysical Journal*, *107*(11), 2559–2566.
Schmidt, H. B., & Görlich, D. (2015). Nup98 FG domains from diverse species spontaneously phase-separate into particles with nuclear pore-like permselectivity. *eLife*, *4*, e04251.
Schuler, B. (2013). Single-molecule FRET of protein structure and dynamics—A primer. *Journal of Nanbiotechnology*, *11*(Suppl. 1), S2.
Schuler, B., & Eaton, W. A. (2008). Protein folding studied by single-molecule FRET. *Current Opinion in Structural Biology*, *18*(1), 16–26.
Sekhar, A., & Kay, L. E. (2013). NMR paves the way for atomic level descriptions of sparsely populated, transiently formed biomolecular conformers. *Proceedings of the National Academy of Sciences of the United States of America*, *110*(32), 12867–12874.
Sezgin, E., Kaiser, H. J., Baumgart, T., Schwille, P., Simons, K., & Levental, I. (2012). Elucidating membrane structure and protein behavior using giant plasma membrane vesicles. *Nature Protocols*, 7(6), 1042–1051.

Shammas, S. L., Garcia, G. A., Kumar, S., Kjaergaard, M., Horrocks, M. H., Shivji, N., et al. (2015). A mechanistic model of tau amyloid aggregation based on direct observation of oligomers. *Nature Communications*, *6*, 7025.
Sharma, R., Bayarjargal, M., Tsunoda, T., Patil, A., & Sharma, A. (2018). MoRFPred-plus: Computational identification of MoRFs in protein sequences using physicochemical properties and HMM profiles. *Journal of Theoretical Biology*, *437*, 9–16.
Shea, J. E., & Levine, Z. A. (2016). Studying the early stages of protein aggregation using replica exchange molecular dynamics simulations. *Methods in Molecular Biology*, *1345*, 225–250.
Shields, P. (1983). An introduction to ergodic-theory—Walters, P. *Bulletin of the American Mathematical Society*, *9*(2), 259–265.
Shoemaker, B. A., Portman, J. J., & Wolynes, P. G. (2000). Speeding molecular recognition by using the folding funnel: The fly-casting mechanism. *Proceedings of the National Academy of Sciences*, *97*(16), 8868–8873.
Srivastava, A., Singh, P. K., Kumbhakar, M., Mukherjee, T., Chattopadyay, S., Pal, H., et al. (2010). Identifying the bond responsible for the fluorescence modulation in an amyloid fibril sensor. *Chemistry*, *16*(30), 9257–9263.
Sugita, Y., & Okamoto, Y. (1999). Replica-exchange molecular dynamics method for protein folding. *Chemical Physics Letters*, *314*(1–2), 141–151.
Tantos, A., & Tompa, P. (2012). Identification of intrinsically disordered proteins by a special 2D electrophoresis. *Methods in Molecular Biology*, *896*, 215–222.
Theillet, F. X., Binolfi, A., Frembgen-Kesner, T., Hingorani, K., Sarkar, M., Kyne, C., et al. (2014). Physicochemical properties of cells and their effects on intrinsically disordered proteins (IDPs). *Chemical Reviews*, *114*(13), 6661–6714.
Toth, G., Gardai, S. J., Zago, W., Bertoncini, C. W., Cremades, N., Roy, S. L., et al. (2014). Targeting the intrinsically disordered structural ensemble of alpha-synuclein by small molecules as a potential therapeutic strategy for Parkinson's disease. *PLoS One*, *9*(2), e87133.
Trexler, A. J., & Rhoades, E. (2009). Alpha-synuclein binds large unilamellar vesicles as an extended helix. *Biochemistry*, *48*(11), 2304–2306.
Uversky, V. N. (2012). Size-exclusion chromatography in structural analysis of intrinsically disordered proteins. In V. N. Uversky & A. K. Dunker (Eds.), *Intrinsically disordered protein analysis: Volume 2, methods and experimental tools* (pp. 179–194). New York, NY: Springer New York.
Uversky, V. N. (2013). Intrinsic disorder-based protein interactions and their modulators. *Current Pharmaceutical Design*, *19*(23), 4191–4213.
Uversky, V. N. (2015). Proteins without unique 3D structures: Biotechnological applications of intrinsically unstable/disordered proteins. *Biotechnology Journal*, *10*(3), 356–366.
Uversky, V. N. (2016). p53 proteoforms and intrinsic disorder: An illustration of the protein structure–function continuum concept. *International Journal of Molecular Sciences*, *17*(11), 1874.
Uversky, V. N. (2017a). Intrinsically disordered proteins in overcrowded milieu: Membrane-less organelles, phase separation, and intrinsic disorder. *Current Opinion in Structural Biology*, *44*, 18–30.
Uversky, V. N. (2017b). Paradoxes and wonders of intrinsic disorder: Stability of instability. *Intrinsically Disordered Proteins*, *5*(1), e1327757.
Uversky, V. N., Gillespie, J. R., & Fink, A. L. (2000). Why are "natively unfolded" proteins unstructured under physiologic conditions? *Proteins: Structure, Function, and Bioinformatics*, *41*(3), 415–427.
Uversky, V. N., Kuznetsova, I. M., Turoverov, K. K., & Zaslavsky, B. (2015). Intrinsically disordered proteins as crucial constituents of cellular aqueous two phase systems and coacervates. *FEBS Letters*, *589*(1), 15–22.

Uversky, V. N., Oldfield, C. J., & Dunker, A. K. (2008). Intrinsically disordered proteins in human diseases: Introducing the D2 concept. *Annual Review of Biophysics*, *37*, 215–246.

van der Lee, R., Buljan, M., Lang, B., Weatheritt, R. J., Daughdrill, G. W., Dunker, A. K., et al. (2014). Classification of intrinsically disordered regions and proteins. *Chemical Reviews*, *114*(13), 6589–6631.

Voropai, E. S., Samtsov, M. P., Kaplevskii, K. N., Maskevich, A. A., Stepuro, V. I., Povarova, O. I., et al. (2003). Spectral properties of thioflavin T and its complexes with amyloid fibrils. *Journal of Applied Spectroscopy*, *70*(6), 868–874.

Wang, J., Cao, Z., Zhao, L., & Li, S. (2011). Novel strategies for drug discovery based on intrinsically disordered proteins (IDPs). *International Journal of Molecular Sciences*, *12*(5), 3205–3219.

Wang, Y., Lomakin, A., McManus, J. J., Ogun, O., & Benedek, G. B. (2010). Phase behavior of mixtures of human lens proteins gamma D and beta B1. *Proceedings of the National Academy of Sciences*, *107*(30), 13282.

Weber, O. C., & Uversky, V. N. (2017). How accurate are your simulations? Effects of confined aqueous volume and AMBER FF99SB and CHARMM22/CMAP force field parameters on structural ensembles of intrinsically disordered proteins: Amyloid-β42 in water. *Intrinsically Disordered Proteins*, *5*(1), e1377813.

Wegmann, S., Eftekharzadeh, B., Tepper, K., Zoltowska, K. M., Bennett, R. E., Dujardin, S., et al. (2018). Tau protein liquid–liquid phase separation can initiate tau aggregation. *The EMBO Journal*, *37*(7), e98049.

Wells, M., Tidow, H., Rutherford, T. J., Markwick, P., Jensen, M. R., Mylonas, E., et al. (2008). Structure of tumor suppressor p53 and its intrinsically disordered N-terminal transactivation domain. *Proceedings of the National Academy of Sciences of the United States of America*, *105*(15), 5762–5767.

Wright, P. E., & Dyson, H. J. (2009). Linking folding and binding. *Current Opinion in Structural Biology*, *19*(1), 31–38.

Wright, P. E., & Dyson, H. J. (2014). Intrinsically disordered proteins in cellular signalling and regulation. *Nature Reviews Molecular Cell Biology*, *16*, 18.

Xue, B., Dunker, A. K., & Uversky, V. N. (2012). Orderly order in protein intrinsic disorder distribution: Disorder in 3500 proteomes from viruses and the three domains of life. *Journal of Biomolecular Structure & Dynamics*, *30*(2), 137–149.

Young, L. M., Saunders, J. C., Mahood, R. A., Revill, C. H., Foster, R. J., Tu, L. H., et al. (2015). Screening and classifying small-molecule inhibitors of amyloid formation using ion mobility spectrometry-mass spectrometry. *Nature Chemistry*, 7(1), 73–81.

Zerze, G. H., Best, R. B., & Mittal, J. (2015). Sequence- and temperature-dependent properties of unfolded and disordered proteins from atomistic simulations. *Journal of Physical Chemistry B*, *119*(46), 14622–14630.

Zerze, G. H., Miller, C. M., Granata, D., & Mittal, J. (2015). Free energy surface of an intrinsically disordered protein: Comparison between temperature replica exchange molecular dynamics and bias-exchange metadynamics. *Journal of Chemical Theory and Computation*, *11*(6), 2776–2782.

Zerze, G. H., Mullen, R. G., Levine, Z. A., Shea, J.-E., & Mittal, J. (2015). To what extent does surface hydrophobicity dictate peptide folding and stability near surfaces? *Langmuir*, *31*(44), 12223–12230.

Zhang, B., Golding, B. T., & Hardcastle, I. R. (2015). Small-molecule MDM2-p53 inhibitors: Recent advances. *Future Medicinal Chemistry*, 7(5), 631–645.

Zhang, M., Ren, B., Liu, Y., Liang, G., Sun, Y., Xu, L., et al. (2017). Membrane interactions of hIAPP monomer and oligomer with lipid membranes by molecular dynamics simulations. *ACS Chemical Neuroscience*, *8*(8), 1789–1800.

FURTHER READING

Aguzzi, A., & Altmeyer, M. (2016). Phase separation: Linking cellular compartmentalization to disease. *Trends in Cell Biology*, *26*(7), 547–558.

Elbaum-Garfinkle, S., & Brangwynne, C. P. (2015). Liquids, fibers, and gels: The many phases of neurodegeneration. *Developmental Cell*, *35*(5), 531–532.

Kaibara, K., Sakai, K., Okamoto, K., Uemura, Y., Miyakawa, K., & Kondo, M. (1992). α-Elastin coacervate as a protein liquid membrane: Effect of pH on transmembrane potential responses. *Biopolymers*, *32*(9), 1173–1180.

Wang, Y., Lomakin, A., Kanai, S., Alex, R., & Benedek, G. B. (2017). Liquid–liquid phase separation in oligomeric peptide solutions. *Langmuir*, *33*(31), 7715–7721.

FSC
www.fsc.org
MIX
Papier | Fördert
gute Waldnutzung
FSC® C083411

Kolibri 360 GmbH
Ferdinand-Jühlke-Straße 7
99095 Erfurt, Deutschland
produktsicherheit@kolibri360.de